Biological Control Systems and Climate Change

CABI Ecostacking Series

The CABI Ecostacking Series is an exciting new set of books focusing on the need to address crop production holistically to ensure balance and minimize damage to the environment.

Ecostacking is a new concept and approach which aims to maximise the benefits of ecosystem service providers in cropping systems to help achieve the goal of long-term sustainable agriculture and food production. The term "ecostacking" means combining synergistically the beneficial services of functional biodiversity from all levels and types. It is a comprehensive approach, where the various ecosystem service providers are fully integrated with the rest of the cropping system including agronomic practices. It is an approach which goes beyond conventional Integrated Pest Management practices, and attempts to take advantage of all the functional biodiversity of a system.

The main focus of ecostacking is on maximising ecosystem services for biological control and pollination from beneficial arthropods, but the approach also utilises beneficial vertebrates, such as bats, birds and small mammals. Microbes also provide invaluable ecosystem services including pest, disease, and weed control, either directly as components of "suppressive soils" or as plant colonisers (as endophytes or as epiphytic microbial flora). The ecostacking approach also aims to maximise other nature-provided services such as maintenance of soil health and nutrient cycling.

Books in the series

- will serve as an inspiration for developing further applications of this breakthrough technology for sustainable agricultural production.
- are a must-read for everyone with an interest in developing sustainable crop protection systems and ecosystem management.
- are written and edited by the world's leading experts in this new and exciting endeavour.

If you have an idea for a new book which you think would be suitable for this series please email the CABI Ecostacking Series Editor-in-Chief, Professor Heikki Hokkanen at heikki.hokkanen@ecostack.eu or Ward Cooper at w.cooper@cabi.org

The ecostacking logo symbolizes the interconnectedness of all ecosystem processes. Affecting one of the processes invariably will affect all other ecosystem components. Conceptual design: Heikki Hokkanen and Ingeborg Menzler-Hokkanen, final design by CABI.

Biological Control Systems and Climate Change

Edited by

Joan van Baaren
UMR-CNRS 6553 ECOBIO, Université de Rennes, France

Cécile Le Lann
UMR-CNRS 6553 ECOBIO, Université de Rennes, France

Chun-Sen Ma
Institute of Plant Protection, Chinese Academy of Agricultural Sciences, Beijing, China and Hebei University, Baoding, China

Gang Ma
Institute of Plant Protection, Chinese Academy of Agricultural Sciences, Beijing

CABI

CABI is a trading name of CAB International

CABI
Nosworthy Way
Wallingford
Oxfordshire OX10 8DE
UK

Tel: +44 (0)1491 832111
E-mail: info@cabi.org
Website: www.cabi.org

CABI
200 Portland Street
Boston
MA 02114
USA

T: +1 (617)682-9015
E-mail: cabi-nao@cabi.org

A catalogue record for this book is available from the British Library, London, UK.

ISBN-13: 9781800625075 (hardback)
9781800625082 (ePDF)
9781800625099 (ePub)

DOI: 10.1079/9781800625099.0000

Commissioning Editor: Ward Cooper
Editorial Assistant: Emma McCann
Production Editor: James Bishop

Typeset by Exeter Premedia Services Pvt Ltd, Chennai, India
Printed and bound in the UK by CPI Group (UK) Ltd, Croydon, CR0 4YY

Contents

Contributors

Alaniz-Zanon, María Silvina: Research Institute on Mycology and Mycotoxicology (IMICO), National Scientific and Technical Research Council – Argentina (CONICET), National University of Río Cuarto (UNRC), Córdoba, Argentina. malaniz@exa.unrc.edu.ar. ORCID 0000-0002-9462-9592.

Alfaro-Tapia, Armando: Centro Regional de Investigación e Innovación para la Sostenibilidad de la Agricultura y los Territorios Rurales, Ceres, Pontificia Universidad Católica de Valparaíso, Quillota Chile. aralfarotapia@gmail.com. ORCID 0000-0001-9166-9874

Auger-Rozenberg, Marie-Anne: INRAE, URZF (Unité de Recherche de Zoologie Forestière), Orléans, France. marie-anne.auger-rozenberg@inrae.fr. ORCID 0000-0002-2455-2438

Barra Bucarei, Lorena: Instituto de Investigaciones Agropecuarias, INIA, Casilla, Chillán, Chile. barra.bucarei@gmail.com. ORCID 0000-0001-8548-2341

Barreto, Carlos: Algoma University, Ste Marie, Ontario, Canada. carlos.barreto@algomau.ca. ORCID 0000-0003-2859-021X

Bonsignore, Carmelo Peter: Dipartimento di Architettura e Design (DAeD) – Laboratorio di Entomologia ed Ecologia Applicata, Università degli studi "Mediterranea" di Reggio Calabria, Reggio Calabria, Italy. cbonsignore@unirc.it. ORCID 0000-0002-9158-1386

Bossa, Marianela: Research Institute on Mycology and Mycotoxicology (IMICO), National Scientific and Technical Research Council – Argentina (CONICET), National University of Río Cuarto (UNRC), Córdoba, Argentina. mbossa@exa.unrc.edu.ar. ORCID 0009-0007-7430-2641

Boutet, Manuel: Université Côte d'Azur, CNRS, UMR GREDEG, Sophia Antipolis, France. manuel.boutet@univ-cotedazur.fr.

Brandl, Stefan: Biodiversity Research Centre, Earth and Life Institute, UCLouvain, ELIV, Louvain-la-Neuve, Belgium. stefan.brandl@uclouvain.be. ORCID 0009-0009-6241-2936

Brodeur, Jacques: Institut de Recherche en Biologie Végétale, Département de Sciences Biologiques, Université de Montréal, Montréal, Canada. jacques.brodeur@umontreal.ca. ORCID 0000-0001-6148-7877

Cáceres Iparraguirre, Hanna: Universidad Autónoma de Ica, Chincha Alta, Perú. hanna.caceres@autonomadeica.edu.pe. ORCID 0000-0001-5040-8950

Celis-Diez, Juan L.: Centro Regional de Investigación e Innovación para la Sostenibilidad de la Agricultura y los Territorios Rurales CERES, La Palma, Chile. juan.celis@pucv.cl. ORCID 0000-0001-6356-264X

Chiotta, Maria Laura: Research Institute on Mycology and Mycotoxicology (IMICO), National Scientific and Technical Research Council – Argentina (CONICET), National University of Río Cuarto (UNRC), Córdoba, Argentina. mchiotta@exa.unrc.edu.ar. ORCID 0000-0003-2601-7233

Chulze, Sofia N.: Research Institute on Mycology and Mycotoxicology (IMICO), National Scientific and Technical Research Council – Argentina (CONICET), National University of Río Cuarto (UNRC), Córdoba, Argentina. schulze@exa.unrc.edu.ar. ORCID 0000-0002-8818-1103

Colmenarez, Yelitza C.: CABI Latin America, FEPAF, Botucatu, Brazil. y.colmenarez@cabi.org. ORCID 0000-0003-4146-5210

Díaz-Siefer, Pablo: Escuela de Agronomía, Facultad de Ciencias Agronómicas y de los Alimentos, Pontificia Universidad Católica de Valparaíso, La Palma, Chile. pablo.diaz@pucv.cl. ORCID 0000-0001-5104-9830

Ding, Jianqing: State Key Laboratory of Crop Stress Adaptation and Improvement, School of Life Sciences, Henan University, Kaifeng, China. jding@henu.edu.cn. ORCID 0000-0002-7399-7388

Edgington, Steve: CABI, Bioscience UK Centre, Egham, UK. s.edgington@cabi.org. ORCID 0000-0002-7551-6402

Flaquer, Carles: BiBio Research Group, Natural Sciences Museum of Granollers, Granollers, Spain. cflaquer@mcng.cat. ORCID 0000-0002-8618-9005

Fontúrbel, Francisco E.: Instituto de Biología, Facultad de Ciencias, Pontificia Universidad Católica de Valparaíso, Valparaíso, Chile. francisco.fonturbel@pucv.cl. ORCID 0000-0001-8585-2816

García, Luis Fernando: Cenur Noreste, Universidad De La República, Ituzaingó 667 Sede Campus, Rivera, Uruguay. luizf.garciah@gmail.com. ORCID 0000-0001-8882-6226

Ghosh, Enakshi: Colorado State University, Fort Collins, USA. Enakshi.Ghosh@colostate.edu. ORCID 0000-0003-4003-5876

Gilbert, François: Biodiversity Research Centre, Earth and Life Institute, UCLouvain, ELIV, Louvain-la-Neuve, Belgium. francois.gilbert@uclouvain.be. ORCID 0009-0002-7318-2880

Hance, Thierry: Biodiversity Research Centre, Earth and Life Institute, UCLouvain, ELIV, Louvain-la-Neuve, Belgium. thierry.hance@uclouvain.be. ORCID 0000-0001-5569-5020

Huang, Shen Nang: College of Resources and Environment, Huazhong Agricultural University, Wuhan, China. sn_huang@webmail.hzau.edu.cn. ORCID 0009-0005-6316-2933

Hubert, Marie-Hélène: CNRS, CREM-UMR, Université de Rennes, France. marie-helene.hubert@univ-rennes.fr. ORCID 0000-0002-6900-1111

Ing Julie: CNRS, CREM-UMR, Université de Rennes, France. julie.ing@univ-rennes.fr. ORCID 0000-0003-1332-8489

Ion Scotta, Michela: Institut Sophia Agrobiotech, INRAE, CNRS, Université Cote d'Azur, Sophia Antipolis, France. michela.ion-scotta@inrae.fr. ORCID 0000-0001-6826-4713

Jacquet, Maxence: Biodiversity Research Centre, Earth and Life Institute, UCLouvain, ELIV, Louvain-la-Neuve, Belgium. maxence.jacquet@uclouvain.be. ORCID 0009-0006-0564-8238

Lago, Clara: Institute for Cross-Disciplinary Physics and Complex Systems (IFISC), Joint Research Institute of the University of the Balearic Islands (UIB) and the Spanish National Research Council (CSIC), Palma de Mallorca, Spain. clara.lago@ifisc.uib-csic.es. ORCID 0000-0002-4074-4747

Latina, Romnick A.: Institute of Weed Science, Entomology and Plant Pathology, College of Agriculture and Food Science, University of the Philippines Los Baños, Laguna, The Philippines. ralatina@up.edu.ph. ORCID 0000-0002-9089-2107

Lavandero, Blas: Laboratorio de Control Biológico, Instituto de Ciencias Biológicas, Universidad de Talca, Talca, Chile. blavandero@utalca.cl. ORCID 0000-0002-2423-7016

Le Lann, Cécile: UMR CNRS ECOBIO (Ecosystèmes, Biodiversité, Evolution), Université de Rennes, Rennes, France. Cecile.lelann@univ-rennes.fr. ORCID 0000-0002-3949-4066

Lemić, Darija: Department of Agricultural Zoology, University of Zagreb Faculty of Agriculture, Zagreb, Croatia. dlemic@agr.hr. ORCID 0000-0002-1175-3716

Lommen, Suzanne T.E.: Institute of Biology, Above-Belowground Interactions Group, Leiden University, Leiden, The Netherlands. s.t.e.lommen@biology.leidenuniv.nl. ORCID 0000-0003-2457-9849

López-Baucells, Adrià: BiBio Research Group, Natural Sciences Museum of Granollers, Granollers, Spain. adria.baucells@gmail.com. ORCID 0000-0001-8446-0108

Ma, Chun-Sen: School of Life Science, Institutes of Life Science and Green Development, Hebei University, Baoding, China. machunsen@caas.cn. ORCID 0000-0002-3253-6921

Ma, Gang: State Key Laboratory for Biology of Plant Diseases and Insect Pests, Institute of Plant Protection, Chinese Academy of Agricultural Sciences, Beijing, China. magang@caas.cn. ORCID 0000-0002-3378-9488

Macasero, John Bethany M.: Institute of Weed Science, Entomology and Plant Pathology, College of Agriculture and Food Science, University of the Philippines Los Baños, Laguna, The Philippines. jmmacasero@up.edu.ph. ORCID 0000-0002-5495-4550

Malausa, Thibaut: Université Côte d'Azur, CNRS, UMR GREDEG, Sophia Antipolis, France. thibaut.malausa@inrae.fr. ORCID 0000-0003-1681-7958

Maldonado-Santos, Enrique: Laboratorio de Control Biológico, Instituto de Ciencias Biológicas, Universidad de Talca, Talca, Chile. ejms15593@gmail.com. ORCID 0009-0001-7477-9518

Mata, Vanessa: CIBIO-InBIO, Centro de Investigação em Biodiversidade de Recursos Genéticos, InBIO Laboratório Associado, Campus de Vairão, Universidade do Porto, Vairão, Portugal. vanessamata@cibio.up.pt. ORCID 0000-0003-3005-9030

Meehan, Matthew L.: University of Manchester, Manchester, UK and Natural Resources Canada, Canadian Forest Service, Laurentian Forestry Centre, Quebec City, Canada. matthew.meehan@nrcan-rncan.gc.ca. ORCID 0000-0001-5427-8241

Moiroux, Joffrey: Avignon Université, Aix Marseille Université, CNRS, IRD, IMBE, Avignon, France. joffrey.moiroux@univ-avignon.fr. ORCID 0000-0002-0132-3763

Müller-Schärer, Heinz: Department of Biology, University of Fribourg, Fribourg, Switzerland. heinz.mueller@unifr.ch. ORCID 0000-0003-0936-1470

Ode, Paul J.: Colorado State University, Fort Collins, USA. paul.ode@colostate.edu. ORCID 0000-0001-7153-1077

Olmos-Moya, Natalia: Escuela de Agronomía, Facultad de Ciencias Agronómicas y de los Alimentos, Pontificia Universidad Católica de Valparaíso, La Palma, Chile. n.olmosmoya@gmail.com. ORCID 0000-0002-6183-7707

Ortiz-Campos, Javiera: Instituto de Investigaciones Agropecuarias, INIA, Casilla, Chile. javiera.ortiz@endofitos.com. ORCID 0000-0003-0666-6252

Pajač Živković, Ivana: Department of Agricultural Zoology, University of Zagreb Faculty of Agriculture, Zagreb, Croatia. ipajac@agr.hr. ORCID 0000-0002-0775-7327

Pangga, Ireneo B.: Institute of Weed Science, Entomology and Plant Pathology, College of Agriculture and Food Science, University of the Philippines Los Baños, Laguna, The Philippines. ibpangga@up.edu.ph. ORCID 0000-0002-9591-6964

Parmentier-Cajaiba, Aura: Université Côte d'Azur, CNRS, UMR GREDEG, Sophia Antipolis, France. aura.parmentier@univ-cotedazur.fr. ORCID 0000-0003-3167-5839

Perrin, Marie: Agroecology and Environment Research Unit, ISARA-Lyon, Avignon, France. mperrin@isara.fr. ORCID 0000-0002-5653-1527

Pétillon, Julien: UMR CNRS ECOBIO (Ecosystèmes, Biodiversité, Evolution), Université de Rennes, Rennes, France. Julien.petillon@univ-rennes.fr. ORCID 0000-0002-7611-5133

Pincebourde, Sylvain: UMR CNRS 7261 Institut de Recherche sur la Biologie de l'Insecte, Université de Tours, Tours, France. sylvain.pincebourde@univ-tours.fr. ORCID 0000-0001-7964-5861

Puig-Montserrat, Xavier: BiBio Research Group, Natural Sciences Museum of Granollers, Granollers, Catalonia, Spain. xpuig@mcng.cat. ORCID 0000-0002-2670-321X

Quijano, Luis: Instituto de Investigaciones en Ecosistemas y Sustentabilidad (IIES), UNAM. Morelia, México. luisquijanocuervo@gmail.com. ORCID 0000-0001-5434-8588

Razgour, Orly: Biosciences, University of Exeter, Hatherly Laboratories, Exeter, UK. O.Razgour@exeter.ac.uk. ORCID 0000-0003-3186-0313

Rebelo, Hugo: cE3c–Centre for Ecology, Evolution and Environmental Changes and CHANGE–Global Change and Sustainability Institute, Departamento de Biologia Animal, Faculdade de Ciências, Universidade de Lisboa, Lisboa, Portugal. herebelo@fc.ul.pt. ORCID 0000-0002-7118-4068

Renoz, François: Biodiversity Research Centre, Earth and Life Institute, UCLouvain, ELIV, Louvain-la-Neuve, Belgium. francois.renoz@uclouvain.be. ORCID 0000-0003-3716-4562

Rhein, Paloma: Research Institute on Mycology and Mycotoxicology (IMICO), National Scientific and Technical Research Council – Argentina (CONICET), National University of Río Cuarto (UNRC), Córdoba, Argentina. hein@exa.unrc.edu.ar. ORCID 0009-0008-2457-7943

Ris, Nicolas: Institut Sophia Agrobiotech, INRAE, CNRS, Université Cote d'Azur, Sophia Antipolis, France. nicolas.ris@inrae.fr. ORCID 0000-0003-1183-234X

Roques, Alain: INRAE, URZF (Unité de Recherche de Zoologie Forestière), Orléans, France. alain.roques@inrae.fr. ORCID 0000-0002-3734-3918

Russo, Danilo: Laboratory of Animal Ecology and Evolution (AnEcoEvo), Dipartimento di Agraria, Università degli Studi di Napoli Federico II, Portici, Italy. danrusso@unina.it. ORCID 0000-0002-1934-7130

Siegwart, Myriam: INRAE, UR 1115 PSH, Plantes et Systèmes de culture Horticoles, Avignon, France. myriam.siegwart@inrae.fr. ORCID 0000-0003-4717-5273

Siemann, Evan: Department of Biosciences, Rice University, Houston, USA. siemann@rice.edu. ORCID 0000-0002-4919-7292

Skendžić, Sandra: Department of Agricultural Zoology, University of Zagreb Faculty of Agriculture, Zagreb, Croatia. sskendzic@agr.hr. ORCID 0000-0002-9115-2899

Sun, Yan: College of Resources and Environment, Huazhong Agricultural University, Wuhan, China. yan.sun@mail.hzau.edu. ORCID 0000-0002-6439-266X

Taylor, Bryony: CABI, Bioscience UK Centre, Egham, UK. b.taylor@cabi.org. ORCID 0000-0002-7676-2824

Tougeron, Kévin: Écologie des Interactions et Changements Globaux, Institut de Recherche en Biosciences, Université de Mons, Mons, Belgium. kevin.tougeron@umons.ac.be. ORCID 0000-0003-4897-3787

Trębicki, Piotr: Applied BioSciences, Macquarie University, Sydney, Australia. piotr.trebicki@mq.edu.au. ORCID 0000-0002-7075-6650

Tuneu-Corral, Carme: BiBio Research Group, Natural Sciences Museum of Granollers, Granollers, Spain. tuneucorral@gmail.com. ORCID 0000-0003-2348-9761

Van Baaren, Joan: UMR CNRS ECOBIO (Ecosystèmes, Biodiversité, Evolution), Université de Rennes, Rennes, France. joan.van-baaren@univ-rennes.fr. ORCID 0000-0002-8552-9645

Vásquez, Carlos: Universidad Técnica de Ambato. Facultad de Ciencias Agropecuarias, Campus Querochaca, Cevallos, Ecuador. ca.vasquez@uta.edu.ec. ORCID 0000-0002-8214-3632

Vercken, Elodie: Institut Sophia Agrobiotech, INRAE, CNRS, Université Cote d'Azur, Sophia Antipolis, France. Elodie.Vercken@inrae.fr.

Vernon, Philippe: Université de Rennes and CNRS, UMR 6553 EcoBio, Rennes, France. philippe.vernon@univ-rennes.fr. ORCID 0000-0002-6237-7511

Viric Gasparic, Helena: Department of Agricultural Zoology, University of Zagreb Faculty of Agriculture, Zagreb, Croatia. hviric@agr.hr. ORCID 0000-0002-4275-4033

Warot, Sylvie: Institut Sophia Agrobiotech, INRAE, CNRS, Université Cote d'Azur, Sophia Antipolis, France. Sylvie.warot@inrae.fr. ORCID 0009-0006-0678-9008

Introduction

Biological communities typically maintain a dynamic balance that supports the coexistence of all their constituent species. Alterations in the system structure or in the density of a given species due to natural or artificial disturbances exceeding a certain threshold can tend towards the breakdown of the system, which will have to adapt to the new conditions, if it has not ceased to exist. During the First Agricultural Revolution, in the Neolithic period, humans started to alter the natural ecosystems to create the first croplands. Complex ecosystems, such as forests or meadows, were changed at a small scale around human settlements to fit the new production needs. Recent evidence suggests that farming did not bring a productivity advantage to the hunter-gatherers who initiated it, unless a previous possession-based private property structure existed, changing the traditional view of farming, because of technological improvements, to an 'institution-first' perspective (Bowles and Choi, 2019). These institutional changes may have predated and provided conditions for subsequent technical changes that allowed for agriculture's birth. Thus, premodern agrarian autocracies benefited most from the advent of agriculture, frequently requiring farmers to grow cereals – easier to harvest and store – rather than root crops, to sustain their dominance.

The Green Revolution (aka the Third Agricultural Revolution) implemented technological improvements to reduce or eliminate hunger. Improved irrigation systems, the use of synthetic pesticides and fertilizers and the selection of high-yielding strains of crops boosted the farmland productivity of a few selected staple crops to unprecedented levels (Soby, 2013). Scaling all these technologies globally was also linked to the birth of a multi-billion industry concentrated in a small number of multinational companies, with a global yearly market of $49 billion on pesticides alone in 2022, according to FAOSTAT (FAO, 2022). But has the Green Revolution proven itself a successful strategy for ending hunger? Increasing production cannot alleviate hunger if it fails to alter the tightly concentrated distribution of economic power, specifically the access to land and the necessary technological products (Rosset *et al.*, 2000). The modern intensification and technological advancement of agriculture have resulted in an agrochemical dependence that represents one of the most pressing challenges to global environmental sustainability and public health (Hu, 2020).

The shift to conventional agriculture, which includes the use of agrochemical pollutants, is the main driver of the decline in insect populations worldwide (Sánchez-Bayo and Wyckhuys, 2019). Agrochemicals have well-known effects on non-target taxa, including alterations in activity, physiology and diversity of terrestrial and aquatic faunal communities (Beringue *et al.*, 2024). The decline in bee populations due to the prevalence of neonicotinoids in farmland is probably one of the best known examples of the negative impact of pesticides on non-target species. This problem is particularly important as bees are a key species needed for the pollination of many crops worldwide (Sgolastra *et al.*, 2020). The detrimental effects of the Green Revolution reached further trophic networks; during the last decades, for instance, up to 70% decline in farmland insect populations and around 50% loss of farmland birds' abundance have been reported in European countries, where solid long-term monitoring schemes exist (Brühl and Zaller, 2019; Geiger *et al.*, 2010). Given the wealth of evidence on the detrimental effects of pesticides on biodiversity, Brühl and Zaller (2019) suggest that the environmental risk assessments enforced in the registration process of agrochemicals fail to measure the detrimental effects of these products on the environment and should therefore be upgraded to a higher level of exigence.

Despite the astounding results in terms of crop yield achieved by modern agricultural intensification, its environmental consequences extend beyond the problems of pest resistance or loss of biodiversity. In what seems to be a Faustian bargain that compromises the sustainability and viability of future human generations, issues linked to agricultural modernisation encompass other key aspects like pollution of critical resources in the face of climate change such as surface and groundwater; dependence on fossil fuels and derivatives corresponding to greenhouse gas emissions; loss of soil fertility and increased soil erosion (Badgley *et al.*, 2007). According to the FAO (2019), 90% of the world's topsoil is likely to be at risk by 2050, partially due to the intensification of agriculture.

Proponents of modern technological agricultural practices argue that no other productive strategy can ensure sufficiently high yields to satisfy the needs of the global food supply chain, even claiming that questioning the technological approach runs counter to scientific evidence (Borlaug, 2000). From this perspective, the assumed lower yield of organic crops would need more land to fulfil the supply needs, thus offsetting any possible environmental benefits of such practices. Badgley *et al.* (2007), using a global dataset from 293 studies on the yield of organic and conventional crops, estimated the viability of organic food production worldwide. Using FAO statistics on global food production and the daily caloric intake of the human population, they conclude that turning global production into organic would yield a ratio between 0.92 and 1.80 times the yield currently achieved by conventional (or Green Revolution) methods. They have also shown that it is also possible to fertilize such organic production using non-synthetic sources, such as crop residues, animal manures, compost and biologically fixed nitrogen from leguminous plants. These results suggest that organic options could play an important role in feeding the current and future human population without increasing the cultivated land, thus alleviating the heavy environmental trade-offs of today's dominant Green Revolution methods and stagnation of some yield crops (Schauberger *et al.*, 2018). Besides striving to improve production, it is generally accepted that the reduction of food losses, mainly attributable to the technological advances in agricultural practices and the food market globalisation boosted by the Green Revolution, would play a significant role in addressing world hunger and inequity (Soria-Lopez *et al.*, 2023).

Novel efficient food systems are required to meet environmental and consumer demands (Soria-Lopez *et al.*, 2023). The Millennium Ecosystem Assessment of 2005 recommended promoting methods that increase food production without harmful trade-offs due to excessive water usage, nutrients or pesticides. For instance, evidence shows that the correct management of the trophic levels of the agroecosystems, either above or below the target pests, can alter the pest population dynamics and reduce crop damage to acceptable thresholds, avoiding the use of pesticides (Matson *et al.*, 1997). Given the current situation, ecological interactions that reinforce biological pest control should become the first option and chemical approaches should only be encouraged when natural solutions prove inefficient in controlling pests.

However, to implement biocontrol instead of chemical pesticides, it is necessary to take into account the major changes of the last 50 years qualified by the term 'global environmental changes', which include changes in both land use (intensification of agriculture mentioned above) and climate change.

Regarding climate change, the main driver is the increase in greenhouse gases emitted into the atmosphere associated with anthropogenic activity such as the extraction and burning of fossil fuels, deforestation and corporate agricultural practices (IPCC, 2023). For example, the International

Panel on Climate Change (IPCC) 6th Assessment Report (AR6) stated that atmospheric carbon dioxide levels have increased to the current level of 410 ppm and will increase to between 730 and 1000 ppm by the end of this century. Rising mean global temperatures as well as changes in precipitation patterns are the direct result of this inexorable rise in concentrations of greenhouse gases, particularly carbon dioxide (CO_2), ozone (O_3) and methane (CH_4). The increase in temperatures has been observed since 1850. The global surface temperature was 1.09°C higher in 2011–2020 than in 1850–1900, with greater increases on land (1.59°C) than over the ocean (0.88°C) (IPCC, 2023) and is projected to increase further as global CO_2 levels are expected to double by the end of the 21st century (IPCC, 2023). Climate change affects both high- and low-income countries, with repercussions for food safety, resulting in fewer opportunities to overcome the crisis for low-income countries, due to the scarcity of resources (FAO, 2020).

At the UN Climate Change Conference in Paris in 2015, the participating countries agreed to limit the temperature increase below 2°C by the end of this century. However, this limit currently seems unachievable, as in practice political decisions have not been followed by the concrete implementation of plans to drastically reduce emissions (Nordhaus, 2018). In this framework, participating countries were required to submit their climate action plans, called nationally determined contributions (NDCs), by 2020. However, these NDCs present a substantial gap in greenhouse gas emissions compared to what would have been necessary to limit warming to 1.5°C or even 2°C assuming immediate action. As a result, a median global warming rate of 2.8 (2.1–3.4) °C is to be expected by 2100. It could even reach 4.4°C for a very high greenhouse gas emissions scenario (IPCC, 2023). This average increase masks significant changes in the frequency, intensity, spatial extent, duration and timing of weather and climate extremes such as droughts or major floods, storms, heat waves or unpredictable cold snaps (Seneviratne *et al.*, 2012). Extreme climate events can be defined as the occurrence of a value of a weather or climate variable above or below a threshold close to the upper or lower end of the range of observed values of the variable (for a discussion of these concepts and the use of extreme indices, see Seneviratne *et al.*, 2012). An increase in extreme temperatures can therefore be expected, especially during summer (Lelieveld *et al.*, 2016), these changes being greater in the northernmost countries. In addition, winters will be milder and vegetation will resume earlier in the spring.

Climate extremes are part of the environmental variability that the living world normally experiences and have therefore always been a selective pressure. Since the industrial era, however, two new elements have emerged: the rapid increase in the amplitude of these extremes and their frequency over time. This raises two questions: how do organisms respond to these increases in amplitude, which go beyond the usual range of conditions they encounter, and what are the consequences of the rapid repetition of these fluctuations for the adaptability of the living world?

For living organisms, a growing number of studies have been able to demonstrate that the modification of abiotic conditions, from the global scale to the local scale, is responsible for changes in the physiology and in all the life-history traits (e.g. fecundity, growth, longevity), in phenology (i.e. periodic events of biological activity determined by seasonal variations), in the relative abundances of individuals of different species, in species geographic distributions and the diversity of communities. As most of the pests (mainly insect herbivores) and most of the biological control agents (BCA) are ectothermic organisms, they are strongly affected by both rises in global temperatures and extremes in temperature and precipitation (Bale *et al.*, 2002; Harvey *et al.*, 2022). These abiotic and biotic changes also alter the interactions between species and trophic levels, leading to modifications in the functioning of these ecosystems and thus in the associated ecosystem services, such as the biological control of crop pests in agroecosystems. This service, supported by biodiversity, has been altered by pesticides as described above, but is also threatened by climate change when a change in practices becomes crucial for both human and environmental health.

The term 'biological control' is generally used to define any suppression or reduction of abundance below an ecological or economical threshold of an organism harmful to humans (pests, such as insects, mites, weeds and pathogens) by another living organism, called a biological control agent or BCA, which includes predators, parasitoids, pathogens, fungi, bacteria, virus etc. Some biocontrol methods rely on active substances or preparations containing one or more active substances of natural origin. This definition includes 'biocontrol products' that use natural mechanisms to combat pests as part of integrated pest management (IPM), such as micro-organisms (bacteria, viruses, fungi but also their extracts), plant defence stimulators and chemical mediators (pheromones, kairomones) and products of natural origin (extracts of plants, animals or minerals). Biological control always involves an active human management role and can be an important component of IPM programmes, even if pesticide application is still possible in IPM.

There are different basic strategies for biological pest control: classical, where a natural enemy of a pest is introduced into its new range; inoculative, when a small number of individuals are introduced in an area (often glasshouses) to establish themselves in this environment and to reproduce before the arrival of the pests; and augmented, in which large numbers of natural enemies are released for rapid pest control; conservational, involving habitat management to favour natural enemy populations against pest populations. The main objective of this book is to synthesize the effects (positive or negative) of climate change on these different methods of biocontrol.

The first part features nine chapters summarizing the impact of climate change on macro-organisms involved in biocontrol strategies (i.e. phytophagous insects against weeds, predators and parasitoids, bats, birds against pest herbivores, etc.). It includes both general concepts and applied studies and details the various impacts of climate change on the life-history traits of macro-organisms and how they vary according to species, inducing phenological modifications and the difference of vulnerability between organisms from temperate and equatorial-tropical areas (Chapters 1 and 2). Chapters of this first section also detail why an increase in pest outbreaks and plant diseases transmitted by vectors is expected following climate change (Chapter 3), and question the relative impacts of climate change on weeds and phytophagous insects to understand if and when the biocontrol will work better or decrease (Chapter 4). They explore the mechanisms of the role of climate change on plant defences and how they are modified by climate change (Chapter 5). Specific chapters present the importance of lesser known but also very important BCAs such as spiders (Chapter 6), birds (Chapter 7) and bats (Chapter 8). Finally, the last chapter of this part (Chapter 9) highlights a case study of community changes and genetic differentiation in parasitoids at a small scale induced by climate change and this knowledge can become a tool to produce BCAs adapted to particular climatic conditions.

The second part of this book focuses on the impact of climate change on pathogens such as fungi, oomycetes, bacteria and viruses, and on the micro-organisms involved in biocontrol strategies (fungi, oomycetes, bacteria and viruses) (Chapters 10 and 11). Chapter 12 deals with the impact of climate change on bacterial–insect symbiosis and the implications for parasitoidism and biological control.

The third part (13 chapters) includes chapters presenting either results of case studies in different continents or agricultural systems or potential future solutions (from the first proof of concept to techniques that still need to be developed). Chapter 13 presents the importance of adapting the biocontrol of aflatoxins in maize under a scenario of climate change, Chapter 14 details a potential adaptation of IPM strategy against the codling moth *Cydia pomonella* to climate change, and Chapter 15 explains how both biocontrol and IPM strategy could be adapted to the effects of climate change in the specific conditions of greenhouses. Chapter 16 provides a tool to determine the threshold of temperature at which the present biocontrol methods will not be efficient any more through a case study with acarine species. Chapter 17 explores how new biocontrol tools could be deployed against plant disease using plant growth-promoting endophytes (specific case of rhizobacteria). Chapter 18 presents how conservation biological control can take opportunities of climate change in some situations in which plant and insect phenology were modified, by developing the management of semi-natural elements as flowering strips or hedges around the fields, to intensify natural enemy pressure on pest populations. Chapter 19 focuses on a tool for early detection of invaders which can become new pests, to determine in advance the future invasive species and control them before invasion and avoid new pest outbreaks. Chapter 20 exposes how microclimate impacts biocontrol, and how managing microclimates can become a tool to improve biocontrol. Chapter 21 introduces a detailed list of new tools, recently used or which can be developed in the future in the area of biocontrol, and Chapter 22 details some case studies in Latin American crops. Chapter 23 reviews the different types of companies involved in biocontrol and the new tools that they could develop to adapt their products to climate change. Chapters 24 and 25 highlight the interest of economic and societal approaches to increase the adoption of biocontrol strategies in the future, either by valuating the ecosystem services using economic tools and methods or by innovation management using living technologies.

The main message of this book is that biocontrol can be improved or adversely affected by climate change, and that the resulting situation is specific to each case (both the system and the localization), with very little possibility of generalization, due to the complexity and the number of effects of climate change both on the pests/diseases and on the biocontrol agents (micro- or macro-organisms).

In this context, this book is timely as over 6000 publications have now provided valuable data on this subject, supplying enough material to produce a synthesis of the current challenges in this research area. This book is intended to provide students, teachers, stakeholders and managers of companies producing BCA or macro-organisms involved in biocontrol with the keys to understanding the impact of climate change on the different biocontrol methods, in order to be able to adapt them to new environmental conditions. Most studies on climate change are focused on the impact of climate change on individual species, whereas biological control methods involve interactions between species at different trophic levels and therefore the necessity to develop an understanding at a large scale. Climate change is a challenge that will require the development of flexible strategies capable of responding to changes in pest distribution and/or food web structure. Biocontrol strategies fit perfectly with Agenda 2030 contributing to the 'zero hunger' goal and particularly to food safety. Likewise, it is in line with the concept of One Health, which is an approach that aims to balance, protect and optimise the health of people, other living organisms such as plants and animals, and ecosystems (Sarrocco, 2023).

References

Badgley, C., Moghtader, J., Quintero, E., Zakem, E., Chappell, M.J. *et al.* (2007) Organic agriculture and the global food supply. *Renewable Agriculture and Food Systems* 22(2), 86–108.

Bale, J.S., Masters, G.J., Hodkinson, I.D., Awmack, C., Bezemer, T.M. *et al.* (2002) Herbivory in global climate change research: Direct effects of rising temperature on insect herbivores. *Global Change Biology* 8, 1–16.

Beringue, A., Queffelec, J., Le Lann, C. and Sulmon, C. (2024) Sublethal pesticide exposure in non-target terrestrial ecosystems: From known effects on individuals to potential consequences on trophic interactions and network functioning. *Environmental Research* 260, 119620.

Borlaug, N.E. (2000). Ending world hunger. The promise of biotechnology and the threat of antiscience zealotry. *Plant Physiology* 124, 487–490.

Bowles, S. and Choi, J.-K. (2019) The Neolithic agricultural revolution and the origins of private property. *Journal of Political Economy* 127(5), 2186–2228.

Brühl, C.A. and Zaller, J.G. (2019) Biodiversity decline as a consequence of an inappropriate environmental risk assessment of pesticides. *Frontiers in Environmental Science* 7.

FAO (2019) *Proceedings of the Global Symposium on Soil Erosion* 2019. FAO, Rome, Italy.

FAO (2022) *Agricultural Production Statistics. 2000–2020.* FAOSTAT Analytical Brief Series 41. FAO, Rome, Italy.

FAO, IFAD, UNICEF, WFP and WHO (2020) *The State of Food Security and Nutrition in the World 2020. Transforming Food Systems for Affordable Healthy Diets*. FAO, Rome, Italy.

Geiger, F., Bengtsson, J., Berendse, F., Weisser, W.W., Emmerson, M. *et al.* (2010) Persistent negative effects of pesticides on biodiversity and biological control potential on European farmland. *Basic and Applied Ecology* 11, 97–105.

Harvey, J., Tougeron, K., Gols, R., Heinen, R., Abarca, M. *et al*. (2022) Scientists' warning on climate change and insects. *Ecological Monographs* 93, e1553.

Hu, Z. (2020) What socio-economic and political factors lead to global pesticide dependence? A critical review from a social science perspective. *International Journal of Environmental Research and Public Health* 17(21), 8119.

IPCC (2023) *Climate Change 2023: Synthesis Report*. IPCC, Geneva, Switzerland.

Lelieveld, J., Proestos, Y., Hadjinicolaou, P. *et al*. (2016) Strongly increasing heat extremes in the Middle East and North Africa (MENA) in the 21st century. *Climatic Change* 137, 245–260.

Matson, P.A., Parton, W.J., Power, A.G. and Swift, M.J. (1997) Agricultural intensification and ecosystem properties. *Science* 277, 504–508.

Nordhaus, W. (2018) Projections and uncertainties about climate change in an era of minimal climate policies. *American Economic Journal- Economic Policy* 10(3), 333–360.

Rosset, P., Collins, J. and Lappé, F.M. (2000) Lessons from the green revolution. *Third World Resurgence* 11–14.

Sánchez-Bayo, F. and Wyckhuys, K.A.G. (2019) Worldwide decline of the entomofauna: A review of its drivers. *Biological Conservation* 232, 8–27.

Sarrocco, S. (2023) Biological disease control by beneficial (micro)organisms: Selected breakthroughs in the past 50 years. *Phytopathology* 113, 732–740.

Schauberger, B., Ben-Ari, T., Makowski, D. *et al*. (2018) Yield trends, variability and stagnation analysis of major crops in France over more than a century. *Science Reports* 8, 16865.

Sgolastra, F., Medrzycki, P., Bortolotti, L., Maini, S., Porrini, C. *et al*. (2020) Bees and pesticide regulation: Lessons from the neonicotinoid experience. *Biological Conservation* 241, 108356.

Soby, S.D. (2013) The end of the Green Revolution. *Journal of Agricultural and Environmental Ethics* 26, 537–546.

Soria-Lopez, A., Garcia-Perez, P., Carpena, M., Garcia-Oliveira, P., Otero, P. *et al*. (2023) Challenges for future food systems: From the Green Revolution to food supply chains with a special focus on sustainability. *Food Frontiers* 4(1), 9–20.

1 Impact of Climate Change on Macro-Organisms Used in Biological Control

Philippe Vernon[1]*, François Renoz[2] and Thierry Hance
[1]Université de Rennes & CNRS, UMR 6553 EcoBio, Rennes, France; [2]Biodiversity Research Centre, Earth and Life Institute, UCLouvain, Louvain-la-Neuve, Belgium

Abstract
Contemporary climate change has many effects on the behaviour, physiology, ecology and evolution of the macro-organisms involved in biological control, as well as on the pests themselves. In temperate zones, for example, global warming may increase the number of generations and improve winter survival, but responses to temperature changes, particularly in ectotherms, are highly variable from species to species, making it very difficult to draw global conclusions, whether about pests or beneficial macro-organisms. From an interspecific point of view, differential rates of reproduction and frequent phenological mismatches are also observed. The ecological impact of climate change varies with latitude: in warm tropical zones, annual thermal variations are less pronounced and ectotherms have generally evolved with little seasonal variability. As a result, in the face of global warming, it is now often postulated that tropical ectotherms are more vulnerable than their temperate counterparts, as they are often already close to their physiological optimum. This chapter attempts to discuss the validity of these assertions.

1.1 Introduction

A wide range of definitions are used for the concepts of climate change and biological control. Two of these seem particularly noteworthy: (i) climate change: 'A change of climate that is attributed directly or indirectly to human activity that alters the composition of the global atmosphere and that is in addition to natural climate variability observed over comparable periods' (Ingram and Hamilton, 2014) (ii) biological control: 'The use of living organisms to suppress the population density or impact of a specific pest organism, making it less abundant or less damaging than it would otherwise be' (Eilenberg *et al.*, 2001). On the other hand, it can paradoxically be difficult to describe what a macro-organism used in biological control is, and it is often the precise definition of what a micro-organism is that makes it possible, as it were by difference, to define a macro-organism. For example, we can consider that micro-organisms are represented by unicellular or multicellular entities, i.e. bacteria, algae, protozoa, fungi and viruses, and that macro-organisms make up the rest of the living world (Amsellem *et al.*, 2017). As a result, the list is long and varied, e.g. nematodes that parasitize insects, beneficial insects (parasitoids, predators), predatory mites and even some species of predatory birds and bats. Parasitoids will be referred to frequently in this chapter. A size criterion can also be used, with macro-organisms larger than 500 µm (Tesson *et al.*, 2015). In this chapter, only the use of

*Corresponding author: philippe.vernon@univ-rennes.fr

DOI: 10.1079/9781800625099.0001

macro-organisms will be considered among living organisms. Macro-organisms are under the control of temperature for their activity, reproduction and development, even if certain environments they occupy (e.g. soils) partially buffer this effect of temperature (Cock *et al.*, 2013).

On another note, it would be a mistake to think that the influence of weather and climate on macro-organisms (particularly insects) used in biological control is a recent field of study. Long before the concerns associated with climate change, the relationship between insects and climate was studied in detail, as in the seminal work by Uvarov (1931). In this 232-page synthesis, Boris Uvarov discussed the various abiotic factors influencing insect life, including the combination of several factors (e.g. temperature and humidity). The final section of this review was already considering the links between weather, climate and economic entomology.

Although we now know that rising temperatures date back to the beginning of the industrial era (see for example Abram *et al.*, 2016), awareness of the consequences of climate change – mainly global warming and changes in climate variability – has been slow and gradual (cf. creation of the United Nations Environment Programme in 1972, IPCC first Assessment Report completed in 1990). Interest in biological control thus preceded in-depth study of the ecological consequences of climate change. Since then, the study of interactions between climate change and biological control, i.e. the responses of macro-organisms to global warming and the limits of these responses – has become a very active field of investigation. The importance of climate change, but also of habitat fragmentation, for the evolution of species interactions and communities functioning is therefore now well documented, and today there is a broad consensus 'that species respond individually to climate change, not as coherent communities' (Lovejoy and Hannah, 2005). Although rising temperatures are the most commonly perceived phenomenon, climate change encompasses several other phenomena that interact with each other, discussed first in this chapter. Moreover, climate change is expressed differently when we consider latitudinal and altitudinal gradients. This is the subject of the second part of this chapter. The aim of this chapter is consequently to briefly describe these features with the help of review articles and multi-authored books, including some very recent ones.

1.2 Climate Change is Not Just About Higher Temperatures

For over 20 years, most textbooks documenting the ecophysiology of ectotherms – and insects in particular – have included a section or chapter devoted to the responses of macro-organisms to current environmental change (e.g.Chown and Nicolson, 2004; Angilletta, 2009; Denlinger and Lee, 2010; Storey and Tanino, 2012; Kang and Banga, 2013; Björkman and Niemelä, 2015; Lehmann *et al.*, 2020; Ma *et al.*, 2021; see also Pareek *et al.*, 2017; Yadav *et al.*, 2019; IPPC Secretariat, 2021). In addition, two recent, well-documented reviews focus on the impact of climate change on agricultural insect pests (Skendžić *et al.*, 2021a, b; Fig. 1.1). Still more recently, a number of papers, book chapters and books have been published about the influence of climate change on pest ecology and the consequences for pest management (Masood *et al.*, 2022; Srivinasa Rao *et al.*, 2022; Eickermann *et al.*, 2023; Gvozdenac *et al.*, 2023; Subedi *et al.*, 2023; Ali *et al.*, 2024; Contosta *et al.*, 2024; González-Tokman and Dáttilo, 2024; Liao *et al.*, 2024; Mandal, 2024; Ramesh *et al.*, 2024; Rosace *et al.*, 2024; Şirin and Taylan, 2024; Siva *et al.*, 2024), reflecting growing concerns in this area. Moreover, until recent years, the historical development of integrated pest management (IPM) had mainly been marked by work carried out by teams from the European Community, Switzerland, UK, Canada and USA. It should be pointed out, however, that of the 14 references cited above for the 2022–2024 period, four were from India, two from Pakistan, and one each from Mexico, Serbia and Turkey. This trend is indicative of the consequences of the globalization of trade and travel (Dara, 2019), but also in raising awareness of environmentally friendly solutions to past and present pest control problems.

From all these studies emerges a complex interweaving of climatic and atmospheric factors which together constitute 'climate

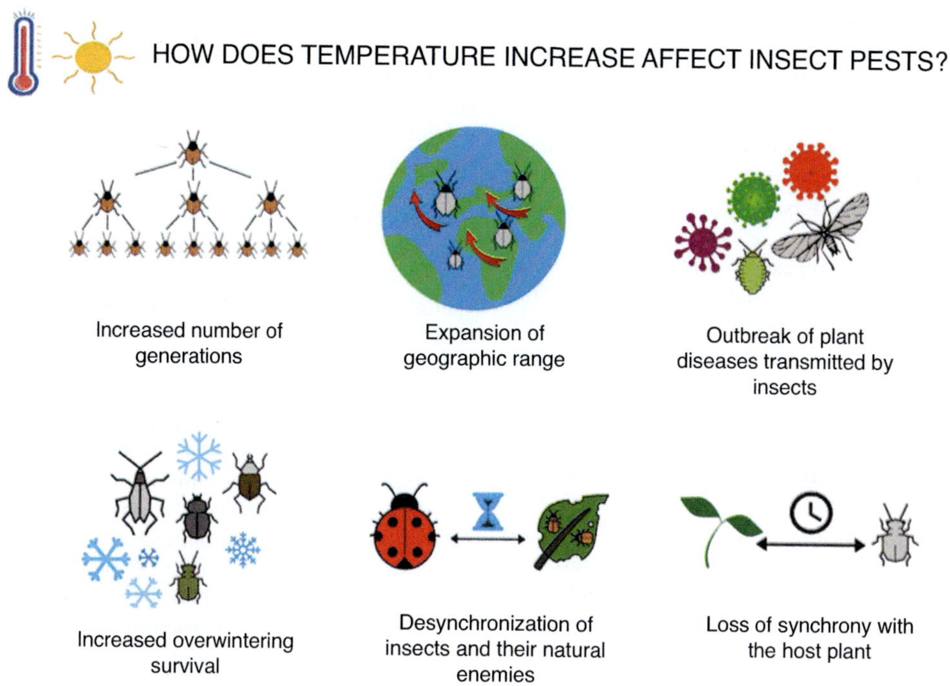

Fig. 1.1. Effects of temperature rise on agricultural insect pests. Reproduced with permission from Skendžić *et al.* (2021a) and Darija Lemić, 2025.

change'. These multiple elements can be grouped into four main components: rising temperatures (e.g. winter warming more pronounced than summers, increased thawing of ice and snow), rising atmospheric carbon dioxide levels (one consequence is the increase in leaf carbon to nitrogen ratio with cascade effects in tritrophic interactions; Dyer *et al.*, 2013), more variable precipitation patterns (changes in water availability and humidity patterns, droughts more pronounced at low latitudes), and an increase in the frequence of extreme climatic events, e.g. increased heavy precipitation periods, increased occurrence of heat waves, with heat stress consequences for host–parasitoid interactions (Wetherington *et al.*, 2017; Malinski *et al.*, 2024). It must be stressed that there is a great deal of geographical variability in the expression of sensitivity to these types of climate change. For example, the Arctic and the Alps are considered as faster warming places (cf. e.g. higher snowmelt rates, lower snow albedo values).

The impact of climate change on the potentialities of macro-organisms used for biological control is both temporal and spatial. On a temporal level, phenological shifts are increasing, affecting, for example, species interactions (Tougeron *et al.*, 2018). These phenological mismatches have been described in detail by Forrest (2016), Buckley (2022) and Ma *et al.* (2024). Spatially, range shifts are now commonly observed, particularly with poleward movements or towards higher altitudes. It is beyond the scope of this chapter to provide exhaustive information in this latter area, as research – at the crossroads of environmental and economic imperatives – is growing rapidly. Information can be found in large reviews and books listed above (e.g. Björkman and Niemelä, 2015; Skendžić *et al.*, 2021a; González-Tokman and Dáttilo, 2024). A recent review (Ramos Aguila *et al.*, 2023) also highlights the particular importance of both phenological shifts and significant population movements to higher altitudes and the consequences of warmer winters. Here, we will focus mainly on the host–parasitoid systems, parasitoids being more and more used in biological control.

1.3 Climate Change Affects the Distribution and Seasonal Ecology of Parasitoids

A parasitoid can be defined as 'an organism which develops on or in another single ('host') organism, extracts nourishment from it, and kills it as a direct or indirect result of that development' (Eggleton and Gaston, 1990). It should be noted that this definition is not 'taxonomically restricted' to the Insect class, but while parasitoids can be nematodes, fungi, protists, bacteria or viruses (Eggleton and Gaston, 1990), the majority of those studied and recorded are insects belonging to the order Hymenoptera (Boivin *et al.*, 2012). Parasitoid wasps are an essential component of food webs in terrestrial ecosystems, with pervasive effects on community structure and diversity. Indeed, these insects have a strong influence on the abundance and dynamics of a range of host populations, exerting top-down control (Godfray and Shimada, 1999; Mills, 2009; Fei *et al.*, 2023). These features make them an effective and well-established tool in biological control programmes. However, the quality of parasitoids may be severely disrupted by climate change, particularly by increasing the amplitude and frequency of temperature variations (Hance *et al.*, 2007; Furlong and Zalucki, 2017). Indeed, each stage of the parasitoid's life cycle is critical and can be influenced by ambient temperature, whether it is host location, host stage selection, egg deposition or larval development through to adult emergence. In addition, the assessment of host quality plays a crucial role in the selection of parasitoids and sex allocation. Offspring development is limited to a finite quantity of resources contained within the host. This can profoundly affect parasitoid traits such as size, maturity, fecundity, sex ratio and longevity (Roitberg *et al.*, 2001; Pekas *et al.*, 2016), with repercussions evident in the context of biological control. This resource limitation can be directly modulated by external factors that influence the interaction between parasitoids and host populations (Fei *et al.*, 2023) but also, as we will discuss later on, by the host microbiota (Ras *et al.*, 2017).

Parasitoid wasps face a heterogeneous environment, including patchy host distribution. As they are usually short-lived, they manage their time and energy efficiently to maximize their fitness (Charnov, 1976; van Alphen *et al.*, 2003). In a complex landscape, a variable range of host species is present, and parasitoid species tend to exhibit two distinct strategies for host exploitation: generalization vs specialization. The adage 'jack of all trades, master of none' suggests that specialist and generalist parasitoid species respond differently to environmental conditions, as host aggregation can potentially mitigate the spatial mismatch caused by warming (Straub *et al.*, 2011; Ramos Aguila *et al.*, 2023). Under increasing and fluctuating conditions, a shift in host distribution towards more suitable areas may be followed by a shift in parasitoid distribution. Recently, Li *et al.* (2023) elegantly tested this hypothesis; they used climate change simulations and knowledge of ecological niches to predict shifts in the distribution of four cereal aphid species (*Sitobion avenae*, *Rhopalosiphum padi*, *Schizaphis graminum* and *Diuraphis noxia*) and their three natural enemies (*Harmonia axyridis*, *Episyrphus balteatus* and *Aphidius gifuensis*) in China. Based on different scenarios, the range of *D. noxia* but also of *E. balteatus* and *A. gifuensis* would be reduced, whereas the range of the other three aphids would be expanded. Furthermore, under the high greenhouse gas emissions scenario for the period 2081–2100, all aphid species except *D. noxia* would move to higher latitudes and altitudes, with an increase in niche overlap between the three aphid species. This spatial expansion to higher altitudes would be less important for *A. gifuensis*. Furthermore, this specialist species shows a reduction in aphid suppressive capacity at higher temperatures, in contrast to the generalist predator *H. axyridis* (Li *et al.*, 2023). This dispersal capacity thus appears to be much greater for a generalist enemy, which can be satisfied with a wider choice of hosts and the benefits of an appropriate microenvironmental temperature.

In addition to the spatial aspect, the temporal synchrony of species during periodic seasonal events provides critical structure and stability for species interactions. Understanding the changing phenology of parasitoid species and their hosts due to climate change is crucial for biological control programmes (Boukal *et al.*, 2019). The close interaction between the parasitoid and its host is based on the optimal overlap of developmental stages, and therefore the range of traits suitable and available for interaction. For example, parasitoids tend to select hosts with specific developmental stages to ensure ideal growth and development of their larvae (Vieira *et al.*, 2019). Changes in temperature, and thus developmental time, alter the host's window of vulnerability to the parasitoid, resulting in a temporal mismatch (Tougeron *et al.*, 2020). Phenological asynchronies between parasitoids and hosts are thus frequently observed, linked in particular to the differential physiological responses of each partner to abiotic conditions, in particular the temperature. A frequently encountered situation is that the host may complete its cycle earlier in the year, before the emergence of adult parasitoids capable of attacking it, with a consequent reduction in the effectiveness of biocontrol. In Canada, this is the case for *Choristoneura fumiferana* (Clemens, 1865) (Lepidoptera: Tortricidae) (Fleming and Candau, 1998; Kocsis and Hufnagel, 2011). However, this pattern is influenced by the variability of parasitoid development times, which may be sufficiently wide to ensure partial or complete synchronization of parasitoid emergence and host presence (see for example Régnière *et al.*, 2020; Tougeron *et al.*, 2020). For *Drosophila sulfurigaster* and *D. birchii* and their parasitoids *Asobara* sp. and *Ganaspis* sp., warming reduces the persistence of the *Drosophila*–parasitoid interaction due to a delay in parasitoid emergence. However, in a community context, the presence of an alternative host may buffer these negative effects (Pardikes *et al.*, 2022). Other studies have highlighted the effect of mild winters on the diapause strategies of parasitoids. Aphid parasitoids with relatively low cold tolerance, such as *Aphidius avenae* and *A. ervi*, must enter diapause in winter, while other species, such as *A. matricariae* or *A. rhopalosiphi*, remain present due to their cold tolerance (Le Lann *et al.*, 2011). With the increasing occurrence of mild winters, diapause is no longer selected for in *A. avenae*, while at the same time an increase in host availability is observed, leading to higher parasitoid activity. Once *A. avenae* is present, it displaces the other species in the winter guild because of the positive effect of higher temperature on this species, while other overwintering species still undergo diapause (Andrade *et al.*, 2016). Within the same season, changes were observed for over 40 years for parasitoids of the genus *Aphidius* with a change in their overwintering strategy that may result in a modification in community structure (Tougeron *et al.*, 2018; Alford *et al.*, 2020). The effect of warming on temporal mismatch in host–parasitoid interactions is often underestimated and poorly supported because it deals with the complexity of community interactions with multiple abiotic and biotic variables. Spatial and temporal mismatch are two responses that determine population dynamics and community structure, and both can destabilize host–parasitoid interactions, reducing top-down control and leading to a relaxation of phytophagous pest pressure. These are crucial aspects to consider for biological control in a context of climate change.

1.4 Climate Change Affects Various Physiological Parameters During the Life Cycle of Natural Enemies

Before they became pests (mainly due to the intensification of agricultural practices), natural enemies were already regulating phytophagous species. Today, the emphasis is on the detailed study of trophic levels, particularly in the context of predator–prey and/or parasitoid–host interactions. While it is true that the host and parasitoid develop more rapidly under higher temperature conditions, the hosts generally have a higher thermal limit than their associated parasitoid species (Furlong and Zalucki, 2017). Extreme temperatures associated with global warming can have a range of physiological impacts that can interfere with the life cycle of parasitoids, with consequences for the effectiveness of biological control programmes. The lifetime reproductive success of parasitoids mainly depends on host availability and the availability of mature eggs, but also on other factors such as sperm quantity and quality. Parasitoid wasps are haplodiploid, meaning that sperm stored by egg-laying females is used only to produce daughters. Thus, the sex ratio of offspring depends on the availability of sperm after mating. Bressac *et al.* (2023) have shown that the male of the parasitoid wasp *Cotesia typhae* produces significantly fewer sperm when exposed to 34°C during the pupal stage, resulting in fewer female offspring. This means that high temperatures could reduce the performance of biological control programmes by interfering with parasitoid reproduction.

Oviposition is a sensitive moment in parasitoid reproduction; its success is dependent on host defences and how the parasitoid can overcome them. Trotta *et al.* (2018) reported that heat shock reduces the success of parasitism by *Aphidius ervi* and therefore increases the chances of survival of the pea aphid *Acyrthosiphon pisum* hosts for each instar. According to the authors, aphid survival can be explained by three hypotheses, one of which is a direct effect and the other two indirect effects on the success of *A. ervi*. (i) The venom and ovarian proteins injected during oviposition lose their efficacy, leading to a reduction in stress for the host (Digilio *et al.*, 2000; Shim *et al.*, 2008); (ii) the immune system, once activated, increases the survival rate of the aphids (Martinez *et al.*, 2016); (iii) there is an adaptation to extreme temperature conditions during the nymphal instars, due to the sedentary moment of this instar compared to the adult stage. The first hypothesis requires further investigation, as it has been shown that under high temperatures, the composition of the venom changes (Cavigliasso *et al.*, 2021), but little is known about the impact of temperature on parasitoid venom functioning. In contrast, the separate application of temperature and drought as abiotic parameters seems to negatively impact the life-history traits of the cabbage aphid *Brevicoryne brassicae*, but not the parasitoid *Diaeretiella rapae*. In line with this, these two parameters induce a decrease in the growth rate of aphids without the presence of parasitoids but decrease the longevity of this one. As for the parasitoids with temperature as a parameter, emergence success is positively impacted despite a low number of offspring for the temperature whereas with drought, the number of offspring is high with no real difference in emergence chances (Romo and Tylianakis, 2013). From the various studies carried out, it is apparent that climate change will directly or indirectly affect host–parasitoid interactions. The response of the two players will depend on the species involved and its biological and ecological functions. It is also important that the application of both abiotic and biotic conditions be adequately considered to provide a clear and concise assessment. This will enable the conditions influencing the interaction between the parasitoid and the host to be reproduced as accurately as possible. Moreover, the combination of adaptive phenotypic plasticity and evolutionary adaptations enables ectotherms to cope with predictably fluctuating temperatures. But when environmental conditions become unpredictable and/or extreme, bet-hedging – i.e. phenotypic diversity in unpredictable environments – can be an indispensable alternative (Le Lann *et al.*, 2021; Costaz *et al.*, 2023; van Baaren *et al.*, 2024).

A brief overview of the impact of warmer temperatures on ectothermic organisms, focusing mainly on insects, allows us to see that these consequences can be variable in their amplitude according to their environment, in particular along latitudinal or elevational gradients, depending on whether the species considered are more or less close to their thermal tolerance limits. A first indication to consider is the rate of change of climatic parameters, and in particular temperature parameters, that vary according to the ecosystems involved. These variables are then compared with the differential capacity of the ectothermic species studied to withstand (or not) these climatic changes. In thermal biology, a dual methodological requirement should also be promoted as far as possible: measurement of heat tolerance and assessing the impact of climate change on life-history traits.

i. To measure heat tolerance, dynamic methods (ramping temperatures giving access to the critical thermal maximum, CTmax) are often to be preferred to static methods which operate at constant temperatures giving access to the survival rate (Herrando-Pérez *et al.*, 2023; see also Ørsted *et al.*, 2022 about the inclusion of both intensity and duration of thermal stress). For both high- and low-temperature tolerance measurements (critical thermal minimum, CTmin), it is important to precisely define the methods used, especially when taking comparative approaches (Terblanche *et al.*, 2011). In addition to this, Gunderson and Stillman (2015) emphasize that ectotherms 'have relatively low acclimation in thermal tolerance' and in insects, Weaving *et al.* (2022) show that CTmax is significantly less plastic than Ctmin and that critical thermal limit plasticity is reduced overall. The latter authors add that, as a result, many insect species will have to rely on changes in their distribution range and behaviour to cope with the effects of climate change.

ii. In order to assess the impact of climate change on life-history traits, it is not possible to simply take into account average values for climatic variables, as is still often the case. Living organisms are subject to fluctuating temperatures, e.g. daily temperature variations of varying intensity, depending on the season and geographical location (Colinet *et al.*, 2015). For example, Paaijmans *et al.* (2013) suggest that fluctuating temperatures may alter sensitivity to climate warming by reducing thermal safety margins. The prevalence of significant thermal heterogeneities is also a plea for the use of microclimatic approaches, in phase with the fine localization of the species studied (Pincebourde *et al.*, 2016). In the same vein, Chevalier *et al.* (2024) warn of the limitations of combining measurements of physiological thermal limits with macroclimatic data. In addition, it is essential to stress that 'the effects of chronic and acute thermal stressors are amplified when they act in combination' (Pilakouta *et al.*, 2024; see also Verheyen and Stoks, 2019).

The consequences of an increase in temperature on the life-history traits of an ectothermic macro-organism are multiple and interrelated. These include acceleration of developmental time and reduction of generation time (e.g. Buckley *et al.*, 2017), the almost general reduction in body size through the phenotypic plastic temperature–size rule (Atkinson, 1994; Ohlberger, 2013), a reduction in life expectancy through accelerated ageing (e.g. Burraco *et al.*, 2020), and increase in metabolic rate (e.g. Dillon *et al.*, 2010). It should be noted that peaks in metabolic performance are observed with this optimal temperature, but also with body mass and latitude (bearing in mind that optimal temperature decreases with latitude) (DeLong *et al.*, 2018). This leads these authors to say that 'hotter is better' at low latitudes and that 'colder is better' at high latitudes. Despite small temperature increases in tropical environments (compared to those observed in temperate environments), Dillon *et al.* (2010) indicate that tropical ectotherms are currently undergoing large increases in their metabolic rates, which may contribute to their vulnerability to climate change (e.g. increased food requirements leading to higher rates of phytophagy or predation).

However, the development of these predictive scenarios should not be focused on the thermobiological approach alone (Kellermann and van Heerwaarden, 2019). Indeed, the complementary integration of traits associated with stress, fitness and phenology is a necessity, so as to better access the plastic and genetic responses of the traits under consideration (see, for example, Kumar *et al.*, 2024). In this context, 'incorporating adaptation into models of climate change resilience is essential for predicting how species will respond under climate change' (Kellermann and van Heerwaarden, 2019). It is therefore essential to simultaneously consider the responses to fluctuating temperatures and abiotic stressors (Harvey *et al.*, 2020). Finally, as convincingly detailed by Kingsolver *et al.* (2011), it is also important to remember that the life cycles of macro-organisms are often complex, with life stages developing in different microhabitats with frequently contrasting abiotic conditions. These life stages can therefore differ in their ability to respond to the microclimates in which they develop, whether in terms of the survival of juvenile stages or the reproductive capacities of adult stages.

1.5 Temperate vs Tropical Species: Is Janzen's Mountain Passes Hypothesis (1967) Still Relevant Today?

The high diversity of species in tropical zones could be *a priori* associated with a greater abundance of parasites, but it should be pointed out that despite the high species richness of tropical lepidopterans and coleopterans, the same is not necessarily true of their parasitoids, as in the case of the underrepresentation of Ichneumonidae in the tropics. An underrepresentation of potential hosts, such as aphids, in tropical zones can also lead to a reduction in the number of parasitoid species (Godfray *et al.*, 1999).

In order to shed light on the diversity of situations observed, it is necessary to understand whether tropical environments could select species differently. In this context, Janzen's mountain passes hypothesis (1967) is an essential tool for combining past and present ideas on ecology, physiology and climate change.

Janzen's hypothesis predicts that tropical habitats with low seasonal variation in temperature would select for species with narrow thermal tolerance breadth (the difference between critical thermal maximum and minimum), which may imply a low acclimation ability and consequently a low dispersal ability along elevational – or, more generally, environmental – gradients. On the other hand, temperate habitats have greater seasonal temperature variations, species with a wider thermal tolerance breadth, a greater acclimation ability and a greater dispersal ability. For over half a century, this comparative approach between tropical and temperate environments has given rise to numerous uses and interpretations, and a number of review papers provide a better understanding of these developments (e.g. Ghalambor *et al.*, 2006; Sheldon *et al.*, 2018).

At first sight, it may seem paradoxical that ectothermic tropical species are particularly sensitive to global warming, even though they are currently subject to slight increases in temperature, but this will especially be the case for tropical species which occupy low altitude, thermally homogeneous environments (variance in spatial temperature broadly similar to variance in seasonal temperature), and which are not very well adapted for even small increases in temperature as they are already close to their optimal temperature (Deutsch *et al.*, 2008; Bonebreak and Deutsch, 2012). It should therefore be emphasized that tropical insects are closer to their lethal temperature than temperate insects, even though global warming is less pronounced in the tropics. This low thermal safety margin can be estimated using a measure of heat tolerance, in the form: [critical thermal maximum – current climatological temperature of the habitat = warming tolerance] (Deutsch *et al.*, 2008). Therefore, and even if mitigation capacities exist (acclimation, behaviour, dispersal), 'the greatest extinction risks from global warming may be in the tropics, where biological diversity is also greatest' (Deutsch *et al.*, 2008). However, this assertion needs to be modulated if we look closely at the spatial variations in temperature that exist within the range of a given species. The existence of these thermal gradients may increase thermal tolerance. This shows that taking into account this spatial dimension (topography) is complementary to the temporal dimension (seasonality) (Bonebreak and Deutsch, 2012).

Among the geographical gradients, latitudinal and altitudinal gradients have been the subject of many studies (much more than longitudinal gradients), and the specific features of each gradient have been described by Jego *et al.* (2023). Janzen's hypothesis combines latitudinal and altitudinal gradients (Janzen, 1967). In this era of global warming, whether we are talking about tropical or temperate environments, it is not enough to deal only with warming tolerance, and it is more informative to consider both upper and lower thermal tolerance (Sunday *et al.*, 2011, 2019). Taking into account the thermal limits of a species as a function of the temperatures to which it is exposed in the field is also an opportunity to establish a link between Janzen's hypothesis and the climate variability hypothesis (CVH) which proposes that the low magnitude of temperatures observed in the tropics is associated with a narrow window of ectotherm thermal tolerance, the opposite being proposed for temperate environments (Stevens, 1989). The CVH was later extended in the approach developed by Pither (2003), which emphasizes the importance of taking into account the most extreme values of the climate variables (climate extremes hypothesis, CEH). A detailed approach to understanding the consequences of extreme weather events has been developed by Kingsolver and Buckley (2017). These extreme climatic events (whose frequency is increasing) now play a major role in the evolution of ectotherms' thermotolerance (Sunday *et al.*, 2019).

When studying thermal tolerance width, i.e. the difference between the critical thermal maximum and minimum, it is important to consider that there is a fundamental dissymmetry between upper and lower temperature data. Indeed, Addo-Bediako *et al.* (2000) have clearly shown that the upper thermal limit varies little with latitude, whereas lower lethal temperatures (and supercooling points, i.e. the lowest temperature to which an ectotherm may be cooled before it freezes) decrease with latitude. This heat-invariant hypothesis (low heat tolerance variability), and its counterpoint the cold variability hypothesis (high cold tolerance variability) have since been the subject of numerous studies, in latitudinal and altitudinal contexts, leading to the same conclusions (e.g. Bozinovic *et al.*, 2014; Bovo *et al.*, 2023).

Climate warming is affecting temperate regions. Interestingly, this is an opportunity to use studies already carried out in tropical regions, where temperatures are naturally higher and where species are closer to their tolerance thresholds (Cock *et al.*, 2013). In this context, it should be noted that the upper temperature limits vary much less than the lower thermal limits, resulting in a limited ability to change thermal limits during seasonal or annual warming phases (Hoffmann *et al.*, 2013). This confirms that the common status of thermal specialists for tropical taxa makes them highly vulnerable to climate change (see also Chan *et al.*, 2016), to the extent that ectotherms often have to resort to subtle behavioural adaptations and search for favourable alternative habitats in order to survive global warming (Sunday *et al.*, 2014). However, this near consensus had been discussed by Kingsolver *et al.* (2013) and convincingly challenged by Johansson *et al.* (2020) who state that 'temperate species have considerably narrower safety margins to warming than suggested by several earlier analyses' due to the frequent failure to consider length and timing of activity periods across latitudes. It must be added that even if seasonal temperature changes are reduced in the tropics, this does not mean that there are no seasonal weather patterns, and the existence of rainy and dry seasons is also subject to global warming (shorter wet seasons, longer dry seasons) (Clusella-Trullas *et al.*, 2011; Guo *et al.*, 2022).

Elevational gradients are often used as a proxy for climate change and in that context, information about species richness and food web structure depending on altitude is useful (Maunsell *et al.*, 2015). These authors show that phytophagous organisms and parasitoids may have different responses depending on altitude, which leads to host–parasitoid food web restructuring. In addition, parasitism rates and the number of parasitoid species decrease with altitude, sometimes in a hump-shaped way for the parasitism rates (Cernecka *et al.*, 2025). This suggests a possible increase in parasitism as temperature rises, although it is not possible to predict with certainty that global warming will lead to a reduction in the density of insect pests. The role of predators and pathogens must of course also be considered (Péré *et al.*, 2013). It has also been shown that the proportion of specialized parasitoids decreases with altitude, which increases the risk of host outbreaks (Di Marco *et al.*, 2023). The same type of approach can be applied to latitudinal gradients (see, for example, Niesenbaum, 2012).

Climate change and the direct impact of human activities have resulted, in particular, in an increase in the number of invasive alien species and in their speed of expansion (McGeoch and Latombe, 2016; Sandvik, 2020). With this in mind, the question of the successful establishment of subtropical or tropical species in temperate zones arises. Are ectothermic species adapted to limited seasonal variations in temperature (these temperatures being high) capable of rapidly colonizing temperate zones where, despite warmer temperatures, the thermal differential remains significant? To our knowledge, the first clue was provided by Nedvěd and Windsor (1994), who showed that a diapausing tropical beetle *Stenotarsus rotundus* has a supercooling ability (an index of resistance to cold; Renault *et al.*, 2002) similar to that of temperate species. It is thus possible to hypothesize that this notable supercooling capacity associated with a diapause syndrome constitutes an 'anticipatory' adaptation for cold resistance in temperate environments. However, recent studies provide contrasting results. For example, Fortuna *et al.* (2023) show that a tropical parasitoid (*Cotesia typhae*) of the corn pest *Sesamia nonagrioides* does not appear to be able to withstand current winter temperatures in temperate regions. In contrast, Leon-Quinto *et al.* (2024) show that the palm borer moth *Paysandisia archon*, native to several South American countries (e.g. Argentina, Paraguay, Uruguay) and accidentally introduced into southern Europe some 30 years ago, presents a resistance to cold considered by the authors as unexpected for a tropical species, even going so far as to be moderately tolerant to freezing. It should be noted that the species has now been reported for less than 10 years in more northern regions of Europe.

1.6 Conclusion

This brief review of the interactions between climate change and macro-organisms involved in biological control shows that the bibliographical references cover a necessarily wide range of complementary fields (behaviour, physiology, ecology, evolution), and that many of the syntheses incorporating an interdisciplinary approach are very recent, and not just from Europe or North America. It is paramount to stress that all life-history traits of a given species are affected by climate change, either positively or negatively. It follows that in order to try to predict the consequences of the rapid intensification of global warming, the only way is to favour field work backed up by comparative experimental approaches. This requirement, which highlights the detailed study of life-history traits in pest–natural enemy models, far from being obsolete, is a convenient and efficient research path provided that geographical specificities are integrated (e.g. temperate zones vs tropical zones, latitudinal and elevational gradients).

References

Abram, N.J., McGregor, H.V., Tierney, J.E., Evans, M.N., McKay, N.P. *et al.* (2016) Early onset of industrial-era warming across the oceans and continents. *Nature* 536, 411–418.

Addo-Bediako, A., Chown, S.L. and Gaston, K.J. (2000) Thermal tolerance, climatic variability and latitude. *Proceedings of the Royal Society B* 267, 739–745.

Alford, L., Louâpre, P., Mougel, F. and van Baaren, J. (2020) Measuring the evolutionary potential of a winter-active parasitic wasp to climate change. *Oecologia* 194, 41–50.

Ali, H., Hou, Y. and Tahir, M.B. (2024) *Climate Change and Insect Biodiversity – Challenges and Implications*. CRC Press, Abingdon, UK.

Amsellem, L., Brouat, C., Duron, O., Porter, S.S., Vilcinskas, A. *et al.* (2017) Importance of microorganisms to macroorganisms invasions: Is the essential invisible to the eye? (The Little Prince, A. de Saint-Exupéry, 1943). *Advances in Ecological Research* 57, 99–146.

Andrade, T.O., Krespi, L., Bonnardot, V., van Baaren, J. and Outreman, Y. (2016) Impact of change in winter strategy of one parasitoid species on the diversity and function of a guild of parasitoids. *Oecologia* 180, 877–888.

Angilletta, M.J., Jr (2009) *Thermal Adaptation: A Theoretical and Empirical Synthesis*. Oxford University Press, Oxford, UK.

Atkinson, D. (1994) Temperature and organism size – a biological law for ectotherms? *Advances in Ecological Research* 25, 1–58.

Björkman, C. and Niemelä, P. (2015) *Climate Change and Insect Pests CABI Climate Change Series 7*. CAB International, Wallingford, UK.

Boivin, G., Hance, T. and Brodeur, J. (2012) Aphid parasitoids in biological control. *Canadian Journal of Plant Science* 92, 1–12.

Bonebreak, T.C. and Deutsch, C.A. (2012) Climate heterogeneity modulates impacts of warming on tropical insects. *Ecology* 93, 449–455.

Boukal, D.S., Bideault, A., Carreira, B.M. and Sentis, A. (2019) Species interactions under climate change: Connecting kinetic effects of temperature on individuals to community dynamics. *Current Opinion in Insect Science* 35, 88–95.

Bovo, R.P., Simon, M.N., Provete, D.B., Lyra, M., Navas, C.A. *et al.* (2023) Beyond Janzen's hypothesis: How amphibians that climb tropical mountains respond to climate variation. *Integrative Organismal Biology* 5, bad009.

Bozinovic, F., Orellana, M.J., Martel, S.I. and Bogdanovich, J.M. (2014) Testing the heat-invariant and cold-variability tolerance hypotheses across geographic gradients. *Comparative Biochemistry and Physiology, Part A* 178, 46–50.

Bressac, C., El Sabrout, A., Kifouche, F., Anne, M., Capdevielle-Dulac, C. *et al*. (2023) Hot and cold waves decrease sperm production and bias sex ratio in the parasitoid wasp *Cotesia typhae* (Hymenoptera, Braconidae). *Journal of Insect Physiology* 149, 104553.

Buckley, L.B. (2022) Temperature-sensitive development shapes insect phenological responses to climate change. *Current Opinion in Insect Science* 52, 100897.

Buckley, L.B., Arakaki, A.J., Cannistra, A.F., Kharouba, H.M. and Kingsolver, J.G. (2017) Insect development, thermal plasticity and fitness implications in changing, seasonal environments. *Integrative and Comparative Biology* 57(5), 988–998.

Burraco, P., Orizaola, G., Monaghan, P. and Metcalfe, N.B. (2020) Climate change and ageing in ectotherms. *Global Change Biology* 26, 5371–5381.

Cavigliasso, F., Gatti, J.-L., Colinet, D. and Poirié, M. (2021) Impact of temperature on the immune interaction between a parasitoid wasp and Drosophila host species. *Insects* 12, 647.

Cernecka, L., Michalko, R., Sykora, J., Gajdos, P., Purgat, P. *et al*. (2025) A holistic approach to parasitoid–host interaction along an elevational gradient revealed coevolution driven by host foraging strategy. *Ecology and Evolution* 15, e71227.

Chan, W.-P., Chen, I.-C., Colwell, R.K., Liu, W.-C., Huang, C.-Y. *et al*. (2016) Seasonal and daily climate variation have opposite effects on species elevational range size. *Science* 1437–1439.

Charnov, E.L. (1976) Optimal foraging, the marginal value theorem. *Theoretical Population Biology* 9, 129–136.

Chevalier, M., Pignard, V., Broennimann, O. and Guisan, A. (2024) A cautionary message on combining physiological thermal limits with macroclimatic data to predict species distribution. *Ecosphere* 15, e4931.

Chown, S.L. and Nicolson, S.W. (2004) *Insect Physiological Ecology: Mechanisms and Patterns*. Oxford University Press, Oxford, UK.

Clusella-Trullas, S., Blackburn, T.M. and Chown, S.L. (2011) Climatic predictors of temperature performance curve parameters in ectotherms imply complex responses to climate change. *American Naturalist* 177, 738–751.

Cock, M.J.W., Biesmeijer, J.C., Cannon, R.J.C., Gerard, P.J., Gillespie, D. *et al*. (2013) The implications of climate change for positive contributions of invertebrates to world agriculture. *CAB Reviews* 8, 1–48.

Colinet, H., Sinclair, B.J., Vernon, P. and Renault, D. (2015) Insects in fluctuating thermal environments. *Annual Review of Entomology* 60, 123–140.

Contosta, A.R., Arndt, K.A., Baulch, H.M., Casson, N.J., Harpold, A. *et al*. (2024) Threshold changes in temperature and precipitation drive threshold responses across nine global climate zones and associated biomes. *Annual Review of Ecology, Evolution, and Systematics* 55, 271–300.

Costaz, T.P.M., de Jong, P.W., van Loon, J.J.A. and Dicke, M. (2023) It is about time: Effects of thermal fluctuations resulting from climate change on trophic interactions with a focus on host-parasitoid interactions. *Evolutionary Ecology* 37, 601–625.

Dara, S.K. (2019) The new integrated pest management paradigm for the modern age. *Journal of Integrated Pest Management* 10(1), 12.

DeLong, J.P., Bachman, G., Gibert, J.P., Luhring, T.M., Montooth, K.L. *et al*. (2018) Habitat, latitude and body mass influence the temperature dependence of metabolic rate. *Biology Letters* 14, 20180442.

Denlinger, D.L. and Lee, R.E., Jr (2010) *Low Temperature Biology in Insects*. Cambridge University Press, Cambridge, UK.

Deutsch, C.A., Tewskbury, J.J., Huey, R.B., Sheldon, K.S., Ghalambor, C.K. *et al*. (2008) Impacts of climate warming on terrestrial ectotherms across latitude. *Proceedings of the National Academy of Sciences* 105, 6668–6672.

Di Marco, M., Santini, L., Corcos, D., Tshorsnig, H.-P. and Cerretti, P. (2023) Elevational homogenization of mountain *parasitoids* across six decades. In: *Proceedings of the National Academy of Sciences 120*, p. e2308273120.

Dyer, L.A., Richards, L.A., Short, S.A. and Dodson, C.D. (2013) Effects of CO2 and temperature on tritrophic interactions. *PLoS ONE* 8(4), e62528.

Digilio, M.C., Isidoro, N., Tremblay, E. and Pennacchio, F. (2000) Host castration by *Aphidius ervi* venom proteins. *Journal of Insect Physiology* 46, 1041–1050.

Dillon, M.E., Wang, G. and Huey, R.B. (2010) Global metabolic impacts of recent climate warming. *Nature* 467, 704–707.

Eggleton, P. and Gaston, K.J. (1990) 'Parasitoid' species and assemblages: Convenient definitions or misleading compromises. *Oikos* 59, 417–421.

Eickermann, M.E., Junk, J. and Rapisarda, C. (2023) Climate change in insects. *Insects* 14, 678.

Eilenberg, J., Hajek, A. and Lomer, C. (2001) Suggestions for unifying the terminology in biological control. *BioControl* 46, 387–400.

Fei, M., Gols, R. and Harvey, J.A. (2023) The biology and ecology of parasitoid wasps of predatory arthropods. *Annual Review of Entomology* 68, 109–128.

Fleming, R.A. and Candau, J.-N. (1998) Influences of climatic change on some ecological processes of an insect outbreak system in Canada's boreal forest and the implications for biodiversity. *Environmental Monitoring and Assessment* 49, 235–249.

Forrest, J.R.K. (2016) Complex responses of insect phenology to climate change. *Current Opinion in Insect Science* 17, 49–54.

Fortuna, T.M., Anne, M., Le Gonnidec, M., Jeannette, R., Bressac, C. *et al.* (2023) Susceptibility to cold suggests low risk of establishment of a tropical parasitoid attacking the corn pest *Sesamia nonagrioides*. *Biological Control* 186, 105359.

Furlong, M.J. and Zalucki, M.P. (2017) Climate change and biological control: The consequences of increasing temperatures on host-parasitoid interactions. *Current Opinion in Insect Science* 20, 39–44.

Ghalambor, C.K., Huey, R.B., Martin, P.R., Tewksbury, J.J. and Wang, G. (2006) Are mountain passes higher in the tropics? Janzen's hypothesis revisited. *Integrative and Comparative Biology* 46, 5–17.

Godfray, H.C.J. and Shimada, M. (1999) Parasitoids as model organisms for ecologists. *Population Ecology* 41, 3–10.

Godfray, H.C.J., Lewis, O.T. and Memmott, J. (1999) Studying insect diversity in the tropics. *Philosophical Transactions of the Royal Society B* 354, 1811–1824.

González-Tokman, D. and Dáttilo, W. (2024) *Effects of Climate Change on Insects – Physiological, Evolutionary, and Ecological Responses*. Oxford University Press, Oxford, UK.

Gunderson, A.R. and Stillman, J.H. (2015) Plasticity in thermal tolerance has limited potential to buffer ectotherms from global warming. *Proceedings of the Royal Society B* 282, 20150401.

Guo, J., Hu, S. and Guan, Y. (2022) Regime shifts of the wet and dry seasons in the tropics under global warming. *Environmental Research Letters* 17, 104028.

Gvozdenac, S., Dedić, B., Mikić, S., Ovuka, J. and Miladinović, D. (2023) Impact of climate change on integrated pest management strategies. In: Benkeblia, N. (ed.) *Climate Change and Agriculture – Perspectives, Sustainability and Resilience*. John Wiley & Sons Ltd, Chichester, UK, pp. 311–372.

Hance, T., van Baaren, J., Vernon, P. and Boivin, G. (2007) Impact of extreme temperatures on parasitoids in a climate change perspective. *Annual Review of Entomology* 52, 107–126.

Harvey, J.A., Heinen, R., Gols, R. and Thakur, M.P. (2020) Climate change-mediated temperature extremes and insects: From outbreaks to breakdowns. *Global Change Biology* 26, 6685–6701.

Herrando-Pérez, S., Vieites, D.R. and Araújo, M.B. (2023) Novel physiological data needed for progress in global change ecology. *Basic and Applied Ecology* 67, 32–47.

Hoffmann, A.A., Chown, S.L. and Clusella-Trullas, S. (2013) Upper thermal limits in terrestrial ectotherms: How constrained are they. *Functional Ecology* 27, 934–949.

Ingram, J. and Hamilton, C. (2014) *Planning for Climate Change*. United Nations Human Settlements Programme (UN-Habitat), Nairobi, Kenya.

IPPC Secretariat (2021) *Scientific Review of the Impact of Climate Change on Plant Pests – A Global Challenge to Prevent and Mitigate Plant Pest Risks in Agriculture, Forestry and Ecosystems*. FAO on Behalf of the IPPC Secretariat, Rome, Italy.

Janzen, D.H. (1967) Why mountain passes are higher in the tropics. *American Naturalist* 101(919), 233–249.

Jego, L., Li, R., Roudine, S., Ma, C.-S., Le Lann, C. *et al.* (2023) Parasitoid ecology along geographic gradients: Lessons from climate change studies. *Current Opinion in Insect Science* 57, 101036.

Johansson, F., Orizaola, G. and Nilsson-Örtman, V. (2020) Temperate insects with narrow seasonal activity periods can be as vulnerable to climate change as tropical insect species. *Scientific Reports* 10, 8822.

Kang, M.J. and Banga, S.S. (2013) *Combating Climate Change: An Agricultural Perspective*. CRC Press/Routledge, Abingdon, UK.

Kingsolver, J.G. and Buckley, L.B. (2017) Quantifying thermal extremes and biological variation to predict evolutionary responses to changing climate. *Philosophical Transactions of the Royal Society* 372, 20160147.

Kocsis, M. and Hufnagel, L. (2011) Impacts of climate change on Lepidoptera species and communities. *Applied Ecology and Environmental Research* 9, 43–72.
Kellermann, V. and Heerwaarden, B. (2019) Terrestrial insects and climate change: Adaptive responses in key traits. *Physiological Entomology* 44(2), 99–115.
Kingsolver, J.G., Woods, H.A., Buckley, L.B., Potter, K.A., MacLean, H.J. *et al.* (2011) Complex life cycles and the responses of insects to climate change. *Integrative and Comparative Biology* 51(5), 719–732.
Kingsolver, J.G., Diamond, S.E. and Buckley, L.B. (2013) Heat stress and the fitness consequences of climate change for terrestrial ectotherms. *Functional Ecology* 27, 1415–1423.
Kumar, H., Doddachowdappa, S., Verma, S., Rajarushi, C.N., Rupali, J.S. *et al.* (2024) Heat stress and thermoregulation in insects: Unveiling molecular and physiological responses in light of climate change. *Journal of Advances in Biology & Biotechnology* 27(9), 820–834.
Le Lann, C., Wardziak, T., van Baaren, J. and van Alphen, J.J.M. (2011) Thermal plasticity of metabolic rates linked to life-history traits and foraging behaviour in a parasitic wasp. *Functional Ecology* 25, 641–651.
Le Lann, C., van Baaren, J. and Visser, B. (2021) Dealing with predictable and unpredictable temperatures in a climate change context: The case of parasitoids and their hosts. *Journal of Experimental Biology* 224, jeb238626.
Lehmann, P., Ammunét, T., Barton, M., Battisti, A., Eigenbrode, S.D. *et al.* (2020) Complex responses of global insect pests to climate warming. *Frontiers in Ecology and the Environment* 18(3), 141–150.
Leon-Quinto, T., Anton-Ruiz, N., Madrigal, R. and Serna, A. (2024) Experimental evidence of a Neotropical pest moderately tolerant to complete freezing. *Journal of Thermal Biology* 123, 103939.
Liao, J.-R., Ho, C.-C. and Ko, C.-C. (2024) Challenges and future directions of predatory mites as biological control agents under climate change. *Journal of Taiwan Agriculture Research* 73(3), 153–168.
Li, D., Li, Z., Liu, Z., Yang, Y., Khos, A.G. *et al.* (2023) Climate change simulations revealed potentially drastic shifts in insect community structure and crop yields in China's farmland. *Journal of Pest Science* 96, 55–69.
Lovejoy, T.E. and Hannah, L. (2005) *Climate Change and Biodiversity*. Yale University Press, New Haven, Connecticut.
Ma, C.-S., Ma, G. and Pincebourde, S. (2021) Survive a warming climate: Insect responses to extreme high temperatures. *Annual Review of Entomology* 66, 163–184.
Ma, G., Ma, C.-S., Le Lann, C. and van Baaren, J. (2024) Effects of climate change on insect phenology. In: González-Tokman, D. and Dáttilo, W. (eds) *Effects of Climate Change on Insects – Physiological, Evolutionary, and Ecological Responses*. Oxford University Press, Oxford, UK, pp. 89–109.
Malinski, K.H., Moore, M.E. and Kingsolver, J.G. (2024) Heat stress and host–parasitoid interactions: Lessons and opportunities in a changing climate. *Current Opinion in Insect Science* 64, 101225.
Mandal, I. (2024) A brief overview of the relationship between global warming and insect outbreaks. *International Journal of Biological Innovations* 6(2), 91–98.
Martinez, A.J., Kim, K.L., Harmon, J.P. and Oliver, K.M. (2016) Specificity of multi-modal aphid defenses against two rival parasitoids. *PLoS ONE* 11(5), e0154670.
Masood, N., Akram, R., Fatima, M., Mubeen, M., Hussain, S. *et al.* (2022) Insect pest management under climate change. In: Nasim Jatoi, W., Mubeen, M., Ahmad, A., Cheema, M.A. and Lin, Z. (eds) *Building Climate Resilience in Agriculture – Theory, Practice and Future Perspective*. Springer Nature, Cham, Switzerland, pp. 225–237.
Maunsell, S.C., Kitching, R.L., Burwell, C.J. and Morris, R.J. (2015) Changes in host–parasitoid food web structure with elevation. *Journal of Animal Ecology* 84, 353–363.
McGeoch, M. and Latombe, G. (2016) Characterizing common and range expanding species. *Journal of Biogeography* 43, 217–228.
Mills, N. (2009) Parasitoids. In: Resh, V.H. and Cardé, R.T. (eds) *Encyclopedia of Insects*, 2nd edn. Academic Press, San Diego, California, pp. 748–751.
Nedvěd, O. and Windsor, D. (1994) Supercooling ability, fat and water contents in a diapausing tropical beetle, *Stenotarsus rotundus* (Coleoptera: Endomychidae). *European Journal of Entomology* 91, 307–312.
Niesenbaum, R.A. (2012) Can latitudinal studies predict potential effects of global climate change on insect herbivore–host plant interactions. *Trends in Entomology* 8, 1–8.
Ohlberger, J. (2013) Climate warming and ectotherm body size – from individual physiology to community ecology. *Functional Ecology* 27, 991–1001.

Ørsted, M., Jørgensen, L.B. and Overgaard, J. (2022) Finding the right thermal limit: A framework to reconcile ecological, physiological and methodological aspects of CTmax in ectotherms. *Journal of Experimental Biology* 225, jeb244514.

Paaijmans, K.P., Heinig, R.L., Seliga, R.A., Blanford, J.I., Blanford, S. *et al.* (2013) Temperature variation makes ectotherms more sensitive to climate change. *Global Change Biology* 19, 2373–2380.

Pardikes, N.A., Revilla, T.A., Lue, C.-H., Thierry, M., Souto-Vilarós, D. *et al.* (2022) Effects of phenological mismatch under warming are modified by community context. *Global Change Biology* 28, 4013–4026.

Pareek, A., Meena, B.M., Sharma, S., Tetarwal, M.L., Kalyan, R.K. *et al.* (2017) Impact of climate change on insect pests and their management strategies. In: Kumar, S., Kanwat, M., Meena, P.D., Kumar, V. and Alone, R.A. (eds) *Climate Change and Sustainable Agriculture*. New India Publishing Agency, New Delhi, India, pp. 253–286.

Pekas, A., Tena, A., Harvey, J.A., Garcia-Marí, F. and Frago, E. (2016) Host size and spatiotemporal patterns mediate the coexistence of specialist parasitoids. *Ecology* 97, 1345–1356.

Péré, C., Jactel, H. and Kenis, M. (2013) Response of insect parasitism to elevation depends on host and parasitoid life-history strategies. *Biology Letters* 9, 20130028.

Pilakouta, N., Allen, D., Moore, E. and Russell, A.A. (2024) Chronic and acute thermal stressors have non-addictive effects on fertility. *Proceedings of the Royal Society B* 291, rspb20241086.

Pincebourde, S., Murdock, C.C., Vickers, M. and Sears, M.W. (2016) Fine-scale microclimatic variation can shape the responses of organisms to global change in both natural and urban environments. *Integrative and Comparative Biology* 56(1), 45–61.

Pither, J. (2003) Climate tolerance and interspecific variation in geographic range size. *Proceedings of the Royal Society B* 270, 475–481.

Ramesh, K.B., Anil, A., Srinivasa, K., Deeksha, M.G., Kallugudi, J. *et al.* (2024) Emerging insect pests under climate change and their management. In: *Research and Technological Advances for Resilient Agriculture*. ICAR-National Academy of Agricultural Research Management, Telangana, India, pp. 59–83.

Ramos Aguila, L.C., Li, X., Akutse, K.S., Bamisile, B.S., Sánchez Moreano, J.P. *et al.* (2023) Host–parasitoid phenology, distribution, and biological control under climate change. *Life* 13, 2290.

Ras, E., Beukeboom, L.W., Cáceres, C. and Bourtzis, K. (2017) Review of the role of gut microbiota in mass rearing of the olive fruit fly, *Actrocera oleae*, and its parasitoids. *Entomologia Experimentalis et Applicata* 164, 237–256.

Régnière, J., Seehausen, L. and Martel, V. (2020) Modeling climatic influences on three parasitoids of low-density spruce budworm populations. Part 1: *Tranosema rostrale* (Hymenoptera: Ichneumonidae). *Forests* 11, 846.

Renault, D., Salin, C., Vannier, G. and Vernon, P. (2002) Survival at low temperatures in insects: What is the ecological significance of the supercooling point. *CryoLetters* 23, 217–228.

Roitberg, B.D., Boivin, G. and Vet, L.E.M. (2001) Fitness, parasitoids, and biological control: An opinion. *Canadian Entomologist* 133, 429–438.

Romo, C.M. and Tylianakis, J.M. (2013) Elevated temperature and drought interact to reduce parasitoid effectiveness in suppressing hosts. *PLoS ONE* 8, e58136.

Rosace, M.C., Björklund, N., Boberg, J., Bradshaw, C.D., Camac, J. *et al.* (2024) Including climate change in pest risk assessment: Current practices and perspectives for future implementation. *European and Mediterranean Plant Protection Organization Bulletin* 54, 52–72.

Sandvik, H. (2020) Expansion speed as a generic measure of spread for alien species. *Acta Biotheoretica* 68, 227–252.

Sheldon, K.S., Huey, R.B., Kaspari, M. and Sanders, N.J. (2018) Fifty years of mountain passes: A perspective on Dan Janzen's classic article. *American Naturalist* 191, 553–565.

Shim, J.-K., Ha, D.-M., Nho, S.-K., Song, K.S. and Lee, K.Y. (2008) Upregulation of heat shock protein genes by envenomation of ectoparasitoid *Bracon hebetor* in larval host of Indian meal moth *Plodia interpunctella*. *Journal of Invertebrate Pathology* 97, 306–309.

Şirin, D. and Taylan, M.S. (2024) The effects of global warming on insects: Current research and approaches. In: Doğruöz Gungör, N. and Arslan Aydöğdu, E.O. (eds) *Ecological Dynamics in the Face of Climate Change – Exploring Organisms, Habitats, and Behavioral Adaptations*. Nobel Tip Kitabevleri Ltd, Istanbul, Turkey, pp. 189–203.

Siva, M., Saravanakumari, K., Nivedha, M., Saranya, P.S., Srimahesvari, D.S. *et al.* (2024) Biological control: A potential tool for pest management under changing climatic conditions. In: Kumar, A., Sah,

R.P. and Gowda, B. (eds) *Enhancing Crop Resilience: Advances in Climate Smart Crop Production Technologies*. Biotica, Tripura, India, pp. 194–216.

Skendžić, S., Zovko, M., Živković, I.P., Lešić, V. and Lemić, D. (2021a) The impact of climate change on agricultural insect pests. *Insects* 12, 440.

Skendžić, S., Zovko, M., Živković, I.P., Lešić, V. and Lemić, D. (2021b) Effect of climate change on introduced and native agricultural invasive insect pests in Europe. *Insects* 12, 985.

Srivinasa Rao, M., Prasad, T.V., Balasubramani, N. and Singh, V.K. (2022) *Adaptation Strategies for Pest Management in Climate Change Scenarios*. ICAR-Central Research Institute for Dryland Agriculture and National Institute of Agricultural Extension Management (MANAGE), Hyderabad, India.

Stevens, G.C. (1989) The latitudinal gradient in geographical range: How so many species coexist in the tropics. *American Naturalist* 133, 240–256.

Storey, K.B. and Tanino, K. (2012) *Temperature Adaptation in a Changing Climate: Nature at Risk*, CABI Climate Change Series 3. CAB International, Wallingford, UK.

Straub, C.S., Ives, A.R. and Gratton, C. (2011) Evidence for a trade-off between host-range breadth and host-use efficiency in aphid parasitoids. *American Naturalist* 177, 389–395.

Subedi, B., Poudel, A. and Aryal, S. (2023) The impact of climate change on insect pest biology and ecology: Implications for pest management strategies, crop production, and food security. *Journal of Agriculture and Food Research* 14, 100733.

Sunday, J., Bates, A.E., Kearney, M.R., Colwell, R.K., Dulvy, N.K. *et al.* (2014) Thermal-safety margins and the necessity of thermoregulatory behavior across latitude and elevation. *Proceedings of the National Academy of Sciences* 111, 5610–5615.

Sunday, J., Bennett, J.M., Calosi, P., Clusella-Trullas, S., Gravel, S. *et al.* (2019) Thermal tolerance patterns across latitude and elevation. *Philosophical Transactions of the Royal Society B* 374, 20190036.

Sunday, J.M., Bates, A.E. and Dulvy, N.K. (2011) Global analysis of thermal tolerance and latitude in ectotherms. *Proceedings of the Royal Society B* 278, 1823–1830.

Terblanche, J.S., Hoffmann, A.A., Mitchell, K.A., Rako, L., le Roux, P.C. *et al.* (2011) Ecologically relevant measures of tolerance to potentially lethal temperatures. *Journal of Experimental Biology* 214, 3713–3725.

Tesson, S.V.M., Okamura, B., Dudaniec, R.Y., Vyverman, W., Löndahl, J. *et al.* (2015) Integrating microorganism and macroorganism dispersal: Modes, techniques and challenges with particular focus on co-dispersal. *Ecoscience* 22(2–4), 109–124.

Tougeron, K., Damien, M., Le Lann, C., Brodeur, J. and van Baaren, J. (2018) Rapid responses of winter aphid-parasitoid communities to climate warming. *Frontiers in Ecology and Evolution* 6, 173.

Tougeron, K., Brodeur, J., Le Lann, C. and van Baaren, J. (2020) How climate change affects the seasonal ecology of insect parasitoids. *Ecological Entomology* 45, 167–181.

Trotta, V., Forlano, P., Falabella, P., Battaglia, D. and Fanti, P. (2018) The aphid *Acyrthosiphon pisum* exhibits a greater survival after a heat shock when parasitized by the wasp *Aphidius ervi*. *Journal of Thermal Biology* 72, 53–58.

Uvarov, B.P. (1931) Insects and climate. *Transactions of the Entomological Society of London* 79, 1–232.

van Alphen, J.J.M., Bernstein, C. and Driessen, G. (2003) Information acquisition and time allocation in insect parasitoids. *Trends in Ecology and Evolution* 18, 81–87.

van Baaren, J., Boivin, G., Visser, B. and Le Lann, C. (2024) Bet-hedging in parasitoids: When optimization is not the best strategy to cope with climatic extremes. *Current Research in Insect Science* 5, 100076.

Verheyen, J. and Stoks, R. (2019) Temperature variation makes an ectotherm more sensitive to global warming unless thermal evolution occurs. *Journal of Animal Ecology* 88, 624–636.

Vieira, L.J.P., Franco, G.M. and Sampaio, M.V. (2019) Host preference and fitness of *Lysiphlebus testaceipes* (Hymenoptera: Braconidae) in different instars of the aphid *Schizaphis graminum*. *Neotropical Entomology* 48, 391–398.

Weaving, H., Terblanche, J.S., Pottier, P. and English, S. (2022) Meta-analysis reveals weak but pervasive plasticity in insect thermal limits. *Nature Communications* 13, 5292.

Wetherington, M.T., Jennings, D.E., Shrewsbury, P.M. and Duan, J.J. (2017) Climate variation alters the synchrony of host–parasitoid interactions. *Ecology and Evolution* 7, 8578–8587.

Yadav, T., Baloda, A.J., Jakhar, B. and Yadav, A. (2019) Impact of climate changes on insect pest and integrated pest management: A review. *Journal of Entomology and Zoology Studies* 7(4), 1150–1156.

2 Change in Diapause Strategy in Insects and Impacts on Pest Populations

Kévin Tougeron[1]* and Jacques Brodeur[2]
[1]Écologie des Interactions et Changements Globaux, Institut de Recherche en Biosciences, Université de Mons, Belgium; [2]Institut de Recherche en Biologie Végétale, Département de Sciences Biologiques, Université de Montréal, QC, Canada

Abstract
Diapause is an ecological and physiological mechanism that enables insects to synchronize their periods of activity with a favourable environment. Climate change has a clear impact on this key event in the phenology of both insect pests and their natural enemies. Changes in diapause strategies have been observed in the face of rising average temperatures and less severe winters. Diapause can start later and end earlier than before, either in direct response to favourable environmental conditions (involving phenotypic plasticity) or via longer-term changes in species-specific critical daylength (involving genetic evolution). The consequences are numerous, ranging from individual to population and community scales. Although predictions are complex, in the agricultural environment, the risk posed is desynchronization between the life cycles of plants, pests and their natural enemies (predators or parasitoids), leading to mismatches between trophic levels and potential disruption of the biological control service.

2.1 Introduction

As temperature imposes significant constraints on physiological, behavioural and ecological traits of ectotherms (Deutsch *et al.*, 2008; Huey *et al.*, 2012; Abram *et al.*, 2017), climate change has already started to impact the occurrence of insects, in both space and time (see Schleuning *et al.*, 2020; Harvey *et al.*, 2023 and references therein). Warmer temperatures have resulted in arthropods expanding their distribution poleward and prolonging their period of activity (Parmesan, 2007). Evidence of phenotypic plasticity and evolutionary adaptation to current global change has emerged for insects and should lead to different strategies to cope with not only the gradual warming but also the increase in the frequency and duration of sporadic, extreme climatic events (e.g. heat waves, unseasonal cold spells, droughts, floods, fires).

The timing of seasonal biological events (phenology) is crucial to the survival and general fitness of all organisms (Forrest and Miller-Rushing, 2010) and climate change is restructuring several elements of the life cycle and year-round ecology of insects. For instance, in some species from temperate regions, warmer temperatures have been shown to speed development and reproduction, provide opportunities for more generations per year (voltinism), advance spring emergence and delay autumn migration or diapause induction (Altermatt,

*Corresponding author: kevin.tougeron@umons.ac.be

DOI: 10.1079/9781800625099.0002

2010; Harvey *et al.*, 2023). These phenological shifts may also modify the nature, frequency and strength of guild and trophic interactions, leading to complex responses at the community level (Davis *et al.*, 1998; Petchey *et al.*, 1999; Suttle *et al.*, 2007; Post *et al.*, 2008). For one thing, phenological changes may modify the synchrony of closely interacting species (e.g. plant vs pollinator; plant vs herbivore; prey or host vs predator or parasitoid). Asynchrony can arise when the type, magnitude and pace of phenological responses to climate change differ between interacting species (Singer and Parmesan, 2010; Gallinat *et al.*, 2015). This issue has particularly become a matter of interest in biological control since temporal asynchrony between pests and their natural enemies may have negative economic impacts (Cock *et al.*, 2013). Understanding the effects of climatic variability on population and community dynamics therefore requires knowledge of how species interact and how these interactions are modulated by temperature variations (Stenseth *et al.*, 2002; Tylianakis *et al.*, 2008; van de Putten *et al.*, 2010; Sentis *et al.*, 2015).

In this chapter we focus on insect diapause in the context of climate change and the potential consequences for biological control of insect pests. We first briefly introduce the role and mechanisms of diapause, then we explore the capacity of insects to develop adapted diapause strategies in response to climate warming, and finally describe how potential diapause-induced asynchrony between pests and natural enemies may impact biological control.

2.2 Diapause

Diapause is a genetically preprogrammed physiological state of dormancy at the heart of the overwintering strategy of most temperate insects (Tauber *et al.*, 1986; Danks, 1987; Koštál *et al.*, 2017; Denlinger, 2022). It is a major adaptation evolved by insects to synchronize their seasonal activity with biotic resources and climatic constraints (Tauber *et al.*, 1986). Diapause is termed 'obligatory', when expressed at a particular time in each generation irrespective of environmental conditions, or 'facultative', when occurring only in the generation that precedes adverse conditions. Diapause can be divided into three main phases: prediapause or induction (insects respond to cues triggering diapause pathways), diapause (period of developmental arrest and insensitivity to environmental cues), and post-diapause (development resumes under favourable conditions) (Tauber and Tauber, 1976; Koštál, 2006; Denlinger *et al.*, 2012) (Fig. 2.1).

The induction phase usually takes place at a given life stage of the insect – egg, larva, pupa or adult – in anticipation of seasonal changes and harsh conditions. Stimuli consist of changes in photoperiod, temperature, moisture, pH, plant secondary compounds and resources such as oxygen, water, food and host. Several cues can be involved in diapause induction, and interactions among stimuli may occur (Tauber *et al.*, 1986). Furthermore, in some cases, responses to diapause inducing cues can be transgenerational, i.e. perceived by the females and extended to their progeny (Saunders, 1965; Rockey *et al.*, 1989; Denlinger, 2022 and references therein). This complexity of response is well illustrated in *Aphidius nigripes* (Hymenoptera: Braconidae), a parasitoid of the potato aphid *Macrosiphum euphorbiae* (Homoptera: Aphididae), which exhibits a facultative prepupal diapause (Brodeur and McNeil, 1989). Abiotic and biotic stimuli are perceived by the female and the egg and first-instar larva. Low temperatures and short daylengths are significant stimuli for both females and immatures, as well as the interactions between these factors. Furthermore, maternal age and host size can also modulate the diapause induction process. The involvement of two generations and the multiplicity of interacting cues are adaptations of a multivoltine species with short developmental time to adequately assess seasonal changes in the environment and programme diapause.

For many temperate insects, a gradual decrease in photoperiod, the most consistent indicator of seasonal change, is the major external signal that initiates molecular, physiological and endocrinological processes associated with diapause induction. Temperature, a less predictable stimulus on a short-term basis, can modify the diapause-regulating role of day length, with high temperatures reinforcing long-day effects and vice versa. Temperature is usually more critical for diapause maintenance and termination, together with other environmental factors

Fig. 2.1. Photograph of a diapausing prepupa of *Aphidius avenaphis* taken out of its mummy (sample from Saskatchewan, Canada, 2015). Credits: Kévin Tougeron (CC BY-NC-SA). Permission received for commercial use.

(e.g. food, host, moisture). Furthermore, a chilling period is often required before an insect can come out of diapause, or a minimum threshold temperature must be attained and sustained for some days to terminate diapause (Lees, 1955; Tauber *et al.*, 1986; Danks, 1987; Koštál, 2006).

For any given insect species, at any given site, 'there is an optimal time to enter diapause that is an adaptive compromise between avoiding winter cold and conserving resources for overwintering and for development and reproduction the following spring' (Bradshaw and Holzapfel, 2008). This optimal timing, which likewise applies for diapause termination and spring emergence, is affected by climate warming and forces insects to respond differently to climatic cues involved in diapause induction and termination. Climate changes will also impact overwintering survival of insects due to decreased severity of winter duration and cold, decreased snow cover and increased environmental stochasticity during the winter (Bale and Hayward, 2010). These conditions will probably affect postwinter fitness components of emerging insects such as body size, fecundity, longevity, dispersal and immune function (Ishihara and Shimada, 1995; Bradshaw *et al.*, 1998; Ellers and van Alphen, 2002). For example, in preparation for winter, most insect species accumulate lipids as energetic fuel for use during dormancy (Hahn and Denliger, 2011; Sinclair and Marshall, 2018; Enriquez and Visser, 2023; Short and Hahn, 2023). CaraDonna *et al.* (2018) showed experimentally for a solitary bee that predicted future climatic conditions would reduce lipid reserves and survival of overwintering individuals.

Insect pest diversity, distribution and abundance are predicted to increase overall with climate change and thus negatively impact crop production (Patterson *et al.*, 1999; Rosenzweig *et al.*, 2014; Deutsch *et al.*, 2018). In this context, from an applied perspective, studying diapause is important to develop and implement pest management strategies. Changes in the phenology, abundance and population cycles (voltinism) of insect pests have the potential to considerably modify crop productivity. For instance, it has been established or anticipated using predictive models that warmer temperatures speed up development and delay diapause in autumn of several pest species, thereby allowing the production of additional generations per year (e.g. Porter *et al.*, 1991; Altermatt, 2010; Bentz *et al.*, 2010). For example, Gagnon and Bourgeois (2024) showed for the carrot weevil *Listronotus oregonensis* (Coleoptera: Curculionidae) that recent increasing temperature in Québec, Canada, promotes the development of a second generation, primarily by disrupting reproductive diapause, resulting in increased pest pressure on carrot crops. Knowledge of diapause is also critical for understanding aspects of overwintering survival and timing of spring emergence of insects. Of significance in high latitudes, climate warming is generally associated with a reduction of the buffering effect of snow cover for insects diapausing in the soil, which has direct impact on their cold-hardiness and energetics (reviewed by Marshall *et al.*, 2020).

Similar considerations apply to biological control agents since diapause and overwintering survival influence their capacity to establish and spread once released in the habitat. Managing diapause can also be an issue when studying or mass-rearing a natural enemy. Gariépy *et al.* (2015) showed that an aphid parasitoid *Binodoxys communis* (Hymenoptera: Braconidae) lost its capacity to enter diapause when reared under non-diapause conditions for extended periods of quarantine and laboratory confinement. Increasing the storage, shelf-life and diapause termination of mass-produced predators and parasitoids through diapause management has also become a challenge for the biological control industry (Colinet and Boivin, 2011).

2.3 Change in Diapause Strategies

This section describes mechanisms allowing insects to cope with and adapt to new climatic conditions. Insect diapause is a classic example of developmental plasticity and polyphenism, intrinsically associated with temperature conditions and other environmental parameters. As diapause is a syndrome selected by environmental variability, it is also liable to change through the action of natural selection (Tauber *et al.*, 1986). This is clearly seen over wide geographic gradients where populations are adapted to local environmental conditions, with greater diapause incidences and earlier initiation timings poleward (Paolucci *et al.*, 2013). In the context of climate change, insect diapause, in terms of both dates of induction and termination, and incidence within populations, is therefore likely to be altered, especially as environmental conditions change rapidly and with great intensity (Bale and Hayward, 2010; Kivelä *et al.*, 2013). Climate change may impact environmental stimuli which act directly on insects or indirectly through food resources or interacting partners (symbionts, prey, predators, competitors) (Tougeron *et al.*, 2019).

Most adaptive responses to climate change are complex and involve a variety of mechanisms acting in combination (Harvey *et al.*, 2023; Edelsparre *et al.*, 2024). Two main routes can lead to changes in diapause expression in insects. The first and most straightforward route is inherent to the definition of the diapause syndrome: phenotypic plasticity in timing and expression. For a given time of year, at a given photoperiod, average temperatures rise with climate warming. As a result, for most insects that rely on both temperature and photoperiod to initiate diapause, the date of entry may be delayed until later in the year through a plastic response to new environmental conditions. This may only be the case for multivoltine species, for which only one generation per year enters diapause. In univoltine species, for which diapause is obligatory, plastic adjustments are much more difficult to achieve (Saunders, 2002; Tougeron, 2019; Lindestad *et al.*, 2020). In some populations exposed to milder climates, only some of the individuals may enter into facultative diapause. For example, parasitoids of the genus *Aphidius* (Hymenoptera: Braconidae) in western France have for most species reduced the expression of their diapause to a small fraction of the population following the onset of increasingly short and warm winters (Tougeron *et al.*, 2017b). Plastic adjustments in diapause strategies may allow insects to adapt to both heterogeneous environments and directional changes in the environment over the short term, but can potentially restrict the action of natural selection by keeping insects away from environmental pressures (Chevin *et al.*, 2013; Sgrò *et al.*, 2016; Abram *et al.*, 2017). Differences in diapause expression over time can also result from maternal effects (transgenerational plasticity), which may enhance or inhibit diapause in the offspring depending on environmental conditions encountered by the mother, and may play a role in the context of climate change (Mousseau and Dingle, 1991; Tougeron *et al.*, 2017a).

The second way in which diapause expression can change over the longer term (i.e. over generations) is through the genetic evolution of thresholds or conditions required to initiate, maintain and terminate diapause. For example, Bean *et al.* (2012) demonstrated an evolution towards a decline in the critical day length inducing diapause in the beetle *Diorhabda carinulata* (Coleoptera: Chrysomelidae), a biological control agent of invasive *Tamarix* sp (Caryophyllales: Tamaricaceae). It resulted in a closer match between the beetle life history and the phenology of the shrub. The change depends on the strength of directional selection on diapause-associated traits and on phenotypic plasticity itself (Bradshaw and Holzapfel, 2006; Forrest, 2016; Sgrò *et al.*, 2016). A genetic alteration affecting seasonal patterns in insects is seen in how the pitcher plant mosquito *Wyeomyia smithii* (Diptera: Culicidae) has adapted to extended growing periods in North America, by modifying the critical photoperiod for inducing larval diapause in the fall, shifting towards shorter daylight thresholds (Bradshaw and Holzapfel, 2001). Most examples of changes in insect phenology in the face of global warming imply strong selection caused by change in winter conditions, such as in the Asian tiger mosquito *Aedes albopictus* (Diptera: Culicidae) which shows genetic divergence in diapause-associated cold tolerance, timing and syndrome (Batz *et al.*, 2020).

Disentangling the relative effects of plasticity and genetic on phenological shifts and local adaptations of insects is a major challenge in evolutionary research on organisms' responses to climate changes (Merilä and Hendry, 2014; Tougeron *et al.*, 2018b). In a warming climate, strong selection pressure for diapause loss could also come from diapause termination requirements for some insects which only resume development if they have first been chilled for a few days or weeks (Tauber and Tauber, 1976; Tougeron, 2019; Toxopeus *et al.*, 2021). For example, the parasitoid *Colpoclypeus florus* (Hymenoptera: Eulophidae) requires a chilling period of at least 5 weeks at 4°C to terminate diapause (Milonas and Savopoulou-Soultani, 2000).

Selection pressures apply to insect diapause expression, whether agricultural pests or their natural enemies, because there are inherent costs, and the cost–benefit balance may alter under climate change when considering timing and ecological costs (e.g. reproductive arrest under mild winters) or metabolic and production costs (e.g. producing unnecessary cryoprotectant molecules) (Hahn and Denlinger, 2007; Lahti *et al.*, 2009; Sgrò *et al.*, 2016; Short and Hahn, 2023). Diapause costs are associated to diapause duration and (mostly thermal) conditions experienced during diapause, by a normally linear consumption of the reserves in function of time, and resources (or time) may be compromised in trade-offs (Nylin and Gotthard, 1998; Danks, 2007; Roff and Fairbairn, 2007). Using *Aphalara itadori* (Hemiptera: Aphalaridae), a biological control agent introduced from Japan to North America and Europe to control an invasive weed *Reynoutria japonica* (Caryophyllales: Polygonaceae), Grevstad *et al.* (2022) suggested that in warmer climates, the insect may be diapausing too early, therefore not fully exploiting the weed throughout the growing season.

The capacity to delay or avert diapause may therefore provide a selective advantage in species that can avoid energetic and metabolic costs while maintaining overwinter activities, including foraging for resources and reproduction (Bale and Hayward, 2010). For example, in the parasitoid *Asobara tabida* (Hymenoptera: Braconidae) an increase in diapause duration led to increased mortality and reduced fecundity (Ellers and van Alphen, 2002). Conversely, when diapause is not expressed in environments that are too cold or variable, this poses serious problems in the context of biological control, for example by introducing species that have evolved in tropical or subtropical environments, or in mass rearing environments (Gariépy *et al.*, 2015).

Depending on the environmental variability and reliability of the diapause-inducing signals, directional selection towards pure and conservative strategies of reduction in diapause expression may be favoured, or bet-hedging strategies can also be selected (i.e. risk minimization strategies) (Hopper, 1999). In many cases, diapause is a pure strategy, with 100% of the individuals in a population entering diapause when environmental and especially winter conditions are harsh. At the opposite end of the spectrum, if winter conditions remain favourable for maintaining activity, a pure strategy of abandoning diapause could theoretically be maintained, as in tropical environments. However, climate change may lead to more variable and unpredictable environments, which can select for bet-hedging strategies (Furness *et al.*, 2015; van Baaren *et al.*, 2024) in insect diapause and overwintering phenotypes to reduce the average fitness in favour of a smaller fitness variance. When a female cannot accurately assess which offspring phenotype would be favoured in the future environment – diapausing offspring or not – the proportion of each offspring phenotype may be adjusted to the probability of encountering a given environment, when biotic or abiotic cues make it at least partially predictable (Uller, 2008). Conditional strategies ('adaptive coin flipping') may therefore be adopted where the probability of expressing a phenotype is based on the probability of encountering a given environment (Cooper and Kaplan, 1982), such as stressful versus mild winter conditions. Alternative bet-hedging strategies may also increase the phenotypic variability produced within a generation or between generations (diversified bet-hedging) or produce the same phenotype (conservative bet-hedging) (Ripa *et al.*, 2010). To cope with a new selection pressure led by increasing environmental stochasticity (variability and unpredictability), mixed strategies such as bet-hedging strategies should become more frequent (Simons, 2011; Kivelä *et al.*, 2013). Selection may therefore act by both lowering diapause induction thresholds and modifying the proportion of diapausing offspring produced by each insect female.

In a nutshell, intrapopulation variability in diapause expression can originate from diversified bet-hedging strategies or from genetic polymorphism of mixed-pure strategies, maintained by balancing selection (Bradshaw and Holzapfel, 2001; Soula and Menu, 2003), which results in different plastic responses to environmental cues within a population. Interpopulation variability results from local adaptation in diapause induction thresholds (i.e. critical day length). Accordingly, climate change is expected to alter the expression of diapause, but a certain degree of unpredictability may be anticipated due to interindividual response elements and environmental variability in relation to global climate change trends. The three forms of adaptations (phenotypic plasticity, genetic evolution, bet-hedging) are not mutually exclusive and depend on climate predictability, variability and rate of change (DeWitt and Scheiner, 2004; Tufto, 2015).

2.4 Matches and Mismatches Across Trophic Levels

Climate change affects insect pest populations and their natural enemies in a variety of ways, which have been extensively reviewed in this book and review articles (see Thomson *et al.*, 2010; Cock *et al.*, 2013; Eigenbrode *et al.*, 2015; Furlong and Zalucki, 2017; Furlong *et al.*, 2021). Many scenarios are possible, but a main pattern is that climate change is expected to increase damage caused by agricultural pests through increasing population growth, accumulative pest generations per year, facilitation of pest dispersion, earlier pest outbreaks and both decreasing natural enemy efficiency and plant resistance (Klapwijk *et al.*, 2012; Björkman and Niemelä, 2015; Pincebourde *et al.*, 2016).

The case of diapause is interesting to explore in detail, as it is a predominant phenological event in the life of insects of temperate environments, and a syndrome with various traits and phenotypes expressed. It is also a polyphenism, i.e. binary situation, which by definition results in either developmental arrest or direct development, with no intermediate situation that might be found in other phenotypes such as heat tolerance (Hance *et al.*, 2007). Different pathways can lead to total or partial phenological desynchronization between interacting organisms, depending on the intensity of selection pressures applied to diapause expression at each trophic level, the response variability of each species and the associated costs for each

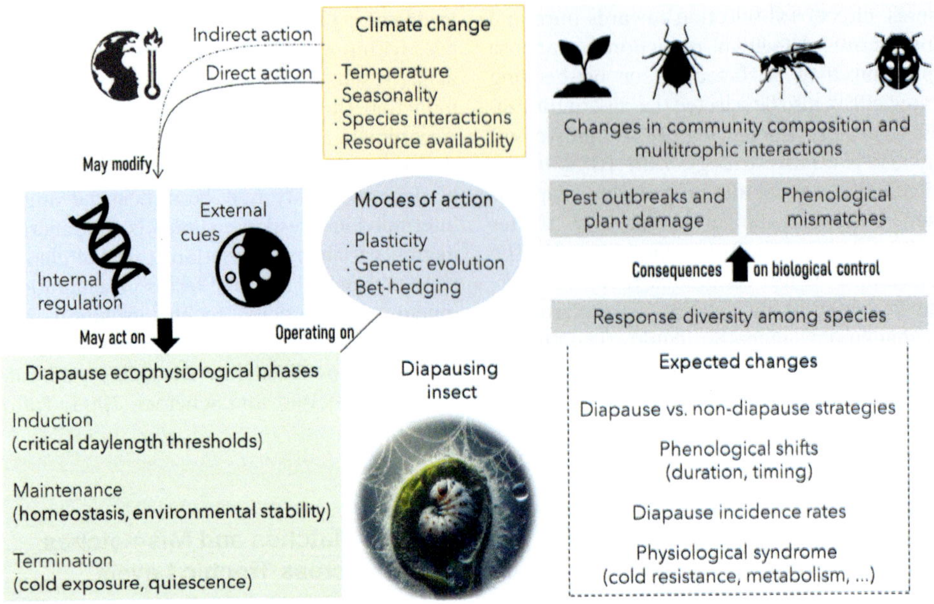

Fig. 2.2. General pathway by which change in the expression of diapause is possible and observed in insects. The consequences are visible on individuals, population dynamics and therefore communities. Biological control services may then be affected.

individual (Renner and Zohner, 2018; Damien and Tougeron, 2019). Therefore, the effects of climate change have to be considered at the species level but also for the entire community because asynchrony and asymmetrical changes in diapause or phenology patterns across trophic levels potentially lead to shift in species interactions (van de Putten *et al.*, 2010; Forrest, 2016; Thackeray *et al.*, 2016; Damien and Tougeron, 2019; Harvey *et al.*, 2023). For example, if diapause persists in natural enemies under future climates but not in pests (primary consumers), warmer winters may favour greater pest population increases during crop auxiliaries diapause and greater damage to crops in spring, as for example expected in orchards infested with aphids in New Zealand (Gerard *et al.*, 2013) (Fig. 2.2).

In any given community, some plant–herbivore, prey–predator or host–parasitoid interactions will reduce in intensity or disappear, and novel interactions will emerge, altering the abundance, distribution and function of each species in a food web (Tylianakis *et al.*, 2008; Gilman *et al.*, 2010). Rapid change in community composition following climate change can be linked to modifications in diapause expression of all or some species of a community (Tougeron *et al.*, 2018a). For example, in parasitoid communities of cereal aphids experiencing warmer winters, the generalist aphid parasitoid *Aphidius avenae* (Hymenoptera: Braconidae) reduced winter diapause expression and became the dominant species in winter, inducing the rarefaction of the previously most abundant host species *A. rhopalosiphi* and *A. matricariae* (Andrade *et al.*, 2016; Tougeron *et al.*, 2017b).

Because early suppression of pests is often crucial for efficient biological control, the loss or decrease of diapause expression in natural enemies could favour biological control, especially if natural enemies can exploit more host or prey for a longer time window. In the parasitoid *Cotesia melitaearum* (Hymenoptera: Braconidae), van Nouhuys and Lei (2004) showed that warmer springs improved the synchrony with its host, the butterfly *Melitaea cinxia*, (Lepidoptera: Nymphalidae) leading to increased parasitoid colonization at the metapopulation scale. However, with climate change, modifications in food web composition may *in fine* decrease biological control efficiency through increasing intraguild competition, intraguild predation (Frances and McCauley, 2018) or activity of new antagonist species. For example, hyper-parasitism levels tend to rise in most cropping systems (Gómez-Marco *et al.*, 2015; Tougeron *et al.*, 2021), suggesting that the loss of diapause expression can occur across different trophic levels and that hyperparasitoid pressure may increase in winter or early spring, leading to poorer pest control by primary parasitoids (Tougeron and Tena, 2018).

Herbivorous pests, in particular sap-sucking insects, are often associated with facultative symbionts that provide nutritional functions and resistance to environmental disturbance (Zytynska *et al.*, 2021). Although diapause as such is not directly affected by these symbionts (Mushegian and Tougeron, 2019), their effects on heat tolerance, and conversely the effects of temperature on their maintenance, can jeopardize interactions with natural enemies present in the environment (Kashima *et al.*, 2006; Cayetano and Vorburger, 2013; Tougeron and Iltis, 2022). Bacterial symbionts are therefore important elements to consider in the context of climate change, as they are partly decisive in the persistence of certain pest populations and in the maintenance of host–parasitoid or prey–predator interactions.

Phenological mismatches and absence of accurate temporal host tracking by natural enemies may result in important yield losses, especially if the natural enemy is highly specialized on a given prey (Eigenbrode *et al.*, 2015). For example, due to warmer winters and springs, parasitoids of the invasive horse chestnut leaf miner (*Cameraria ohridella* (Lepidoptera: Gracillariidae)) broke diapause too early, resulting in poor synchronization between species and reduced control of the host populations (Grabenweger *et al.*, 2007). For the biocontrol agent *Torymus sinensis* (Hymenoptera: Torymidae), a lower parasitism pressure due to phenological mismatch with its host *Dryocosmus kuriphilus* (Hymenoptera: Cynipidae) increases the probability of pest outbreaks at warmer temperatures (Ferracini *et al.*, 2022). Natural enemies that rely on their hosts or prey to initiate their own diapause (Tougeron *et al.*, 2021), or in which there is strong selection pressure to follow the same diapause-inducing signals, will be less likely to desynchronize and reduce the effectiveness of the biological control service (Furlong and Zalucki, 2017; Renner and Zohner, 2018). In addition, diapause asynchrony may not be an issue for biological control when there is enough functional redundancy within a guild of natural enemies to control a host (Hance *et al.*, 2007; Damien and Tougeron, 2019). However, it is necessary to include potential changes in diapause in population and community dynamic models destined to inform pest control decisions (Lalonde, 2004; Kalinkat and Rall, 2015).

References

Abram, P.K., Boivin, G., Moiroux, J. and Brodeur, J. (2017) Behavioural effects of temperature on ectotherm animals: Unifying thermal physiology and behavioural plasticity. *Biological Reviews* 92, 1859–1876.

Altermatt, F. (2010) Climatic warming increases voltinism in European butterflies and moths. *Proceedings of the Royal Society B: Biological Sciences* 277, 1281–1287.

Andrade, T.O., Krespi, L., Bonnardot, V., van Baaren, J. and Outreman, Y. (2016) Impact of change in winter strategy of one parasitoid species on the diversity and function of a guild of parasitoids. *Oecologia* 180, 877–888.

Bale, J.S. and Hayward, S.A.L. (2010) Insect overwintering in a changing climate. *Journal of Experimental Biology* 213, 980–994.

Batz, Z.A., Clemento, A.J., Fritzenwanker, J., Ring, T.J., Garza, J.C. *et al.* (2020) Rapid adaptive evolution of the diapause program during range expansion of an invasive mosquito. *Evolution* 74, 1451–1465.

Bean, D.W., Dalin, P. and Dudley, T.L. (2012) Evolution of critical day length for diapause induction enables range expansion of *Diorhabda carinulata*, a biological control agent against tamarisk (*Tamarix spp.*): Evolution and range expansion of a biocontrol agent. *Evolutionary Applications* 5, 511–523.

Bentz, B.J., Régnière, J., Fettig, C.F., Hansen, M.E., Hayes, J.L. *et al.* (2010) Climate change and bark beetles of the Western United States and Canada: Direct and indirect effects. *BioScience* 60, 602–613.

Björkman, C. and Niemelä, P. (eds) (2015) *Climate Change and Insect Pests*. CAB International, Wallingford, UK.

Bradshaw, W.E. and Holzapfel, C.M. (2001) Genetic shift in photoperiodic response correlated with global warming. *Proceedings of the National Academy of Sciences* 98, 14509–14511.

Bradshaw, W.E. and Holzapfel, C.M. (2008) Genetic response to rapid climate change: It's seasonal timing that matters. *Molecular Ecology* 17, 157–166.

Bradshaw, W.E., Armbruster, P.A. and Holzapfel, C.M. (1998) Fitness consequences of hibernal diapause in the pitcher-plant mosquito, *Wyeomyia Smithii*. *Ecology* 79, 1458–1462.

Bradshaw, W.E. and Holzapfel, C.M. (2006) Evolutionary response to rapid climate change. *Science* 312, 1477–1478.

Brodeur, J. and McNeil, J.N. (1989) Biotic and abiotic factors involved in diapause induction of the parasitoid, *Aphidius nigripes* (Hymenoptera: Aphidiidae). *Journal of Insect Physiology* 35, 969–974.

CaraDonna, P.J., Cunningham, J.L. and Iler, A.M. (2018) Experimental warming in the field delays phenology and reduces body mass, fat content and survival: Implications for the persistence of a pollinator under climate change. *Functional Ecology* 32, 2345–2356.

Cayetano, L. and Vorburger, C. (2013) Effects of heat shock on resistance to parasitoids and on life history traits in an aphid/endosymbiont system. *PLoS ONE* 8, e75966.

Chevin, L.-M., Gallet, R., Gomulkiewicz, R., Holt, R.D. and Fellous, S. (2013) Phenotypic plasticity in evolutionary rescue experiments. *Philosophical Transactions of the Royal Society of London B: Biological Sciences* 368, 20120089.

Cock, M.J., Biesmeijer, J.C., Cannon, R.J., Gerard, P.J., Gillespie, D. *et al.* (2013) The implications of climate change for positive contributions of invertebrates to world agriculture. *CABI Reviews* 8(28), 1–48.

Colinet, H. and Boivin, G. (2011) Insect parasitoids cold storage: A comprehensive review of factors of variability and consequences. *Biological Control* 58, 83–95.

Cooper, W.S. and Kaplan, R.H. (1982) Adaptive 'coin-flipping': A decision-theoretic examination of natural selection for random individual variation. *Journal of Theoretical Biology* 94, 135–151.

Damien, M. and Tougeron, K. (2019) Prey–predator phenological mismatch under climate change. *Current Opinion in Insect Science* 35, 60–68.

Danks, H.V. (1987) *Insect Dormancy: An Ecological Perspective. Biological Survey of Canada (Terrestrial Arthropods)*. National Museum of Natural Sciences, Ottawa, Canada.

Danks, H.V. (2007) The elements of seasonal adaptations in insects. *Canadian Entomologist* 139, 1–44.

Davis, A.J., Jenkinson, L.S., Lawton, J.H., Shorrocks, B. and Wood, S. (1998) Making mistakes when predicting shifts in species range in response to global warming. *Nature* 391, 783–785.

Denlinger, D.L. (2022) Regulation of diapause. *Annual Review of Entomology* 47, 93–12.

Denlinger, D.L., Yocum, G.D. and Rinehart, J.P. (2012) Hormonal control of diapause. In: Gilbert, L.I. (ed.) *Insect Endocrinology*. Academic Press, Cambridge, Massachussetts, pp. 430–463.

Deutsch, C.A., Tewksbury, J.J., Huey, R.B., Sheldon, K.S., Ghalambor, C.K. *et al.* (2008) Impacts of climate warming on terrestrial ectotherms across latitude. *Proceedings of the National Academy of Sciences of the United States of America* 105, 6668–6672.

Deutsch, C.A., Tewksbury, J.J., Tigchelaar, M., Battisti, D., Merrill, S.C. *et al.* (2018) Increase in crop losses to insect pests in a warming climate. *Science* 361, 916–919.

DeWitt, T.J. and Scheiner, S.M. (eds) (2004) *Phenotypic Plasticity: Functional and Conceptual Approaches*. Oxford University Press, Oxford, UK.

Edelsparre, A.H., Fitzpatrick, M.J., Saastamoinen, M. and Teplitsky, C. (2024) Evolutionary adaptation to climate change. *Evolution Letters* 8, 1–7.

Eigenbrode, S.D., Davis, T.S. and Crowder, D.W. (2015) Climate change and biological control in agricultural systems: Principles and examples from North America. *Climate Change and Insect Pests* 7, 119–135.

Ellers, J. and van Alphen, J.J. (2002) A trade-off between diapause duration and fitness in female parasitoids. *Ecological Entomology* 27, 279–284.

Enriquez, T. and Visser, B. (2023) The importance of fat accumulation and reserves for insect overwintering. *Current Opinion in Insect Science* 60, 101118.

Forrest, J.R.K. (2016) Complex responses of insect phenology to climate change. *Current Opinion in Insect Science* 17, 49–54.

Ferracini, C., Pogolotti, C. and Alma, A. (2022) A mismatch in the emergence of *Torymus sinensis* may affect the effectiveness of this biocontrol agent. *Biological Control* 174, 105029.

Forrest, J. and Miller-Rushing, A.J. (2010) Toward a synthetic understanding of the role of phenology in ecology and evolution. *Philosophical Transactions of the Royal Society B* 365, 3101–3112.

Frances, D.N. and McCauley, S.J. (2018) Warming drives higher rates of prey consumption and increases rates of intraguild predation. *Oecologia* 187, 585–596.

Furlong, M.J. and Zalucki, M.P. (2017) Climate change and biological control: The consequences of increasing temperatures on host–parasitoid interactions. *Current Opinion in Insect Science* 20, 39–44.

Furlong, M.J., Day, M.D. and Zalucki, M.P. (2021) Biological control and climate change. In: Mason, P.G. (ed.) *Biological Control Global Impacts Challenges Future Directions of Pest Management*. CSIRO Publishing, Clayton, Australia, pp. 220–235.

Furness, A.I., Lee, K. and Reznick, D.N. (2015) Adaptation in a variable environment: Phenotypic plasticity and bet-hedging during egg diapause and hatching in an annual killifish. *Evolution* 69, 1461–1475.

Gagnon, A.-E. and Bourgeois, G. (2024) Impact of climate change on the reproductive doiapause and voltinism of the carrot weevil, *Listronotus oregonensis*. *Journal of Insect Physiology* 155, 104653.

Gallinat, A.S., Primack, R.B. and Wagner, D.L. (2015) Autumn, the neglected season in climate change research. *Trends in Ecology & Evolution* 30, 169–176.

Gariépy, V., Boivin, G. and Brodeur, J. (2015) Why two species of parasitoids showed promise in the laboratory but failed to control the soybean aphid under field conditions. *Biological Control* 80, 1–7.

Gerard, P., Barringer, J., Charles, J., Fowler, S., Kean, J. *et al*. (2013) Potential effects of climate change on biological control systems: Case studies from New Zealand. *BioControl* 58, 149–162.

Gilman, S.E., Urban, M.C., Tewksbury, J., Gilchrist, G.W. and Holt, R.D. (2010) A framework for community interactions under climate change. *Trends in Ecology & Evolution* 25, 325–331.

Gómez-Marco, F., Urbaneja, A., Jaques, J.A., Rugman-Jones, P.F., Stouthamer, R. *et al*. (2015) Untangling the aphid-parasitoid food web in citrus: can hyperparasitoids disrupt biological control? *Biological Control* 81, 111–121.

Grabenweger, G., Hopp, H., Jäckel, B., Balder, H., Koch, T. *et al*. (2007) Impact of poor host-parasitoid synchronisation on the parasitism of *Cameraria ohridella* (Lepidoptera: Gracillariidae). *European Journal of Entomology* 104, 153.

Grevstad, P.S., Wepprich, T., Barker, B., Coop, L.B., Shaw, R. *et al*. (2022) Combining photoperiod and thermal responses to predict phenological mismatch for introduced insects. *Ecological Applications* 32, e2557.

Hahn, D.A. and Denlinger, D.L. (2007) Meeting the energetic demands of insect diapause: Nutrient storage and utilization. *Journal of Insect Physiology* 53, 760–773.

Hahn, D.A. and Denliger, D.L. (2011) Energetics of insect diapause. *Annual Review of Entomology* 56, 103–121.

Hance, T., van Baaren, J., Vernon, P. and Boivin, G. (2007) Impact of extreme temperatures on parasitoids in a climate change perspective. *Annual Review of Entomology* 52, 107–126.

Harvey, J.A., Tougeron, K., Gols, R., Heinen, R., Abarca, M. *et al*. (2023) Scientists' warning on climate change and insects. *Ecological Monographs* 93, e1553.

Hopper, K.R. (1999) Risk-spreading and bet-hedging in insect population biology. *Annual Review of Entomology* 44, 535–560.

Huey, R.B., Kearney, M.R., Krockenberger, A., Holtum, J.A., Jess, M. *et al*. (2012) Predicting organismal vulnerability to climate warming: Roles of behaviour, physiology and adaptation. *Philosophical Transactions of the Royal Society B: Biological Sciences* 367, 1665–1679.

Ishihara, M. and Shimada, M. (1995) Trade-off in allocation of metabolic reserves: Effects of diapause on egg production and adult longevity in a multivoltine bruchid, *Kytorhinus sharpianus*. *Functional Ecology* 9, 618–624.

Kalinkat, G. and Rall, B.C. (2015) Effects of climate change on the interactions between insect pests and their natural enemies. In: Bjorkman, C. and Niemela, P. (eds) *Climate Change and Insect Pests*. CAB International, Wallingford, UK, pp. 74–91.

Kashima, T., Nakamura, T. and Tojo, S. (2006) Uric acid recycling in the shield bug, *Parastrachia japonensis* (Hemiptera: Parastrachiidae), during diapause. *Journal of Insect Physiology* 52, 816–825.

Kivelä, S.M., Välimäki, P. and Gotthard, K. (2013) Seasonality maintains alternative life-history phenotypes: Evolution of life-history polyphenisms. *Evolution* 67, 3145–3160.

Klapwijk, M.J., Ayres, M.P., Battisti, A. and Larsson, S. (2012) Assessing the impact of climate change on outbreak potential. In: Barbosa, P., Letourneau, D.K. and Agrawal, A.A. (eds) *Insect Outbreaks Revisited*. Blackwell Publishing, Hoboken, New Jersey, pp. 429–450.

Koštál, V. (2006) Eco-physiological phases of insect diapause. *Journal of Insect Physiology* 52, 113–127.

Koštál, V., Stettin, T., Poupardin, R., Korbelová, J. and Bruce, A.W. (2017) Conceptual framework of the eco-physiological phases of insect diapause development justified by transcriptomic profiling. *Proceedings of the National Academy of Sciences of the United States of America* 32, 8532–8537.

Lahti, D.C., Johnson, N.A., Ajie, B.C., Otto, S.P., Hendry, A.P. *et al.* (2009) Relaxed selection in the wild. *Trends in Ecology & Evolution* 24, 487–496.

Lalonde, R.G. (2004) Some dynamical consequences of parasitoid diapause. *Oikos* 107, 338–344.

Lees, A.D. (1955) *The Physiology of Diapause in Arthropods*. Cambridge University Press, New York.

Lindestad, O., von Schmalensee, L., Lehmann, P. and Gotthard, K. (2020) Variation in butterfly diapause duration in relation to voltinism suggests adaptation to autumn warmth, not winter cold. *Functional Ecology* 34, 1029–1040.

Marshall, K.E., Gotthard, K. and Williams, C.M. (2020) Evolutionary impacts of winter climate change on insects. *Current Opinion in Insect Science* 41, 54–62.

Merilä, J. and Hendry, A.P. (2014) Climate change, adaptation, and phenotypic plasticity: The problem and the evidence. *Evolutionary Applications* 7, 1–14.

Milonas, P.G. and Savopoulou-Soultani, M. (2000) Diapause induction and termination in the parasitoid *Colpoclypeus florus* (Hymenoptera: Eulophidae): Role of photoperiod and temperature. *Annals of the Entomological Society of America* 93, 512–518.

Mousseau, T.A. and Dingle, H. (1991) Maternal effects in insect life histories. *Annual Review of Entomology* 36, 511–534.

Mushegian, A.A. and Tougeron, K. (2019) Host-microbe interactions in the context of diapause. *Biological Bulletin* 237, 180–191.

Nylin, S. and Gotthard, K. (1998) Plasticity in life-history traits. *Annual Review of Entomology* 43, 63–83.

Paolucci, S., van de Zande, L. and Beukeboom, L.W. (2013) Adaptive latitudinal cline of photoperiodic diapause induction in the parasitoid *Nasonia vitripennis* in Europe. *Journal of Evolutionary Biology* 26, 705–718.

Parmesan, C. (2007) Influences of species, latitudes and methodologies on estimates of phenological response to global warming. *Global Change Biology* 13, 1860–1872.

Patterson, D.T., Westbrook, J.K., Joyce, R.J.V., Lingren, P.D. and Rogasik, J. (1999) Weeds, insects, and diseases. *Climatic Change* 43, 711–727.

Petchey, O.L., Mcphearson, P.T., Casey, T.M. and Morin, P.J. (1999) Environmental warming alters food-web structure and ecosystem function. *Nature* 402, 69–72.

Pincebourde, S., van Baaren, J., Rasmann, S., Rasmont, P., Rodet, G. *et al.* (2016) Plant–insect interactions in a changing world. In: Sauvion, N., Thiery, D. and Calatavud, P.A. (eds) *Advances in Botanical Research*. Academic Press, Cambridge.

Porter, J.H., Parry, M.L. and Carter, T.R. (1991) The potential effects of climatic change on agricultural insect pests. *Agricultural and Forest Meteorology* 57, 221–240.

Post, E., Pedersen, C., Wilmers, C.C. and Forchhammer, M.C. (2008) Warming, plant phenology and the spatial dimension of trophic mismatch for large herbivores. *Proceedings of the Royal Society B: Biological Sciences* 275, 2005–2013.

Renner, S.S. and Zohner, C.M. (2018) Climate change and phenological mismatch in trophic interactions among plants, insects, and vertebrates. *Annual Review of Ecology, Evolution, and Systematics* 49, 165–182.

Ripa, J., Olofsson, H. and Jonzen, N. (2010) What is bet-hedging, really? *Proceedings of the Royal Society B: Biological Sciences* 277, 1153–1154.

Rockey, S.J., Miller, B.B. and Denlinger, D.L. (1989) A diapause maternal effect in the flesh fly, *Sarcophaga bullata*: Transfer of information from mother to progeny. *Journal of Insect Physiology* 35, 553–558.

Roff, D. and Fairbairn, D. (2007) The evolution of trade-offs: where are we? *Journal of Evolutionary Biology* 20, 433–447.

Rosenzweig, C., Elliott, J., Deryng, D., Ruane, A.C., Muller, C. *et al.* (2014) Assessing agricultural risks of climate change in the 21st century in a global gridded crop model intercomparison. *Proceedings of the National Academy of Sciences of the United States of America* 111, 3268–3273.

Saunders, D.S. (1965) Larval diapause of maternal origin: Induction of diapause in *Nasonia vitripennis* (Walk.) (Hymenoptera: Pteromalidae). *Journal of Experimental Biology* 42, 495–508.

Saunders, D.S. (2002) *Insect Clocks*, 3rd edn. Elsevier, Amsterdam, The Netherlands.

Schleuning, M., Neuschulz, E.L., Albrecht, J., Bender, I.M., Bowler, D.E. *et al.* (2020) Trait-based assessments of climate-change impacts on interacting species. *Trends in Ecology & Evolution* 35, 319–328.

Sentis, A., Ramon-Portugal, F., Brodeur, J. and Hemptinne, J.L. (2015) The smell of change: Warming affects species interactions mediated by chemical information. *Global Change Biology* 21, 3586–3594.

Sgrò, C.M., Terblanche, J.S. and Hoffmann, A.A. (2016) What can plasticity contribute to insect responses to climate change. *Annual Review of Entomology* 61, 433–451.

Short, C.A. and Hahn, D.A. (2023) Fat enough for the winter? does nutritional status affect diapause? *Journal of Insect Physiology* 145, 104488.

Simons, A.M. (2011) Modes of response to environmental change and the elusive empirical evidence for bet hedging. *Proceedings of the Royal Society B: Biological Sciences* 278, 1601–1609.

Sinclair, B.J. and Marshall, K.E. (2018) The many roles of fats in overwintering insects. *Journal of Experimental Biology* 221, 161836.

Singer, M.C. and Parmesan, C. (2010) Phenological asynchrony between herbivorous insects and their hosts: signal of climate change or pre-existing adaptive strategy? *Philosophical Transactions of the Royal Society B: Biological Sciences* 365, 3161–3176.

Soula, B. and Menu, F. (2003) Variability in diapause duration in the chestnut weevil: mixed ESS, genetic polymorphism or bet-hedging? *Oikos* 100, 574–580.

Stenseth, N.C., Mysterud, A., Ottersen, G., Hurrell, J.W., Chan, K.S. *et al.* (2002) Ecological effects of climate fluctuations. *Science* 297, 1292–1296.

Suttle, K.B., Thomsen, M.A. and Power, M.E. (2007) Species interactions reverse grassland responses to changing climate. *Science* 315, 640–642.

Tauber, M.J. and Tauber, C.A. (1976) Insect seasonality: Diapause maintenance, termination, and *postdiapause* development. *Annual Review of Entomology* 21, 81–107.

Tauber, M.J., Tauber, C.A. and Masaki, S. (1986) *Seasonal Adaptations of Insects*. Oxford University Press, New York.

Thackeray, S.J., Henrys, P.A., Hemming, D., Bell, J.R., Botham, M.S. *et al.* (2016) Phenological sensitivity to climate across taxa and trophic levels. *Nature* 535, 241–245.

Thomson, L.J., Macfadyen, S. and Hoffmann, A.A. (2010) Predicting the effects of climate change on natural enemies of agricultural pests. *Biological Control* 52, 296–306.

Tougeron, K. (2019) Diapause research in insects: Historical review and recent work perspectives. *Entomologia Experimentalis et Applicata* 167, 27–36.

Tougeron, K. and Iltis, C. (2022) Impact of heat stress on the fitness outcomes of symbiotic infection in aphids: A meta-analysis. *Proceedings of the Royal Society B: Biological Sciences* 289, 20212660.

Tougeron, K. and Tena, A. (2018) Hyperparasitoids as new targets in biological control in a global change context? *Biological Control* 130, 164–171.

Tougeron, K., Hraoui, G., Le Lann, C., van Baaren, J. and Brodeur, J. (2017a) Intraspecific maternal competition induces summer diapause in insect parasitoids. *Insect Science* 25, 1080–1088.

Tougeron, K., Le Lann, C., Brodeur, J. and van Baaren, J. (2017b) Are aphid parasitoids from mild winter climates losing their winter diapause. *Oecologia* 183, 619–629.

Tougeron, K., Damien, M., Le Lann, C., Brodeur, J. and van Baaren, J. (2018a) Rapid responses of winter aphid-parasitoid communities to climate warming. *Frontiers in Ecology and Evolution* 6, 173.

Tougeron, K., Brodeur, J., van Baaren, J., Renault, D. and Le Lann, C. (2021) Phenology under bottom-up control: Change in host quality induces diapause in parasitic wasps. *Peer Community Journal* 1, e12.

Tougeron, K., van Baaren, J., Llopis, S., Ridel, A., Doyon, J. *et al.* (2018b) Disentangling plasticity from local adaptation in diapause expression in parasitoid wasps from contrasting thermal environments: A reciprocal translocation experiment. *Biological Journal of the Linnean Society* 124, 756–764.

Tougeron, K., Brodeur, J., Le Lann, C. and van Baaren, J. (2019) How climate change affects the seasonal ecology of insect parasitoids. *Ecological Entomology* 45, 167–181.

Toxopeus, J., Gadey, L., Andaloori, L., Sanaei, M. and Ragland, G.J. (2021) Costs of averting or prematurely terminating diapause associated with slow decline of metabolic rates at low temperature. *Comparative Biochemistry and Physiology Part A: Molecular & Integrative Physiology* 255, 110920.

Tufto, J. (2015) Genetic evolution, plasticity, and bet-hedging as adaptive responses to temporally autocorrelated fluctuating selection: A quantitative genetic model: Evolutionary responses to fluctuating. *Evolution* 69, 2034–2049.

Tylianakis, J.M., Didham, R.K., Bascompte, J. and Wardle, D.A. (2008) Global change and species interactions in terrestrial ecosystems. *Ecology Letters* 11, 1351–1363.

Uller, T. (2008) Developmental plasticity and the evolution of parental effects. *Trends in Ecology & Evolution* 23, 432–438.

van Baaren, J., Boivin, G., Visser, B. and Le Lann, C. (2024) Bet-hedging in parasitoids: When optimization is not the best strategy to cope with climatic extremes. *Current Research in Insect Science* 5, 100076.

van de Putten, W.H., Macel, M. and Visser, M.E. (2010) Predicting species distribution and abundance responses to climate change: Why it is essential to include biotic interactions across trophic levels. *Philosophical Transactions of the Royal Society B: Biological Sciences* 365, 2025–2034.

van Nouhuys, S. and Lei, G. (2004) Parasitoid–host metapopulation dynamics: The causes and consequences of phenological asynchrony. *Journal of Animal Ecology* 73, 526–535.

Zytynska, S.E., Tighiouart, K. and Frago, E. (2021) The benefits and costs of hosting facultative symbionts in plant-sucking insects: A meta-analysis. *Molecular Ecology* 30, 2483–2494.

3 Biocontrol, Climate Change and Population Dynamics: Why is an Increase of Pest Outbreaks and Plant Diseases Transmitted by Vectors Expected Following Climate Changes?

Gang Ma[1]* and Chun-Sen Ma[2]

[1]*State Key Laboratory for Biology of Plant Diseases and Insect Pests, Institute of Plant Protection, Chinese Academy of Agricultural Sciences, Beijing, China;*
[2]*School of Life Science, Institutes of Life Science and Green Development, Hebei University, Baoding, China*

Abstract

Climate change has profoundly affected insect pests and their natural enemies. Changes in climate may contribute to pest outbreaks and insect vector-borne plant disease epidemics by modifying abiotic and biotic environmental factors and thereby altering spatial-temporal synchronizations of insect pests and their natural enemies. The mismatches between interacting species can also alter the structure and dynamics of insect communities. Generally, the impact of climate change on species at higher trophic levels is expected to be greater than that at lower trophic levels, which is likely to hinder biocontrol efficacy. Moreover, insect pests are expected to present better plastic and/or evolutionary adaptation capacities due to their shorter generation time and greater niche breadth than their natural enemies. All of these factors can scale up to facilitate higher potential of pest outbreaks and disease epidemics under climate change, challenging the biocontrol functions in protecting plant and food security worldwide.

3.1 Introduction

Under global climate change, the reduction of insect diversity and abundance has received widespread attention, and many studies have reported substantial reductions at the global and regional scales (Outhwaite *et al.*, 2022). Outbreaks of plant pests and diseases transmitted by insect vectors are also expected and observed to be closely related to climate change (Deutsch *et al.*, 2018; Ma *et al.*, 2021a; b), leading to increases in global crop yield loss (Deutsch *et al.*, 2018) and damage to global forests by pests and diseases (Marini *et al.*, 2017). Meanwhile, the natural biological control services of pests in many ecosystems are gradually decreasing (Wu *et al.*, 2016; Furlong and Zalucki,

*Corresponding author: magang@caas.cn

DOI: 10.1079/9781800625099.0003

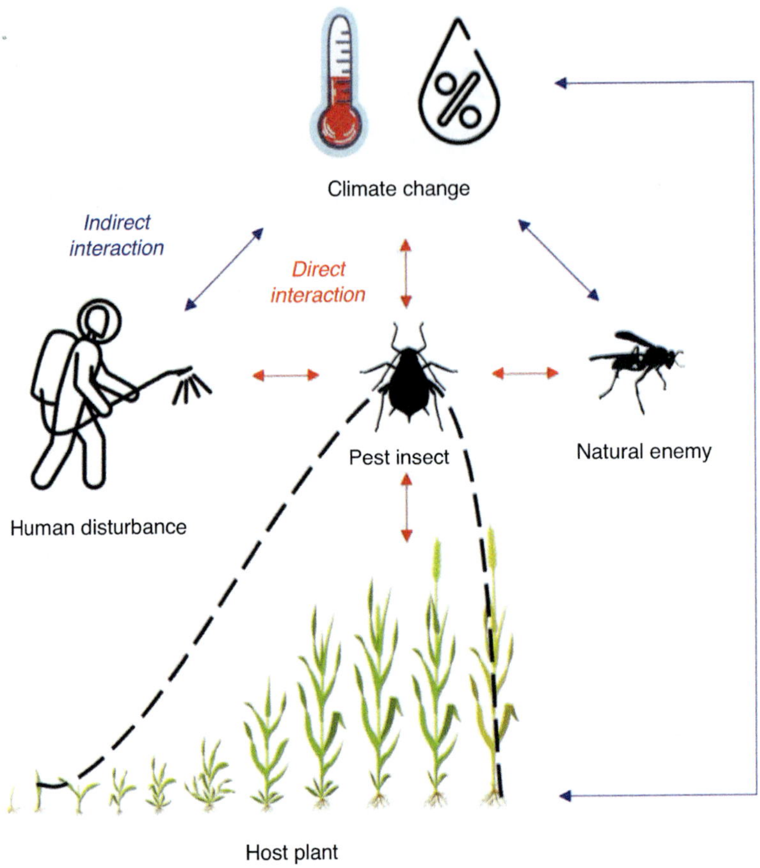

Fig. 3.1. A conceptual framework illustrating the abiotic and biotic factors potentially regulating the population dynamics (dashed line) of pest insects (e.g. aphid, vectoring plant viral diseases). On the one hand, the insect population growth can be affected directly by climate change (temperature, humidity, etc.), host plant (e.g. phenology), natural enemy (e.g. parasitoid) and human disturbance (e.g. pesticide use). On the other hand, climate change may influence insect population growth indirectly by modifying the conditions of the host plant, natural enemy and human disturbance. The b-directional arrows represent the interactions between pest insects and environmental factors via direct (red) and indirect (blue) pathways. Thus, climate change may facilitate pests directly by providing favourable climate conditions or indirectly by promoting their adaptability to abiotic/biotic factors and weakening biological control efficacy.

2017). Unfortunately, we do not currently have a clear understanding about the pathways by which climate change is causing the outbreaks of plant pests and diseases.

Generally, the population dynamics of pest insects, and thereby the diseases transmitted by insect vectors, are mainly regulated by abiotic and biotic environmental factors (Fig. 3.1). Abiotic factors mainly include climate conditions such as temperature and humidity (precipitation/drought) and human activity disturbances such as land use, pesticides, fertilization, irrigation, etc. Biological factors involve the components of the food web closely related to the ecosystems, including the availability of host plant resources and the natural biological control of pests by predatory/parasitic enemies. Insects, as ectotherms, are significantly affected by global temperature rise, reflected in their earlier and longer activity timing, faster development, more generations per year and constantly changing geographical distributions (Ma *et al.*, 2021a). An increase in the number of annual generations is expected to accelerate the evolutionary responses of insects to selection pressures. In addition, climate change also affects the effectiveness of biological control by altering species interactions. Together, these pathways can scale up to potentially cause pest outbreaks and disease epidemics under climate change.

In this chapter, we discuss the potential causes of the outbreaks of pests and insect-vectoring diseases caused by climate change. First, we explore how changes in abiotic climate factors such as temperature and drought can alter the fitness of insects. Then, we discuss how insects can improve their adaptability to environments in the context of climate change. Finally, by focusing on the interspecific interactions between insect pests and natural enemies, we discuss the changing trends of biocontrol service functions under climate change. This chapter may allow us to understand the widely debated scientific question of why pest insects and insect vector disease outbreaks are becoming increasingly severe with climate change.

3.2 Favourable Climates Promoting Pest Fitness

Climate conditions are the most important abiotic factor affecting the population growth of insects in both natural and agricultural ecosystems. As such, climate change can reduce or even eliminate the negative effects of climate factors on insect population growth, which may be beneficial for the population growth of pest insects (Ma and Ma, 2022). For example, an increase in temperature can lead to earlier phenology of pest emergence and delayed phenology of winter overwintering, resulting in an extended occurrence period and an increase in the number of annual generations. Changes in drought or precipitation can directly influence insects or affect insects indirectly through changes to host plants (Hamann *et al.*, 2021). Importantly, warming and drought can combine to increase the fitness of pest insects (Dale and Frank, 2017). Suitable climate conditions also help insects expand their distribution ranges, leading to the risk of emerging pests in new distribution areas (Osawa *et al.*, 2018).

3.2.1 Temperature

The increase in average temperature is one of the most significant features of global climate change, which has substantial impacts on insects. In theory, insect performance exhibits a single peak bell-shaped curve with temperature increase (Fig. 3.2A). When the temperature is below optimal, an increase is beneficial for insect population growth. By contrast, when the temperature is above optimal, an increase can depress insect population growth. Overall, for insects distributed in cold and temperate regions, because their current average ambient temperatures are expected to be much lower than optimal temperatures, temperature increase may be beneficial for their population growth. In constrast, for insects in tropical areas, an increase in temperature is likely to be detrimental due to their average ambient temperatures approaching optimal temperatures (Deutsch *et al.*, 2008). This may explain why pest outbreaks are more common in mid to high latitudes (Schneider *et al.*, 2022). For example, in Bavaria, Germany, a long-term observation during 1800–2011 showed that the voltinism of a forest pest, the common pine sawfly *Diprion pini*, is influenced by temperature, with more outbreaks in hot summers (Haynes *et al.*, 2014). Importantly, according to the thermal performance curve, the relationship between temperature and insect fitness is non-linear, suggesting a sharp increase in insect fitness with warming in the areas with cooler basal temperatures. For example, in the past 40 years, the distribution of the pine beauty moth *Panolis flammea* in Finland has expanded from 64°N to 68°51'N at a speed of nearly 10.2 kilometres per year, which is much faster than previous predictions (Pulgarin Díaz *et al.*, 2022). As the temperature rises, ecosystem stability weakens, making it easier for pests to break out (Nelson *et al.*, 2013). However, in the Mediterranean region with hot summers, heat waves can offset the increase in pine processionary moth *Thaumetopoea pityocampa* diseases caused by winter warming (Domingo *et al.*, 2024).

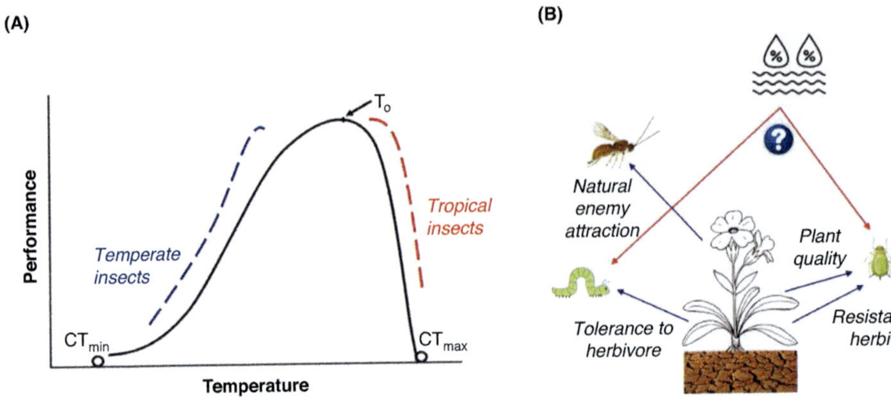

Fig. 3.2. Effects of (A)temperature and (B)humidity on insects in the context of climate change. A. Insect performance shows a unimodal curve with temperature theoretically. An increase in temperature below the optimal temperature (T_o) is beneficial for insect fitness while temperature rise approaching or above T_o would depress insect population growth. Warming can benefit temperate insects because their average ambient temperatures are expected to be much lower than T_o whereas temperature increase may be detrimental for tropical insects because their average ambient temperatures are already close to T_o. However, microclimate heterogeneity and thermoregulation may help insects to buffer the impacts of extreme high temperatures. Thus, warming will potentially facilitate the population growth of insect pests. B. Main direct (red arrows) and indirect (blue arrows) effects of drought on insect pests under climate change. The direct effects are uncertain depending on the sensitivity of focal insect species or populations to the changes in humidity. Nevertheless, drought is expected to influence insects indirectly via modifying the biochemical conditions of host plants, which is likely to facilitate insect pests by increased plant quality, declined ability to herbivore resistance and tolerance and decreased natural enemy attraction of the host plants.

3.2.2 Drought

Changes in precipitation patterns have led to increases in the intensity and frequency of drought in many regions, which may also have significant impacts on insect population growth. Drought can directly affect insects by altering the ambient air or soil humidity (Fig. 3.2B). However, such effects may be uncertain, depending on the sensitivity of focal species in response to the changes in humidity. Drought can also affect insects indirectly by modifying the chemical defence, nutritional quality, digestibility and palatability of plants (Jamieson *et al.*, 2017). Water stress is generally thought to increase the vulnerability of leaves to defoliators by increasing solid concentrations (Jactel *et al.*, 2012). Outbreaks of pine insect pests tend to follow drought. After long periods of low precipitation, the increase in leaf fall caused by pine caterpillars may be a result of drought-induced changes in host tree suitability (Haynes *et al.*, 2014). A meta-analysis from 2012 showed that water stress often causes greater damage by deciduous insects compared to unstressed trees (Jactel *et al.*, 2012). By analysing the spatiotemporal pattern of the outbreak of the main invasive pest of pine trees, the woodwasp *Sirex noctilio*, in South America, the effect of the pest in 29 pine forests in Argentina was reconstructed over a 16-year period. Outbreaks of the pest are caused by a sudden increase in drought, which may increase the availability of susceptible host trees (Lantschner *et al.*, 2019). Herbivorous insects have been considered to perform better on plants under water stress (Huberty and Denno, 2004). During drought, the accumulation of N compounds in plant tissues may lead to outbreaks of herbivores, thereby promoting the growth and reproduction of herbivores. Drought stress increases herbivorous feeding in agricultural systems, and the likelihood of reduced plant resistance to herbivores is expected to rise (Hamann *et al.*, 2021). In addition, the increase in nutrient concentration in plant tissues enhances the effectiveness of soil nutrients through frass and litter deposition, and herbivorous feeding leads to positive feedback on the severity of outbreaks (De Grandpré *et al.*, 2022).

3.2.3 Combination of temperature and drought

The additive effects of warming and drought often lead to more severe pest outbreaks. For example, a field experiment showed that the concurrent increase of urban warming and drought stress led to increased abundance in the scale pest insect (*Melaaspis tenebricosa*), which manifested as the additive effect of temperature and drought stress as increases of female production and body size with temperature were greater on dry stressed than watered trees (Dale and Frank, 2017). In Europe, a 30-year multinational observation showed that rising temperatures and insufficient summer precipitation resulted in the outbreak of the bark beetle (*Ips typographus*) while such effects can be dampened by the occurrence of heavy rains (Marini *et al.*, 2017). In East Africa, six serious locust outbreaks were reported between 1912 and 1989, all of which were closely related to long-term drought, warm winters and high precipitation in spring and summer (Peng *et al.*, 2020). A recent survey in South Africa showed that warm winters, dryness and heat can affect the prevalence of insect pests (Phophi *et al.*, 2020).

The increase in pest populations caused by the combined effects of warming and drought has led to the prevalence of insect-borne diseases. For example, museum specimen data suggest that the emergence of plant diseases may be related to an increase in the number of insect vectors. Since the mid-1990s, potato growers in the western United States, Mexico and Central America have experienced severe yield losses to the zebra chip disease and have had to suppress the population of the insect vector, the potato psyllid *Bactericera cockerelli*, by increasing insecticide use (Zeilinger *et al.*, 2017). In southern Africa, climate change has led to short winters, high temperatures and long periods of drought, resulting in an increase in the number of vector pests carrying destructive viral pathogens on various vegetables and the emergence of new viral diseases caused by whitefly-transmitted viruses (Mafongoya *et al.*, 2019).

3.3 Enhanced Environmental Adaptability of Pests

Changes in climate and its associated environments have impacted insects significantly. However, at the same time, insects can improve their adaptability to environmental changes through plasticity and evolution. On the one hand, although previous studies indicate that the ability of insects to improve their heat tolerance through thermal plasticity and adaptive evolution is relatively limited (Gunderson and Stillman, 2015; Bennett *et al.*, 2021), the evidence for thermal plasticity and evolution is found to be pervasive in insects (Hoffmann and Sgrò, 2011; Weaving *et al.*, 2022). On the other hand, to cope with climate variability and extremes, as well as the changes in their food resources, insects can respond rapidly to environmental changes by modifying their life history (van Asch *et al.*, 2013; Kearney *et al.*, 2018; Martin *et al.*, 2022). Moreover, the interaction between climate change and artificial chemical input into ecosystems may facilitate pest insects via reducing pesticide efficacy and promoting pesticide resistance (Pu *et al.*, 2020; Crossley *et al.*, 2024). Furthermore, climate warming may promote pesticide resistance via expanding the overwintering distribution range of pest insects (Ma *et al.*, 2021b).

3.3.1 Plasticity and rapid evolution in heat tolerance

Evidence from experimental selections

The increase in average temperature and the intensity and frequency of extreme heat events are the challenges that insects face under climate change (Deutsch *et al.*, 2008; Ma *et al.*, 2021a). In this case, there is evidence in laboratory and field conditions that some herbivorous insects can rapidly respond to climate change via plasticity and evolution (Muñoz-Valencia *et al.*, 2016; Zhu *et al.*, 2021). For instance, the whitefly *Bemisia tabaci* (MED and MEAM1) can enhance its heat tolerance after selections for only a few generations (Ma *et al.*, 2014; Muñoz-Valencia *et al.*, 2016). Although a fitness cost of decreased fertility is expected, recent studies concerning experimental selection showed that the promoted capacity of heat tolerance is not necessarily related to reduced reproductivity, which suggests independent pathways of thermal evolution (Muñoz-Valencia *et al.*, 2016; Mesas *et al.*, 2021). For example, the aphid *Rhopalosiphum padi* can maintain both high heritability of heat tolerance (critical thermal maximum, CT_{max}) and reproductive capacity simultaneously after short-term selection, which is in line with its robust adaptation to subtropical climates (Zhu *et al.*, 2021). Some species, such as the invasive fruit fly *Drosophila subobscura*, can even increase both CT_{max} and thermal optimum when selected in an ecologically relevant low-ramping rate (Mesas *et al.*, 2021).

Field evidence in natural populations

Some insects in natural ecosystems have improved their heat tolerance after a period of hot years. For example, the harlequin bug *Murgantia histrionica* is an important herbivorous insect feeding on the plant species *Isomeris arborea* in California, USA. In 2009, it was found that no eggs could develop to the second nymphal stage at 33°C. In the following decade, there were some of the hottest years on record, with average temperatures continuously rising, accompanied by an increase in the frequency and duration of extreme high temperatures. A repeated experiment in 2021 showed that eggs could now develop into reproductive adults at 33°C. The upper thermal limit for maturity and survival increases, and as age and temperature increase, the risk of death decreases. All observed traits are developing towards greater heat tolerance, which suggests that sustained exposure to heat stress may be at least part of the reason for these changes (Ardelan *et al.*, 2023).

A recent study found significant lineage changes in the main pest of alfalfa, the pea aphid *Acyrthosiphon pisum*, on a latitudinal gradient of approximately 600 km in central Chile. After 10 years of unprecedented climate warming, a new clone was found to almost completely replace the original aphid super-colonies. The new clone exhibits higher population growth rates at higher temperatures. These results support the idea of supercolonial transformation mediated by differential reproductive performance at changing temperatures (Martel *et al.*, 2024).

3.3.2 Plasticity and evolution in life history

Evolutionary phenological responses to host plants

The synchronization of phenology between insect pests and their host plants determines the availability of their food resources, which is crucial for the persistence and growth of pest populations. Many studies have shown that, on average, climate warming can lead to a smaller advance in the phenology of herbivorous insects compared to the plant resources they rely on, resulting in a mismatch between pests and their resources (Thackeray *et al.*, 2016). However, there are also many herbivorous insects that can rapidly respond in advance to the phenology of their host plants through evolution, buffering their mismatch with changes in host plant phenology (Weir and Phillimore, 2024).

On the other hand, when thermal responses of herbivorous insects are more sensitive to climate warming than their host plants, the insects may delay their phenology to maintain synchronization with the host plant. For example, the winter moth *Operophtera brumata* is the main leaf-chewing pest insect of oak *Quercus robur*. The leaves of oak trees are only suitable for short-term growth, which is critical for the hatching eggs and newly emerged caterpillars as food resources. Even the difference of a few days between egg hatching and oak bud opening has significant fitness consequences. The opening of oak buds and the hatching of winter moth eggs both depend on temperature, and as temperature increases, both phenologies have advanced in recent decades. However, the hatching of winter moth eggs showed a stronger response to temperature rise, resulting in a decrease in the synchronization between oak trees and winter moths (van Asch *et al.*, 2013). The mismatch of phenology caused by climate change can lead to serious individual fitness consequences, including significantly increasing the mortality rate of newly hatched caterpillars and affecting their growth and development time. At the population level, phenological mismatch has a significant impact on population density, and in years with significant phenological mismatch, population growth rates are lower (van Dis *et al.*, 2023). However, long-term observations have shown that the hatching date of winter moth eggs has undergone genetic changes (delays) to ensure greater synchronization with oak bud burst. Therefore, changes in selection pressure under climate change can lead to rapid evolutionary adaptation responses of insects to match host phenological changes caused by climate change (van Asch *et al.*, 2013).

Bet-hedging strategies coping with climate variability

Pest insects may evolve bet-hedging strategies to cope with the changes in interannual climates or the alternating of growth–dormancy seasons. These strategies can enable a certain proportion of individuals to track the synchronization with their foraging/mating resources when missing the previous phenological window. For example, the spruce beetle *Dendroctonus rufipennis* diapauses obligately as an adult, but low temperatures can cause a facultative prepupal diapause, resulting in a semivoltine life cycle. However, warm temperatures may enable this species to complete a univoltine life cycle to avoid a facultative prepupal diapause (Schebeck *et al.*, 2017). The semivoltine individuals of the pine processionary moth *Thaumetopoea pityocampa* can emerge 1 year later than their own cohort and were found to be synchronized with the univoltine individuals that appeared in the same year, rather than with their own cohort, in order to optimize mating probability, which means an adaptive response to new climate conditions (Martin *et al.*, 2022).

The European cherry fruit fly *Rhagoletis cerasi* undergoes obligatory pupal diapause during the autumn and winter seasons, requiring sufficient chilling to terminate diapause (Moraiti *et al.*, 2014). When the chilling requirements for diapause termination cannot be met (insufficient or prolonged chilling), individuals can deal with specific climates by adopting two different pathways. A cold winter may lead to prolonged coldness, resulting in another facultative diapause to allow adults to emerge 1 year later. In contrast, warm winters may result in insufficient low temperatures, causing prolonged dormancy to delay diapause termination.

Summer diapause buffering extreme high temperatures

With ongoing climate warming, insects have to face frequent high temperature events. Under the strong selection pressure of hot climates, some insects may adopt summer diapause to change their life history and avoid the adverse effects of high temperatures. In hot summers in China and Africa, the cotton bollworm *Helicoverpa armigera* enters diapause as pupae (Nibouche, 1998; Liu *et al.*, 2006). When the third instar larvae were exposed to 37°C, a large proportion of individuals (males 94.7%; females 60.9%) went into diapause. After 60 days of storage at this temperature, 64.3% of dormant pupae survived, with most adults emerging subsequently (Nibouche, 1998). Under high temperature conditions (33–39°C), the weight and energy storage capacity of summer diapause pupae are much higher than those of non-diapause pupae (Liu *et al.*, 2006).

The onion maggot *Delia antiqua* is one of the most important insect pests for onions and garlic worldwide. This species exhibits overlapping or slightly separated peaks in the second and third generations during summer in higher latitude regions such as the United States, Canada and Germany (42°N to 53°N). However, in lower latitude regions such as South Korea (33°N to 35°N), the population ceases activity for more than 2 months during summer, and the third generation only begins to appear after summer (Shin *et al.*, 2023). This indicates that this species responds to the hot summer in South Korea through summer diapause.

In Australia, the matchstick grasshopper *Warramaba virgo* responds to extreme high temperatures through egg diapause when the diurnal temperature cycle exceeds 36°C in order to maintain a univoltine lifecycle (Kearney *et al.*, 2018).

3.3.3 Reduced insecticide efficacy and promoted pesticide resistance

Reduced insecticide efficacy

The efficacy of pesticides is closely related to environmental conditions. As such, changes in climate conditions can lead to a decrease in the insecticide sensitivity of pest insects. First, the use of chemicals to control pests may become less effective as climate changes (Pu *et al.*, 2020) because existing studies have shown that changes in temperature, humidity and CO_2 concentration can affect the toxicity of many insecticides used in pest management, resulting in reduced efficacy. The elevated temperature and CO_2 level can lead to a decrease in pesticide efficacy either directly by modifying pesticide metabolism and transportation or indirectly by increasing pesticide detoxification in plants, thereby reducing the pesticide effectiveness to target pests (Matzrafi, 2019). Under laboratory conditions, the reduced sensitivity of pests to pesticides has been widely detected under conditions of elevated temperature and CO_2 concentration. For example, an increase in temperature can reduce the biological control effect of avermectins on the western flower thrip *Frankliniella occidentalis* (Li *et al.*, 2017a). As the concentration of CO_2 increases, the control effect of triazophos on the brown planthopper *Nilaparvata lugens* declines (Ge *et al.*, 2013). Similar situations have also been found in field pest management. In the south-eastern United States, the whitefly *Bemisia tabaci* has become a major threat to vegetable production. Under hotter conditions, whiteflies are more abundant in landscapes with higher levels of insecticide use (Crossley *et al.*, 2024).

Promoted insecticide resistance

Evolutionary responses to climate change may lead to rapid changes in the frequency of insecticide resistance gene alleles. This is because these insecticide resistance genes have a pleiotropic effect on other traits, such as heat or dry tolerance. Therefore, insecticide resistance and climate adaptation may evolve together. A change in one trait can have a negative or positive impact on another by balancing or promoting it (Kliot and Ghanim, 2012). For example, climate adaptation will promote the development of insecticide resistance, leading to an increase in the alleles of insecticide resistance. This promoting effect is due to the shared molecular basis of climate adaptation and insecticide resistance. Therefore, the selection of one trait will also benefit the other. For example, there is a positive correlation between the resistance of the small brown planthopper *Laodelphax striatellus* to thiazide and its heat tolerance (Li *et al.*, 2017b).

In many regions of Australia, analysis of pyrethroid resistance to the main agricultural pest mite *Halloydeus destructor* showed that regional differences in drought, temperature seasonality and precipitation patterns affect spatial patterns of resistance. Climate regions with increased voltinisms are positively correlated with pesticide resistance, indicating that increased voltinisms accelerate evolutionary responses to selection pressure (Maino *et al.*, 2018). The level of insecticide resistance of pest insects is related to the overwintering range. Compared to locations that only occur seasonally, the average insecticide resistance in overwintering locations is 158 times higher in the diamondback moth *Plutella xylostella*, an important pest of cruciferous vegetables worldwide. By promoting local sustainability throughout the year, climate change can increase and expand the resistance of this destructive species globally, potentially causing significant economic losses (Ma *et al.*, 2021b).

3.4 Weakened Efficacy of Biological Control

Climate change can affect the effectiveness and function of biological control by altering interactions between herbivorous insects and their natural enemies. For example, changes in winter climates may disrupt the seasonal phenological patterns of parasitoid wasps and modify the synchronization of host–parasitoid interactions. In this case, decreases in winter cold intensity and duration can alter the seasonal ecological evolution of parasitoid wasps although their responses to climate change are greatly limited by their host insects (Tougeron *et al.*, 2020). The differences in response of pests and natural enemies to temperature changes can also impact on biological control functions through changes in species phenology and distribution. Meanwhile, climate change can also affect the function and efficiency of micro-organisms as biological control agents by altering the interactions between insect hosts and microbial natural enemies. In addition, under the trend of global climate change, the reduction of insect biodiversity may further reduce the natural biological control functions mediated by insect diversity.

3.4.1 Desynchronization between herbivorous insects and their natural enemies

Differences in thermal tolerance

The differences in heat tolerance among insects at different nutritional levels can alter the relative fitness and abundance of interacting species, causing significant impacts on ecosystems. Generally, if the predator/parasitoid can tolerate higher temperatures than its prey, then extreme high temperatures are beneficial for biological control. By contrast, if prey can tolerate higher temperatures than their predators/parasitoids, extreme temperatures are not conducive to biological control (Ma *et al.*, 2021a).

In the herbivorous insect and parasitoid wasp system, a study based on published data analysed the relative heat demand and tolerance of Hymenopteran parasitoid wasps and their hosts. The optimal temperature for the development of parasitoids (T_o) is significantly lower than that of the hosts. Considering the limited plasticity of insects in their response to high temperatures and the proximity of T_o to the critical thermal maximum, this suggests that host–parasitoid interactions may be negatively affected by global temperature increases (Furlong and Zalucki, 2017). In addition, extreme high temperature events can also reduce parasitism in the system of Lepidopteran pests and *Trichogramma* as demonstrated by a meta-analysis (Wu *et al.*, 2016). Therefore, in the context of continuously rising global temperatures, herbivorous pests seem to be more advantaged than parasitoid wasps (Furlong and Zalucki, 2017). Studies in different regions around the world, such as North America, Africa and Australia, have found that populations of the diamondback moth *P. xylostella* have higher heat tolerance than its local parasitoids such as *Cotesia vestalis*, *Diadema semalausum* and *Diadema insulin* (Bahar *et al.*, 2013; Machekano *et al.*, 2018; Wang *et al.*, 2022).

Similarly, in systems involving herbivorous insects and predators, the heat tolerance of pests is often higher than that of their natural enemies. For example, the tolerance of the predatory mites *Euseius stimulatus* and *Neoseiulus californicus* to extreme temperatures is significantly lower than that of their prey *Oligonychus perseae*. The sustained high temperatures in the summer of southern Spain significantly prolonged the negative effect of high temperatures on the populations of these two predatory mites, leading to a decrease in the control effect of predatory mites on their prey (Montserrat *et al.*, 2013). Also, high temperatures can decrease the hoverfly *Epipsychus baltaneus* fitness and their biocontrol potential (Noel *et al.*, 2022).

Mismatch in phenology

The phenology of insects is influenced by climate change, mainly driven by phenotypic plasticity and evolutionary changes. Any changes in the seasonal activity of a species can affect the interactions within and between trophic levels. The occurrence of asynchrony is mainly due to the different responses of species to climate change. A meta-analysis applied the climate sensitivity profile method to 10,003 terrestrial and aquatic phenolic datasets, which were spatially matched with temperature and precipitation data to quantify changes in climate sensitivity. It was found that the timing of phenological events in primary consumers, including herbivorous pests, varies more than that of secondary consumers such as natural enemies, and thus is more sensitive to climate change (Thackeray *et al.*, 2016). Insect species change their seasonal activity by altering their life cycle or their developmental rate. The relative changes in phenology can generate strong bottom-up effects, thereby affecting consequences at the community and ecosystem levels (Damien and Tougeron, 2019).

The weather in different years or locations can affect the biological control of pests by altering phenological matching between local specific parasitoid wasps and their host insects. A 10 year study in Utah, USA, has shown that the life cycle timing of the cereal leaf beetle *Oulema melanopus* varies more with warm rather than cold springs than its natural enemy, the parasitoid wasp *Tetrastichus julis*. As such, in a warm spring, the larval phenology of the beetle is relatively delayed compared to the activity of parasitoid wasp adults, resulting in a decrease in parasitism (Evans *et al.*, 2013). A survey conducted in six chestnut orchards in northern Italy from 2018 to 2022 found that as temperatures rise in winter and spring, the parasitoid wasp *Torymus sinensis* exhibits earlier phenology than the Asian chestnut gall wasp *Dryocosmus kuriphilus*, leading to a mismatch in phenology and a decrease in biological control efficiency. In the orchards with relatively lower winter temperatures, the parasitism rate by *T. sinensis* was high (77%~99%). Conversely, in the orchards with relatively higher winter temperatures (+2.3–2.7°C), the parasitism rate was low (44%~50%). Warmer winter temperatures led to an early emergence of the biocontrol agent in the spring when the hosts were not available (Ferracini *et al.*, 2022).

Mismatch in distribution

If pest distribution shifts to the areas outside the distribution of natural enemies, the effectiveness of natural enemies in controlling pests will decrease, leading to potential pest outbreaks (Thomson *et al.*, 2010; Ramos Aguila *et al.*, 2023). For example, a model study on the interaction between the Australian diamondback moth and its parasitoid wasp *Diadegma seminaum* suggests that predicted temperature increase will have a greater negative impact on the distribution of parasitoid wasps than on the diamondback moth and may lead to their exclusion from some currently important agricultural areas (Furlong and Zalucki, 2017). A species distribution model (Maxent) was used to predict the current and future distribution of three important lepidopteran stem borer pests (*Busseola fusca*, *Chilo partellus* and *Semia calamitis*) in East African maize as well as two parasitoid wasps (*Cotesia flavipes* and *C. sesamiae*) currently used in biocontrol. Based on these predictions and field surveys by local farmers in Kenya and Tanzania, future corn yield losses were also predicted. The results indicate that under more severe climate change, biocontrol will be less effective, thereby reducing the ability of biological control to suppress corn yield losses caused by pests (Jendritzki *et al.*, 2023).

Climate change may also help pests expand their distribution range by invading new climate zones (Ma and Ma, 2022). The spotted wing drosophila *Drosophila suzukii* is a highly harmful fruit pest. The potential of the parasitoid wasp *Leptopilina japonica* as a biocontrol agent against *D. suzukii* was evaluated using a species distribution model. The results indicated that globally, only half of countries estimate that biological control coverage is greater than 50%. Therefore, in the release plan for specific locations, the niche specificity of pests and parasitoid wasps must be considered in order to effectively control pests (Nair and Peterson, 2023). In addition to being influenced by climate through host plants and related herbivores, adaptive management strategies adopted by farmers to address climate change will also alter the abundance and activity of natural enemies. These strategies may lead to spatial and temporal mismatches between pests and enemies, reducing the effectiveness of their biological control (Thomson *et al.*, 2010; Jendritzki *et al.*, 2023).

3.4.2 Reduced microorganism functioning as biocontrol services

Climate change can also affect the function and efficiency of micro-organisms as biocontrol agents by altering the interactions between insect hosts and microbial natural enemies (such as entomopathogenic fungi, entomopathogenic nematode, etc.). For example, *Bacillus thuringiensis* is a biocontrol microbial agent used as a substitute for chemical pesticides to control the armyworm on grasslands. The increase in temperature is beneficial for the tolerance of the armyworm on corn to *B. thuringiensis*-based bioproducts, which in turn reduces the effectiveness of biological control (Carrizo *et al.*, 2023). Entomopathogenic nematodes are effective biological control agents for fruit flies. However, nematodes are more sensitive to temperature change, leading to the nematode *Steinernema carpocapsae* losing its efficacy at low diurnal temperature ranges (Mastore *et al.*, 2023). Drought and changes in precipitation can also affect the effectiveness of microbial biocontrol agents in controlling pests. Water stress can reduce the biocontrol effect of nucleopolyhedrovirus on the fall armyworm *Spodoptera frugiperda* (Real-Santillán *et al.*, 2023). Water stress and plant nutrition can affect the growth of viruses, thereby affecting the efficiency of biological control and pest management, especially in situations where water and nutrient limitations may increase under climate change (Real-Santillán *et al.*, 2023). Reduced precipitation can lead to increased pest damage by decreasing the biocontrol effect of enteropathogenic nematodes such as *Heterorhabditis bacteriophora* against the larvae of the Western corn rootworm *Diabrotica virgifera virgifera* (Toepfer *et al.*, 2010).

The interaction between multiple factors related to climate change will reduce the biological control function of microbial agents. By manipulating the three major abiotic factors associated with climate change (atmospheric CO_2, temperature and soil moisture), the individual and joint effects of these factors on the interaction between maize, the banded cucumber beetle *Diabrotica balteata* and the entomopathogenic nematode *Heterorhabditis bacteriophora* (EPN) were investigated. Elevated temperature and CO_2 directly hinder the infectivity potential of EPN, while lower water levels improve it. In the RCP 8.5 scenario, temperature and CO_2 showed an interactive effect on the infectivity of EPN, with an overall decrease of 40%. Therefore, the problem of root pests may worsen with climate change due to increased performance of herbivores and reduced top-down control of biocontrol agents (Guyer *et al.*, 2021).

3.4.3 Declining insect diversity weakens biodiversity-mediated biocontrol

Insect diversity plays a crucial role in directly supporting ecosystem services through pollination and natural pest control. Due to the mismatch between pests and natural enemies in space and time, climate change may increase the risk of pest outbreaks (Steffan-Dewenter *et al.*, 2024). In recent years, the abundance of terrestrial insects has generally declined (Raven and Wagner, 2021). A recent study investigated the long-term trends of 923 terrestrial insect communities monitored in 106 studies and found that their abundance and species richness also decreased accordingly. The observed decrease in total insect abundance can be mainly explained by a general decrease in previously abundant species (van Klink *et al.*, 2024). Using distribution data of over 600 arthropod species collected from Italy over the past 150 years, it was found that rapid changes in precipitation make communities unstable and maximize colonization and extinction rates (Marta *et al.*, 2021).

Declining insect diversity is most likely to impact natural biological control. By using 18 year radar and light trap datasets from East Asia (2003–2020) to evaluate the temporal population trends of 98 aerial insect species and characterize relevant H-NE interactions, it was found that over time, the summer abundance of natural enemy insects decreased by 19.3%. In a food web composed of 124 pest–enemy pairs, the abundance of NE decreases by 0.7% annually. The study revealed how the gradual decrease in insect populations weakens pest nutrition regulation and ecosystem stability at the macro scale (Zhou *et al.*, 2023). In New Hampshire, USA, the abundance and diversity of Coleoptera insects sharply declined from the mid-1970s to the mid-2010s, and the reduction in winter snow depth and duration was identified as the main reason for the decline in beetle diversity. Compared to the mid-1970s, the abundance of beetles in all taxa decreased by 83%, the number of taxa decreased by 39%, and the number of predatory beetles decreased the most (Harris *et al.*, 2019).

3.5 Conclusion

Although a global decline in insect diversity and abundance has been reported in recent years (Outhwaite *et al.*, 2022), the outbreaks of pest insects and their transmitted diseases (Deutsch *et al.*, 2018; Ma *et al.*, 2021a, b) as well as weakened natural biocontrol services (Wu *et al.*, 2016; Furlong and Zalucki, 2017) are also widely expected and frequently detected under climate change, challenging both natural and agricultural ecosystems worldwide (Marini *et al.*, 2017; Deutsch *et al.*, 2018). In this chapter, we have discussed the potential mechanisms, such as favourable climates, promoted environmental adaptability and weakened biocontrol efficacy, which may underlie why climate change can facilitate pest outbreaks and insect vector-borne plant disease epidemics. However, the impacts of climate change on fitness and population dynamics are multifaceted and complex. Many other possible pathways may potentially contribute to insect outbreaks, such as avoiding climate extremes through behavioural thermoregulation and microclimate heterogeneity exploration (Ma *et al.*, 2021a), warming-induced changes in pest dominance and community structure (Ma *et al.*, 2015), exacerbated plant disease through enhancing interactive facilitation of coinfested insect vectors (Zhu *et al.*, 2023), emerging pests due to climate change-associated biological invasion, insect range expansion and consequent potential insect hybridization (Corrêa *et al.*, 2019; Schneider *et al.*, 2022). Moreover, natural biocontrol functioning may further decrease owing to interspecific competition between endoparasitoids (Costaz *et al.*, 2023) or the strengthened hyperparasitoids disrupting biocontrol services (Tougeron and Tena, 2019) under climate change. Therefore, from the perspective of sustainable pest management using biological control, more attention should be paid to how to enhance the response of natural enemies to stress resistance and adaptability while inhibiting that of pest insects in the context of ongoing climate change.

References

Ardelan, A., Tsai, A., Will, S., McGuire, R. and Amarasekare, P. (2023) Increase in heat tolerance following a period of heat stress in a naturally occurring insect species. *Journal of Animal Ecology* 92, 2039–2051.

Bahar, M.H., Hegedus, D., Soroka, J., Coutu, C., Bekkaoui, D. *et al.* (2013) Survival and Hsp70 gene expression in *Plutella xylostella* and its larval parasitoid *Diadegma insulare* varied between slowly ramping and abrupt extreme temperature regimes. *PLOS One* 8(9), e73901.

Bennett, J.M., Sunday, J., Calosi, P., Vilalobos, F., Martinez, B. *et al.* (2021) The evolution of critical thermal limits of life on earth. *Nature Communications* 12, 1198.

Carrizo, A.E., de Valle Loto, F., Baigori, M.D. and Pera, L.M. (2023) *Bacillus thuringiensis*-based bioproduct: Characterization and performance against *Spodoptera frugiperda* strains in maize under different environmental temperatures. *Neotropical Entomology* 52, 283–291.

Corrêa, A.S., Cordeiro, E.M. and Omoto, C. (2019) Agricultural insect hybridization and implications for pest management. *Pest Management Science* 75, 2857–2864.

Costaz, T.P.M., de Jong, P.W., Harvey, J.A., van Loon, J.J.A., Dicke, M. *et al.* (2023) Temperature affects the outcome of competition between two sympatric endoparasitoids. *Animal Behaviour* 203, 11–20.

Crossley, M.S., Smith, O.M., Barman, A.K., Croy, J.R., Schmidt, J.M. *et al.* (2024) Warmer temperatures trigger insecticide-associated pest outbreaks. *Pest Management Science* 80, 1008–1015.

Dale, A.G. and Frank, S.D. (2017) Warming and drought combine to increase pest insect fitness on urban trees. *PLOS One* 12(3), e0173844.

Damien, M. and Tougeron, K. (2019) Prey–predator phenological mismatch under climate change. *Current Opinion in Insect Science* 35, 60–68.

De Grandpré, L., Marchand, M., Kneeshaw, D.D., Paré, D., Boucher, D. *et al.* (2022) Defoliation-induced changes in foliage quality may trigger broad-scale insect outbreaks. *Communications Biology* 5, 463.

Deutsch, C., Tewksbury, J.J., Huey, R.B., Sheldon, K.S., Ghalambor, C.K. *et al.* (2008) Impacts of climate warming on terrestrial ectotherms across latitude. *Proceedings of the National Academy of Sciences USA* 105, 6668–6672.

Deutsch, C.A., Tewksbury, J.J., Tigchelaar, M., Battisti, D., Merrill, S.C. *et al.* (2018) Increase in crop losses to insect pests in a warming climate. *Science* 361, 916–919.

Domingo, D., Vicente-Serrano, S.M., Gómez, C., Olano, J. and Sanguessa-Barreda, G. (2024) Summer heat waves could counterbalance the increasing incidence of pine processionary due to warmer winters in Mediterranean pine forests. *Forest Ecology and Management* 555, 121695.

Evans, E.W., Carlile, N.R., Innes, M.B. and Pitigala, N. (2013) Warm springs reduce parasitism of the cereal leaf beetle through phenological mismatch. *Journal of Applied Entomology* 137, 383–391.

Ferracini, C., Pogolotti, C. and Alma, A. (2022) A mismatch in the emergence of *Torymus sinensis* may affect the effectiveness of this biocontrol agent. *Biological Control* 174, 105029.

Furlong, M.J. and Zalucki, M.P. (2017) Climate change and biological control: The consequences of increasing temperatures on host-parasitoid interactions. *Current Opinion in Insect Science* 20, 39–44.

Ge, L.Q., Wu, J.C., Sun, Y.C., Ouyang, F. and Ge, F. (2013) Effects of triazophos on biochemical substances of transgenic Bt rice and its nontarget pest *Nilaparvata lugens* Stål under elevated CO_2. *Pesticide Biochemistry and Physiology* 107, 188–199.

Gunderson, A.R. and Stillman, J.H. (2015) Plasticity in thermal tolerance has limited potential to buffer ectotherms from global warming. *Proceedings of the Royal Society B* 282, 20150401.

Guyer, A., Van Doan, C., Maurer, C., Machado, R.A., Mateo, P. *et al.* (2021) Climate change modulates multitrophic interactions between maize, a root herbivore, and its enemies. *Journal of Chemical Ecology* 47, 889–906.

Hamann, E., Blevins, C., Franks, S.J., Jameel, M.I. and Anderson, J.T. (2021) Climate change alters plant–herbivore interactions. *New Phytologist* 229, 1894–1910.

Harris, J.E., Rodenhouse, N.L. and Holmes, R.T. (2019) Decline in beetle abundance and diversity in an intact temperate forest linked to climate warming. *Biological Conservation* 240, 108219.

Haynes, K.J., Allstadt, A.J. and Klimetzek, D. (2014) Forest defoliator outbreaks under climate change: Effects on the frequency and severity of outbreaks of five pine insect pests. *Global Change Biology* 20, 2004–2018.

Hoffmann, A. and Sgrò, C. (2011) Climate change and evolutionary adaptation. *Nature* 470, 479–485.

Huberty, A.F. and Denno, R.F. (2004) Water stress and its consequences for herbivorous insects: A new synthesis. *Ecology* 85, 1383–1398.

Jactel, H., Petit, J., Desprez-Loustau, M.-L., Delzon, S., Piou, D. *et al.* (2012) Drought effects on damage by forest insects and pathogens: A meta-analysis. *Global Change Biology* 18, 267–276.

Jamieson, M.A., Burkle, L.A., Manson, J.S., Runyon, J.B., Trowbridge, A.M. *et al.* (2017) Global change effects on plant–insect interactions: The role of phytochemistry. *Current Opinion in Insect Science* 23, 70–80.

Jendritzki, I., Tonnang, H.E.Z., Calatayud, P.A., Borgemeister, C., Johansson, T. *et al.* (2023) Uncertainties in the effectiveness of biological control of stem borers under different climate change scenarios in Eastern Africa. *Climatic Change* 176, 56.

Kearney, M.R., Deutscher, J., Kong, J.D. and Hoffmann, A.A. (2018) Summer egg diapause in a matchstick grasshopper synchronizes the life cycle and buffers thermal extremes. *Integrative Zoology* 13, 437–449.

Kliot, A. and Ghanim, M. (2012) Fitness costs associated with insecticide resistance. *Pest Management Science* 68, 1431–1437.

Lantschner, M.V., Aukema, B.H. and Corley, J.C. (2019) Droughts drive outbreak dynamics of an invasive forest insect on an exotic host. *Forest Ecology and Management* 433, 762–770.

Li, H.B., Zheng, Y.T., Sun, D.D., Wang, J.J. and Du, Y.Z. (2017a) Combined effects of temperature and avermectins on life history and stress response of the western flower thrips, *Frankliniella occidentalis*. *Pesticide Biochemistry and Physiology* 108, 42–48.

Li, Y., Zhang, Y., Liu, X. and Guo, H. (2017b) Does resistance to buprofezin improve heat and cold tolerance of *Laodelphax striatellus* (Hemiptera: Delphacidae)? *Environmental Entomology* 46, 988–994.

Liu, Z., Gong, P., Wu, K., Sun, J. and Li, D. (2006) A true summer diapause induced by high temperatures in the cotton bollworm, *Helicoverpa armigera* (Lepidoptera: Noctuidae). *Journal of Insect Physiology* 52, 1012–1020.

Ma, G. and Ma, C.S. (2022) Potential distribution of invasive crop pests under climate change: Incorporating mitigation responses of insects into prediction models. *Current Opinion in Insect Science* 49, 15–21.

Ma, C.S., Ma, G. and Pincebourde, S. (2021a) Survive a warming climate: Insect responses to extreme high temperatures. *Annual Review of Entomology* 66, 163–184.

Ma, C.S., Zhang, W., Peng, Y., Zhao, F., Chang, X.Q. *et al.* (2021b) Climate warming promotes pesticide resistance through expanding overwintering range of a global pest. *Nature Communications* 12, 5351.

Ma, F.-Z., Lü, Z.-C., Wang, R. and Wan, F.-H. (2014) Heritability and evolutionary potential in thermal tolerance traits in the invasive Mediterranean cryptic species of *Bemisia tabaci* (Hemiptera: Aleyrodidae). *PLOS One* 9, e103279.

Ma, G., Rudolf, V. and Ma, C.S. (2015) Extreme temperature events alter demographic rates, relative fitness, and community structure. *Global Change Biology* 21, 1794–1808.

Machekano, H., Mvumi, B.M. and Nyamukondiwa, C. (2018) Loss of coevolved basal and plastic responses to temperature may underlie trophic level host-parasitoid interactions under global change. *Biological Control* 118, 44–54.

Mafongoya, P., Gubba, A., Moodley, V., Chapoto, D., Kisten, L. *et al.* (2019) Climate change and rapidly evolving pests and diseases in Southern Africa. In: Ayuk, E. and Unuigbe, N. (eds) *New Frontiers in Natural Resources Management in Africa*, Vol. 53. Natural Resource Management and Policy, Springer, Cham, Switzerland.

Maino, J.L., Umina, P.A. and Hoffmann, A.A. (2018) Climate contributes to the evolution of pesticide resistance. *Global Ecology and Biogeography* 27, 223–232.

Marini, L., Økland, B., Jönsson, A.M., Bentz, B., Carroll, A. *et al.* (2017) Climate drivers of bark beetle outbreak dynamics in Norway spruce forests. *Ecography* 40, 1426–1435.

Marta, S., Brunetti, M., Manenti, R., Provenzala, A. and Ficetola, G.F. (2021) Climate and land-use changes drive biodiversity turnover in arthropod assemblages over 150 years. *Nature Ecology and Evolution* 5, 1291–1300.

Martel, S.I., Zamora, C.A., Ricote, N., Sepúlveda, D.A., Mahéo, F. *et al.* (2024) Rapid turnover of a pea aphid superclone mediated by thermal endurance in Central Chile. *Proceedings of the Royal Society B* 291, 20232462.

Martin, J.-C., Mesmin, X., Buradino, M., Rossi, J.-P. and Kerdelhué, C. (2022) Complex drivers of phenology in the pine processionary moth: Lessons from the past. *Agricultural and Forest Entomology* 24, 247–259.

Mastore, M., Quadroni, S., Rezzonico, A. and Brivio, M.F. (2023) The influence of daily temperature fluctuation on the efficacy of bioinsecticides on spotted wing Drosophila larvae. *Insects* 14, 43.

Matzrafi, M. (2019) Climate change exacerbates pest damage through reduced pesticide efficacy. *Pest Management Science* 75, 9–13.

Mesas, A., Jaramillo, A. and Castañeda, L.E. (2021) Experimental evolution on heat tolerance and thermal performance curves under contrasting thermal selection in *Drosophila subobscura*. *Journal of Evolutionary Biology* 34, 767–778.

Montserrat, M., Guzman, C., Sahun, R.M., Belda, J. and Hormaza, J. (2013) Pollen supply promotes, but high temperatures demote, predatory mite abundance in avocado orchards. *Agriculture, Ecosystems and Environment* 164, 155–161.

Moraiti, C.A., Nakas, C.T. and Papadopoulos, N.T. (2014) Diapause termination of *Rhagoletis cerasi* pupae is regulated by local adaptation and phenotypic plasticity: Escape in time through bet-hedging strategies. *Journal of Evolutionary Biology* 27, 43–54.

Muñoz-Valencia, V., González-Obando, R. and Díaz, F. (2016) Rapid responses to a strong experimental selection for heat hardening in the invasive whitefly *Bemisia tabaci* MEAM 1. *Entomologia Experimentalis et Applicata* 160, 147–155.

Nair, R.R. and Peterson, A.T. (2023) Mapping the global distribution of invasive pest *Drosophila suzukii* and parasitoid *Leptopilina japonica*: Implications for biological control. *PeerJ* 11, e15222.

Nelson, W.A., Bjornstad, O. and Yamanaka, T. (2013) Recurrent insect outbreaks caused by temperature-driven changes in system stability. *Science* 341, 796–799.

Nibouche, S. (1998) High temperature induced diapause in the cotton bollworm *Helicoverpa armigera*. *Entomologia Experimentalis et Applicata* 87, 271–274.

Noel, G., Caetano, J., Blanchard, S., Boullis, A. and Francis, F. (2022) High temperatures adversely affect the hoverfly *Episyrphus balteatus* (Diptera: Syrphidae) fitness and aphid prey consumption. *Turkish Journal of Zoology* 46, 186–193.

Osawa, T., Yamasaki, K., Tabuchi, K., Yoshioka, A., Ishigooka, Y. *et al.* (2018) Climate-mediated population dynamics enhance distribution range expansion in a rice pest insect. *Basic and Applied Ecology* 30, 41–51.

Outhwaite, C.L., McCann, P. and Newbold, T. (2022) Agriculture and climate change are reshaping insect biodiversity worldwide. *Nature* 605, 97–102.

Peng, W., Ma, N.L., Zhang, D., Zhou, Q., Yue, X. *et al.* (2020) A review of historical and recent locust outbreaks: Links to global warming, food security and mitigation strategies. *Environmental Research* 191, 110046.

Phophi, M.M., Mafongoya, P. and Lottering, S. (2020) Perceptions of climate change and drivers of insect pest outbreaks in vegetable crops in limpopo province of South Africa. *Climate* 8, 27.

Pu, J., Wang, Z. and Chung, H. (2020) Climate change and the genetics of insecticide resistance. *Pest Management Science* 76, 846–852.

Pulgarin Díaz, J.A., Melin, M. and Tikkanen, O.P. (2022) Thermal sum drives abundance and distribution range shift of *Panolis flammea* in Finland. *Scandinavian Journal of Forest Research* 37(2), 93–98.

Ramos Aguila, L.C., Li, X., Akutse, K.S., Bamisile, B.S., Sánchez Moreano, J.P. *et al.* (2023) Host–parasitoid phenology, distribution, and biological control under climate change. *Life* 13, 2290.

Raven, P.H. and Wagner, D.L. (2021) Agricultural intensification and climate change are rapidly decreasing insect biodiversity. *Proceedings of the National Academy of Sciences USA* 118(2), e2002548117.

Real-Santillán, R.O., del-Val, E., Williams, T., Martinez-Castillo, A., Gavito, M. *et al.* (2023) Water stress decreases the biocontrol efficacy of a nucleopolyhedrovirus against the fall armyworm on maize. *Journal of Pest Science* 97, 1315–1324.

Schebeck, M., Hansen, E.M., Schopf, A., Ragland, G.J., Stauffer, C. *et al.* (2017) Diapause and overwintering of two spruce bark beetle species. *Physiological Entomology* 42, 200–210.

Schneider, L., Rebetez, M. and Rasmann, S. (2022) The effect of climate change on invasive crop pests across biomes. *Current Opinion in Insect Science* 50, 100895.

Shin, Y.K., Kim, S., Park, C.G. and Kim, D.S. (2023) A new perspective to understand the late season abundance of *Delia antiqua* (Diptera: Anthomyiidae): A modeling approach for the hot summer effect. *Insects* 14, 816.

Steffan-Dewenter, I., Kerr, R.B. and Peters, M.K. (2024) Insect diversity for agroecosystem resilience in a changing climate. *One Earth* 7, 541–544.

Thackeray, S.J., Henrys, P.A., Hemming, D., Bell, J., Botham, M. *et al.* (2016) Phenological sensitivity to climate across taxa and trophic levels. *Nature* 535, 241–245.

Thomson, L.J., Macfadyen, S. and Hoffmann, A.A. (2010) Predicting the effects of climate change on natural enemies of agricultural pests. *Biological Control* 52, 296–306.

Toepfer, S., Hatala Zseller, I., Ehlers, R.U., Peters, A. and Kuhlmann, U. (2010) The effect of application techniques on field-scale efficacy: Can the use of entomopathogenic nematodes reduce damage by western corn rootworm larvae? *Agricultural and Forest Entomology* 12, 389–402.

Tougeron, K. and Tena, A. (2019) Hyperparasitoids as new targets in biological control in a global change context. *Biological Control* 130, 164–171.

Tougeron, K., Brodeur, J., Le Lann, C. and van Baaren, J. (2020) How climate change affects the seasonal ecology of insect parasitoids. *Ecological Entomology* 45, 167–181.

van Asch, M., Salis, L., Holleman, L.J.M., Lith, B. and Visser, M.E. (2013) Evolutionary response of the egg hatching date of a herbivorous insect under climate change. *Nature Climate Change* 3, 244–248.

van Dis, N.E., Sieperda, G.J., Bansal, V., Lith, B., Wertheim, B. *et al.* (2023) Phenological mismatch affects individual fitness and population growth in the winter moth. *Proceedings. Biological Sciences* 290, 20230414.

van Klink, R., Bowler, D.E., Gongalsky, K.B., Shen, M., Swengel, S. *et al.* (2024) Disproportionate declines of formerly abundant species underlie insect loss. *Nature* 628, 359–364.

Wang, L., Etebari, K., Zhao, Z., Walter, G.H. and Furlong, M.J. (2022) Differential temperature responses between *Plutella xylostella* and its specialist endo-larval parasitoid *Diadegma semiclausum* – implications for biological control. *Insect Science* 29, 855–864.

Weaving, H., Terblanche, J.S., Pottier, P. and English, S. (2022) Meta-analysis reveals weak but pervasive plasticity in insect thermal limits. *Nature Communications* 13, 5292.

Weir, J.C. and Phillimore, A.B. (2024) Buffering and phenological mismatch: A change of perspective. *Global Change Biology* 30, e17294.

Wu, L.H., Hoffmann, A.A. and Thomson, L.J. (2016) Potential impact of climate change on parasitism efficiency of egg parasitoids: A meta-analysis of *Trichogramma* under variable climate conditions. *Agriculture, Ecosystems and Environment* 231, 143–155.

Zeilinger, A.R., Rapacciuolo, G., Turek, D., Oboyski, P.T., Almeida, R.P.P. *et al.* (2017) Museum specimen data reveal emergence of a plant disease may be linked to increases in the insect vector population. *Ecological Applications* 27, 1827–1837.

Zhou, Y., Zhang, H., Liu, D., Khashaveh, A., Li, Q. *et al.* (2023) Long-term insect censuses capture progressive loss of ecosystem functioning in East Asia. *Science Advances* 9, eade9341.

Zhu, L., Hoffmann, A.A., Li, S.M. and Ma, C.S. (2021) Extreme climate shifts pest dominance hierarchy through thermal evolution and transgenerational plasticity. *Functional Ecology* 35, 1524–1537.

Zhu, L., Xue, Q., Ma, G. and Ma, C.S. (2023) Climate warming exacerbates plant disease through enhancing commensal interaction of co-infested insect vectors. *Journal of Pest Science* 96, 945–959.

4 Weed Biocontrol with Introduced Insects Under Climate Change

Yan Sun[1]*, Shen Nang Huang[1], Suzanne T.E. Lommen[2], Evan Siemann[3], Jianqing Ding[4] and Heinz Müller-Schärer[1,5]

[1]College of Resources and Environment, Huazhong Agricultural University, Wuhan, China; [2]Institute of Biology, Above-Belowground Interactions Group, Leiden University, Leiden, The Netherlands; [3]Department of Biosciences, Rice University, Houston, USA; [4]State Key Laboratory of Crop Stress Adaptation and Improvement, School of Life Sciences, Henan University, Kaifeng, China; [5]Department of Biology, University of Fribourg, Fribourg, Switzerland

Abstract

This chapter investigates the complex challenges of using introduced insects for weed biocontrol in the context of climate change. It explores how changing climatic factors, such as temperature, precipitation and CO_2 levels, affect the physiology, behaviour and ecological interactions of both invasive plants and their biological control agents (BCAs). These changes can greatly impact the effectiveness and safety of biocontrol efforts. This chapter emphasizes the need to integrate evolutionary biology with ecological modelling to accurately predict and manage these impacts. Additionally, it addresses the potential for non-target impacts resulting from altered habitats and phenological overlaps caused by climate change. By proposing interdisciplinary strategies, this chapter offers a framework for developing sustainable, eco-friendly biocontrol methods as traditional chemical approaches become less viable. This synthesis of current research is crucial for those seeking to improve weed management practices in the context of global climate change and the One Health concept.

4.1 Introduction

In this chapter, we delve into the intricate trophic interplay of plant–herbivore dynamics, a complexity climate change is anticipated to amplify. The profound effects of climate change, such as alterations in temperature, precipitation, CO_2 levels, insolation and the frequency of extreme climatic events, are shaping all stages of the plant invasion process, from introduction and establishment up to spread and impact. Concomitantly, they also affect growth and development of associated herbivores as well as antagonists at higher trophic levels. Together, these changes subsequently impact species interactions, ecosystem services and human livelihoods. The chapter asserts that classical biological control has historically been an effective and cost-efficient strategy to mitigate plant invasions (Müller-Schärer and Schaffner, 2008, 2020). However, understanding how climate change alters the dynamics between host plants and biocontrol agents (BCAs) remains largely unexplored.

*Corresponding author: yan.sun@mail.hzau.edu

DOI: 10.1079/9781800625099.0004

The complexity of interactions involving invasive plants, biotic and abiotic components under varying climatic scenarios presents policy and management challenges. These include assessing effectiveness and biosafety of biocontrol methods under different climate change scenarios. Specifically, climate change is expected to affect invasive weeds and BCAs through impacts on metabolism, phenology, physiology and their interactions, as well as effects on non-target attack of native plant species.

As we scrutinize the interplay between climate change and weed biocontrol, we observe that climate change may severely alter interactions of invasive weeds and BCAs by influencing phenotypic traits such as physiology, biochemistry, life history and phenology, or by changing the abiotic and/or biotic environments in which these interactions occur. This can modify the frequency, timing, intensity and duration of interactions between plant invaders and BCAs, thereby enhancing or reducing effects on plant invaders. Because plants differ in their susceptibility to damage at different life stages, changes in timing of insect and weed emergences under climate change could modify the consequences of insect feeding. Geographic distribution changes, especially with warming and more frequent extreme events, can impact interactions between BCAs and weeds as differential migration rates of insects, host plants, higher trophic level agents and other species create novel biotic interactions and new above- and below-ground communities. The net effect of climate change on weed biocontrol will depend on the relative strengths of these various responses to multiple climate change factors.

Understanding evolutionary aspects is also crucial as invaders and BCAs rapidly evolve in response to environmental shifts. Many cases of rapid adaptive evolution have been reported for invasive plants, including shifts in resource allocation from defence to growth, local adaptation to new habitats and climates, evolution of phenotypic clines along climatic gradients, evolution of greater dispersal ability,and increases in rates of population growth and expansion (Helliwell *et al.*, 2018; Sotka *et al.*, 2018). For instance, van Boheemen *et al.* (2019) found a shift to warmer and wetter climates in the introduced European and Australian ranges of the North American native *Ambrosia artemisiifolia* L. (Asteraceae). Despite the recent introductions into Australia and a bottleneck limiting genetic variation, their results indicate rapid and repeated adaptation to more productive environments, potentially contributing to trait divergence between the ranges.

Evolutionary adaptation is also expected for BCAs when they encounter novel environments or changes in climatic conditions, especially for species with short generation times. Integrating evolutionary studies into different stages of weed biocontrol programmes can help predict and adapt management strategies to these changes. For instance, using reciprocal transplants or common garden experiments coupled with population modelling, researchers could match climatically adapted BCA candidates with their target release environments or evaluate their potential evolvability under changing selection pressures (Müller-Schärer *et al.*, 2020). High genetic variance in life history and abiotic stress tolerance traits may allow researchers to screen for strains or populations that could adapt to future climates.

In conclusion, the intricate trophic relationships among plants, herbivores and their natural enemies under the lens of climate change reveal a complex, dynamic and challenging field of study, necessitating an interdisciplinary approach that integrates ecological, evolutionary and modelling perspectives to foresee and mitigate the impacts of climate change on weed biocontrol. Practitioners of classical biological control of invasive weeds are confronted with a dual expectation: successfully control plant invaders while avoiding damage to non-target plants and adverse indirect effects. In the following, we specifically address (1) effectiveness and (2) biosafety of weed biocontrol under climate change, and then (3) future opportunities for weed biocontrol in the light of increasing plant invasions due to global change and decreasing availability of alternative management tools (herbicide bans and absence of new molecules). Thus, eco-friendly management methods are becoming increasingly important as an integral part of the One Health concept that emphasizes biological control (Essack, 2018).

4.2 Classical Biological Control under Climate Change

Classical biological control involves intentional introductions of natural enemies from the invasive plant's native region to suppress its populations in the new environment (Müller-Schärer and Schaffner, 2008). Climate change introduces variables that can affect the biology and ecology of invasive plants and their BCAs. The complexity of these interactions under varying climatic conditions presents unique challenges for policy and management strategies in biocontrol programmes. Integrating considerations of climate change into biocontrol strategies is crucial for ensuring their effectiveness and environmental responsibility in the long term.

With growing concern about climate change effects and plant invasions, publications and their citations of plant invasions under climate change have sharply increased over the past 20 years. This is paralleled by increases in studies on climate change and weed biocontrol, highlighting the interest in exploring biological control measures as an integrated and adaptable tool to successfully manage plant invasions under climate change conditions (Fig. 4.1). However, our understanding of biocontrol outcomes under the full complement of climate-driven changes remains rudimentary (Fig. 4.2). Multiple reports suggest that climate change may promote plant invasions and increase their impact (Dukes and Mooney, 1999; Bradley *et al.*, 2010; Gioria *et al.*, 2023), while studies of climate change effects on performance and impact of BCAs report both positive and negative outcomes, with variable consequences for their efficacy and safety (Sun *et al.*, 2020a).

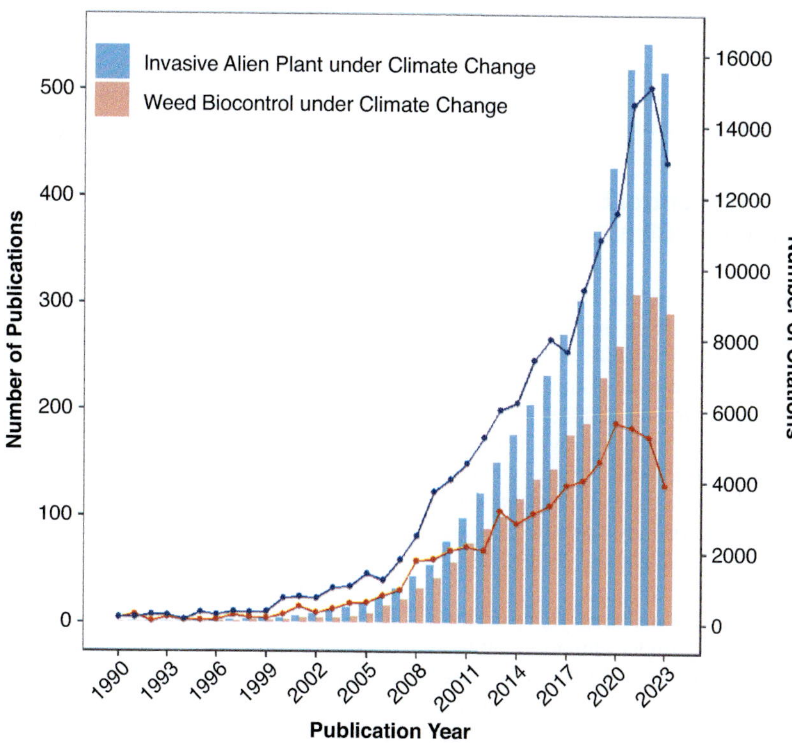

Fig. 4.1. Literature survey since 1990 on increase of publications on invasive weeds under climate change compared to increase of weed biocontrol under climate change. Bars represent the number of publications per year and the lines represent the sums of citations per year.

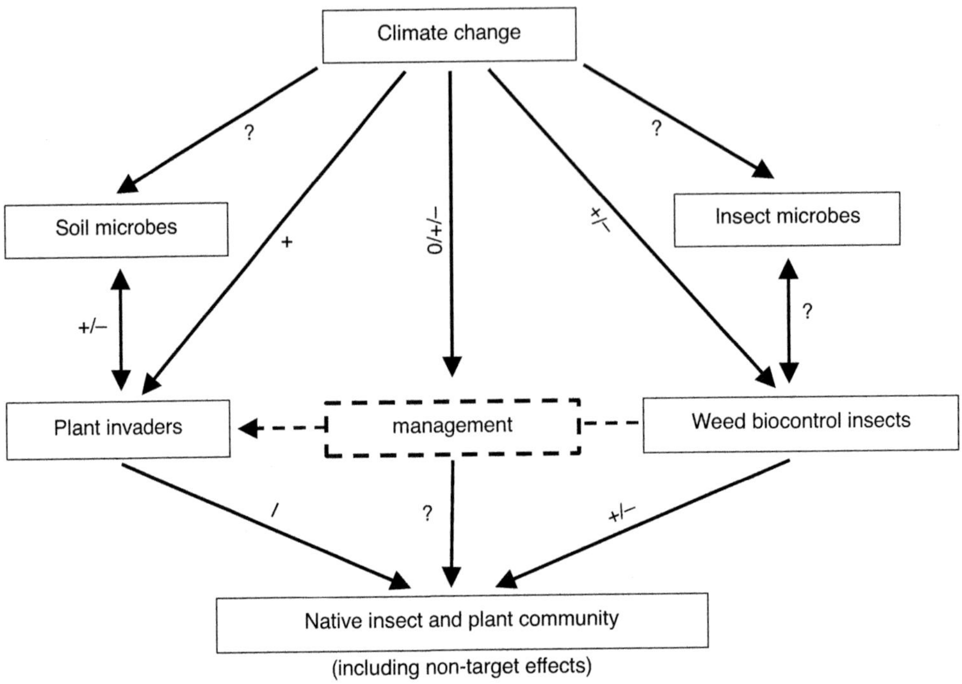

Fig. 4.2. Impact of climate change on plant invasions, weed biocontrol insects and their interactions (i.e. management; dashed lines) as evidenced from literature. 0/+/– represent no, positive and negative effects. Modified from Sun *et al.* (2020a).

4.3 Weed Biocontrol Efficiency and Impact under Climate Change

Climate change is expected to affect performance, population build-up and spread of both plant invaders and their BCAs independently, and therefore the frequency, timing, intensity and duration of their interactions. Thus, predicting agent impacts on plant invaders under climate change conditions is complex and highly demanding and net effects of climate change on weed biocontrol will depend on relative strengths of responses to multiple climate change factors (see Fig. 4.2).

In the following, we address sharp declines in insects worldwide under climate change vs increases in insect pests and ask: (i) are weed biocontrol insect impacts like natural insects (decreasing) or crop pests (increasing) under climate change? (ii) How will various climate change factors affect BCA impacts? (iii) Climate change will affect invader demography and population genetics, thus how to best match genotypes and phenotypes of invaders and BCA to optimize biocontrol effectiveness. (iv) Will recent advances in predictive modelling tools including life-history traits and climate-dependent vital rates allow spatially explicit long-term forecasts of BCA success and impacts?

4.3.1 Are weed biological control agents winners or losers under climate change?

A recent review reported dramatic declines of insect species worldwide that may lead to extinction of 40% of insect species over the next few decades, with Lepidoptera, Hymenoptera and Coleoptera being most affected in terrestrial ecosystems (Sánchez-Bayo and Wyckhuys, 2019). Coleoptera and Lepidoptera are also the taxa most often used in weed biological control projects (Schwarzländer *et al.*, 2018). The main drivers of species decline in order of importance appear to be habitat change, pollution (mainly synthetic pesticides and fertilizers), biological traits and climate change, with climate warming accounting for 12.6% of the decline, mainly in tropical areas. Concurrently, a small number of species are becoming more abundant; these are adaptable, mobile generalist species occupying newly vacant niches (Thomas *et al.*, 2004; Sánchez-Bayo and Wyckhuys, 2019).

In contrast, a recent report released by the Intergovernmental Panel on Climate Change (IPCC, the UN's authority on climate change) (Legg, 2021) addressed how climate changes will affect crop losses worldwide, especially from insects. For instance, Deutsch *et al.* (2018) estimate grain (rice, maize and wheat) yield losses of 10–25% per 1°C of global warming, especially in temperate regions that produce most grain. This reflects rising temperatures boosting insect digestion rates and warming temperatures in temperate regions increasing insect activity and reproduction. Furthermore, warming is expected to increase crop losses to insect pests through geographic range expansions, human-assisted introductions, biological invasions of pest species, effects on natural enemies of pests and changes in plant nutrient content and defence. Some responses may be compounded by interactive effects of warming with other factors, such as changing rainfall patterns and increasing atmospheric CO_2 concentrations. Thus, how will insect BCAs influence invasions from establishment, performance and population build-up, up to spread and impact?

First, BCAs must be host plant specific and reared free of antagonists. They often originate from one or few populations from a limited eco-geographic range (Sun *et al.*, 2017). Habitat loss and conversion to intensive agriculture and urbanization will favour the target host and so the BCA. Biological factors, including pathogens and introduced species, are not expected to negatively impact BCAs, at least in the short term. Pollution (mainly synthetic pesticides and fertilizers) will only negatively affect BCAs targeted on crop weeds, but not the 90% of BCAs that are used against environmental weeds (Müller-Schärer *et al.*, 2018). Like many insect pests, BCAs rely on a single host plant in a given area, but crops, noxious weeds and plant invaders are abundant and predictable, and thus less affected by the main drivers of insect decline. Based on this, BCAs are rather like insect pests and thus their performance and impacts are expected to increase under rising temperatures and CO_2 levels, especially in temperate regions. In biological control against insect pests, the BCAs are predators and parasitoids, i.e. at the top of the trophic chains, while in the control of invasive plants, the BCAs are phytophagous insects, so at a lower level of the trophic chain. Based on this, the weed BCAs should be less affected by climate change than the BCAs of insect pests, with either positive or negative outcomes for the population increase of the weed BCA and thus for the impacts on the plant invaders.

4.3.2 Drivers of climate change affecting herbivorous insects

Insects are poikilotherms so temperature is probably the critical environmental factor affecting their behaviour, distribution, development and reproduction (Damos and Savopoulou-Soultani, 2012; see also Chapter 1 of this book). Warming is expected to increase overwintering survival, increasing the number of generations but with lower body weight and thus fecundity, although with overall increasing abundance. Postintroduction expansion to newly suitable areas under climate change is expected, but impacts will greatly depend on parallel spread and phenological synchrony with the target host, which remains difficult to predict.

Theoretically, in temperate regions, temperature increase is expected to result in optimal metabolic activity and thus population increase and range expansion, but not in the tropics, where temperatures are already optimal for many insects for rapid growth (Deutsch *et al.*, 2018; Skendžić *et al.*, 2021). As outlined above, this response to climate warming has been well documented for insects in general (population decline mainly in the tropics) and insect pests in particular (increasing crop losses due to insects mainly in temperate regions). Unfortunately, based on our bibliographic review (see Fig. 4.1), this has not yet been explored for outcomes of weed BCAs.

Effects of elevated CO_2 on insect performance are mainly mediated through changes in host plant quality. For instance, Baso *et al.* (2021) investigated the effects of elevated atmospheric CO_2 on the biological control of four invasive aquatic weeds, *Azolla filiculoides* Lam. (Salviniaceae), *Salvinia molesta* Mitch. (Salviniaceae), *Pistia stratiotes* L. (Araceae) and *Myriophyllum aquaticum* (Vell.) Verdc. (Haloragaceae), which presently are under successful control by their respective BCAs *Stenopelmus rufinasus* Gyllenhal (Coleoptera: Curculionidae), *Cyrtobagous salviniae* Calder and Sands (Coleoptera: Curculionidae), *Neohydronomus affinis* Hustache (Coleoptera: Curculionidae) and *Lysathia sp.* (Coleoptera: Chrysomelidae), respectively, in South Africa. At elevated CO_2, the authors found a general overall increase in biomass production across all weed species, and a reduced insect feeding damage, except for *S. molesta*. They conclude that although plants will respond differently to CO_2 increase, the general trend suggests that these species will become more challenging to manage through biological control in future. In another experimental study, Paper *et al.* (2023) examined the interaction between water hyacinth *Pontederia crassipes* Mart. (Pontederiaceae), one of the world's worst aquatic weeds and a target for biological control, and two insect biocontrol agents, the leaf-chewing *Cornops aquaticum* Brüner (Orthoptera: Acrididae) and the phloem-feeding *Megamelus scutellaris* Berg (Hemiptera: Delphacidae) under elevated CO_2. *Pontederia crassipes* (Mart.) Solms (Pontederiaceae) growth responses increased when exposed to insect herbivory, with chewing herbivory by *C. aquaticum* being consistent across CO_2 conditions, whereas the feeding by *M. scutellaris* increased substantially at elevated CO_2. This indicates that successful biological control of water hyacinth under elevated CO_2 conditions might rely on phloem-feeding insects, with chewers playing a lesser role.

Similarly, changing precipitation patterns may result in more frequent and intense drought and flooding. On the one hand, heavy rainfall can wash away insect larvae and eggs from plants and decrease overwintering probability through flooding and prolonged stagnation of water, while drought-stressed plants may become more susceptible to insect attack because of decreased chemical defences. Henrikson *et al.* (2018) studied the interaction between the plant invader *Alternanthera philoxeroides* (Mart.) Griseb. (Amaranthaceae) and the introduced biocontrol beetle *Agasicles hygrophila* Selman and Vogt (Coleoptera: Chrysomelidae: Halticinae) in terrestrial (well-watered) and flooded environments in Texas. They found that the hydrological environment may affect invasive plant performance by altering herbivore oviposition and feeding preferences. Under controlled conditions, flooding increased leaf toughness and decreased beetle consumption, and beetles preferred to feed and oviposit on terrestrial plants in choice trials.

In summary, a multitude of direct and indirect host plant-mediated effects of various climate change factors can alter the frequency, timing, intensity and duration of interactions between plant invaders and BCAs and, as the above examples show, impacts of BCAs on plant invaders may be enhanced or reduced. Furthermore, changes in geographic distributions, especially as the climate continues to warm and extreme events become more common, can impact interactions between BCAs and weeds as differential migration rates of insects, host plants, higher trophic levels and other species create novel biotic interactions. Sun *et al.* (2017) used a biogeographic approach and species distribution models to select among six BCA candidates and their combinations to potentially cover a large European range of the North American invader *A. artemisiifolia*, under both current and future bioclimatic conditions. Besides a considerable discrepancy in geographic range overlap between Europe (31.4%) and North America (83.3%), which might be partially due to the expansion of ragweed in Europe and the fact that habitats with high ragweed occurrences in Europe are rare in North America, and predicted to be unsuitable for the candidates, they found total geographic range of all candidates combined to decrease under climate change in both ranges, but that the species will respond differently. Complex modelling tools are therefore urgently needed to make realistic predictions of biological control outcomes under future climate conditions (see below).

4.3.3 How to best match genotypes and phenotypes between plant invaders and BCAs to optimize biocontrol effectiveness

Many cases of rapid adaptive evolution have been reported for invasive plants, including shifts in resource allocation from defence to growth (Coley, *et al.*, 1985; Orians and Ward,

2010), local adaptation to new habitats (Colautti and Barrett, 2013) and climates such as by the evolution of phenotypic clines along climatic gradients (Yang *et al.*, 2021), evolution of greater dispersal ability (Alex Perkins *et al.*, 2013) and increases in rates of population growth and expansion (Lee, 2010; Hodgins *et al.*, 2018). According to the evolution of increased competitive ability (EICA) hypothesis (Blossey and Nötzold, 1995), mechanical and/or chemical defensive characters may be adaptive in the presence of enemies, but they become unnecessary and costly when these enemies are absent, such as when introduced into new areas. This has been widely tested, for instance in *Silene latifolia* Poiret (Caryophyllaceae), which was introduced from Europe to North America c.200 years ago. A common garden experiment in Europe revealed that plants from North America have evolved a two-to-threefold higher reproductive potential, at the expense of higher susceptibility to fungal infection, fruit predation and aphid infestation, compared to plants from Europe (Wolfe *et al.*, 2004). Rapid evolutionary change can also occur in BCAs post introduction in response to new climatic and environmental conditions (Suarez and Tsutsui, 2008; Roderick *et al.*, 2012), especially for species with short generation times (Van Asch *et al.*, 2013; Szűcs *et al.*, 2017; Wright and Bennett, 2018). A recent review of experiments and theory suggests BCA changes following introduction may be larger than previously considered (Szűcs *et al.*, 2019). As outlined above (see Fig. 4.2), *a priori* predictions of interactions between a plant invader and its BCAs are difficult, even more so when including a parallel change in climate conditions for both a plant invader and BCA has an ecological and evolutionary dimension.

Populations of both a plant invader in the introduced range and its potential BCAs in the native range may be genetically differentiated among geographically distinct regions. A first pragmatic approach to test the outcome of their interaction when brought together has recently been tested by Sun *et al.* (2020b) by challenging various plant genotypes of the target invader *A. artemisiifolia* with genotypes of the accidentally introduced and now successful BCA in China, the ragweed leaf beetle *Ophraella communa* L. (Coleoptera: Chrysomelidae). Both the invader and its antagonist were collected over three continents, i.e. in their native North American and introduced European and Chinese ranges and across large environmental amplitudes. The authors found the BCA genotype to be the critical factor, as some were effective against most plant genotypes, indicating the potential for at least short-term control efficacy. This approach will also indicate where to find the most efficient BCA genotypes, and it could be extended to set up these genotype-by-genotype interactions under different temperatures mimicking climate warming. An alternative approach for understanding and predicting population dynamics under climate change is the incorporation of climate-dependent vital rates of the BCA into species distribution models (cf. below).

4.3.4 Recent advances in predictive demographic modelling tools

In addressing the escalating complexities of weed biocontrol under climate change, it becomes evident that traditional modelling approaches may not suffice. To effectively predict and manage the future interplay of plants and herbivores, we must develop more sophisticated predictive modelling tools that incorporate a detailed understanding of life-history traits and climate-dependent vital rates (Sun *et al.*, 2020a). Life-history traits, including reproduction, growth and survival rates, are fundamental to understanding how species respond to environmental changes. These traits are shaped by evolutionary forces and current environmental conditions and can provide insights into resilience and adaptability of both invasive plants and their BCAs. Models incorporating these traits can help predict how species will respond to various climate scenarios, including changes in temperature, precipitation patterns and the frequency of extreme events (Kearney and Porter, 2009), such as integral projection models (IPMs) (Easterling *et al.*, 2000) and DYMEX (Hearne Scientific Software) (Maywald *et al.*, 2007). For example, a process-based population dynamics model predicted that invasive *Buddleja davidii* Franchet (Buddlejaceae) would succumb rapidly to damage by its biocontrol weevil *Cleopus japonicus* Wingelmüller

(Coleoptera: Curculionidae) under warming conditions (Kriticos *et al.*, 2009).

Incorporating these factors into predictive models requires a detailed understanding of the species in question, including their physiology, behaviour and interactions with other species and the environment. It also requires sophisticated modelling techniques capable of capturing the complexity of biological systems and their responses to a changing climate. This might include mechanistic models based on underlying biological processes that can make detailed predictions about how species will respond to specific environmental changes. Moreover, these models must be dynamic, capable of adapting to new information and changing conditions. This requires ongoing research and data collection, as well as a willingness to revise and improve models as new information becomes available. Combining species distribution models (SDMs) and mechanistic (process-based) models by integrating physiological models of insect development into SDMs based on habitat suitability may enable more robust predictions of both range shifts (Kearney and Porter, 2009) and population abundances (Keith *et al.*, 2008; Gallien *et al.*, 2010). Such a combined approach identified climatic factors limiting populations of *O. communa*, a biocontrol candidate of invasive *A. artemisiifolia*, and predicted its potential population density across its suitable European range and the relative importance of those climatic factors on the population growth (Augustinus *et al.*, 2020).

In conclusion, as the interplay among plants and herbivores becomes increasingly complex under climate change, we need more detailed predictive modelling tools. These tools must incorporate life-history traits and climate-dependent vital rates to provide accurate predictions about weed biocontrol. Further studies would benefit from integrating theoretical modelling, physiological/behavioural experiments and experimental population studies under climate change conditions into a biocontrol programme. By embracing complexity and striving for a deeper understanding of the systems we seek to manage, we can develop models that not only predict the future but also guide our actions.

4.4 Non-target Effects under Climate Change

Climate change can modify non-target plant effects of insect BCAs via ecological mechanisms by increasing geographical or phenological overlap of a BCA and a plant in its potential host range (examples for alligator weed flea beetle detailed below) or by changing the realized host range (Acacia bud-galling wasp example below). Evolution can enhance these ecological mechanisms through evolved expansions in climatic niches or fundamental host ranges. In the short term, ecological effects are likely to be more important despite the potentially rapid evolutionary rates of herbivorous insects.

Non-target effects can increase if plants potentially within the BCA's host range that formerly had ranges distinct from the agent become sympatric with it. For instance, alligator weed flea beetle *A. hygrophila*, a BCA introduced in 1986 to China (from USA populations originally sourced from South America) to control alligator weed *A. philoxeroides*, has shifted its distribution to overlap with the non-target native plant *Alternanthera sessilis* (L.) R. Br. ex DC. (Amaranthaceae) and is now damaging this plant (up to 40% leaf area removal) (Lu *et al.*, 2015). Because, on average, insects shift their elevational and latitudinal range limits more rapidly than plants do (Lenoir *et al.*, 2020), potential increases in non-target effects with climate change are likely to be driven primarily by BCAs expanding their ranges to overlap those of non-target plants rather than non-target plants expanding their ranges to overlap with BCAs. Evolved increases in climatic niches of BCAs could magnify their range overlap with non-target plants but insects appear to have little evolutionary response to warming (Weaving *et al.*, 2022).

Non-target effects can increase if plants potentially within the BCA's host range that formerly had phenologies distinct from the agent begin to have overlap of the non-target plant's vulnerable stages with a period of active feeding by the BCA. For instance, *A. hygrophila* increased its non-target effects on *A. sessilis* in China in part through a shift in the plant's life history from perennial to annual with increasing latitude around 26°N (and from no emergence holes to ~10/0.25 m^2 of cover) (Lu *et al.*, 2015). Because both insects (Bale *et al.*, 2002) and plants (Parmesan and Hanley, 2015) change their phenology with climate, non-target effects may either increase or decrease depending on their relative changes (Forkner *et al.*, 2008). The net impact of climate change on non-target effects depends on phenology of plant defences and plant nutrition but these also respond directly to climate change (Pincebourde *et al.*, 2017) which makes it difficult to predict the overall contributions of phenology. Insects have been shown to evolve new phenologies in response to climate change, such as *Rhinocyllus conicus* (Froel.) (Coleoptera: Curculionidae) introduced to control invasive thistles (especially *Carduus nutans* L., Asteraceae) in North America (Russell and Louda, 2004), which indicates that evolved changes in the phenology of non-target plants and/or BCAs could magnify or dampen impacts of climate change.

Climate change can impact realized host ranges of BCAs through a phenotypic response (i.e. plants that already co-occur geographically and phenologically with the agent are fed upon). Non-target effects of BCAs may increase or decrease as their population densities vary in response to changing climate conditions. For instance, one study with *Trichilogaster acaciaelongifoliae* (Frogatt) (Hymenoptera: Pteromalidae) introduced to control *Acacia longifolia* (Andrews) Willd. (Fabaceae) in South Africa found that a greater number of exotic woody plants were attacked when its population densities were higher (Dennill *et al.*, 1993). Although insect populations and herbivory increase when temperatures increase (Currano *et al.*, 2008; Lemoine *et al.*, 2013), which would be expected to increase non-target effects by spillover, such increases may be offset by direct effects of temperature on insect diet breadth. For example, *Popillia japonica* (Coleoptera: Scarabaeidae; not a BCA) has a narrower diet breadth at higher temperatures (Lemoine *et al.*, 2013). So, the net impact of climate change on non-target effects via phenotypic changes in host ranges will depend on agent densities together with the indirect and direct effects of warming on agent diet breadth.

Biological control agents could also evolve a broader fundamental host range, putting additional non-target plants at risk of attack. However, reviews have found little evidence of such expansions of fundamental host ranges (van Klinken and Edwards, 2002; Wright and Bennett, 2018). More common, though still rare, is evolved higher performance of a BCA on a non-target plant but there does not appear to be a role of climate change in this evolved higher performance (van Klinken and Edwards, 2002). Even more commonly observed are evolved changes in quantitative traits that enhance performance of insects on their new hosts (van Klinken and Edwards, 2002). Such changes may increase non-target effects if an expansion of fundamental host range occurs.

A special case is accidental introductions of potential biological weed control agents, i.e. unintentionally introduced natural enemies, which are already used for biological control of plant invaders in another part of the invaded range (Müller-Schärer *et al.*, 2020). As for all non-native species, such introductions have increased with global exchange of goods and human travel. They offer opportunities to assess host use of agents with a potentially broader fundamental host range than those approved for field release directly in target areas, allowing comparison with records from literature surveys or, if available, with predictions from pre-release studies that rejected their introduction (Müller-Schärer *et al.*, 2020). An example, presently well studied by an international consortium, is the accidental introduction of *O. communa* into Europe as a potential BCA against ragweed. It was first recorded in 2013 in southern Switzerland and northern Italy, from where it has subsequently spread eastward up to Romania and recently also westwards to France (Müller-Schärer *et al.*, 2018; Müller-Schärer *et al.*, 2024; Sun *et al.*, 2022). Insights gained from detailed postintroduction studies of such accidentally introduced species will help to critically review and extend pre-release studies and thus re-evaluate the presently strict release criteria, as they might exclude safe and efficient agents. Müller-Schärer *et al.* (2020) proposed full development on non-target species under no-choice laboratory conditions should indicate only which test plant species should be used in further studies on biosafety. The focus should be on population growth rate rather than on survival on individual non-target plants, as a negative population growth rate on a non-target plant would indicate no sustainable population on this non-target plant, although spillover effects may occur (Hinz *et al.*, 2019).

4.5 Biological Control Opportunities Will Increase

4.5.1 Continued plant invasions under climate change

The introduction of alien species in new regions is an old phenomenon and their accumulation shows no signs of saturation (Seebens *et al.*, 2017). In 2020, ~ 14,000 plant species with established alien populations in at least one region were known, representing 4% of the world flora (Pyšek *et al.*, 2020). The first stages of biological invasion (transport, introduction) are mainly driven by economic activities, and vascular plants will continue to be spread to new areas through increased air travel (main vector for low-income countries) and trade of cultivated plants (main vector for high-income countries) (Early *et al.*, 2016). A global model projects an average increase of new vascular plant species of 18% in 2050, with a maximum of 41% in temperate Asia, and most new species in Australasia (1065), followed by Europe (997) (Seebens *et al.*, 2021).

Climate change is proportionally more important in facilitating later stages of biological invasion (establishment, spread). Interacting with land and sea use change, climate change is predicted to amplify the future threat from invasive species, especially in disturbed habitats and in nearby natural habitats (IPBES, 2023; www.ipbes.net). A global analysis of climate change effects on invading taxa and receptivity of habitats showed overall increased performance of invasive species amongst plants, and in terrestrial habitats. Specifically, altered precipitation and CO_2 levels would enhance invasive woody plants and forbs in outcompeting native plant species (Stephens *et al.*, 2019). Severe plant invasions are expected in coastal and high-latitude ecoregions, such as coastal rivers, temperate forests and alpine vegetation (Wang *et al.*, 2019). In line with these global predictions, local studies forecasted northward range expansion of tropical invasive plants in the south-eastern USA due to future winter warming (Osland *et al.*, 2023) and increased chances of establishment for invasive plants in Poland (Solarz *et al.*, 2023).

In agriculture, climate change is generally expected to aggravate weed problems by positively impacting weed growth, but imposing more biotic and abiotic stress on crops (Varanasi *et al.*, 2016). Invasive plants have been identified as an increased risk for agro-ecosystems under climate change, for example in Sri Lanka (Marambe and Wijesundara, 2021). In addition, climate change will probably decrease the current efficacy of most herbicides, making weed management an even greater challenge for sustainable crop production (Varanasi *et al.*, 2016). As addressed earlier in this chapter, adaptation of invasive plants to climate change may exacerbate problems. As climate change and invasive species are drivers influencing each other, policy and management should thus consider their combined threats (Bradley *et al.*, 2022).

4.5.2 Political landscape – chances and threats for weed biocontrol

In recent decades, the need to take action globally against invasive species to protect biodiversity has been widely recognized, as reflected in several global agreements. In 2015, the United Nations' 2030 Agenda for Sustainable Development set 17 goals, including ecosystem conservation and measures against invasive alien species. The CBD 2022 (Convention on Biological Diversity, www.cbd.int) urges action against invasive species. The 2022 Kunming-Montreal Global Biodiversity Framework (www.cbd.int/gbf) aims to stimulate national efforts on biodiversity goals, including invasive species prevention and control. A 2023 IPBES report highlights that, with sufficient resources, political will and long-term commitment, preventing and controlling invasive species are achievable and beneficial for people and nature. This need for action can enhance development and/or adoption of (sustainable) methods like biological control.

In an agricultural context, recent policies are also promoting herbicide-free alternatives for weed control. The EU Environmental Agency acknowledges that widely used chemical pesticides in food production systems are a major source of pollution and biodiversity loss (www.eea.europa.eu/publications/how-pesticides-impact-human-health). In the context of the mission-driven European Green Deal to achieve a climate-neutral continent by 2025 (Boix-Fayos and de Vente, 2023), the Farm2Fork strategy includes a specific goal to reduce the use and risk of chemical pesticides by 50% by 2030 compared to the (EC, 2020). Herbicides made up 34% of the mass of chemical pesticides used in the EU from 2010 to 2019 (Triantafyllidis *et al.*, 2023). In the past few decades several important and commonly used hazardous chemical herbicides have already been banned, including atrazine in 2004 and paraquat in 2007. Currently, a ban on glyphosate is being debated (but rejected by the European Commission in 2023). In addition, the EU's mission 'A Soil Deal for Europe' stimulates research into innovative solutions contributing to healthier soils. Together, these policies and strategies favour the inclusion of non-chemical alternatives for integrated weed management, including the use of biological control agents (Tataridas *et al.*, 2022).

Although it is suggested that reducing herbicides may be more challenging than other groups (e.g. fungicides, insecticides), integrated weed control is feasible (Triantafyllidis *et al.*, 2023). Indeed, eliminating paraquat has saved lives without reducing agricultural productivity, thanks to the use of less hazardous and more sustainable alternatives (Stuart *et al.*, 2023). As farmers are often confronted with a multitude of weed species, it is unlikely that classical weed control will be the solution for all of these species. Agriculture may nevertheless benefit greatly from classical weed biocontrol programmes. Examples include cases where specific exotic weed species were widespread and very dominant in specific agricultural sectors or landscapes, such as the very cost-effective control of common ragwort on dairy farms in New Zealand using flea beetles (Fowler *et al.*, 2016). In other cases, invasive plant species are weeds in both crop and non-crop land (Müller-Schärer *et al.*, 2018) and are not exclusively targeted for agricultural purposes. Finally, classical BCAs may be augmented by mass releases to accelerate spread or increase local densities, as is being practised with the leaf beetle *O. communa* in China to control the highly allergenic common ragweed that is both an environmental and a crop weed (Müller-Schärer *et al.*, 2018).

Implementation of these (global) policies that support integrated weed control and biological weed control can be challenging. Implementation of the CBD recommendations to act against invasive alien plants is limited by resources, capacity, a lack of legal frameworks and uncertainties associated with future trajectories of societies and global change (Early *et al.*, 2016; Pyšek *et al.*, 2020; Roura-Pascual *et al.*, 2024). In the context of the EU Green Deal, scientists recently warned that EU member states should improve their National Action Plans to realise integrated pest management goals (Helepciuc and Todor, 2022a; 2022b). The EEA (2023) also acknowledges that more action is needed to achieve the goals.

Biological control is a critical tool to realise these environmental goals. It is an important method for controlling invasive plants in terrestrial and closed water systems as, next to adaptive management, it is most effective in the long term (IPBES, 2023). Impact has been demonstrated for a large majority of invasive plant species targeted, and for a substantial part of these, need for other control methods was greatly reduced or eliminated (Hinz *et al.*, 2020; Sun *et al.*, 2022). Compared to adaptive management, biological control is currently more available and easier to use. Scientists warn, however, that recent global (e.g. Nagoya protocol), continental and national regulations that restrict or lengthen the sourcing, importation or use of exotics may slow down or limit possibilities for biological weed control, even though this method is critical for the management of invasive alien plants (Paynter *et al.*, 2020; Sun *et al.*, 2022: Müller-Schärer *et al.*, 2024).

References

Alex Perkins, T., Phillips, B.L., Baskett, M.L. and Hastings, A. (2013) Evolution of dispersal and life history interact to drive accelerating spread of an invasive species. *Ecology Letters* 16(8), 1079–1087.

Augustinus, B., Sun, Y., Beuchat, C. and Schaffner, U. (2020) Predicting impact of a biocontrol agent: Integrating distribution modelling with climate-dependent vital rates. *Ecological Applications* 30(1), e02003.

Bale, J.S., Masters, G.J., Hodkinson, I.D., Awmack, C., Bezemer, T.M. *et al.* (2002) Herbivory in global climate change research: Direct effects of rising temperature on insect herbivores. *Global Change Biology* 8(1), 1–16.

Baso, N.C., Coetzee, J.A., Ripley, B.S. and Hill, M.P. (2021) The effects of elevated atmospheric CO2 concentration on the biological control of invasive aquatic weeds. *Aquatic Botany* 170, 103348.

Blossey, B. and Nötzold, R. (1995) Evolution of increased competitive ability in invasive nonindigenous plants: A hypothesis. *Journal of Ecology* 83, 887.

Boix-Fayos, C. and de Vente, J. (2023) Challenges and potential pathways towards sustainable agriculture within the European Green Deal. *Agricultural Systems* 207, 103634.

Bradley, B.A., Blumenthal, D.M., Wilcove, D.S. and Ziska, L.H. (2010) Predicting plant invasions in an era of global change. *Trends in Ecology & Evolution* 25(5), 310–318.

Bradley, B.A., Beaury, E.M., Fusco, E.J. and Lopez, B.E. (2022) Invasive species policy must embrace a changing climate. *Bioscience* 73(2), 124–133.

Colautti, R.I. and Barrett, S.C. (2013) Rapid adaptation to climate facilitates range expansion of an invasive plant. *Science* 342(6156), 364–366.

Coley, P.D., Bryant, J.P. and Chapin III, F.S. (1985) Resource availability and plant antiherbivore defense. *Science* 230(4728), 895–899.

Currano, E.D., Wilf, P., Wing, S.L., Labandeira, C.C., Lovelock, E.C. *et al.* (2008) Sharply increased insect herbivory during the Paleocene-Eocene thermal maximum. *Proceedings of the National Academy of Sciences USA* 105(6), 1960–1964.

Damos, P. and Savopoulou-Soultani, M. (2012) Temperature-driven models for insect development and vital thermal requirements. *Psyche* 2012(1), 123405.

Dennill, G.B., Donnelly, D. and Chown, S.L. (1993) Expansion of host-plant range of a biocontrol agent *Trichilogaster acaciaelongifoliae* (Pteromalidae) released against the weed *Acacia longifolia* in South Africa. *Agriculture, Ecosystems & Environment* 43(1), 1–10.

Deutsch, C.A., Tewksbury, J.J., Tigchelaar, M., Battisti, D.S., Merrill, S.C. *et al.* (2018) Increase in crop losses to insect pests in a warming climate. *Science* 361(6405), 916–919.

Dukes, J.S. and Mooney, H.A. (1999) Does global change increase the success of biological invaders? *Trends in Ecology & Evolution* 14(4), 135–139.

Early, R., Bradley, B.A., Dukes, J.S., Lawler, J.J., Olden, J.D. *et al.* (2016) Global threats from invasive alien species in the twenty-first century and national response capacities. *Nature Communications* 7(1), 12485.

Easterling, M.R., Ellner, S.P. and Dixon, P.M. (2000) Size-specific sensitivity: Applying a new structured population model. *Ecology* 81(3), 694–708.

EC (2020) Farm to Fork targets – Progress. Available at: https://food.ec.europa.eu/plants/pesticides/sustainable-use-pesticides/farm-fork-targets-progress_en#:~:text=Further%20information-,About%20the%20targets,use%20of%20more%20hazardous%20pesticides (accessed 5 June 2025).

EEA (2023) Environmental statement. Available at: www.eea.europa.eu/en/analysis/publications/environmental-statement-2023#:~:text=The%20EEA's%20environmental%20performance%20in,an%20increase%20compared%20to%202022 (accessed 5 June 2025).

Essack, S. (2018) Antibiotic resistance and One Health: A mapping project. *Lancet Global Health* 6, S27.

Forkner, R.E., Marquis, R.J., Lill, J.T. and Le Corff, J. (2008) Timing is everything: Phenological synchrony and population variability in leaf-chewing herbivores of *Quercus*. *Ecological Entomology* 33(2), 276–285.

Fowler, S., Gourlay, A. and Hill, R. (2016) Biological control of ragwort in the New Zealand dairy sector: An ex-post economic analysis. *New Zealand Journal of Agricultural Research* 59(3), 205–215.

Gallien, L., Münkemüller, T., Albert, C.H., Boulangeat, I. and Thuiller, W. (2010) Predicting potential distributions of invasive species: Where to go from here? *Diversity and Distributions* 16(3), 331–342.

Gioria, M., Hulme, P.E., Richardson, D.M. and Pyšek, P. (2023) Why are invasive plants successful? *Annual Review of Plant Biology* 74, 635–670.

Helepciuc, F.-E. and Todor, A. (2022a) Evaluating the EU's efforts to improve resilience to health and environmental risks associated with pesticide use by analyzing the national action plans of EU Member States from 2009 to 2019. *International Journal of Environmental Research and Public Health* 19(9), 5446.

Helepciuc, F.-E. and Todor, A. (2022b) Greener European agriculture? Evaluating EU Member States' transition efforts to integrated pest management through their national action plans. *Agronomy* 12(10), 2438.

Helliwell, E.E., Faber-Hammond, J., Lopez, Z.C., Garoutte, A., von Wettberg, E. *et al.* (2018) Rapid establishment of a flowering cline in Medicago polymorpha after invasion of North America. *Molecular Ecolog* 27, 4758–4774.

Henrikson, J.W., Lim, D.S., Lu, X., Ding, J. and Siemann, E. (2018) Strong effects of hydrologic environment and weak effects of elevated CO2 on the invasive weed *Alternanthera philoxeroides* and the biocontrol beetle *Agasicles hygrophila*. *Arthropod Plant Interaction* 12, 691–700.

Hinz, H.L., Winston, R.L. and Schwarzländer, M. (2019) How safe is weed biological control? A global review of direct nontarget attack. *Quarterly Review of Biology 94*(1), 1–27.

Hinz, H.L., Winston, R.L. and Schwarzländer, M. (2020) A global review of target impact and direct nontarget effects of classical weed biological control. *Current Opinion in insect Science* 38, 48–54.

Hodgins, K.A., Bock, D.G. and Rieseberg, L.H. (2018) Trait evolution in invasive species. *Annual Plant Reviews Online* 459–496.

IPBES (2023) Summary for policymakers of the thematic assessment report on invasive alien species and their control of the intergovernmental science-policy platform on biodiversity and ecosystem services. In: Roy, H.E., Pauchard, A., Stoett, P., Renard Truong, T. and Bacher, S., et al (eds) *IPBES Secretariat*. Bonn, Germany. Available at: https://doi.org/10.5281/zenodo.7430692

Kearney, M. and Porter, W. (2009) Mechanistic niche modelling: Combining physiological and spatial data to predict species' ranges. *Ecology Letters* 12(4), 334–350.

Keith, D.A., Akçakaya, H.R., Thuiller, W., Midgley, G.F., Pearson, R.G. *et al.* (2008) Predicting extinction risks under climate change: Coupling stochastic population models with dynamic bioclimatic habitat models. *Biology Letters* 4(5), 560–563.

Kriticos, D.J., Watt, M.S., Withers, T.M., Leriche, A. and Watson, M.C. (2009) A process-based population dynamics model to explore target and non-target impacts of a biological control agent. *Ecological Modelling* 220(17), 2035–2050.

Lee, C.E. (2010) Evolution of invasive populations. In: *Proceedings of the National Academy of Sciences USA*, pp. 15793–15798 (Vol. 104).

Legg, S. (2021) IPCC, 2021: Climate change 2021 – The physical science basis. *International Panel on Climate Change* 49, 44–45.

Lemoine, N.P., Drews, W., Burkepile, D.E. and Parker, J.D. (2013) Increased temperature alters feeding behavior of a generalist herbivore. *Oikos* 122(12), 1669–1678.

Lenoir, J., Bertrand, R., Comte, L., Bourgeaud, L., Hattab, T. *et al.* (2020) Species better track climate warming in the oceans than on land. *Nature Ecology & Evolution* 4(8), 1044–1059.

Lu, X.M., Siemann, E., He, M.Y., Wei, H., Shao, X. *et al.* (2015) Climate warming increases biological control agent impact on a non-target species. *Ecology Letters* 18(1), 48–56.

Marambe, B. and Wijesundara, S. (2021) Effects of climate change on weeds and invasive alien plants in Sri Lankan agro-ecosystems: Policy and management implications. *Frontiers in Agronomy* 3, 641006.

Maywald, G., Kriticos, D., Sutherst, R. and Bottomley, W. (2007) *DYMEX Model Builder Version 3: User's Guide*. Hearne Scientific Software, Melbourne, Australia, p. 172.

Müller-Schärer, H., Sun, Y. and Schaffner, U. (2024) When a plant invader meets its old enemy abroad: What can be learnt from accidental introductions of biological control agents. *Pest Management Science* 80(1), 19–27.

Müller-Schärer, H. and Schaffner, U. (2008) Classical biological control: Exploiting enemy escape to manage plant invasions. *Biological Invasions* 10(6), 859–874.

Müller-Schärer, H. and Schaffner, U. (2020) Editorial overview: Biological control of plant invaders: A continued stimulus and yet untapped potential to link and advance applied and basic research. *Current Opinion in Insect Science* 38, v–viii.

Müller-Schärer, H., Sun, Y., Chauvel, B., Karrer, G., Kazinczi, G. *et al.* (2018) Cross-fertilizing weed science and plant invasion science to improve efficient management: A European challenge. *Basic and Applied Ecology* 33, 1–13.

Müller-Schärer, H., Bouchemousse, S., Litto, M., McEvoy, P.B., Roderick, G.K. *et al.* (2020) How to better predict long-term benefits and risks in weed biocontrol: An evolutionary perspective. *Current Opinion in Insect Science* 38, 84–91.

Orians, C.M. and Ward, D. (2010) Evolution of plant defenses in nonindigenous environments. *Annual Review of Entomology* 55, 439–459.

Osland, M.J., Chivoiu, B., Feher, L.C., Dale, L.L., Lieurance, D. *et al.* (2023) Plant migration due to winter climate change: Range expansion of tropical invasive plants in response to warming winters. *Biological Invasions* 25(9), 2813–2830.

Paper, M.K., Righetti, T., Raubenheimer, S.L., Coetzee, J.A., Sosa, A.J. *et al.* (2023) Effects of elevated CO_2 on feeding responses of biological control agents of Pontederia crassipes. *Entomologia Experimentalis et Applicata* 171(12), 998–1008.

Parmesan, C. and Hanley, M.E. (2015) Plants and climate change: Complexities and surprises. Review, *Annals of Botany* 116(6), 849–864.

Paynter, Q., Paterson, I.D. and Kwong, R.M. (2020) Predicting non-target impacts. *Current Opinion in Insect Science* 38, 79–83.

Pincebourde, S., van Baaren, J., Rasmann, S., Rasmont, P., Rodet, G. *et al.* (2017) Plant–insect interactions in a changing world. In: Sauvion, N., Thiéry, D. and P.A, Calatayud. (eds) *Advances in Botanical Research*, Vol. 81. Academic Press, New York, pp. 289–332.

Pyšek, P., Hulme, P.E., Simberloff, D., Bacher, S., Blackburn, T.M. *et al.* (2020) Scientists' warning on invasive alien species. *Biological Reviews* 95(6), 1511–1534.

Roderick, G.K., Hufbauer, R. and Navajas, M. (2012) Evolution and biological control. *Evolutionary Applications* 5(5), 419–423.

Roura-Pascual, N., Saul, W.C., Pérez-Granados, C., Rutting, L., Peterson, G.D. *et al.* (2024) A scenario-guided strategy for the future management of biological invasions. *Frontiers in Ecology and the Environment* 22(4), e2725.

Russell, F.L. and Louda, S.M. (2004) Phenological synchrony affects interaction strength of an exotic weevil with *Platte thistle*, a native host plant. *Oecologia* 139, 525–534.

Sánchez-Bayo, F. and Wyckhuys, K.A. (2019) Worldwide decline of the entomofauna: A review of its drivers. *Biological Conservation* 232, 8–27.

Schwarzländer, M., Hinz, H.L., Winston, R. and Day, M. (2018) Biological control of weeds: An analysis of introductions, rates of establishment and estimates of success, worldwide. *BioControl* 63(3), 319–331.

Seebens, H., Blackburn, T.M., Dyer, E.E., Genovesi, P., Hulme, P.E. *et al.* (2017) No saturation in the accumulation of alien species worldwide. *Nature Communications* 8, 14435.

Seebens, H., Bacher, S., Blackburn, T.M., Capinha, C., Dawson, W. *et al.* (2021) Projecting the continental accumulation of alien species through to 2050. *Global Change Biology* 27(5), 970–982.

Skendžić, S., Zovko, M., Živković, I.P., Lešić, V. and Lemić, D. (2021) The impact of climate change on agricultural insect pests. *Insects* 12(5), 440.

Solarz, W., Najberek, K., Tokarska-Guzik, B. and Pietrzyk-Kaszyńska, A. (2023) Climate change as a factor enhancing the invasiveness of alien species. *Environmental & Socio-Economic Studies* 11(4), 36–48.

Sotka, E.E., Baumgardner, A.W., Bippus, P.M., Destombe, C., Duermit, E.A. *et al.* (2018) Combining niche shift and population genetic analyses predicts rapid phenotypic evolution during invasion. *Evolutionary Applications* 11, 781–793.

Stephens, K.L., Dantzler-Kyer, M.E., Patten, M.A. and Souza, L. (2019) Differential responses to global change of aquatic and terrestrial invasive species: Evidences from a meta-analysis. *Ecosphere* 10(4), e02680.

Stuart, A.M., Merfield, C.N., Horgan, F.G., Willis, S., Watts, M.A. *et al.* (2023) Agriculture without paraquat is feasible without loss of productivity – lessons learned from phasing out a highly hazardous herbicide. *Environmental Science and Pollution Research* 30(7), 16984–17008.

Suarez, A.V. and Tsutsui, N.D. (2008) The evolutionary consequences of biological invasions. *Molecular Ecology* 17(1), 351–360.

Sun, Y., Brönnimann, O., Roderick, G.K., Poltavsky, A., Lommen, S.T. and Müller-Schärer, H. (2017) Climatic suitability ranking of biological control candidates: A biogeographic approach for ragweed management in Europe. *Ecosphere* 8(4), e01731.

Sun, Y., Ding, J., Siemann, E. and Keller, S.R. (2020a) Biocontrol of invasive weeds under climate change: Progress, challenges and management implications. *Current Opinion in Insect Science* 38, 72–78.

Sun, Y., Beuchat, C. and Müller-Schärer, H. (2020b) Is biocontrol efficacy rather driven by the plant or the antagonist genotypes? A conceptual bioassay approach. *NeoBiota* 63, 81–100.

Sun, Y., Müller-Schärer, H. and Schaffner, U. (2022) Fighting neobiota with neobiota: Consider it more often and do it more rigorously. *Biological Conservation* 268, 109506.

Szűcs, M., Vahsen, M., Melbourne, B., Hoover, C., Weiss-Lehman, C. and Hufbauer, R. (2017) Rapid adaptive evolution in novel environments acts as an architect of population range expansion. *Proceedings of the National Academy of Sciences USA* 114(51), 13501–13506.

Szűcs, M., Vercken, E., Bitume, E.V. and Hufbauer, R.A. (2019) The implications of rapid eco-evolutionary processes for biological control-A review. *Entomologia Experimentalis et Applicata* 167(7), 598–615.

Tataridas, A., Kanatas, P., Chatzigeorgiou, A., Zannopoulos, S. and Travlos, I. (2022) Sustainable crop and weed management in the era of the EU Green Deal: A survival guide. *Agronomy* 12(3), 589.

Thomas, J.A., Telfer, M.G., Roy, D.B., Preston, C.D., Greenwood, J.J.D. *et al.* (2004) Comparative losses of British butterflies, birds, and plants and the global extinction crisis. *Science* 303(5665), 1879–1881.

Triantafyllidis, V., Mavroeidis, A., Kosma, C., Karabagias, I.K., Zotos, A. *et al.* (2023) Herbicide use in the era of farm to fork: Strengths, weaknesses, and future implications. *Water, Air and Soil Pollution* 234(2), 94.

Van Asch, M., Salis, L., Holleman, L.J., Van Lith, B. and Visser, M.E. (2013) Evolutionary response of the egg hatching date of a herbivorous insect under climate change. *Nature Climate Change* 3(3), 244.

van Boheemen, L.A., Atwater, D.Z. and Hodgins, K.A. (2019) Rapid and repeated local adaptation to climate in an invasive plant. *New Phytologist* 222, 614–627.

van Klinken, R.D. and Edwards, O.R. (2002) Is host-specificity of weed biological control agents likely to evolve rapidly following establishment? *Ecology Letters* 5(4), 590–596.

Varanasi, A., Prasad, P.V. and Jugulam, M. (2016) Impact of climate change factors on weeds and herbicide efficacy. *Advances in Agronomy* 135, 107–146.

Wang, C.-J., Li, Q.-F. and Wan, J.-Z. (2019) Potential invasive plant expansion in global ecoregions under climate change. *PeerJ* 7, e6479.

Weaving, H., Terblanche, J.S., Pottier, P. and English, S. (2022) Meta-analysis reveals weak but pervasive plasticity in insect thermal limits. *Nature Communications* 13(1), 5292.

Wolfe, L.M., Elzinga, J.A. and Biere, A. (2004), Increased susceptibility to enemies following introduction in the invasive plant Silene latifolia. *Ecology Letters* 7, 813–820.

Wright, M.G. and Bennett, G.M. (2018) Evolution of biological control agents following introduction to new environments [Review]. *Biocontrol* 63(1), 105–116.

Yang, Y., Liu, M., Pan, Y., Huang, H., Pan, X. *et al.* (2021) Rapid evolution of latitudinal clines in growth and defence of an invasive weed. *New Phytologist* 230(2), 845–856.

5 Effects of Climate Change on Plant Defences and Its Consequences for Biocontrol

Paul J. Ode* and Enakshi Ghosh
Colorado State University, Fort Collins, USA

Abstract
Climate change is widely appreciated to severely disrupt species interactions, including those between plants, their insect herbivores and their natural enemies. Central to plant–insect interactions are chemical defences against herbivory. Understanding how trophic interactions are structured by plant defence chemistry is important for gauging how biological control efforts to manage invasive plants as well as herbivorous pests in agricultural crops will respond to changing climates. This chapter explores the complex ways in which climate change can alter the expression of plant chemical defences and consequently alter interactions between plants, insect herbivores, and their parasitoids and predators. In particular, the effects of three climate change variables (elevated CO_2, elevated O_3 and elevated temperatures) known to influence the expression of plant chemical defences are examined. This chapter reviews studies that explore the effects of these climate change variables on species interactions and their implications for biological control.

5.1 Introduction

Reciprocal selection between insect herbivores and their host plants is widely thought to be responsible for the high species numbers within both groups (Mitter *et al.*, 1988; Futuyma and Agrawal, 2009). Central components of most, if not all, plant–insect herbivore interactions are plant chemical defences (Richards *et al.*, 2015). Herbivory has selected for a diversity of plant chemical defences (Speed *et al.*, 2015). In turn, diverse plant chemistries and other defensive traits have selected for an array of physiological and behavioural mechanisms within herbivores to cope with these plant defences (Schoonhoven *et al.*, 2005). Furthermore, plant defence chemistry can directly and indirectly affect how herbivores interact with their own natural enemies, namely parasitoids and predators (Price *et al.*, 1980; Ode, 2006; Kessler and Heil, 2011; Kaplan *et al.*, 2016; Turlings and Erb, 2018; Pearse *et al.*, 2020).

Insect herbivores are important biological control agents against invasive plants in forests and rangelands. Likewise, insect parasitoids and predators are widely used as biological control agents against insect herbivore pests (typically in production agriculture). However, climate change threatens to alter the trophic relationships that form the foundation of biological

*Corresponding author: paul.ode@colostate.edu

 Biological Control Systems and Climate Change (eds Joan van Baaren *et al.*)
DOI: 10.1079/9781800625099.0005

control programmes. As reviewed elsewhere in this volume (see Chapter 1), climate change (especially rising temperatures and changes in precipitation patterns) is causing changes in species phenologies and distribution patterns and, consequently, interactions among species. Elevated temperatures are also known to affect the expression of plant chemical defences. Furthermore, elevated concentrations of greenhouse gases themselves, especially CO_2 and O_3, can directly affect the production of plant defensive chemicals, thereby potentially altering the outcome of plant–insect and herbivore–natural enemy interactions, including those at the centre of biological control programmes. Consequently, consideration of how climate change can alter the expression of plant defence chemistry is of great importance to understanding how biological control programmes will be affected by our rapidly changing climate (Ode, 2022).

This chapter explores the effects of climate change and greenhouse gas concentrations on the expression of plant defensive chemistry and how these are expected to affect interactions with higher trophic levels. While relatively few studies have directly examined the effects of climate change on defence chemistry and how this affects the outcome of biological control efforts, we explore the potential implications for biological control. Much of this discussion is focused on the effects of elevated concentrations of greenhouse gases (e.g. CO_2, O_3) and the consequential increase in temperatures on the expression of plant signalling pathways that are responsible for the production of plant antiherbivore toxins as well as herbivore-induced plant volatiles (HIPVs) that serve as indirect defences, attracting parasitoids of many insect herbivores. We primarily focus on the separate effects of rising temperatures and elevated greenhouse gas concentrations on the expression of plant defence chemistry as most studies have considered these separately. Nevertheless, it is important to recognize that these factors and others change concurrently in nature and the induction of plant defences is also driven by the interaction between these factors (Zvereva and Koslov, 2006; Jamieson *et al.*, 2017).

5.1.1 Plant defence chemistry

Plants produce a truly astounding diversity of defensive agents, with estimates of over 200,000 compounds (Mithöfer and Boland, 2012), and we refer the reader to any number of excellent reviews that discuss the different classes of these defensive compounds (e.g. Rosenthal and Janzen, 1979; Rosenthal and Berenbaum, 1991; Harborne, 1993; Schoonhoven *et al.*, 2005; Mithöfer and Boland, 2012). Examples of different classes of chemical defences include alkaloids, cyanogenic glycosides, glucosinolates, terpenoids and phenolics (including tannins) (Schoonhoven *et al.*, 2005; Mithöfer and Boland, 2012). Some classes are characteristic of a single plant family; for example, glucosinolates are produced primarily by the Brassicaceae and closely related families. Some of these classes are extremely diverse, notably the terpenoids and phenolics (which include tannins), and are found ubiquitously in plants. Many terpenoids (especially monoterpenoids) are volatile compounds which are produced upon tissue damage and play roles in attracting predators and parasitoids of insect herbivores (Dicke *et al.*, 1990; De Moraes *et al.*, 1998; Kessler and Baldwin, 2001). Subclasses of terpenoids and phenolics can be characteristic of individual plant families. Cucurbitacins (a class of triterpenoids) are characteristic of the Cucurbitaceae and cardenolides (another class of triterpenoids) are found predominantly within the Apocynaceae. Furanocoumarins (a class of phenolics) are found primarily in the Apiaceae and Rutaceae (Berenbaum, 1991). Finally, inorganic compounds such as silica (SiO_2) and calcium oxalate (CaC_2O_4) can provide antiherbivore functions when accumulated by plants (McNaughton and Tarrants, 1983; Reynolds *et al.*, 2009; Cooke and Leishman, 2010).

Plants produce a diverse mix of chemical defences against attacking herbivores and pathogens. Such defences can be expressed constitutively, that is, produced irrespective of whether they are attacked or not by herbivores and pathogens. Alternatively, higher concentrations of chemical defences can be induced in response to attack or perception of imminent attack by herbivores or pathogens. In general, constitutive defences are more commonly

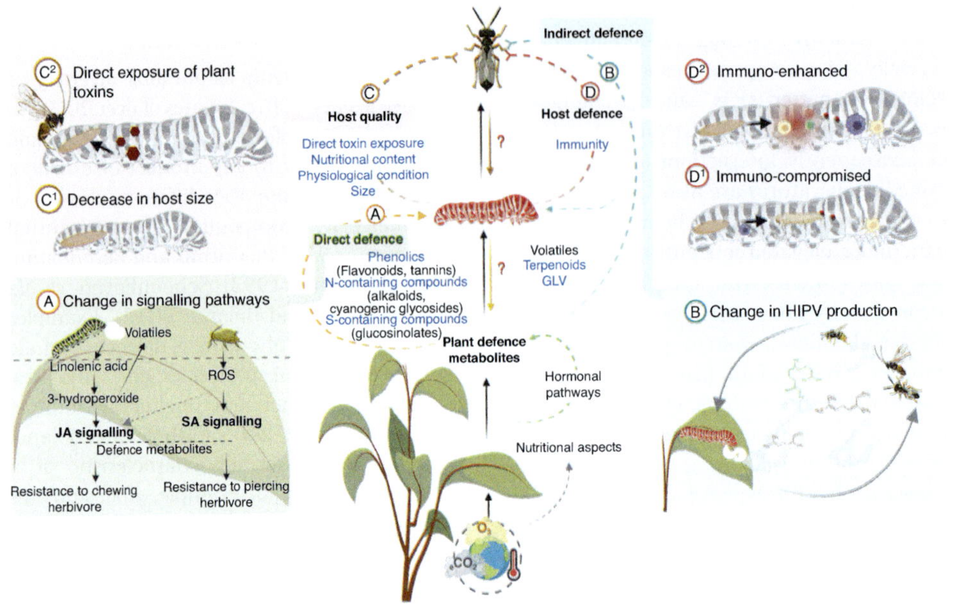

Fig. 5.1. Impact of climate change variables (elevated CO_2, elevated O_3 and increased temperature) on plant chemical defence chemistry. These variables affect hormonal pathways, which in turn influence the production of direct (plant toxins) (A) and indirect (volatile compounds that attract parasitoids and predators) (B) defence chemical production. Plant volatiles may also attract specialist herbivores or repel generalist herbivores. Specifically, temperature, e[CO_2] and e[O_3] modulate jasmonic acid and salicylic acid pathways (A), crucial for plant defence against chewing and sucking herbivores. Direct defence chemicals influence these interactions by modifying host quality (C) and altering the immune status of the host (D).

produced when the threat of herbivory is constant and/or predictable. Induced defences are more indicative of herbivore pressure that is less predictable and probably serve to conserve resources that could be rediverted to growth and reproduction when herbivory is absent (Strauss *et al.*, 2002; Cipollini *et al.*, 2003a,b).

Production of induced defences is primarily mediated by two plant hormone signalling pathways: the jasmonic acid (JA) pathway and the salicylic acid (SA) pathway (Fig. 5.1a). The JA signalling pathway is typically induced by feeding damage by chewing herbivores and, along with the ethylene pathway, ultimately produces many chemical defences including toxins, antinutrients and digestion inhibitors as well as volatile compounds that deter herbivory (see Fig. 5.1a) (Zavala *et al.*, 2013). Such defences are referred to as direct defences in that they directly act against attacking herbivores and/or pathogens. The JA signalling pathway also results in the production of volatile organic compounds (VOCs; also known as herbivore-induced plant volatiles – HIPVs – when induced by chewing damage and oral secretions of herbivores) that are attractive to parasitoids and predators of herbivores. These volatile compounds are often referred to as indirect defences in that they do not directly affect herbivores but nonetheless provide defence for the plant by attracting natural enemies of the herbivores (Fig. 5.1b). The SA signalling pathway induces defences against plant pathogens and many piercing-sucking insects. Significantly, the SA pathway is known to suppress the expression of the JA pathway in many (but not all) systems, a phenomenon known as cross-talk (Schweiger *et al.*, 2014).

5.2 Climate Change and Plant Defence Chemistry

5.2.1 Effects of increasing temperature

In addition to the effects of rising mean temperatures and more extreme temperature events on plant and insect physiology, phenology and trophic interactions, increasing temperatures have been shown to alter the expression of plant defence chemistry (Monticelli *et al.*, 2021). Compared to the effects of elevated carbon dioxide concentrations (e[CO_2]) and elevated ozone concentrations (e[O_3]) on plant defence chemistry discussed below, relatively few studies have focused on the effects of elevated temperature on the production of plant defence chemistry, making generalizations difficult (Zvereva and Koslov, 2006). Some studies, including a meta-analysis (Zvereva and Koslov, 2006), have suggested that warmer temperatures are correlated with higher levels of terpenoids but decreased levels of phenolics, but these trends are by no means universal. As pointed out elsewhere, interactions (ranging from additive to synergistic, to antagonistic) among climate change factors can obscure the main effects of a single factor (Zvereva and Koslov, 2006; Jamieson *et al.*, 2017). Consequently, the effects of elevated temperatures on plant defence chemistry depend greatly on plant species and the specific defence chemical (Bidart-Bouzat and Imeh-Nathaniel, 2008; Jamieson *et al.*, 2012). On the other hand, increasing temperatures have been associated with increases in the production of VOCs, including isothiocyanates, the hydrolysis products of glucosinolates that mediate many herbivore and parasitoid interactions with glucosinolate containing plants such as *Brassica* spp. (Brassicaceae) (Matusheski *et al.*, 2004; van Poecke and Dicke, 2004).

5.2.2 Effects of greenhouse gases on hormone signalling pathways

Both JA and SA signalling pathways are known to be modulated by greenhouse gases, notably e[CO_2] and e[O_3], thereby influencing the production of plant antiherbivore toxins (direct defences) and the production of HIPVs that function as indirect defences insofar as they attract parasitoids and predators of herbivores (Ode *et al.*, 2014). In general, e[CO_2] suppresses gene expression in the JA signalling pathway but induces the SA signalling pathway, further exacerbating the intensity of cross-talk (Casteel *et al.*, 2008; Sun *et al.*, 2013; Zavala *et al.*, 2013). Elevated [O_3] induces the production of reactive oxygen species, thus increasing synthesis of SA (Mittler, 2006), similar to the response to damage by plant pathogens and piercing-sucking insects such as aphids and whiteflies (Cui *et al.*, 2012a,b).

As CO_2 is a key substrate for photosynthesis, e[CO_2] is generally positively correlated with increased photosynthetic rates and net primary productivity (Drake *et al.*, 1997; Long *et al.*, 2004; Ainsworth and Long, 2005; Stiling and Cornelessen, 2007; Robinson *et al.*, 2012). In general, the foliar concentrations of nitrogen decrease under e[CO_2] as carbohydrates increase (Bezemer and Jones, 1998; Körner, 2003). This tends to reduce the nutritional quality of a plant for herbivores, which in some cases may induce increased feeding by herbivores to obtain sufficient nitrogen for their growth and development. On the other hand, e[O_3] tends to reduce rates of photosynthesis and consequently plant biomass (Morgan *et al.*, 2003; Fiscus *et al.*, 2005; Ainsworth *et al.*, 2012; Moreno-Delafuente *et al.*, 2021a). Once O_3 enters the leaf stomata, it reacts with the cell walls to form reactive oxygen species that activate SA defence signalling pathways (Valkama *et al.*, 2007). Elevated [O_3] is well known to result in crop damage and reduced yields (McGrath *et al.*, 2015).

5.2.3 Climate change and plant defence chemistry: e[CO_2], e[O_3] and plant toxins (direct defences)

Given the effects of e[CO_2] and e[O_3] on the JA and SA signalling pathways described above, it is tempting to predict that e[CO_2] will decrease the production of plant antiherbivore toxins (direct defences) and HIPVs which attract parasitoids and predators (indirect defences). If true, herbivory by chewing herbivores should increase under e[CO_2] and the success of biological control efforts of herbivores using parasitoids should be diminished as these natural enemies would be less efficient in locating their hosts with the reduced production of HIPVs. Conversely, e[CO_2] might be expected to enhance biological control efforts of invasive plants as herbivore biological control agents would be able to feed on less chemically defended plants.

Similarly, e[O_3] may also be expected to increase feeding damage by chewing insect herbivores. Elevated [O_3] induces the production of reactive oxygen species (similar to damage by plant pathogens and piercing-sucking insects), stimulating SA production (Mittler, 2006; Casteel *et al.*, 2008; Cui *et al.*, 2012a), which ultimately suppresses JA signalling (resulting in the same effect as e[CO_2]). Elevated ozone is also known to stimulate antioxidant defence systems resulting in increased production of certain alkaloids, glucosinolates, coumarins and cyanogenic glycosides (Valkama *et al.*, 2007; Lindroth, 2010). With regard to predictions for the success of weed and insect biological control programmes, we can make similar predictions for the effects of e[O_3] as we can for e[CO_2].

The following two sections consider the effects of elevated greenhouse gases on herbivores and their parasitoids. The relationship between e[CO_2], e[O_3] and the induction of specific plant antiherbivore toxins is very idiosyncratic, depending on plant species and the class of chemical defensive compounds involved (Jackson *et al.*, 2000; Holopainen, 2002; Jøndrup *et al.*, 2002; Menéndez *et al.*, 2009, 2010). Furthermore, this relationship can depend on the identity of the herbivore attacking the plant inducing specific chemical defences. This makes it extremely challenging to make broad predictions about how climate change will influence induced plant defences and their effects on herbivores and their natural enemies.

e[CO_2], e[O_3] and direct defence effects on herbivores

Much of the early work of the effects of e[CO_2] and e[O_3] on insect herbivores involved woody plants in which defences are dominated by phenolics, tannins and terpenoids (e.g. Koricheva *et al.*, 1998). Furthermore, these studies often measured the effects of total phenolics or tannins rather than individual compounds, potentially obscuring the effects of specific compounds. One such study showed that aspen trees grown under e[CO_2] produced higher levels of condensed tannins that resulted in decreased growth of the forest tent caterpillar *Malacosoma disstria* (Lasiocampidae) (Roth *et al.*, 1998), although this pattern was not consistent across all tree species nor classes of tannins (Lindroth and Kinney, 1998; Roth *et al.*, 1998). The effects of e[CO_2] have also been well studied in plants in the Brassicales which are characterized by the production of glucosinolates. The relationship between e[CO_2] and glucosinolates depends on the specific glucosinolate as well as the plant species in question (Karowe *et al.*, 1997; Schonhof *et al.*, 2007). Furthermore, herbivory by the diamondback moth *Plutella xylostella* (Plutellidae) induced higher levels of glucosinolates in certain genotypes of *Arabidopsis thaliana* (Brassicaceae) under e[CO_2] (Bidart-Bouzat *et al.*, 2005). Elevated [CO_2] was also found to significantly increase the production of glucosinolates (both aliphatic and indole) and suppress population growth rates of the specialist cabbage aphid *Brevicoryne brassicae* (Aphididae) as well as the specialist large cabbage white butterfly *Pieris brassicae* (Pieridae) (Klaiber *et al.*, 2013a).

Similarly, the relationship between e[O_3] and herbivory is highly variable (Menéndez *et al.*, 2009). Some studies have shown that e[O_3] results in increased herbivory by aphids (Percy *et al.*, 2002; Kopper and Lindroth, 2003) whereas others found a negative relationship (Docherty *et al.*, 1997; Holopainen, 2002). Rather than being driven by substrate availability (like net primary productivity), most plant chemical defences are strongly regulated by differential gene expression after attack by herbivores and pathogens (Hamilton *et al.*, 2001; Moore *et al.*, 2014).

A small but growing number of studies generally support the notion that e[CO_2] suppresses the JA signalling pathway, resulting in the suppression of direct defences against herbivores. Rarer still are studies that demonstrate that herbivore damage increases as a consequence of e[CO_2]. Elevated [O_3], on the other hand, induces the SA signalling pathway similarly to piercing-sucking insects and many plant pathogens. This can increase the production of direct defences against many piercing-sucking insect herbivores (see Fig. 5.1a).

Recent work on silicon defences in grasses supports this relationship. Silicon (Si) is important for tolerance to abiotic stresses as well as resistance to herbivory, especially in grasses. Feeding damage by caterpillars of *Helicoverpa armigera* (Noctuidae) activate the JA signalling pathway in the grass *Brachypodium distachyon* (Poaceae) which stimulates uptake of silicon (Hall *et al.*, 2020). Silicon generally reduces *H. armigera* growth rates. Elevated [CO_2] suppressed JA production by 70%, which reduced silicon uptake by 19%. However, despite the decrease in silicon uptake, herbivory levels by *H. armigera* were unaffected (Hall *et al.*, 2020). In another example, Japanese beetles *Popillia japonica* (Scarabaeidae) feeding on soybean *Glycine max* (Fabaceae) under e[CO_2] lived longer and had higher fecundity (O'Neill *et al.*, 2008). Rather than stimulating an increase sugar content, the increased fecundity and longevity of beetles appears to be the result of the reduced abundance of JA defence-related transcripts as well as reduced amounts of protease inhibitors in plants grown under e[CO_2], suggesting that such plants are less defended against herbivory (Casteel *et al.*, 2008; Zavala *et al.*, 2008). Similarly, e[CO_2] suppresses JA-based glucosinolate defences in *Arabidopsis thaliana*, allowing green peach aphid *Myzus persicae* (Aphididae) colonies to increase (Sun *et al.*, 2013). In other systems, however, aphids suffer reduced body weights when feeding on plants exposed to e[CO_2]. Body weights of both the corn leaf aphid *Rhopalosiphum maidis* (Aphididae) feeding on barley (Chen *et al.*, 2019) and the cotton aphid *Aphis gossypii* (Aphididae) feeding on melon *Cucumis melo* (Cucurbitaceae) (Moreno-Delafuente *et al.*, 2021b) were reduced under e[CO_2], presumably due to changes in nutritional quality of their host plants.

Elevated [O_3] has been shown to induce the SA signalling pathway, increasing chemical defences and reducing herbivore fitness. Increased [O_3] reduced both sugar and amino acid content as well as total phenolics and condensed tannins in three genotypes of tomato (wild, JA-overexpressed, JA-deficient). Correspondingly, the whitefly *Bemisia tabaci* (Aleyrodidae) suffered increased development times, reduced fecundity and reduced intrinsic rates of increase on plants grown under increased [O_3] (Cui *et al.*, 2012b). That the same trends were seen across the three genotypes suggests that e[O_3] did not affect the JA signalling pathway with regard to the production of plant toxins.

e[CO_2], e[O_3] and direct defence effects on parasitoids

Plant defensive toxins can affect parasitoids, and by extension the outcome of biocontrol efforts, in one of three ways. The overwhelming majority of studies of the multitrophic effects of plant antiherbivore toxins on parasitoids have focused on one of two mechanisms (Kaplan *et al.*, 2016), both of which have negative effects on parasitoid fitness. First, ingestion and metabolism of plant toxins frequently lead to small body sizes of herbivorous hosts, resulting in reduced body sizes and brood sizes as well as reduced survivorship for many parasitoids (see Fig. 5.1c). Second, parasitoids may directly encounter unmetabolized toxins or their toxic metabolites within the bodies of their hosts. This includes situations where hosts sequester plant toxins in defence against parasitoids and especially predators (Nishida, 2002; Ode, 2006; Beran and Petschenka, 2022).

To date, relatively few studies have explicitly linked the effects of e[CO_2] or e[O_3] on plant toxin production to parasitoid performance. One of the earliest studies to do so showed that e[CO_2] resulted in the increased production of phenolics (as well as carbohydrates) of four species of deciduous trees with variable effects on the performance of the polyphagous spongy moth *Lymantria dispar* (Erebidae) (Roth and Lindroth, 1995). Effects on the parasitoid *Cotesia melanoscela* (Braconidae) were weak and variable across the four tree species. A study of the effects of e[CO_2] and e[O_3] on the plant defensive chemistry of quaking aspen *Populus tremuloides* (Salicaceae) found similarly weak effects on the dipteran parasitoid *Compsilura concinnata* (Tachinidae) of the forest tent caterpillar *Malacosoma disstria* (Holton *et al.*, 2003). More recent studies on non-tree systems have shown stronger, and generally negative, effects of e[CO_2] on parasitoid and predator performance. Elevated [CO_2] increased glucosinolate content and changed plant nutritional quality in cultivated *Brassica oleracea*, with negative effects on the body size and development time of the cabbage aphid *Brevicoryne brassicae* (Aphididae) as well as the development time, longevity and overall parasitism efficiency of its parasitoid *Diaretiella rapae* (Braconidae) (Klaiber *et al.*, 2013a,b). Aphids that fed on plants grown under e[CO_2] were 50% lighter and contained significantly lower carbohydrate concentrations than aphids that fed on plants grown under ambient CO_2 levels (Najar-Rodriguez *et al.*, 2024), suggesting that reduced performance of *D. rapae* was explained by the smaller body size and reduced nutritional quality of *B. brassicae* that fed on plants grown under e[CO_2] (Klaiber *et al.*, 2013a). On the other hand, laboratory studies of the effects of elevated temperatures and CO_2 on defensive chemistry production in alfalfa *Medicago sativa* (Fabaceae) showed decreased plant quality and decreased development time of the herbivore *Spodoptera exigua* (Noctuidae), leading to developmental asynchrony for the parasitoid *Cotesia marginiventris* (Braconidae) (Dyer *et al.*, 2013).

A far smaller but growing number of studies have explored the third mechanism by which plant antiherbivore toxins may affect host–parasitoid interactions: the effect of plant secondary chemistry on the expression of immune responses against parasitoids (Smilanich and Muchoney, 2022). In general, herbivores feeding on plants containing higher concentrations of defensive toxins suffer compromised immune systems and are more susceptible to attack by parasitoids (Smilanich and Muchoney, 2022) (see Fig. 5.1d[1]). However, some studies have shown that herbivores feeding on plants with higher defensive compound concentrations can have enhanced immunity against their own pathogens (e.g. Ojala *et al.*, 2005; Gassmann *et al.*, 2010). Recent studies have shown that at least in some specialist herbivores (e.g. the small cabbage white *Pieris rapae* and the diamondback moth *Plutella xylostella*), consumption of *Brassica* spp. plants containing higher levels of glucosinolates results in enhanced immune responses against their respective *Cotesia* spp. parasitoids (Ghosh *et al.*, 2022; 2023a) (see Fig. 5.1d[2]). On the other hand, the generalist cabbage looper *Trichoplusia ni* (Noctuidae) suffered reduced immunocompetence when fed a diet containing higher concentrations of isothiocyanates, the bioactive hydrolysis product of glucosinolates (Ghosh *et al.*, 2023b).

Whether the positive relationship between plant toxins and immunocompetence of specialist herbivores and the negative relationship in generalist herbivores holds warrants further study. One of the few studies of which we are aware that has explicitly explored the relationship between e[CO_2] and herbivore immunocompetence against their own natural enemies involves the chrysomelid beetle *Paropsis atomaria* (Chrysomelidae) feeding on *Eucalyptus tereticornis* (Myrtaceae) (Gherlenda *et al.*, 2016). Cellular immunity (formation of cellular capsules around parasitoid eggs) of the beetle increased but, interestingly, humoral (production of phenol oxidase, involved in melanization) immune responses decreased under conditions of e[CO_2]. Total phenolics increased under e[CO_2], which possibly explains the positive relationship between e[CO_2] and cellular immunity (Gherlenda *et al.*, 2016). In another study of plant-mediated influence of e[CO_2] on herbivore immune responses, milkweed plants (*Asclepias* spp.; Apocynaceae) produced lower concentrations of cardenolides under e[CO_2] which enhanced both cellular and humoral immunity of the monarch butterfly *Danaus plexippus* (Nymphalidae) against the protozoan parasite *Ophryocystis elektroscirrha* (Ophryocystidae) (Decker *et al.*, 2021).

5.2.4 Climate change and plant defence chemistry: e[CO_2], e[O_3] and HIPVs (indirect defences)

Production of volatile organic compounds, particularly those induced in response to herbivore damage (HIPVs), are broadly appreciated to play roles in the indirect defence of the plant against insect herbivores in that HIPVs are well known to be attractive to parasitoids and predators (Heil, 2008; Turlings and Erb, 2018). In terms of indirect defences, elevated temperatures have long been known to result in the increased release of VOCs, including HIPVs that are attractive to parasitoids and predators of insect herbivores (Holopainen and Gershenzon, 2010). This is partly due to increased vapour pressure at higher temperatures, but VOC release is often at rates higher than vapour pressure alone can explain, suggesting that biosynthesis is higher at elevated temperatures. One generalized prediction is that biocontrol of insect pests may benefit if indirect plant defences in the form of increased HIPV production enhance attractiveness to parasitoids.

However, the effects of e[CO_2] on HIPV production are highly idiosyncratic (like the effect of e[CO_2] on antiherbivore toxins), depending strongly on the plant species, specific volatile compound and the herbivore species involved (e.g. Vuorinen *et al.*, 2004a,b; Yuan *et al.*, 2009). Not surprisingly, the effects of e[CO_2] on volatile production and whether parasitoids are attracted or not is also highly system specific, with examples of increased attractiveness, decreased attractiveness or no change in attractiveness to parasitoids and predators in response in e[CO_2] (Vuorinen *et al.*, 2004a; Himanen *et al.*, 2009). Three factors serve to complicate the relationship between e[CO_2] and indirect defences in the form of attraction of parasitoids and predators of insect herbivores. First, as noted above, e[CO_2] tends to suppress JA signalling and consequently the production of HIPVs, possibly making it more difficult for parasitoids and predators to respond to the activities of their herbivorous hosts and prey. Second, and possibly counteracting the first factor, suppression of JA signalling is expected to result in reduced induction of plant toxins, allowing herbivores to feed more and thus increase HIPV production. This may result in increased attractiveness to parasitoids and predators. Third, HIPVs are typically complex blends of several volatile compounds. Parasitoids and predators often respond to specific combinations of compounds in specific proportions rather than simply the most abundant volatile in a particular blend or even the total amount of the blend (Moore *et al.*, 2014; Dicke, 2016). Taken together, these three factors help explain the idiosyncratic effects of e[CO_2] on the production of HIPVs and the attractiveness to predators and parasitoids.

While plant volatile production, including HIPV production, is widely known to increase under conditions of e[O_3], relatively little is known about how e[O_3] affects plant volatile production. Elevated [O_3] is known to stimulate the increased production of volatile terpenoid emissions, particularly when the plant is under attack by a herbivore (Beauchamp *et al.*, 2005; Himanen *et al.*, 2009; Holopainen and Gershenzon, 2010; Souza *et al.*, 2013). Many insect herbivores, in particular aphids and whiteflies, along with elevated temperature, high light and elevated ozone all induce VOCs through similar mechanisms that result in production of reactive oxygen species such as superoxide and hydrogen peroxide (Holopainen and Gershenzon, 2010). Injury by ozone exposure is known to increase JA concentrations which increase volatile production, but the mechanisms by which this occurs are unclear (Cui *et al.*, 2016).

5.3 Climate Change, Plant Chemical Defences and Biological Control of Weedy Plants

To the extent that e[CO_2] is expected to suppress the JA signalling pathway and decrease the induction of chemical defences against chewing herbivores, our simplistic expectation is that climate change involving e[CO_2] would improve biological control efforts using chewing insect herbivores to control invasive plants. Conversely, e[CO_2] and e[O_3] are expected to induce the SA signalling pathway (and further suppress the JA signalling pathway) and enhance induction of chemical defences against many piercing-sucking insects and some gall-forming insects. Therefore, our expectation is that climate change may hinder efforts of weed biological control using agents such as aphids and gall-forming insects.

Plant–herbivore population dynamics may be disrupted if the development of plants and insect herbivores responds differently to increasing temperatures. A recent study examined the effects of simulated warming on the effectiveness of biological control of the invasive common ragweed *Ambrosia artemisiifolia* (Asteraceae) using the chrysomelid beetle *Ophraella communa*

(Chrysomelidae). Where these two species co-occur, the beetle can provide effective control. However, under climate warming the effectiveness of the beetle to control *A. artemisiifolia* populations is expected to decline, possibly because of changes in the production of sesquiterpene lactones in warmer conditions (Sun *et al.*, 2022). In another study of temperature effects on biological control effectiveness, Liu *et al.* (2022) explored the effects of elevated temperature on the plant chemistry of two species of alligatorweed: *Alternanthera philoxeroides* (Amaranthaceae), which is invasive (to China), and the native congener *A. sessilis*. Both plants are fed upon by several native herbivores including the noctuid moth *Spodoptera litura* (Noctuidae) and the chrysomelid beetle *Cassida piperata* (Chrysomelidae). Under simulated warming, total phenol concentration increased (although only marginally for *A. philoxeroides*) for both species of alligatorweed. Larval weight of *S. litura* was reduced and development time was increased under conditions of elevated temperature (Liu *et al.*, 2022). While these native insect species are not used as biological control agents in China (indeed, they attack a native congener), these results add to the body of evidence that increasing temperatures will probably disrupt efforts to maintain effective biological control programmes.

Similar to the effects of increased temperature on the effectiveness of biological control efforts, e[CO_2] is also predicted to disrupt efforts directed at the biological control of invasive plants. Four species of invasive aquatic plants in South Africa are currently under effective control by their respective introduced biological control agents: three species of weevils and one chrysomelid beetle (Baso *et al.*, 2021). Under conditions of e[CO_2], plant biomass increased with a corresponding drop in C:N ratio. Percent damage by three of the four herbivores declined under conditions of e[CO_2], suggesting that biological control efforts directed at these aquatic plants are likely to be disrupted in the future (Baso *et al.*, 2021). While not directly measured in this study, it is possible that plants grown under e[CO_2] were less nutritious or more toxic.

5.4 Climate Change, Plant Chemical Defences and Biological Control of Herbivorous Insects

While our predictions for the effects of e[CO_2] and e[O_3] on weed biological control efforts are straightforward, the predicted outcome of e[CO_2] and e[O_3] on biocontrol efforts against chewing herbivorous insects is more complicated and depends on various traits specific to the herbivore, the parasitoid and the host–parasitoid relationship. For instance, chewing herbivore populations may outgrow parasitoid populations when feeding on less defended plants, resulting in reduced efficiency of biological control (Fig. 5.2a). On the other hand, parasitoids may encounter reduced concentrations of plant toxins (or their metabolites) in the body of their hosts that feed on plants with reduced defensive toxins. We expect the same predictions to hold in cases where hosts sequester plant toxins in defence against parasitoids and predators (see Fig. 5.2b). In either case, parasitoids should benefit from developing within hosts that contain lower levels of plant toxins, because such hosts are either larger and/or contain lower levels of toxins, resulting in more efficient biological control (see Fig. 5.2b). Finally, herbivorous hosts feeding on plants whose ability to increase the production of toxins has been suppressed by e[CO_2] may experience either suppressed or enhanced immune response capabilities depending on the herbivore species and how immunity is affected by dietary plant toxins (see Fig. 5.2c).

Several studies have examined the effect of e[CO_2] on plant chemistry, insect herbivores and their natural enemies. An early study by Roth and Lindroth (1995) showed that e[CO_2] increased phenolic content, which reduced the larval weight of the spongy moth *Lymantria dispar*. This led to reduced body size and increased mortality of the parasitoid *Cotesia melanoscela*. Cotton (*Gossypium hirsutum*; Malvaceae) grown under conditions of e[CO_2] produced decreased foliar nitrogen and increased concentrations of gossypol, a toxin active against several cotton herbivores including the cotton aphid *Aphis gossypii* (Aphididae) which exhibited reduced survivorship, longevity and reproductive output (Gao *et al.*, 2008). In turn, adult females of the predatory coccinellid beetle *Propylaea japonica*

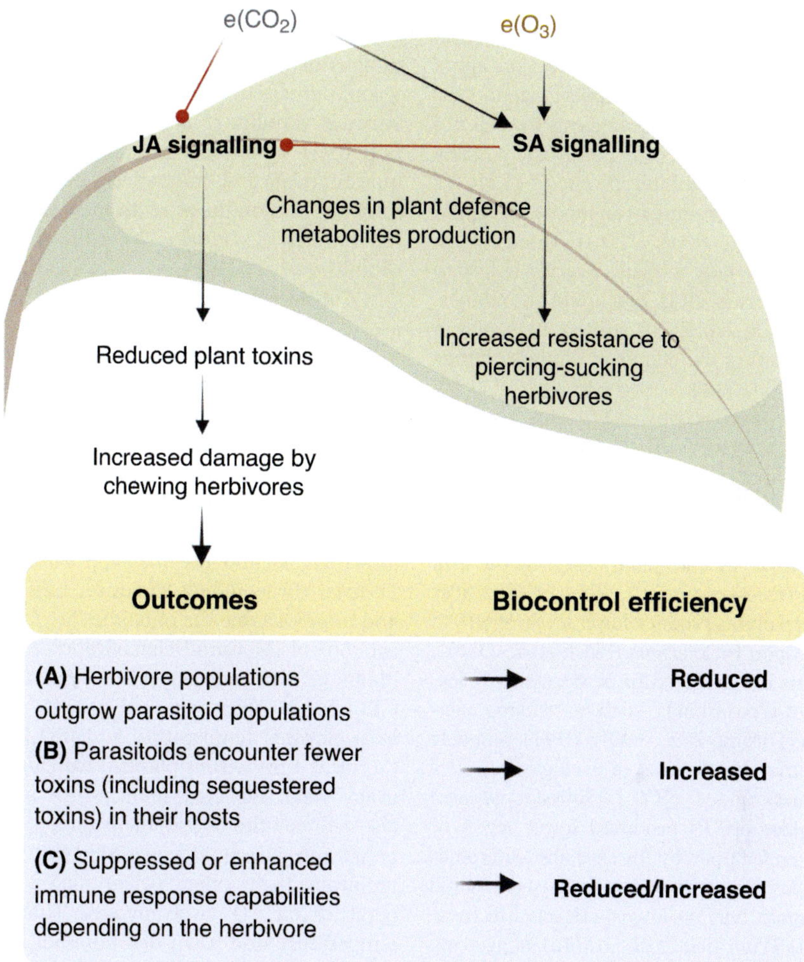

Fig. 5.2. Predicted outcomes of e[CO_2] and e[O_3] for plant, herbivore and natural enemy interactions and their consequences for biological control. Lines ending in arrows indicate positive effects; lines ending in circles represent suppressive or inhibitory effects.

(Coccinellidae) fed aphids from plants exposed to e[CO_2] were heavier and had shorter development times, two traits that the authors argued would enhance efforts using this biological control agent (Gao *et al.*, 2008). In a separate study of the cotton–cotton aphid system, Chen *et al.* (2005) found that the multicolored Asian lady beetle *Leis* (=*Harmonia*) *axyridis* (Coccinellidae) preferred to feed on aphids that fed on cotton plants grown under e[CO_2], suggesting that biological control of the cotton aphid should be enhanced under rising CO_2 levels. In other biological control systems, e[CO_2] and temperatures are expected to result in less effective control of insect pests. In a phytotron study of maize (*Zea mays*; Poaceae), both elevated temperatures and e[CO_2] improved performance of the banded cucumber beetle *Diabrotica balteata* (Chrysomelidae), but hindered biocontrol efforts using the entomopathogenic nematode *Heterorhabditis bacteriophora* (Rhabditida) (Guyer *et al.*, 2021).

Plant genotypes and their interactions with herbivores and their natural enemies can vary in their responses to climate change. Elevated [CO_2] resulted in a 70% increase in raspberry aphid

Amphorophora idaei (Aphididae) abundance on resistant raspberry *Rubus idaeus* (Rosaceae) genotypes but e[CO_2] had little effect on aphid abundance on susceptible raspberry genotypes (Hentley *et al.*, 2014). This suggests that e[CO_2] disrupts plant resistance to herbivory, possibly due to unmeasured inhibitory effects of e[CO_2] on JA signalling, resulting in an increase in aphid populations (Sun *et al.*, 2013). Interestingly, predation by the lady beetle *H. axyridis* helped to reduce the effect of e[CO_2] on aphid abundance, suggesting that biological control may play a role in mitigating the effects of climate change (Hentley *et al.*, 2014).

As discussed above, e[CO_2] can alter the production of HIPVs that are attractive to parasitoids and predators. Corn (a.k.a. maize; *Zea mays*) is well known to produce HIPVs such as (*E*)-β-caryophyllene that attract parasitoids when fed upon by caterpillars such as the beet armyworm *Spodoptera exigua*. When grown under e[CO_2], corn plants produce lower levels of HIPVs when fed upon by *S. exigua* (Block *et al.*, 2017). Such plants are expected to be significantly less attractive to parasitoids such as *Cotesia marginiventris* (Turlings *et al.*, 1990; 1991), probably reducing the effectiveness of biological control programmes under e[CO_2]. *Brassica oleracea* grown under e[CO_2] produced lower levels of HIPVs when fed upon by the specialist caterpillar *Plutella xylostella* such that its specialist parasitoid *Cotesia vestalis* was no longer attracted to these caterpillars (Vuorinen *et al.*, 2004a), suggesting that biocontrol would be disrupted in this system. Interestingly, with a different plant but with the same herbivore and parasitoid, e[CO_2] increased HIPVs produced by *Brassica napus* when fed on by *P. xylostella*, but this had no effect on the attractiveness to *C. vestalis* (Himanen *et al.*, 2009). Clearly the effects of e[CO_2] can vary between very similar systems.

Elevated [CO_2] can also alter the composition of volatiles released by aphid honeydew, potentially affecting the attractiveness to aphid natural enemies. While not directly involving plant volatile chemistry, many aphid parasitoids and predators are attracted to volatiles released by aphid honeydew, the excess sugars and liquids excreted as aphids feed on plant phloem. In some aphid–plant systems, bacteria colonize honeydew and can alter the volatile composition. Honeydew excreted by the pea aphid *Acyrthosiphon pisum* (Aphididae) is often colonized by bacteria in the genus *Staphylococcus* (Staphylococcaceae) which can contribute to volatile production that attracts aphid natural enemies (Boullis *et al.*, 2018). Under e[CO_2] honeydew volatile composition decreased both quantitatively and qualitatively, but this did not have an effect on the attractiveness to the predator hoverfly *Episyrphus balteatus* (Syrphidae) (Boullis *et al.*, 2018).

Consideration of the joint effects of multiple aspects of climate change (e.g. CO_2, O_3, elevated temperature, drought) is arguably more realistic than considering the effects of either climate change variable alone (Dyer *et al.*, 2013; Guyer *et al.*, 2021). A study of the effects of elevated temperature and CO_2 on wheat *Triticum aestivum* (Poaceae), the aphid *Rhopalosiphum padi* (Aphididae) (which is a vector of barley yellow dwarf virus) and the parasitic wasp *Aphidius colemani* (Braconidae) illustrates how complex and unpredictable the outcomes are for biological control (Moreno-Delafuente *et al.*, 2021b). Plants grown under elevated temperature and [CO_2] experienced increased growth, although both elevated temperature and [CO_2] increase the virus titre within plants, severely reducing grain yield. Elevated temperature and [CO_2] also reduced the development time of both the aphid and parasitoid, suggesting that both will maintain their phenological match as temperature and CO_2 levels increase. That elevated temperature and [CO_2] had minimal effects on any other parasitoid performance parameters (survivorship, body size) suggests that biological control efforts using this parasitoid may be unaffected by climate change (Moreno-Delafuente *et al.*, 2021b) (see Fig. 5.2c).

Not all studies of the effects of e[CO_2] suggest that the outcome for biological control efforts will be improved. Cabbage aphids *Brevicoryne brassicae* that fed on *Brassica oleracea* plants grown under e[CO_2] were significantly smaller and nutritionally poorer (reduced concentration of carbohydrates), which resulted in reduced survivorship and parasitism rates by the specialist parasitoid *Diaretiella rapae* (Najar-Rodriguez *et al.*, 2024).

Increasing frequency and intensity of drought conditions are another outcome of climate change that can alter plant defence chemistry and the success of biological control programmes. Drought stress reduces plant growth in sugar beet *Beta vulgaris* which benefits population growth rates of the aphid *Aphis fabae*. However, sugar beets produce lower quantities of volatiles under drought stress which reduces the indirect defence from the parasitoid *Aphidius colemani* (Rahman *et al*., 2024). Similar findings have been shown in drought-stressed Brussels sprouts *Brassica oleracea* (Tariq *et al*., 2013) that are attacked by two species of aphids: the green peach aphid *Myzus persicae* and the cabbage aphid *Brevicoryne brassicae*. The combination of drought stress and root herbivory by the cabbage root fly *Delia radicum* (Anthomyiidae) generally reduced the HIPV profiles emitted by Brussels sprouts plants. Parasitism rates and survivorship of *Aphidius colemani* and *Diaretiella rapae* (the parasitoids of the two aphid species) were both reduced under drought conditions. These studies indicate that a combination of increased aphid population growth rates and reduced mortality from the parasitoid make it likely that the effectiveness of biological control will be greatly reduced in the future.

Elevated $[O_3]$ can increase the production of volatile compounds that are attractive to parasitoids. Tomato plants grown under $e[O_3]$ which experienced whitefly *Bemisia tabaci* herbivory produced significantly higher quantities of volatile terpenoids. A study of three tomato genotypes (wild type, JA-overexpressed mutant, JA-deficient mutant) showed that under $e[O_3]$, volatile production was highest in the JA-overexpression mutant, intermediate in the wild type and lowest in the JA-deficient mutant. Correspondingly, under $e[O_3]$ but not ambient $[O_3]$, the parasitoid *Encarsia formosa* (Aphelinidae) was most strongly attracted to JA-overexpressed plants attacked by *B. tabaci* and least attracted to JA-deficient plants (Cui *et al*., 2014). Interestingly, whiteflies were least attracted to the JA-overexpressed plants and most attracted to the JA-deficient plants, providing another mechanism of resistance to whiteflies under $e[O_3]$ (Cui *et al*., 2016). Furthermore, under $e[O_3]$, the JA signalling pathway upregulates the production of HIPVs that are attractive to parasitoids (Cui *et al*., 2014; 2016). Previous studies have suggested that $e[O_3]$ results in increased HIPV production that is attractive to parasitoids (Ellis *et al*., 2002; Pinto *et al*., 2007a,b). Together, these studies suggest that $e[O_3]$ should enhance indirect defences and therefore should improve the success of biological control programmes that rely on attraction of parasitoids to herbivore-infested plants.

5.5 Conclusion

That atmospheric gas concentrations (including carbon dioxide and ozone), along with mean global temperatures, will continue to rise through the rest of this century is a virtual certainty. While such climate changes will undoubtedly change patterns of induction of plant defence chemistry (both direct and indirect), the direction and magnitude of these changes are difficult to predict. Furthermore, the impacts on the interactions between plants, insect herbivores and their parasitoids and predators are even more complex and difficult to predict. The implications for future effectiveness of biological control programmes are also complex and difficult to predict. Nevertheless, it is clear from many of the examples discussed in this chapter that many biological control programmes against invasive plants and against herbivorous pests of crops will probably be disrupted. More studies on a wider range of systems will undoubtedly enhance our understanding of how such relationships will probably be changed.

We also need more studies conducted under realistic field conditions that involve multiple herbivore species and natural enemy species. Most plants are attacked by multiple species of herbivores, which in turn are attacked by multiple species of natural enemies. Yet most studies have, understandably, focused on single species plant, herbivore and natural enemy systems as these are already complex. Nevertheless, we encourage more studies that incorporate greater degrees of complexity so that phenomena such as apparent competition and apparent predation can be studied. Whether such phenomena act to mitigate or exacerbate the effects of climate change will be very interesting from an ecological perspective and vitally important from the perspective of designing resilient biological control programmes.

References

Ainsworth, E.A. and Long, S.P. (2005) What have we learned from Free-Air CO_2 Enrichment (FACE)? A meta-analytic review of the responses of photosynthesis, canopy properties and plant production to rising CO_2. *New Phytologist* 165, 351–371.

Ainsworth, E.A., Yendrek, C.R., Sitch, S., Collins, W.J. and Emberson, L.D. (2012) The effects of tropospheric ozone on net primary productivity and implications for climate change. *Annual Review of Plant Biology* 63, 637–661.

Baso, N.C., Coetzee, J.A., Ripley, B.S. and Hill, M.P. (2021) The effects of elevated atmospheric CO_2 concentration on the biological control of invasive aquatic weeds. *Aquatic Botany* 170, 103348.

Beauchamp, J., Wisthaler, A., Hansel, A., Kleist, E., Miebach, M. *et al.* (2005) Ozone induced emissions of biogenic VOC from tobacco: Relationships between ozone uptake and emission of LOX products. *Plant, Cell and Environment* 28, 1334–1343.

Beran, F. and Petschenka, G. (2022) Sequestration of plant defense compounds by insects: From mechanisms to insect-plant coevolution. *Annual Review of Entomology* 67, 163–180.

Berenbaum, M.R. (1991) Coumarins. In: Rosenthal, G.A. and Berenbaum, M.R. (eds) *Herbivores: Their Interactions with Secondary Plant Metabolites, Volume 1: The Chemical Participants*, 2nd edn. Academic Press, San Diego, California, pp. 221–249.

Bezemer, T.M. and Jones, T.H. (1998) Plant-insect herbivore interactions in elevated atmospheric CO_2: Quantitative analyses and guild effects. *Oikos* 82, 212–222.

Bidart-Bouzat, M.G. and Imeh-Nathaniel, A. (2008) Global change effects on plant chemical defenses against insect herbivores. *Journal of Integrative Plant Biology* 50, 1339–1354.

Bidart-Bouzat, M.G., Mithen, R. and Berenbaum, M.R. (2005) Elevated CO_2 influences herbivory-induced defense responses of *Arabidopsis thaliana*. *Oecologia* 145, 415–424.

Block, A., Vaughan, M.M., Christensen, S.A., Alborn, H.T. and Tumlinson, J.H. (2017) Elevated carbon dioxide reduces emission of herbivore-induced volatiles in *Zea mays*. *Plant, Cell and Environment* 40, 1725–1734.

Boullis, A., Blanchard, S., Francis, F. and Verheggen, F. (2018) Elevated CO_2 concentrations impact the semiochemistry of aphid honeydew without having a cascading effect on an aphid predator. *Insects* 9, 47.

Casteel, C.L., O'Neill, B.F., Zavala, J.A., Bilgin, D.D., Berenbaum, M.R. *et al.* (2008) Transcriptional profiling reveals elevated CO_2 and elevated O_3 alter resistance of soybean (*Glycine max*) to Japanese beetles (*Popillia japonica*). *Plant, Cell and Environment* 31, 419–434.

Chen, F., Ge, F. and Parajulee, M.N. (2005) Impact of elevated CO_2 on tri-trophic interaction of *Gossypium hirsutum*, *Aphis gossypii*, and *Leis axyridis*. *Environmental Entomology* 34, 37–46.

Chen, Y., Serteyn, L., Wang, Z., He, K.L. and Francis, F. (2019) Reduction of plant suitability for corn leaf aphid (Hemiptera: Aphididae) under elevated carbon dioxide conditions. *Environmental Entomology* 48, 935–944.

Cipollini, D.F., Busch, J.W., Stowe, K.A., Simms, E.L. and Bergelson, J. (2003a) Genetic variation and relationships of constitutive and herbivore-induced glucosinolates, trypsin inhibitors, and herbivore resistance in *Brassica rapa*. *Journal of Chemical Ecology* 29, 285–302.

Cipollini, D.F., Purrington, C.B. and Bergelson, J. (2003b) Costs of induced responses in plants. *Basic and Applied Ecology* 4, 79–85.

Cooke, J. and Leishman, M.R. (2010) Is plant ecology more siliceous than we realize? *Trends in Plant Science* 16, 61–68.

Cui, H., Sun, Y., Su, J., Li, C. and Ge, F. (2012a) Reduction in the fitness of *Bemisia tabaci* fed on three previously infested tomato genotypes differing in the jasmonic acid pathway. *Environmental Entomology* 41, 1443–1453.

Cui, H., Sun, Y., Su, J., Ren, Q., Li, C. *et al.* (2012b) Elevated O_3 reduces the fitness of *Bemisia tabaci* via enhancement of the SA-dependent defense of the tomato plant. *Arthropod–Plant Interactions* 6, 425–437.

Cui, H., Su, J., Wei, J., Hu, Y. and Ge, F. (2014) Elevated O_3 enhances the attraction of whitefly-infested tomato plants to *Encarsia formosa*. *Scientific Reports* 4, 5350.

Cui, H., Wei, J., Su, J., Li, C. and Ge, F. (2016) Elevated O_3 increases volatile organic compounds via jasmonic acid pathway that promote the preference of parasitoid *Encarsia formosa* for tomato plants. *Plant Science* 253, 243–250.

De Moraes, C.M., Lewis, W.J., Paré, P.W., Alborn, H.T. and Tumlinson, J.H. (1998) Herbivore-infested plants selectively attract parasitoids. *Nature* 393, 570–573.

Decker, L.E., Jeffrey, C.S., Ochsenrider, K.M., Potts, A.S., de Roode, J.C. *et al.* (2021) Elevated atmospheric concentrations of CO_2 increase endogenous immune function in a specialist herbivore. *Journal of Animal Ecology* 90, 628–640.

Dicke, M. (2016) Induced plant volatiles: Plant body odours structuring ecological networks. *New Phytologist* 110, 10–12.

Dicke, M., Vanbeek, T.A., Posthumus, M.A., Bendom, N., Vanbokhoven, H. *et al.* (1990) Isolation and identification of volatile kairomone that affects acarine predator-prey interactions – involvement of host plant in its production. *Journal of Chemical Ecology* 16, 381–396.

Docherty, M., Wade, F.A., Hurst, D.K., Whittaker, J.B. and Lea, P.J. (1997) Responses of tree sap-feeding herbivores to elevated CO_2. *Global Change Biology* 3, 51–59.

Drake, B.G., Gonzàlez-Meler, M.A. and Long, S.P. (1997) More efficient plants: a consequence of rising atmospheric CO_2? *Annual Review of Plant Physiology and Plant Molecular Biology* 48, 609–639.

Dyer, L.A., Richards, L.A., Short, S.A. and Dodson, C.D. (2013) Effects of CO_2 and temperature on tritrophic interactions. *PLOS One* 8(4), e62528.

Ellis, C., Karafylldis, I. and Turner, J.G. (2002) Constitutive activation of jasmonate signaling in an *Arabidopsis* mutant correlates with enhanced resistance to *Erysiphe cichoracearum*, *Pseudomonas syringae*, and *Myzus persicae*. *Molecular Plant Microbe Interactions* 15, 1025–1030.

Fiscus, E.L., Booker, F.L. and Burkey, K.O. (2005) Crop responses to ozone: Uptake, modes of action, carbon assimilation and partitioning. *Plant, Cell and Environment* 28, 997–1011.

Futuyma, D.J. and Agrawal, A.A. (2009) Macroevolution and the biological diversity of plants and herbivores. In: *Proceedings of the National Academy of Sciences*, pp. 18054–18061 (Vol. 106).

Gao, F., Zhu, S.-R., Sun, Y.-C., Du, L., Parajulee, M. *et al.* (2008) Interactive effects of elevated CO_2 and cotton cultivar on tri-trophic interaction of *Gossypium hirsutum*, *Aphis gossypii*, and *Propylaea japonica*. *Environmental Entomology* 37, 29–37.

Gassmann, A.J., Stock, S.P., Tabashnik, B.E. and Singer, M.S. (2010) Tritrophic effects of host plants on an herbivore-pathogen interaction. *Annals of the Entomological Society of America* 103, 371–378.

Gherlenda, A.N., Haigh, A.M., Moore, B.D., Johnson, S.N. and Riegler, M. (2016) Climate change, nutrition and immunity: Effects of elevated CO_2 and temperature on the immune function of an insect herbivore. *Journal of Insect Physiology* 85, 57–64.

Ghosh, E., Sasidharan, A., Ode, P.J. and Venkatesan, R. (2022) Oviposition preference and performance of a specialist herbivore is modulated by natural enemies, larval odors, and immune status. *Journal of Chemical Ecology* 48, 670–682.

Ghosh, E., Paul, R.L. and Ode, P.J. (2023a) Plant species with higher chemical defences enhance herbivore cellular immunity with differential effectiveness against two parasitoid species. *Functional Ecology* 37, 1492–1503.

Ghosh, E., Tafesh-Edwards, G.S.Y., Eleftherianos, I., Goldin, S.L. and Ode, P.J. (2023b) The plant toxin 4-methylsulfinylbutyl isothiocyanate decreases herbivore performance and modulates cellular and humoral immunity. *PLOS One* 18(8), e0289205.

Guyer, A., van Doan, C., Maurer, C., Machado, R.A.R., Mateo, P. *et al.* (2021) Climate change modulates multitrophic interactions between maize, a root herbivore, and its enemies. *Journal of Chemical Ecology* 47, 889–906.

Hall, C.R., Mikhael, M., Hartley, S.E. and Johnson, S.N. (2020) Elevated atmospheric CO_2 suppresses jasmonate and silicon-based defences without affecting herbivores. *Functional Ecology* 34, 993–1002.

Hamilton, J.G., Zangerl, A.R., DeLucia, E.H. and Berenbaum, M.R. (2001) The carbon-nutrient balance hypothesis: Its rise and fall. *Ecology Letters* 4, 86–95.

Harborne, J.B. (1993) *Introduction to Ecological Biochemistry*, 4th edn. Academic Press, London, UK.

Heil, M. (2008) Indirect defence via tritrophic interactions. *New Phytologist* 178, 41–61.

Hentley, W.T., Hails, R.S., Johnson, S.N., Jones, T.H. and Vanbergen, A.J. (2014) Top-down control by *Harmonia axyridis* mitigates the impact of elevated atmospheric CO_2 on a plant-aphid interaction. *Agricultural and Forest Entomology* 16, 350–358.

Himanen, S.J., Nerg, A.M., Nissinen, A., Pinto, D.M., Stewart, C.N., Jr *et al.* (2009) Effects of elevated carbon dioxide and ozone on volatile terpenoid emissions and multitrophic communication of transgenic insecticidal oilseed rape (*Brassica napus*). *New Phytologist* 181, 174–186.

Holopainen, J.K. (2002) Aphid response to elevated ozone and CO_2. *Entomologia Experimentalis et Applicata* 104, 137–142.

Holopainen, J.K. and Gershenzon, J. (2010) Multiple stress factors and the emission of plant VOCs. *Trends in Plant Science* 15, 176–183.

Holton, M.K., Lindroth, R.L. and Nordheim, E.V. (2003) Foliar quality influences tree-herbivore-parasitoid interactions: effects of elevated CO_2, O_3, and plant genotype. *Oecologia* 137, 233–244.

Jackson, D.M., Rufty, T.W., Heagle, A.S., Severson, R.F. and Eckel, R.V.W. (2000) Survival and development of tobacco hornworm larvae on tobacco plants grown under elevated levels of ozone. *Journal of Chemical Ecology* 26, 1–19.

Jamieson, M.A., Trowbridge, A.M., Raffa, K.F. and Lindroth, R.L. (2012) Consequences of climate warming and altered precipitation patterns for plant-insect and multitrophic interactions. *Plant Physiology* 160, 1719–1727.

Jamieson, M.A., Burkle, L.A., Manson, J.S., Runyon, J.B., Trowbridge, A.M. *et al.* (2017) Global change effects on plant-insect interactions: The role of phytochemistry. *Current Opinion in Insect Science* 23, 70–80.

Jøndrup, P.M., Barnes, J.D. and Port, G.R. (2002) The effect of ozone fumigation and different *Brassica rapa* lines on the feeding behaviour of *Pieris brassicae* larvae. *Entomologia Experimentalis et Applicata* 104, 143–151.

Kaplan, I., Carrillo, J., Garvey, M. and Ode, P.J. (2016) Indirect plant-parasitoid interactions mediated by changes in herbivore physiology. *Current Opinion in Insect Science* 14, 112–119.

Karowe, D.N., Seimens, D.H. and Mitchell-Olds, T. (1997) Species-specific response of glucosinolate content to elevated atmospheric CO_2. *Journal of Chemical Ecology* 23, 2569–2582.

Kessler, A. and Baldwin, I.T. (2001) Defensive function of herbivore-induced plant volatile emissions in nature. *Science* 291, 2141–2144.

Kessler, A. and Heil, M. (2011) The multiple faces of indirect defences and their agents of natural selection. *Functional Ecology* 25, 348–357.

Klaiber, J., Dorn, S. and Najar-Rodriguez, A.J. (2013a) Acclimation to elevated CO_2 increases constitutive glucosinolate levels of *Brassica* plants and affects the performance of specialized herbivores from contrasting feeding guilds. *Journal of Chemical Ecology* 39, 653–665.

Klaiber, J., Najar-Rodriguez, A.J., Dialer, E. and Dorn, S. (2013b) Elevated carbon dioxide impairs the performance of a specialized parasitoid of an aphid host feeding on *Brassica* plants. *Biological Control* 66, 49–55.

Kopper, B.J. and Lindroth, R.L. (2003) Effects of elevated carbon dioxide and ozone on the phytochemistry of aspen and performance of an herbivore. *Oecologia* 134, 95–103.

Koricheva, J., Larsson, S., Haukioja, E. and Keinänen, M. (1998) Regulation of woody plant secondary metabolism by resource availability: Hypothesis testing by means of meta-analysis. *Oikos* 83, 212–226.

Körner, C. (2003) Ecological impacts of atmospheric CO_2 enrichment on terrestrial ecosystems. *Proceedings of the Royal Society A* 361, 2023–2041.

Lindroth, R.L. (2010) Impacts of elevated atmospheric CO_2 and O_3 on forests: phytochemistry, trophic interactions, and ecosystem dynamics. *Journal of Chemical Ecology* 26, 2–21.

Lindroth, R.L. and Kinney, K.K. (1998) Consequences of enriched atmospheric CO_2 and defoliation for foliar chemistry and gypsy moth performance. *Journal of Chemical Ecology* 24, 1677–1695.

Liu, Z., Yu, H., Sun, X. and Ding, J. (2022) Effects of elevated temperature on chemistry of an invasive plant, its native congener and their herbivores. *Journal of Plant Ecology* 15, 450–460.

Long, S.P., Ainsworth, E.A., Rogers, A. and Ort, D.R. (2004) Rising atmospheric carbon dioxide: Plants FACE the future. *Annual Review of Plant Biology* 55, 591–628.

Matusheski, N.V., Juvik, J.A. and Jeffery, J.H. (2004) Heating decreases epithiospecifier protein activity and increases sulforaphane formation in broccoli. *Phytochemistry* 65, 1273–1281.

McGrath, J.M., Betzelberger, A.M., Wang, S., Shook, E., Zhu, X.-G. *et al.* (2015) An analysis of ozone damage to historical maize and soybean yields in the United States. *Proceedings of the National Academy of Sciences USA* 112, 14390–14395.

McNaughton, S.J. and Tarrants, J.L. (1983) Grass leaf silicification: Natural selection for an inducible defense against herbivores. *Proceedings of the National Academy of Sciences USA* 80, 790–791.

Menéndez, A.I., Romero, A.M., Folcia, A.M. and Martínez-Ghersa, M.A. (2009) Getting the interactions right: Will higher O_3 interfere with induced defenses to aphid feeding? *Basic and Applied Ecology* 10, 255–264.

Menéndez, A.I., Romero, A.M., Folcia, A.M. and Martínez-Ghersa, M.A. (2010) Aphid and episodic O_3 injury in arugula plants (*Eruca sativa* Mill) grown in open-top field chambers. *Agriculture, Ecosystems and Environment* 135, 10–14.

Mithöfer, A. and Boland, W. (2012) Plant defense against herbivores: Chemical aspects. *Annual Review of Plant Biology* 63, 431–450.

Mitter, C., Farrell, B. and Wiegmann, B. (1988) The phylogenetic study of adaptive zones: Has phytophagy promoted insect diversification? *American Naturalist* 132, 107–128.

Mittler, R. (2006) Abiotic stress, the field environment and stress combination. *Trends in Plant Science* 11, 15–19.

Monticelli, L.S., Bishop, J., Desneux, N., Gurr, G.M., Jaworski, C.C. *et al.* (2021) Multiple global change impacts on parasitism and biocontrol services in future agricultural landscapes. *Advances in Ecological Research* 65, 245–304.

Moore, B.D., Andrew, R.L., Külheim, C. and Foley, W.J. (2014) Explaining intraspecific diversity in plant secondary metabolites in an ecological context. *New Phytologist* 201, 733–750.

Moreno-Delafuente, A., Morales, I., Garzo, E., Fereres, A., Viñuela, E. *et al.* (2021a) Changes in melon plant phytochemistry impairs *Aphis gossypii* growth and weight under elevated CO_2. *Scientific Reports* 11, 2186.

Moreno-Delafuente, A., Viñuela, E., Fereres, A., Medina, P. and Trębicki, P. (2021b) Combined effects of elevated CO_2 and temperature on multitrophic interactions involving a parasitoid of plant virus vectors. *BioControl* 66, 307–319.

Morgan, P.B., Ainsworth, E.A. and Long, S.P. (2003) How does elevated ozone impact soybean? A meta-analysis of photosynthesis, growth and yield. *Plant, Cell and Environment* 26, 1317–1328.

Najar-Rodriguez, A.J., Lacorazza, S., Klaiber, J., Avila, G.A., Zhang, J. *et al.* (2024) Long-term effects of elevated CO_2 on the nutrition provided to parasitoids by their herbivorous hosts. *Physiological Entomology* 49, 244–252.

Nishida, R. (2002) Sequestration of defensive substances from plants by Lepidoptera. *Annual Review of Entomology* 47, 57–92.

Ode, P.J. (2006) Plant chemistry and natural enemy fitness: Effects on herbivore and natural enemy interactions. *Annual Review of Entomology* 51, 163–185.

Ode, P.J., Johnson, S.N. and Moore, B.D. (2014) Atmospheric change and induced plant secondary metabolites – are we reshaping the building blocks of multi-trophic interactions? *Current Opinion in Insect Science* 5, 57–65.

Ode, P.J. (2022) Caterpillars, plant chemistry, and parasitoids in natural vs. agroecosystems. In: Marquis, R.J. and Koptur, S. (eds) *Caterpillars in the Middle*. Springer Nature, Cham, Switzerland, pp. 395–423.

Ojala, K., Julkunen-Tiitto, R., Lindström, L. and Mappes, J. (2005) Diet affects the immune defence and life-history traits of an arctiid moth *Parasemia plantaginis*. *Evolutionary Ecology Research* 7, 1153–1170.

O'Neill, B.F., Zangerl, A.R., DeLucia, E.H. and Berenbaum, M.R. (2008) Longevity and fecundity of Japanese beetle (*Popillea japonica*) on foliage grown under elevated carbon dioxide. *Environmental Entomology* 37, 601–607.

Pearse, I.S., LoPresti, E., Schaeffer, R.N., Wetzel, W.C., Mooney, K.A. *et al.* (2020) Generalising indirect defence and resistance of plants. *Ecology Letters* 23, 1137–1152.

Percy, K.E., Awmack, C.S., Lindroth, R.L., Kubiske, M.E., Kopper, B.J. *et al.* (2002) Altered performance of forest pests under atmospheres enriched by CO_2 and O_3. *Nature* 420, 403–407.

Pinto, D.M., Blande, J.D., Nykänen, R., Dong, W.X., Nerg, A.M. *et al.* (2007a) Ozone degrades common herbivore-induced plant volatiles: Does this affect herbivore prey location by predators and parasitoids? *Journal of Chemical Ecology* 33, 683–694.

Pinto, D.M., Nerg, A.M. and Holopainen, J.K. (2007b) The role of ozone-reactive compounds, terpenes, and Green Leaf Volatiles (GLVs), in the orientation of *Cotesia plutellae*. *Journal of Chemical Ecology* 33, 2218–2228.

Price, P.W., Bouton, C.E., Gross, P., McPheron, B.A., Thompson, J.N. *et al.* (1980) Interactions among three trophic levels: Influence of plants on interactions between insect herbivores and natural enemies. *Annual Review of Ecology and Systematics* 11, 41–65.

Rahman, S., Rostás, M. and Vosteen, I. (2024) Drought aggravates plant stress by favouring aphids and weakening indirect defense in a sugar beet tritrophic system. *Journal of Pest Science* 98, 549–564.

Reynolds, O.L., Keeoing, M.G. and Meyer, J.H. (2009) Silicon-augmented resistance of plants to herbivorous insects: A review. *Annals of Applied Biology* 1655, 171–186.

Richards, L.A., Dyer, L.A., Forister, M.L., Smilanich, A.M., Dodson, C.D. *et al.* (2015) Phytochemical diversity drives plant-insect community diversity. In: *Proceedings of the National Academy of Sciences*, pp. 10973–10978 (Vol. 112).

Robinson, E.A., Ryan, G.D. and Newman, J.A. (2012) A meta-analytical review of the effects of elevated CO_2 on plant-arthropod interactions highlights the importance of interacting environmental and biological variables. *New Phytologist* 194, 321–336.

Rosenthal, G.A. and Janzen, D.H. (eds) (1979) *Herbivores: Their Interaction with Secondary Plant Metabolites*. Academic Press, New York.

Rosenthal, G.A. and Berenbaum, M.R. (eds) (1991) *Herbivores: Their Interactions with Secondary Plant Metabolites, Volume 1: The Chemical Participants*, 2nd edn. Academic Press, New York.

Roth, S. and Lindroth, R.L. (1995) Elevated atmospheric CO_2: Effects on phytochemistry, insect performance and insect-parasitoid interactions. *Global Change Biology* 1, 173–182.

Roth, S., Lindroth, R.L., Volin, J.C. and Kruger, E.L. (1998) Enriched atmospheric CO_2 and defoliation: Effects on tree chemistry and insect performance. *Global Change Biology* 4, 419–430.

Schonhof, I., Kläring, H.P., Krumbein, A. and Schreiner, M. (2007) Interaction between atmospheric CO_2 and glucosinolates in broccoli. *Journal of Chemical Ecology* 33, 105–114.

Schoonhoven, L.M., Loon, J.J.A. and Dicke, M. (2005) *Insect Plant Biology*, 2nd edn. Oxford University Press, Oxford, UK.

Schweiger, R., Heise, A.-M., Persicke, M. and Müller, C. (2014) Interactions between the jasmonic and salicylic acid pathway modulate the plant metabolome and affect herbivores of different feeding types. *Plant, Cell and Environment* 37, 1574–1585.

Smilanich, A.M. and Muchoney, M.D. (2022) Host plant effects on the caterpillar immune response. In: Marguis, R.J. and Koptur, S. (eds) *Caterpillars in the Middle*. Springer, Cham, Switzerland, pp. 449–484.

Souza, S.R., Blande, J.D. and Holopainen, J.K. (2013) Pre-exposure to nitric oxide modulates the effect of ozone on oxidative defenses and volatile emissions in lima bean. *Environmental Pollution* 179, 111–119.

Speed, M.P., Fenton, A., Jones, M.G., Ruxton, G.D. and Brockhurst, M.A. (2015) Coevolution can explain defensive secondary metabolite diversity in plants. *New Phytologist* 208, 1251–1263.

Stiling, P. and Cornelessen, T. (2007) How does elevated carbon dioxide (CO_2) affect plant-herbivore interactions? A field experiment and meta-analysis of CO_2-mediated changes on plant chemistry and herbivore performance. *Global Change Biology* 13, 1823–1842.

Strauss, S.Y., Rudgers, J.A., Lau, J.A. and Irwin, R.E. (2002) Direct and ecological costs of resistance to herbivory. *Trends in Ecology and Evolution* 17, 278–285.

Sun, Y., Guo, H., Zhu-Salzman, K. and Ge, F. (2013) Elevated CO_2 increases the abundance of the peach aphid on *Arabidopsis* by reducing jasmonic acid defenses. *Plant Science* 210, 128–140.

Sun, Y., Züst, T., Silvestro, D., Erb, M., Bossdorf, O. *et al.* (2022) Climate warming can reduce biocontrol efficacy and promote plant invasion due to both genetic and transient metabolomic changes. *Ecology Letters* 25, 1387–1400.

Tariq, M., Wright, D.J., Bruce, T.J.A. and Staley, J.T. (2013) Drought and root herbivory interact to alter the response of above-ground parasitoids to aphid-infested plants and associated plant volatile signals. *PLOS One* 8(7), e69013.

Turlings, T.C.J. and Erb, M. (2018) Tritrophic interactions mediated by herbivore-induced plant volatiles: Mechanisms, ecological relevance, and application potential. *Annual Review of Entomology* 63, 433–452.

Turlings, T.C.J., Tumlinson, J.H. and Lewis, W.J. (1990) Exploitation of herbivore-induced plant odors by host-seeking parasitic wasps. *Science* 250, 1251–1253.

Turlings, T.C.J., Tumlinson, J.H., Heath, R.R., Proveaux, A.T. and Doolittle, R.E. (1991) Isolation and identification of allelochemicals that attract the larval parasitoid, *Cotesia marginiventris* (Cresson), to the microhabitat of one of its hosts. *Journal of Chemical Ecology* 17, 2235–2250.

Valkama, E., Koricheva, J. and Oksanen, E. (2007) Effects of elevated O_3, alone and in combination with elevated CO_2, on tree leaf chemistry and insect herbivore performance: A meta-analysis. *Global Change Biology* 13, 184–201.

van Poecke, R.M.P. and Dicke, M. (2004) Indirect defence of plants against herbivores: Using *Arabidopsis thaliana* as a model plant. *Plant Biology* 6, 387–401.

Vuorinen, T., Nerg, A.M., Ibrahim, M.A., Reddy, G.V.P. and Holopainen, J.K. (2004a) Emission of *Plutella xylostella*-induced from cabbages grown at elevated CO_2 and orientation behavior of the natural enemies. *Plant Physiology* 135, 1984–1992.

Vuorinen, T., Reddy, G.V.P., Nerg, A.-M. and Holopainen, J.K. (2004b) Monoterpene and herbivore-induced emissions from cabbage plants grown at elevated atmospheric CO_2 concentration. *Atmospheric Environment* 38, 675–682.

Yuan, J.S., Himanen, S.J., Holopainen, J.K., Chen, F. and Stewart, C.N., Jr (2009) Smelling global climate change: Mitigation of function for plant volatile organic compounds. *Trends in Ecology and Evolution* 24, 323–330.

Zavala, J.A., Casteel, C.L., DeLucia, E.H. and Berenbaum, M.R. (2008) Anthropogenic increase in carbon dioxide compromises plant defense against invasive insects. In: *Proceedings of the National Academy of Sciences*, pp. 5129–5133 (Vol. 105).

Zavala, J.A., Nabity, P.D. and DeLucia, E.H. (2013) An emerging understanding of mechanisms governing insect herbivory under elevated CO_2. *Annual Review of Entomology* 58, 79–97.

Zvereva, E.L. and Koslov, M.V. (2006) Consequences of simultaneous elevation of carbon dioxide and temperature for plant-herbivore interactions: A metaanalysis. *Global Change Biology* 12, 27–41.

6 What Do We Know about the Role of Spiders in Biological Control under Climate Change? A Bibliometric Approach with Insights on the Neotropical Region

Luis Fernando García[1]*, Luis Quijano[2] and Julien Pétillon[3,4]
[1]Cenur Noreste, Universidad De La República, Rivera, Uruguay; [2]Instituto de Investigaciones en Ecosistemas y Sustentabilidad (IIES), UNAM, Morelia, Mexico; [3]UMR CNRS ECOBIO (Ecosystèmes, Biodiversité, Evolution), University of Rennes, Rennes, France; [4]Institute for Coastal and Marine Research, Nelson Mandela University, Gqeberha, South Africa

Abstract
Climate change affects the abundance, distribution and interactions of arthropods, impacting ecosystem services such as pollination and pest control. Spiders, as generalist predators, are crucial for insect regulation, yet their role as biological pest control agents, especially in tropical regions, is underresearched. Using a bibliometric approach, this study examined the extent of research on spiders in agroecosystems within the context of climate change and biological control. Results indicate a low number of studies on climate change and agriculture compared to other topics, with research predominantly conducted in North America and Europe, leaving significant gaps in tropical regions. Given the importance of agriculture in tropical areas, it is strongly recommended to increase research on the effects of spiders as biocontrol agents in these zones to better understand and utilize their potential in pest management, especially under current changing climatic conditions.

6.1 Introduction

Climate change is considered a significant local threat to the diversity of various taxonomic groups. While the impact is more obvious in some groups, such as vertebrates, which are facing massive extinctions, other groups like arthropods are also highly affected by climate change due to their specific ecophysiological conditions, but evidence remains scarce for these taxa (Harvey *et al.*, 2023). Although in some cases climate change is positively related to the population growth of certain groups, particularly insects from temperate zones, in other regions like the tropics, which are highly diverse, climate will probably negatively affect insect populations due to abrupt changes in the usually stable climatic conditions of these regions (Halsch *et al.*, 2021).

Another possible consequence of climate change is an alteration of the well-known biotic to abiotic control of community structure at different levels, including ecological aspects such as trophic interactions, one of the most

*Corresponding author: luizf.garciah@gmail.com

DOI: 10.1079/9781800625099.0006

relevant. For example, in the case of insects it has been shown that temperature changes may affect several community and population levels, including interactions between prey and predators by altering, for example, the behaviour and nutritional ecology of the latter, while non-consumptive effects are affected too since temperature can alter the intra- and interspecific properties of chemical signals (Laws, 2017). Changes in trophic interactions can affect the role of beneficial organisms in agriculture, particularly in the case of programmes which depend on predator–prey interactions such as conservation biological control, where native generalist predators play a key role (Schmitz and Barton, 2014).

Spiders are among the primary groups of insect predators within arthropods (Pekár and Toft, 2014). It is estimated that spiders can consume up to 800 million tonnes of insects annually, making them one of the leading regulators of insect populations (Nyffeler and Birkhofer, 2017). Their diet mainly consists of other arthropods, primarily insects, although some species can prey on small vertebrates (Pekár and Toft, 2014; Valdez, 2020; Nyffeler and Gibbons, 2022). This makes spiders an important group of predators across various microhabitats (Nyffeler and Birkhofer, 2017). While most spider species are generalists with adaptations to consume a wide array of prey, some are strict specialists, preying specifically on certain arthropods such as ants, lepidopterans, isopods, dipterans, termites and even other spiders (Pekár and Toft, 2014). The control of insect populations by spiders extends beyond natural habitats, providing practical benefits for humans. Nevertheless, most research on spiders as biological control agents has focused on agriculture using generalist spider species (Michalko *et al.*, 2019a), considering they are found in a wide variety of crops and have been reported as one of the most abundant predators in these environments (Benamú *et al.*, 2017a). Certain spider groups dominate crop ecosystems and are classified as agrobionts due to traits such as periodic colonization and strong competitive abilities compared to other species (Samu and Szinetár, 2002). This adaptability to agricultural settings, coupled with their synchrony with local pest populations (González *et al.*, 2009) and their ability to persist even when major pest populations decline, makes them highly suitable for biological control (Cuff *et al.*, 2021).

Despite being common predators in crops, for many years the role of spiders as insect regulators in agroecosystems was controversial, as occurs with some other generalist predators, due to the consumption of non-specific prey (Symondson *et al.*, 2002). However, recent evidence has shown that spiders are a promising group for pest control in crops. In a recent meta-analysis, Michalko *et al.* (2019b) studied the beneficial effects of spiders in 32 crops, including apple, avocado, barley, cabbage, rice and wheat, and found that spiders could contribute to reduce pest infestations in up to 70% of all the studied cases. These authors found that spiders had positive effects on several parameters related to pest control, such as reducing pest densities, decreasing crop damage and improving crop performance.

The effectiveness of spiders in reducing pest populations can be partially attributed to their diverse spider guilds and generalist habits. These traits enable them to target prey with various morphologies and defensive strategies (Sunderland, 1999; García *et al.*, 2021). The diversity among spider guilds not only minimizes intraguild competition in crops but also allows access to prey with different behaviours and ecological niches. For instance, Butt and Tahir (2010) found that hunting spiders consumed a higher proportion of homopteran prey compared to web-building spiders, whereas the opposite trend was observed for lepidopterans. Additionally, García *et al.* (2021) discovered that adult females of the generalist spider *Lycosa thorelli* can prey on various pest species and developmental stages due to their generalist habits, even consuming prey with strong chemical defences, such as stink bugs, and different stages like larvae and adults of the fall armyworm. Moreover, spiders possess other behavioural traits that make them suitable for biological control, such as overkilling, where spiders kill more prey than they consume (Michalko *et al.*, 2019a), influencing their functional response and, consequently, their impact on pest populations (López-Mercadal *et al.*, 2024). This overkilling behaviour has been observed in generalist agrobiont spiders such as linyphiids and lycosids (Mansour and Heimbach, 1993). Furthermore, several spiders potentially useful for pest control can persist even when the crop is no longer available by feeding on non-pest alternative prey, such as collembolans, thanks to their generalist habits (Sigsgaard, 2002 ; Kuusk and Ekbom, 2010).

Conservation biological control aims to utilize native natural enemies in agricultural environments by boosting their populations through practices such as habitat management (Holland *et al.*, 2016; Peñalver-Cruz *et al.*, 2019). Given the local abundance of spiders in various crops, they are considered ideal candidates for conservation biological control programmes. For example, habitat modifications in crops like soybean and alfalfa have been shown to increase spider populations, thereby reducing pest numbers and crop damage or delaying pest establishment (Halaj *et al.*, 2000; Schmidt *et al.*, 2007). Although there is extensive information on this topic from recent reviews, most studies have focused on data from North American and European crops (Michalko *et al.*, 2019a, b). Despite the wealth of research from temperate regions, studies on spiders in the Neotropics are relatively scarce, and there is limited information on conservation biological control for Neotropical species compared to insects (Vargas *et al.*, 2023). Nonetheless, it has been observed that the positive effects of spiders on crops increase towards the tropics, which could be partially attributed to a rise in the annual mean temperature (Michalko *et al.*, 2019a). This suggests that temperature significantly influences the effectiveness of spiders as biological control agents.

6.2 Temperature and Feeding Behaviour in Spiders

In recent years, several studies have linked temperature with the predatory capacity of spiders, showing that they are efficient predators under different thermal scenarios. For example, certain studies have shown that spiders exposed to high temperatures exhibit an increase in predation rate and prey biomass (Walker *et al.*, 2020). Similarly, it has been shown that the functional response of wolf spiders is affected by temperature, displaying unimodal behaviour optimal values at specific temperatures which coincides with temperatures that predators are expected to experience (DeLong *et al.*, 2023). For example, Pekár *et al.* (2015) demonstrated that in pear orchards, spiders could significantly reduce the populations of the psyllid *Cacopsylla pyri* during winter, which is considered one of the most important pests in this system. Although other mortality factors besides spiders might reduce *Cacopsylla* populations, early predation by spiders can prevent pest outbreaks.

Considering that temperatures affect the feeding behaviour of spiders, it is expected that variations in temperature due to climate change will influence the trophic interactions involving spiders in crops. For example, in barley fields, it has been shown that certain parameters, such as prey selectivity and web properties, change with the weather (Cuff *et al.*, 2023). Similarly, Birkhofer *et al.* (2022) demonstrated that climate change might affect the trophic interactions of spiders in crops, given the higher susceptibility of hemipterans to climate fluctuations. This could lead to an increased consumption of alternative prey with higher resistance, such as collembolans, due to climatic instability. Nevertheless, some studies suggest that while warm temperatures might favour nocturnally active spiders by increasing their foraging behaviour, the opposite pattern might occur in diurnally active spiders, potentially leading to changes in the temporal niches of spiders and niche overlap between different species (DeLong *et al.*, 2023). Therefore, temperature might have mixed effects depending not only on spider species and foraging habits but also on the climatic zone where the species is present.

Temperature can also affect the role of spiders in crops alongside the action of some insecticides commonly used in agriculture. For example, studies have shown a differential effect of the insecticides Spinosad and Lambda-cyhalothrin on philodromid (vegetation-dwelling) spiders under different temperatures, suggesting that climate change might influence the interaction between spiders present in crops and other agricultural practices such as pesticide use (Michalko and Košulič, 2016; see also Chapter 14 in this book).

Despite the relevance of spiders as study models in the context of crops and climate change, the number of studies is limited and often restricted to certain countries, including the USA, Germany and Switzerland together with some other European countries, suggesting a lack of research focus on some zones, such as the Neotropical region which has a lower number of studies (see Birkhofer *et al.*, 2022). Furthermore, given the recent surge in studies on climate change, it is unclear how this topic has evolved in relation to other aspects of climate change and spiders. Therefore, the aim of the following section is to provide an updated bibliographic review of the knowledge on spiders as biological control agents in the context of climate change. This includes examining how this knowledge has evolved over the years and identifying the most studied aspects related to biological control in spiders. Finally, the production and number of papers per country and over time are also analysed.

We evaluated the bibliographic production in the Scopus database without any word or time limits, using the following search algorithms: 'spiders AND climate AND change', 'spiders AND global and warming', 'spiders AND climate AND change AND behavioural AND changes', 'spiders AND climate AND physiological AND responses', 'spiders AND climate AND change AND distribution', 'spiders AND climate AND change AND behavioural AND changes', 'spiders AND climate AND change AND behavioural AND change', 'spiders AND climate AND change AND crops', 'spiders AND climate AND change AND species AND distribution', in the time period between 1970 and 2024. Once the data were extracted, we conducted a bibliometric analysis using the R package Bibliometrix (Aria and Cuccurullo, 2017), and generated plots using the ggplot2 package (Valero-Mora, 2010). We filtered out all the papers that did not focus on spiders or were not related to climate change by checking their abstracts. In addition to the bibliometric analyses, we evaluated the yearly production and identified the most productive countries regarding authors' affiliation on this topic. Furthermore, we contrasted studies focused on agriculture-related fields against other subjects. All topics and papers were evaluated in the context of climate change.

We classified the topics found into the following categories, all of which were selected in the context of climate change by including papers which included the term 'climate change' in the keywords or evaluated the effect of temperature on the parameters considered.

1. *Agriculture-related fields*: this category included all subjects related to crops in different fields such as biological control, crop diversity and ecotoxicology.
2. *Biodiversity and conservation*: this included all topics related to alpha and beta diversity, as well as conservation status, including local threats.
3. *Ecophysiology*: this category included topics related to physiology and ecology, such as thermal tolerance and reproductive rates, among other ecophysiological parameters.
4. *Phylogeography and dispersion*: this included studies related to the dispersal of spiders, including biogeographical and phylogeographical patterns.
5. *Trophic ecology*: this category included studies evaluating the feeding behaviour of spiders, as well as other parameters like trophic webs.
6. *Others*: this included topics that did not fit into the other categories, including molecular biology and medically important spiders.

Once the information was filtered, we evaluated all information regarding agriculture-focused papers by analysing the yearly production for this topic, as well as the most productive countries. We further divided the agricultural papers into those focused on biological control topics and those focused on other fields of knowledge. In this category, we defined the following topics.

1. *Abundance and diversity of natural enemies in crops*: this included studies on diversity, phenology and abundance in crops.
2. *Biocontrol-related studies*: this included papers where direct or indirect evidence of pest control by spiders was evaluated.
3. *Ecotoxicology-related topics*: this included all topics related to the effects of insecticides and herbicides on spiders.

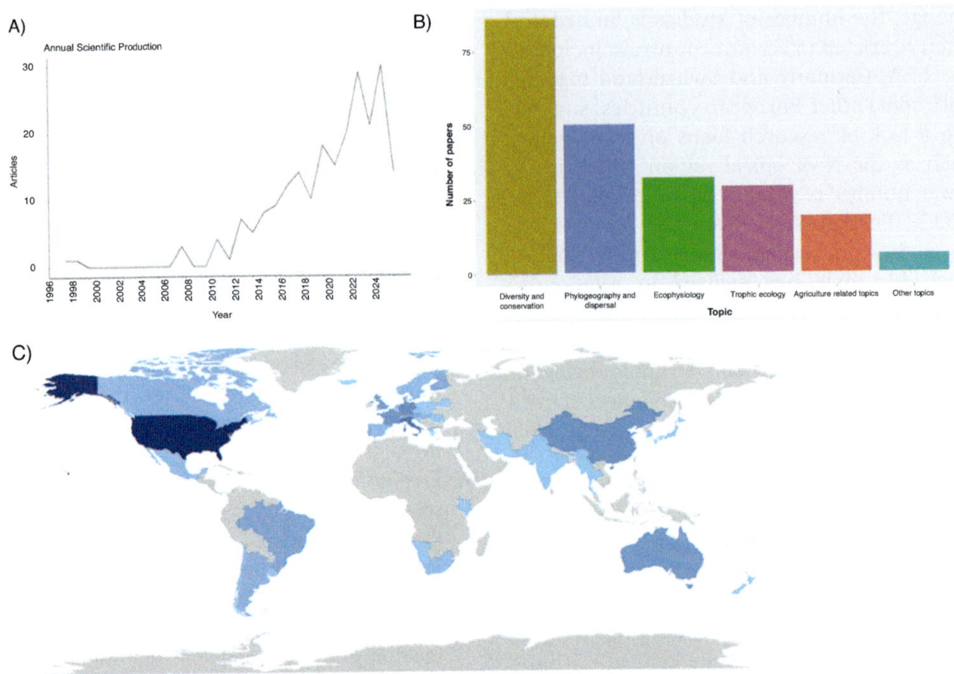

Fig. 6.1. A. Bibliographic production over time of studies focused on spiders and climate change in the Scopus database. B. Number of published studies in various fields of knowledge related to spiders and climate change (see text for details). C. Scientific production per country on climate change and spider research.

Similar to the general classification, we included only papers related to climate change. We compared the number of papers published for each topic and per year, as well as the countries with the highest productivity on this topic. After performing the search and removing all papers that did not fit our criteria, we found a total of 222 papers which were used in the following sections. A list of the papers used for bibliometric analysis can be requested from the corresponding author (see email address at the start of the chapter).

6.3 What is the Overall Knowledge of the Biocontrol Potential of Spiders under Climate Change?

Upon evaluating the yearly production, a clear trend was found of a marked increase in topics related to climate change after the year 2005, reaching an almost exponential trend in the last 6 years (Fig. 6.1a). When comparing countries, as expected, the highest productivity was found in the USA (Fig. 6.1c) and China as well as in several European countries, particularly in Italy, Germany, the UK and France; as shown by Birkhofer *et al.* (2022) in their review, there is a significant lack of knowledge in the Global South, with almost a complete lack of production in several countries, such as most African and Asian countries. It is particularly interesting that in the case of Latin America, there is a strong lack of studies in megadiverse countries such as Colombia, Panama, Ecuador and Venezuela. This trend contrasts with other groups such as pollinators which have been studied in more detail regarding climate change in the Neotropical region (Galetto *et al.*, 2022). This is particularly important since temperatures severely affect arachnids, and in consequence, changes in trophic networks and local extinction processes might be happening without being noticed.

When comparing the topics related to climate change (see Fig. 6.1b), a higher proportion of studies focused on biodiversity was found, followed by phylogeography and distribution, ecophysiology, trophic ecology, agriculture-related topics and finally other topics. The literature shows a significant trend towards diversity and conservation topics, which agrees with previous studies in other groups such as vertebrates and insects (Quintero and Wiens, 2013; Halsch *et al.*, 2021; Raven and Wagner, 2021), where it has been shown that processes like local extinction or threats to local diversity caused by climate change are the main focus of recent research. Second, the high number of studies focused on phylogeography might be explained by the recent interest sparked by the rapid distributional changes of several spider groups. For example, it has been shown that several spider species have recently expanded their current distribution and, in some cases, might become a threat to local diversity. Nevertheless, in our search, we found that some studies regarding phylogeography have modelled the effects of climate change on spiders, intending to predict the impact of temperature changes on local spider populations, suggesting that they represent a suitable model to evaluate the effects of climate change in the future.

Regarding ecophysiology, most studies evaluated aspects such as reproduction, demography, thermal stress and the natural history of various species. In trophic ecology, most studies focused on the feeding behaviour of spiders, trophic webs and changes in feeding parameters like functional response and prey biomass under different thermal scenarios. Agricultural-related topics and studies of medical importance occupy the last places with fewer than 25 published studies. The low number of studies on climate change and spiders' roles in agricultural systems contrasts with studies on other predatory arthropods like carabids which, although still understudied, receive more attention regarding climate change.

Given the local importance of arachnids as models for studying climate change, top predators and pest regulators, we strongly suggest increasing research on this group, especially in regions with high arachnid diversity such as the tropics, where many species may face extinction.

6.4 Climate Change and Spiders as Biological Control Agents

Knowledge about climate change and spiders' roles as biological control agents is limited. Similar to the overall knowledge of spiders in the context of climate change, studies on spiders in agriculture, particularly concerning biological control and climate change, are scarce. In the analysis described in this chapter, it was found that scientific studies on spiders in crops and climate change have seen interrupted growth, with a slight increase since 2019, indicating its appearance as a new topic (see Fig. 6.2a). This production is concentrated in North America and Europe, with no contributions from tropical countries (Fig. 6.2b).

For example, in Latin America, studies on spiders in areas like conservative biological control are scarce compared to other predator groups such as insects (Vargas *et al.*, 2023). The lack of knowledge about local fauna probably explains the low number of studies on spiders and biological control in crops. Therefore, it is necessary to increase local diversity studies, not only in natural areas but also in agroecosystems, considering the importance of agriculture in tropical regions, which include several megadiverse areas. It was found that studies on spiders as biological control agents are fewer compared to other aspects (see Fig. 6.2b), which is consistent with other studies. For instance, Benamú *et al.* (2017a) noted that most studies on spiders as biological control agents in the Neotropical region focus on spider surveys in crops, with scarce data on their direct effects on pests.

The role of spiders as natural enemies is affected by pesticides like herbicides and insecticides by lethal and sublethal means (Pekár, 2012). For example, glyphosate can affect spiders' functional responses to pests (Lacava *et al.*, 2021), and similar trends are observed with various insecticides (Řezáč *et al.*, 2010; Boyd *et al.*, 2022). Pesticides can cause physiological changes in spiders, altering silk properties and reproductive rates (Benamú *et al.*, 2010, 2017b). Considering that climate change and insecticides can have negative and synergistic effects on spiders, this should be a highly explored topic. However, our search found very few studies combining climate change

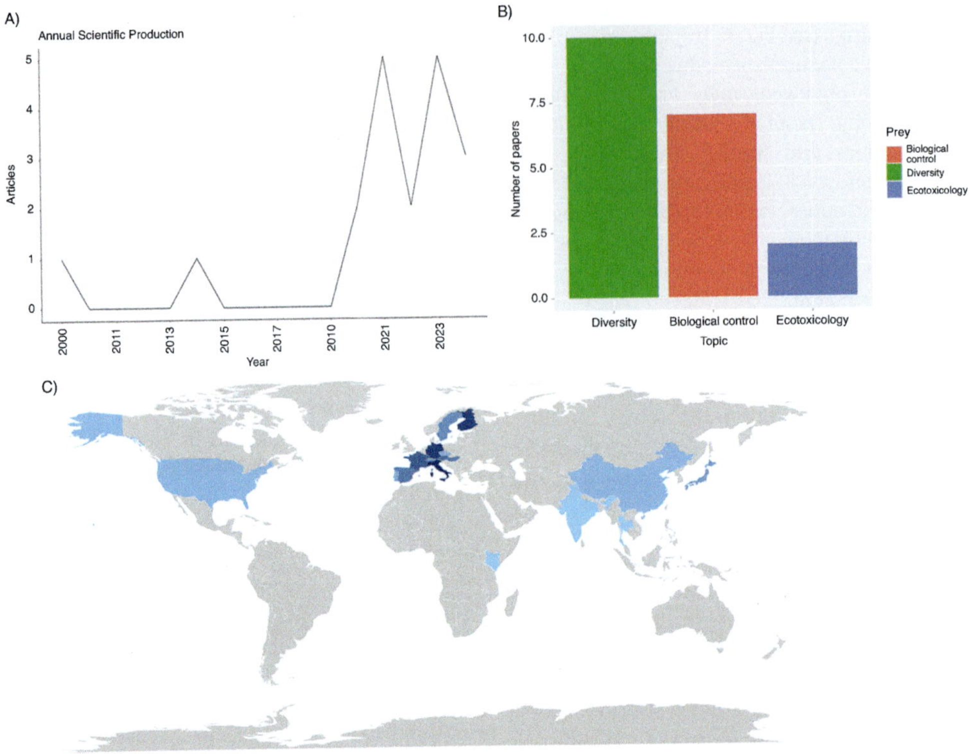

Fig. 6.2. A. Bibliographic production over time of studies focused on spiders, climate change and agroecosystems in the Scopus database. B. Number of published studies in various fields of knowledge related to spiders, climate change and agroecosystems (see text for details). C. Scientific production per country on agroecosystems, climate change and spider research.

and pesticide use, especially in tropical regions where temperature changes might negatively impact local spider diversity, exacerbated by pesticide use.

Given that studies in trophic ecology have shown the effects of temperature on foraging behaviour and functional response, we suggest these studies should be conducted on tropical species and local pests, considering the variation between the ecophysiological conditions of pests and predators in agroecosystems.

6.5 Conclusion

Spiders are important predators and useful models in various fields, potentially valuable for evaluating the impact of climate change on local diversity. Their emerging role as recognized biological control agents makes them suitable for studying how changing environments affect local fauna. The results of this chapter summarize the current knowledge of spiders as biological control agents in the context of climate change. Despite limitations such as using a single database and not analysing the study site on each country, the findings of this chapter basically align with previous studies highlighting similar issues, such as the lack of knowledge in tropical regions and the mixed effects of pesticides on spiders. Further studies should explore the role of spiders as biological control agents and their relationship with climate change, especially in Neotropical regions where their role has been poorly explored.

References

Aria, M. and Cuccurullo, C. (2017) *Bibliometrix*: An R-tool for comprehensive science mapping analysis. *Journal of Informetrics* 11(4), 959–975.

Benamú, M., Lacava, M., García, L.F., Santana, M., Fang, J. *et al.* (2017b) Nanostructural and mechanical property changes to spider silk as a consequence of insecticide exposure. *Chemosphere* 181, 241–249.

Benamú, M.A., Schneider, M.I. and Sánchez, N.E. (2010) Effects of the herbicide glyphosate on biological attributes of *Alpaida veniliae* (Araneae, Araneidae), in laboratory. *Chemosphere* 78(7), 871–876.

Benamú, M.A., Lacava, M., García, L.F., Santana, M. and Viera, C. (2017a) Spiders associated with agroecosystems: Roles and perspectives. In: Viera, C. and Gonzaga, M.O. (eds) *Behavior and Ecology of Spiders: Contributions from the Neotropical Region*. Springer International, New York, pp. 275–302.

Birkhofer, K., Djoudi, E.A., Schnerch, B. and Michalko, R. (2022) Climatic conditions and functional traits affect spider diets in agricultural and non-agricultural habitats worldwide. *Ecography* 2022(3), e06090.

Boyd, K., Hesselberg, T. and Alexander, M. (2022) Determination of the functional response in the orb-weaving spider *Araneus diadematus* (Araneae: Araneidae) according to insecticide type. *Ecological Entomology* 47, 791–800.

Butt, A. and Tahir, H.M. (2010) Resource partitioning among five agrobiont spiders of a rice ecosystem. *Zoological Studies* 49(4), 470–480.

Cuff, J.P., Drake, L.E., Tercel, M.P.T.G., Stockdale, J.E., Orozco-terWengel, P. *et al.* (2021) Money spider dietary choice in pre- and post-harvest cereal crops using metabarcoding. *Ecological Entomology* 46(2), 249–261.

Cuff, J.P., Windsor, F.M., Tercel, M.P.T.G., Bell, J.R., Symondson, W.O.C. *et al.* (2023) Temporal variation in spider trophic interactions is explained by the influence of weather on prey communities, web building and prey choice. *Ecography* 7, e06737.

DeLong, J., Uiterwaal, S.F. and Magallanes, A. (2023) Temperature has a unimodal effect on the functional response of wolf spiders. *Climate Change Ecology* 4, 100063.

Galetto, L., Aizen, M.A., Arizmendi, M.C., Freitas, B.M., Garibaldi, L.A. *et al.* (2022) Risks and opportunities associated with pollinators' conservation and management of pollination services in Latin America. *Ecología Austral* 32(1).

García, L.F., Núñez, E., Lacava, M., Silva, H., Martínez, S. *et al.* (2021) Experimental assessment of trophic ecology in a generalist spider predator: Implications for biocontrol in Uruguayan crops. *Journal of Applied Entomology* 145(1–2), 82–91.

González, A., Liljesthröm, G., Castro, D.C. and Armendano, A. (2009) Development and recruitment of *Misumenops pallidus* (Keyserling) (Araneae: Thomisidae), and its synchronicity with three potential prey species in soybean cultures from Argentina. *Entomological News* 120(1), 41–52.

Halaj, J., Cady, A.B. and Uetz, G.W. (2000) Modular habitat refugia enhance generalist predators and lower plant damage in soybeans. *Environmental Entomology* 29(2), 383–393.

Halsch, C.A., Shapiro, A.M., Fordyce, J.A., Nice, C.C., Thorne, J.H. *et al.* (2021) Insects and recent climate change. *Proceedings of the National Academy of Sciences* 118(2), e2002543117.

Harvey, J.A., Tougeron, K., Gols, R., Heinen, R., Abarca, M. *et al.* (2023) Scientists' warning on climate change and insects. *Ecological Monographs* 93(1), e1553.

Holland, J.M., Bianchi, F.J., Entling, M.H., Moonen, A., Smith, B.M. *et al.* (2016) Structure, function and management of semi-natural habitats for conservation biological control: A review of European studies. *Pest Management Science* 72(9), 1638–1651.

Kuusk, A.K. and Ekbom, B. (2010) Lycosid spiders and alternative food: Feeding behavior and implications for biological control. *Biological Control* 55(1), 20–26.

Lacava, M., García, L.F., Viera, C. and Michalko, R. (2021) The pest-specific effects of glyphosate on functional response of a wolf spider. *Chemosphere* 262, 127785.

Laws, A.N. (2017) Climate change effects on predator-prey interactions. *Current Opinion in Insect Science* 23, 28–34.

López-Mercadal, J., Benhadi-Marín, J., Rodrigues, I., Miranda, M.Á. and Pereira, J.A. (2024) Does the superfluous killing exhibited by spiders mask the functional response parameters? A case study from the perspective of biological control. *Journal of Applied Entomology* 148(4), 364–370.

Mansour, F. and Heimbach, U. (1993) Evaluation of lycosid, micryphantid and linyphiid spiders as predators of *Rhopalosiphum padi* (Hom.: Aphididae) and their functional response to prey density-laboratory experiments. *Entomophaga* 38(1), 79–87.

Michalko, R. and Košulič, O. (2016) Temperature-dependent effect of two neurotoxic insecticides on predatory potential of Philodromus spiders. *Journal of Pest Science* 89(2), 517–527.

Michalko, R., Pekár, S. and Entling, M.H. (2019a) An updated perspective on spiders as generalist predators in biological control. *Oecologia* 189(1), 21–36.

Michalko, R., Pekár, S., Dul'a, M. and Entling, M.H. (2019b) Global patterns in the biocontrol efficacy of spiders: A meta-analysis. *Global Ecology and Biogeography* 28(9), 1366–1378.

Nyffeler, M. and Birkhofer, K. (2017) An estimated 400–800 million tons of prey are annually killed by the global spider community. *Science of Nature* 104(3–4), 30.

Nyffeler, M. and Gibbons, J.W. (2022) Spiders feeding on vertebrates is more common and widespread than previously thought, geographically and taxonomically. *Journal of Arachnology* 50(2), 121–134.

Pekár, S. (2012) Spiders (Araneae) in the pesticide world: An ecotoxicological review. *Pest Management Science* 68(11), 1438–1446.

Pekár, S. and Toft, S. (2014) Trophic specialisation in a predatory group: The case of prey-specialised spiders (Araneae). *Biological Reviews* 90, 744–761.

Pekár, S., Michalko, R., Loverre, P., Líznarová, E. and Černecká, L. (2015) Biological control in winter: Novel evidence for the importance of generalist predators. *Journal of Applied Ecology* 52(1), 270–279.

Peñalver-Cruz, A., Alvarez-Baca, J.K., Alfaro-Tapia, A., Gontijo, L. and Lavandero, B. (2019) Manipulation of agricultural habitats to improve conservation biological control in South America. *Neotropical Entomology* 48(6), 875–898.

Quintero, I. and Wiens, J.J. (2013) Rates of projected climate change dramatically exceed past rates of climatic niche evolution among vertebrate species. *Ecology Letters* 16(8), 1095–1103.

Raven, P.H. and Wagner, D.L. (2021) Agricultural intensification and climate change are rapidly decreasing insect biodiversity. *Proceedings of the National Academy of Sciences* 118(2), e2002548117.

Samu, F. and Szinetár, C. (2002) On the nature of agrobiont spiders. *Journal of Arachnology* 30(2), 389–402.

Schmidt, N.P., O'Neal, M.E. and Singer, J.W. (2007) Alfalfa living mulch advances biological control of soybean aphid. *Environmental Entomology* 36(2), 416–424.

Schmitz, O.J. and Barton, B.T. (2014) Climate change effects on behavioral and physiological ecology of predator–prey interactions: Implications for conservation biological control. *Biological Control* 75, 87–96.

Sigsgaard, L. (2002) Early season natural biological control of insect pests in rice by spiders and some factors in the management of the cropping system that may affect this control. In: Toft, S. and Scharff, N. (eds) *European arachnology 2000: Proceedings of the 19th European Colloquium of Arachnology, Århus*, Aarhus Universitetsforlag, pp. 57–74.

Sunderland, K. (1999) Mechanisms underlying the effects of spiders on pest populations. *Journal of Arachnology* 27(1), 308–316.

Symondson, W.O.C., Sunderland, K.D. and Greenstone, M.H. (2002) Can generalist predators be effective biocontrol agents? *Annual Review of Entomology* 47, 561–594.

Valdez, J.W. (2020) Arthropods as vertebrate predators: A review of global patterns. *Global Ecology and Biogeography* 29(10), 1691–1703.

Valero-Mora, P.M. (2010) Ggplot2: Elegant graphics for data analysis. *Journal of Statistical Software* 35, 1–3.

Vargas, G., Rivera-Pedroza, L.F., García, L.F. and Jahnke, S.M. (2023) Conservation biological control as an important tool in the neotropical region. *Neotropical Entomology* 52(2), 134–151.

Walker, R., Wilder, S.M. and González, A.L. (2020) Temperature dependency of predation: Increased killing rates and prey mass consumption by predators with warming. *Ecology and Evolution* 10(18), 9696–9706.

Řezáč, M., Pekár, S. and Stará, J. (2010) The negative effect of some selective insecticides on the functional response of a potential biological control agent, the spider *Philodromus cespitum*. *BioControl* 55(4), 503–510.

7 Bird-Mediated Effects of Biological Control under Climate Change: Opportunities and Challenges

Blas Lavandero[1,2*], Pablo Díaz-Siefer[3], Natalia Olmos-Moya[3], Francisco E. Fontúrbel[4] and Juan L. Celis-Diez[2,3]

[1]Laboratorio de Control Biológico, Instituto de Ciencias Biológicas, Universidad de Talca, Talca, Chile; [2]Centro Regional de Investigación e Innovación para la Sostenibilidad de la Agricultura y los Territorios Rurales CERES, Quillota, Chile; [3]Escuela de Agronomía, Facultad de Ciencias Agronómicas y de los Alimentos, Pontificia Universidad Católica de Valparaíso, Quillota, Chile; [4]Instituto de Biología, Facultad de Ciencias, Pontificia Universidad Católica de Valparaíso, Valparaíso, Chile

Abstract

This chapter explores the role of birds in biological pest control, emphasizing their potential as natural enemies of insect pests in agriculture. It highlights how climate change can both enhance and challenge bird-mediated pest control. While warmer temperatures and changing precipitation patterns may boost predation rates, extreme weather events could disrupt the balance between birds and pests, potentially reducing effectiveness. The chapter calls for more research on bird–pest dynamics under different agricultural systems and climate scenarios, especially in vulnerable regions. It also advocates for biodiversity-friendly agricultural practices, such as maintaining natural areas within productive lands, to support bird populations and enhance their pest control services. Finally, we underscore the need for innovative strategies to mitigate climate change impacts on bird populations, which are crucial for sustainable food production and conservation efforts.

7.1 Introduction: Bird-Mediated Effects on Agricultural Ecosystems – Ecological Intensification

Biological control aims to use natural enemies to reduce pest populations, being a promising alternative to conventional agricultural practices and an ally in the integrated production management of insect pests. Among the natural enemies of agricultural pests, insects, both predators and parasitoids, have historically attracted the most attention and research. However, recent studies have highlighted the role of birds in biological pest control (Díaz-Siefer *et al.*, 2022). Birds are abundant and diverse predators, which can play a critical role in the biological control of insect pests in agricultural and natural ecosystems. In the Neotropics, >60% of all bird species are insectivorous (Sherry, 2021). However, the effectiveness of bird-mediated biological control can

*Corresponding author: blavandero@utalca.cl

DOI: 10.1079/9781800625099.0007

be influenced by climate change (Sherry, 2021), posing both opportunities and challenges for the understanding of this ecosystem service delivery. Climate change can create favourable conditions for bird-mediated biological control by influencing the abundance, distribution and behaviour of both birds and their insect prey. For instance, warmer temperatures can increase reproductive rates and activity levels, potentially leading to enhanced predation rates. Additionally, changes in precipitation patterns can alter food and habitat availability, which may influence foraging behaviour and prey selection.

Climatic change also poses challenges to bird-mediated biological control by disrupting the natural balance between predators and prey. Extreme weather events, such as droughts and floods, can lead to declines in bird populations or alter their foraging patterns, potentially reducing predation effectiveness on insect pests (Sherry, 2021). Furthermore, shifts in temperature and precipitation can affect the distribution and abundance of certain insect species, making it more difficult for birds to locate and consume them. In addition to this, several studies have documented a significant increase in insect pest populations under future climate change scenarios (Estay *et al.*, 2009), so understanding the complex interactions between bird-mediated biological control and climatic variation is essential for developing effective and sustainable pest management strategies. However, there is a clear knowledge gap to better quantify the overall effect of bird-mediated pest control, as well as the effect on crop production, particularly under different agroecosystems managements, in different geographical areas. Furthermore, it is important to identify the main drivers that explain the positive, neutral or negative effects of bird predation on pest populations. By carefully considering the potential impacts of climate change on bird populations and insect pest dynamics, we can harness the opportunities presented by climate change while mitigating the challenges it poses.

This chapter will explore these opportunities and challenges in detail, providing a comprehensive overview of the current state of knowledge on bird-mediated biological control under climate change scenarios. It will also discuss future research directions and management implications for enhancing the effectiveness of bird-mediated biological control in the face of a changing climate, not only as an ally to the sustainable production of food but also for bird conservation.

7.2 Pest Control in Modern Agriculture

Pest control is one of the biggest challenges of modern agriculture (Thrupp, 2000; Tilman *et al.*, 2002; Manosathiyadevan *et al.*, 2017; Isman, 2019). Intensive agriculture, which typically harbours low biodiversity (Tilman *et al.*, 2002), is highly vulnerable to animal pest attacks (including vertebrates and invertebrates), leading to an average yield loss of 9–37% when crops are not protected (Oerke, 2006; Culliney, 2014). Pest control in modern monocrop-based agriculture relies on chemical control, which is not ecologically sustainable (Pimentel, 2005; Chaplin-Kramer *et al.*, 2011; Isman, 2019). Moreover, pesticides are one of the factors that most negatively affect farmland birds (Stanton *et al.*, 2018), but also pose a risk to human health (Ansari *et al.*, 2021). However, in the last decades, alternative strategies have become more common, including biological control or the replacement of toxic pesticides by more environmentally friendly molecules, which may reduce pesticide-related mortality of natural enemies (Stenberg, 2017; Dara, 2019; Isman, 2019; Snyder, 2019).

Birds have important roles in different ecosystems (Whelan *et al.*, 2015; Şekercioğlu *et al.*, 2016; Pejchar *et al.*, 2018). Thus, the nature of pest control and insect–insect interactions is highly context dependent, and any conclusion must consider system-specific factors that may override any general benefits in ecosystem services (Karp *et al.*, 2018; Etienne *et al.*, 2024). Ecological intensification practices, such as habitat manipulation or crop diversification, have promoted the integration of ecosystem services mediated by biodiversity into agroecosystems in order to maintain crop productivity while minimizing negative environmental impacts (Kleijn *et al.*, 2019). Within ecological intensification practices, the role of biodiversity and natural areas in agricultural landscapes has been highlighted for its value in sustaining insect and bird populations that benefit crop production (Garibaldi *et al.*, 2011; Karp *et al.*, 2013).

While ecological intensification practices aim to assess the quantitative relationship between pest control and crop production, there is limited availability of studies that have undertaken a quantitative evaluation of this connection, despite being discussed in reviews and previous research (Karp *et al.*, 2013). On the other hand, how these links are modified due to climate change remains unclear, as climatic variability has been shown to decrease insect populations overall but to increase pest insect populations (Wagner *et al.*, 2021; Harvey *et al.*, 2023) and also decrease insectivorous birds, particularly dietary specialists (Bowler *et al.*, 2019). Therefore, an increase of prey species in agricultural environments could potentially favour certain well-adapted species, as long as the management in these agroecosystems keeps favouring bird populations (Bowler *et al.*, 2019).

7.3 Birds under Climate Change Scenarios

Bird abundance, behaviour and migration patterns are largely determined by climate. Therefore, under a climate change scenario we also expect changes in bird populations (Lees *et al.*, 2022). Previous results suggest that models describing associations between climatic variables and abundance patterns can predict changes through time, for some species. For example, bird populations in western North America appear to have already shifted their geographic patterns of abundance in response to changes in precipitation and winter temperature (Illán *et al.*, 2014). Among birds, insectivorous species are the most vulnerable to climate change (Lees *et al.*, 2022). Adaptation may be achieved by phenotypic adjustment or by changes in the genetic composition of populations. Both processes can promote the survival of populations in changing environments, but at different time scales and at different costs.

Recent studies indicate that the mechanisms underlying adaptive phenotypic changes in birds may be complex, involving both plastic responses and genetic changes. Changes in breeding periods, for instance, seem to be predominantly caused by phenotypic adjustment to environmental conditions. Shifts in the genetic composition of populations have been demonstrated to be involved in recent changes in morphology and migratory behaviour of birds. Additive genetic variation within and among avian populations, along with rapid evolutionary responses to rare climatic events, suggest that birds have a high potential for adaptive evolutionary change. However, it remains unclear whether this is a general pattern, and which factors limit the adaptability of avian populations. Antagonistic genetic correlations and maladaptive phenotypic responses (i.e. evolutionary traps) are probably the most important constraints to microevolutionary change. Furthermore, the loss of genetic variation due to population declines and gene flow in the presence of among-population variation in the response to climate change may limit the rate of adaptive evolution. Future research should try to identify the targets of selection and gauge the importance of constraints to microevolutionary change (Pulido and Berthold, 2004).

7.4 Birds as Biological Pest Control Agents

Birds have been documented to induce strong cascading effects by suppressing different types of crop herbivores, such as insects, mammals or even other bird species in some natural and human-modified ecosystems (Mazia *et al.*, 2009; Mäntylä *et al.*, 2011; Karp *et al.*, 2013; Whelan *et al.*, 2015; Kross *et al.*, 2016; Labuschagne *et al.*, 2016; Chain-Guadarrama *et al.*, 2019; Díaz-Siefer *et al.*, 2022; Monteagudo *et al.*, 2023). For example, in southern American temperate rainforests, birds showed strong suppression of herbivory in *Nothofagus pumilio* (Fagales:Nothofagacea) (a native forest species of South America) saplings (Mazia *et al.*, 2009), while in Australian vineyards, management to attract raptors significantly reduced bird peck damage to grapes (Peisley *et al.* 2017) and rodent abundance (Labuschagne *et al.*, 2016). There is increasing evidence highlighting the role of birds as biological control agents in agroecosystems, aiming to change the traditional biological control paradigms in agroecosystems which have overlooked the potential contribution of vertebrate predators over parasitoids and predatory insects (García *et al.*, 2021; Díaz-Siefer *et al.*, 2022; Monteagudo *et al.*, 2023).

At a global scale, it has been estimated that birds consume between 400 and 500 million tonnes of arthropods per year in many biomes, including cropland ecosystems, thus contributing significantly towards preventing crop damage and loss (Nyffeler *et al.*, 2018). Furthermore, birds can reduce crop pest damage and pest abundance, and increase crop yields across crop types, climatic regions and agricultural management (Díaz-Siefer *et al.*, 2022), but the relevance of this may have been largely underestimated (García *et al.*, 2021). Moreover, the literature has shown some contrasting results related to wild bird pest control in agricultural landscapes.

On one hand, birds in agricultural landscapes are generally associated with environmental disservices by farmers (Lindell *et al.*, 2012), since they can act as pests in some cases, such as fruit-eating birds (Shave *et al.*, 2018). For instance, bird damage to crops can cause losses up to US$189 million in five crops (Honeycrisp apples, wine grapes, blueberries, and sweet and tart cherries) (Anderson *et al.*, 2013) and US$1.3 million in corn (Klosterman *et al.*, 2013) in the USA, or they can have a neutral effect on yield as arthropod suppression can equal the damage inflicted to those crops by these birds (Gonthier *et al.*, 2019). In addition, birds can suppress agricultural pest predators' suppression (i.e. intraguild predation (IGP); Rosenheim *et al.*, 1993; Letourneau *et al.*, 2009; Grass *et al.*, 2017; Olimpi *et al.*, 2020). For example, birds can prey on certain spider species, which are generalist predators that consume agricultural pests, thereby providing a disservice to farmers (Greenberg *et al.*, 2000; Recher and Majer, 2006). But this can also be the opposite as many spider species are actually more prejudicial to natural enemies than pests (Lang, 2003).

On the other hand, several recent reviews have highlighted the importance of many wild birds in suppressing pests in agricultural landscapes, reducing crop damage (Greenberg *et al.*, 2000; Pejchar *et al.*, 2018; Boesing *et al.*, 2018; García *et al.*, 2021), consuming sentinel prey species (Jedlicka *et al.*, 2011; García *et al.*, 2021; Olmos-Moya *et al.*, 2022), reducing pest abundance (Mols and Vissr, 2007; Karp *et al.*, 2013) and increasing yield (Mols and Vissr, 2007; Grass *et al.*, 2016). These studies have shown that bird-mediated biocontrol has positive impacts on herbivore abundance, crop damage and local economies. For instance, Shave *et al.* (2018) found that for every US dollar spent on a raptor's nest box to control birds, a loss of up to US$220 in sweet cherry crops could be prevented. Hence, pests are diluted in the herbivore community and subject to a wider diversity of potential natural enemies (Letourneau and Goldstein, 2001). These nest boxes can increase the abundance of wild birds and therefore lessen crop damage by reducing pest abundance in organic farms (Muneret *et al.*, 2018). Based on our results from a recent meta-analysis (Díaz-Siefer *et al.*, 2022) and previous work, there seems to be a clear link between the ecological context of agricultural land and its resilience to biodiversity loss. In this context, agriculture within highly diverse wild bird communities also seems to be successful at natural pest control and leads to lower pest abundance (Letourneau and Goldstein, 2001; Muneret *et al.*, 2018).

Finally, many studies have documented that most farmers associate birds with environmental disservices (e.g. damage to fruit crops) and that bird-friendly practices are not associated with economic incentives (Jacobson *et al.*, 2003). However, our results suggest that ecological intensification practices can complement and, in some cases, replace the use of pesticides or exotic enemy releases for biological control, with ecological processes to sustain agricultural production (Bommarco *et al.*, 2013; Kross *et al.*, 2018; Smith *et al.*, 2022).

7.4.1 Effects depending on crop type

Most of the published studies included in a recent meta-analysis by Díaz-Siefer *et al.* (2022), including data from 57 papers that followed the selection criteria, showed a bias in crop type. Coffee, cereal and apple

Table 7.1. Examples of studies per crop of the effect of the presence of insectivorous birds on pest abundance, crop damage and yields. Symbols: (+) decrease in pests and crop damage, increased yield; (-) increase in pest presence and crop damage, decreased yield; (0) no effect.

Crop	Abundance of pests	Examples	Crop damage by pests	Examples	Crop yield	Examples
Coffee	+	Karp *et al.*, (2013) Philpott *et al.*, 2004	+	Kellerman *et al.*, 2008 Peters and Greenberg, 2012	+	Classen *et al.*, 2014
Cereals	+	Tremblay *et al.*, 2001				
	-					
	0	Borkhataria *et al.* (2012)			0	Borkhataria *et al.* (2012)
Cacao	+	Maas *et al.*, 2013			Grass *et al.* (2017)+	Grass *et al.*, 2017
			0	Maas *et al.*, 2013		
Vineyards	+	Rey-Benayas *et al.*, 2017 Jedlicka *et al.*, 2011				

crops represent almost 60% of the published research, so it is not easy to conclude much. In any case, annual agroecosystems including crop rotations are a highly changing environment for herbivores, so prey ability for birds may vary seasonally to a greater extent than in more permanent systems (orchard fruit production, for example). As the diet preference of most species involved in biological control is not known, the effect of crop rotation on the abundance and composition of different bird assemblages and their effect on herbivores control are also a mystery. However, more permanent crops will offer more stable resources and probably can be managed in the long run. It would be easier to understand the effect of climate change in these systems, and easier to develop management strategies that favour bird populations, and hence bird-driven biological control (García *et al.*, 2021). The reduced variety of crops included in the international literature highlights an urgent need to expand research to other crops. For examples of studies in different crops see Table 7.1.

7.4.2 Effects depending on agricultural management

The consequences of agronomic management on bird-mediated biological control have been discussed in the literature. The meta-analysis of Díaz-Siefer *et al.* (2022) showed that the majority of positive cases of bird-mediated herbivore suppression were recorded under conventional management. Surprisingly, no significant effect was observed under organic management. This result could be attributed to the fact that organic farming supports a greater biodiversity, including many natural enemies, than conventional agriculture (Tuck *et al.*, 2014; Muneret *et al.*, 2018). Therefore, the net effect of pest biological control in the agroecosystem is not observable by the presence/absence of a particular natural enemy (Mols and Vissr, 2007). Consequently, the relative impact of birds on pest abundance and crop pest damage is diluted in these communities by the greater diversity of natural enemies (Letourneau and Goldstein, 2001; Muneret *et al.*, 2018).

Conventional agricultural practices have led to a significant decrease in insectivorous birds (Şekercioğlu *et al.*, 2002; Muñoz-Sáez *et al.*, 2020) and a reduction in nesting sites for cavity-nesting birds (Cockle *et al.*, 2011). Additionally, the configuration of agricultural landscapes can act as barrier for the movement of insectivorous birds (Şekercioğlu *et al.*, 2016; Assandri *et al.*, 2017; Jerrentrup *et al.*, 2017; Muñoz-Sáez *et al.*, 2020; Stanton *et al.*, 2018). Consequently, the expansion of agriculture into natural habitats (Armesto *et al.*, 2010) has diminished the presence of large, aged and decaying trees, adversely affecting populations of secondary cavity-nesting birds (Newton, 1994; Cockle *et al.*, 2011; Bonaparte *et al.*, 2020), which are mainly insectivorous (Ibarra *et al.*, 2017). Thus, placing nesting boxes in crops can increase the abundance of insectivorous birds (Jedlicka *et al.*, 2017; Olmos-Moya *et al.*, 2022).

Only 16% of the selected articles in Díaz-Siefer *et al.* (2022) documented increased bird diversity through habitat manipulation practices (e.g. perches or nest boxes). We found that attracting wild birds, an ecological intensification tool that is easily implemented by farmers, could significantly reduce crop damage and pest abundance (García *et al.*, 2021; Olmos-Moya *et al.*, 2022). For instance, nest boxes for insectivorous birds or raptors provide a highly cost-effective ecosystem service to control fruit-eating birds or herbivore arthropods (Suhonen *et al.*, 1994; Shave *et al.*, 2018; García *et al.*, 2021; Olmos-Moya *et al.*, 2022). In-farm evaluation could be a way to improve farmer attitudes towards the conservation of biodiversity (Jacobson *et al.*, 2003; Kross *et al.*, 2018), and a bridge to advance towards other ecological intensification practices (Kleijn *et al.*, 2019). In fact, as proposed by García *et al.* (2021), by attracting birds to farms via ecological innovations, farmers should receive direct benefits as reported in this study, as well as indirect benefits through reduction of pesticides with the associated consequences of environmental damage. These innovations could be the focus of public subsidies from (i) environmental policy focused on biodiversity conservation in rural landscapes, and (ii) agricultural policy focused to increase productivity in a sustainable way. Therefore, enhancing crops with ecological innovation practices can promote both biological control and biodiversity conservation within productive landscapes. For examples of studies in different agricultural management types see Table 7.2.

7.4.3 Effects depending on pest type

There is a limited amount of studies considering different pest types, but when considering arthropods and birds as groups of pests, the effect of exclusion is significant for pest abundance (Díaz-Siefer *et al.*, 2022). There is no clear information of how the variability may affect these interactions, but the response of climatic variability will change depending on the pest type and the repercussions for bird populations will need to be quantified, as well as other important indirect effects on pest populations, which have not been quantified in the literature. One example where the effect of bird suppression may have been underestimated, as

Table 7.2. Examples of studies per agricultural type of the effect of the presence of insectivorous birds on pest abundance, crop damage and yields. Symbols: (+) decrease in pests and crop damage, increased yield; (-) increase in pest presence and crop damage, decreased yield; (0) no effect.

Agriculture type	Abundance of pests	Examples	Crop damage by pests	Examples	Crop yield	Examples
Conventional	+	Tremblay *et al.*, 2001	+	Libran-Embid *et al.*, 2017		
		Kross *et al.*, 2016				
Organic	+	Garfinkel *et al.*, 2015	+	Gonthier *et al.*, 2019	-	Gonthier *et al.*, 2019

it did not provide evidence of direct pest suppression when birds were present, was a study on the rodent *Microtus arvalis* (vole). Wolff *et al.* (2011) investigated the effect of added perch sites (attraction) and did not find a direct effect on the pest populations, although an 11-fold increase on the visitation by American kestrels (*Falco sparverius*) was observed. Interestingly, however, in this same study, a decrease in male voles' home range and activity was measured, which on closer inspection could be interpreted as an avoidance response to the predation risk. More long-term studies may in the future find a decrease of the overall pest population (in this case the voles) arising from this reduced activity.

In regard to climate change, species of rodents have been shown to be highly sensitive to precipitation changes, due to changes in specific plant composition, which for the case of voles has been shown to reduce the content of fructose and fatty acids, which could lead to body mass and fitness changes mediated by gut microbiota shifts (Li *et al.*, 2020). These changes could eventually increase vole populations and species such as *Falco sparverius* could be key to their regulation in sites with greater precipitation under climate change.

7.4.4 Effects depending on geographical location

There is a geographic bias in the current literature. Most studies on avian-mediated pest control have been conducted in tropical (49%) rather than in temperate (35%) areas or Mediterranean agroecosystems (16%). For examples of studies in different geographical locations see Table 7.3. The overrepresentation of tropical areas has been changing slowly, with a recent increase in the number of studies focused on other geographical areas (Boesing *et al.*, 2018). Mediterranean regions are considered as global biodiversity hotspots for conservation (Myers *et al.*, 2000) but they also represent one of the most vulnerable ecosystems on the planet because of the negative effects of land use change and intensive agriculture (Sala *et al.*, 2000; Armesto *et al.*, 2010). Thus, identifying the contribution of biodiversity to crop production in Mediterranean regions is key to promoting sustainable land use management and conservation of productive and natural areas (Cox and Underwood, 2011), as climate projections show more severe effects in Mediterranean areas (Rojas *et al.*, 2019).

Conservation of natural areas or biodiversity-friendly landscapes in the agroecosystem can benefit pest regulation, even beyond organic farming (Tscharntke *et al.*, 2008, 2021). A larger variety of wild bird-mediated pest control species have been frequently reported in woodland landscapes (Boesing *et al.*, 2018).

It is important to acknowledge important biases in published studies. For example, in Díaz-Siefer *et al.* (2022), 78% of the studies reviewed were conducted in the northern hemisphere. This knowledge gap results from a lack of data reported from the southern hemisphere in terms of research and monitoring of global ecosystems (Rozzi *et al.*, 2012). This lack of representation of the Global South is an important issue to consider for agricultural and conservation management purposes.

7.4.5 Other effects of birds in agricultural landscapes

Previous work on the role of birds in agroecosystems highlights that landscape structure has an important effect on species diversity (Boesing *et al.*, 2018 Muñoz-Sáez *et al.*, 2020). Bird taxonomic and functional diversity increases avian-mediated insect pest control services in agroecosystems (Martínez-Salinas *et al.*, 2016). Landscape configuration may constrain the functional diversity of birds in tropical agroecosystems (Tscharntke *et al.*, 2008; Barbaro *et al.*, 2017; Mariano-Neto and Santos, 2023). The remaining vegetation in agroecosystems allows the transfer of controlling agents from natural areas to crops, so it can host predators such as birds and arthropods, which interact in complex ways (Pejchar *et al.*, 2018; García *et al.*, 2021). Pest suppressive landscapes have been shown to positively affect pest control, ut birds have not been considered. These could not only suppress pests but act as reservoirs of bird populations, even species with conservation problems, where land-sparing schemes will not be enough to promote bird biodiversity.

Table 7.3. Examples of studies per geographical area of the effect of the presence of insectivorous birds on pest abundance, crop damage and yields. Symbols: (+) decrease in pests and crop damage, increased yield; (-) increase in pest presence and crop damage, decreased yield; (0) no effect.

Geographical area	Abundance of pests	Examples	Crop damage by pests	Examples	Crop yield	Examples
Tropical	+	Karp *et al.*, 2013	+	Martinez-Salinas *et al.*, 2016	+	Classen *et al.*, 2014
		Maas *et al.*, 2013		Koh, 2008		Maas *et al.*, 2013
		Peters and Greenberg, 2012		Johnson *et al.*, 2009		Linden *et al.*, 2019
		Guenat *et al.*, 2017		Kellerman *et al.*, 2008		Grass *et al.*, 2016
		Ndang'ang'a *et al.*, 2013		Libran-Embid *et al.*, 2017		
	0		0	Maas *et al.* (2013)		
				Guenat *et al.*, 2017		
Temperate	+	Kross *et al.*, 2016	+	Gonthier *et al.*, 2019	+	Mols and Visser, 2007
		Tremblay *et al.*, 2001		Peisley *et al.*, 2016		
	-	Grass *et al.*, 2017			-	Gonthier *et al.*, 2019
		Martin *et al.*, 2013				
	0	Borkhataria *et al.* (2012)			0	Borkhataria *et al.*, 2012
		Martin *et al.* (2013)				

Management ecosystems will have a major role in the conservation of species, as land use pressure will increase to meet food security.

Structural diversity within an agricultural landscape allows for greater diversity of habitats and therefore increases the arrival of different functional bird guilds (i.e. other than insectivores), which increases the resilience of agroecosystems as there are species that play different roles (Boesing *et al.* 2018; Pejchar *et al.*, 2018). For example, there are records that the house wren (*Troglodytes aedon*), common in agricultural systems and a skilled consumer of insects, is a minority part of the diet of the Magellanic horned owl (*Bubo magellanicus*) (Trejo and Grigera, 1998), a key species in maintaining the balance of biodiversity that lives in forest remnants surrounding crops and a predator of micromammals that are harmful to crops. In this way, it is essential to consider the ecosystem services and disservices that birds provide to determine the best agricultural management in pursuit of birdlife conservation and food production.

7.5 Intraguild Predation (Birds Eating Natural Enemies) or Birds Eating Birds That Eat Pests

To encourage the presence and activity of natural enemies, it is critical to determine feeding patterns as well as potential negative effects of general predators (Gurr and You, 2016). The development of molecular techniques has allowed for the relatively simple elucidation of trophic linkages within a food chain, and a better understanding of many natural enemy guilds in agroecosystems, but information for birds is still relatively scarce (Agustí *et al.*, 2003; Clare, 2014; Ekbom *et al.*, 2014; Birkhofer *et al.*, 2017; Alhmedi *et al.*, 2018).

As the relationship between bird abundance and pest control could be modulated by intraguild predation (i.e. birds predating other birds), more information is needed to explain different outcomes when an increase in natural enemy diversity, due to management or landscape characteristics, may intensify competition between natural enemies and/or direct or coincidental intraguild predation (Letourneau *et al.*, 2009). For example, in a study by Garfinkel *et al.* (2022), birds feeding on natural enemies of soybean defoliators indirectly increased defoliation. Although in this specific study there were no negative effects on yield, it exemplifies the need for more knowledge on insectivorous bird diets. Therefore, more field studies that address the effect of diverse factors affecting interactions between birds and other natural enemies are lacking.

However, these negative interactions have been observed indirectly for bird species. Studies of the molecular diet of insectivorous birds have found DNA from carabids (Jedlicka *et al.*, 2017) and spiders, which are considered natural enemies of agricultural pests (Crisol-Martínez *et al.*, 2016; Trevelline *et al.*, 2016). Another study on feeding of nestlings, based on direct observations, found that hoverflies and ladybird beetles accounted for 77% of the food for nestlings during peak aphid density (Grass *et al.*, 2017), which is surprising considering that there is evidence of coccinellids being slightly toxic and distasteful to some bird species (Marples, 1993). Similarly in another study carried out in three sympatric songbirds foraging on small invertebrates in Phragmites reed bed ecosystems (*Acrocephalus scirpaceus*, *A. schoenobaenus* and *Cettia cetti*), amongst the prey found were major pests (including the tomato moth *Lacanobia oleracea*) as well as many potentially valuable natural enemies including aphidophagous hoverflies and braconid wasps (King *et al.*, 2015).

Another study found evidence through metabarcoding of individuals from the families of natural enemies Syrphidae, Miridae and Carabidae (Forsman *et al.*, 2022). However, it may be worth mentioning that the frequency of diet items in metagenomic studies and molecular detection in general cannot be used as a proxy for prey biomass. Frequently detected prey items may be a very small portion of the total biomass ingestion of a predator, so conclusions on prey detection must not be generalized in terms of the relative importance for a predatory species.

If landscape complexity and other factors such as management promote negative interactions among these natural enemies, then overall pest suppression could be reduced. Nevertheless, in cases where the most effective natural enemies may act as an intraguild predator, the resulting pest suppression may not be diminished (e.g. for arthropod IGP in: Cardinale *et al.* (2003); Finke and Denno (2005); Gardiner and Landis (2007); Traugott *et al.* (2012)). Therefore, not only do we need more studies looking for prey items of birds but also we need to look at the relative importance of the removal of these prey for birds compared to their arthropod counterpart natural enemies.

7.6 Climate Change and Bird-Mediated Pest Control

As climatic variability increases as a result of climate change, insect and bird populations are expected to decline. Climate change is expected to significantly influence bird communities mainly through changes in the composition and structure of their habitats (Cadieux *et al.*, 2020). Climate change drives phenological mismatches between birds and their insect prey (Martin and Maron, 2012), altering ecosystem function and services which is enhanced when landscape heterogeneity decreases (Boesing *et al.*, 2018). Likewise, this could affect bird-mediated pest control in agroecosystems.

Skendžić *et al.* (2021) showed that the effects of climate change on insects are complex since their diversity, abundance and phenology are affected, inhibiting some species and enhancing others, suggesting a general increase in the number of pest outbreaks that involve a wider range of insect pests. Climate change has multiple complementary and synergistic consequences with other drivers of global change, such as land use change and forest loss, agricultural intensification and agrochemicals, and invasive species, which impact the decline of insectivorous birds (Sherry, 2021). Several hypotheses have been proposed to explain the abrupt decline or vulnerability of insectivorous birds compared to other trophic guilds, among which the biotic challenge hypothesis (BCH) stands out, based on the evolutionary food specialization of these birds and their low dispersal capacity (Sherry, 2021). More detail on the effects of climate change on macro-organisms involved in biocontrol is explained in Chapter 1.

Changes in the populations or distributions of birds and insects can be attributed to changes in average temperatures or precipitation, and also to changes in land use, such as expansion of agriculture. Therefore, designing diverse agroecosystems is vital for the conservation of insectivorous bird communities. Maintaining habitat heterogeneity and hedgerows may indirectly benefit bird species impacted by climate change by providing refuge, food and nesting sites (Pejchar *et al.*, 2018; Rollan *et al.*, 2019).

7.7 Challenges

It is necessary to generate more quantitative data on the relationship between bird and pest species and agricultural management under a climate change scenario, especially in areas more susceptible to the effects of climate change such as Mediterranean climate regions.

Future research efforts should focus on understanding the effects of the landscape on the provision of ecosystem services such as pest control under climate change scenarios, as well as developing agricultural crop management practices as diversified systems to improve climate resilience and conservation of bird species.

Studies that provide baseline dietary data on avian insectivores are crucial for evaluating how community dynamics change as a result of the global decline in terrestrial insect populations (van Klink *et al.*, 2020). Although diet studies are complicated and logistically difficult, they are crucial for monitoring and conserving bird populations in the future as well as assessing their importance as biological control agents under climatic variation.

7.8 Opportunities

Bird conservation and ecological intensification will promote sustainable agriculture (Watson and Venter, 2017). However, it is necessary to design and implement better strategies that involve bird conservation and ecosystem service delivery, including trophic interactions. As well as birds being pests and natural enemies, there are species which can act as important seed dispersers of invasive plants (Cordero *et al.*, 2023; Juncosa-Polzella *et al.*, 2023; Zhu *et al.*, 2023). Wild birds play a major role in pest control in agroecosystems, as we have seen here, by reducing crop damage and pest abundance and increasing crop yield. Maintaining natural areas within productive lands enhances these natural pest control services, which will be under more threat due to climatic variability. However, these areas will be more and more important for many species' survival. Ecological intensification practices are a win–win goal for farmers and the environment. More strategies are therefore needed to enhance the survival and attraction of birds in agricultural landscapes. The use of nest boxes, perching sites and natural vegetation needs to be studied more. Creative strategies to mitigate extreme heat and cold for the survival of birds and to reduce nest predation by invasive species should also be incorporated in the toolbox for sustainable food production. A concise understanding of the diet of insectivorous and omnivorous bird species will also be pivotal to the design of resource supplementation strategies in conservation biological programmes. Other food items, access to seeds or alternative prey for the continued survival of birds in agricultural landscapes will also need to be considered in the future.

7.9 Conclusion

As human activities continue to expand at the expense of biodiversity, we urgently need to reconcile production with conservation. In this regard, ecological intensification principles can provide a bridge between production and conservation, taking advantage of the ecosystem services provided by neighbouring natural/semi-natural vegetation without compromising production yield. This is particularly relevant for controlling crop pests, which become a major problem in large monoculture fields where they can generate significant losses. Therefore, implementing biodiversity-friendly agricultural practices and keeping remnants of natural vegetation near to crop fields can attract wild birds that can provide biocontrol services, while contributing to minimizing the effect of global change that harms biodiversity. Management innovation promoting bird populations within crop fields can increase pest removal, reduce the use of agrochemicals and open up conservation opportunities. In a climate change scenario, we expect that greater climatic variability (e.g. extreme meteorological events) may result in crop losses and decrease of native species populations. Therefore, bird-mediated biocontrol may become more relevant as crop pests can benefit from warmer and drier conditions, increasing their abundance and impacting crop species that may also be suffering from water stress. Consequently, promoting biodiversity-friendly landscapes may be more resilient to climate change, even beyond conventional or organic farming, generating benefits for producers while sustaining wild bird populations at the same time.

Acknowledgements

The authors would like to thank the continual support by the projects ANID/PIA/ACT192027, ANID/CERES/R23F0003 and PIA/BASAL FB210006 during the writing of this chapter.

References

Agustí, N., Shayler, S.P., Harwood, J.D., Vaughan, I.P., Sunderland, K.D. *et al.* (2003) Collembola as alternative prey sustaining spiders in arable ecosystems: Prey detection within predators using molecular markers. *Molecular Ecology* 12, 3467–3475.

Alhmedi, A., Raymaekers, S., Tomanović, Ž., Bylemans, D. and Beliën, T. (2018) Food web structure of aphids and their parasitoids in Belgian fruit agroecosystems. *Entomological Science* 21, 279–291.

Anderson, A., Lindell, C.A., Moxcey, K.M., Siemer, W.F., Linz, G.M. *et al.* (2013) Bird damage to select fruit crops: The cost of damage and the benefits of control in five states. *Crop Protection* 52, 103–109.

Ansari, I., El-Kady, M.M., Arora, C., Sundararajan, M., Maiti, D. *et al.* (2021) A review on the fatal impact of pesticide toxicity on environment and human health. *Global Climate Change* 361–391.

Armesto, J.J., Manuschevich, D., Mora, A., Smith-Ramirez, C., Rozzi, R. *et al.* (2010) From the Holocene to the Anthropocene: A historical framework for land cover change in southwestern South America in the past 15,000 years. *Land Use Policy* 27, 148–160.

Assandri, G., Bogliani, G., Pedrini, P. and Brambilla, M. (2017) Insectivorous birds as 'non-traditional' flagship species in vineyards: Applying a neglected conservation paradigm to agricultural systems. *Ecological Indicators* 80, 275–285.

Barbaro, L., Rusch, A., Muiruri, E.W., Gravellier, B., Thiery, D. *et al.* (2017) Avian pest control in vineyards is driven by interactions between bird functional diversity and landscape heterogeneity. *Journal of Applied Ecology* 54, 500–508.

Birkhofer, K., Gossner, M.M., Diekötter, T., Drees, C., Ferlian, O. *et al.* (2017) Land-use type and intensity differentially filter traits in above- and below-ground arthropod communities. *Journal of Animal Ecology* 86, 511–520.

Boesing, A.L., Nichols, E. and Metzger, J.P. (2018) Land use type, forest cover and forest edges modulate avian cross-habitat spillover. *Journal of Applied Ecology* 55, 1252–1264.

Bommarco, R., Kleijn, D. and Potts, S.G. (2013) Ecological intensification: Harnessing ecosystem services for food security. *Trends in Ecology and Evolution* 28, 230–238.

Bonaparte, E.B., Ibarra, J.T. and Cockle, K.L. (2020) Conserving nest trees used by cavity-nesting birds from endangered primary Atlantic forest to open farmland: Increased relevance of excavated cavities in large dead trees on farms. *Forest Ecology and Management* 475, 118440.

Borkhataria, R.R., Nuessly, G.S., Pearlstine, E. and Cherry, R.H. (2012) Effects of blackbirds (*Agelaius phoenicius*) on stink bug (hemiptera: pentatomidae). *Florida Entomologist* 95(1), 143–149. DOI: 10.1653/024.095.0122.

Bowler, D.E., Heldbjerg, H., Fox, A.D., de Jong, M. and Böhning-Gaese, K. (2019) Long-term declines of European insectivorous bird populations and potential causes. *Conservation Biology* 33, 1120–1130.

Cadieux, P., Boulanger, Y., Cyr, D., Taylor, A.R., Price, D.T. *et al.* (2020) Projected effects of climate change on boreal bird community accentuated by anthropogenic disturbances in western boreal forest, Canada. *Diversity and Distributions* 26, 668–682.

Cardinale, B.J., Harvey, C.T., Gross, K. and Ives, A.R. (2003) Biodiversity and biocontrol: Emergent impacts of a multi-enemy assemblage on pest suppression and crop yield in an agroecosystem. *Ecology Letters* 6, 857–865.

Chain-Guadarrama, A., Martínez-Salinas, A., Aristizábal, N. and Ricketts, T.H. (2019) Ecosystem services by birds and bees to coffee in a changing climate: A review of coffee berry borer control and pollination. *Agriculture, Ecosystems & Environment* 280, 53–67.

Chaplin-Kramer, R., O'Rourke, M.E., Blitzer, E.J. and Kremen, C. (2011) A meta-analysis of crop pest and natural enemy response to landscape complexity. *Ecology Letters* 14, 922–932.

Clare, E.L. (2014) Molecular detection of trophic interactions: Emerging trends, distinct advantages, significant considerations and conservation applications. *Evolutionary Applications* 7, 1144–1157.

Classen, A., Peters, M.K., Ferger, S.W., Helbig-Bonitz, M., Schmack, J.M, *et al.* (2014) Complementary ecosystem services provided by pest predators and pollinators increase quantity and quality of coffee yields. *Proceedings of the Royal Society B: Biological Sciences* 281(1779). Available at: https://doi.org/10.1098/rspb.2013.3148

Cockle, K.L., Martin, K. and Wesołowski, T. (2011) Woodpeckers, decay, and the future of cavity-nesting vertebrate communities worldwide. *Frontiers in Ecology and the Environment* 9, 377–382.

Cordero, S., Gálvez, F. and Fontúrbel, F.E. (2023) Ecological impacts of exotic species on native seed dispersal systems: A systematic review. *Plants* 12(2), 261.

Cox, R.L. and Underwood, E.C. (2011) The importance of conserving biodiversity outside of protected areas in Mediterranean ecosystems. *PLoS ONE* 6, e14508.

Crisol-Martínez, E., Moreno-Moyano, L.T., Wormington, K.R., Brown, P.H. and Stanley, D. (2016) Using next-generation sequencing to contrast the diet and explore pest-reduction services of sympatric bird species in macadamia orchards in Australia. *PLoS ONE* 11(3), e0150159.

Culliney, T. (2014) Crop losses to arthropods. In: Pimentel, D. and Peshin, R. (eds) *Integrated Pest Management*. Springer, Dordrecht, The Netherlands.

Dara, S.K. (2019) The new integrated pest management paradigm for the modern age. *Journal of Integrated Pest Management* 10(1), 12.

Díaz-Siefer, P., Olmos-Moya, N., Fontúrbel, F.E., Lavandero, B., Pozo, R.A. *et al.* (2022) Bird-mediated effects of pest control services on crop productivity: A global synthesis. *Journal of Pest Science* 1, 3.

Ekbom, B., Kuusk, A.-K., Malsher, G., Åström, S. and Cassel-Lundhagen, A. (2014) Consumption of flea beetles (*Phyllotreta, coleoptera*: Chrysomelidae) by spiders in field habitats detected by molecular analysis. *Canadian Entomologist* 146(6), 639–651.

Estay, S.A., Lima, M. and Labra, F.A. (2009) Predicting insect pest status under climate change scenarios: Combining experimental data and population dynamics modelling. *Journal of Applied Entomology* 133, 491–499.

Etienne, L., Franck, P., Rusch, A. and Lavigne, C. (2024) Apple pest and pathogen reduction in landscapes with large patch size and small area of orchards: A national-scale analysis. *Landscape Ecology* 39, article 8.

Finke, D.L. and Denno, R.F. (2005) Predator diversity and the functioning of ecosystems: The role of intraguild predation in dampening trophic cascades. *Ecology Letters* 8, 1299–1306.

Forsman, A.M., Hoenig, B.D., Gaspar, S.A., Fischer, J.D., Siegrist, J. *et al.* (2022) Evaluating the impacts of metabarcoding primer selection on DNA characterization of diet in an aerial insectivore, the Purple Martin. *Ornithology* 139, ukab075.

García, D., Miñarro, M. and Martínez-Sastre, R. (2021) Enhancing ecosystem services in apple orchards: Nest boxes increase pest control by insectivorous birds. *Journal of Applied Ecology* 58, 465–475.
Gardiner, M.M. and Landis, D.A. (2007) Impact of intraguild predation by adult *Harmonia axyridis* (Coleoptera: Coccinellidae) on *Aphis glycines* (Hemiptera: Aphididae) biological control in cage studies. *Biological Control* 40(3), 386–395.
Garfinkel, M. and Johnson, M. (2015) Pest-removal services provided by birds on small organic farms in northern california. *Agriculture, Ecosystems & Environment* 211, 24–31. Available at: https://doi.org/10.1016/j.agee.2015.04.023
Garfinkel, M.B., Fuka, M.E., Minor, E. and Whelan, C.J. (2022) When a pest is not a pest: Birds indirectly increase defoliation but have no effect on yield of soybean crops. *Ecological Applications* 32, e2527.
Garibaldi, L.A., Steffan-Dewenter, I., Kremen, C., Morales, J.M., Bommarco, R. *et al.* (2011) Stability of pollination services decreases with isolation from natural areas despite honey bee visits. *Ecology Letters* 14, 1062–1072.
Gonthier, D.J., Sciligo, A.R., Karp, D.S., Lu, A., Garcia, K. *et al.* (2019) Bird services and disservices to strawberry farming in Californian agricultural landscapes. *Journal of Applied Ecology* 56, 1948–1959.
Gras, P., Tscharntke, T., Maas, B., Tjoa, A., Hafsah, A, *et al.* (2016) How ants, birds and bats affect crop yield along shade gradients in tropical cacao agroforestry. *Journal of Apply Ecology* 53, 953–963. Available at: https://doi.org/10.1111/1365-2664.1262
Grass, I., Lehmann, K., Thies, C. and Tscharntke, T. (2017) Insectivorous birds disrupt biological control of cereal aphids. *Ecological Society of America* 98, 1583–1590.
Greenberg, R., Bichier, P. and Angón, A.C. (2000) The conservation value for birds of cacao plantations with diverse planted shade in Tabasco, Mexico. *Animal Conservation* 3, 105–112.
Guenat, S., Kaartinen, R. and Jonsson, M. (2017) Shade trees decrease pest abundances on brassica crops in Kenya. *Agroforestry Systems* 93(2), 1–12. Available at: https://doi.org/10.1007/s10457-017-0159-5
Gurr, G.M. and You, M. (2016) Conservation biological control of pests in the molecular era: New opportunities to address old constraints. *Frontiers in Plant Science* 6, 1255.
Harvey, J.A., Tougeron, K., Gols, R., Heinen, R., Abarca, M. *et al.* (2023) Scientists' warning on climate change and insects. *Ecological Monographs* 93(1), e1553.
Ibarra, J.T., Martin, M., Cockle, K.L. and Martin, K. (2017) Maintaining ecosystem resilience: Functional responses of tree cavity nesters to logging in temperate forests of the Americas. *Science Reports* 7, 4467.
Illán, J.G., Thomas, C.D., Jones, J.A., Wong, W.K., Shirley, S.M. *et al.* (2014) Precipitation and winter temperature predict long-term range-scale abundance changes in Western North American birds. *Global Change Biology* 20, 3351–3364.
Isman, M.B. (2019) Challenges of pest management in the Twenty First Century: New tools and strategies to combat old and new foes alike. *Frontiers in Agronomy* 1, 2.
Jacobson, S.K., Sieving, K.E., Jones, G.A. and Van Doorn, A. (2003) Assessment of farmer attitudes and behavioral intentions toward bird conservation on organic and conventional Florida farms. *Conservation Biology* 17, 595–606.
Jedlicka, J.A., Greenberg, R. and Letourneau, D.K. (2011) Avian conservation practices strengthen ecosystem services in California vineyards. *PLoS ONE* 6, e27347.
Jedlicka J., A., Vo A. T., E. and Almeida R., P. (2017) Molecular scatology and high-throughput sequencing reveal predominately herbivorous insects in the diets of adult and nestling western bluebirds (*Sialia mexicana*) in California vineyards. *Auk: Ornithological Adv* 134, 116–127. DOI: 10.1642/AUK-16-103.1.
Jerrentrup, J.A., Dauber, J., Strohbach, M.W., Mecke, S., Mitschke, A. *et al.* (2017) Impact of recent changes in agricultural land use on farmland bird trends. *Agriculture, Ecosystems & Environment* 239, 334–341.
Johnson, M.D., Levy, N.J., Kellermann, J.L. and Robinson, D.E. (2009) Effects of shade and bird exclusion on arthropods and leaf damage on coffee farms in jamaica's blue mountains. *Agroforestry Systems* 76, 139–148.
Juncosa-Polzella, A.S., Tecco, P.A. and Vergara-Tabares, D.L. (2023) Seed dispersal of a fleshy-fruited invasive shrub is affected by changes in the frugivorous bird assemblage along an elevational gradient. *Austral Ecology* 48, 1797–1814.
Karp, D.S., Mendenhall, C.D., Sandí, R.F., Chaumont, N., Ehrlich, P.R. *et al.* (2013) Forest bolsters bird abundance, pest control and coffee yield. *Ecology Letters* 16, 1339–1347.

Karp, D.S., Chaplin-Kramer, R., Meehan, T.D., Martin, E.A., DeClerck, F. *et al.* (2018) Crop pests and predators exhibit inconsistent responses to surrounding landscape composition. *Proceedings of the National Academy of Sciences USA* 115, E7863–E7870.

Kellerman, J.L., Johnson, M.D., Stercho, A.M. and Hackett, S.C. (2008) Ecological and economic services provided by birds on jamaican blue mountain coffee farms. *Conservation Biology* 22, 1177–1185. Available at: https://doi.org/10.1111/j.1523-1739.2008.00968.x

King, R.A., Symondson, W.O.C. and Thomas, R.J. (2015) Molecular analysis of faecal samples from birds to identify potential crop pests and useful biocontrol agents in natural areas. *Bulletin of Entomological Research* 105, 261–272.

Kleijn, D., Bommarco, R., Fijen, T.P.M., Garibaldi, L.A., Potts, S.G. *et al.* (2019) Ecological intensification: Bridging the gap between science and practice. *Trends in Ecology & Evolution* 34(2), 154–166.

Klosterman, M.E., Linz, G.M., Slowik, A.A. and Homan, H.J. (2013) Comparisons between blackbird damage to corn and sunflower in North Dakota. *Crop Protection* 53, 1–5.

Koh, L.P. (2008) Birds defend oil palms from herbivorous insects. *Ecological Applications* 18, 821–825. Available at: https://doi.org/10.1890/07-1650.1

Kross, S.M., Kelsey, T.R., McColl, C.J. and Townsend, J.M. (2016) Field-scale habitat complexity enhances avian conservation and avian-mediated pest-control services in an intensive agricultural crop. *Agriculture, Ecosystems & Environment* 225, 140–149.

Kross, S.M., Ingram, K.P., Long, R.F. and Niles, M.T. (2018) Farmer perceptions and behaviors related to wildlife and on-farm conservation actions. *Conservation Letters* 11, e12364.

Labuschagne, L., Swanepoel, L.H., Taylor, P.J., Belmain, S.R. and Keith, M. (2016) Are avian predators effective biological control agents for rodent pest management in agricultural systems? *Biological Control* 101, 94–102.

Lang, A. (2003) Intraguild interference and biocontrol effects of generalist predators in a winter wheat field. *Oecologia* 134, 144–153.

Lees, A.C., Haskell, L., Allinson, T., Bezeng, S.B., Burfield, I.J. *et al.* (2022) State of the world's birds. *Annual Review of Environment and Resources* 47, 231–260.

Letourneau, D.K. and Goldstein, B. (2001) Pest damage and arthropod community structure in organic vs. conventional tomato production in California. *Journal of Applied Ecology* 38, 557–570.

Letourneau, D.K., Jedlicka, J.A., Bothwell, S.G. and Moreno, C.R. (2009) Effects of natural enemy biodiversity on the suppression of arthropod herbivores in terrestrial ecosystems. *Annual Review of Ecology, Evolution, and Systematics* 40, 573–592.

Librán-Embid, F., De Coster, G. and Metzger, J.P. (2017) Effects of bird and bat exclusion on coffee pest control at multiple spatial scales. *Landscape Ecology* 32, 1907–1920. Available at: https://doi.org/10.1007/s10980-017-0555-2

Li, G., Yin, B., Li, J., Wang, J., Wei, W. *et al.* (2020) Host-microbiota interaction helps to explain the bottom-up effects of climate change on a small rodent species. *ISME Journal* 14, 1795–1808.

Lindell, C.A., Eaton, R.A., Lizotte, E.M. and Rothwell, N.L. (2012) Bird consumption of sweet and tart cherries and implications for bird control techniques. *Human–Wildlife Interactions* 6(2), 283–290.

Linden, V.M.G., Grass, I., Joubert, E., Tscharntke, T., Weier, S.M, *et al.* (2019) Ecosystem services and disservices by birds, bats and monkeys change with macadamia.

Maas, B., Clough, Y. and Tscharntke, T. (2013) Bats and birds increase crop yield in tropical agroforestry landscapes. *Ecology Letters* 16(12), 1480–1487. Available at: https://doi.org/10.1111/ele.12194

Manosathiyadevan, M., Bhuvaneshwari, V. and Latha, R. (2017) Impact of insects and pests in loss of crop production: A review. In: Dhanarajan, A. (ed.) *Sustainable Agriculture towards Food Security*. Springer, Singapore, pp. 57–67.

Mäntylä, E., Klemola, T. and Laaksonen, T. (2011) Birds help plants: A meta-analysis of top-down trophic cascades caused by avian predators. *Oecologia* 165, 143–151.

Mariano-Neto, E. and Santos, R.A.S. (2023) Changes in the functional diversity of birds due to habitat loss in the Brazil Atlantic Forest. *Frontiers in Forests and Global Change* 6, 1041268.

Marples, N.M. (1993) Toxicity assays of ladybirds using natural predators. *Chemoecology* 4, 33–38.

Martin, E.A., Reineking, B., Seo, B. and Steffan-Dewenter, I. (2013) Natural enemy interactions constrain pest control in complex agricultural landscapes. *Proceedings of the National Academy of Sciences of the United States of America* 110(14), 5534–5539. Available at: https://doi.org/10.1073/pnas.1215725110

Martínez-Salinas, A., DeClerck, F., Vierling, K., Vierling, L., Legal, L. *et al.* (2016) Bird functional diversity supports pest control services in a Costa Rican coffee farm. *Agriculture, Ecosystems & Environment* 235, 277–288.

Martin, T.E. and Maron, J.L. (2012) Climate impacts on bird and plant communities from altered animal-plant interactions. *Nature Climate Change* 2, 195–200.

Mazia, C.N., Chaneton, E.J., Kitzberger, T. and Garibaldi, L.A. (2009) Variable strength of top–down effects in Nothofagus forests: Bird predation and insect herbivory during an ENSO event. *Austral Ecology* 34, 359–367.

Mols, C.M.M. and Vissr, M.E. (2007) Great tits (Parus major) reduce caterpillar damage in commercial apple orchards. *PLoS ONE* 2(2), e202.

Monteagudo, N., Benayas, J.M.R., Andivia, E. and Rebollo, S. (2023) Avian regulation of crop and forest pests, a meta-analysis. *Pest Management Science* 79, 2380–2389.

Muneret, L., Mitchell, M., Seufert, V., Aviron, S., Djoudi, E.A. *et al.* (2018) Evidence that organic farming promotes pest control. *Nature Sustainability* 1, 361–368.

Muñoz-Sáez, A., Heaton, E.E., Reynolds, M. and Merenlender, A.M. (2020) Agricultural adapters from the vineyard landscape impact native oak woodland bird. *Agriculture, Ecosystems & Environment* 300, 106960.

Myers, N., Mittermier, R.A., Mittermier, C.G., da Fonseca, G.A.B. and Kent, J. (2000) Biodiversity hotspots for conservation priorities. *Nature* 403, 853–858.

Ndang'ang'a, P.K., Njoroge, J.B.M. and Vickery, J. (2013) Quantifying the contribution of birds to the control of arthropod pests on kale, brassica oleracea acephala, a key crop in east african highland farmland. *International Journal of Pest Management* 59(3), 211–216. Available at: https://doi.org/10.1080/09670874.2013.820005

Newton, I. (1994) The role of nest sites in limiting the numbers of hole-nesting birds: A review. *Biological Conservation* 70, 265–276.

Nyffeler, M., Şekercioğlu, Ç.H. and Whelan, C.J. (2018) Insectivorous birds consume an estimated 400–500 million tons of prey annually. *Science of Nature* 105, 47.

Oerke, E.C. (2006) Crop losses to pests. *Journal of Agricultural Science* 144(1), 31–43.

Olimpi, E.M., Garcia, K., Gonthier, D.J., De Master, K.T., Echeverri, A. *et al.* (2020) Shifts in species interactions and farming contexts mediate net effects of birds in agroecosystems. *Ecological Applications* 30(5), e02115.

Olmos-Moya, N., Díaz-Siefer, P., Pozo, R.A., Fontúrbel, F.E., Lavandero, B. *et al.* (2022) The use of cavity-nesting wild birds as agents of biological control in vineyards of Central Chile. *Agriculture, Ecosystems & Environment* 334, 107975.

Peisley, R.K., Saunders, M.E. and Luck, G.W. (2016) Cost-benefit trade-offs of bird activity in apple orchards. *PeerJ* 4, e2179. Available at: https://doi.org/10.7717/peerj.2179

Peisley, R.K., Saunders, M.E. and Luck, G.W. (2017) LuckProviding perches for predatory and aggressive birds appears to reduce the negative impact of frugivorous birds in vineyardswildl. *Res* 44, 334–342.

Pejchar, L., Clough, Y., Ekroos, J., Nicholas, K.A., Olsson, O. *et al.* (2018) Net effects of birds in agroecosystems. *BioScience* 68, 896–904.

Peters, V.E. and Greenberg, R. (2012) Fruit supplementation affects birds but not arthropod predation by birds in costa rican agroforestry systems. *Biotropica* 1–9. Available at: https://doi.org/10.1111/j.1744-7429.2012.00891.x

Philpott, S.M., Greenberg, R., Bichier, P. and Perfecto, I. (2004) Impacts of major predators on tropical agroforest arthropods: Comparisons within and across taxa. *Oecologia* 140(1), 140–149. Available at: https://doi.org/10.1007/s00442-004-1561-z

Pimentel, D. (2005) Environmental and economic costs of the application of pesticides primarily in the United States. *Environment, Development and Sustainability* 7, 229–252.

Pulido, F. and Berthold, P. (2004) Microevolutionary response to climatic change. In: Moller, A.P., Fielder, W. and Berthold, P. (eds) *Birds and Climate Change*. Academic Press, London, UK, pp. 151–183.

Recher, H.F. and Majer, J.D. (2006) Effects of bird predation on canopy arthropods in *wandoo* Eucalyptus *wandoo* woodland. *Austral Ecology* 31, 349–360.

Rey Benayas, J.M., Meltzer, J., De Las Heras-Bravo, D. and Cayuela, L. (2017) Potential of pest regulation by insectivorous birds in mediterranean woody crops. *PLoS One* 12, 1–19. Available at: https://doi.org/10.1371/journal.pone.0180702

Rojas, M., Lambert, F., Ramirez-Villegas, J. and Challinor, A.J. (2019) Emergence of robust precipitation changes across crop production areas in the 21st century. *Proceedings of the National Academy of Sciences USA* 116, 6673–6678.

Rollan, À., Hernández-Matías, A. and Real, J. (2019) Organic farming favours bird communities and their resilience to climate change in Mediterranean vineyards. *Agriculture, Ecosystems & Environment* 269, 107–115.

Rosenheim, J.A., Wilhoit, L.R. and Armer, C.A. (1993) Influence of intraguild predation among generalist insect predators on the suppression of an herbivore population. *Oecologia* 96, 439–449.

Rozzi, R., Armesto, J.J., Gutierrez, J.R., Massardo, F., Likens, G.E. *et al.* (2012) Integrating ecology and environmental ethics: Earth stewardship in the southern end of the Americas. *BioScience* 62, 226–236.

Sala, O.E., Chapin, F.S., Armesto, J.J., Berlow, E., Bloomfield, J. *et al.* (2000) Biodiversity – global biodiversity scenarios for the year 2100. *Science* 287, 1770–1774.

Shave, M.E., Shwiff, S.A., Elser, J.L., Lindell, C.A. and Megan Shave, C.E. (2018) Falcons using orchard nest boxes reduce fruit-eating bird abundances and provide economic benefits for a fruit-growing region. *Journal of Applied Ecology* 55, 2451–2460.

Sherry, T.W. (2021) Sensitivity of tropical insectivorous birds to the anthropocene: A review of multiple mechanisms and conservation implications. *Frontiers in Ecology and Evolution* 9, 662873.

Skendžić, S., Zovko, M., Živković, I.P., Lešić, V. and Lemić, D. (2021) The impact of climate change on agricultural insect pests. *Insects* 12(5), 440.

Smith, O.M., Olimpi, E.M., Navarro-Gonzalez, N., Cornell, K.A., Frishkoff, L.O. *et al.* (2022) A trait-based framework for predicting foodborne pathogen risk from wild birds. *Ecological Applications* 32(2), e2523.

Snyder, W.E. (2019) Give predators a complement: Conserving natural enemy biodiversity to improve biocontrol. *Biological Control* 135, 73–82.

Stanton, R.L., Morrissey, C.A. and Clark, R.G. (2018) Analysis of trends and agricultural drivers of farmland bird declines in North America: A review. *Agriculture, Ecosystems & Environment* 254, 244–254.

Stenberg, J.A. (2017) A conceptual framework for integrated pest management. *Trends in Plant Science* 22(9), 759–769.

Suhonen, J., Norrdahl, K. and Korpimäki, E. (1994) Avian predation risk modifies breeding bird community on a farmland area. *Ecology* 75(6), 1626–1634.

Şekercioğlu, C., Wenny, D.G. and Whelan, C.J. (2016) *Why Birds Matter: Avian Ecological Function and Ecosystem Services*. University of Chicago Press, Chicago, Illinois.

Thrupp, L.A. (2000) Linking agricultural biodiversity and food security: The valuable role of agrobiodiversity for sustainable agriculture. *International Affairs* 76, 283–297.

Tilman, D., Cassman, K.G., Matson, P.A., Naylor, R. and Polasky, S. (2002) Agricultural sustainability and intensive production practices. *Nature* 418, 671–677.

Traugott, M., Bell, J.R., Raso, L., Sint, D. and Symondson, W.O.C. (2012) Generalist predators disrupt parasitoid aphid control by direct and coincidental intraguild predation. *Bulletin of Entomological Research* 102(2), 239–247.

Trejo, A. and Grigera, D.E. (1998) Food habits of the great horned owl (Bubo virginianus) in a Patagonian steppe in Argentina. *Journal of Raptor Research* 32, 306–311.

Tremblay, A., Mineau, P. and Stewart, R. (2001) Effects of bird predation on some pest insect populations in corn.agriculture. *Ecosystems & Environment* 83, 143–152. Available at: https://doi.org/10.1016/S0167-8809(00)00247-4

Trevelline, B.K., Latta, S.C., Marshall, L.C., Nuttle, T. and Porter, B.A. (2016) Molecular analysis of nestling diet in a long-distance Neotropical migrant, the Louisiana waterthrush (Parkesia Motacilla). *The Auk* 133, 415–428.

Tscharntke, T., Sekercioglu, C.H., Dietsch, T.V., Sodhi, N.S. and Hoehn, P. (2008) Landscape constraints on functional diversity of birds and insects in tropical agroecosystems. *Ecology* 89, 944–951.

Tscharntke, T., Grass, I., Wanger, T.C., Westphal, C. and Batáry, P. (2021) Beyond organic farming, harnessing biodiversity-friendly landscapes. *Trends in Ecology & Evolution* 36(10), 919–930.

Tuck, S.L., Winqvist, C., Mota, F., Ahnström, J., Turnbull, L.A. *et al.* (2014) Land-use intensity and the effects of organic farming on biodiversity: A hierarchical meta-analysis. *Journal of Applied Ecology* 51, 746–755.

van Klink, R., Bowler, D.E., Gongalsky, K.B., Swengel, A.B., Gentile, A. *et al.* (2020) Meta-analysis reveals declines in terrestrial but increases in freshwater insect abundances. *Science* 368, 417–420.

Wagner, D.L., Grames, E.M., Forister, M.L., Berenbaum, M.R. and Stopak, D. (2021) Insect decline in the Anthropocene: Death by a thousand cuts. *Proceedings of the National Academy of Sciences USA* 118, e2023989118.

Watson, J.E.M. and Venter, O. (2017) A global plan for nature conservation. *Nature* 550, 48–49.

Whelan, C.J., Şekercioğlu, Ç.H. and Wenny, D.G. (2015) Why birds matter: From economic ornithology to ecosystem services. *Journal of Ornithology* 156(1), 227–238.

Wolff, J.O., Fox, T., Skillen, R.R. and Wang, G. (2011) The effects of supplemental perch sites on avian predation and demography of vole populations. *Canadian Journal of Zoology* 77, 535–541.

Zhu, C., Li, W., Campos-Arceiz, A., Dalsgaard, B., Ren, P. *et al*. (2023) The reliability of regional ecological knowledge to build local interaction networks: A test using seed-dispersal networks across land-bridge islands. *Proceedings of the Royal Society B* 290, 20231221.

8 Impact of Climate Change on Bats Involved in Biological Control

Danilo Russo[1]*, Adrià López-Baucells[2], Carles Flaquer[2], Vanessa Mata[3,4], Orly Razgour[5], Carme Tuneu-Corral[2,3,4], Xavier Puig-Montserrat[2] and Hugo Rebelo[6]

[1]Laboratory of Animal Ecology and Evolution (AnEcoEvo), Dipartimento di Agraria, Università degli Studi di Napoli Federico II, Portici, Italy; [2]BiBio Research Group, Natural Sciences Museum of Granollers, Granollers, Spain; [3]CIBIO-InBIO, Centro de Investigação em Biodiversidade de Recursos Genéticos, InBIO Laboratório Associado, Campus de Vairão, Universidade do Porto, Vairão, Portugal; [4]BIOPOLIS Program in Genomics, Biodiversity and Land Planning, CIBIO, Campus de Vairão, Vairão, Portugal; [5]Biosciences, Hatherly Laboratories, Streatham Campus, Exeter, UK; [6]cE3c – Centre for Ecology, Evolution and Environmental Changes and CHANGE – Global Change and Sustainability Institute, Departamento de Biologia Animal, Faculdade de Ciências, Universidade de Lisboa, Lisboa, Portugal

Abstract

Climate change can disrupt predator–prey interactions in agricultural systems by altering species' physiology, distribution and behaviour. Temperature and precipitation changes may cause mismatches in resource timing, destabilising these dynamics. Such shifts can reduce natural pest control effectiveness, increasing reliance on chemical methods. Many bat species prey on arthropods and play a crucial role in pest control. Bats help to reduce pesticide use, saving farmers costs. However, climate change impacts bat distribution, reproduction and behaviour. Some species have shifted their ranges, while others face challenges like heat stress and drought, affecting survival and reproduction. These shifts may disrupt bat–pest interactions, creating mismatches between pest outbreaks and bat presence. Climate change also affects insect emergence and bat reproduction timing, potentially leading to pest outbreaks. Future research should monitor bat and pest dynamics, and predict climate change's impact on the important ecosystem services bats deliver.

8.1 How Climate Change Can Alter Prey–Predator Interactions in Agricultural Systems

Empirical evidence and modelling work show that climate change may profoundly affect predator–prey dynamics, particularly within agricultural systems. The physiological and distributional changes induced by global warming can potentially destabilize these interactions, impacting both predation success and the distribution of predators and prey. Alterations in temperature and precipitation patterns can affect the growth rates, reproductive cycles and habitat ranges of predators and their prey. These shifts may imply mismatches in

*Corresponding author: danrusso@unina.it

 Biological Control Systems and Climate Change (eds Joan van Baaren *et al.*)
DOI: 10.1079/9781800625099.0008

the timing and availability of resources, further complicating the stability of such ecological interactions. Changes in the abundance and behaviour of prey species and predators' range shifts can create new challenges within agricultural systems, potentially leading to cascading effects throughout the ecosystem.

One critical way in which climate change affects predator–prey interactions is through alterations in physiology and metabolic rates. Climate-induced changes in prey vulnerability due to changes in body condition or habitat significantly influence predator responses (Morin *et al.*, 2021). For instance, changes in the prey's body condition might make the prey vulnerable to predation. These shifts may in turn affect predator populations, depending on their hunting strategies. In African savannah, cursorial predators such as spotted hyenas are predicted to benefit from more stable prey populations under arid conditions. However, ambush predators like lions could face challenges due to reduced prey availability or altered prey behaviours (Morin *et al.*, 2021).

Phenotypic plasticity might, to some extent, buffer predator–prey dynamics from the effects of climate change. Phenotypic plasticity is an organism's capacity to alter its phenotype in response to varying environmental conditions within its lifetime. Therefore, the same genotype can lead to different phenotypes under different environmental conditions. Thermal acclimation, a form of phenotypic plasticity, can buffer the impacts of warming on predator–prey interactions as seen in a dyad of freshwater invertebrates (Sohlström *et al.*, 2021). Acclimation to higher temperatures can stabilize predator–prey dynamics by reducing the sensitivity of metabolic and feeding rates to temperature changes. This stabilization results in increased biomass of predator and prey populations, suggesting that some predator–prey systems might be less affected by warming than previously anticipated (Sohlström *et al.*, 2021).

Climate change also alters species distribution, leading to potential mismatches between predators and prey. Mathematical models predict that global warming and wind flow can affect prey and predator populations, potentially leading to instability in predator–prey interactions (Thirthar *et al.*, 2024). Specifically, changes in prey distribution due to warming are predicted to disrupt established interactions with potential implications for ecosystem stability. Modelling including three-species interactions (crop, pest, predator) in agroecosystems shows that increasing global temperatures can shift the equilibrium points in predator–prey models (Panja, 2023). Such shifts might lead to situations where prey and predator populations no longer align spatially or temporally, leading to mismatches. For example, if predators cannot adapt quickly enough to changes in prey distribution, their effectiveness in controlling prey populations could diminish, leading to increased pest pressures in agricultural systems.

Phenological mismatches – discrepancies in the timing of biological events – are another critical concern (see Chapter 2). Such mismatches can impact population dynamics and the provision of ecosystem services, such as biological pest control. This process can result in increased pest populations and reduced crop yields in agricultural systems. For instance, if predators do not adjust their breeding times to match the peak abundance of their prey, their effectiveness in controlling pest populations may be compromised, leading to increased reliance on chemical pest control methods.

Adaptive management strategies are crucial to mitigate the adverse effects of climate change on predator–prey interactions in agricultural systems. This includes modifying integrated pest management tactics to account for shifts in pest and predator distributions and employing predictive models to anticipate changes in these dynamics (Skendžić *et al.*, 2021). Long-term monitoring and mechanistic modelling will also be essential for understanding and managing the complex interactions between climate change, predator–prey dynamics and agricultural productivity (Panja, 2023). In summary, climate change may alter predator–prey interactions in agroecosystems through physiological and distributional changes, implying mismatches and alteration of ecological interactions. This affects the effectiveness of biological control measures and farmland productivity. Understanding these dynamics and implementing adaptive management strategies are critical for mitigating the impacts of climate change and ensuring the resilience of agricultural ecosystems.

With about 1500 described species, bats display a wide range of foraging behaviours and dietary habits, including the consumption of fruits, nectar, seeds, small vertebrates, blood and, in most species, insects (Frick *et al.*, 2020; Simmons and Cirranello, 2025). This wide variety of feeding strategies makes bats crucial providers of ecosystem services such as regulation of arthropod populations (Kunz *et al.*, 2011). Bats are also potentially sensitive to climate change (Festa *et al.*, 2023), raising concerns about how this might affect their pest control performance, as discussed below.

8.2 Bats as Natural Enemies in Agricultural Systems

About 70% of bat species base their diet on arthropods and are equipped with highly developed and complex echolocation systems that make them among the world's most voracious predators (Kunz and Stern, 1995; Kalka and Kalko, 2006; Kunz *et al.*, 2011). Insectivorous bats employ different hunting strategies depending on their foraging guilds. Some are specialized in capturing flying insects (aerial-hawking bats), while others seize prey directly from foliage or the ground (gleaning bats), and some skim it from the water surface (trawling bats). These varied hunting strategies make bats versatile and effective predators in different environments and on a wide variety of prey. However, a recent review of the impact of bats as agricultural pest suppressors by Tuneu-Corral *et al.* (2023) listed 2308 documented interactions between 81 bat species and their respective insect pest prey (760 spp.), underpinning the generalized lack of studies for many species in most countries worldwide.

Estimating the exact quantity of prey consumed by bats is challenging. Various methods have been employed to quantify prey consumption, including direct observation, comparisons of pre- and postflight body mass and analysis of faecal samples. After several decades of different studies, many authors have demonstrated the significant magnitude of arthropod consumption of bats, some of them suggesting that daily prey intake can reach almost their entire body mass (Kurta *et al.*, 1989; Kunz and Stern, 1995; Kalka and Kalko, 2006; Kunz *et al.*, 2011), some suggesting between 25–50%, whereas others report that insectivorous bats can consume 600–1200 mosquitos nightly or up to 200 moths (Gonsalves *et al.*, 2013). Such estimations highlight bats' great potential as pest suppressors and their fundamental role in regulating arthropod populations.

In agricultural landscapes, bats act as natural enemies of many harmful insects, providing invaluable ecosystem services as pest suppressors (Tuneu-Corral *et al.*, 2023). Numerous studies highlight the ecological importance of bats in agroecosystems, emphasizing their ability to reduce crop damage and yield loss. This in turn decreases the need for agrochemical pesticide application, which not only reduces costs for farmers but also mitigates the negative impacts of these chemicals on ecosystems and human health (Maine and Boyles, 2015; Taylor *et al.*, 2018; Kolkert *et al.*, 2021). Ultimately, by lowering pest levels and promoting more sustainable farming practices, bats contribute to the long-term viability of agricultural systems and save substantial amounts of money for farmers and land managers, reinforcing their crucial role in the farming economy.

Bats have been proven to provide substantial economic benefits across a wide range of crops, including cereals such as rice (Puig-Montserrat *et al.*, 2015; Tuneu-Corral *et al.*, 2023) or corn (Maine and Boyles, 2015), fruit trees like vineyards (Rodríguez-San Pedro *et al.*, 2020; Charbonnier *et al.*, 2021) and apple orchards (Ancillotto *et al.*, 2024), oil palm plantations (Denan *et al.*, 2020), plantations of coffee or sugarcane (Schmitt *et al.*, 2021), and non-food crops such as cotton (Cleveland *et al.*, 2006) or rubber trees (Denan *et al.*, 2020). Researchers have sometimes quantified these ecosystem services to provide more tangible economic value to farmers and policymakers. These assessments have been generally performed using exclusion structures in crops and agricultural fields, excluding bats at specific places to create controlled treatment comparative experiments, and have been valued, for instance, at more than $200 per hectare per year in vineyards in Chile (Rodríguez-San Pedro *et al.*, 2020), €551 per hectare per year in apple orchards in Italy (Ancillotto *et al.*, 2024) and up to €135 per hectare per year in Spanish rice fields (Tuneu-Corral *et al.*, 2023).

Despite the challenges and difficulties of extrapolating these quantifications at broader scales, some efforts have been carried out at national and global levels to estimate the overall economic impacts of bats on farmers. Bats have been suggested to save more than $1 billion annually on corn plantations at a global scale (Maine and Boyles, 2015), with values exceeding $390 million in Brazil alone (Aguiar *et al.*, 2021). In rice paddies, bat activity results in annual savings of over €6 million in Spain (Tuneu-Corral *et al.*, 2023) and more than $1 million in Thailand (Wanger *et al.*, 2014). For macadamia orchards, benefits are estimated to exceed US$3 million in South Africa (Taylor *et al.*, 2018). In cotton plantations in Australia, this value can reach up to $63 million annually (Kolkert *et al.*, 2021). Moreover, in cacao farms across the Global South, bats and birds help mitigate crop damage, preventing annual losses estimated at $730 per hectare in Indonesia, $478 per hectare in Cameroon and $959 per hectare in Perú (Maas *et al.*, 2013; Ferreira *et al.*, 2023; Ocampo-Ariza *et al.*, 2023).

Apart from the impacts on agricultural pests in crops, bats significantly influence forest insect dynamics by preying on numerous herbivorous insects like moths and beetles, such as the well-known processionary moth, which can cause defoliation and other damage to trees (Charbonnier *et al.*, 2014; Garin *et al.*, 2019; Ancillotto *et al.*, 2022; Tuneu-Corral *et al.*, 2023; Augusto *et al.*, 2024). These and other studies have shown that bats also suppress pest insect populations, mitigating the risk of pest outbreaks that threaten forest health, critical in pine plantations, among many other forest structures. By consuming large quantities of these insects, bats help maintain ecosystem balance and biodiversity in agro-forestry areas.

Predator–prey interactions are shaped by direct predation but also by the prey's perception of predatory risk, significantly influencing their behaviour and spatial distribution (Laundré *et al.*, 2010; Gallagher *et al.*, 2017; Gaynor *et al.*, 2019; MacKay *et al.*, 2021). The published evidence indicates that bats exert non-consumptive effects on certain insect species, particularly those capable of detecting ultrasounds (Huang *et al.*, 2003; Huang and Subramanyam, 2004; Gillam *et al.*, 2011; Nakano *et al.*, 2022). Several nocturnal lepidopterans, which are primary prey for insectivorous bats, have evolved tympanic organs or high-frequency 'ears' sensitive to ultrasounds (Ratcliffe *et al.*, 2009; Conner and Corcoran, 2012). These auditory systems enable them to detect echolocating bats and alter their normal behaviour to reduce predation risk, offering a survival advantage over non-tympanate species (Yack *et al.*, 2007; Ratcliffe *et al.*, 2009).

However, while these behavioural changes favour survival probability, they often negatively impact reproductive success. Evidence shows that in the presence of synthesized ultrasounds, some male moths reduce their response towards pheromones, and both male and female moths may cease sexual calling or spermatophore transfer, decreasing mating success (Acharya and Mcneil, 1998; Huang *et al.*, 2003; Huang and Subramanyam, 2004; Skals *et al.*, 2005). Ultrasound exposure can also adversely affect egg production and larval weight (Huang *et al.*, 2003; Huang and Subramanyam, 2004) and decrease egg viability (Akinneye and Ashamo, 2009). These findings suggest that the acoustic dimension significantly impacts agriculture, as some of the most harmful crop pests are tympanate nocturnal Lepidoptera (Wagner, 2013).

Bats are crucial in managing agricultural ecosystems sustainably, acting as natural regulators of numerous pests. By reducing reliance on chemical pesticides, bats foster healthier ecosystems and minimize environmental impact. As invaluable allies in sustainable farming practices, bats globally contribute to the natural balance of agroecosystems. Therefore, the conservation of bat populations is paramount for preserving the stability, health and resilience of agricultural environments worldwide, which, in parallel, enhances long-term food production (Fig. 8.1). However, although scientific evidence on the role of bats in pest suppression continues to grow (Tuneu-Corral *et al.*, 2023), further research is necessary to unravel the complexities of predator–prey interactions within changing and adaptative landscapes and environments. Direct observations and quantifications of pest suppression in specific crops have been documented globally. Large-scale implementation among farmers, including comprehensive social-ecological assessments and in-depth societal analyses, is essential for advancing integrated pest management strategies.

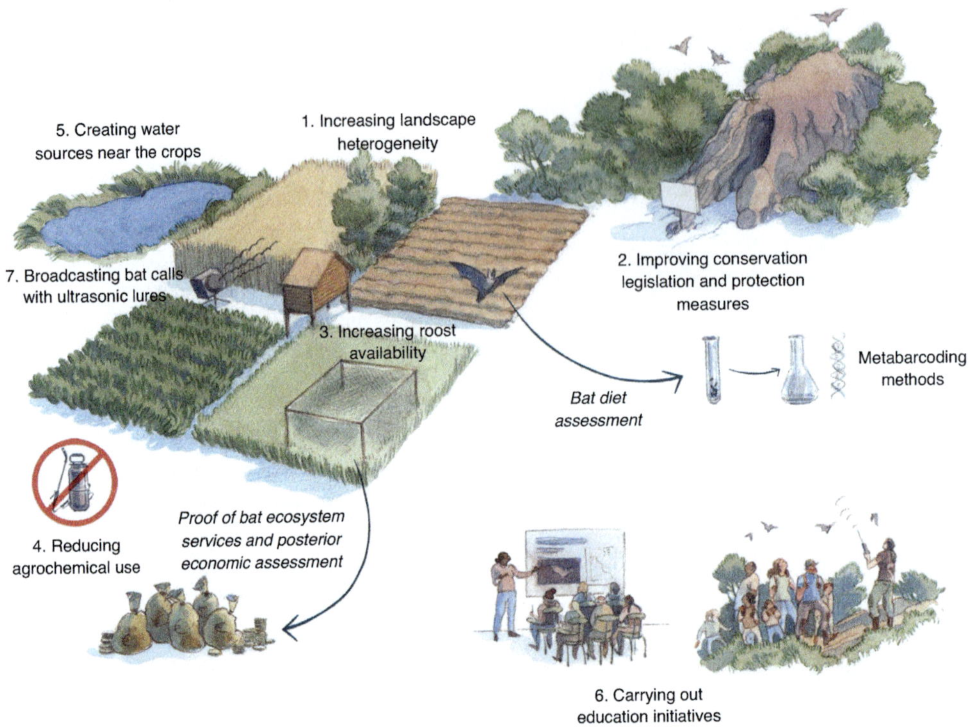

Fig. 8.1. Proposed recommendations from the scientific literature to improve the effectiveness of bats as pest suppressors and their associated ecosystem services. The figure emphasizes the need for future research on bat diet analysis, evidence of their ecosystem service, and the economic evaluation of pest suppression. From Tuneu-Corral *et al.* (2023) and used with permission of the authors. Published under a CC BY-NC-ND 4.0 licence: https://creativecommons.org/licenses/by-nc-nd/4.0/deed.en.

8.3 Bat Responses to Climate Change

Bats are sensitive to climate change because their large non-insulated wing and tail membranes relative to their generally small body mass mean they have high surface-to-volume ratios and are therefore prone to dehydration (Webb, 1995). Indeed, bats in arid environments have specific physiological adaptations, such as reduced evaporative water loss rates, which enable them to cope with higher temperatures and reduced water availability (Muñoz-Garcia *et al.*, 2016). In addition, bats are vulnerable to climate change because their low reproductive output limits population recovery after extreme events, and their ecological specialization and high trophic position make them sensitive to changes in food webs (Jones and Rebelo, 2013).

Evidence over the past few decades shows that bats have responded to climate change through changes in their range, demography, morphology, reproduction, physiology, migration, hibernation, habitat use and mortality (Festa *et al.*, 2023).

Ecologically plastic, generalist species are expanding their ranges to higher latitudes in temperate regions. For example, in Europe, Kuhl's pipistrelle *Pipistrellus kuhlii* (Kuhl, 1817), a Mediterranean vespertilionid species, has expanded its geographical range into central and eastern Europe by ~394% since 1980 (Sachanowicz *et al.*, 2006; Ancillotto *et al.*, 2016), while in North America, the largely tropical molossid Mexican free-tailed bat *Tadarida brasiliensis* (Geoffroy

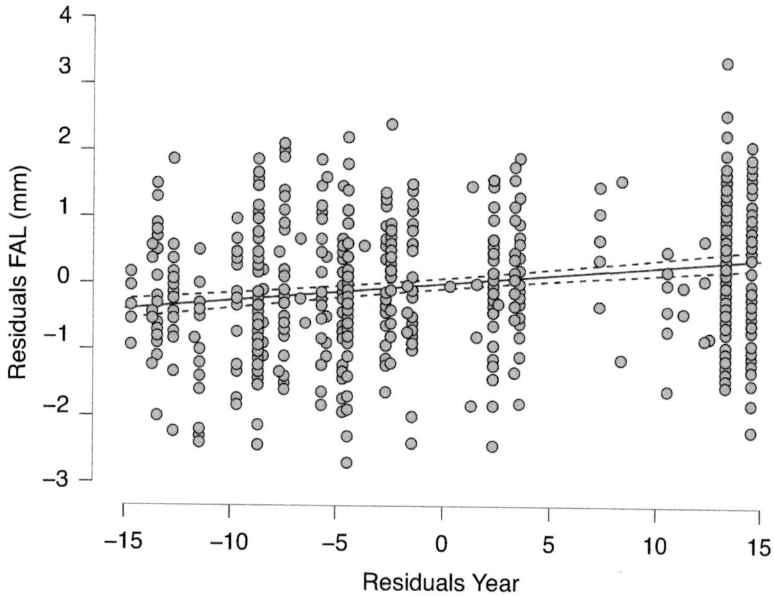

Fig. 8.2. Partial regression plots from a multiple linear regression with categorical predictors (using dummy coding) illustrating the relationship between forearm length (FAL, in mm) and year of record (top row) in Italian Daubenton's bats (*Myotis daubentonii*). The observed increase in body size is interpreted as a phenotypically plastic response to rising temperatures, with warmer roosts reducing thermoregulatory energy costs for mothers, allowing greater energy investment in the biomass of their offspring. Adapted from Russo *et al.* (2024), published under open access (licence CC BY 4.0).

Saint-Hilaire, 1824) has expanded its range along the eastern US coast since 2007 (McCracken *et al.*, 2018). *T. brasiliensis* has also experienced phenological changes over the past two decades in the form of advanced spring migration and summer reproduction cycle (Stepanian and Wainwright, 2018). In certain species, such as Daubenton's bat *Myotis daubentonii* (Kuhl, 1817), family Vespertilionidae, phenotypically plastic responses to warmer nursery roosts enable reproducing females to conserve energy, potentially allocating more resources to increase the biomass of their offspring. In Italian populations,this process has caused a gradualincrease in body size over time (Russo et al., 2024) (Fig. 8.2). Likewise, contrary to expectations based on thermoregulation under warmer temperatures (Bergmann's Rule), the body sizes of some European bat species, including a rhinolophid, the greater horseshoe bat *Rhinolophus ferrumequinum* (Schreber, 1774) in Italy and a vespertilionid, Natterer's bat *Myotis nattereri* (Kuhl, 1817) in Germany, increased in the past century, probably linked to the longer growing period (Salinas-Ramos *et al.*, 2021; Stapelfeldt *et al.*, 2023). Such increases in body size in another vespertilionid, the Bechstein's bat *Myotis bechsteinii* (Kuhl, 1817), have been linked to increased mortality rates, raising concerns about the conservation of this species under climate change (Mundinger *et al.*, 2021). Larger individuals need to store more fat for hibernation, which increases their foraging demands. When prey availability is low, larger females may struggle to meet these demands.

Bats are sensitive to extreme events such as heatwaves, droughts, wildfires and storms. Mass mortality events of bats due to heat stress have been reported in tree-roosting Old World fruit bats (*Pteropus* spp., family Pteropodidae) in Australia during recurring heatwaves since 1994 (Welbergen *et al.*, 2008) and in cave-dwelling insectivorous bats during a heatwave in Cambodia (Pruvot *et al.*, 2019). In southern Europe heatwaves are associated with higher rates of admittance of bats to wildlife rehabilitation centres (Salinas-Ramos *et al.*, 2023). Droughts have been associated with up to 50%

reduction in reproductive output of several *Myotis* bat species in the US (Adams, 2010). Demographic simulations predict population decreases of up to 90% across the range of the fringed myotis *Myotis thysanodes* (Miller, 1897) under future climate change scenarios due to reduced reproductive rates (Adams and Hayes, 2021).

Species distribution modelling studies predict changes in range suitability for bats under projected climate change scenarios by the end of the century, including severe losses in range suitability for forest specialist frugivorous bat species in the Neotropics (Gonçalves *et al.*, 2021) and Mediterranean bat species in Europe (Rebelo *et al.*, 2010), and substantial losses in bat species richness in South-East Asia (Hughes *et al.*, 2012). Predicted range losses can result in considerable reduction of genetic diversity in Mediterranean bat species, limiting the evolutionary potential of these species and their ability to respond to future changes (Razgour *et al.*, 2013). Limited landscape connectivity due to anthropogenic habitat loss and fragmentation can impede the ability of bats to shift their ranges to track shifting suitable climatic conditions (Razgour, 2015) and the spread of adaptive genetic variation between populations (Razgour *et al.*, 2019).

8.4 Climate Change and Bat–Pest Interactions: Impacts and Future Studies

Climate change can significantly influence bat predation on pest insects, affecting ecosystems and agricultural systems. Beyond habitat suitability, climate exerts a powerful influence on the occurrence of bats and pests by shaping energy and water dynamics in an area. These changes can lead to spatial mismatches, where pests thrive but bats cannot due to unsuitable conditions for foraging or roosting. This imbalance can disrupt the natural pest control functions that bats provide.

As the climate warms and affects precipitation levels, the life cycles of both bats and insects may be disrupted. There is growing evidence that climate change may be causing bats to give birth earlier in the season in the northern hemisphere (e.g. Ransome and McOwat, 1994; Lučan *et al.*, 2013; Linton and Macdonald, 2018; Matthäus *et al.*, 2023), while some insects may be having earlier flight periods, increased winter survival rates and more generations per year (e.g. Robinet and Roques, 2010; Subedi *et al.*, 2023). These shifts can create ecological disturbances, particularly in agriculture where a lack of natural predators can lead to pest outbreaks. Unstable weather conditions may also reduce bat pup survival (Adams, 2010; Hayes and Adams, 2017), further diminishing the population of pest predators. In turn, insects may emerge when bats are absent, potentially leading to uncontrolled pest proliferation.

Additionally, novel climates offer opportunities for new species to colonize previously unsuitable regions. Vespertilionids such as the above-mentioned *P. kuhlii* and *Hypsugo savii* (Bonaparte, 1837) are expanding in eastern Europe (Ancillotto *et al.*, 2016; Uhrin *et al.*, 2016), and pest species like the pine processionary moth *Thaumetopoea pityocampa* (Denis and Schiffermüller, 1775), from the family Notodontidae, and winter moth (*Operophtera brumata* (L.), family Geometridae) are moving northwards in Europe (Battisti *et al.*, 2005; Jepsen *et al.*, 2008). Similar responses have been recorded for the curculionid mountain and southern pine beetles (*Dendroctonus ponderosae* Hopkins, 1902, and *Dendroctonus frontalis* Zimmermann, 1868, respectively) in North America (Cudmore *et al.*, 2010; Lesk *et al.*, 2017). Human trade has also contributed to the spread of exotic species, and with climate change accelerating the emergence of new climates, pests may spread more rapidly through human-mediated pathways rather than natural migration (Gippet *et al.*, 2019). In summary, climate-induced temporal and spatial shifts combined with new species introduction could severely disrupt the ecosystem services bats provide as pest predators. These changes pose unpredictable consequences for both natural ecosystems and agriculture.

Studying the impact of climate change on bats and insects often relies on species distribution models (SDMs), which predict species occurrence based primarily on climatic variables. However, these models rarely incorporate future land cover scenarios, partly due to uncertainty in highly populated regions like Europe and parts of

Asia and the Americas (Razgour *et al.*, 2016). In these areas, land use is driven by socio-economic factors rather than ecological suitability, making it difficult to predict future habitat changes (Festa *et al.*, 2023). Consequently, SDMs tend to focus on identifying suitable climatic conditions for species, often neglecting the habitat and roosting needs of bats and the plant preferences of insects (Razgour *et al.*, 2016). Despite these limitations, SDMs remain valuable for identifying regions which may be resilient to climate change and forecasting potential areas where species may shift, enabling proactive strategies to manage these changes (Rebelo *et al.*, 2010).

However, the co-occurrence of bats and pest species does not necessarily imply predation (Blanchet *et al.*, 2020). Various factors influence predation levels, including prey availability, interspecies interactions and landscape structure. For example, studies have shown that bat predation on pine processionary moths increases in areas with greater bat diversity, though this predation is not constant across pest populations (Augusto *et al.*, 2024). These findings suggest that bat–pest interactions are more complex than what current models capture, highlighting the need for more refined approaches.

In recent years, advancements in diet analysis through metabarcoding have provided valuable insights into bat ecology. These studies have revealed that many bat species feed on a broad range of pest insects, underscoring their role as natural pest suppressors (e.g. Aizpurua *et al.*, 2018; Mata *et al.*, 2021). Monitoring bat diets over time and across different regions could offer crucial insights into how pest suppression may change in response to ongoing climate shifts. Additionally, molecular diet analysis can serve as an early warning system for detecting new pests in a given area (Maslo *et al.*, 2017; Montauban *et al.*, 2021). The vespertilionid big brown bat *Eptesicus fuscus* (Palisot de Beauvois, 1796) can detect invasive pests like the brown marmorated stink bug *Halyomorpha halys* (Stål, 1855), family Pentatomidae, weeks earlier than conventional blacklight traps (Maslo *et al.*, 2017). Likewise, molecular analysis of dropping contents of the vespertilionid soprano pipistrelle *Pipistrellus pygmaeus* (Leach, 1825) first revealed the occurrence in the Iberian peninsula of the rice water weevil (*Lissorhoptrus oryzophilus* Kuschel, 1950).

Since bats consume a diverse range of prey, they are likely to ingest newly arrived pests, allowing detection of invasive species before they become established.

Despite its potential, the widespread use of metabarcoding is limited by cost, which remains a barrier, particularly in lower-income countries (Cruz-Cano *et al.*, 2024). However, the cost of this technology has been declining in recent years, making it increasingly accessible. Researchers are also developing alternative methods for assessing pest suppression, such as using ecological indicators or proxies that correlate with pest control levels (Tougeron *et al.*, 2022). Integrating artificial intelligence with environmental data, such as satellite imagery and citizen science data, may improve our ability to monitor bat–pest interactions cost-effectively (Emery *et al.*, 2024). Artificial intelligence could aid in exploring molecular dietary analysis and link bat occurrence, population trends and their impact on pest arthropods with environmental variables. This would enhance our ability to assess the delivery of ecosystem services under different conditions and forecast trends. Citizens can participate in monitoring bat presence using simple bat detectors and may collect bat droppings from roost buildings, provided it does not disturb the bats. These droppings can then be sent to specialized laboratories for dietary analysis. These combined approaches could allow researchers to develop indicators that track changes in pest suppression over time and space.

In conclusion, understanding how climate change affects bat predation on pests requires integrating biological and environmental data. While species distribution models and metabarcoding offer valuable insights, future advances in artificial intelligence and ecological monitoring hold promise for more accurately predicting and managing the impacts of climate change on bat–pest interactions. By combining these tools, researchers and policymakers can better prepare for the potential disruptions to pest control services caused by global environmental shifts.

8.5 Conclusion

We have seen that bats play an indispensable role in agricultural systems worldwide, serving as natural enemies of arthropod pests and significantly reducing the use of pesticides. Bats' diverse hunting strategies, which target various insect pest species, make them remarkably versatile and effective in pest suppression across many crop types, from rice and corn to fruit orchards and cotton plantations. The contributions of bats to agriculture, as demonstrated by exclusion experiments and economic valuations, highlight their substantial cost-saving benefits. By mitigating pest populations, bats enhance crop yield and quality and contribute to healthier ecosystems, promoting sustainable agricultural practices. These contributions highlight the value of bat conservation as a strategic component in sustainable agriculture and integrated pest management.

However, climate change poses a multifaceted threat to bats, especially due to their physiological sensitivity to dehydration, low reproductive output and ecological specializations. Climate-driven changes in distribution, morphology, phenology and demography have already been observed in bat populations. Generalist species are expanding into previously cooler regions, and specialist species face habitat loss and population declines. Extreme weather events, such as heatwaves and droughts, have led to mass mortality and reproductive declines, intensifying the vulnerability of bat populations. These changes underline the urgency of incorporating bats into climate resilience planning, emphasizing habitat protection and addressing the specific needs of climate-sensitive species.

Bats are critical providers of ecosystem services and key components of biodiversity. Bat conservation links with broader goals such as food security, environmental health and climate resilience. To fully exploit these benefits, future research should focus on refining the quantification of bats' pest suppression effects, exploring non-consumptive influences on pest behaviour, and investigating large-scale impacts on agricultural practices. Additionally, comprehensive socio-ecological studies are essential to embed bats within integrated pest management frameworks globally. Ensuring the stability and resilience of bat populations is key for agricultural productivity and maintaining the ecological balance in rapidly changing environments. Therefore, sustained efforts in bat conservation will serve as a proactive approach toward fostering sustainable agricultural landscapes and building resilience against climate change impacts.

References

Acharya, L. and Mcneil, J.N. (1998) Predation risk and mating behavior: The responses of moths to bat-like ultrasound. *Behavioral Ecology* 9, 552–558.

Adams, R. (2010) Bat reproduction declines when conditions mimic climate change projections for western north america. *Ecology* 918, 2437–2445.

Adams, R.A. and Hayes, M.A. (2021) The importance of water availability to bats: Climate warming and increasing global aridity. In: Lim, B.K., Fenton, M.B., Brigham, R.M., Mistry, S. and Kurta, A. (eds) *50 Years of Bat Research: Foundations and New Frontiers. Fascinating Life Sciences*. Springer, Cham, Switzerland, pp. 105–120.

Aguiar, L.M.S., Bueno-Rocha, I.D., Oliveira, G., Pires, E.S., Vasconcelos, S. *et al.* (2021) Going out for dinner – the consumption of agriculture pests by bats in urban areas. *PLoS ONE* 16, 1–23.

Aizpurua, O., Budinski, I., Georgiakakis, P., Gopalakrishnan, S., Ibañez, C. *et al.* (2018) Agriculture shapes the trophic niche of a bat preying on multiple pest arthropods across Europe: Evidence from DNA metabarcoding. *Molecular Ecology* 27(3), 815–825.

Akinneye, J.O. and Ashamo, M.O. (2009) Effects of delayed mating on reproductive performance of *Ephestia Cautella* [Walker] [Lepidoptera: Pyralidae]. *Pakistan Journal of Scientific and Industrial Research* 6, 322–327.

Ancillotto, L., Santini, L., Ranc, N., Maiorano, L. and Russo, D. (2016) Extraordinary range expansion in a common bat: The potential roles of climate change and urbanisation. *Science of Nature* 103, 1–8.

Ancillotto, L., Rummo, R., Agostinetto, G., Tommasi, N., Garonna, A.P. *et al.* (2022) Bats as suppressors of agroforestry pests in beech forests. *Forest Ecology and Management* 522, 1–7.

Ancillotto, L., Borrello, M., Caracciolo, F., Dartora, F., Ruberto, M. *et al.* (2024) A bat a day keeps the pest away: Bats provide valuable protection from pests in organic apple orchards. *Journal for Nature Conservation* 78, 126558.

Augusto, A.M., Raposeira, H., Horta, P., Mata, V.A., Aizpurua, O. *et al.* (2024) Bat diversity boosts ecosystem services: Evidence from pine processionary moth predation. *Science of the Total Environment* 912, 169387.

Battisti, A., Stastny, M., Netherer, S., Robinet, C., Schopf, A. *et al.* (2005) Expansion of geographic range in the pine processionary moth caused by increased winter temperatures. *Ecological Applications* 15(6), 2084–2096.

Blanchet, F.G., Cazelles, K. and Gravel, D. (2020) Co-occurrence is not evidence of ecological interactions. *Ecology Letters* 23(7), 1050–1063.

Charbonnier, Y., Barbaro, L., Theillout, A. and Jactel, H. (2014) Numerical and functional responses of forest bats to a major insect pest in pine plantations e109488. *PLoS ONE* 9, 0109488.

Charbonnier, Y., Papura, D., Touzot, O., Rhouy, N., Sentenac, G. *et al.* (2021) Pest control services provided by bats in vineyard landscapes. *Agriculture Ecosystems and Environment* 306, 107207.

Cleveland, C.J., Betke, M., Federico, P., Frank, J.D., Hallam, T.G. *et al.* (2006) Economic value of the pest control service provided by Brazilian free-tailed bats in south-central Texas. *Frontiers in Ecology and the Environment* 4, 238–243.

Conner, W.E. and Corcoran, A.J. (2012) Sound strategies: The 65-million-year-old battle between bats and insects. *Annual Reviews in Entomology* 57, 21–39.

Cruz-Cano, R., Kolb, M., Saldaña-Vázquez, R.A., Bretón-Deval, L., Cruz-Cano, N. *et al.* (2024) Existing evidence on the use of environmental DNA as an operational method for studying rivers: A systematic map and thematic synthesis. *Environmental Evidence* 13(1), 2.

Cudmore, T., Björklund, N., Carroll, A. and Lindgren, B. (2010) Climate change and range expansion of an aggressive bark beetle: Evidence of higher beetle reproduction in naïve host tree populations. *Journal of Applied Ecology* 47, 1036–1043.

Denan, N., Wan Zaki, W.M., Norhisham, A.R., Sanusi, R., Nasir, D.M. *et al.* (2020) Predation of potential insect pests in oil palm plantations, rubber tree plantations, and fruit orchards. *Ecology and Evolution* 10, 654–661.

Emery, S.E., Rosenheim, J.A., Chaplin-Kramer, R., Sharp, R. and Karp, D.S. (2024) Leveraging satellite observations to reveal ecological drivers of pest densities across landscapes. *Science of the Total Environment* 924, 171591.

Ferreira, D.F., Jarrett, C., Wandji, A.C., Atagana, P.J., Rebelo, H. *et al.* (2023) Birds and bats enhance yields in Afrotropical cacao agroforests only under high tree-level shade cover. *Agriculture Ecosystems and Environment* 345, 108325.

Festa, F., Ancillotto, L., Santini, L., Pacifici, M., Rocha, R. *et al.* (2023) Bat responses to climate change: A systematic review. *Biological Reviews* 98(1), 19–33.

Frick, W.F., Kingston, T. and Flanders, J. (2020) A review of the major threats and challenges to global bat conservation. *Annals of the New York Academy of Sciences* 1469, 5–25.

Gallagher, A.J., Creel, S., Wilson, R.P. and Cooke, S.J. (2017) Energy landscapes and the landscape of fear. *Trends in Ecology and Evolution* 32, 88–96.

Garin, I., Aihartza, J., Goiti, U. and Arrizabalaga-Escudero, A. (2019) Bats from different foraging guilds prey upon the pine processionary moth. *PeerJ* 7, e7169.

Gaynor, K.M., Brown, J.S., Middleton, A.D., Power, M.E. and Brashares, J.S. (2019) Landscapes of fear: Spatial patterns of risk perception and response. *Trends in Ecology and Evolution* 34, 355–368.

Gillam, E.H., Westbrook, J.K., Schleider, P.G. and McCracken, G.F. (2011) Virtual bats and real insects: Effects of echolocation on pheromone-tracking behavior of male corn earworm moths, *Helicoverpa zea*. *Southern Naturalist* 56, 103–133.

Gippet, J., Liebhold, A., Fenn-Moltu, G. and Bertelsmeier, C. (2019) Human-mediated dispersal in insects. *Current Opinion in Insect Science* 35, 96–102.

Gonçalves, F., Sales, L.P., Galetti, M. and Pires, M.M. (2021) Combined impacts of climate and land use change and the future restructuring of Neotropical bat biodiversity. *Perspectives in Ecology and Conservation* 19(4), 454–463.

Gonsalves, L., Bicknell, B., Law, B., Webb, C. and Monamy, V. (2013) Mosquito consumption by insectivorous bats: Does size matter? *PLoS ONE* 8, e77183.

Hayes, M. and Adams, R. (2017) Simulated bat populations erode when exposed to climate change projections for western North America. *PLoS ONE* 12, 0180693.

Huang, F. and Subramanyam, B. (2004) Behavioral and reproductive effects of ultrasound on the Indian meal moth, *Plodia interpunctella*. *Entomologia Experimentalis et Applicata* 113, 157–164.

Huang, F., Subramanyam, B. and Taylor, R. (2003) Ultrasound affects spermatophore transfer, larval numbers, and larval weight of *Plodia interpunctella* (Hübner) (Lepidoptera: Pyralidae). *Journal of Stored Products Research* 39, 413–422.

Hughes, A.C., Satasook, C., Bates, P.J., Bumrungsri, S. and Jones, G. (2012) The projected effects of climatic and vegetation changes on the distribution and diversity of Southeast Asian bats. *Global Change Biology* 18(6), 1854–1865.

Jepsen, J.U., Hagen, S.B., Ims, R.A. and Yoccoz, N.G. (2008) Climate change and outbreaks of the geometrids *Operophtera brumata* and *Epirrita autumnata* in subarctic birch forest: Evidence of a recent outbreak range expansion. *Journal of Animal Ecology* 77(2), 257–264.

Jones, G. and Rebelo, H. (2013) Responses of bats to climate change: Learning from the past and predicting the future. In: Adams, R.A. and Pedersen, S.C. (eds) *Bat Evolution, Ecology, and Conservation*. Springer, New York, pp. 457–478.

Kalka, M. and Kalko, E.K.V. (2006) Gleaning bats as underestimated predators of herbivorous insects: Diet of *Micronycteris microtis* (Phyllostomidae) in Panama. *Journal of Tropical Ecology* 22, 1–10.

Kolkert, H., Smith, R., Rader, R. and Reid, N. (2021) Insectivorous bats provide significant economic value to the Australian cotton industry. *Ecosystem Services* 49, 101280.

Kunz, T.H. and Stern, A.A. (1995) Maternal investment and post-natal growth in bats. *Symposia of the Zoological Society of London* 67, 123–138.

Kunz, T.H., Torrez, E.B., Bauer, D., Lobova, T. and Fleming, T.H. (2011) Ecosystem services provided by bats. *Annals of the New York Academy of Sciences* 1223, 1–38.

Kurta, A., Bell, G.P., Nagy, K.A. and Kunz, T.H. (1989) Energetics of pregnancy and lactation in free-ranging little brown bats (*Myotis lucifugus*). *Physiological Zoology* 62, 804–818.

Laundré, J.W., Hernández, L. and Ripple, W.J. (2010) The landscape of fear: Ecological implications of being afraid. *Open Ecology Journal* 3, 1–7.

Lesk, C., Coffel, E., D'Amato, A.W., Dodds, K. and Horton, R. (2017) Threats to North American forests from southern pine beetle with warming winters. *Nature Climate Change* 7(10), 713–717.

Linton, D. and Macdonald, D. (2018) Spring weather conditions influence breeding phenology and reproductive success in sympatric bat populations. *Journal of Animal Ecology* 87, 1080–1090.

Lučan, R., Weiser, M. and Hanák, V. (2013) Contrasting effects of climate change on the timing of reproduction and reproductive success of a temperate insectivorous bat. *Journal of Zoology* 290, 151–159.

Maas, B., Clough, Y. and Tscharntke, T. (2013) Bats and birds increase crop yield in tropical agroforestry landscapes. *Ecology Letters* 16, 1480–1487.

MacKay, R.N., Wood, T.C. and Moore, P.A. (2021) Running away or running to? Do prey make decisions solely based on the landscape of fear or do they also include stimuli from a landscape of safety? *Journal of Experimental Biology* 224, jeb.242687.

Maine, J.J. and Boyles, J.G. (2015) Bats initiate vital agroecological interactions in corn. *Proceedings of the National Academy of Sciences* 112, 12438–12443.

Maslo, B., Valentin, R., Leu, K., Kerwin, K., Hamilton, G.C. *et al.* (2017) Chiro surveillance: The use of native bats to detect invasive agricultural pests. *PLoS ONE* 12(3), e0173321.

Mata, V.A., Silva, L.P., Veríssimo, J., Horta, P., Raposeira, H. *et al.* (2021) Combining DNA metabarcoding and ecological networks to inform conservation biocontrol by small vertebrate predators. *Ecological Applications* 31(8), e02457.

Matthäus, L., Kugelschafter, K. and Fietz, J. (2023) Influence of ambient temperature on the phenology of the greater mouse-eared bat (*Myotis myotis*). *Ecology and Evolution* 13, 10081.

McCracken, G.F., Bernard, R.F., Gamba-Rios, M., Wolfe, R., Krauel, J.J. *et al.* (2018) Rapid range expansion of the Brazilian free-tailed bat in the southeastern United States, 2008–2016. *Journal of Mammalogy* 99(2), 312–320.

Montauban, C., Mas, M., Wangensteen, O.S., Monteys, V.S., Fornós, D.G. *et al.* (2021) Bats as natural samplers: First record of the invasive pest rice water weevil *Lissorhoptrus oryzophilus* in the Iberian Peninsula. *Crop Protection* 141, 105427.

Morin, A., Chamaillé-Jammes, S. and Valeix, M. (2021) Climate effects on prey vulnerability modify expectations of predator responses to short-and long-term climate fluctuations. *Frontiers in Ecology and Evolution* 8, 601202.

Mundinger, C., Scheuerlein, A. and Kerth, G. (2021) Long-term study shows that increasing body size in response to warmer summers is associated with a higher mortality risk in a long-lived bat species. *Proceedings of the Royal Society B* 288(1952), 20210508.

Muñoz-Garcia, A., Larraín, P., Ben-Hamo, M., Cruz-Neto, A., Williams, J.B. *et al.* (2016) Metabolic rate, evaporative water loss and thermoregulatory state in four species of bats in the negev desert. *Comparative Biochemistry and Physiology Part A: Molecular and Integrative* 156–165.

Nakano, R., Ito, A. and Tokumaru, S. (2022) Sustainable pest control inspired by prey-predator ultrasound interactions. *Proceedings of the National Academy of Sciences* 119, e2211007119.

Ocampo-Ariza, C., Vansynghel, J., Bertleff, D., Maas, B., Schumacher, N. *et al.* (2023) Birds and bats enhance cacao yield despite suppressing arthropod *mesopredation*. *Ecological Applications* 33, e2886.

Panja, P. (2023) Dynamics of a crop, pest and predator model in an agricultural system. *Results in Control and Optimization* 12, 100274.

Pruvot, M., Cappelle, J., Furey, N., Hul, V., Heng, H.S. *et al.* (2019) Extreme temperature event and mass mortality of insectivorous bats. *European Journal of Wildlife Research* 65(3), 41.

Puig-Montserrat, X., Torre, I., López-Baucells, A., Guerrieri, E., Monti, M.M. *et al.* (2015) Pest control service provided by bats in Mediterranean rice paddies: Linking agroecosystems structure to ecological functions. *Mammalian Biology* 80, 237–245.

Ransome, R. and McOwat, T. (1994) Birth timing and population changes in greater horseshoe bat colonies (*Rhinolophus ferrumequinum*) are synchronized by climatic temperature. *Zoological Journal of the Linnean Society* 112, 337–351.

Ratcliffe, J.M., Fullard, J.H., Arthur, B.J. and Hoy, R.R. (2009) Tiger moths and the threat of bats: Decision-making based on the activity of a single sensory neuron. *Biology Letters* 5(3), 368–371.

Razgour, O. (2015) Beyond species distribution modeling: A landscape genetics approach to investigating range shifts under future climate change. *Ecological Informatics* 30, 250–256.

Razgour, O., Juste, J., Ibáñez, C., Kiefer, A., Rebelo, H. *et al.* (2013) The shaping of genetic variation in edge-of-range populations under past and future climate change. *Ecology Letters* 16(10), 1258–1266.

Razgour, O., Rebelo, H., Di Febbraro, M. and Russo, D. (2016) Painting maps with bats: Species distribution modelling in bat research and conservation. *Hystrix* 27(1), 30–37.

Razgour, O., Forester, B., Taggart, J.B., Bekaert, M., Juste, J. *et al.* (2019) Considering adaptive genetic variation in climate change vulnerability assessment reduces species range loss projections. *Proceedings of the National Academy of Sciences* 116(21), 10418–10423.

Rebelo, H., Tarroso, P. and Jones, G. (2010) Predicted impact of climate change on European bats in relation to their biogeographic patterns. *Global Change Biology* 16(2), 561–576.

Robinet, C. and Roques, A. (2010) Direct impacts of recent climate warming on insect populations. *Integrative Zoology* 5, 132–142.

Rodríguez-San Pedro, A., Allendes, J.L., Beltrán, C.A., Chaperon, P.N., Saldarriaga-Córdoba, M.M. *et al.* (2020) Quantifying ecological and economic value of pest control services provided by bats in a vineyard landscape of central chile. *Agriculture, Ecosystems & Environment* 302, 107063.

Russo, D., Jones, G., Martinoli, A., Preatoni, D.G., Spada, M. *et al.* (2024) Climate is changing, are European bats too? A multispecies analysis of trends in body size. *Ecology and Evolution* 14(2), e10872.

Sachanowicz, K., Wower, A. and Bashta, A.T. (2006) Further range extension of Pipistrellus kuhlii (Kuhl, 1817) in central and eastern Europe. *Acta Chiropterologica* 8(2), 543–548.

Salinas-Ramos, V.B., Agnelli, P., Bosso, L., Ancillotto, L. and Russo, D. (2021) Body size of Italian greater horseshoe bats (*Rhinolophus ferrumequinum*) increased over one century and a half: A response to climate change? *Mammalian Biology* 101, 3.

Salinas-Ramos, V.B., Tomassini, A., Ferrari, F., Boga, R. and Russo, D. (2023) Admittance to wildlife rehabilitation centres points to adverse effects of climate change on insectivorous bats. *Biology* 12(4), 543.

Schmitt, L., Greenberg, R., Ibarra-Núñez, G., Bichier, P., Gordon, C.E. *et al.* (2021) Cascading effects of birds and bats in a shaded coffee agroforestry system. *Frontiers in Sustainable Food Systems* 5, 512998.

Simmons, N.B. and Cirranello, A.L. (2025) Bat Species of the World: A Taxonomic and Geographic Database. Available at: https://batnames.org/ (accessed 16 June 2025).

Skals, N., Anderson, P., Kanneworff, M. and Surlykke, A. (2005) Her odours make him deaf: Crossmodal modulation of olfaction and hearing in a male moth. *Journal of Experimental Biology* 208(4), 595–601.

Skendžić, S., Zovko, M., Živković, I.P., Lešić, V. and Lemić, D. (2021) The impact of climate change on agricultural insect pests. *Insects* 12(5), 440.

Sohlström, E.H., Archer, L.C., Gallo, B., Jochum, M., Kordas, R.L. *et al.* (2021) Thermal acclimation increases the stability of a predator–prey interaction in warmer environments. *Global Change Biology* 27(16), 3765–3778.

Stapelfeldt, B., Tress, C., Koch, R., Tress, J., Kerth, G. *et al.* (2023) Long-term field study reveals that warmer summers lead to larger and longer-lived females only in northern populations of Natterer's bats. *Oecologia* 201(3), 853–861.

Stepanian, P.M. and Wainwright, C.E. (2018) Ongoing changes in migration phenology and winter residency at Bracken Bat Cave. *Global Change Biology* 24(7), 3266–3275.

Subedi, B., Poudel, A. and Aryal, S. (2023) The impact of climate change on insect pest biology and ecology: Implications for pest management strategies, crop production, and food security. *Journal of Agriculture and Food Research* 14, 100733.

Taylor, R.A.J., Herms, D.A., Cardina, J. and Moore, R.H. (2018) Climate change and pest management: Unanticipated consequences of trophic dislocation. *Agronomy* 8(1), 7.

Thirthar, A.A., Jawad, S., Majeed, S.J. and Nisar, K.S. (2024) Impact of wind flow and global warming in the dynamics of prey–predator model. *Results in Control and Optimization* 15, 100424.

Tougeron, K., Couthouis, E., Marrec, R., Barascou, L., Baudry, J. *et al.* (2022) Multi-scale approach to biodiversity proxies of biological control service in European farmlands. *Science of the Total Environment* 822, 153569.

Tuneu-Corral, C., Puig-Montserrat, X., Riba-Bertolín, D., Russo, D., Rebelo, H. *et al.* (2023) Pest suppression by bats and management strategies to favour it: A global review. *Biological Reviews* 98(5), 1564–1582.

Uhrin, M., Hüttmeir, U., Kipson, M., Estók, P., Sachanowicz, K. *et al.* (2016) Status of Savi's pipistrelle *Hypsugo savii* (Chiroptera) and range expansion in Central and south-eastern Europe: A review. *Mammal Review* 46, 1–16.

Wagner, D.L. (2013) Moth. In: Levin, S. (ed.) *Encyclopedia of Biodiversity*. Academic Press, San Diego, California, pp. 384–403.

Wanger, T.C., Darras, K., Bumrungsri, S., Tscharntke, T. and Klein, A.M. (2014) Bat pest control contributes to food security in Thailand. *Biological Conservation* 171, 220–223.

Webb, P.I. (1995) The comparative ecophysiology of water balance in microchiropteran bats. *Symposia of the Zoological Society of London* 67, 203–218.

Welbergen, J.A., Klose, S.M., Markus, N. and Eby, P. (2008) Climate change and the effects of temperature extremes on Australian flying-foxes. *Proceedings of the Royal Society B: Biological Sciences* 275(1633), 419–425.

Yack, J.E., Kalko, E.K. and Surlykke, A. (2007) Neuroethology of ultrasonic hearing in nocturnal butterflies (Hedyloidea). *Journal of Comparative Physiology A* 193, 577–590.

9 Community Change and Genetic Differentiation in a Small But Contrasted Area: Insight from *Trichogramma* in South-Eastern France

Michela Ion Scotta, Sylvie Warot, Elodie Vercken and Nicolas Ris*
Institut Sophia Agrobiotech, INRAE, CNRS, Université Cote d'Azur, Nice, France

Abstract
Because of its direct and indirect impacts on species, climate change impacts their distribution areas. To what extent the intraspecific genetic variability for tolerance to abiotic factors is able to buffer the evolution of the communities is, however, unclear. Some answers may be provided by investigating small-scale contrasted areas such as latitudinal and/or altitudinal transects. Within this frame, this chapter investigated the distribution of *Trichogramma* species (hymenopteran egg parasitoids) in south-eastern France and the extent to which local adaptations to climatic conditions may exist in one of the most cosmopolitan species (*T. cacoeciae*). These results are discussed in the frame of the possible consequences of climate change on *Trichogramma* geographical distribution but also as possible opportunities or risks for the use of these species as biological control agents.

9.1 Context

9.1.1 Eco-evolutionary background

Biological control is commonly described as the use of living organisms – so-called biocontrol agents – to decrease the populations of and/or damage caused by other species considered as noxious (Eilenberg *et al.*, 2001). Underneath this general definition, there exist three different strategies whose ins and outs actually differ: introduction (a.k.a. classical) biological control, augmentation biological control (including inoculation and inundation substrategies) and conservation biological control. All these strategies can be implemented in different contexts (agriculture, animal rearing, human health, conservation biology, etc). In agriculture, most of the biocontrol agents used against phytophagous arthropods are other arthropods, in particular predatory or parasitoid insects and predatory mites (Cock *et al.*, 2010).

As ectotherms, both biocontrol agents and pests are sensitive to various abiotic factors, in particular (but not exclusively) temperature. Temperature impacts them at two different levels: phenotypic and evolutionary. At the phenotypic level, several traits are well known to be modified by the external temperature: development time, adult size, potential fecundity, etc. (see Chapter 1 of this book; Angilletta *et al.*, 2002; Trudgill *et al.*, 2005; Angilletta, 2009; Rezende *et al.*, 2014). This kind of phenotypic plasticity can be observed not only in the individual itself but also

*Corresponding author: nicolas.ris@inrae.fr

DOI: 10.1079/9781800625099.0009

its offspring (see for instance Ferrer *et al.* (2013) and Stelgenga and Fischer (2007)).

Beyond the individual phenotype, this temperature-induced phenotypic plasticity impacts key parameters of population dynamics, in particular the population growth rate and the number of generations per year. Because genetic variability for temperature-dependent traits usually exists within a species, temperature also acts as a selective factor and thus tends to promote the best genotypes according to local conditions (Partridge *et al.*, 1995; Trudgill *et al.*, 2005; Shelomi, 2012). Genetic differentiation between populations is thus likely to occur in widely distributed species, especially if 'between-environments trade-offs' occur (Huey and Hertz, 1984; Partridge *et al.*, 1995; Shirley and Sibly, 1999). Under such constraints, a genotype well adapted to one particular environment will perform comparatively worse than others in a different environment. Within a community of arthropods, each species is thus exposed to these two types of temperature-induced impacts (phenotypic and evolutionary). As consequence, most if not all interspecific interactions are modulated by temperature change, be they between species of the same trophic levels (competitors) or between two trophic levels (prey/predators or host/parasitoid). An additional layer of complexity is the interaction between the phytophagous arthropods and their own resource, *i.e.* their host plants. Indeed, plants may also be impacted by temperature or other abiotic factors, both at the phenotypic and evolutionary levels. This is known to have significant repercussions on higher trophic levels (Pratt *et al.*, 2017). Predicting how such communities can evolve in response to temperature and, more generally, climate change is challenging.

A possible approach to understanding the evolution of arthropod communities, at least in the short and medium terms, is the study of climatic gradients such as latitudinal, altitudinal or longitudinal transects (Körner, 2007; Adrion *et al.*, 2015; Jego *et al.*, 2023). Of course, some biases do exist. For instance, abiotic factors other than temperature (precipitation, UV radiation or partial pressure of the air) change along an altitudinal transect (Hodkinson, 2005; Körner, 2007) and can thus act concomitantly as a source of phenotypic plasticity and/or a selective factor. The same holds true for latitudinal transects with possible confounding effects linked to progressive variations of photoperiod, for instance. Despite these limits, this approach may be still relevant, in particular if both latitudinal and altitudinal transects are *a posteriori* providing similar patterns. We thus surveyed the scientific literature, seeking for case studies that would simultaneously document:

- how arthropod communities evolve along a climatic gradient
- how some widely distributed resident species adapt locally to thermal conditions
- how climate change favours some species through time and alters the initial pattern
- to what extent pre-existing local adaptations in resident species may buffer this change.

This survey was, however, quite disappointing, the sole consistent corpus being the investigations on *Drosophila* species and their associated hymenopteran larval parasitoids along a latitudinal transect (see, for instance, Fleury *et al.*, 2009; Vuarin *et al.*, 2012; Delava *et al.*, 2014).

9.1.2 Focus on *Trichogramma* communities

Long-term investigations on the previously described eco-evolutionary framework were initiated in a context particularly relevant for agriculture and biological control: the community of *Trichogramma* species and their hosts (Ion Scotta, 2019). *Trichogramma*, which has about 200 species worldwide and probably about 60 species in the West Palearctic areas (Pinto and Stouthamer, 1994; Pintureau, 2008; Querino *et al.*, 2010), are micro-hymenopteran egg parasitoids, the females ovipositing their eggs in eggs of other species (mainly Lepidoptera) (Nagarkatti and Nagaraja, 1977; Smith, 1996; Romeis *et al.*, 2005). The host egg is quickly killed and thus does not hatch, the pre-imaginal *Trichogramma* developing inside it until its emergence as an adult. If this biological feature is shared by all *Trichogramma* species, interspecific variability exists for other important traits, in particular three that are relevant for this chapter:

- *Reproduction*: most species are arrhenotokous with haploid males and diploid females while some are thelytokous (diploid females only), this thelytoky being either an intrinsic property, as in *T. cacoeciae* (Vavre *et al.*, 2004), or induced by bacterial endosymbionts, as in *T. dendrolimi* (Liu *et al.*, 2019) or *T. pretiosum* (Xiao *et al.*, 2024).

- *Host specificity*: while most of the *Trichogramma* species are described as generalist and indeed able to develop in different host species in laboratory conditions, *Trichogramma* species could be more specialized or even actual specialists *in natura* (Romeis *et al.*, 2005).
- *Winter diapause* in response to low temperatures, this feature varying for both inter- and intraspecific species (Rahimi-Kaldeh *et al.*, 2018; Viscarret *et al.*, 2023; Mashal *et al.*, 2024; Reznik and Voinovich, 2024).

Because of their biology and ecology, *Trichogramma* species contribute naturally to the regulation of Lepidopteran pests, this action being sometimes completed by augmentation biological control practices (Smith, 1996; Parra and Zucchi, 2004; Zang *et al.*, 2021). This value as biocontrol agents as well as the relevance of *Trichogramma* as 'model organisms' for laboratory experiments has led, through the years, to a large set of published studies (more than 2800 references found in Web of Science with *Trichogramma* as the topic).

Despite this long-term interest, several aspects of *Trichogramma* remain surprisingly poorly documented. Their taxonomy itself is still subject to debate because of their minute size, the paucity of reliable morphological traits and the lack of experts. The modalities through which *Trichogramma* species are able to maintain themselves in some areas also remain intriguing. Indeed, even assuming that most *Trichogramma* species are generalist and thus exploit a diversity of resources, the presence of hosts all year round is not ensured and long period of host shortage may occur. Moreover, *Trichogramma* are usually described as active short-distance dispersers (some centimetres or metres by walking or jumping) and/or passive long-distance dispersers (Wright *et al.*, 2001; Kuske *et al.*, 2003; Fatouros and Huigens, 2012; Zhou *et al.*, 2019), two behaviours that *a priori* do not warrant the optimal exploration and exploitation of their local environment.

With regard to this situation, two opposite visions of the (meta-)population biology of *Trichogramma* species can be suggested.

- On one side, local functioning in which *Trichogramma* are able to maintain themselves locally in the long term through diverse (and currently poorly documented) adaptations (pre-imaginal diapause, adult persistence, phoresy, etc.), despite temporary adverse conditions (lack of hosts, unfavourable temperatures, etc.).
- On the other, global circulation of *Trichogramma* within wide geographic areas, mainly through passive methods with local 'sources' (habitats with more or less high and/or perennial presence of hosts) and 'sinks' (ephemeral favourable habitats).

With regard to the issue of population differentiation, these two scenarios led to different predictions. Under the 'local functioning' scenario, it is likely that widespread species may exhibit local adaptations in response to strong and contrasted local conditions, and experience reduced gene flows between populations. Under the 'global functioning' scenario, local adaptions are less likely to occur, in particular if gene flows are strong with regard to local selection.

9.1.3 Objectives

Focusing on *Trichogramma* species, we thus address here the two first questions mentioned earlier:

- how do arthropod communities evolve a climatic gradient?
- how do some widely distributed resident species locally adapt to thermal conditions?

In the first instance, the *Trichogramma* communities were investigated in a small geographic area characterized by a steep altitudinal gradient. In the second instance, complementary experimental approaches were performed to investigate population differentiation for traits linked to cold temperature and/or winter conditions in the most widespread species and to try to evidence 'adaptive syndromes', *i.e.* a genetic differentiation observed on a set of traits, this differentiation being reliably linked to a particular driver (Sinnott-Armstrong *et al.*, 2022).

9.2 *Trichogramma* Diversity in South-Eastern France

9.2.1 Sampling design

We selected an altitudinal transect in the Alpes-Maritimes department in south-eastern France to explore the spatial distribution of *Trichogramma* diversity. This area shows a diversity of climatic conditions, with six of the eight 'bioclimatic zones' described by Blondel and Aronson (2011). The altitudinal transect (30 km as the crow flies) covers an area located between Cannes (sea level) and Gréolières (1400 m altitude) and two bioclimatic zones, the Meso- and Supra-Mediterranean stages. Within this area, we focused on four habitat types likely to host *Trichogramma*: (i) cultivated areas (vegetables, horticulture, arboriculture; only in Meso-Mediterranean level), (ii) natural grasslands, (ii) shrubland ('maquis') and (iv) forests. Although some plant species are common to several levels and/or habitats (e.g. the European box *Buxus sempervirens* (Buxaceae), the common heather *Calluna vulgaris* (Ericaceae), the tree heather *Erica arborera* (Ericaceae), the spiny spurge *Euphorbia spinosa* (Euphorniaceae), the cade juniper *Juniperus oxycedrus* (Cupressaceae), the pubescent oak *Quercus pubescens* (Fagaceae) or the common thyme *Thymus vulgaris* (Lamiaceae)), each natural habitat is characterized by its own plant associations depending on altitude (Noble *et al.*, 2011).

A total of 42 sites (six replicates for each of the seven 'bioclimatic zone × natural habitat' combinations) were selected in the study area and each site was sampled monthly from April until November 2016. For each site and date, the sampling relied on the exposure of sentinel eggs of *Ephestia kuehniella* Zeller (Lepidoptera: Pyralidae) on ten plants. Thus, about 500 sterile host eggs/cm² were stuck on a piece of rigid paper (6 × 1 cm) and these 'patches' were stapled onto plant organs (mainly leaves but also flowers and stems and even trunks – three patches on a single plant). Plant species were chosen according to different criteria including (i) their physiological state (presence of fresh leaves and flowers were privileged), (ii) their potential attractiveness with regard to *Trichogramma* or their hosts, and (iii) the ability to identify them, at least to the genus level. *A posteriori*, sampling included 57 plant family, 128 plant genus and 101 plant species. Some of these were sampled more than 100 times: the European box *B. sempervirens*, the spiny broom *Calicotome spinosa* (Fabaceae), the scorpion senna *Coronilla emerus* (Fabaceae), the common hawthorn *Cratageus monogyna* (Rosaceae), the tree heather *E. arborea*, the silvery broom *Genista cinerea* (Fabaceae), the pubescent oak *Q. pubescens*, the dog rose *Rosa canina* (Rosaceae) and thymes *Thymus* sp.

After 48 hour outdoors, each patch was recovered and inserted into a tube containing small drops of honey and closed by a cotton cap. All tubes were then stored in a climatic chamber at 22 +/- 1°C, with a photoperiod (light/dark) of 16 hour/8 hour and a relative humidity of 70 +/- 20%. They were regularly checked to detect as early as possible signs of parasitism (parasitized eggs turning black) and then the emerging *Trichogramma* adults. All *Trichogramma* emerging from a single tube were collected, allowed to reproduce to found laboratory strains and finally counted and stored in 100% ethanol in a freezer for further molecular characterization. Molecular characterization was realized using part of the mitochondrial gene cytochrome c oxidase subunit 1 (COI), all technical details being given in Warot (2018) and Ion Scotta (2019).

9.2.2 Main results

A total of 9924 patches were set in the field and 9397 were recovered; 53% of recovered patches had suffered from strong predation pressure (less than 25% of the initial quantity of eggs left on the patch). Predation was observed directly from Hymenoptera (ants, Vespoidae), Mecoptera and Coleoptera (ladybirds) and suspected from Orthoptera and even molluscs (snail and/or slugs). The highest rates of predation (almost 90%) were observed in July, August and September while the lowest (around 10%) were observed in November. Among the weakly or non-predated patches (predation lower than

50%), only 1.8% were parasitized for a total of 784 identified *Trichogramma*. The *Trichogramma* genus nevertheless appeared to be widely distributed insofar as at least one individual was found in 67% of the sites.

Although differences between such low values (parasitism rates, occurrences, numbers of individuals) have to be treated cautiously, the presence of *Trichogramma* tended to be higher in the Meso- than in the Supra-Mediterranean stage since, for the same habitats (forest, shrubland and grassland), the first provided 63% of the total of individuals. The temporal dynamics of *Trichogramma* recoveries also appeared to vary between these two bioclimatic levels. In the Meso-Mediterranean, two peaks were observed, the first during May–June (26% of the individuals) and a second during October and early November (52% of the individuals). In the Supra-Mediterranean stage, the temporal dynamics was 'flatter' and most of the individuals (48%) were collected in May–June. Among natural habitats, shrublands appeared to be more favourable than others, both in the Meso- and the Supra-Mediterranean stages (respectively 48% and 52% of the individuals) followed by grasslands (respectively 36% and 29%). In the Meso-Mediterranean stage, cultivated areas provided quite similar results to the most favourable natural habitats (shrublands).

At the plant level, no reproducible pattern was observed, a list of plant species where *Trichogramma* were collected being provided in Ion Scotta (2019).

A total of 692 COI sequences were obtained from the *Trichogramma* sampling. They corresponded to 38 different haplotypes (Fig. 9.1). Among them, 11 (*i.e.* 31% of the haplotypes, less than 2% of the individuals) were observed only once and thus considered as technical artefacts. Moreover, seven haplotypes (18% of the haplotypes and 13% of the individuals) exhibited non-synonymous variations in the amino acid sequence and were thus putative pseudogenes and/or technical artefacts, the amino acid sequence being conserved, to our knowledge, at the *Trichogramma* genus level. Other COI haplotypes were considered as valid and compared with our reference database (see Warot, 2018 and Ion Scotta, 2019 for details). This allows us to identify eleven 'molecular entities', nine of them being associated to a species names (*T. bourarachae*, *T. cacoeciae*, *T. cordubensis*, *T. dendrolimi*, *T. evanescens*, *T. oleae* and *T. principium*) or to a species complex ('brassicae-euproctidis' and 'cacoeciae-embryophagum'). Finally, two molecular entities (arbitrarily named 'misB' and 'mis D') presented no strong similarities with previously identified species or species complex.

More precisely, regarding the complex 'cacoeciae – embryophagum':

- investigations on strains initiated from field-caught individuals highly suggest that the individuals correspond to *T. cacoeciae sensu stricto*
- a subclustering was, however, observed, with two subclusters (subcluster 1, 47% of the individuals, and subcluster 2, 45%) and one isolated haplotype ('Hap_076', 8% of the individuals) observed once in the Meso-Mediterranean level
- the two subclusters presented quite similar distributions along the transect and were mainly found in the Meso-Mediterranean level (respectively 61% and 78% of the occurrences and 65% and 82% of the individuals).

Based on this molecular characterization, we *a posteriori* estimated that 85% of parasitized host patches were parasitized by only one *Trichogramma* species, others being parasitized by two (12%) or even three species (3%). As shown in Table 9.1, the most abundant taxon was *T. cacoeciae*, be it in terms of number of sites (40% of the sites), occurrences or total number of individuals. Then, *T. cordubensis* and *T. evanescens* were present in a similar number of sites and occurrences, the first one being more abundant. In decreasing order, we then observed the four following taxa: 'misB', 'brassicae-euproctidis', *T. bourarachae* and *T. principium*. Finally, 'misD' was observed only once. Taken as a whole, the species' diversity of *Trichogramma* qualitatively (presence/absence) varied according to bioclimatic zone. Among the more abundant species, all 'brassicae-euproctidis', almost all *T. cordubensis* and 93% of *T. evanescens* were observed in the Meso-Mediterranean level. Even if the numbers of occurrences are lower and have thus to be treated with caution, an opposite pattern

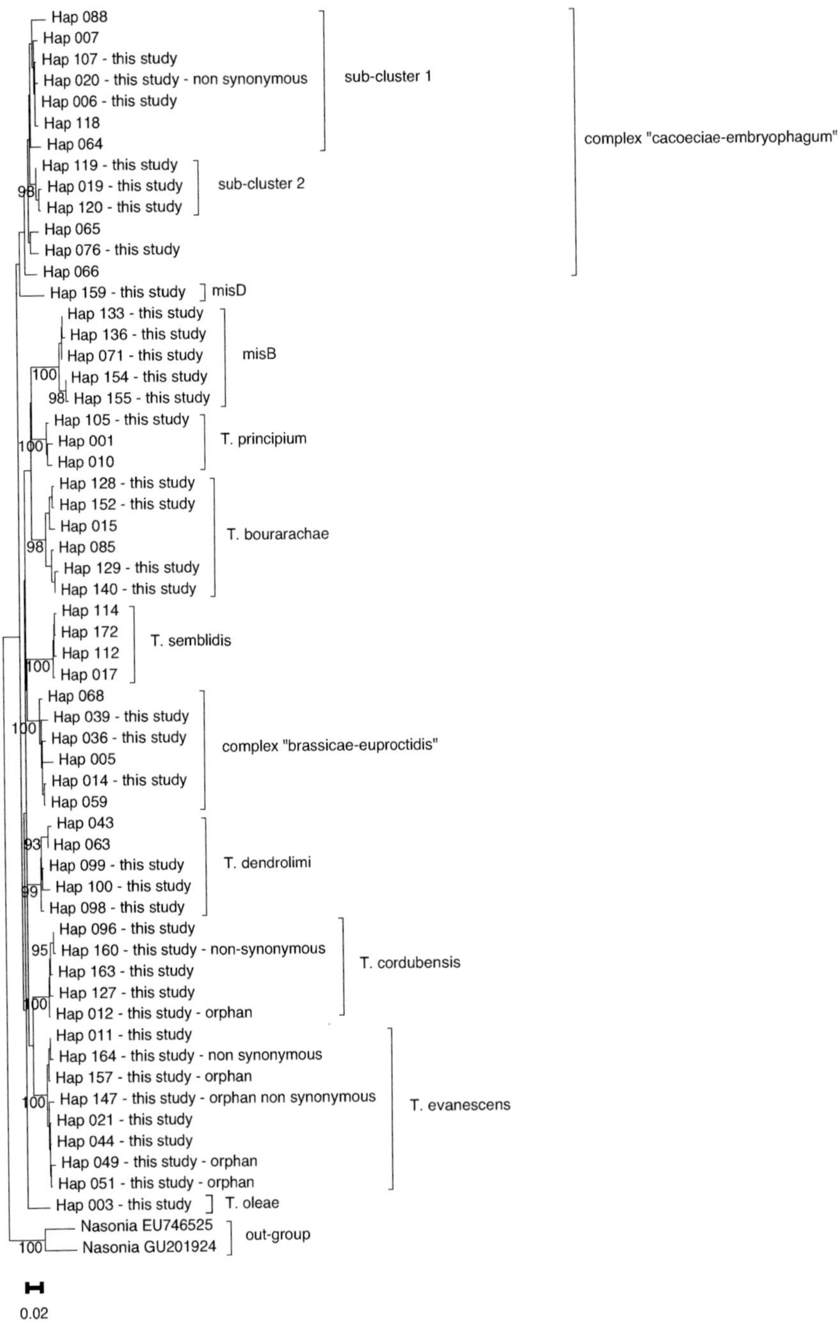

Fig. 9.1. Neighbour-joining tree inferred from the COI haplotypes. Details on the COI haplotypes and related reference strains are given in Warot (2018) and Ion Scotta (2019).The label 'this study' refers to haplotypes found during the field sampling, putative artefacts or pseudogenes being indicated by the labels 'orphan' (haplotypes observed only once) and 'non-synonymous' (nucleotidic mutations associated with changes in the amino-acid sequence). Adapted from Ion Scotta *et al*. (2021). Used under licence CC BY 4.0; www.mdpi.com/2075-4450/12/11/1013.

Table 9.1. Summary of *Trichogramma* occurrences and abundancies in the different bioclimatic zones (Meso- vs Supra- Mediterranean) and habitats (cultivated areas, grassland, shrubland and forest). For each environmental combination (bioclimatic zone × habitats) and each *Trichogramma* species, three parameters are indicated: (i) the total number of sites where parasitism was observed (left), the total number of parasitized patches (middle) and the total number of collected individuals (right).

	Bioclimatic zone - Habitats							
	Meso-				Supra			
	Cultivated areas	Forest	Shrubland	Grassland	Forest	Shrubland	Grassland	Total
Species' levels								
T. bourarachae	0/0/0	0/0/0	1/1/3	0/0/0	0/0/0	1/4/23	0/0/0	2/5/26
«*brassicae – euproctidis*»	2/2/8	1/1/12	1/2/19	1/1/18	0/0/0	0/0/0	0/0/0	5/6/57
T. cacoeciae	2/2/18	2/3/51	3/9/75	4/8/51	2/5/18	2/3/26	2/3/23	17/33/262
T. cordubensis	4/11/83	0/0/0	2/4/25	4/5/53	1/1/01	0/0/0	0/0/0	11/21/162
T. evanescens	4/10/64	0/0/0	3/6/20	2/2/07	0/0/0	2/4/07	0/0/0	11/22/98
T. principium	2/2/8	0/0/0	0/0/0	0/0/0	0/0/0	0/0/0	0/0/0	2/2/8
misB	0/0/0	0/0/0	1/2/10	0/0/0	0/0/0	4/6/42	2/2/18	7/10/70
misD	0/0/0	0/0/0	1/1/09	0/0/0	0/0/0	0/0/0	0/0/0	1/1/9
Genus' level								
Locations	5	3	4	5	2	5	4	
Parasitized patches	23	4	23	15	5	15	6	
Total of individuals	200	66	194	144	20	107	53	

was observed for *T. bourarachae* and 'misB' for which most of the individuals were found in the Supra-Mediterranean level.

9.2.3 Intermediate take-home message

At a short geographical scale, we evidenced a noteworthy species diversity with local co-occurrences of different species, even on the same plants, and thus competitive interactions between species with quite similar biology. We also documented that the bioclimatic consequences of higher altitude negatively impact the *Trichogramma* community both quantitatively (more individuals at lower altitude) and qualitatively (higher species diversity at lower altitude). Bioclimatic conditions in the Supra-Mediterranean level thus undoubtedly constitute a challenge for most of the *Trichogramma* species.

Among the identified species, two types can be distinguished in terms of distribution areas (Pintureau, 2008; Warot, 2018):

- on one side, widely distributed species at, at least, European level (including septentrional countries like Germany and Great Britain): *T. cacoeciae*, *T. dendrolimi*, *T. evanescens*, *T. principium* and the species complex 'brassicae-euproctidis'
- on the other, more septentrional species (only reported in Mediterranean countries): *T. bourarachae*, *T. cordubensis* or *T. oleae*.

Only one species (the so-called 'misD') may be specialized on conditions present in the Supra-Mediterranean level. The extent to which this specialization is linked to climatic and/or ecological (host insect, host plant, habitat, etc.) conditions has to be further investigated but this persistence of this species could be at risk in the future.

Trichogramma cacoeciae was the sole species noted to be present in the two climatic levels, this species being known to be widely distributed (at least) in Europe and around the Mediterranean Sea. This species was thus chosen to investigate further possible local adaptations to temperature.

9.3 Population Differentiation of *T. cacoeciae* in South-Eastern France

9.3.1 Experimental designs

Three experiments were done on a common set of 40 *T. cacoeciae* strains (see details in Ion Scotta *et al.*, 2021). Most of these strains originated from the sampling described above, either in the Meso-Mediterranean level or the Supra-Mediterranean level. This set was completed by other more septentrional strains (referred as to 'Northern' here). All the strains were maintained by the Biological Resource Centre egg Parasitoids Collection (https://crb-eggparasitoids-coll.hub.inrae.fr/) prior to their use. The conditions of rearing (temperature, photoperiod, hosts, etc.) prior to each of the experiments were standardized for all strains, preventing possible biases.

The three designed experiments were:

- measure of cold thermal tolerance indices in a laboratory setting
- measure of diapause behaviour with regard to low temperatures in a laboratory setting
- measure of overwintering abilities in a common-garden experiment.

All these experiments are detailed elsewhere (Ion Scotta, 2019; Ion Scotta *et al.*, 2021) and we only report below the main information. All the results are provided here: https://entrepot.recherche.data.gouv.fr/dataset.xhtml?persistentId=doi:10.57745/XLCCED.

Cold thermal tolerance indices

Thermal tolerance indices describe the ability of individuals to maintain locomotion in stressful (either cold or warm) conditions (Cooper *et al.*, 2008; Angilletta, 2009; Andersen *et al.*, 2015). With regard to cold conditions, the critical thermal minimum (CTmin) is the temperature at which individuals lose their locomotor activity while the chill coma temperature (CCT) is the temperature at which individuals lose all movement. After this step, a progressive warming allows estimation of activity recovery (AR), *i.e.* the temperature at which individuals recover locomotor activity. Several authors emphasize the relevance of these thermal tolerance indices to experimentally investigate interspecific and, to a lesser extent, intraspecific variabilities (see, for instance, Gibert *et al.*, 2001; Castañeda *et al.*, 2005; Andersen *et al.*, 2015).

Because of their small size, investigating the thermal tolerance indices in *Trichogramma* was challenging and a specific prototype, the so-called Thermo-arena (licence granted by INRAE to the private company INNOFENSO), was created for this purpose. This prototype allows a progressive decrease of the temperature from 17°C to −2°C or −4°C with a controlled rate of 0.3 °C/min in a small cell where about 50 *Trichogramma* young adults were previously inserted. Once the minimum was reached, a progressive warming started with a controlled rate of 0.3 °C/min until the temperature reached 14°C. During this time, the behaviours of *Trichogramma* were recorded using a digital video camera (Nikon D800 digital camera equipped with a 60 mm 1:2.8 G ED AF-S Micro Nikkor lens), the light being provided from the bottom. All the video sequences were then carefully checked and only the behaviour of individuals (about 15–20) located on the floor of the arena and well visible were taken into account to estimate the thermal tolerance indices. All the *Trichogramma* strains were tested only once.

Diapause induction

Diapause is an adaptive strategy frequently observed in species living in climates with marked and predictable seasons (Denlinger, 2002; Hodek, 2002; Danks, 2007; Denlinger, 2022). This arrest of development or activity enables the insects to survive the winter period and sometimes even a summer period when abiotic conditions, such as temperature, are unfavourable and/or resources are scarce or even unavailable. Diapause may be obligatory or facultative. In the second case, diapause occurs in response to environmental signals. If a certain environmental signal is received during a sensitive period, the insect will enter diapause. This type of diapause is typical of multivoltine species and is induced only in the generation preceding the onset of winter (Denlinger 2002, 2022). Diapause has been observed and even exploited in *Trichogramma* but this ability seems to exhibit both inter- and intraspecific variabilities (see, for instance, Garcia *et al.*, 2002; Reznik *et al.*, 2008; Rahimi-Kaldeh *et al.*, 2018; Zhang *et al.*, 2018).

This experiment was carried out in collaboration with BIOLINE Agrosciences (Benoit Frank, Corinne Frenoy and Paloma Martinez-Rodriguez), a biocontrol agent manufacturer that has developed an expertise in the diapause of *Trichogramma*. However, the different strains were provided to BIOLINE Agrosciences in a pre-imaginal state. BIOLINE Agrosciences stored them in conditions close to those at INRAE. The strains were synchronized by making minor differences in development temperatures based on degree-days. The induction of diapause is a trade secret of BIOLINE Agrosciences and thus kept confidential. After induction, the strains were stored at low temperature (around 3°C) for varying storage periods. For the first 6 months, a fraction of the individuals from each strain was returned each month to conditions compatible with the termination of diapause and adult emergence (22°C and L16:D8). After the first 6 months, the tests of emergence were conducted every 2 months until 12 months. For each strain and each storage period, the corresponding fraction of parasitized *E. kuehniella* eggs was photographed with a Nikon D750 camera so that the parasitized host could be counted later using the 'Cell Counter' manual counting option in the Image J software. The emergence kinetics of the adult *Trichogramma* were then observed daily. Seven days after the first emergences, all the adult individuals were frozen at −20°C and counted.

Overwintering abilities

In regions with marked seasonality (temperate, mountain or even polar climates), the ability to survive the winter (hereafter referred to as 'overwintering') is a major issue for insects at both individual and population levels, as low temperatures have a negative impact not only on several of their phenotypic traits but also on the resources provided by lower trophic levels (Boivin, 1994; Hodkinson, 2005). If laboratory investigations tend to indicate a variability in diapause potential across populations of *T. cacoeciae* (see previous paragraph), its actual expression in the field is not certain. Moreover, other types of resistance to cold may exist.

Following past experiments on overwintering (see, for instance, Rossi and Pizzol, 1997; Babendreier *et al.*, 2003; Özder and Saglam,

2005), we thus designed a 'common-garden' experiment during the winter of 2017–2018 to evaluate the intraspecific variability of this traits. The first 'common garden' was located in the Meso-Mediterranean level at an altitude of 200m (Sophia-Antipolis) while the second was placed in the Supra-Mediterranean zone at an altitude of 800m (Cipières). One Logtag® temperature recorder was placed in each garden at the beginning of the experiment. As expected, this recording evidenced a higher temperature of 4–5°C throughout the survey in the Meso-Mediterranean garden in contrast to the Supra-Mediterranean garden where the coldest mean temperature was observed in February (about 4°C). Within this context, the *T. cacoeciae* strains were placed at three different dates (November 2017, December 2017 and January 2018) in each of the two gardens as follows. A patch (0.8 mm ×2 cm) covered with eggs of *E. kuehniella* was offered to the females of each strain in a glass tube sealed with a piece of cotton. After 24 hour, the females were removed and the patch was divided into three equal parts. Each part was placed in a new glass tube. Of the three new tubes, one was kept as a laboratory control at 23°C, 75 ± 10% relative humidity and a photoperiod of L16:D8. The other two tubes were placed under cover, one in each of the common gardens. For each strain, each exposure date and each garden, the tubes kept outdoors were checked every week in autumn and spring, and every fortnight in winter. The dates of *Trichogramma* emergence were recorded and the development time was calculated.

9.3.2 Main results

Trait-by-trait analysis

Results on thermal tolerance indices were extensively presented in Ion Scotta *et al.* (2021). No significant differences were found between the three geographical origins (Meso-Mediterranean, Supra-Mediterranean and Northern) of the *T. cacoeciae* strains for the CTmin and the AR. A significant but quite counterintuitive difference was, however, seen for the CCT. Indeed, strains from the Supra-Mediterranean origin exhibited higher CCT than strains from the Northern origin. Another significant result of this study was the positive correlation observed between CTmin and CCT.

Regarding the 'diapause induction' experiment, three types of relationships between the storage duration at low temperature and the percentage of emergence of *Trichogramma* adults were evidenced.

- A 'logistic' response characterized by a low rate of emergence during the first weeks/months of storage followed by high and relatively constant emergence rates. This pattern was unambiguously attributed to diapausing strains.
- A rather 'linearly decreasing' response characterized by a maximal rate of emergence for short storage duration followed by a regular decrease of the rate of emergence with increasing storage duration. This pattern was attributed to non-diapausing strains.
- An intermediate response whose biological meaning remains unclear, a diversified bet-hedging (Hopper, 1999) being a possible explanation.

The proportion of these three qualitative responses (diapause, intermediate or quiescence) varied according to the geographical origins (Meso-Mediterranean, Supra-Mediterranean and Northern) of the strains (Fisher's exact test on a contingency table: p=0.009). Diapausing strains were more frequent in the Supra-Mediterranean and Northern origins (respectively 50 and 100%) than in the Meso-Mediterranean origin (15%).

Quite surprisingly, all the strains were able to overwinter in both common gardens, regardless of the date of exposure and the origin of the strain. This seems to indicate that cold temperatures do not constitute a challenge for *T. cacoeciae* overwinter survival, even in the Supra-Mediterranean level. More subtle information related to the observed development time was, however, acquired during this experiment. The first interesting point was the observation of a linear relationships between the mean temperature observed in the field and the development time. This means that, although diapause was possible for some of the strains (see previous section), it was not expressed here. Moreover, it appears that the two parameters (the slope and the intercept) of the 'mean temperature – observed development time' regression were highly correlated (Spearman's rank correlation test: r = −0981 with $p<10^{-3}$). Finally the values of intercepts and slopes also varied between geographical origins (Kruskal Wallis tests: intercept XXII^2_{2df} = 7.26 with p=0.027; slope XXII^2_{2df} = 7.55 with p=0.023). Meso-Mediterranean strains tend to have higher intercepts and correlatively lower slopes, the only significant *post hoc* test being observed between the slopes of the Meso-Mediterranean strains and those of the Supra-Mediterranean strains (*post hoc* Wilcoxon test adjusted with Bonferroni corrections: p=0.044). Biologically speaking, Supra-Mediterranean strains thus tend to develop more slowly than others at low temperatures, this difference disappearing or even being inverted as the mean temperature increases. Because this pattern was observed under very variable temperatures, the extrapolation of classical parameters used in degree-day models (Hagstrum and Milliken, 1991) is not straightforward but undoubtedly deserves more attention.

Multi-traits analysis

The deliberate choice to perform the three experiments with the same set of strains allows aggregation of results from independent experiments and different traits at the strain level. This was the case for 27 strains for which all the information was obtained. In order to take into account the different natures (qualitative or quantitative) of the response variables in the three experiments, all the quantitative variables (AR, CC and CTmin for the 'thermal tolerance indices' experiment and the intercept for the 'overwintering' experiment) were transformed into qualitative variables. For each, three categories were distinguished (A, B and C) respectively including to the first quartile, the interquartile and the fourth quartile. A multiple correspondence analysis was then performed using R software ('ade' package). The cumulative projected inertia related to the two first axes reached 35%. As shown in Fig. 9.2, this approach evidenced a clustering of the different strains according to their geographical origins (Meso-Mediterranean, Supra-Mediterranean and Northern). In particular, the diapause ability and lower intercepts ('overwintering' experiment) seemed to characterize the Northern strains.

Intermediate take-home message

As shown in the previous paragraph, trait-by-trait analysis evidenced some genetic differentiations according to geographic origin. These were rather intuitive (diapause), more complex but finally meaningful (overwintering) or rather counterintuitive (thermal tolerance indices). With regard to the two bioclimatic levels along the altitudinal transect, population differentiation does exist for at least two of the three (types of) traits (diapause and overwintering). With regard to the two scenarios about the (meta-)population biology of *Trichogramma* species, this at least suggests that local selective pressure in the supposedly more 'hostile' Supra-Mediterranean level is strong enough to maintain a differentiation even with possible gene flows. Although preliminary, our meta-analysis of the different experiments on a single set of strains tends to suggest the existence of an adaptive syndrome to low temperature at least between the propensity to diapause and the ability to develop fast at low temperature.

9.4 Conclusion

The results presented here constitute a first contribution in order to understand eco-evolutionary processes at work in insect communities of agricultural relevance in response to climate change. For such a risky long-term initiative to be successful, three technical/methodological criteria were required: (i) a feasible sampling design, (ii) a relevant environmental gradient, and (iii) sufficiently accurate phenotyping approaches. Luckily, all these three criteria were met *a posteriori*. Indeed, even if quite time-consuming and undoubtedly not exhaustive, the study allowed sampling of several *Trichogramma* species with different geographical distributions. Among them, one species (*T. cacoeciae*) was widely distributed along the altitudinal gradient and elsewhere.

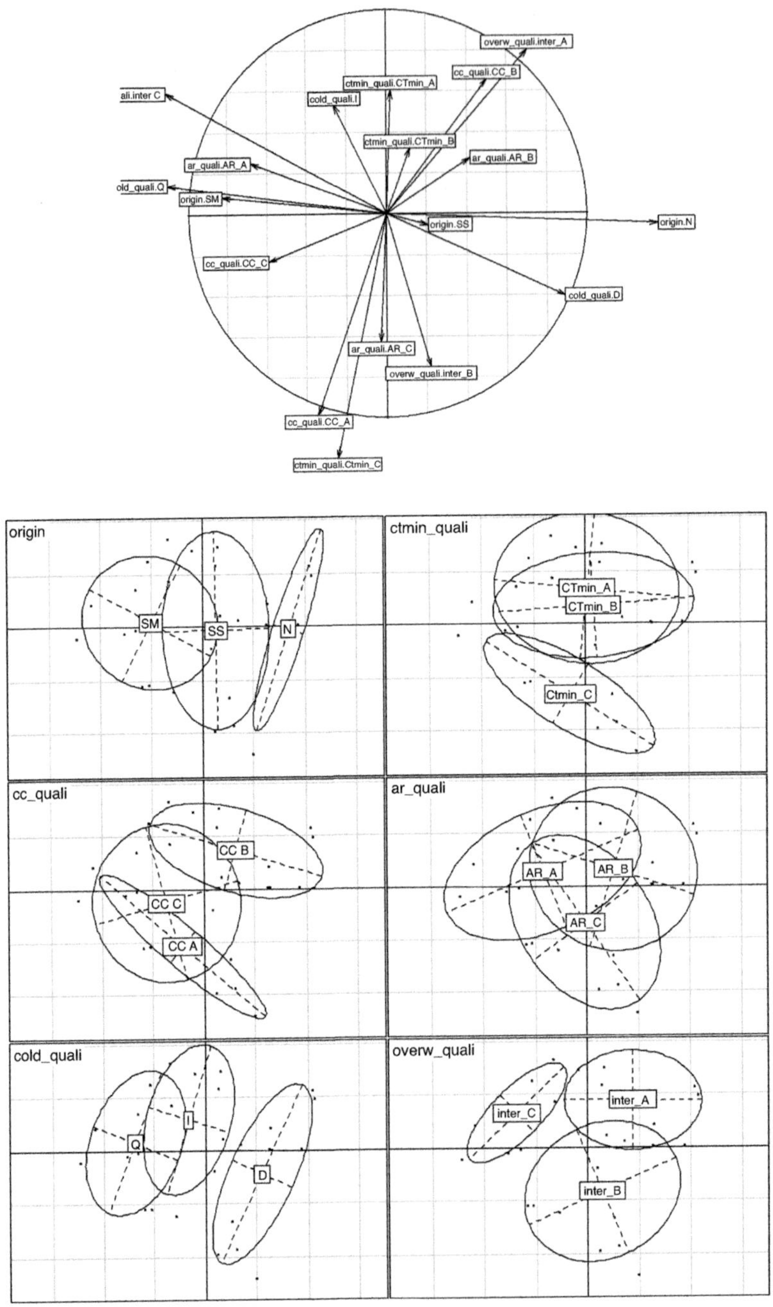

Fig. 9.2. Graphical outputs obtained from the multiple correspondence analysis (MCA) which was performed using four variables: (i) the geographic origin: Meso-Mediterranean (SM), Supra-Mediterranean (SS) and Northern (N); (ii) the cold thermal tolerance indices AR, CC and CTmin; (iii) the qualitative profile – diapause (D), intermediate (I) and quiescence (Q) – obtained from the 'diapause induction' experiment; (iv) the intercept (inter) extrapolated from the linear regression between mean temperature and development time from the 'overwintering' experiment. All quantitative variables were categorized as follow: A = first quartile; B = second and third quartiles; C = fourth quartile.

Finally, the phenotyping approaches used here (thermal tolerance, diapause/quiescence and overwintering) were accurate enough to discriminate the three strains' geographical origins (Meso-Mediteranean, Supra-Mediterranean and Northern). Surprisingly, patterns of population differentiation were not always as intuitive as expected (more similarities between Supra-Mediterranean and Northern origins). Why the local altitudinal gradient does not exactly mimic the more global latitudinal gradient is quite intriguing. One hypothesis relies on undetected subtle climatic differences between the two bioclimatic zones (harsher winter conditions in the Northern bioclimatic area) which could explain why diapause is pervasive in Northern strains and less frequent in Supra-Mediterranean strains. This pattern could, however, also be explained by gene flows from the Meso-Mediterranean towards the close Supra-Mediterranean.

Despite being potentially blurred by statistical analysis (transformation of quantitative continuous variables into qualitative variables), the research on some adaptive syndromes to cold conditions was quite encouraging. In particular, the responses to diapause induction and parameters (intercepts and slopes) of the regression between the mean temperature and the observed development time appear to covary. The physiological and/or genetic (possible pleiotropy) bases of this pattern remain to be understood. In understanding the patterns of population differentiation and covariations between traits, the extent to which adaption to cold temperatures can be traded off by a maladaptation at higher temperatures is of course central. Within this frame, investigations are ongoing to understand to what extent thermal tolerance indices at cold and hot temperatures are negatively correlated.

References

Adrion, J.R., Hahn, M.W. and Cooper, B.S. (2015) Revisiting classic clines in *Drosophila melanogaster* in the age of genomics. *Trends in Genetics* 31, 434–444.

Andersen, J.L., Manenti, T., Sorensen, J.G., Macmillan, H.A., Loeschcke, V. *et al.* (2015) How to assess *Drosophila* cold tolerance: Chill coma temperature and lower lethal temperature are the best predictors of cold distribution limits. *Functional Ecology* 29, 55–65.

Angilletta, M.J. (2009) *Thermal Adaptation: A Theoretical and Empirical Synthesis*. Oxford University Press, New York.

Angilletta, M.J., Niewiarowski, P.H. and Navas, C.A. (2002) The evolution of thermal physiology in ectotherms. *Journal of Thermal Biology* 27, 249–268.

Babendreier, D., Kuske, S. and Bigler, F. (2003) Overwintering of the egg parasitoid *Trichogramma brassicae* in Northern Switzerland. *Biocontrol* 48, 261–273.

Blondel, J. and Aronson, J. (2011) *Biodiversity and Ecosystem Function in the Mediterranean Basin: Human and Non-Human Determinants*. Oxford University Press, New York.

Boivin, G. (1994) Overwintering strategies of egg parasitoids. In: Wajnberg, E. and Hassan, S.A. (eds) *Biological Control with Egg Parasitoids*. CABI, Wallingford, UK, pp. 219–244.

Castañeda, L.E., Lardies, M.A. and Bozinovic, F. (2005) Interpopulational variation in recovery time from chill coma along a geographic gradient: A study in the common woodlouse, *Porcellio laevis*. *Journal of Insect Physiology* 51, 1346–1351.

Cock, M.J.W., Van Lenteren, J.C., Brodeur, J., Barratt, B.I.P., Bigler, F. *et al.* (2010) Do new access and benefit sharing procedures under the Convention on Biological Diversity threaten the future of biological control? *Biocontrol* 55, 199–218.

Cooper, B.S., Williams, B.H. and Angilletta, M.J. (2008) Unifying indices of heat tolerance in ectotherms. *Journal of Thermal Biology* 33, 320–323.

Danks, H.V. (2007) The elements of seasonal adaptations in insects. *Canadian Entomologist* 139, 1–44.

Delava, E., Allemand, R., Léger, L., Fleury, F. and Gibert, P. (2014) The rapid northward shift of the range margin of a *Mediterranean parasitoid* insect (Hymenoptera) associated with regional climate warming. *Journal of Biogeography* 41, 1379–1389.

Denlinger, D.L. (2002) Regulation of diapause. *Annual Review of Entomology* 47, 93–122.

Denlinger, D.L. (2022) *Insect Diapause*. Cambridge University Press, Cambridge, UK.

Eilenberg, J., Hajek, A. and Lomer, C. (2001) Suggestions for unifying the terminology in biological control. *Biocontrol* 46, 387–400.

Fatouros, N.E. and Huigens, M.E. (2012) Phoresy in the field: Natural occurrence of *Trichogramma* egg parasitoids on butterflies and moths. *Biocontrol* 57, 493–502.

Ferrer, A., Dorn, S. and Mazzi, D. (2013) Cross-generational effects of temperature on flight performance, and associated life-history traits in an insect. *Journal of Evolutionary Biology* 26, 2321–2330.

Fleury, F., Gibert, P., Ris, N. and Allemand, R. (2009) Ecology and life history evolution of frugivorous *Drosophila* parasitoids. *Advances in Parasitology* 70, 3–44.

Garcia, P.V., Wajnberg, E., Pizzol, J. and Oliveira, M.L.M. (2002) Diapause in the egg parasitoid *Trichogramma cordubensis*: Role of temperature. *Journal of Insect Physiology* 48, 349–355.

Gibert, P., Moreteau, B., Pétavy, G., Karan, D. and David, J.R. (2001) Chill-coma tolerance, a major climatic adaptation among *Drosophila* species. *Evolution* 55, 1063–1068.

Hagstrum, D.W. and Milliken, G.A. (1991) Modeling differences in insect developmental times between constant and fluctuating temperatures. *Annals of the Entomological Society of America* 84, 369–379.

Hodek, I. (2002) Controversial aspects of diapause development. *European Journal of Entomology* 99, 163–173.

Hodkinson, I.D. (2005) Terrestrial insects along elevation gradients: Species and community responses to altitude. *Biological Reviews* 80, 489–513.

Hopper, K.R. (1999) Risk-spreading and bet-hedging in insect population biology. *Annual Review of Entomology* 44, 535–560.

Huey, R.B. and Hertz, P.E. (1984) Is a jack-of-all-temperatures a master of none? *Evolution* 38, 441–444.

Ion Scotta, M. (2019) Distribution des espèces du genre *Trichogramma* le long d'un gradient altitudinal et adaptations locales aux basses températures chez l'espèce Trichogramma cacoeciae. PhD thesis, Université Côte d'Azur, Nice, France.

Ion Scotta, M., Margris, L., Sellier, N., Warot, S., Gatti, F. *et al.* (2021) Genetic variability, population differentiation, and correlations for thermal tolerance indices in the minute wasp, *Trichogramma cacoeciae*. *Insects* 12, 17.

Jego, L., Li, R.N., Roudine, S., Ma, C.S., Le Lann, C. *et al.* (2023) Parasitoid ecology along geographic gradients: Lessons for climate change studies. *Current Opinion in Insect Science* 57, 8.

Körner, C. (2007) The use of altitude in ecological research. *Trends in Ecology & Evolution* 22, 569–574.

Kuske, S., Widmer, F., Edwards, P.J., Turlings, T.C.J., Babendreier, D. *et al.* (2003) Dispersal and persistence of mass released *Trichogramma brassicae* (Hymenoptera: Trichogrammatidae) in non-target habitats. *Biological Control* 27, 181–193.

Liu, Q.Q., Zhou, J.C., Zhang, C., Ning, S.F., Duan, L.J. *et al.* (2019) Co-occurrence of thelytokous and bisexual *Trichogramma dendrolimi* Matsumura (Hymenoptera: Trichogrammatidae) in a natural population. *Scientific Reports* 9, 14.

Mashal, S., Agamy, E., El-Husseini, M., Abd El-Wahab, T. and El Behery, H. (2024) Improving quality of stored *Trichogramma evanescens* (Hymenoptera: Trichogrammatidae) by inducing diapause or quiescence. *Egyptian Journal of Biological Pest Control* 34, 10.

Nagarkatti, S. and Nagaraja, H. (1977) Biosystematics of *Trichogramma* and *Trichogrammatoidea* species. *Annual Review of Entomology* 22, 157–176.

Noble, V., Diadema, K. and Offerhaus, B. (2011) *La flore des Alpes-Maritimes et de la principauté de Monaco: Originalité et diversité*. Naturalia Publications, Turriers, France.

Özder, N. and Saglam, Ö. (2005) Overwintering of the egg parasitoids *Trichogramma brassicae* and *T. cacoeciae* (Hymenoptera: Trichogrammatidae) in the Thrace region of Turkey. *Journal of Pest Science* 78, 129–132.

Parra, J.R.P. and Zucchi, R.A. (2004) *Trichogramma* in Brazil: Feasibility of use after twenty years of research. *Neotropical Entomology* 33, 271–281.

Partridge, L., Barrie, B., Barton, N.H., Fowler, K. and French, V. (1995) Rapid laboratory evolution of adult life-history traits in *Drosophila melanogaster* in response to temperature. *Evolution* 49, 538–544.

Pinto, J.D. and Stouthamer, R. (1994) Systematics of the Trichogrammatidae with emphasis on *Trichogramma*. In: Wajnberg, E. and Hassan, S.A. (eds) *Biological Control with Egg Parasitoids*. CABI, Wallingford, UK, pp. 1–36.

Pintureau, B. (2008) *Les espèces européennes de Trichogrammes*. In Libro Veritas, pp. 978–2. ISBN: 978-2-35209-181-3

Pratt, J.D., Datu, A., Tran, T., Sheng, D.C. and Mooney, K.A. (2017) Genetically based latitudinal clines in *Artemisia californica* drive parallel clines in arthropod communities. *Ecology* 98, 79–91.
Querino, R.B., Zucchi, R.A. and Pinto, J.D. (2010) Systematics of the *Trichogrammatidae* (Hymenoptera: Chalcidoidea) with a focus on the genera attacking Lepidoptera. In: Consoli, F.L., Parra, J.R.P. and Zucchi, R.A. (eds) *Egg Parasitoids in Agroecosystems with Emphasis on Trichogramma*. Springer, Dordrecht, The Netherlands, pp. 191–218.
Rahimi-Kaldeh, S., Ashouri, A., Bandani, A. and Ris, N. (2018) Abiotic and biotic factors influence diapause induction in sexual and asexual strains of *Trichogramma brassicae* (Hym: Trichogrammatidae). *Scientific Reports* 8, 6.
Rezende, E.L., Castañeda, L.E. and Santos, M. (2014) Tolerance landscapes in thermal ecology. *Functional Ecology* 28, 799–809.
Reznik, S.Y. and Voinovich, N.D. (2024) Influence of temperature on the photoperiodic time measurement and on the maternal induction of diapause in *Trichogramma telengai*: The separation of the two effects. *Journal of Insect Physiology* 155, 7.
Reznik, S.Y., Vaghina, N.P. and Voinovich, N.D. (2008) Diapause induction in *Trichogramma embryophagum* Htg. (Hym., Trichogrammatidae): The dynamics of thermosensitivity. *Journal of Applied Entomology* 132, 502–509.
Romeis, J., Babendreier, D., Wäckers, F.L. and Shanower, T.G. (2005) Habitat and plant specificity of *Trichogramma* egg parasitoids: Underlying mechanisms and implications. *Basic and Applied Ecology* 6, 215–236.
Rossi, M.M. and Pizzol, J. (1997) Development of *Trichogramma cacoeciae* and *T. evanescens* during autumn and winter under natural conditions in southern France. *Journal of Applied Entomology-Zeitschrift Fur Angewandte Entomologie* 121, 29–36.
Shelomi, M. (2012) Where are we now? Bergmann's rule *sensu lato* in insects. *American Naturalist* 180, 511–519.
Shirley, M.D.F. and Sibly, R.M. (1999) Genetic basis of a between-environment trade-off involving resistance to cadmium in *Drosophila melanogaster*. *Evolution* 53, 826–836.
Sinnott-Armstrong, M.A., Deanna, R., Pretz, C., Liu, S.K., Harris, J.C. *et al.* (2022) How to approach the study of syndromes in macroevolution and ecology. *Ecology and Evolution* 12, 10.
Smith, S.M. (1996) Biological control with *Trichogramma*: Advances, successes, and potential of their use. *Annual Review of Entomology* 41, 375–406.
Stelgenga, M.J. and Fischer, K. (2007) Within- and between-generation effects of temperature on life-history traits in a butterfly. *Journal of Thermal Biology* 32, 396–405.
Trudgill, D.L., Honek, A., Li, D. and Van Straalen, N.M. (2005) Thermal time – concepts and utility. *Annals of Applied Biology* 146, 1–14.
Vavre, F., De Jong, J.H. and Stouthamer, R. (2004) Cytogenetic mechanism and genetic consequences of thelytoky in the wasp *Trichogramma cacoeciae*. *Heredity* 93, 592–596.
Viscarret, M.M., Cagnotti, C.L. and López, S.N. (2023) Evaluation of the maternal photoperiodic effect as a storage method on *Trichogramma nerudai*. *Bulletin of Insectology* 76, 95–100.
Vuarin, P., Allemand, R., Moiroux, J., Van Baaren, J. and Gibert, P. (2012) Geographic variations of life history traits and potential trade-offs in different populations of the parasitoid *Leptopilina heterotoma*. *Naturwissenschaften* 99, 903–912.
Warot, S. (2018) Caractérisation moléculaire et isolements reproducteurs chez des auxiliaires de lutte biologique. EPHE thesis, Ecole Pratique des Hautes Etudes, Paris, France.
Wright, M.G., Hoffmann, M.P., Chenus, S.A. and Gardner, J. (2001) Dispersal behavior of *Trichogramma ostriniae* (Hymenoptera: Trichogrammatidae) in sweet corn fields: Implications for augmentative releases against *Ostrinia nubilalis* (Lepidoptera: Crambidae). *Biological Control* 22, 29–37.
Xiao, Z.T., Xu, X., Liang, Q.L., Zhu, C.Y., He, X.Y. *et al.* (2024) *Wolbachia* titers are related to sex determination of *Trichogramma pretiosum*. *Journal of Pest Science* 97, 1879–1888.
Zang, L.S., Wang, S., Zhang, F. and Desneux, N. (2021) Biological control with *Trichogramma* in China: History, present status, and perspectives. *Annual Review of Entomology* 66, 463–484.
Zhang, J.J., Zhang, X., Zang, L.S., Du, W.M., Hou, Y.Y. *et al.* (2018) Advantages of diapause in *Trichogramma dendrolimi* mass production on eggs of the Chinese silkworm, *Antheraea pernyi*. *Pest Management Science* 74, 959–965.
Zhou, J.C., Dong, Q.J., Zhang, T.S., Duan, L.J., Ning, S.F. *et al.* (2019) Effect of wind time on the dispersal capacity of *Trichogramma dendrolimi* Matsumura (Hymenoptera Trichogrammatidae). *Journal of Asia-Pacific Entomology* 22, 742–749.

10 Dynamics and Evolution of Fungal and Oomycete Plant Pathogens and Biological Control Agents under Climate Change

Ireneo B. Pangga*, John Bethany M. Macasero and Romnick A. Latina
Institute of Weed Science, Entomology and Plant Pathology, College of Agriculture and Food Science, University of the Philippines Los Baños, Laguna, The Philippines

Abstract
This chapter explores the effects of climate change on the growth, development and evolutionary changes of fungal and oomycete plant pathogens and biological control agents (BCAs). The different stages of fungal and oomycete plant pathogen development from germination, latency or incubation, to sporulation are influenced by elevated CO_2, increased temperature and altered moisture conditions (either under drought or increased precipitation) and their combinations, with variable effects resulting in an increase, decrease or no change. Fungal and oomycete plant pathogen evolution under climate change showed either an increase or decrease in aggressiveness, fitness and adaptation or survival under elevated CO_2 and increased temperature. Studies on the effect of elevated CO_2, increased temperature and their combination with moisture on fungal and oomycete BCAs showed an increase or decrease in effectiveness. Coevolution of the target fungal and oomycete pathogens and their corresponding BCAs needs to be studied under climate change conditions.

10.1 Introduction

This chapter focuses on the impact of climate change on, first, plant pathogenic fungi and oomycetes and, second, their fungal and oomycete biological control agents (BCAs). For each of these parts, both the dynamics and the evolution, together with prospects for research and management of BCAs, will be presented. This chapter will not discuss the interacting effects of plant-associated factors such as changes in plant disease resistance, plant physiological processes, plant canopy architecture and microclimate which are reviewed extensively (Manning and Tiedemann, 1995; Eastburn *et al.*, 2011; Pangga *et al.*, 2011, 2013). This chapter aims to gain an understanding on climate change–plant fungal and oomycete disease and BCA interactions.

True fungi are eukaryotic, mostly microscopic, usually filamentous and spore-bearing heterotrophic organisms with branched mycelia that lack chlorophyll in the Kingdom Fungi (Eumycota) (Agrios, 2005; Oliver, 2024). Oomycetes are fungal-like, eukaryotic, filamentous heterotrophic micro-organisms with aseptate hyphae that were classified as true fungi until about 1990 but then moved to Phylum Oomycota under Kingdom Chromista (Agrios, 2005) and they belong to the

*Corresponding author: ibpangga@up.edu.ph

DOI: 10.1079/9781800625099.0010

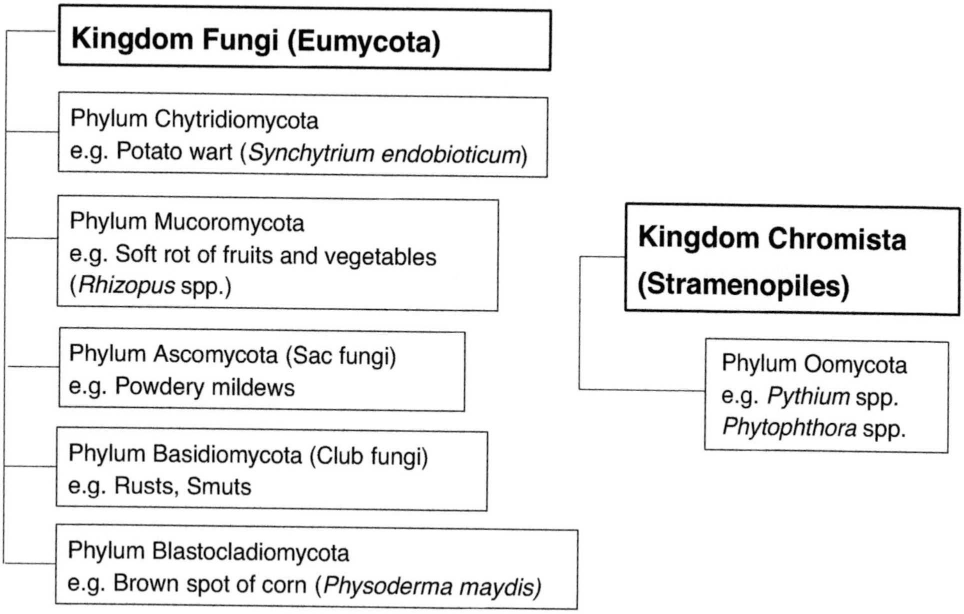

Fig. 10.1. Classification of plant pathogenic fungi and oomycetes with examples (Agrios, 2005; Tronsmo *et al*., 2020; Oliver, 2024).

Stramenopila lineage in the supergroup TSAR (Telonemia, Stramenopila, Alveolata and Rhizaria) (Govers, 2024). Most fungi have cell walls that contain chitin and glucans without cellulose while the majority of oomycetes have cell walls composed of β-glucans and cellulose, but not chitin. Biotrophic fungi and oomycetes are obligate parasites, completing their life cycle on the host plant, while necrotrophic fungi and oomycetes kill host cells before hyphal growth. Hemibiotrophs are biotrophic in the initial phase followed by necrotrophy (Agrios, 2005; Tronsmo *et al*., 2020). The classification of plant pathogenic fungi and oomycetes is shown in Fig. 10.1.

10.2 Climate Change Effects on Fungal and Oomycete Dynamics

Plant fungal and oomycete disease development under elevated CO_2 (Luck *et al*., 2011; Debela and Tola, 2018; Roy and Mathur, 2021; Maurya *et al*., 2022; Wang *et al*., 2022), increased temperature (Luck *et al*., 2011; Debela and Tola, 2018; Maurya *et al*., 2022; Wang *et al*., 2022) and altered moisture conditions (Wang *et al*., 2022) showed increase, decrease or no effects (Juroszek *et al*., 2020).Variable results were observed for interaction studies of these climate change variables (Gullino *et al*., 2018). These variable results are influenced by plant factors such as disease resistance or immunity (Velasquez *et al*., 2018; Smith and Luna, 2023), plant–pathogen (biotic) interactions (Kazan, 2018) and canopy microclimate (Manning and Tiedemann, 1995; Eastburn *et al*., 2011; Pangga *et al*., 2013). Furthermore, global warming and altered moisture conditions also affect the distribution and dispersal of fungal and oomycete plant pathogens.

10.2.1 Elevated CO_2

Reviews on the effect of elevated CO_2 on the growth and development of plant pathogenic fungi and oomycetes indicate a range of results: positive, negative or no effect (Manning and Tiedemann, 1995; Grover *et al*., 2015; Debela and Tola, 2018; Roy and Mathur, 2021; Maurya *et al*., 2022; Smith and Luna, 2023). The studies reviewed in Table 10.1 show negative effects of elevated CO_2 on spore

Table 10.1. Effects of elevated CO_2 and temperature on spore germination and latent or incubation period of plant fungal and oomycete pathogens.

Crop	Pathogen	Treatment	Effect	Reference
Spore germination				
Barley	*Erysiphe graminis*	700 ppm CO_2	Reduced rates of primary penetration	Hibberd *et al.*, 1996
Stylo	*Colletotrichum gloeosporioides*	700 ppm CO_2	Reduced spore germination Delay in germ tube growth and production of appressoria	Chakraborty *et al.*, 2000
Stylo	*Colletotrichum gloeosporioides*	700 ppm CO_2	Reduced infection efficiency	Pangga *et al.*, 2004
Eucalyptus	*Puccinia psidii*	700 ppm CO_2	Decrease in rust pustules per leaf	Ghini *et al.*, 2014
Potato	*Phytophthora infestans*	Large oscillations in temperature	Decrease in infection efficiency	Shakya *et al.*, 2015
Latent/incubation period				
Stylo	*Colletotrichum gloeosporioides*	700 ppm CO_2	Longer latent and incubation periods in resistant cv. Seca and not in susceptible cv. Fitzroy	Chakraborty *et al.*, 2000
Red oak	*Cronarium quercuum*	700 ppm CO_2	Longer latent period	Runion *et al.*, 2010
Wheat	*Puccinia striiformis* f. sp. *tritici*	800 ppm CO_2	Shorter latent period Higher lesion expansion	Han *et al.*, 2011
Wheat	*Puccinia striiformis* f. sp. *tritici*	700 ppm CO_2	No change in latent period	Chakraborty *et al.*, 2011
Coffee	*Hemileia vastatrix*	Shorter diurnal thermal amplitude	Shorter latent period	Avelino *et al.*, 2015
Triticale	*Puccinia recondita* f. sp. *tritici*	Elevated temperature	Shorter latent period	Wójtowicz *et al.*, 2020

germination. The latent period of fungal plant pathogens has been observed to shorten in several rust diseases (see Table 10.1). Short generation time combined with large numbers of spores can enhance the rust fungi's adaptive and evolutionary potential under climate change (Helfer, 2014). Variable results in the effect of elevated CO_2 on fungal sporulation are shown in Table 10.2 indicating increase, decrease or no effect. However, increased sporulation or fecundity can also be related to a decrease in aggressiveness of *Colletotrichum gloeosporioides* causing anthracnose on *Stylosanthes* (Chakraborty and Datta, 2003), which is an example of a trade-off, a decrease in one function due to an increase in another, in plant and pathogen interactions under climate change (Yang *et al.*, 2022).

Table 10.2. Effects of elevated CO_2 and temperature on sporulation of plant pathogenic fungi and oomycetes.

Crop	Pathogen	Treatment	Effect	Reference
(soil)	*Aspergillus* *Penicillium* *Cladosporium* *Fusarium* *Alternaria*	700 ppm CO_2	222–499% increase 30–195% increase	Klironomos *et al.*, 1997
Stylo	*Colletotrichum gloeosporioides*	700 ppm CO_2	Twice the number of spores produced at ambient CO_2	Chakraborty and Datta, 2003
Wheat	*Puccinia striiformis* f. sp. *tritici*	700 ppm CO_2	No change in fecundity	Chakraborty *et al.*, 2011
Timothy	*Alternaria alternata*	500–600 ppm CO_2	Three times the number of spores produced	Wolf *et al.*, 2010
Rubber vine	*Maravalia cryptostegiae*	700 ppm CO_2	Increased fecundity	Chakraborty *et al.*, unpublished
Chilli pepper fruits	*Colletotrichum acutatum*	700 ppm CO_2	Reduced fecundity	Koo *et al.*, 2016
Eucalyptus clonal plants	*Puccinia psidii*	700 ppm CO_2	Reduced number of spores per uredinia	Ghini *et al.*, 2014
Five *Cucurbita* species	*Sphaerotheca fuliginea*	500–600 ppm CO_2	Higher number of conidia per cm^2 leaf surface	Khan and Rizvi, 2020
Potato	*Phytophthora infestans*	Small oscillations in temperature Large oscillations in temperature	Increase in sporulation Decrease in sporulation	Shakya *et al.*, 2015

Elevated CO_2 influences host–pathogen interactions but there is no unifying framework in its effects on plant diseases (Velasquez *et al.*, 2018; Singh *et al.*, 2023). In general, plant diseases caused by necrotrophic fungal pathogens increase at elevated CO_2 while diseases caused by biotrophic and hemibiotrophic fungal pathogens decrease. These reactions are caused by effects of elevated CO_2 on plant defence pathways, in which elevated CO_2 promotes the salicylic acid (SA) and suppresses the jasmonic acid (JA) pathways (Li *et al.*, 2023). SA signalling positively affects plant immunity against biotrophic plant pathogens while JA signalling positively influences resistance against necrotrophic plant pathogens (Yang *et al.*, 2019). However, variable effects of elevated CO_2 on plant disease resistance were observed in C3 plants in a meta-analysis, where 41.7% and 48.6% of studies on biotrophic and necrotrophic fungi and oomycetes, respectively, showed an increase in disease while 45.5% of studies on hemibiotrophic fungi and oomycetes showed increased infection but another 45.5% decreased infection (Bazinet *et al.*, 2022). The variability in reactions can be due to other exogenous environmental factors such as light and temperature influencing phytohormone levels (Li and Ahammed, 2023).

10.2.2 Elevated temperature

Several review articles have surveyed the effects of elevated temperature on plant fungal and oomycete diseases and showed positive, negative and no associated effects (Debela and Tola, 2018; Maurya *et al.*,

2022). An increase in disease due to elevated temperature was observed in 62.5% (Debela and Tola, 2018) and 70% (Maurya *et al.*, 2022) of studies. A decrease in disease due to elevated temperature was observed in 4% (Debela and Tola, 2018) and 20% (Maurya *et al.*, 2022) of studies. An increase in night temperature of 3°C influenced *Fusarium* head blight on wheat and barley grains in field conditions from 2016 to 2017, which showed that warm nights favoured disease development of *Fusarium graminearum* over *F. poae*, reduced grain quality and increased mycotoxin concentration (Martinez *et al.*, 2022). Infection and sporulation of *Phytophthora infestans* increased under small temperature oscillations (±5°C) compared with constant temperatures but further decreased when temperature oscillations were large (±10°C). At a daily amplitude of ±5°C and ±10°C, night temperature is closer to and below, respectively, the optimum constant temperature for sporulation and infection (Shakya *et al.*, 2015).

Temperature plays a significant role in host–pathogen interactions and can affect the stages in fungal and oomycete disease development (Pathak *et al.*, 2018; Maurya *et al.*, 2022). Average temperature may not always be a good predictor of infection potential, and fluctuating temperature in the field causing infection highlights that temperature extremes can be more important than small increments in average temperature, but both are projected in future climate change scenarios (Velasquez *et al.*, 2018). The latent period of fungal plant pathogens has been observed to shorten for coffee rust in Central America (Avelino *et al.*, 2015) and wheat stripe rust in Poland (Wójtowicz *et al.*, 2020) due to elevated temperature (see Table 10.1), leading to a possible increase in disease incidence. When exposed to average decadal increase in air temperature, mycelia of six fungal species, e.g. *Alternaria alternata*, grew faster at higher temperatures that was clearly pronounced at the 2100 temperature projection in the United Kingdom (Damialis *et al.*, 2015).

10.2.3 Altered moisture conditions

Climate change can affect plant fungal and oomycete diseases through extreme drought. Climate change between 1996 and 2016 moved the climate optimum of pine blister rust in Sierra Nevada, USA into a higher elevation. There was a decrease of 33% mean disease prevalence that could have been due to drought–host–pathogen interactions. In an aridity gradient, there was decreasing infection at lower elevations and increasing infection at higher elevations (Dudney *et al.*, 2021). A recent survey of fungal plant diseases affected by drought stress showed a reduction in infection for *Botrytis cinerea* and *Oidium neolycopersici* on tomato, and an increase in infection for *Rhizoctonia bataticola* on chickpea. These results show fungal species-specific disease development under drought stress and that host susceptibility depends on the pathogen species, drought intensity and plant species (Zahra *et al.*, 2023).

Increased precipitation has influenced diseases caused by plant pathogenic fungi and oomycetes. The unprecedented pine needle blight epidemic caused by *Dothistroma septosporum* in north-western British Columbia, Canada, starting in the 1980s was caused by increased summer precipitation and not climate warming (Woods *et al.*, 2005). The abundance of plant pathogenic fungal pathogens in the *Eucalyptus* phyllosphere in Australia showed a positive linear relationship with precipitation. Abundance of plant pathogenic fungal pathogens is projected to increase in the coastal regions of Australia up to 100-fold by 2050 compared to the current abundance (Chen *et al.*, 2021).

10.2.4 Interaction of climate change factors

Based on an analysis of controlled environment studies on the effect of elevated CO_2 and temperature on fungal and oomycete plant diseases, variable effects were obtained when these two factors were considered separately. However, positive effects of the interaction of these two climate change variables were observed based on their overall effect. An increase in disease severity was observed in combined elevated CO_2 and temperature for necrotrophic fungi, e.g. *Fusarium equiseti* and *Alternaria* sp., hemibiotroph *C. gloeosporioides* and biotroph oomycete *Plasmopara viticola* (Gullino *et al.*,

2018). Sporulation of *Alternaria alternata* increased at elevated CO_2 (Wolf *et al.*, 2010) but its spore production decreased from 1980 to 2000 influenced by increasing average decadal temperature (Damialis *et al.*, 2015).

The interaction of altered moisture and temperature conditions has variable effects on fungal and oomycete plant pathogens. A survey showed that favourable temperature ranges and high humidity influence diseases caused by destructive fungal pathogens (Waheed *et al.*, 2023). In an experiment on the seasonal abundance of *Phytophthora cinnamomi* in a Mediterranean oak forest from 2016 to 2020, rainfall caused a reduction of 30% spring inoculum density of *P. cinnamomi* in wet years, but warming caused an increase of 30% in inoculum density in spring and early summer (Serrano *et al.*, 2022). Humidity and temperature provide vital information to understand fungal disease epidemics worldwide (Romero *et al.*, 2022) and are significant variables in predicting plant disease epidemic risk in future climate change scenarios.

10.2.5 Distribution and dispersal

The global distribution of plant fungal and oomycete pathogens has been influenced by global warming. Fungi and oomycete in crops are moving 6–7 km poleward per year on average (Bebber *et al.*, 2013). Expansion of overwintering and oversummering ranges of the wheat stripe rust fungal pathogen *Puccinia striiformis* f. sp. *tritici* resulted in earlier and more severe epidemics in central and eastern USA and Canada in 2000 compared to pre-2000 due to changing climate and the introduction of more virulent strains (Lyon and Broders, 2017). More aggressive and thermotolerant novel strains of the wheat stripe rust pathogen are replacing old strains and moving to new regions due to their thermal generalist behaviour invading warm and cold areas of Europe (de Vallavieille-Pope *et al.*, 2018). A peak in species richness of plant pathogenic fungi was shown in mid-latitude regions, and phytopathogenic fungal α-diversity will increase in forests and cropland ecosystems that will become more pronounced in fossil fuel-dependent climate change scenarios (Li et al., 2023).

The dispersal of plant pathogenic fungi and oomycetes has been affected by climate change in several studies. Temperature is the most important determinant of plant disease risk, and in areas where plants will benefit from climate change, their pathogens will also benefit (Garrett *et al.*, 2022). In Europe, increases in air temperature and precipitation by 0.1°C and 6.33 mm per year, respectively, from 1998 to 2014 were associated with changes in spore detection in the atmosphere. The spores of *Leptosphaeria maculans* and *L. biglobosa* causing stem canker of oilseed rape were detected 22 days earlier and the occurrence of the maximum number of spores detected was delayed by 50 days (Kaczmarek *et al.*, 2016). In a spore trap monitoring study in Pistoia, Italy, *Phytophthora* DNA detected using qPCR in the air increased with relative humidity and decreased with increasing mean temperature. There was a positive correlation between *Phytophthora* DNA detected and rainfall (Migliorini *et al.*, 2019).

10.3 Climate Change Effects on Fungal and Oomycete Pathogen Evolution

The major responses of plant pathogens to climate change are to exploit phenotypic plasticity, migrate to new areas or evolve new attributes (Chakraborty, 2013; Santini and Ghelardini, 2015). Forces of evolution such as mutation and selection can act on large pathogen populations to accelerate evolution, leading to more aggressive races which can be influenced by elevated CO_2, elevated temperature and altered precipitation, and mediated by host plant attributes (Eastburn *et al.*, 2011; Pangga *et al.*, 2011; Chakraborty, 2013). The fitness of biotrophic plant pathogens is based solely on pathogenicity while the fitness of necrotrophic plant pathogens consists of pathogenic and saprophytic fitness. Changing geographical distributions of plant pathogens can bring together diverse lineages which can potentially increase diversity. A large population size, rapid growth, short generation time and genetic variation can assist in the evolution of plant pathogens (Chakraborty, 2013).

10.3.1 Aggressiveness

Aggressiveness can be defined as the quantitative component of pathogenicity. It has great sensitivity to environmental conditions such as climate (Pariaud *et al.*, 2009). Aggressiveness as a quantitative trait is highly affected by temperature, rainfall and relative humidity, and variability in aggressiveness among and within populations partly determines adaptation and fitness of the pathogen (Chakraborty and Datta, 2003). The emergence of more aggressive fungal pathogens that were previously less aggressive can be due to climate change (Gullino *et al.*, 2018). For example, *Fusarium equiseti*, causing leaf spots on several host plants, has become more invasive despite being considered as a 'weak' pathogen before (Gilardi *et al.*, 2017).

The effect of elevated CO_2 on pathogen aggressiveness can be positive or negative. In a controlled environment (CE) study on the effect of 25 infection cycles on anthracnose caused by *C. gloeosporioides* on *Stylosanthes* (stylo), a tropical pasture legume, under elevated CO_2, the aggressiveness of two isolates of *C. gloeosporioides* on a resistant stylo cv. Seca and a susceptible stylo cv. Fitzroy decreased after an initial lag period of increase in the first ten infection cycles. In comparison, aggressiveness increased for both isolates and cultivars over 25 infection cycles at ambient CO_2 (Chakraborty and Datta, 2003). In a CE study on the effect of elevated CO_2 and temperature on *C. gloeosporioides* causing anthracnose on hot pepper, the aggressiveness of *C. gloeosporioides* decreased significantly at elevated CO_2 and high temperature of 30°C compared to ambient CO_2 and 25°C (Koo *et al.*, 2016). In a study on *Fusarium* head blight of wheat, the aggressiveness of *F. culmorum* and *F. pseudograminearum* increased by 110% and weakly pathogenic *F. equiseti* and *F. oxysporum* increased by 50% in partially resistant and resistant wheat cultivars after five continuous cropping cycles at elevated CO_2 (Khudhair *et al.*, 2014).

Elevated temperature can cause an increase in aggressiveness in plant pathogenic fungi and oomycetes. The more aggressive *F. graminearum* has caused epidemics in temperate regions in the past 20 years due to warm and humid conditions, indicating a shift from the less aggressive *F. culmorum* occurring in cool and wet conditions (Jennings *et al.*, 2004). *Puccinia striiformis* f. sp. *tritici* isolates obtained post-2000 showed increased aggressiveness at a high temperature range of 12–18°C and these isolates have caused severe epidemics in the USA (Milus *et al.*, 2006). Temperature influenced two *Monilinia* species causing brown rot of peach, in which *M. fructicola* was more aggressive at a high temperature of 35°C while *M. laxa* was more aggressive at low temperatures of 4°C and 11°C (Fidanoglu *et al.*, 2023). Elevated temperature enhanced the variation in aggressiveness of *P. infestans*. At a high acclimation temperature of 25°C, the aggressiveness of *P. infestans* was positively correlated with annual mean temperature, indicating that isolates from warm places produced more disease (Wu *et al.*, 2022).

10.3.2 Fitness

Fitness is defined as the ability of an organism to survive and reproduce (Pringle and Taylor, 2002). The fitness of necrotrophs consists of pathogenic fitness (based on pathogenicity) and saprophytic fitness (based on feeding on dead matter) (Chakraborty, 2013). Limited studies have been done on the effect of elevated CO_2 on plant pathogen fitness, and a few studies showed conflicting results. In a free-air CO_2 enrichment (FACE) facility and controlled environment studies, the saprophytic fitness of necrotroph *F. graminearum* was measured as colonized straw length at 4 days after incubation at 23°C under elevated CO_2. The results showed that there were no significant differences in the saprophytic fitness of *F. graminearum* isolates under both ambient and elevated CO_2 but fungal biomass, a measure of pathogen fitness, was significantly higher at elevated CO_2 than at ambient CO_2 in two out of three studies (Melloy *et al.*, 2010).

Fitness of plant pathogenic fungi and oomycetes can be affected by elevated temperature. In a study on the pathogen fitness of *F. pseudograminearum* causing wheat crown rot on different wheat lines with different resistance levels, the lines with superior crown rot resistance were generally associated with reduced overall pathogen fitness regardless of temperature. Pathogen fitness decreased among wheat lines with increasing temperature (Sabburg *et al.*, 2015). In a study on the effect of temperature on lineages of *P. infestans* on tomato, US-22 and US-23 clonal lineages grew faster at nearly 2°C above the optimum temperature of EC-1 default lineage. The increase in *P. infestans* infection rate on tomato and potato in Wisconsin, USA, at warmer temperatures in 2011 may indicate that they have a fitness advantage at warmer temperatures (Seidl Johnson *et al.*, 2015).

10.3.3 Abundance and genetic diversity

An increase in abundance and diversity of plant pathogenic fungi is observed at elevated temperature and CO_2. In a study using a global field survey and 9-year experiment, an increment of 2°C increased the relative abundance of soil-borne potential fungal pathogens three times worldwide. Relative abundance is measured as a percentage of the total fungal communities in soils (Delgado-Baquerizo *et al.*, 2020). Across an elevated CO_2 experiment, no change or an increase in biomass and diversity of soil-borne fungi were found worldwide. All studies > 10 years and any study of any length with warming of >2°C showed significant changes in fungal community composition (Baldrian *et al.*, 2022).

The interaction of elevated temperature and altered precipitation affects plant pathogenic fungal and oomycete abundance and diversity. In a field experiment that studied fungal communities at a future climate scenario of 20% reduction in precipitation and 1.5°C increase in soil temperature, plant pathogenic fungi dominated 87% of the total sequences within the wheat residue mycobiome. New potential phytopathogenic fungi were detected in the future climate change plots and not in ambient climate plots (Wahdan *et al.*, 2020). Average temperature and precipitation especially in the spring months from March to May influenced the differences in the spectrum of plant pathogenic fungi in wheat from 2009 to 2013 in the Czech Republic; different species were dominant in different years. *Alternaria triticina* was dominant in the warmest year, 2009 (Hysek *et al.*, 2017).

10.3.4 Phenotypic plasticity

Plant pathogenic fungi and oomycetes can adapt to climate change via phenotypic plasticity, which is the capacity of a genotype to demonstrate variable phenotypes in different environments (Fusco and Minelli, 2010). Respiration rates can reflect the general activity of *Heterobasidion parviporum* causing *Heterobasidion* root rot in trees because respiration rate and temperature response curves are similar (Müller *et al.*, 2014). In a study on the respiration rate of *H. parviporum* subpopulations from climatically driven environments at different temperatures, older isolates from 2 to 18 years before which were stored at 5°C showed lower respiration rates at 20°C independent of geographic origin, which indicated the phenotypic plasticity of *H. parviporum* in response to elevated temperature (Müller *et al.*, 2015). *Aspergillus flavus*, *A. niger*, *A. tubiengensis* and *Fusarium strains* obtained from tropical regions tolerated heatwaves based on a climate change scenario of a maximum temperature of 40°C (Tischner *et al.*, 2022).

Thermal adaptation has been observed in several studies on *Phytophthora infestans* with significant implications on its potential for adaptation to climate change. Thermal adaptation was observed by Wu *et al.* (2020) in China for 21 *P. infestans* isolates acclimatized to high temperature for 200 days that adapted quickly to high temperature as shown by growth inhibition at 30°C, which was close to the maximum growth temperature of 31.5°C (Yang *et al.*, 2016). However, all *P. infestans* isolates from different temperate and subtropical environments showed significant ability to respond to thermal stress. Another study showed thermal adaptation of 140 of *P. infestans* isolates under five temperature regimes. At a high acclimation temperature (25°C), isolates from warmer regions produced more disease than those from cooler regions, indicating geographic variation in the epidemic risk of *P. infestans* under global warming in China (Wu *et al.*, 2022).

10.4 Climate Change Effects on Fungal and Oomycete Biological Control Agent Dynamics

Fungal and oomycete BCAs against fungal and oomycete plant pathogens used for the past 50 years were reviewed recently, identifying 300 fungal BCAs belonging to 13 classes and 113 genera. For the biological control of plant fungal and oomycete pathogens by fungi, *Trichoderma* has the greatest potential with 25 species while *Pythium oligandrum* has the greatest potential for oomycetes (Thambugala *et al.*, 2020). Broad-spectrum BCAs are effective over a wide range of target fungal and/or oomycete pathogens, e.g. *Trichoderma* spp. (Thambugala *et al.*, 2020), while narrow-spectrum BCAs are specific to a target fungal and/or oomycete pathogen, e.g. *Phlebiopsis gigantea* against *Heterobasidion annosum* s.l. species complex causing root rot of conifer trees (Prospero *et al.*, 2021).

The efficacy of fungal and oomycete BCAs can be influenced by different environmental factors such as temperature, pH, water availability, soil metal ions and climate-change related factors such as elevated CO_2 and UV-B radiation (Ghorbanpour *et al.*, 2018). Temperature and humidity are two key environmental factors influencing fungal and oomycete BCA efficacy (Elad and Stewart, 2004; He *et al.*, 2021). The effects of climate change on bacterial and viral BCAs are discussed in Chapter 11.

The success of fungal and oomycete BCAs is dependent on ecological processes including host establishment, environmental persistence and trait expression that are also greatly affected by the microbiome (Sullam and Musa, 2021). The interaction of the different components of the fungal and oomycete BCA–target fungal and oomycete pathogen under climate change is depicted in a biological control pyramid in Fig. 10.2 where the interaction of the plant host, BCA and target pathogen is affected by the abiotic environment (Alfiky and Weisskopf, 2021). Climate change will affect host susceptibility, pathogen infectivity and biotic and abiotic environments which are all interconnected (Yang *et al.*, 2022). The biological control pyramid considers climate change variables in the environment component, and the fungal or oomycete BCA as another equally important component of the pyramid.

10.4.1 Elevated CO_2

The effects of elevated CO_2 singly and in combination with elevated temperature and other climate change variables on fungal BCAs are limited or unclear (Gilardi *et al.*, 2017; Desai *et al.*, 2020), and various generalizations can be deduced from these limited studies. In a study on the biological control of *Fusarium oxysporum* f. sp. *lactucae* Race 1 on lettuce, *Beauveria bassiana* showed consistent disease control of 53–56% efficacy at 20–24°C at ambient and elevated CO_2 but disease control of *Trichoderma asperellum* was inconsistent across all CO_2 and temperature treatments (Gilardi *et al.*, 2023). An interaction experiment on the effect of elevated CO_2 and temperature on the biological control of zucchini powdery mildew by *Amphelomyces quisqualis* showed improved disease control at 800–850 ppm CO_2 at 26–30°C by 23.3% for disease incidence and 22.8% for disease severity (Gilardi *et al.*, 2017). In a study of the interaction of elevated temperature, relative humidity (RH) and elevated CO_2 on the resilience of different formulations of *Candida sake* CPA-1, a BCA of fungal pathogens of fruits, high temperature at 35°C and lower RH at 40% reduced the number of viable cells of *C. sake* on grapes but the resilience of the three formulations of these treatments at elevated CO_2 (1000 ppm) improved with an increase in the population of viable cells (Carbo *et al.*, 2018).

10.4.2 Increased temperature and altered moisture conditions

Elevated temperature and its interaction with RH have variable results, positive or negative, on their effect on fungal and oomycete BCAs. The suppression of grey mould of cucumber by *Trichoderma harzianum* T39 was more pronounced at higher temperature (>20°C) and lower RH between 80% and 97% (Elad *et al.*, 1993). A positive result was found in the evaluation of different *Trichoderma* native strains in grapevine pruning wounds at different

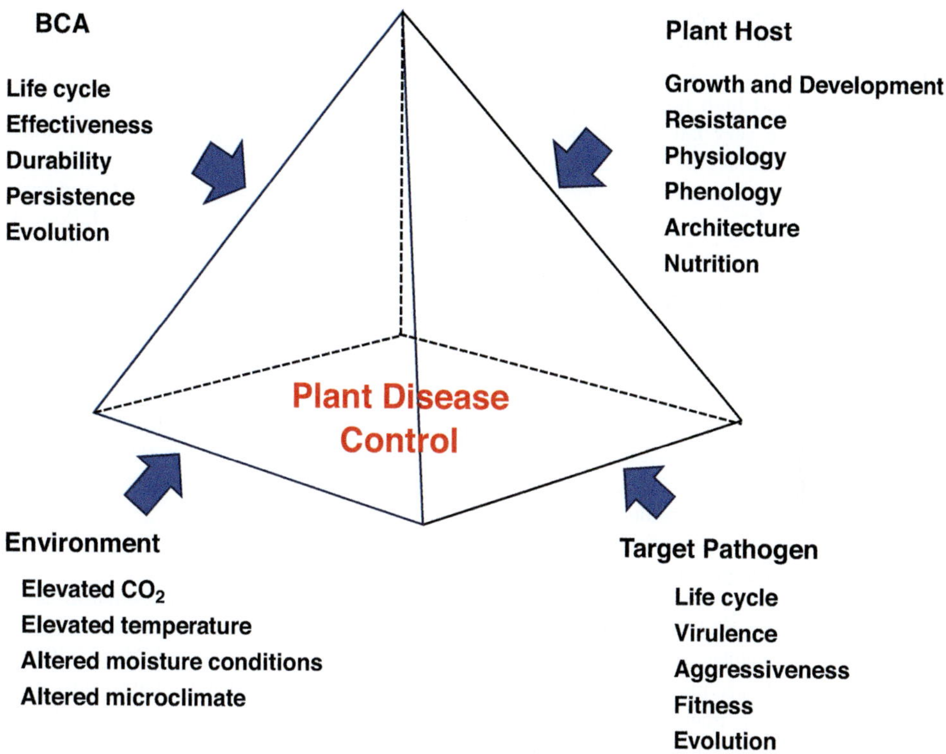

Fig. 10.2. Biological control pyramid under climate change using fungal and oomycete biological control agents (BCA) showing the interaction of the different components representing each pyramid side, namely BCA, plant host, environment and target fungal or oomycete pathogen resulting in plant disease control. The characteristics of each component are listed.

temperatures in Spain, which showed that strains from clade Harzianum/Virens (HV) grew at higher temperatures. *Trichoderma* strains T154 and T24 that belong to clade HV showed high mycelial growth rates at 35°C (Carro-Huerga *et al.*, 2021). Mycoparasitism of *Sclerotinia sclerotiorum* sclerotia by *T. harzianum* was higher at 25°C than at 15°C on pea in the USA (Knudsen *et al.*, 1991) but *Trichoderma* sp. did not suppress *Sclerotium rolfsii* at temperature >30°C on ginger in India (Mukherjee and Raghu, 1977).

In a study on the effect of temperature and RH combinations on the growth and mycotoxigenesis of three commercial BCAs on *Fusarium graminearum* on wheat in France, *Trichoderma asperellum* was ineffective at 30°C and 33°C at all RH levels to limit *F. graminearum* growth and deoxylivenol production but the oomycete *Pythium oligandrum* significantly limited *F. graminearum* growth and deoxylivenol production at 33°C (Dieye *et al.*, 2024).

10.5 Climate Change Effects on Fungal and Oomycete Biological Control Agent Evolution

Evolution of pathogen strains and adaptation to BCAs can change the dynamics of host–pathogen interaction (Bardin *et al.*, 2015). Sporulation of *Trichoderma* was increased considerably under elevated CO_2 after 30 generations (Grover *et al.*, 2015). A BCA of *Botrytis*, *Chlonotachys rosea*, and an entomopathogen, *Metarhizium anisopliae*, were found to be associated with the cover crop *Lolium repens* under elevated CO_2 in a FACE experiment. This could indicate increased soil suppressiveness to plant pathogenic fungi under a cover crop vegetation (Rezacova *et al.*, 2005). The evolutionary potential of fungal and oomycete BCAs is high due to their short generation time and high population number. However, there are very few studies on long-term evolutionary studies on BCAs (Sentis *et al.*, 2021).

10.6 Research and Management Outlook

The selection of the BCA should consider stable effectiveness under various environmental conditions such as temperature, moisture and interaction with biotic and abiotic factors (He *et al.*, 2021). There is a need to consider temperature fluctuations in field conditions on biological control. Climate change studies on biological control have focused on instantaneous changes in temperature but subtle changes accumulate over many years (Crowder and Harwood, 2014). Stress-tolerant BCAs are needed that should be released at the proper times to synchronize the phenology of both target host pathogen and BCA to increase BCA efficacy under extreme climate change events (Wang *et al.*, 2022).

The coevolution of the target pathogen and BCA should be studied further as BCA durability is associated negatively with the evolutionary potential of the target pathogen. The durability of the BCA will be shorter for the target pathogen with high genetic variation from sexual reproduction, long-distance dispersal ability and large effective population size (Zhan and McDonald, 2013). Climate change may complicate the current interaction of BCAs and target pathogen which could lead to ineffectiveness of the BCA due to the evolution of target pathogens, as erosion of BCAs in laboratory and greenhouse conditions has been observed (He *et al.*, 2021), for example at high constant (Mukherjee and Raghu, 1977) and high day (Dik and Elad, 1999) temperatures for *Trichoderma* sp. and *T. harzianum* T39, respectively.

Phenotypic plasticity and genotype by environment (G × E) interactions can influence the BCA's efficacy, so testing different environments and genotype combinations is important before BCA deployment (Sullam and Musa, 2021). Evolution of the plant host–pathogen– BCA interaction should be considered in formulating the biological control strategy in time and space to achieve BCA durability and efficacy (He *et al.*, 2021).

References

Agrios, G. (2005) *Plant Pathology*, 5th edn. Elsevier Academic Press, Amsterdam.

Alfiky, A. and Weisskopf, L. (2021) Deciphering *Trichoderma*–plant–pathogen interactions for better development of biocontrol applications. *Journal of Fungi* 7, 61.

Avelino, J., Cristancho, M., Georgiou, S., Imbach, P., Aguilar, L. *et al.* (2015) The coffee rust crises in Colombia and Central America (2008–2013): Impacts, plausible causes and proposed solutions. *Food Security* 7, 303–321.

Baldrian, P., Bell-Dereske, L., Lepinay, C., Větrovský, T. and Kohou, P. (2022) Fungal communities in soils under global change. *Studies in Mycology* 103, 1–24.

Bardin, M., Ajouz, S., Comby, M., Lopez-Ferber, M., Graillot, B. *et al.* (2015) Is the efficacy of biological control against plant diseases likely to be more durable than that of chemical pesticides? *Frontiers in Plant Science* 6, 566.

Bazinet, B., Tang, L. and Bede, J.C. (2022) Impact of future elevated carbon dioxide on C_3 plant stresses. *Molecular Plant-Microbe Interaction* 5, 527–539.

Bebber, D.P., Ramotowski, M.A.T. and Gurr, S.J. (2013) Crop pests and pathogens move polewards in a warming world. *Nature Climate Change* 3, 985–988.

Carbo, A., Torres, R., Teixido, N., Usali, J., Medina, A. *et al.* (2018) Impact of climate change environmental conditions on the resilience of different formulations of the biocontrol agent *Candida* sake CPA-1 on grapes. *Letters in Applied Microbiology* 67, 2–8.

Carro-Huerga, G., Mayo-Prieto, S., Rodríguez-González, Á., Álvarez-García, S., Gutiérrez, S. *et al.* (2021) The influence of temperature on the growth, sporulation, colonization, and survival of *Trichoderma* spp. in grapevine pruning wounds. *Agronomy* 11, 1771.

Chakraborty, S. and Datta, S. (2003) How will plant pathogens adapt to host plant resistance at 2 elevated CO2 under a changing climate? *New Phytologist* 159, 733–742.

Chakraborty, S., Pangga, I.B., Lupton, J., Hart, L., Room, P.M. *et al.* (2000) Production and dispersal of *Colletotrichum gloeosporioides* spores on *Stylosanthes scabra* under elevated CO_2. *Environmental Pollution* 108, 381–387.

Chakraborty, S., Luck, J., Hollaway, G., Fitzgerald, G. and White, N. (2011) Rust-proofing wheat for a changing climate. *Euphytica* 179, 19–32.

Chakraborty, S. (2013) Migrate or evolve: Options for plant pathogens under climate change. *Global Change Biology* 19, 1985–2000.

Chen, Q.L., Hu, H.W., Yan, Z.Z., Li, C.Y., Nguyen, B.A.T. *et al.* (2021) Precipitation increases the abundance of fungal plant pathogens in Eucalyptus phyllosphere. *Environmental Microbiology* 23(12), 7688–7700.

Crowder, G.W. and Harwood, J.D. (2014) Promoting biological control in a rapidly changing world. *Biological Control* 75, 1–7.

Damialis, A., Mohammad, A.B., Halley, J.M. and Gange, A.C. (2015) Fungi in a changing world: Growth rates will be elevated, but spore production may decrease in future climates. *International Journal of Biometeorology* 59, 1157–1167.

de Vallavieille-Pope, C., Bahn, B., Leconte, M., Zurfluh, O., Belaid, Y. *et al.* (2018) Thermal generalist behaviour of invasive *Puccinia graminis* f. sp. *tritici* under current and future climate conditions. *Plant Pathology* 67, 1307–1320.

Debela, C. and Tola, M. (2018) Effects of elevated CO_2 and temperature on crop-disease interactions under rapid climate change. *International Journal of Environmental Sciences and Natural Resources* 13(1), 555851.

Delgado-Baquerizo, M., Guerra, C.A., Cano-Díaz, C., Egidi, E., Wang, J.-T. *et al.* (2020) The proportion of soil-borne pathogens increases with warming at the global scale. *Nature Climate Change* 10, 550–554.

Desai, S., Dubey, S.C. and Prasad, R.D. (2020) Impacts of climate change on *Fusarium* species vis-à-vis adaptation strategies. *Indian Phytopathology* 73, 593–603.

Dieye, C.A.T., Durand, N., Schorr-Galindo, S., Strub, C. and Fontana, A. (2024) Impacts of abiotic factors on the growth of three commercial biological control agents, on the growth and mycotoxinogenesis of *Fusarium graminearum* and on their interaction. *Journal of Science, Food and Agriculture* 104, 932–941.

Dik, A.J. and Elad, Y. (1999) Comparison of antagonists of *Botrytis cinerea* in greenhouse-grown cucumber and tomato under different climatic conditions. *European Journal of Plant Pathology* 105, 123–137.

Dudney, J., Willing, C.E., Das, A.J., Latimer, A.M., Nesmith, J.C. *et al.* (2021) Nonlinear shifts in infectious rust disease due to climate change. *Nature Communications* 12, 5102.

Eastburn, D.M., McElrone, A.J. and Bilgin, D.D. (2011) Influence of atmospheric and climatic change on plant pathogen interactions. *Plant Pathology* 60, 54–69.

Elad, Y., Zimand, G., Zaqs, Y., Zuriel, S. and Chet, I. (1993) Use of *Trichoderma harzianum* in combination or alternation with fungicides to control cucumber grey mould (*Botrytis cinerea*) under commercial greenhouse conditions. *Plant Pathology* 42, 324–332.

Elad, Y. and Stewart, A. (2004) Microbial control of *Botrytis* spp. In: Elad, Y., Williamson, B., Tudzynski, P. and Delen, N. (eds) *Botrytis: Biology, Pathology and Control*. Kluwer Academic Publishers, Amsterdam, The Netherlands, pp. 223–241.

Fidanoglu, B.T., Mestav, B. and Ozkilinc, H. (2023) A case study of the effect of temperature on aggressiveness in the *Monilinia*-peach pathosystem. *European Journal of Phytopathology* 1–10.

Fusco, G. and Minelli, A. (2010) Phenotypic plasticity in development and evolution: Facts and concepts. *Philosophical Transactions of the Royal Society B: Biological Sciences* 365, 547–556.

Garrett, K.A., Bebber, D.P., Etherton, B.A., Gold, K.M., Plex Sula, A.I. *et al.* (2022) Climate change effects on pathogen emergence: Artificial intelligence to translate big data for mitigation. *Annual Review of Phytopathology* 60, 357–378.

Ghini, R., MacLeod, R.E. de O., Torre Neto, A., Cardoso, D.C., Bettiol, W. *et al.* (2014) Increased atmospheric carbon dioxide concentration: Effects on eucalypt rust (*Puccinia psidii*), C:N ratio and essential oils in eucalypt clonal plantlets. *Forest Pathology* 44, 409–416.

Ghorbanpour, M., Omidvarib, M., Abbaszadeh-Dahajic, P., Omidvard, R. and Karimane, K. (2018) Mechanisms underlying the protective effects of beneficial fungi against plant diseases. *Biological Control* 117, 147–157.

Gilardi, G., Gisi, U., Garibaldi, A. and Gullino, M.L. (2017) Effect of elevated atmospheric CO_2 and temperature on the chemical and biological control of powdery mildew of zucchini and the Phoma leaf spot of leaf beet. *European Journal of Plant Pathology* 148, 229–236.

Gilardi, G., Tabone, G., Gullino, M.L. and Garibaldi, A. (2023) Effect of the use of biocontrol agents and resistance inducers against race 1 of *Fusarium oxysporum* f. sp. *lactucae* on lettuce in a simulated climate change scenario. *Journal of Plant Pathology* 106.

Govers, F. (2024) Diseases caused by oomycetes. In: Oliver, R.P. (ed.) *Agrios' Plant Pathology*. Elsevier Academic Press, London, pp. 435–463.

Grover, M., Maheswaria, M., Desaia, M., Gopinatha, K.A. and Venkateswarlu, B. (2015) Elevated CO_2: Plant associated microorganisms and carbon sequestration. *Applied Soil Ecology* 95, 73–85.

Gullino, M.L., Pugliese, M., Gilardi, G. and Garibaldi, A. (2018) Effect of increased CO_2 and temperature on plant diseases: A critical appraisal of results obtained in studies carried out under controlled environment facilities. *Journal of Plant Pathology* 100, 371–389.

Han, G.F., Wang, H.G. and Ma, Z.H. (2011) Effects of elevated CO_2 and enhanced UV-b on epidemic components of wheat stripe rust. *Scientia Agricultura Sinica* 44, 4326–4332.

He, D.C., He, M.H., Amalin, D.M., Liu, W., Alvindia, D.G. *et al.* (2021) Biological control of plant diseases: An evolutionary and eco-economic consideration. *Pathogens* 10(10), 1311.

Helfer, S. (2014) Rust fungi and global change. *New Phytologist* 201, 770–780.

Hibberd, J.M., Whitbread, R. and Farrar, J.F. (1996) Effect of elevated concentrations of CO_2 on infection of barley by *Erysiphe graminis*. *Physiological and Molecular Plant Pathology* 48, 37–53.

Hysek, J., Vavera, R. and Ruzek, P. (2017) Influence of temperature, precipitation, and cultivar characteristics on changes in the spectrum of pathogenic fungi in wither wheat. *International Journal of Biometeorology* 61, 967–975.

Jennings, P., Coates, M.E., Walsh, K., Turner, J.A. and Nicholson, P. (2004) Determination of deoxynivalenol- and nivalenol-producing chemotypes of *Fusarium graminearum* isolated from wheat crops in England and Wales. *Plant Pathology* 53, 643–652.

Juroszek, P., Racca, R., Linka, S., Farhumand, J. and Kleinhenza, B. (2020) Overview on the review articles published during the past 30 years relating to the potential climate change effects on plant pathogens and crop disease risks. *Plant Pathology* 69, 179–193.

Kaczmarek, J., Kedziora, A., Brachaczek, A., Latunde-Dada, A.O., Dakowska, S. *et al.* (2016) Effect of climate change on sporulation of the teleomorphs of *Leptosphaeria* species causing stem canker of brassicas. *Aerobiologia* 32, 39–51.

Kazan, K. (2018) Plant-biotic interactions under elevated CO_2: A molecular perspective. *Environmental and Experimental Botany* 153, 249–261.

Khan, M.R. and Rizvi, T.F. (2020) Effect of elevated levels of CO_2 on powdery mildew development in five cucurbit species. *Scientific Reports* 10, 4986.

Khudhair, M., Melloy, P., Lorenz, D.J., Obanor, F., Aitken, E. *et al.* (2014) Fusarium crown rot under continuous cropping of susceptible and partially resistant wheat in microcosms at elevated CO_2. *Plant Pathology* 63, 1033–1043.

Klironomos, J.K., Rillig, M.C., Allen, M.F., Zak, D.R., Pregitzer, K.S. *et al.* (1997) Increased levels of airborne fungal spores in response to *Populus tremuloides* grown under elevated atmospheric CO_2. *Canadian Journal of Botany* 75, 1670–1673.

Knudsen, G.R., Eschen, D.J., Dandurand, L.M. and Bin, L. (1991) Potential for biocontrol of *Sclerotinia sclerotiorum* through colonization of sclerotia by *Trichoderma harzianum*. *Plant Disease* 75, 466–470.

Koo, T.H., Hong, S.J. and Yun, S.C. (2016) Changes in the aggressiveness and fecundity of hot pepper anthracnose pathogen (*Colletotrichum acutatum*) under elevated CO_2 and temperature over 100 infection cycles. *Plant Pathology Journal* 32, 260–265.

Li, Z. and Ahammed, G.J. (2023) Salicylic acid and jasmonic acid in elevated CO_2-induced plant defense response to pathogens. *Journal of Plant Physiology* 286, 154019.

Li, P.F., Wang, B.Z., Tedersoo, L., Crowther, T.W., Shi, Y. *et al.* (2023) Global diversity and biogeography of potential phytopathogenic fungi in a changing world. *Nature Communications* 14, 6482.

Luck, J., Spackman, M., Freeman, A., Trebicki, P., Griffiths, W. *et al.* (2011) Climate change and diseases of food crops. *Plant Pathology* 60, 113–121.

Lyon, B. and Broders, K. (2017) Impact of climate change and race evolution on the epidemiology and ecology of stripe rust in Central and Eastern USA and Canada. *Canadian Journal of Plant Pathology* 39, 385–392.

Manning, W.J. and Tiedemann, A.V. (1995) Climate change: Potential effects of increased atmospheric carbon dioxide (CO_2), ozone (O_3), and ultraviolet-B (UV-b) radiation on plant diseases. *Environmental Pollution* 88, 219–245.

Martinez, M., Biganzoli, F., Arata, A., Dinolfo, M.I., Rojas, D. *et al.* (2022) Warm nights increase fusarium head blight negative impact on barley and wheat grains. *Agricultural and Forest Meteorology* 318, 108909.

Maurya, M.K., Yadav, V.K., Singh, S.P., Jatoth, R., Singh, H.K. *et al.* (2022) Impact of climate change n diseases of crops and their management – a review. *Journal of Agricultural Science and Technology B* 12, 1–15.

Melloy, P., Hollaway, G., Luck, J.O., Norton, R.O.B., Aitken, E. *et al.* (2010) Production and fitness of *Fusarium pseudograminearum* inoculum at elevated carbon dioxide in FACE. *Global Change Biology* 16, 3363–3373.

Migliorini, D., Ghelardini, L., Luchi, N., Capretti, P., Onorari, M. *et al.* (2019) Temporal patterns of airborne *Phytophthora* spp. in a woody plant nursery area detected using real-time PCR. *Aerobiologia* 35, 201–214.

Milus, E.A., Seyran, E. and McNew, R. (2006) Aggressiveness of *Puccinia striiformis* f. sp. *tritici* isolates in the South-Central United States. *Plant Disease* 90, 847–852.

Mukherjee, P.K. and Raghu, K. (1977) Effect of temperature on antagonistic and biocontrol potential of *Trichoderma* sp on *Sclerotium rolfsii*. *Mycopathologia* 139, 151–155.

Müller, M.M., Sievanen, R., Beuker, E., Meesenburg, H., Kuuskeri, J. *et al.* (2014) Predicting the activity of *Heterobasidion parviporum* on Norway spruce in warming climate from its respiration rate at different temperatures. *Forest Pathology* 44, 325–336.

Müller, M.M., Hamberg, L., Kuuskeri, J., LaPorta, N., Pavlov, I. *et al.* (2015) Respiration rate determinations suggest *Heterobasidion parviporum* subpopulations have potential to adapt to global warming. *Forest Pathology* 45, 515–524.

Oliver, R.P. (2024) Diseases caused by fungi. In: Oliver, R.P. (ed.) *Agrios' Plant Pathology*. Elsevier Academic Press, London, pp. 339–427.

Pangga, I.B., Chakraborty, S. and Yates, D. (2004) Canopy size and induced resistance in *Stylosanthes scabra* determine anthracnose severity at high CO_2. *Phytopathology* 94, 221–227.

Pangga, I.B., Hanan, J. and Chakraborty, S. (2011) Pathogen dynamics in a crop canopy and their evolution under changing climate. *Plant Pathology* 60, 70–81.

Pangga, I.B., Hanan, J. and Chakraborty, S. (2013) Climate change impacts on plant canopy architecture: Implications for pest and pathogen management. *European Journal of Plant Pathology* 135, 595–610.

Pariaud, B., Ravigné, V.F., Halkett, F.H., Goyeau, H., Carlier, J. *et al.* (2009) Aggressiveness and its role in the adaptation of plant pathogens. *Plant Pathology* 58, 409–424.

Pathak, R., Singh, S.K., Tak, A. and Gehlot, P. (2018) Impact of climate change on host, pathogen and plant disease adaptation regime: A review. *Biosciences Biotechnology Research Asia* 15, 529–540.

Pringle, A. and Taylor, J.W. (2002) The fitness of filamentous fungi. *Trends in Microbiology* 10, 474–481.

Prospero, S., Botella, L., Santini, A. and Robin, C. (2021) How can we move from dreams to reality? *Forest Ecology and Management* 496, 119377.

Rezacova, V., Blum, H., Hrselova, H., Gamper, H. and Gryndler, M. (2005) Saprobic microfungi under *Lolium perenne* and *Trifolium repens* at different fertilization intensities and elevated atmospheric CO_2 concentration. *Global Change Biology* 11, 224–230.

Romero, F., Cazzato, S., Walder, F., Vogelgsang, S., Franz Bender, S. *et al.* (2022) Humidity and high temperature are important for predicting fungal disease outbreaks worldwide. *New Phytologist* 234, 1553–1556.

Roy, S. and Mathur, P. (2021) Delineating the mechanisms of elevated CO_2 mediated growth, stress tolerance and phytohormonal regulation in plants. *Plant Cell Reports* 40, 1345–1365.

Runion, G.B., Prior, S.A., Rogers, H.H. and Mitchel, R.J. (2010) Effects of elevated atmospheric CO_2 on two southern forest diseases. *New Forests* 39, 275–285.
Sabburg, R., Obanor, F., Aitken, E. and Chakraborty, S. (2015) Changing fitness of a necrotrophic plant pathogen under increasing temperature. *Global Change Biology* 21, 3126–3137.
Santini, A. and Ghelardini, L. (2015) Plant pathogen evolution and climate change. *CAB Reviews* 10, 1–8.
Seidl Johnson, A.C., Frost, K.E., Rouse, D.I. and Gevens, A.J. (2015) Effect of temperature on growth and sporulation of US-22, US-23, and US-24 clonal lineages of *Phytophthora infestans* and implications for late blight epidemiology. *Phytopathology* 105, 445–459.
Sentis, A., Hemptinne, J.-L., Magro, A. and Outreman, Y. (2021) Biological control needs evolutionary perspectives of ecological interactions. *Evolutionary Applications* 15, 1537–1554.
Serrano, M.S., Romero, M.A., Homet, P. and Aparicio, L. (2022) Climate change impact on the population dynamics of exotic pathogens: The case of the worldwide pathogen *Phytophthora cinnamomi*. *Agricultural and Forest Meteorology* 322, 109002.
Shakya, S.K., Goss, E.M., Dufault, N.S. and van Bruggen, A.H.C. (2015) Potential effects of diurnal temperature oscillations on potato late blight with special reference to climate change. *Phytopathology* 105, 230–238.
Singh, B.K., Delgado-Baquerizo, M., Egidi, E., Guirado, E., Leach, J.E. *et al.* (2023) Climate change impacts on plant pathogens, food security and paths forward. *Nature Reviews Microbiology* 21, 640–656.
Smith, F. and Luna, E. (2023) Elevated atmospheric carbon dioxide and plant immunity to fungal pathogens: Do the risks outweigh the benefits? *Biochemical Journal* 480, 1791–1804.
Sullam, K.E. and Musa, T. (2021) Ecological dynamics and microbial treatments against oomycete plant pathogens. *Plants* 10, 2697.
Thambugala, K.M., Daranagama, D.A., Phillips, A.J.L., Kannangara, S.D. and Promputtha, I. (2020) Fungi vs. fungi in biocontrol: An overview of fungal antagonists applied against fungal plant pathogens. *Frontiers in Cellular Infection Microbiology* 10, 604923.
Tischner, Z., Páldy, A., Kocsubé, S., Kredics, L., Dobolyi, C. *et al.* (2022) Survival and growth of microscopic fungi derived from tropical regions under future heat waves in the Pannonian Biogeographical Region. *Fungal Biology* 126, 511–520.
Tronsmo, A.M., Collinge, D.B., Djurle, A., Munk, L., Yuen, J. *et al.* (2020) *Plant Pathology and Plant Diseases*. CAB International, Boston, Massachusetts.
Velasquez, A.C., Castroverde, C.D.M. and He, S.Y. (2018) Plant and pathogen warfare under changing climate conditions. *Current Biology* 28, R619–R6.
Wahdan, S.F.M., Hossen, S., Tanunchai, B., Schädler, M., Buscot, F. *et al.* (2020) Future climate significantly alters fungal plant pathogen dynamics during the early phase of wheat litter decomposition. *Microorganisms* 8, 908.
Waheed, A., Haxim, Y., Islam, W., Ahmad, M., Muhammad, M. *et al.* (2023) Climate change reshaping plant-fungal interaction. *Environmental Research* 238, Part 2, 117282.
Wang, B.X., Hof, A.R. and Ma, C.S. (2022) Impacts of climate change on crop production, pests and pathogens of wheat and rice. *Frontiers of Agricultural Science and Engineering* 9, 4–18.
Wójtowicz, A., Wójtowicz, M., Sigvald, R., Czernecki, B., Ratajkiewicz, H. *et al.* (2020) Assessment of the impact of climate change on the latency period of leaf rust on triticale in Poland. *Acta Agriculturae Scandinavica, Section B – Soil & Plant Science* 70, 195–207.
Wolf, J., O'Neill, N.R., Rogers, C., Muilenberg, M.L. and Ziska, L.H. (2010) Elevated atmospheric carbon dioxide concentrations amplify *Alternaria alternata* sporulation and total antigen production. *Environmental Health Perspectives* 118, 1223–1228.
Woods, A., Coates, K.D. and Hamann, A. (2005) Is an unprecedented Dothistroma needle blight epidemic related to climate change? *Bioscience* 55, 761–769.
Wu, E.J., Wang, Y.P., Yahuza, L., He, M.H., Sun, D.L. *et al.* (2020) Rapid adaptation of the Irish potato famine pathogen *Phytophthora infestans* to changing temperature. *Evolutionary Application* 13, 768–780.
Wu, E.J., Wang, Y.P., Yang, L.N., Zhao, M.Z. and Zhan, J. (2022) Elevating air temperature may enhance future epidemic risk of the plant pathogen *Phytophthora infestans*. *Journal of Fungi* 8, 808.
Yang, J., Duan, G., Li, C., Liu, L., Han, G. *et al.* (2019) The crosstalks between jasmonic acid and other plant hormone signaling highlight the involvement of jasmonic acid as a core component in plant response to biotic and abiotic stresses. *Frontiers in Plant Science* 10, 1349.

Yang, L.N., Ren, M. and Zhan, J. (2022) Modeling plant diseases under climate change: Evolutionary perspectives. *Trends in Plant Science* 28, 519–526.
Yang, L.N., Zhu, W., Wu, E.J., Yang, C., Thrall, P.H. *et al.* (2016) Trade-offs and evolution of thermal adaptation in the Irish potato famine pathogen *Phytophthora infestans*. *Molecular Ecology* 25, 4047–4058.
Zahra, N., Hafeez, M.B., Shukaily, M.A., Al-Sadi, A.M., Siddique, K.H.M. *et al.* (2023) Influence of abiotic stress on disease infestation in plants. *Physiological and Molecular Plant Pathology* 127, 102125.
Zhan, J. and McDonald, B. (2013) Field-based experimental evolution of three cereal pathogens using a mark-release-capture strategy. *Plant Pathology* 62, 106–114.

11 Impact of Climate Change on Bacteria and Viruses in Biocontrol

Clara Lago[1,2*] and Piotr Trębicki[3,4]

[1]Institute of Agricultural Sciences, Spanish National Research Council (ICA-CSIC), Madrid, Spain; [2]Institute for Cross-Disciplinary Physics and Complex Systems, Spanish National Research Council and University of the Balearic Islands (IFISC-CSIC-UIB), Palma de Mallorca, Spain; [3]Applied BioSciences, Macquarie University, Sydney, Australia; [4]School of Agriculture, Food andEcosystem Sciences, University of Melbourne, Parkville, Australia

Abstract

Biological control agents, such as bacteria and viruses, offer viable alternatives to agrochemicals for controlling pests and diseases, playing a critical role in integrated pest management (IPM) and integrated plant protection (IPP). However, their use is not without challenges, particularly in the context of climate change. Therefore, this chapter examines how climate change affects the effectiveness of bacterial and viral biocontrol agents (BCAs). Research indicates that climate change can both enhance and diminish the efficacy of plant growth-promoting bacteria (PGPBs) and rhizobacteria (PGPRs) based on specific environmental conditions. Similarly, viral BCAs face significant challenges due to temperature fluctuations and UV radiation, which can alter their effectiveness and impact insect pest dynamics. Overall, this chapter highlights the importance of incorporating these BCAs into IPM and IPP strategies to mitigate climate change effects on agriculture, while also calling for more field-based and long-term research to address existing limitations.

11.1 Introduction

Climate change and its impact on agriculture, particularly in relation to pests and diseases, can lead to reduced crop yields and quality, further exacerbating the challenge of meeting the increasing demand for food. Therefore, understanding the impact of climate change on pests and plant diseases has become paramount for developing effective control strategies.

Since the Industrial Revolution, greenhouse gas concentrations have escalated to alarming levels, triggering a cascade of repercussions impacting temperature, carbon dioxide (CO_2), precipitation and UV radiation, as well as the incidence and frequency of extreme weather events (i.e. droughts, heat waves, floods) (IPCC, 2023). These disturbances can lead to changes in pest and disease dynamics in agriculture, with potential for both positive and negative effects (Das *et al.*, 2011).

Temperature is a critical factor impacting the prevalence and distribution of agricultural pests and diseases, leading to an expansion of their geographic range, increasing

*Corresponding author: clara.lago@ifisc.uib-csic.es

 Biological Control Systems and Climate Change (eds Joan van Baaren *et al.*)
DOI: 10.1079/9781800625099.0011

overwintering survival and affecting interaction with host plants and natural enemies (Das *et al.*, 2011). Elevated CO_2 affects plant biochemistry, which may in turn impact insect fecundity, feeding behaviour, survival and population dynamics (Trębicki *et al.*, 2017). Additionally, UV radiation can have both positive and negative effects on agriculture. It can enhance plant responses by inducing alterations in secondary metabolism, enhancing photoprotection and resistance to pests and diseases (Wargent and Jordan, 2013). However, some studies show that altered UV radiation can also negatively affect plant growth, development and yield (Liaqat *et al.*, 2024). Finally, extreme weather events can directly damage crops and create conditions that favour the proliferation of pests and diseases (Sivakumar, 2020).

The use of agrochemicals has been a predominant strategy to control pests and diseases. The frequent application of these products has led to the development of resistance in various organisms to insecticides, herbicides and fungicides (Sjut, 1997) and has resulted in environmental contamination and the infiltration of these compounds into the food chain, causing health, social and economic issues (Song *et al.*, 2017).

Consequently, crop protection research has shifted its focus toward integrated plant protection (IPP) and integrated pest management (IPM). IPP and IPM are sustainable strategies aimed at managing pests in agriculture, combining biological, cultural, physical and chemical tools. These methods aim to minimize economic, health and environmental risks, with a focus on reducing the use of synthetic pesticides. In this context, the use of biological control agents (BCAs) has become a key component of IPP and IPM strategies, which when combined with other management methods can provide more effective pest control and disease management (Galli *et al.*, 2024).

Several organisms have been proven to be effective as BCAs. Natural enemies, including predators, parasitoids and micro-organisms, play a crucial role in biological control, helping to suppress crop pests and maintain a stable pest equilibrium (Hajek and Eilenberg, 2018). Additionally, fungi such as *Trichoderma* spp., bacterial agents (e.g. *Pseudomonas chlororaphis*, *Bacillus thuringiensis*) (Thilagam *et al.*, 2023) and insect viruses (mainly baculoviruses) (Sun, 2015) have been proven to be effective for controlling plant diseases and pests. In this chapter, we focus on bacteria and viruses used as BCAs (Fig. 11.1).

In crop protection, interactions between plants and pathogens are of paramount importance since this relationship is highly specific and differs between pathosystems. When

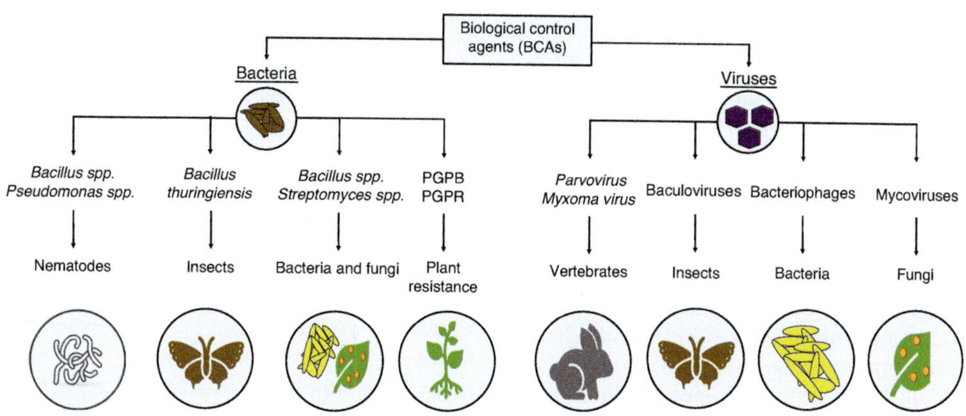

Fig. 11.1. Overview of bacteria and viruses as biological control agents, including BCA examples and the pathogen or pest. PGPB, plant growth promoting bacteria; PGPR, plant growth-promoting rhizobacteria.

integrating BCAs into IPP and IPM programmes, the plant–pathogen–BCA complex becomes the central focus. This depends on the biocontrol agent implemented (mechanism of action, conditioning, dose, methods of application), the pests/pathogens target (sensitivity) and the host (cultivar type, physical properties), and is shaped by the environmental conditions (Montesinos and Bonaterra, 2009). Therefore, climate change adds another layer of complexity and uncertainty to crop protection. Consequently, besides choosing the optimal BCAs, understanding pathogen dynamics and interaction with the host plant in a specific environment is crucial to developing effective IPP and IPM and considering possible disturbances induced by climate change.

This highlights the need for a deep understanding of the effects of climate change on BCAs, which is critical for addressing the challenges of biocontrol under a climate change context. In this chapter, we will describe bacteria and viruses used in biocontrol, explain their advantages and limitations, provide insights into the interactions between the environment and the plant–pathogen–BCA complex, and highlight challenges imposed by climate change.

11.2 Climate Change Interactions with Biocontrol Strategies Based on the Use of Bacteria as Biocontrol Agents

11.2.1 Bacteria as biological control agents

In agriculture, bacteria-based BCAs have the potential to suppress the activity of several pests and pathogens. They are particularly effective for controlling plant parasitic nematodes (PPN) (Migunova and Sasanelli, 2021) and insect pests (Bravo *et al.*, 2011). Furthermore, some bacterial groups, specifically *Bacillus* spp., have demonstrated the ability to produce secondary metabolites with antimicrobial activity, making them good candidates for controlling bacterial and fungal pathogens (Raaijmakers *et al.*, 2010). Additionaly, plant growth-promoting bacteria (PGPBs) are soil bacteria that enhance plant growth and provide biocontrol against plant pathogens. Specifically, the plant growth-promoting rhizobacteria (PGPR) are the most well studied among them (Ayaz *et al.*, 2023).

Despite bacteria-based BCAs presenting a series of advantages, their effectiveness is often not equal to synthetic agrochemicals, requiring their integration with other management tools (Jacobsen *et al.*, 2004). Furthermore, using bacteria-based BCAs presents some limitations, including the need for a better understanding of their interactions with plant hosts and microbiomes, as well as with the environment. Moreover, although bacterial BCAs have demonstrated significant control in laboratory and greenhouse studies, field trial results have been inconsistent (Akhtar and Siddiqui, 2011). Despite these challenges and limitations, bacteria-based BCAs have great potential in IPP and IPM systems, when used in combination with other disease management tools.

Bacteria-based BCAs against nematodes

Bacillaceae spp. and Pseudomonadaceae spp. are efficient against PPNs, as they produce nematicidal and antimicrobial metabolites, compete for ecological niches against PPNs and enhance plant defence (Migunova and Sasanelli, 2021). The nematicidal commercial bioproduct containing the bio-agent *Bacillus megaterium* (Bioarc®) has been proven to reduce root galling in tomato, caused by the root-knot nematode *Meloidogyne incognita*, by 89.20%. Similarly, field applications of *B. thuringiensis* var. kurstaki (LBT-3 strain) resulted in average nematode population reductions of 87%, mainly targeting *Radopholus similis* (Fernández *et al.*, 2005). Furthermore, *B. subtilis* has proven to be efficient in controlling root knot and lesion nematodes on sugarcane and common bean (Mazzuchelli *et al.*, 2020), while *B. aryabhattai* exhibited suppressive effects against *Heterodera glycines* on soybean (Zhao *et al.*, 2011). Regarding the Pseudomonadaceae family, *P. fluorescens* isolates induce defence enzymes in tomatoes against root-knot nematodes (Nikoo *et al.*, 2014).

Bacteria-based BCAs against insects

More than 100 bacteria have been identified as pathogenic against insects, emerging as promising biopesticides against insect pests. *Bacillus thuringensis* (*Bt*) is the most widely used for insect biocontrol (Bravo *et al.*, 2011). *Bt* targets insects from different orders at larval stage and kills them by disrupting the midgut tissue followed by septicaemia (Raymond *et al.*, 2010). The insecticidal activity of *Bt* comes from Cry toxins, which disrupt cell membranes upon ingestion, causing insect mortality (Khachatourians, 1986). Most research on the toxic properties of *Bt* has been performed on Lepidoptera and, to a lesser extend, on Diptera and other groups of insects (Bravo *et al.*, 2011). Similarly, *Bacillus sphaericus* produces an intracellular toxin during the sporulation phase with insecticide properties, which are extensively used against mosquitoes (Park *et al.*, 2010). Moreover, many strains of the *Pseudomonas* group have exhibited potential as BCAs due to their ability to colonize insects, even inducing lethal infections. Specifically, the *P. chlororaphsis* subgroup displays insecticidal activities when injected and also when ingested by insect larvae (Flury *et al.*, 2016).

Bacteria-based BCAs against fungal and bacterial diseases

Several bacteria groups have been tested as candidates for controlling bacterial and fungal diseases. Among these groups, *Bacillus* spp. can produce a wide range of bioactive compounds that play a crucial role against plant pathogens. These compounds include metabolites with antimicrobial activities, surface-actives, that are involved in the induction of plant defence responses. For example, certain *Bacillus* spp. produce bacteriocins with demonstrated antimicrobial properties against plant pathogens with a wide host range such as *Agrobacterium tumefaciens* (Mouloud *et al.*, 2013) or *Pseudomonas savastanoi* and *P. syringae* (Ugras *et al.*, 2013), when tested on an agar spot assay. Moreover, particular *Bacillus* spp. and *Pseudomonas* spp. produce cyclic lipopeptides (CLPs), which have demonstrated antifungal activity under laboratory conditions (Raaijmakers *et al.*, 2010). These studies suggest that these bacterial secondary metabolites have biocontrol potential, although further studies on plants and under field conditions need to be performed.

Beyond the antimicrobial activity, some *Bacillus* spp. produce metabolites that trigger plant defence response. For example, *B. amyloliquefaciens* FZB42 produces secondary metabolites that promote gene expression involved in induced plant defence, efficiently reducing lettuce bottom rot (Chowdhury *et al.*, 2015).

Plant growth-promoting bacteria (PGPBs)

Plant growth-promoting bacteria, specifically PGPRs, enhance plant performance by root colonization, provide biocontrol against pathogens, enhance plant defence and stimulate plant growth (Ayaz *et al.*, 2023). PGPRs enable plants to perform efficient nitrogen fixation (Srivastava *et al.*, 2016) and phosphate solubilization, access iron due to siderophore generation (Beck *et al.*, 2020) or produce hydrogen cyanide (HCH) (Rivera-Méndez *et al.*, 2020). PGPRs also produce phytohormones such as auxins, gibberellins and cytokinins which are involved in several physiological processes critical for plant development, plant defence and resistance to abiotic stress (such as drought) (Issazadeh *et al.*, 2012; Prasanth *et al.*, 2015). PGPRs also produces auxins which augment root growth and structure, promote plant growth and regulate the expression of genes involved in plant defence systems (Srivastava *et al.*, 2016).

11.2.2 Impact of climate change on bacteria-based biocontrol strategies

Climate change impacts several environmental factors which may affect bacteria-based BCAs. Environmental disturbances impact pests, pathogens, plants and the interactions between them, and affect the frequency, intensity and duration of pests and diseases. Indeed, one of the major challenges in biocontrol is to transfer the efficacy of BCAs from the laboratory to the field, as BCAs are highly sensitive to environmental conditions (Barratt *et al.*, 2018). However, knowledge on how climate change affects bacteria-based BCAs is still limited.

Most research focuses on how climate change impacts rhizobacteria communities, although some research has been also performed on its impact on *Bt*-transgenic plants. Some climate change disturbances such as elevated CO_2 (eCO_2) or increased temperature affect plant physiology, and these physiological changes alter the microbial communities present in the soil (Bei *et al.*, 2019) via root exudates (Rangel-Castro *et al.*, 2005) and plant metabolites (Rasche *et al.*, 2009). Besides, several PGPBs enhance plant defence against climate change-induced stresses, such as drought and elevated temperatures through various mechanisms, including enhancing plant growth, nutrient uptake and stress tolerance (Khan and Mehmood, 2023).

In this section, we will explain how climate change impacts different bacteria-based biocontrol strategies and the role of bacteria in mitigating the effects of climate change, and discuss limitations and future perspectives.

Climate change and Bacillus thuringensis

Environmental conditions affect the efficacy of *Bt*-transgenic plants against insect pests (Selvaraj *et al.*, 2013). The expression of Cry toxins in *Bt*-transgenic plants is highly dependent on the environmental conditions. Kaiser (1996) reported that *Bt*-cottons were destroyed by the cotton bollworm *Heliothis virescens* since high temperatures reduced the production of *Bt*-toxins. Elevated CO_2 also impacts the efficacy of *Bt*-transgenic crops. Changes in the nitrogen and carbon proportion in plants caused by eCO_2 reduced the expression of *Bt*-toxin in cotton leaves (Coviella *et al.*, 2002). However, these changes may occur by increasing herbivory and thus a higher intake of *Bt* toxins (Coviella and Trumble, 2000). Since the environmental conditions are critical to the efficacy of *Bt*-transgenic crops, it is of the utmost importance to understand the effects imposed by climate change.

Impact of climate change on plant growth-promoting bacteria

One critical factor that impacts the structure of the rhizosphere is the CO_2 level, which affects the dominance of specific bacterial groups and the abundance ratio between bacteria and fungi. Many studies have revealed that eCO_2 has a positive effect on PGPBs present in the soil, although this may be conditioned by other factors such as temperature (Compant *et al.*, 2010).

Bei et al. (2019) observed in a field trial that eCO_2, in combination with sustained summer heat, notably modified the composition of the rhizosphere in European grassland. In this study, it was observed that Bacteria abundances significantly increased relative to Eukarya, due to an enrichment of Actinobacteria and Proteobacteria and a significant decline in fungal populations. Furthermore, analyses suggested an increment in plant production of secondary metabolites, which had a positive impact on plant performance against phytopathogenic fungi such as Leotiomycetes. Similarly, *Pseudomonas* communities, which inhibit growth of parasitic fungi and are used as PGPRs, are also affected by eCO_2 (Marilley *et al.*, 1999). In contrast, a study conducted in *Lolium perenne* and *Medicago coerula* showed that eCO_2 decreased the frequencies of *Pseudomonas* strains which are able to produce HCN (Tarnawski *et al.*, 2006). However, in the same study the authors observed that the proportion of siderophore producers and nitrate-dissimilating strains increased. Drigo *et al.* (2009) showed different responses of eCO_2 effects on *Pseudomonas* spp. and *Burkholderia* spp. and on production of antimicrobial compounds, depending on plant genotype and the soil, while *Actinomycetes* spp. and *Bacillus* spp., which are also antibiotic producers, were not affected. These variable research outcomes show that eCO_2 exhibits both positive and negative effects on bacteria used as BCAs.

In addition, some PGPB strains which grow better at higher temperatures can be of particular agricultural interest in warmer climates, although bacterial performance depends on temperature and soil type. For example, rhizobia isolated from nodules of the desert woody legume *Prosopis glandulosa* exhibited better growth at 36°C compared to 26°C (Waldon *et al.*, 1989).

Beneficial bacteria and their role in mitigating climate change effects on plants

Several PGPBs used as BCAs directly enhance the resilience of plants against climate change, protecting crops from several abiotic stressors. *Pseudomonas mendocina* strain enhanced the growth of lettuce plants under eCO_2 (Kohler *et al.*, 2009). Plants inoculated with *P. mendocina* exhibited the highest plant biomass production and the highest foliar potassium levels and leaf relative water content under eCO_2 and drought conditions. In contrast, the effects of elevated temperature vary widely depending on studied system.

Some PGPRs and endophytes have the capacity to mitigate thermal stress in plants (Bilal *et al.*, 1993; Bashan, 1998). A study performed with PGPBs belonging to different species and strains isolated from different geographic areas with different climates revealed that bacterial efficacy in protecting plants from temperature stress is dependent on the bacterial isolate (Egamberdiyeva and Höflich, 2003). In this study, *Mycobacterium* spp., *Pseudomonas fluorescens* and *Pantoea agglomerans* isolated from a semi-continental climate enhanced root and shoot growth of winter wheat at 16°C compared to growth at 26°C in loamy sand (Waldon *et al.*, 1989). Furthermore, bacteria which colonize different niches may exhibit different responses to specific environmental conditions. Some rhizosphere bacteria and endophytes have been shown to reduce temperature stress on plants, and these strains may induce growth promotion of different crops at different climates, soils and temperatures (Bilal *et al.*, 1993; Javed and Arshad, 1997). In addition, rhizosphere and endophytic bacteria are able to enhance plant resistance to temperature or drought stress by inducing plant systemic responses (Aroca and Ruiz-Lozano, 2009).

Rhizobacteria also have several mechanisms that enhance plant response to drought. PGPRs impact phytohormone synthesis, osmotic tuning, antioxidant machinery, stomatal conductivity, nutrient uptake, the release of volatile organic compounds (VOCs) and exopolysaccharide production (Sati *et al.*, 2023). Moreover, some strains of Actinobacteria induce plant resistance to drought by accelerating the lignification of sieve elements and inducing callose accumulation (Hasegawa *et al.*, 2005). In addition, wheat, maize and sorghum seedlings inoculated with *Azospirillum* strains exhibited better water status than control plants and an increase in crop production when plants were under drought conditions (Sarig *et al.*, 1988; Creus *et al.*, 1998). It has also been reported that maize plants inoculated with *A. brasilense* accumulated higher levels of trehalose, exhibiting an enhanced drought tolerance and a significant increase in leaf and root biomass (Rodríguez-Salazar *et al.*, 2009). Similarly, *Burkholderia phytofirmans* strain PsJN demonstrated positive effects on potato and Cucurbitaceae plants under drought stress (Nowak *et al.*, 1995).

Biological control agents also play a crucial role in improving plant performance by interfering with the phytohormone plant system. Auxin-producing PGPRs have been shown to regulate the expression of various genes involved in plant growth and defence (Issazadeh *et al.*, 2012). Indole-3-acetic acid (IAA), an auxin produced by PGPR, has gained much attention for plant growth promotion under drought stress; IAA induces root initiation and favours plant searching for moisture, conferring improved drought resistance (Gong *et al.*, 2015). Karimi *et al.* (2021) reported that wheat plants inoculated with *Azospirillium* improved water status in leaves by modulating IAA. In line with these results, other authors reported that different PGPR consortiums, i.e. of *Bacillus megatherium* and *Pseudomonas putida*, improved IAA level and thus improved water status in clover plants (Marulanda *et al.*, 2009), *Phyllobacterium brassicacearum* in *A. thaliana* plants (Bresson *et al.*, 2014) and *B. thuringiensis* in lavender plants (Armada *et al.*, 2014).

Auxin-producing *Bacillus amyloliquefaciens* enhanced grain production under drought stress (Raheem *et al.*, 2018) and its application in wheat plants triggered elongation of root hairs and increased hair density, improving plant water uptake. Other authors reported that inoculation with bacteria which produce aminocyclopropane-1-carboxylic acid deaminase (ACCd) improved water use efficiency in several crops. For example, it was observed that when chickpea and wheat plants were inoculated simultaneously with nitrogen-fixing bacteria and ACCd-producing bacteria, they exhibited enhanced drought tolerance (Shakir *et al.*, 2012).

Bacteria used as BCAs under climate change: conclusions and outcomes

Available research indicates that climate change, especially eCO_2, impacts PGPBs and PGPRs. However, effects vary, depending on the bacterial species, plant type and soil conditions. Although to a lesser extent, some research also shows that temperatures may also impact the efficacy of certain beneficial bacteria. Moreover, several studies show that some PGPRs enhance plant resilience to climate change-induced stresses, such as drought and elevated temperatures, through various mechanisms. These include enhancing plant growth and nutrient uptake, and influencing phytohormone synthesis, osmotic adjustment and antioxidant activity.

Given the variability in how different bacterial species respond to climate change, it is crucial to conduct further research on the comprehensive effects of climate change on bacterial BCAs, including exploring the potential of specific bacterial strains for use in agriculture and their effectiveness in mitigating climate change effects on plants.

11.3 Climate Change Interactions with Biocontrol Strategies Based on the Use of Viruses as Biocontrol Agents

11.3.1 Viruses as biological control agents

Viruses are obligate parasites, which need a host to reproduce. As they can infect cellular organisms of all domains of life, they have emerged as good candidates as BCAs against different diseases and pests in agriculture. Their specificity, environmental safety and compatibility with other management practices make them attractive for IPP and IPM programmes (Wagemans *et al.*, 2022). Insect viruses are the most extensively studied and used viral BCAs in agriculture, especially the group of baculoviruses (BV). Baculoviruses are arthropod-specific viruses that offer significant advantages for insect pest control. Their high host specificity and environmental compatibility make them ideal for pest management, as they do not harm beneficial insects or non-target organisms (McCutchen and Flexner, 1999). To a lesser extent, some viruses have been used in controlling vertebrates. The feline panleukopenia virus (parvovirus) was used to control cats on Marion Island and the myxoma virus (MYXV) and rabbit haemorrhagic disease virus (RHDV) were released in Australia and New Zealand to eliminate rabbits (Di Giallonardo and Holmes, 2015). However, the use of viruses that cause disease to vertebrate pests can be concerning due to the risk associated with affecting non-target organisms, including humans.

Concerning other diseases, some studies have explored the potential of using mycoviruses against phytopathogenic fungi (Kotta-Loizou, 2021) and bacteriophages against plant pathogenic bacteria (Holtappels *et al.*, 2021) although more research is needed before they can be proposed as an effective tool.

Viral-based BCAs against insect pests

The use of viruses for controlling insect and mite pests in agriculture has grown in recent decades (Lacey *et al.*, 2015). More than 200 insect viruses have been isolated from several families including Baculoviridae, Reoviridae, Densovirinae and Entomopoxvirinae (Sun, 2015). Baculoviruses are the most extensively used, specifically nucleopolyhedroviruses (NPV; *Alphabaculovirus* spp.) and granuloviruses (GV; *Betabaculovirus* spp.), which have been developed for commercial use (Eberle *et al.*, 2012). During their infective stage, BVs produce protective occlusion bodies (OB) formed by crystalline protein, which varies depending on the group, that allow them to reproduce within an insect host (Sajjan and Hinchigeri, 2016). Their robust nature allows their long-term storage and formulation, making them favourable for commercial development (Rowley *et al.*, 2011). Infection by BVs generally occurs during host ingestion. After ingestion, BV virions bind to the receptors in the insect's gut, penetrate epithelial cells and initiate infection. Subsequently, infection often spreads to the haemocoel and then to organs and tissues, particularly to fat bodies, inducing host mortality within 5–14 days (Arrizubieta *et al.*, 2014).

Baculoviruses have a restricted host range, within the order or the family of the host.

In the Insecta order, BV targets Lepidoptera, Hymenoptera, Diptera, Coleoptera and Neuroptera (Wagemans *et al.*, 2022). These viruses have been successfully commercialized for controlling various pests in agriculture and forestry worldwide (Gelaye and Negash, 2023). They are particularly effective against lepidopteran caterpillars, with the potential to trigger epizootics in insect populations. To provide some examples, the codling moth granulovirus (CpGV) has been proven to effectively control *Cydia pomonella* caterpillars (Lacey *et al.*, 2008). Also, BVs have been used as efficient BCAs against *Spodoptera* spp. (Sun, 2015) and *Helicoverpa* spp. (Rowley *et al.*, 2011).

Baculoviruses also present some limitations such as slow speed of action and field instability due to their sensitivity to UV light (McCutchen and Flexner, 1999). To address these limitations, genetic engineering approaches have been developed to improve their insecticidal activity, infectivity and UV resistance (Gelaye and Negash, 2023).

Some research has been performed on non-BV entomopathogenic viruses, although their development for use in crop protection is still limited. These studies include the use of tetraviruses under field conditions against heliothines (Lepidoptera) in Australia (Christian and Scotti, 2005) and cypoviruses against oil palm pests in South America (Zeddam *et al.*, 2003). However, none of those viruses is close to commercialization.

Viral-based BCAs against pathogenic bacteria

The use of bacterial viruses or bacteriophages as BCAs in agriculture could be promising for crop protection, although currently only a limited number of products have been commercialized. In the USA, a few phage cocktails have been authorized so far, all from the OmniLytics™ company (Wagemans *et al.*, 2022). The AgriPhage® cocktail contains bacteriophages for controlling *Xanthomonas campestris* pv. *vesicatoria* and *Pseudomonas syringae* pv. *tomato*, in tomato and pepper plants. Cocktails against bacterial canker disease on tomatoes and citrus, and against *Erwinia amylovora* fire blight on apple and pear trees have also been designed. A cocktail against *Xanthomonas citri* subsp. *citri* reduced citrus canker disease severity by 59% under glasshouse conditions (Balogh *et al.*, 2008). Unfortunately, no specific information on the phage contents of the mentioned cocktails was described in the literature. Finally, the recently approved cocktail XylPhi-PD®, developed by the Otsuka Pharmaceutical subsidiary A&P Inphatec, according to the company reduces Pierce's disease in grapevines in California. In Europe, no phages have been registered as pesticides, except in Hungary where the government briefly authorized the use of a cocktail containing two phage species, PhiEaH1 (Meczker *et al.*, 2014) and PhiEaH2 (Dömötör *et al.*, 2012), for reducing fire blight cases in field experiments without artificial infection.

The use of phages in crop protection, although promising, still requires further research, since it is critical to deeply understand the pathosystem to select effective application methods and cocktails, and perform environmental assessments to ensure safe application.

Viral-based BCAs against pathogenic fungi

Mycoviruses are obligate parasites, which are often associated with latent infections in fungi. Despite some successful cases of using mycoviruses as BCAs, to date no mycovirus-based products have been registered for commercial use. The first case was the use of mycovirus *Cryphonectria hypovirus* 1 (CHV1) against the pathogenic fungus *Cryphonectria parasitica* for the control of chestnut blight (*Cryphonectria parasitica*) in orchards and forests (Anagnostakis, 1982). In Europe, the process of using CHV1 for controlling chestnut blight has achieved some therapeutic success, although the long-term persistence is more limited (Wagemans *et al.*, 2022). In contrast, attempts in the USA to introduce a biocontrol strategy based on CHV1 have failed, mainly due to the higher susceptibility of American chestnut (Meynadier *et al.*, 1977). Another successful case was published by Zhang *et al.* (2020) in which the authors showed successful biocontrol using SsHADV-1 against *Sclerotinia sclerotiorum*, exhibiting results similar to chemical fungicides and improving seed yield in rapeseed crops.

These approaches highlight both the successes and limitations of using mycoviruses in the biological control of fungal diseases, emphasizing the need for strategies tailored to local conditions and the potential of biotechnological innovations.

11.3.2 Impact of climate change on viruses as biocontrol agents

There are many environmental factors that can be affected by climate change, including temperature, UV radiation, rainfall and wind patterns that can influence viral-based BCAs, pathogens, host plants and the interactions between the three elements. After reviewing the literature, available research exploring the impact of climate change on viral-based BCAs focuses on BV since they are the most researched and used in crop protection.

One critical factor that affects biocontrol strategies against insect pests is temperature. Viruses used as BCAs are not particularly affected by temperature directly but the development of insects, being ectothermic organisms, depends on temperature (Herms, 2004). Temperature affects the developmental rate of insects and number of generations, as well as egg development and hatching (Gilbert and Raworth, 1996; Lago *et al.*, 2023). Therefore, fluctuations in temperature will impact the population dynamics of insect pests. As stated in Chapter 1, when mean temperatures increase, insects developmental rates rise, leading to more generations per year. Thus, a greater number of treatments need to be applied for pest control, affecting control management costs and the times for the release of viruses used as BCAs.

Baculoviruses are particularly sensitive to UV radiation. Indeed, UV radiation is one of the limiting factors for field application of these products. Nowadays, commercial products incorporate UV protectors to improve virus survival. In their extensive review, Anggraini *et al.* (2022) described several methods to prevent UV inactivation of BV, including the use of UV protectants, the selection of insect BVs resistant to UV irradiation and enhancing photostability of BVs by genetic engineering. However, future expected disturbances in UV radiation may compromise the viability of commercial products.

Other factors that impact virus persistence in the field are changes in rainfall patterns and wind intensity. Heavy rainfall increases virus wash-off from leaves, while strong winds may reduce the presence of insect viruses in the soil (Jaques, 1977). Therefore, virus-based biocontrol products need to be regularly supplemented with substances that enhance the adhesion of viral particles to plant surfaces.

Complex interactions between viruses, insects and host plants may also be affected by disturbed environmental factors, affecting virus transmission dynamics. MacDonald *et al.* (2023) studied how the use of specific NPVs against the western tent caterpillar (*Malacosoma californicum pluviale*) may be affected by high temperatures. Specifically, the authors analysed the impact of high temperatures on viral transmission and the mortality rate of the pest. They observed that caterpillars died more rapidly from virus infection as temperature increased, but no difference in the production of viral transmission stages was reported. It was observed that, despite larval movement and feeding increasing at warmer temperatures, and thus higher viral transmission and higher larval mortality rates are expected, the likelihood of preinfected larvae dying was double under low temperature conditions. This resulted in higher virus mortality and a higher transmission rate at lower temperatures. Thus, the reduced movement of infected larvae at lower temperatures raised the risk of infection between these social insects. This effect was more relevant for virus transmission than the increase in larval activity under warmer conditions. In addition, in response to different stresses (i.e.higher temperatures, drought), plants secrete components that may interact with viral BCAs affecting their effectiveness. In response to stress, many plants produce sclerotized and thickened leaves, reducing ingestion by insect larvae and reducing virus uptake by target insects.

Viruses used as BCAs under climate change: conclusions and outcomes

The complex interactions between viral BCAs, insect pests and plants under climate change pose significant challenges for the effectiveness of biocontrol strategies. One critical factor for BV application is their sensitivity to UV radiation, and so improving UV resistance has become of paramount importance for the use of BV-based products in the field. In addition, higher temperatures may increase virus transmission and insect mortality rates, yet lower temperatures might enhance virus transmission due to reduced insect movement. Furthermore, stress responses in plants can reduce feeding activity of insects, thus reducing virus uptake and diminishing the efficacy of viral BCAs. Finally, since knowledge on the use of phages and mycoviruses in biocontrol is more limited, future studies should also consider the effect of climate change for future implementation of these agents as effective BCAs.

11.4 Conclusion

Climate change affects several environmental factors that shape the interactions between BCAs, target pests or pathogens and host plants. Understanding the responses of pests, pathogens and their BCAs to climate change is crucial for effective biocontrol management in the future. After reviewing the existing literature, it is evident that further research is necessary to understand the effects of climate change on bacteria and viruses used as BCAs. Some studies explore how temperature, CO_2, rainfall and UV radiation impact specific BCAs, pathogens/pests or host plants under controlled conditions. However, research exploring the effect of climate change on the interactions between these three elements is still limited. In addition, most research is based on laboratory or glasshouse experiments, while field-based experiments, long-term studies and modelling based on future climate change predictions are scarce. Furthermore, most research regarding climate change primarily focuses on bacteria-based BCAs while little is known about the effects on viral-based BCAs.

The use of bacteria and viruses in biocontrol has grown in recent years, since they are environmentally safe and highly specific, avoiding potential harm to non-target organisms. Therefore, they have been implemented as a crucial element of IPP and IPM strategies. However, the use of bacteria and viruses in biocontrol also presents several challenges. They are less effective in crop protection when compared to agrochemical compounds. Also, there are inconsistencies in performance depending on the pathosystems, host plant and environmental conditions, and their stability sometimes compromises their use in the field. Therefore, further research is needed to enhance their effectiveness and extend their persistence.

Available research shows that eCO_2 and temperature fluctuations impact PGPBs and PGPRs, altering rhizosphere composition and potentially increasing plant performance and defence. However, the effects vary depending on bacterial species, strains, host plant and soil conditions. Additionally, PGPBs and PGPRs have been proven to enhance plant resistance to abiotic stress caused by climate change, such as drought. Therefore, they are emerging as a promising tool to be implemented in crop protection to mitigate the negative impacts of climate change on agriculture. In addition, research shows that viral-based biocontrol strategies against insect pests can be affected by fluctuations in temperature, since the pests' life cycle is highly dependent on temperature. Moreover, UV radiation and altered rainfall and wind patterns have also been proven to compromise the effectiveness of viral-based BCAs in the field.

Overall, bacterial and viral BCAs emerge as powerful tools in crop protection, capable of being implemented against phytopathogenic organisms and pests and mitigating the effects of abiotic stress caused by climate change to plants. However, their implementation still faces several limitations that need to be addressed. Therefore, further research should be conducted to improve biocontrol-based strategies, under a changing climate context.

References

Akhtar, M.S. and Siddiqui, Z.A. (2011) Role of plant growth promoting rhizobacteria in biocontrol of plant diseases and sustainable agriculture. *Plant Growth and Health Promoting Bacteria* 18, 157–195.

Anagnostakis, S.L. (1982) Biological control of chestnut blight. *Science* 215(4532), 466–471.

Anggraini, E., Hong, L.W., Vadamalai, G., Kong, L.L. and Mat, M. (2022) The effect and possible mitigation of UV radiation on baculoviruses as bioinsecticides. *Biodiversitas Journal of Biological Diversity* 23(7), 3721–3735.

Armada, E., Roldán, A. and Azcón, R. (2014) Differential activity of autochthonous bacteria in controlling drought stress in native *Lavandula* and *Salvia* plants species under drought conditions in natural arid soil. *Microbial Ecology* 67(2), 410–420.

Aroca, R. and Ruiz-Lozano, J.M. (2009) Induction of plant tolerance to semi-arid environments by beneficial soil microorganisms–a review. *Environmental Science Agricultural and Food Sciences Biology* 2, 121–135.

Arrizubieta, M., Williams, T., Caballero, P. and Simon, O. (2014) Selection of a nucleopolyhedrovirus isolate from *Helicoverpa armigera* as the basis for a biological insecticide. *Pest Management Science* 70(6), 967–976.

Ayaz, M., Li, C.-H., Ali, Q., Zhao, W., Chi, Y.-K. *et al*. (2023) Bacterial and fungal biocontrol agents for plant disease protection: Journey from lab to field, current status, challenges, and global perspectives. *Molecules* 28, 18.

Balogh, B., Canteros, B.I., Stall, R.E. and Jones, J. (2008) Control of citrus canker and citrus bacterial spot with bacteriophages. *Plant Disease* 92(7), 1048–1052.

Barratt, B.I.P., Moran, V.C., Bigler, F. and Van Lenteren, J.C. (2018) The status of biological control and recommendations for improving uptake for the future. *BioControl* 63, 155–167.

Bashan, Y. (1998) Inoculants of plant growth-promoting bacteria for use in agriculture. *Biotechnology Advances* 16(4), 729–770.

Beck, C., Garzón, J.F.G. and Weber, T. (2020) Recent advances in Re-engineering Modular PKS and NRPS Assembly Lines. *Biotechnology and Bioprocess Engineering* 25(6), 886–894.

Bei, Q., Moser, G., Wu, X., Müller, C. and Liesack, W. (2019) Metatranscriptomics reveals climate change effects on the rhizosphere microbiomes in European grassland. *Soil Biology and Biochemistry* 138, 107604.

Bilal, R., Rasul, G., Arshad, M. and Malik, K.A. (1993) Attachment, colonization and proliferation of *Azospirillum brasilense* and *Enterobacter* spp. On root surface of grasses. *World Journal of Microbiology and Biotechnology* 9, 63–69.

Bravo, A., Likitvivatanavong, S., Gill, S.S. and Soberón, M. (2011) *Bacillus thuringiensis*: A story of a successful bioinsecticide. *Insect Biochemistry and Molecular Biology* 41(7), 423–431.

Bresson, J., Vasseur, F., Dauzat, M., Labadie, M., Varoquaux, F. *et al*. (2014) Interact to survive: *Phyllobacterium brassicacearum* improves *Arabidopsis* tolerance to severe water deficit and growth recovery. *PLOS One* 9(9), e107607.

Chowdhury, S.P., Uhl, J., Grosch, R., Alquéres, S., Pittroff, S. *et al*. (2015) Cyclic lipopeptides of *Bacillus amyloliquefaciens* subsp. *plantarum* colonizing the lettuce rhizosphere enhance plant defense responses toward the bottom rot pathogen *Rhizoctonia solani*. *Molecular Plant–Microbe Interactions* 28(9), 984–995.

Christian, P.D. and Scotti, P.D. (2005) The invertebrate small RNA viruses; changing perceptions and 40 years of evolving taxonomy. *Bee Research and Virus in Europe (BRAVE) Symposium Proceedings* 11–43.

Compant, S., Van Der Heijden, M.G. and Sessitsch, A. (2010) Climate change effects on beneficial plant–microorganism interactions. *FEMS Microbiology Ecology* 73(2), 197–214.

Coviella, C.E. and Trumble, J.T. (2000) Effect of elevated atmospheric carbon dioxide on the use of foliar application of *Bacillus thuringiensis*. *Biocontrol* 45(3), 325–336.

Coviella, C.E., Stipanovic, R.D. and Trumble, J.T. (2002) Plant allocation to defensive compounds: Interactions between elevated CO_2 and nitrogen in transgenic cotton plants. *Journal of Experimental Botany* 53(367), 323–331.

Creus, C.M., Sueldo, R.J. and Barassi, C.A. (1998) Water relations in *Azospirillum*-inoculated wheat seedlings under osmotic stress. *Canadian Journal of Botany* 76(2), 238–244.

Das, D.K., Singh, J. and Vennila, S. (2011) Emerging crop pest scenario under the impact of climate change – a brief review. *Journal of Agricultural Physics* 11, 13–20.

Di Giallonardo, F. and Holmes, E.C. (2015) Viral biocontrol: Grand experiments in disease emergence and evolution. *Trends in Microbiology* 23(2), 83–90.

Dömötör, D., Becságh, P., Rákhely, G., Schneider, G. and Kovács, T. (2012) Complete genomic sequence of *Erwinia amylovora* phage *PhiEaH2*. *Journal of Virology* 86(19), 10899.

Drigo, B., Veen, J.A. and Kowalchuk, G.A. (2009) Specific rhizosphere bacterial and fungal groups respond differently to elevated atmospheric CO2. *ISME Journal* 3(10), 1204–1217.

Eberle, K.E., Jehle, J.A. and Huber, J. (2012) Microbial control of crop pests using insect viruses. In: Singh, D.P. (ed.) *Integrated Pest Management: Principles and Practice*. CABI, Wallingford, UK, pp. 281–298.

Egamberdiyeva, D. and Höflich, G. (2003) Influence of growth-promoting bacteria on the growth of wheat in different soils and temperatures. *Soil Biology and Biochemistry* 35(7), 973–978.

Fernández, E., Mena, J., González, J., Márquez, M.E., Turner, D.W. *et al.* (2005) Biological control of nematodes in banana. In: Turner, D.W. and Rosales, F.E. (eds) *Banana Root System: Towards a Better Understanding for Its Management. Proceedings of an International Symposium, San José, Costa Rica, 2003*, INIBAP, Montpellier, France, pp. 193–200.

Flury, P., Aellen, N., Ruffner, B., Péchy-Tarr, M., Fataar, S. *et al.* (2016) Insect pathogenicity in plant-beneficial pseudomonads: Phylogenetic distribution and comparative genomics. *ISME Journal* 10(10), 2527–2542.

Galli, M., Feldmann, F., Vogler, U.K. and Kogel, K.-H. (2024) Can biocontrol be the game-changer in integrated pest management? A review of definitions, methods and strategies. *Journal of Plant Diseases and Protection* 131(2), 265–291.

Gelaye, Y. and Negash, B. (2023) The role of baculoviruses in controlling insect pests: A review. *Cogent Food and Agriculture* 9(1), 2254139.

Gilbert, N. and Raworth, D.A. (1996) Insects and temperature – a general theory. *Canadian Entomologist* 128(1), 1–13.

Gong, A.-D., Li, H.-P., Yuan, Q.-S., Song, X.-S., Yao, W. *et al.* (2015) Antagonistic mechanism of iturin A and plipastatin A from *Bacillus amyloliquefaciens* S76-3 from wheat spikes against *Fusarium graminearum*. *PLOS One* 10(2), e0116871.

Hajek, A.E. and Eilenberg, J. (2018) *Natural Enemies: An Introduction to Biological Control*. Cambridge University Press, New York.

Hasegawa, S., Meguro, A., Toyoda, K., Nishimura, T. and Kunoh, H. (2005) Drought tolerance of tissue-cultured seedlings of mountain laurel (*Kalmia latifolia* L.) induced by an endophytic actinomycete II. acceleration of callose accumulation and lignification. *Actinomycetologica* 19(1), 13–17.

Herms, D. (2004) Using degree-days and plant phenology to predict pest activity. In: Krischik, V. and Davidson, J. (eds) *IPM (Integrated Pest Management) of Midwest Landscapes*. Minnesota Agricultural Experiment Station, St Paul, Minnesota, pp. 49–59.

Holtappels, D., Fortuna, K., Lavigne, R. and Wagemans, J. (2021) The future of phage biocontrol in integrated plant protection for sustainable crop production. *Current Opinion in Biotechnology* 68, 60–71.

IPCC (2023) *Climate Change 2023: Synthesis Report*. Contribution of Working Groups I, II and III to the Sixth Assessment Report of the Intergovernmental Panel on Climate Change. Available at: www.ipcc.ch/report/ar6/syr/

Issazadeh, K., Rad, S.K., Zarrabi, S. and Rahimibashar, M.R. (2012) Antagonism of *Bacillus* species against *Xanthomonas campestris* pv. *Campestris* and *Pectobacterium carotovorum* subsp. *Carotovorum*. *African Journal of Microbiology Research* 6(7), 1615–1620.

Jacobsen, B.J., Zidack, N.K. and Larson, B.J. (2004) The role of *Bacillus*-based biological control agents in integrated pest management systems: Plant diseases. *Phytopathology* 94(11), 1272–1275.

Jaques, R.P. (1977) Stability of entomopathogenic viruses. In: Hostetter, D.L. and Ignoffo, C.M. (eds) *Environmental Stability of Microbial Insecticides*, Vol. 10. Entomological Society of America, College Park, Maryland, p. 99.

Javed, M. and Arshad, M. (1997) Growth promotion of two wheat cultivars by plant growth-promoting rhizobacteria. *Pakiston Journal of Botany* 29, 243–248.

Kaiser, J. (1996) Pests overwhelm Bt cotton crop. *Science* 273(5274), 423–0423.

Karimi, N., Goltapeh, E.M., Amini, J., Mehnaz, S. and Zarea, M.J. (2021) Effect of *Azospirillum zeae* and seed priming with zinc, manganese and auxin on growth and yield parameters of wheat, under dryland farming. *Agricultural Research* 10, 44–55.

Khachatourians, G.G. (1986) Production and use of biological pest control agents. *Trends in Biotechnology* 4(5), 120–124.

Khan, N. and Mehmood, A. (2023) Revisiting climate change impacts on plant growth and its mitigation with plant growth promoting rhizobacteria. *South African Journal of Botany* 160, 586–601.

Kohler, J., Caravaca, F., del Mar Alguacil, M. and Roldán, A. (2009) Elevated CO_2 increases the effect of an arbuscular mycorrhizal fungus and a plant-growth-promoting rhizobacterium on structural stability of a semiarid agricultural soil under drought conditions. *Soil Biology and Biochemistry* 41(8), 1710–1716.

Kotta-Loizou, I. (2021) Mycoviruses and their role in fungal pathogenesis. *Current Opinion in Microbiology* 63, 10–18.

Lacey, L.A., Thomson, D., Vincent, C. and Arthurs, S.P. (2008) Codling moth granulovirus: A comprehensive review. *Biocontrol Science and Technology* 18(7), 639–663.

Lacey, L.A., Grzywacz, D., Shapiro-Ilan, D.I., Frutos, R., Brownbridge, M. *et al.* (2015) Insect pathogens as biological control agents: Back to the future. *Journal of Invertebrate Pathology* 132, 1–41.

Lago, C., Giménez-Romero, À., Morente, M., Matías, M.A., Moreno, A. *et al.* (2023) Degree-day-based model to predict egg hatching of *Philaenus spumarius* (Hemiptera: Aphrophoridae), the main vector of *Xylella fastidiosa* in Europe. *Environmental Entomology* 52(3), 350–359.

Liaqat, W., Altaf, M.T., Barutçular, C., Nawaz, H., Ullah, I. *et al.* (2024) Ultraviolet-B radiation in relation to agriculture in the context of climate change: A review. *Cereal Research* 1–24.

MacDonald, P., Myers, J.H. and Cory, J.S. (2023) Warmer temperatures reduce the transmission of a virus in a gregarious forest insect. *Ecology* 104(10), e4159.

Marilley, L., Hartwig, U.A. and Aragno, M. (1999) Influence of an elevated atmospheric CO_2 content on soil and rhizosphere bacterial communities beneath *Lolium perenne* and *Trifolium repens* under field conditions. *Microbial Ecology* 38, 39–49.

Marulanda, A., Barea, J.-M. and Azcón, R. (2009) Stimulation of plant growth and drought tolerance by native microorganisms (AM fungi and bacteria) from dry environments: Mechanisms related to bacterial effectiveness. *Journal of Plant Growth Regulation* 28, 115–124.

Mazzuchelli, R. de C.L., Mazzuchelli, E.H.L. and de Araujo, F.F. (2020) Efficiency of *Bacillus subtilis* for root-knot and lesion nematodes management in sugarcane. *Biological Control* 143, 104185.

McCutchen, W.F. and Flexner, L. (1999) *Joint actions of baculoviruses and other control agents.* Use and Delivery, Biopesticides, pp. 341–355.

Meczker, K., Dömötör, D., Vass, J., Rákhely, G., Schneider, G. *et al.* (2014) The genome of the *Erwinia amylovora* phage PhiEaH1 reveals greater diversity and broadens the applicability of phages for the treatment of fire blight. *FEMS Microbiology Letters* 354(1), 27–33.

Meynadier, G., Amargier, A. and Genty, P. (1977) Une virose de type densonucleose chez le lépidoptère *Sibine fusca* Stoll. *Oléagineux* 32, 357–361.

Migunova, V.D. and Sasanelli, N. (2021) Bacteria as biocontrol tool against phytoparasitic nematodes. *Plants* 10(2), 389.

Montesinos, E. and Bonaterra, A. (2009) Microbial pesticides. In: Schaechter, M. (ed.) *Encyclopedia of Microbiology*, 3rd edn. Elsevier, St Louis, Missouri, pp. 110–120.

Mouloud, G., Daoud, H., Bassem, J., Laribi Atef, I. and Hani, B. (2013) New bacteriocin from *Bacillus clausii* strain GM17: Purification, characterization, and biological activity. *Applied Biochemistry and Biotechnology* 171, 2186–2200.

Nikoo, F.S., Sahebani, N., Aminian, H., Mokhtarnejad, L. and Ghaderi, R. (2014) Induction of systemic resistance and defense-related enzymes in tomato plants using *Pseudomonas fluorescens* CHAO and salicylic acid against root-knot nematode *Meloidogyne javanica*. *Journal of Plant Protection Research* 54(4), 383–389.

Nowak, J., Asiedu, S. and Lazarovits, G. (1995) Enhancement of in vitro growth and transplant stress tolerance of potato and vegetable plantlets co-cultured with a plant growth promoting *Pseudomonad* bacterium. *Ecophysiology and Photosynthetic in Vitro Cultures* 173–179.

Park, H.-W., Bideshi, D.K. and Federici, B.A. (2010) Properties and applied use of the mosquitocidal bacterium, *Bacillus sphaericus*. *Journal of Asia-Pacific Entomology* 13(3), 159–168.

Prasanth, M., Ramesh, N., Gothandam, K.M., Sivamangala, K. and Shanthini, T. (2015) Pseudomonas syringae: An overview and its future as a “rain making bacteria”. *International Research Journal of Biological Sciences* 4(2), 70–77.

Raaijmakers, J.M., Bruijn, I., Nybroe, O. and Ongena, M. (2010) Natural functions of lipopeptides from *Bacillus* and *Pseudomonas*: More than surfactants and antibiotics. *FEMS Microbiology Reviews* 34(6), 1037–1062.

Raheem, A., Shaposhnikov, A., Belimov, A.A., Dodd, I.C. and Ali, B. (2018) Auxin production by rhizobacteria was associated with improved yield of wheat (*Triticum aestivum* L.) under drought stress. *Archives of Agronomy and Soil Science* 64(4), 574–587.

Rangel-Castro, J.I., Killham, K., Ostle, N., Nicol, G.W., Anderson, I.C. *et al.* (2005) Stable isotope probing analysis of the influence of liming on root exudate utilization by soil microorganisms. *Environmental Microbiology* 7(6), 828–838.

Rasche, F., Lueders, T., Schloter, M., Schaefer, S., Buegger, F. *et al.* (2009) DNA-based stable isotope probing enables the identification of active bacterial endophytes in potatoes. *New Phytologist* 181(4), 802–807.

Raymond, B., Johnston, P.R., Nielsen-LeRoux, C., Lereclus, D. and Crickmore, N. (2010) *Bacillus thuringiensis*: An impotent pathogen? *Trends in Microbiology* 18(5), 189–194.

Rivera-Méndez, W., Obregón, M., Morán-Diez, M.E., Hermosa, R. and Monte, E. (2020) *Trichoderma asperellum* biocontrol activity and induction of systemic defenses against *Sclerotium cepivorum* in onion plants under tropical climate conditions. *Biological Control* 141, 104145.

Rodríguez-Salazar, J., Suarez, R., Caballero-Mellado, J. and Iturriaga, G. (2009) Trehalose accumulation in *Azospirillum brasilense* improves drought tolerance and biomass in maize plants. *FEMS Microbiology Letters* 296(1), 52–59.

Rowley, D.L., Popham, H.J. and Harrison, R.L. (2011) Genetic variation and virulence of nucleopolyhedroviruses isolated worldwide from the heliothine pests *Helicoverpa armigera*, *Helicoverpa zea*, and *Heliothis virescens*. *Journal of Invertebrate Pathology* 107(2), 112–126.

Sajjan, D.B. and Hinchigeri, S.B. (2016) Structural organization of baculovirus occlusion bodies and protective role of multilayered polyhedron envelope protein. *Food and Environmental Virology* 8, 86–100.

Sarig, S., Blum, A. and Okon, Y. (1988) Improvement of the water status and yield of field-grown grain sorghum (*Sorghum bicolor*) by inoculation with *Azospirillum brasilense*. *Journal of Agricultural Science* 110(2), 271–277.

Sati, D., Pande, V., Pandey, S.C. and Samant, M. (2023) Recent advances in PGPR and molecular mechanisms involved in drought stress resistance. *Journal of Soil Science and Plant Nutrition* 23(1), 106–124.

Selvaraj, S., Ganeshamoorthi, P. and Pandiaraj, T. (2013) Potential impacts of recent climate change on biological control agents in agro-ecosystem: A review. *International Journal of Biodiversity and Conservation* 5(12), 845–852.

Shakir, M.A., Bano, A. and Arshad, M. (2012) Rhizosphere bacteria containing ACC-deaminase conferred drought tolerance in wheat grown under semi-arid climate. *Soil and Environment* 31(1), 108–112.

Sivakumar, M.V. (2020) Climate extremes and impacts on agriculture. *Agroclimatology: Linking Agriculture to Climate* 60, 621–647.

Sjut, V. (ed.) (1997) *Molecular Mechanisms of Resistance to Agrochemicals*, Vol. 13. Springer, Cham, Switzerland.

Song, Q., Zhang, Y., Yan, L., Wang, J., Lu, C. *et al.* (2017) Risk assessment of the endocrine-disrupting effects of nine chiral pesticides. *Journal of Hazardous Materials* 338, 57–65.

Srivastava, S., Bist, V., Srivastava, S., Singh, P.C., Trivedi, P.K. *et al.* (2016) Unraveling aspects of *Bacillus amyloliquefaciens* mediated enhanced production of rice under biotic stress of *Rhizoctonia solani*. *Frontiers in Plant Science* 7, 587.

Sun, X. (2015) History and current status of development and use of viral insecticides in China. *Viruses* 7(1), 306–319.

Tarnawski, S., Hamelin, J., Jossi, M., Aragno, M. and Fromin, N. (2006) Phenotypic structure of *Pseudomonas* populations is altered under elevated pCO_2 in the rhizosphere of perennial grasses. *Soil Biology and Biochemistry* 38(6), 1193–1201.

Thilagam, P., Sharanappa, C.H., Roy, S., Deb, L., Padhan, S. *et al.* (2023) A review on advances in biocontrol techniques for managing insect pests in sustainable agriculture. *International Journal of Environment and Climate Change* 13(11), 2114–2125.

Trębicki, P., Dáder, B., Vassiliadis, S. and Fereres, A. (2017) Insect–plant–pathogen interactions as shaped by future climate: Effects on biology, distribution, and implications for agriculture. *Insect Science* 24(6), 975–989.

Ugras, S., Sezen, K., Kati, H. and Demirbag, Z. (2013) Purification and characterization of the bacteriocin Thuricin Bn1 produced by *Bacillus thuringiensis* subsp *kurstaki* Bn1 isolated from a hazelnut pest. *Journal of Microbiology and Biotechnology* 23(2), 167–176.

Wagemans, J., Holtappels, D., Vainio, E., Rabiey, M., Marzachì, C. *et al.* (2022) Going viral: Virus-based biological control agents for plant protection. *Annual Review of Phytopathology* 60(1), 21–42.

Waldon, H.B., Jenkins, M.B., Virginia, R.A. and Harding, E.E. (1989) Characteristics of woodland rhizobial populations from surface-and deep-soil environments of the Sonoran Desert. *Applied and Environmental Microbiology* 55(12), 3058–3064.

Wargent, J.J. and Jordan, B.R. (2013) From ozone depletion to agriculture: Understanding the role of UV radiation in sustainable crop production. *New Phytologist* 197(4), 1058–1076.

Zeddam, J.-L., Arroyo Cruzado, J., Luna Rodriguez, J., Ravallec, M. and Candiotti Subilete, E. (2003) A cypovirus from the South American oil-palm pest *Norape argyrrhorea* and its potential as a microbial control agent. *BioControl* 48(1), 101–112.

Zhang, H., Xie, J., Fu, Y., Cheng, J., Qu, Z. *et al.* (2020) A 2-kb mycovirus converts a pathogenic fungus into a beneficial endophyte for *Brassica* protection and yield enhancement. *Molecular Plant* 13(10), 1420–1433.

Zhao, Q., Shen, Q., Ran, W., Xiao, T., Xu, D. *et al.* (2011) Inoculation of soil by *Bacillus subtilis* Y-IVI improves plant growth and colonization of the rhizosphere and interior tissues of muskmelon (*Cucumis melo* L.). *Biology and Fertility of Soils* 47(5), 507–514.

12 Bacterial–Insect Symbiosis in a Context of Climate Change: Implications for Parasitoidism and Biological Control

Thierry Hance*, François Gilbert, Stefan Brandl, Maxence Jacquet and François Renoz

Biodiversity Research Centre, Earth and Life Institute, UCLouvain, Louvain-la-Neuve, Belgium

Abstract

Insects host a wide range of bacteria, some of which are heritable and have major impacts on their evolutionary ecology, with consequences for host–parasitoid relationships. Climate change is expected to have a significant impact on host–parasitoid and host–symbiont relationships, but these two axes are rarely considered together. We address this issue by focusing primarily on aphids, aphid parasitoids and the diversity of their heritable symbionts. We outline the diversity of bacterial symbionts that coevolve with crop pests and parasitoid wasps and discuss the influence of temperature on bacteria–insect symbiosis. Along the way, we discuss various hypotheses and predictions on how increasing temperature in a context of climate change, by influencing endosymbionts, may affect host–parasitoid relationships and biological control programmes. This critical synthesis suggests new avenues to delve into and should serve as a roadmap for further exploration of the complexity of the host–microbiota–parasitoid network in a context of global warming.

12.1 Setting the Scene: Insects are a Food Resource for Parasitoids and Symbiotic Bacteria

Insects represent the most diverse group of animals on the planet and are the food source for many species (Scudder, 2017). In food webs, insects are also the target of a multitude of natural enemies, either as prey or as hosts. Parasitoidism is a type of biological relationship, particularly common in insects, that has been exploited in crop pest control programmes. A parasitoid is an organism that lives on or in another organism (the host) and eventually causes the death of the host in which it develops (Blaimer *et al.*, 2023). Parasitoidism has contributed to the diversification of the Hymenoptera order, and today forms the basis of many biological control strategies against insect pests in various agricultural systems (Waage and Hassell, 1982).

Insects also interact closely with diverse micro-organisms (bacteria, fungi, viruses, etc.) that form their microbiota and find resources in insects to proliferate (Engel and Moran, 2013). The bacterial dimension of the insect microbiota

*Corresponding author: thierry.hance@uclouvain.be

DOI: 10.1079/9781800625099.0012

has received increasing attention in recent years, as it has been shown to play an important role in the evolutionary ecology of these invertebrates, affecting host fitness and phenotypes in a variety of ways (Zientz *et al.*, 2004; Feldhaar, 2011; Engel and Moran, 2013; Sudakaran *et al.*, 2017). The nature of relationships between insects and prokaryotes can evolve along the parasitism–mutualism continuum as a function of genetic and ecological parameters (Drew *et al.*, 2021). Insect-associated bacteria can colonize diverse parts of the host, including the surface (ectosymbionts), the digestive tract (gut symbionts) and the host body cavity (endosymbionts) where they can sometimes live intracellularly in specialized cells called bacteriocytes (Baumann, 2005; Engel and Moran, 2013; Ganesan *et al.*, 2023). While some insect symbionts are transient associates, others can hardly live outside their host, which provides them with a habitat with all the resources they need to thrive. These bacteria, which maintain a highly intimate relationship with the host, are faithfully transmitted vertically from one generation to the next (Frago *et al.*, 2020).

Depending on the degree of dependence insects have evolved with heritable symbionts, the latter are usually classified as obligate or facultative symbionts. Obligate symbionts are pervasive in insect taxa that feed on a diet poor in certain nutrients (e.g. blood, plant sap and wood). These taxa generally include species that are agronomically important as crop pests. Without these nutritional symbionts, these insects can hardly develop and reproduce. In many cases, obligate symbioses appeared tens of millions of years ago, and have fostered the diversification of insects, enabling them to exploit otherwise inaccessible ecological niches (Fisher *et al.*, 2017; Sudakaran *et al.*, 2017; Cornwallis, *et al.*, 2023). Insects are also hosts to facultative symbionts, which are heritable associates that are not essential to their host's survival but can affect the insect's adaptive phenotype by producing beneficial or detrimental effects, and/or by selfishly manipulating host reproduction (Feldhaar, 2011; Kaur *et al.*, 2021).

Interestingly, the last two decades have seen a growing awareness of the influence of insect microbiota on host–parasitoid interactions. This is particularly relevant for sap-feeding insects, which can carry a diversity of heritable bacterial symbionts that can confer direct protective effects on their hosts, induce behavioural changes and indirectly influence ecological dynamics (Oliver *et al.*, 2003; Dion *et al.*, 2011; McLean *et al.*, 2016). Although receiving more limited attention, the evolutionary ecology of parasitoids targeting sap-feeding pests may also be influenced by the heritable symbionts with which they are associated, in particular the widespread reproductive manipulator *Wolbachia pipientis* (Dicke *et al.*, 2020). Several studies have highlighted the importance of taking the microbial axis into account in parasitoidism-based pest control (Douglas, 2007; Berasategui *et al.*, 2016; Vorburger, 2018). However, all these interactions are influenced by temperature, and are being challenged by the thermal constraints imposed by global warming. In this context, it is therefore essential to understand how thermal stress influences prey–predator and parasitoid–host relationships through bacterial symbiosis.

A variety of studies have examined the impact of temperature on parasitoidism (Hance *et al.*, 2007; van Baaren *et al.*, 2010; Furlong and Zalucki, 2017). Also, numerous studies have investigated the influence of temperature on microbial symbiosis and its consequences for insect development and the expression of certain phenotypes (Fan and Wernegreen, 2013; Corbin *et al.*, 2017; Hussain *et al.*, 2017; Renoz *et al.*, 2019). However, few studies have focused on the effects of temperature on the host–microbiota–parasitoid network. The aim of this chapter is to examine how temperature, through its influence on heritable symbiosis, can impact parasitoidism. We address this issue in the context of heat stress and global warming, highlighting gaps to offer a fresh perspective on these aspects in the context of biological control. Biological control using parasitoids and predators is a vast field of study. The same is true of the insect microbiota, which comprises an incredible diversity of micro-organisms associated with a variety of effects. For this reason, this chapter deliberately overlooks predators, concentrating instead on parasitoidism, particularly that involving parasitic wasps. It also focuses on the bacterial component of the insect microbiota, in particular heritable symbionts that have evolved intimate relationships with insects, especially sap-feeding insect taxa that include important crop pests. Through this framework, we aim to provide the community with a source of guidance for addressing the diversity of biological systems relevant to biological control in a warming climate.

First, we will examine the nutritional dependence of many insect pests on ancestral symbionts and discuss how their heat sensitivity can impact host–parasitoid relationships. We will then take a closer look at the case of facultative symbiosis, with aphids as study models. Indeed, some heritable facultative symbionts can interfere with interspecific interactions, including parasitoidism, and this influence tends to be temperature dependent. Finally, we will look at the diversity of parasitoid microbiota, a rather neglected area. The prevalent symbiont *W. pipientis* will serve as a case study to understand how an increase in temperature can affect the course of biological control by influencing the interaction between the bacterium and its parasitoid host.

12.2 Many Insect Pests Depend on Degenerate and Heat-sensitive Symbionts

Host quality shapes the life-history traits of parasitoids and is modulated by many factors, including host symbionts (Heimpel *et al.*, 1996; Attia *et al.*, 2022). In the following paragraphs, we examine how symbionts can influence host quality in a context of climate change and increasing extreme heat events. Many insect crop pests have their biology inextricably linked to the presence of heritable symbiotic bacteria in their tissues and cells, which can have a variety of effects on the host (Moran *et al.*, 2008; Łukasik and Kolasa, 2024). Sap-feeding insects have generally evolved a dependence on nutritional endosymbionts that provide them with nutrients essential for their development that are insufficiently present in their diet (e.g. amino acids and vitamins) (Fig. 12.1) (Zientz *et al.*, 2004). Obligate symbionts are pervasive in hemipterans such as aphids, adelgids, psyllids, cicadas, leafhoppers, planthoppers and

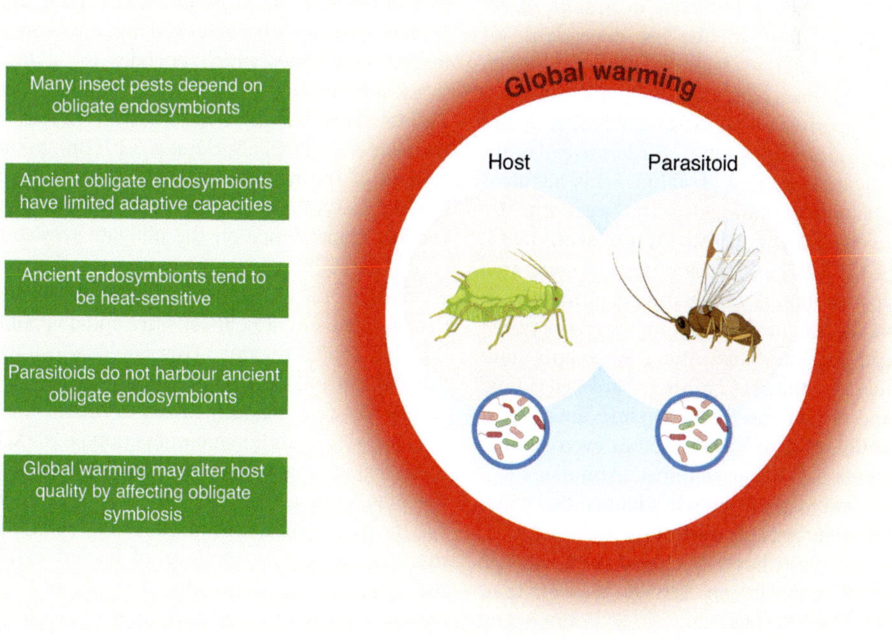

Fig. 12.1. Global warming may modify host quality by affecting ancient obligate symbioses, leading to altered host–parasitoid relationships.

treehoppers (Sudakaran *et al.*, 2017). All these taxa are associated, with one – sometimes several – nutritional symbionts that they often acquired tens of millions of years ago.

A typical feature of these ancient associations is that obligate symbionts are internalized within specific cells that insects have evolved: the bacteriocytes. These cells perform several functions: they mediate metabolic exchanges between the host and its bacterial partners, promote the transmission of symbionts from mother to offspring and regulate symbiont populations according to the host's nutritional needs over its lifetime (Koga *et al.*, 2012; Simonet *et al.*, 2018; Whittle *et al.*, 2021). However, this confined and stable intracellular lifestyle exposes symbionts to reductive genome evolution because it severely limits genetic exchange between nutritional symbionts and other bacteria, promotes asexuality and causes Muller's ratchet. This, coupled with severe population bottlenecks during vertical transmission and the relaxation of purifying selection on genes no longer needed in the context of interdependent association, leads to the emergence of bacterial associates that resemble organelles (Bennett and Moran, 2015). Consequently, after tens of millions of years of reductive evolution, insect host species find themselves dependent on nutritional symbionts with tiny genomes (with a loss of sometimes more than 95% of the original genetic repertoire) and extremely limited adaptive capacities. For example, *Buchnera aphidicola*, the obligate symbiont of aphids, was acquired over 100 million years ago and has a genome size of between 0.4 and 0.6 Mb (Moran *et al.*, 1993; Chong *et al.*, 2019). This is very small compared with 'conventional' free-living bacteria such as *Escherichia coli*, whose genome size is between 4.5 and 5.5 Mb. Some insect symbionts have genomes as small as 0.2 Mb (e.g.*Nardonella* found in some weevil species), encoding fewer than 200 genes, and so have genomes even smaller than some organellar genomes (Anbutsu *et al.*, 2017). Due to their limited adaptive capacities and low tolerance to environmental variation, it has been postulated that these organelle-like symbionts represent an Achilles heel for their host species (Wernegreen, 2012; Bennett and Moran, 2015; Renoz *et al.*, 2019). In the context of global warming, the functioning of the symbiosis could be significantly altered, with consequences for the biology of the host and its quality for higher trophic levels.

The thermosensitivity of obligate insect symbionts has been highlighted and discussed by studies which, in most cases, have used aphids as models (Dunbar *et al.*, 2007; Burke *et al.*, 2010a; b; Fan and Wernegreen, 2013; Moran and Yun, 2015; Hussain *et al.*, 2017; Shan *et al.*, 2017; Zhang *et al.*, 2019; Heyworth *et al.*, 2020). For example, pea aphids *Acyrthosiphon pisum* (Hemiptera: Aphididae) exposed for a few hours at temperatures up to 38°C have a reduced density of the bacteriocyte-associated symbiont *B. aphidicola*, which can even be completely lost by being lysed (Burke *et al.*, 2010a). However, heat stress experiments conducted on different aphid species have shown that *Buchnera*'s heat tolerance depends on the strain's genotype (Zhang *et al.*, 2019). For example, the cotton aphid *Aphis gossypii* (Hemiptera: Aphididae) and its associate *Buchnera* are little affected by the heat stress applied, while the black bean aphid *Aphis fabae* (Hemiptera: Aphididae) and its associate *Buchnera* are very sensitive, with a significant reduction in aphid fitness and *Buchnera* symbiont numbers. These differences in heat tolerance are not yet well understood. However, previous studies in *A. pisum* have shown that *Buchnera* strains characterized by a mutation in the *ibpA* gene, which encodes for a heat shock protein, are more sensitive to heat than *Buchnera* strains that do not have this mutation (Dunbar *et al.*, 2007; Burke *et al.*, 2010b). In the context of the reductive evolution experienced by obligate symbionts, gene loss is virtually irreversible. Dependence on the obligate symbiont could therefore impose rigidity on the thermal adaptation of host aphids and eventually lead to the displacement of heat-sensitive aphid populations to cooler regions. This could affect the distribution and seasonal ecology of parasitoids, whose life cycle is closely linked to that of their host. However, while symbiont genotype is a key factor explaining symbiont heat tolerance and thermotolerance of hosts, it is also possible that specific host-level mechanisms are at work, for example, perhaps at the level of bacteriocytes, the cells in which nutritional symbionts are confined and which regulate their populations via various cellular and molecular processes (Shigenobu and Stern, 2013; Simonet *et al.*, 2018; Whittle *et al.*, 2021). The study by Shan *et al.* (2017) suggests that bacteriocytes are also thermosensitive, as heat stress tends to severely impact their proliferation and morphology. The genetic variability of obligate symbionts linked to their thermosensitivity, and the compensatory mechanisms deployed by the host to deal with it, are issues that should be addressed in future research.

The thermosensitivity of obligate symbiosis is not limited to aphids; experiments conducted on *Bemisia tabaci* (Hemiptera: Aleyrodidae) show that continuous exposure to heat stress at 35°C almost completely depletes the nutritional symbionts *Portiera* and *Hamiltonella* and inhibits bacteriocyte proliferation (Shan *et al.*, 2017). In most studies, insects are exposed to acute heat stress but even if the frequency of such stresses increases with global warming, this is only one of the many types of heat stress that insects can experience in nature. However, a study that examined the impact of a slight and prolonged increase in temperature on the phytophagous green stink bug *Nezara viridula* (Hemiptera: Pentatomidae) shows that the obligate gut symbiont is severely affected, impairing the insect's normal development (Kikuchi *et al.*, 2016). This study suggests that even a slight increase in temperature can have a dramatic effect on nutritional symbiosis and then on insect development. This study confirms that heat-sensitive symbionts, whether residing in bacteriocytes or in specific crypts of the digestive tract, can be the Achilles heel of insect hosts under climate change conditions, impairing their quality as a food source for other organisms.

Data on the thermosensitivity of obligate symbioses is still limited and comes mainly from aphids living in temperate regions. Further studies are therefore needed to determine whether this high thermosensitivity of obligate symbiosis can be generalized to most symbiotic systems (e.g. insects evolving in tropical regions). However, in the light of current knowledge, we can expect heat stress, by destabilizing the nutritional mutualisms born of millions of years of coevolution, to affect the health of insect hosts in the wild. While it is certain that this will affect higher trophic levels, the exact consequences for parasitoids are difficult to predict. However, as their life cycle is intimately linked to that of the hosts they parasitize, we can assume that, depending on the thermal tolerance of the host's nutritional symbionts, climate change will modify the physiology and feeding habits of parasitoid insects. The quality and geographical distribution of hosts may change, with important consequences for interspecific interactions such as host–parasitoid and prey–predator interactions. Today, these aspects remain little studied, even though they are relevant to agriculture and the orientation of biological control methods based on parasitoidism.

Finally, it is important to note that unlike many of their phytophagous hosts, parasitoid wasps do not harbour ancient obligate symbionts with highly reduced genomes (Dicke *et al.*, 2020). However, the parasitic wasp *Asobara tabida* (Hymenoptera: Braconidae) seems to have evolved a dependence on *Wolbachia* for the proper conduct of oogenesis (Dedeine *et al.*, 2001). It is not known whether this mutualism extends to other parasitoid species, or how temperature sensitive it is. These are aspects that merit further study. Many viruses also form specific obligate mutualistic associations with parasitoids (Dicke *et al.*, 2020), but this aspect is not covered in this review.

12.3 Insect Pests Host Facultative Symbionts Associated with Important Ecological Traits

12.3.1 The diversity of facultative symbionts and their effects: Aphids in the spotlight

Many insect species, whether associated with obligate symbionts or not, are infected by heritable facultative symbionts that are not necessary for host development but which can affect the insect's adaptive phenotype in different ways (Oliver *et al.*, 2010; Feldhaar, 2011; Kaur *et al.*, 2021). These bacteria can produce beneficial or detrimental effects, the expression of which is conditioned by the genotype of both partners and by the ecological context, including the thermal environment (Fig. 12.2). Facultative symbionts also include bacteria that selfishly manipulate insect reproduction to propagate within populations (e.g. *Wolbachia* symbionts). As these bacteria have been acquired more recently than obligate symbionts, they have

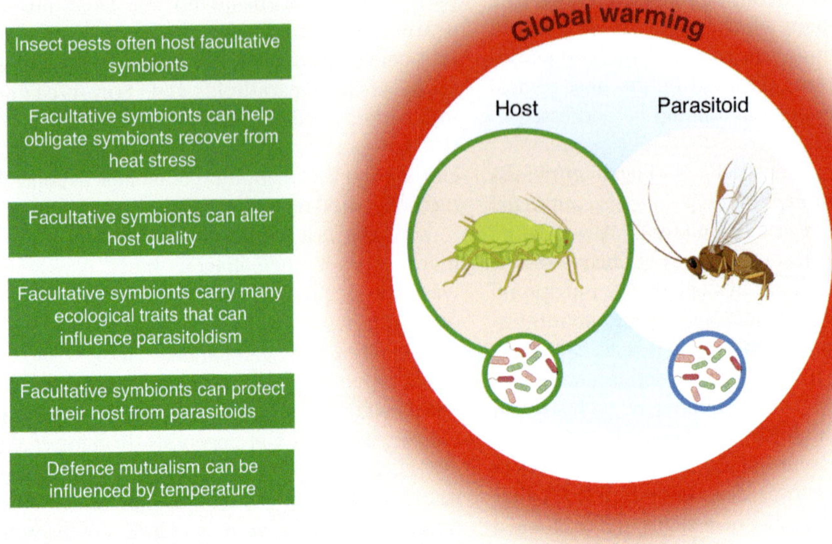

Fig. 12.2. Facultative symbiosis is an important dimension to consider in the evolutionary ecology of insect pests in a context of global warming.

less reduced genomes, can infect a wider range of tissues and cells (bacteriocytes, sheath cells, hemolymph, digestive tract, etc.) and can occasionally circulate in the environment and infect new hosts.

Facultative symbiosis has evolved in many insect species but aphids are historical models for its study as they can host a dozen or so facultative symbionts associated with diverse properties (Oliver *et al.*, 2010). For example, *Spiroplasma* bacteria can manipulate aphid reproduction (Simon *et al.*, 2011), *Rickettsiella viridis* can modify the body colour of infected insects (Tsuchida *et al.*, 2010) and *Regiella* and *Arsenophonus* symbionts can play a nutritional role by influencing aphid adaptability to host plants (Tsuchida *et al.*, 2004; Wagner *et al.*, 2015; Tian *et al.*, 2023; Renoz, 2024). *Serratia symbiotica* and other facultative symbionts species are known to induce protection against acute thermal stress (Burke *et al.*, 2010a). Finally, certain facultative symbionts confer protection to aphids against pathogenic fungi and/or parasitoids (Oliver *et al.*, 2003; Scarborough *et al.*, 2005; Vorburger *et al.*, 2010; Ballinger and Perlman, 2019). Although faithfully transmitted vertically, facultative symbionts can also undergo horizontal transmission, so that the important ecological traits with which they are associated can be passed on within insect populations or even acquired by other taxa (Oliver *et al.*, 2010).

It is important to note that while *Wolbachia* (Alphaproteobacteria) is widely distributed in many insect taxa (e.g. Hymenoptera and Diptera), the symbiont seems rarer in aphids and other sap-feeding insects, and its effects on these insects remain puzzling (Augustinos *et al.*, 2011). On the other hand, it is widespread in other insects targeted by biological control with parasitoids, such as certain *Drosophila* species (Diptera: Drosophilidae). In these insects, *Wolbachia* can have a wide range of effects. While some strains are pathogenic, others are nutritional mutualists (Min and Benzer, 1997; Brownlie *et al.*, 2009). Some strains protect hosts against pathogenic viruses and parasites (Bruner-Montero and Jiggins, 2023; Perlmutter *et al.*, 2023). *Wolbachia* can also modify the behaviour of *Drosophila* hosts (Bi and Wang,

2020). In the context of *Drosophila*–parasitoid relationships, *Wolbachia* can interfere with parasitoidism by interacting with parasitoid viruses. For example, Martinez *et al.* (2012) showed that *Wolbachia* hosted by *Drosophila* flies could either modify the oviposition preferences of the parasitoid *Leptopilina boulardi* (Hymenoptera: Figitidae), depending on the presence of LbFV virus in the latter, or help the host fly escape parasitism from LbFV-infected parasitoids by increasing the rate of encapsulation, a typical defence mechanism against parasitoid eggs. Our study deliberately focuses on sap-feeding insect hosts but given that the effects of *Wolbachia* on aphids remain unknown, the interaction between *Drosophila* and *Wolbachia* is a source of inspiration to fill this gap.

12.3.2 Facultative symbionts are also heat sensitive (but maybe less than obligate symbionts!)

In general, facultative symbionts appear to be less sensitive to heat stress than obligate symbionts. For example, exposure of *B. tabaci* to heat stress severely affects the populations of obligate symbionts, whereas facultative symbionts show an ability to resist more effectively (Shan *et al.*, 2014; 2017). This could be explained by the fact that, although these heritable symbionts possess many of the same genomic features as obligate mutualists (e.g. in terms of genome reduction), these are generally less pronounced, and they retain a generally complete wall and membrane (Wernegreen, 2012). However, a number of studies also show that high temperatures can deplete populations of facultative symbionts and reduce the efficiency of their vertical transmission (Anbutsu *et al.*, 2008; Burke *et al.*, 2010a; Corbin *et al.*, 2017; Liu *et al.*, 2019). For example, temperatures above 30°C suppress *Wolbachia* infection in the spider mite *Tetranychus urticae* (Arachnida: Tetranychidae) (van Opijnen and Breeuwer, 1999). The aphid symbiont *R. viridis* can be readily transmitted horizontally by plants, but not at temperatures above 25°C (Gu *et al.*, 2023). This is important because horizontal transmission drives inter- and intraspecific transfer of facultative symbionts and the important ecological traits they are associated with. All this suggests that the increased frequency of heat stress due to climate change could have an impact on the persistence and dynamics of facultative symbioses in natural insect populations, and on how the associated traits are transmitted and expressed (Martins *et al.*, 2023).

12.3.3 Facultative symbionts can help obligate symbionts recover from heat stress

Facultative symbionts can be a game changer when it comes to insect resistance to extreme thermal events. Indeed, some facultative symbionts have the ability to protect their hosts from the deleterious effects of high temperatures, thereby increasing their thermal tolerance. For example, certain strains of *S. symbiotica* increase the survival and reproduction of aphids exposed to heat shock (Burke *et al.*, 2010a). Furthermore, the heat protection delivered by *S. symbiotica* tends to be resilient to repeated thermal stress (Tougeron *et al.*, 2023). Subsequent studies have shown that these protective effects extend to other facultative symbionts (e.g. *Regiella*, *Hamiltonella* or *Fukatsuia*) (Russell and Moran, 2006; Heyworth *et al.*, 2020). However, while a recent meta-analysis shows that facultative symbionts tend to improve or maintain aphid fitness under heat stress, these protective effects could vary greatly depending on the strain, as some symbiont strains tend to exacerbate host thermosensitivity (Russell and Moran, 2006).

The mechanisms behind this heat protection are largely unknown, but one clearly observed trend is that the presence of certain facultative symbionts attenuates the collapse of *Buchnera* cell density under heat stress, while the density of facultative symbionts falls rapidly but recovers after the stress episode (Burke *et al.*, 2010a; Heyworth *et al.*, 2020). A study conducted on *S. symbiotica* in the pea aphid *A. pisum* suggests that it is the metabolites released after cell lysis of the facultative symbionts that help protect the integrity of the obligate symbiosis with *B. aphidicola* (Burke *et al.*, 2010a). More recently, it has been shown that *S. symbiotica* can complement the *Buchnera* DNA repair system to form a complete system that would preserve the integrity of the *Buchnera* genome and thus improve the thermostability of the nutritional symbiosis (Ling *et al.*, 2024). It is not known whether this mechanism extends to other facultative symbionts. However, in the field, this increase in heat tolerance provided by facultative symbionts could result in greater prevalence in populations evolving in warm areas (Montllor *et al.*, 2002).

There is no longer any doubt that many facultative symbionts are associated with protective effects against heat stress. However, the other aspects of these bacterial associates should not be overlooked: despite their advantages in specific ecological contexts, facultative symbionts are often moderate pathogens that impose costs on their host's fitness (Zytynska *et al.*, 2021; Clavé *et al.*, 2022). For example, aphids infected with *S. symbiotica* have a reduced survival rate, reproduce less and are smaller (Perreau *et al.*, 2021; Attia *et al.*, 2022; Tougeron *et al.*, 2023). The costs associated with facultative symbionts may explain why, in some cases, their presence exacerbates the host's thermosensitivity (Russell and Moran, 2006). Thus, the extreme versatility of facultative symbionts on the parasitism–mutualism evolutionary continuum makes it difficult to predict the impact of these bacteria on host–parasitoid relationships in a warming climate. One hypothesis is that an increase in the prevalence of facultative symbionts with heat-protective effects in a warming climate could lead to lower-quality hosts and hinder parasitoid development, with overall impacts on their populations and their effectiveness in pest regulation. This is supported experimentally by the study of Attia *et al.* (2022), which showed that a *S. symbiotica* strain with heat-protective effects alters the quality of aphid hosts, which in turn influences the foraging behaviour and life-history traits of the parasitoid *Aphidius ervi* (Hymenoptera: Braconidae).

12.3.4 Defensive mutualism under heat stress

As discussed above, facultative heritable symbionts can provide aphids with a whole range of advantages. One of the most notable is defensive mutualism: some facultative symbionts effectively protect their host from fungal pathogens or parasitoids (Oliver *et al.*, 2003; Scarborough *et al.*, 2005; Oliver *et al.*, 2014; Ballinger and Perlman, 2019; McLean *et al.*, 2020). Defensive mutualism has been reported in aphids but also in other insect groups (e.g. *Drosophila* flies) (Vorburger, 2022).

In aphids, the ability of the facultative symbiont *H. defensa* to protect these insects from parasitoids has received significant attention. This protection is due to a phage called APSE, which infects the symbiont and produces toxins that prevent the development of parasitoids in the host, with the level of protection depending on the phage variant (Oliver *et al.*, 2009). The thermal sensitivity of the defensive symbiosis involving *Hamiltonella* has been examined in a few studies. There is experimental evidence that heat stress can undermine the protection provided by *H. defensa* against parasitoids (Bensadia *et al.*, 2006; Guay *et al.*, 2009; Cayetano and Vorburger, 2013). In their study, Bensadia *et al.* (2006) hypothesized that the resistance provided by *H. defensa* to the pea aphid *A. pisum* against *A. ervi* might be temperature dependent and tested this on several resistant and non-resistant aphid clonal lines under controlled conditions. Three clones showing complete immunity to *A. ervi* at 20°C and containing *H. defensa* lost resistance to the parasitoids at 25°C and especially at 30°C, a temperature normally reached during the agricultural season. These differences were not related to host acceptance but were clearly related to the development of the first parasitoid larvae. The authors concluded that the temperature-dependent susceptibility of the resistant clones could be explained by a strong inhibitory effect of temperature on the physiological immunity of pea aphids.

The results of the Bensadia *et al.* (2006) study were later confirmed by that of Guay *et al.* (2009) conducted on other *A. pisum* clones. In addition, they showed that clones coinfected with *H. defensa* and *Fukatsuia symbiotica* (previously named PAXS) (Patel *et al.*, 2019)) remained resistant to *A. ervi* at high temperature compared with clones containing *H. defensa* alone. In this study, the effect of the *F. symbiotica* symbiont alone was not investigated (Guay *et al.*, 2009). Heyworth and Ferrari (2015) also showed that the aphid endosymbiont *F. symbiotica* (again formerly 'X-type' or PAXS symbiont) improves both tolerance to thermal shock and resistance to natural enemies, particularly against the parasitoid *A. ervi* and the fungal pathogen *Pandora neoaphidis*, but the authors did not test the combined effect of high temperature and natural enemies. Heyworth and Ferrari (2016) examined the effect of *F. symbiotica* alone on protection against parasitoids under conditions of heat stress. They found that under a benign temperature regime, aphids carrying *F. symbiotica* tend to be better protected against the parasitoid than cured aphids. However, when aphids underwent heat shock before being parasitized, aphids carrying *F. symbiotica* were more susceptible than cured aphids. All these studies revealed that symbionts other than *H. defensa* can play an important role in host–parasitoid interaction under heat stress, and that the thermosensitivity of the defensive symbiosis may depend on a combination of symbionts within the same host.

Several field studies have been conducted to examine the influence of heat stress on defensive mutualism. One of these studies showed a negative correlation between *H. defensa* frequencies and mean temperature in various North American localities, confirming the trends reported by the experimental approaches (Doremus *et al.*, 2018). However, Gimmi *et al.* (2023) reported a positive correlation between the presence of *H. defensa* in the aphid *Aphis fabae* and the number of hot days (temperature ≥30°C) in the preceding 8 to 4 weeks, while its frequency dynamics was only weakly correlated with parasitism risk. Quite similar results were reported by Smith *et al.* (2021), who found no relationship between *H. defensa* prevalence and parasitism levels, while showing that temperature was the strongest predictor of endosymbiont presence, with a marked effect of seasonality. These two field studies suggest that rapid seasonal adaptation in multivoltine organisms may be mediated by the multiple effects of heritable bacterial endosymbionts. These somewhat contradictory results between field and experimental studies call for further investigation of the role of *H. defensa* under stressful temperatures and the additional effects associated with the symbiont (e.g. costs, benefits under high temperatures) (Hudson *et al.*, 2024). Although there is evidence that resistance to parasitoid wasps conferred by facultative symbionts is highly temperature dependent, further studies are needed to better assess how rising global temperatures influence defensive mutualisms, what the consequences are for host–parasitoid coevolution, and at what level this may interfere with parasitoid-based pest management strategies.

12.3.5 Other ecological traits carried by facultative symbionts that could influence parasitoidism

Beyond defensive mutualism, facultative symbionts may carry other important ecological traits that can influence parasitoidism. The expression of these effects should undoubtedly be studied from a thermal perspective. For example, the bacterial endosymbiont *R. viridis* can change the body colour of aphids from red to green, which is ecologically important as parasitoid wasps tend to prefer to lay their eggs in green aphids (Losey *et al.*, 1997; Libbrecht *et al.*, 2007). Interestingly, Gu *et al.* (2023) found that this body colour change tends to disappear at higher temperatures. By modulating aphid colour through bacterial symbiosis, temperature could therefore influence the behaviour of parasitoids towards their host.

Facultative symbionts can also influence wing polyphenism. For example, it has been reported that *S. symbiotica* can inhibit the apterization of pea aphids and favour the development of winged forms (Kang *et al.*, 2022). In the context of parasitoidism, this is important because winged forms can disperse to colonize new plants and escape parasitoids. Infected winged aphids can also contribute to parasitoid dispersal (Walton *et al.*, 2011). Winged forms are also reputed to be of lesser quality to parasitoids, although this does not always appear to be true (Pirotte *et al.*, 2018). At the same time, studies have reported that the production of winged forms tends to decrease with increasing temperature (Müller *et al.*, 2001). The combined effect of temperature and the presence of symbionts in wing polyphenism should undoubtedly be the subject of further study.

Facultative symbionts can also influence host behaviour (Hosokawa and Fukatsu, 2020). For example, by analysing the behaviour of aphids in the presence of parasitoids, it was found that aphids infected with *H. defensa* tend to show reduced aggressiveness and escape reactions compared with uninfected aphids (Dion *et al.*, 2011). However, this influence on behaviour seems to vary from one symbiont species to another and from one strain to another (Sochard *et al.*, 2021). It is therefore conceivable that temperature, by influencing defensive symbiosis, could modify aphid behaviour towards parasitoids.

Finally, as mentioned above, facultative symbionts are often associated with fitness costs for the host. For example, a costly strain of the symbiont *S. symbiotica* hosted by *A. pisum* has an impact on the behavioural response and development of the parasitoid *A. ervi*. Parasitoids tend to choose uninfected hosts and when they develop in infected hosts, they show reduced physical quality, lower emergence rates and a higher proportion of male offspring when developing in infected aphids (Attia *et al.*, 2022). This bias in sex ratio is an important consideration in biological control with parasitoids, as it can lead to cascading effects on populations and even at community level (Monticelli *et al.*, 2019). If rising temperatures increase the frequency of certain symbionts and this combination degrades host quality for parasitoids, the result could be a skewed sex ratio among parasitoids, favouring males and thus compromising the reproductive fitness of parasitoid populations.

12.4 Parasitoid Microbiota Diversity and Thermal Considerations for Biological Control

12.4.1 Bacterial symbionts associated with parasitoids are mainly reproductive manipulators

By regulating host populations and their dynamics, parasitoids have a strong influence on ecosystems and are key players in crop pest management (Fei *et al.*, 2023). Most of them are Hymenoptera, with a parasitic larval stage dependent on a host insect (Mills, 2009). Like most insects, parasitoids host a microbial community (viruses, fungi and bacteria) and these microbial partners influence their life-history traits and more generally their evolutionary ecology (Dicke *et al.*, 2020). While the virobiota of parasitoids has been well studied (Beckage and Drezen, 2011; Dicke *et al.*, 2020), the bacterial dimension of their microbiota remains poorly documented (Zouache *et al.*, 2009; Brucker and Bordenstein, 2012; Nedoluzhko *et al.*, 2017). The main heritable symbionts identified in parasitoids to date belong to genera known to have the ability to manipulate the host reproduction, such as *Spiroplasma*, *Cardinium*, *Rickettsia*, *Arsenophonus* and *Wolbachia*, the most frequent symbiont in these insects (Fig. 12.3) (Werren, 1997; Engelstädter and Hurst, 2009; Vavre *et al.*, 2009; Taylor *et al.*, 2011; Pollmann *et al.*, 2022).

12.4.2 *Wolbachia pipientis* is the most common heritable symbiont among parasitoids

Wolbachia pipientis (Alphaproteobacteria) is the most widespread insect symbiont, present in around two-thirds of all insect species, including many parasitoid species (Hilgenboecker *et al.*, 2008). This symbiont species is particularly well known for manipulating host reproduction to promote its own spread within insect populations by various strategies such as feminization, male-killing, induction of pathogenesis or cytoplasmic incompatibility (Werren *et al.*, 2008; Kaur *et al.*, 2021). In parasitoids, two of these strategies have been reported: induction of pathogenesis or cytoplasmic incompatibility. Some strains have been reported to favour thelytokous parthenogenesis, thus preventing the production of males. This was first reported in *Trichogramma* spp. (Hymenoptera: Trichogrammatidae), egg parasitoids widely used for biocontrol of certain lepidopteran pests (Stouthamer *et al.*, 1990; 1993), but also in species of the families Pteromalidae (Stouthamer *et al.*, 1993), Aphelinidae (Zchori-Fein *et al.*, 1995) and Scelionidae (Arakaki *et al.*, 2000). Despite lower fecundity, thelytokous *Trichogramma* are often considered to be more effective biological control agents than uninfected arrhenotokous strains because all offspring are female, thus accelerating population growth and reducing production costs (Silva *et al.*, 2000). It is therefore in the

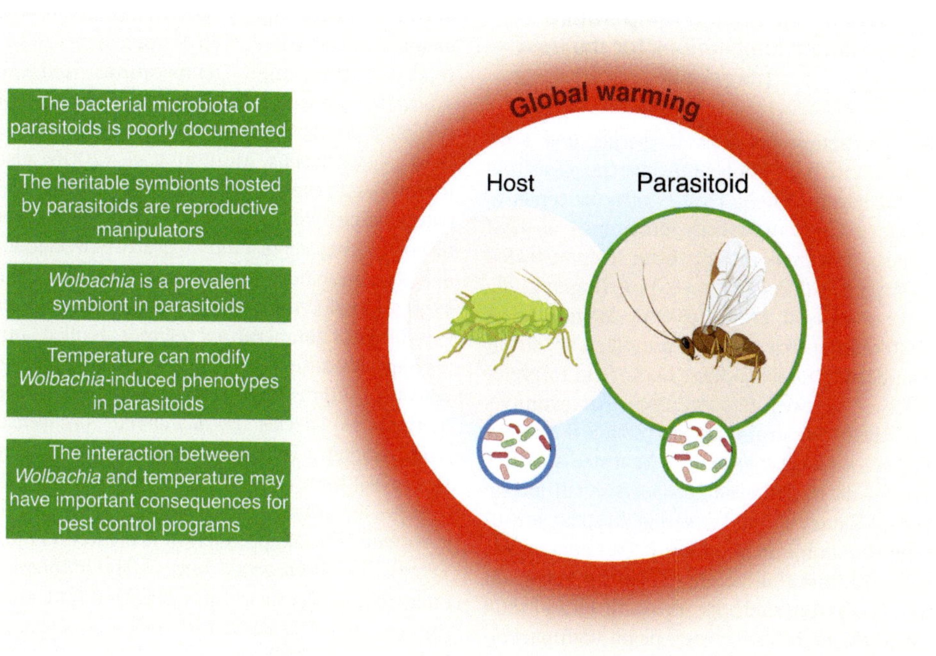

Fig. 12.3. The diversity of parasitoids' bacterial microbiota must be considered in biological control programmes.

interest of biological control to have *Wolbachia*-infected control agents.

In parasitoids, *Wolbachia* can also cause cytoplasmic incompatibility (CI), a phenomenon resulting in high embryonic mortality when mating occurs between infected males and uninfected females due to sperm modifications, as reported in the parasitoid *Encarsia inaron* (Hymenoptera: Aphelinidae) (White *et al.*, 2009). *Wolbachia* maintains a high prevalence in insect populations thanks to its efficient vertical transmission rates (typically 100%), its ability to manipulate host reproduction and its propensity for horizontal transmission between species that are sometimes phylogenetically far apart. This is particularly true for parasitoids, due to the close interaction maintained with the host, as first demonstrated by Werren *et al.* (1997) with *Nasonia vitripennis* (Hymenoptera: Pteromalidae) and its fly host. Superparasitism, i.e. the oviposition behaviour of parasitoid females that lay their eggs in an already parasitized host, also appears to be a cause of horizontal transmission when infected and uninfected larvae are in the same host egg, as reported by Huigens *et al.* (2000).

Beyond the effects of manipulating parasitoid reproduction, *Wolbachia* can also have an impact on the fitness of these insects. In the *Drosophila* parasitoid wasp *Leptopilina heterotoma* (Hymenoptera: Figitidae), *Wolbachia* can significantly decrease certain fitness traits such as fecundity, host survival and locomotor performance (Fleury *et al.*, 2000). Recent evidence suggests that *Wolbachia* may have an impact on mating preferences in parasitoid wasps, by making infected females less selective in their choice of mating, but the actual implications of decrease in female selectivity by *Wolbachia* are not yet clear (Amini *et al.*, 2024). In very rare cases, the *Wolbachia*–parasitoid association may have evolved into a strictly mutualistic relationship, as shown by Dedeine *et al.* (2001) in *A. tabida*, where the presence of *Wolbachia* strain *w*Atab3 became necessary for host oogenesis. This is the only reported case, and the extent of *Wolbachia*-based obligate mutualism in parasitoids remains unknown.

12.4.3 *Wolbachia*-temperature interaction: Implications for the host parasitoid

The interaction between *Wolbachia* and temperature can have implications for the parasitoid host at several levels. First, *Wolbachia* tends to influence the temperature preference of host insects. This has mainly been demonstrated in *Drosophila* (Truitt *et al.*, 2019; Hague *et al.*, 2020; Ferguson *et al.*, 2024). Hague *et al.* (2020) demonstrated that these effects are strain dependent. *Drosophila* flies' hosts tend to prefer cooler or warmer temperatures. For example, some *Wolbachia* strains appear to have the ability to increase insect tolerance to heat stress, as has been found in *Drosophila melanogaster* (Diptera: Drosophilidae) where *Wolbachia* do so by stimulating dopamine metabolism (Gruntenko *et al.*, 2017; Burdina *et al.*, 2021). However, Ferguson *et al.* (2024) reported opposite results: they found that *Wolbachia* infection has a negative impact on heat tolerance in *Drosophila simulans* (Diptera: Drosophilidae). The impact of *Wolbachia* on host heat tolerance undoubtedly depends on the genotype of both host and symbiont. Results concerning the impact of *Wolbachia* on temperature preference and heat tolerance come mainly from the *Drosophila* model but no data are available for parasitoids. Yet this is a crucial issue to address in the context of biological control of pests under global warming. Testing the influence of *Wolbachia* on the temperature preference and heat tolerance of parasitoids is an area for future study.

12.4.4 Effect of temperature on *Wolbachia*-induced phenotypes

Numerous studies have examined the impact of the thermal environment on the expression of parasitic reproductive phenotypes due to *Wolbachia*. While most results have been obtained on *Drosophila* flies (Corbin *et al.*, 2017), some studies have been conducted on parasitoid wasps (Girin and Boulétreau, 1995; Pintureau *et al.*, 1999; Mouton *et al.*, 2006; 2007; Bordenstein and Bordenstein, 2011; Zhou *et al.*, 2019; Power *et al.*, 2022). In general, high temperatures tend to reduce the expression of the parasitic reproductive phenotypes. For example, Bordenstein and Bordenstein (2011) reported that high temperatures reduced *Wolbachia* densities and the penetrance of cytoplasmic incompatibility in the parasitoid wasp *N. vitripennis*. Mouton *et al.* (2006) also investigated the impact of temperature on cytoplasmic incompatibility in *L. heterotoma*, but although they reported temperature-dependent density variability, they found no change in the expression of cytoplasmic incompatibility under the different temperature regimes applied, but perhaps because they did not test the critical warm conditions for the *Wolbachia* strain. In the thelytokous species *Trichogramma cordubensis* (Hymenoptera: Trichogrammatidae) infected by *Wolbachia*, high-temperature treatments (30°C) tend to 'inactivate' the symbiont, resulting in induction of male production (Pintureau *et al.*, 1999). In *Ooencyrtus mirus* (Hymenoptera: Encyrtidae), Power *et al.* (2022) found that the sex ratio was skewed in favour of males in offspring from *Wolbachia*-infected parents reared at 34°C and 36°C (high temperatures). In this species, the population growth rate without male generation is highest at 30°C.

These studies teach us some important lessons for biological control programmes. First, *Wolbachia* can be hijacked to produce more female parasitoids, the sex that enables pest control in the field. Second, temperature conditions are essential for the expression of *Wolbachia*-induced phenotypes. Understanding how temperature influences *Wolbachia*-induced phenotypes is therefore essential for optimizing the mass rearing of parasitoids dedicated for pest control.

One thing to bear in mind is that symbiont within-host density is an important parameter in the interaction between the host insect and its symbiont because it can significantly influence transmission and the amplitude of expression of the symbiont-induced phenotypes (Martinez *et al.*, 2017; López-Madrigal and Duarte, 2019). For the insect host, the higher the density of symbionts, the greater the fitness costs can be (Chrostek and Teixeira, 2015; Martinez *et al.*, 2017; Parker *et al.*, 2021). However, at the same time, a high symbiont density can also mean greater benefits for the host, as it correlates with the degree of expression of induced phenotypes (Martinez *et al.*, 2017; López-Madrigal and

Duarte, 2019; Parker *et al.*, 2021; Drew and King, 2022). And temperature is a highly influential parameter when it comes to within-host density symbiont (Mouton *et al.*, 2006; Anbutsu *et al.*, 2008; Doremus *et al.*, 2018). Thus, understanding the thermal niche of heritable symbionts and the influence of temperature on their within-host density is crucial for grasping how the costs, benefits and parasitic reproductive phenotypes are expressed in parasitoid wasps and their hosts, and finally for understanding how this modulates parasitoidism with the aim of optimizing biological control in a context of climate change.

12.5 Concluding Remarks and Perspectives

Global change is causing a general rise in temperatures and a greater incidence of extreme climatic events. In this context, insects, which are ectotherms and therefore particularly sensitive to thermal conditions, are exposed to a greater frequency of heat stress events, which could destabilize the coevolutionary processes in which they are involved and which have been shaped over millions of years in a relatively stable climatic context (van Baaren *et al.*, 2010; Harvey *et al.*, 2023). In this context, host–parasitoid relationships could be turned upside down: the life-history traits, development times and metabolism of hosts and parasitoids could be affected. On a broader scale, the geographical distribution of species involved in these interactions and phenological synchronization between trophic levels could be disrupted. The aim of this chapter is to highlight the importance of invisible players that have a key role in the evolutionary ecology of their hosts and can influence parasitoidism: heritable symbionts. Indeed, temperature highly influences the effects these microbes can have on their host, and therefore host–parasitoid interactions.

The question that motivates our discussion is: 'How can temperature influence parasitoidism through heritable symbiosis?'. Our review highlights the difficulty of predicting this. Indeed, the diversity of the microbial players involved is great, and the nature of the interactions they can have with their host on the parasitism–mutualism evolutionary continuum can be extremely variable, depending on various genetic and environmental factors. Heat tolerance of insects can be limited by their endosymbionts. This is especially true for insects harbouring obligate symbionts. In a warming environment, the thermosensitivity of the obligate symbionts of many pests could lead to a reduction in their quality for parasitoids and/or lead to their geographical redistribution. At the same time, facultative symbionts can change this paradigm. Some symbionts can rescue old, heat-sensitive symbioses while others, by being costly to their host, inflict on them a double penalty in a stressful environment (Renoz *et al.*, 2019). Defensive symbiosis can also be disrupted, perhaps to the advantage of parasitoids. Reproductive manipulators, such as *Wolbachia*, which are widespread in parasitoids, could also be impacted by thermal changes. This could have consequences for the sex ratio of parasitoid populations, depending on the species and the thermal niche of their symbionts. From an applied point of view, this is a crucial aspect that needs to be further studied; a better understanding of the thermal niches of reproductive manipulators is needed for optimized, more effective biological control. *Wolbachia* is undoubtedly an important tool to be considered in biological control (and still too neglected) to increase yields in terms of mass production of parasitoid females.

On the other hand, both the host and the parasitoid may carry symbionts that react in opposite directions to temperature increases and extremes. Determining the fate of such systems with the information we currently have is virtually impossible. To resolve these questions, complex interaction experiments need to be set up, for example with aphids carrying *S. symbiotica* facing a parasitoid carrying *Wolbachia*, under fluctuating temperatures, with all the necessary positive and negative controls. This type of experiment is difficult but not impossible to carry out in a microcosm, for example, and it is also essential if we want to make predictions that will hold up in the future.

In the same way, one aspect that is sorely lacking is the absence of long-term data on changes in the various components of trophic levels in the field. Field studies that consider thermal or seasonal variables and evaluate the success of biological control programmes in the context of aphid–parasitoid–microbiota interactions are still lacking (Gimmi *et al.*, 2023). Moreover, among the sample of field studies conducted, the trend appears to be surprisingly reversed from interpretations supported by laboratory results (Smith *et al.*, 2015; Doremus *et al.*, 2018; Hudson *et al.*, 2024). It is imperative to go out into the field and monitor changes in pest populations and their natural enemies, as well as the symbionts they contain. Metagenomic/metabarcoding approaches currently provide us with the tools we need to do this but require us to set up sampling protocols involving different types of organism that may be involved. In addition, studies should focus more on the transmission of endosymbionts under natural conditions to better understand the dynamics of infection within host populations.

One of the key factors in the future evolution of climate is the speed of change. This speed is such that it generates selection pressures that are too severe to enable adaptation, and selection could therefore lead to extinction. Phenotypic plasticity, especially thermal plasticity, should allow faster adaptation to these changes and may even be heritable (Svensson *et al.*, 2020). Future populations should therefore be characterized by greater phenotypic plasticity in response to temperature. In this case, however, it is clear that selection pressures are exerted on both the host and its obligate and facultative symbionts. Selection should therefore favour groups of organisms that allow the host and its partners to develop, survive and be as fertile as possible under stress conditions; in other words, greater phenotypic plasticity of the holobiont (i.e. the assemblage of a host and the many other species living in or around it). Similarly, parasitoids also exert a selection pressure on the composition of their host populations by favouring those that harbour protective symbionts, leading to dynamic coevolution between these species. In an environment where warmer temperatures become the norm, symbiont-induced genotypes conferring better heat resistance are likely to be selected. By a cascade effect, associations of strains that are both protective against parasitoids and heat tolerant will become more frequent. These coinfections should have an impact on the physical quality and fecundity of the pests. In short, we could be witnessing a subtle interplay of the effects of high temperatures, which can result in faster pest population growth because the age of first reproduction is earlier and the host is protected by the endosymbionts it carries.

Finally, we have dealt here with only a small part of the possible interactions, and only with certain organisms. We did not cover the case of prey–predator interactions, although many predators used in biological control are infected with *Wolbachia* (Sokolova *et al.*, 2002; Li *et al.*, 2021), and there is evidence that prey symbionts have an impact on predator fitness (Kovacs *et al.*, 2017). Also, the question of the virobiota is only just beginning to be addressed, although it has a considerable influence on the life-history traits and behaviour of the insects concerned (Murhububa *et al.*, 2024). Moreover, at the level of insect vectors, there seems to be a clear potential interaction between the viruses transmitted and the presence of symbionts. We lack data to understand these interactions and the influence that temperature fluctuations or an increase in mean temperature could have on them.

Fortunately, more and more data are accumulating that will eventually enable us to make predictions and perhaps even envisage new biological control techniques directly involving micro-organisms that live in close interaction with their hosts and thus modify their plastic response capacity to sudden variations in environmental conditions.

Acknowledgement

This paper was made possible thanks to the grant that Thierry Hance received from the SPWFeder-Interreg TransPest project (grant number 0100226).

References

Amini, S., Fathipour, Y., Hoffmann, A. and Mehrabadi, M. (2024) Effect of female mate preference and offspring fitness in a parasitoid wasp. *Pest Management Science* 80(10), 5432–5439.

Anbutsu, H., Goto, S. and Fukatsu, T. (2008) High and low temperatures differently affect infection density and vertical transmission of male-killing *Spiroplasma* symbionts in *Drosophila* hosts. *Applied and Environmental Microbiology* 74(19), 6053–6059.

Anbutsu, H., Moriyama, M., Nikoh, N., Hosokawa, T., Futahashi, R. *et al.* (2017) Small genome symbiont underlies cuticle hardness in beetles. *Proceedings of the National Academy of Sciences USA* 114(40), E8382–E8391.

Arakaki, N., Noda, H. and Yamagishi, K. (2000) Wolbachia-induced parthenogenesis in the egg parasitoid Telenomus nawai. *Entomologia Experimentalis et Applicata* 96(2), 177–184.

Attia, S., Renoz, F., Pons, I., Louâpre, P., Foray, V. *et al.* (2022) The aphid facultative symbiont *Serratia symbiotica* influences the foraging behaviors and the life-history traits of the parasitoid Aphidius ervi. *Entomologia Generalis* 42(1), 21–33.

Augustinos, A.A., Santos-Garcia, D., Dionyssopoulou, E., Moreira, M., Papapanagiotou, A. *et al.* (2011) Detection and characterization of *Wolbachia* infections in natural populations of aphids: Is the hidden diversity fully unraveled? *PLOS One* 6(12), e28695.

Ballinger, M.J. and Perlman, S.J. (2019) The defensive *Spiroplasma*. *Current Opinion in Insect Science* 32, 36–41.

Baumann, P. (2005) Biology of bacteriocyte-associated endosymbionts of plant sap-sucking insects. *Annual Review of Microbiology* 59(1), 155–189.

Beckage, N.E. and Drezen, J.-M. (eds) (2011) *Parasitoid Viruses: Symbionts and Pathogens*. Academic Press, New York.

Bennett, G.M. and Moran, N.A. (2015) Heritable symbiosis: The advantages and perils of an evolutionary rabbit hole. *Proceedings of the National Academy of Sciences USA* 112(33), 10169–10176.

Bensadia, F., Boudreault, S., Guay, J.F., Michaud, D. and Cloutier, C. (2006) Aphid clonal resistance to a parasitoid fails under heat stress. *Journal of Insect Physiology* 52(2), 146–157.

Berasategui, A., Shukla, S., Salem, H. and Kaltenpoth, M. (2016) Potential applications of insect symbionts in biotechnology. *Applied Microbiology and Biotechnology* 100(4), 1567–1577.

Bi, J. and Wang, Y. (2020) The effect of the endosymbiont *Wolbachia* on the behavior of insect hosts. *Insect Science* 27(5), 846–858.

Blaimer, B.B., Santos, B.F., Cruaud, A., Gates, M.W., Kula, R.R, *et al.* (2023) Key innovations and the diversification of hymenoptera. *Nat Commun* 14(1), 1212.

Bordenstein, S.R. and Bordenstein, S.R. (2011) Temperature affects the tripartite interactions between Bacteriophage WO, *Wolbachia*, and cytoplasmic incompatibility. *PLOS One* 6(12), e29106.

Brownlie, J.C., Cass, B.N., Riegler, M., Witsenburg, J.J., Iturbe-Ormaetxe, I. *et al.* (2009) Evidence for metabolic provisioning by a common invertebrate endosymbiont, *Wolbachia* pipientis, during periods of nutritional stress. *PLOS Pathogens* 5(4), e1000368.

Brucker, R.M. and Bordenstein, S.R. (2012) The roles of host evolutionary relationships (Genus: Nasonia) and development in structuring microbial communities. *Evolution* 66(2), 49–362.

Bruner-Montero, G. and Jiggins, F.M. (2023) *Wolbachia* protects *Drosophila melanogaster* against two naturally occurring and virulent viral pathogens. *Scientific Reports* 13(1), 8518.

Burdina, E.V., Bykov, R.A., Menshanov, P.N., Ilinsky, Y.Y. and Gruntenko, N.E. (2021) Unique *Wolbachia* strain wMelPlus increases heat stress resistance in *Drosophila melanogaster*. *Archives of Insect Biochemistry and Physiology* 106(4), e21776.

Burke, G., Fiehn, O. and Moran, N. (2010a) Effects of facultative symbionts and heat stress on the metabolome of pea aphids. *ISME Journal* 4(2), 242–252.

Burke, G.R., McLaughlin, H.J., Simon, J.C. and Moran, N.A. (2010b) Dynamics of a recurrent *Buchnera* mutation that affects thermal tolerance of pea aphid hosts. *Genetics* 186(1), 367–372.

Cayetano, L. and Vorburger, C. (2013) Genotype-by-genotype specificity remains robust to average temperature variation in an aphid/endosymbiont/parasitoid system. *Journal of Evolutionary Biology* 26(7), 1603–1610.

Chong, R.A., Park, H. and Moran, N.A. (2019) Genome evolution of the obligate endosymbiont *Buchnera aphidicola*. *Molecular Biology and Evolution* 36(7), 1481–1489.

Chrostek, E. and Teixeira, L. (2015) Mutualism breakdown by amplification of *Wolbachia* genes. *PLOS Biology* 13(2), e1002065.

Clavé, C., Sugio, A., Morlière, S., Pincebourde, S., Simon, J.C. *et al.* (2022) Physiological costs of facultative endosymbionts in aphids assessed from energy metabolism. *Functional Ecology* 36(10), 2580–2592.

Corbin, C., Heyworth, E.R., Ferrari, J. and Hurst, G.D.D. (2017) Heritable symbionts in a world of varying temperature. *Heredity* 118(1), 10–20.

Cornwallis, C.K., van't Padje, A., Ellers, J., Klein, M. and Jackson, R. (2023) Symbioses shape feeding niches and diversification across insects. *Nature Ecology & Evolution* 7(7), 1022–1044.

Dedeine, F., Vavre, F., Fleury, F., Loppin, B., Hochberg, M.E. *et al.* (2001) Removing symbiotic *Wolbachia* bacteria specifically inhibits oogenesis in a parasitic wasp. *Proceedings of the National Academy of Sciences USA* 98(11), 6247–6252.

Dicke, M., Cusumano, A. and Poelman, E.H. (2020) Microbial symbionts of parasitoids. *Annual Review of Entomology* 65(65), 171–190.

Dion, E., Polin, S.E., Simon, J.C. and Outreman, Y. (2011) Symbiont infection affects aphid defensive behaviours. *Biology Letters* 7(5), 743–746.

Doremus, M.R., Smith, A.H., Kim, K.L., Holder, A.J., Russell, J.A. *et al.* (2018) Breakdown of a defensive symbiosis, but not endogenous defences, at elevated temperatures. *Molecular Ecology* 27(8), 2138–2151.

Douglas, A.E. (2007) Symbiotic microorganisms: Untapped resources for insect pest control. *Trends in Biotechnology* 25(8), 338–342.

Drew, G.C. and King, K.C. (2022) More or less? The effect of symbiont density in protective mutualisms. *American Naturalist* 199(4), 443–454.

Drew, G.C., Stevens, E.J. and King, K.C. (2021) Microbial evolution and transitions along the parasite–mutualist continuum. *Nature Reviews Microbiology* 19(10), 623–638.

Dunbar, H.E., Wilson, A.C.C., Ferguson, N.R. and Moran, N.A. (2007) Aphid thermal tolerance is governed by a point mutation in bacterial symbionts. *PLOS Biology* 5(5), e96.

Engel, P. and Moran, N.A. (2013) The gut microbiota of insects – diversity in structure and function. *FEMS Microbiology Reviews* 37(5), 699–735.

Engelstädter, J. and Hurst, G.D.D. (2009) The ecology and evolution of microbes that manipulate host reproduction. *Annual Review of Ecology, Evolution, and Systematics* 40(1), 127–149.

Fan, Y. and Wernegreen, J.J. (2013) Can't take the heat: High temperature depletes bacterial endosymbionts of ants. *Microbial Ecology* 66(3), 727–733.

Fei, M., Gols, R. and Harvey, J.A. (2023) The biology and ecology of parasitoid wasps of predatory arthropods. *Annual Review of Entomology* 68, 109–128.

Feldhaar, H. (2011) Bacterial symbionts as mediators of ecologically important traits of insect hosts. *Ecological Entomology* 36(5), 533–543.

Ferguson, L.F., Ross, P.A. and van Heerwaarden, B. (2024) *Wolbachia* infection negatively impacts *Drosophila* simulans heat tolerance in a strain- and trait-specific manner. *Environmental Microbiology* 26(4), e16609.

Fisher, R.M., Henry, L.M., Cornwallis, C.K., Kiers, E.T. and West, S.A. (2017) The evolution of host-symbiont dependence. *Nature Communications* 8, 15973.

Fleury, F., Vavre, F., Ris, N., Fouillet, P. and Boulétreau, M. (2000) Physiological cost induced by the maternally-transmitted endosymbiont *Wolbachia* in the *Drosophila* parasitoid *Leptopilina heterotoma*. *Parasitology* 121(5), 493–500.

Frago, E., Zytynska, S.E. and Fatouros, N.E. (2020) Microbial symbionts of herbivorous species across the insect tree. In: Oliver, K.M. and Russell, J.A. (eds) *Advances in Insect Physiology*. Academic Press, New York, pp. 111–159.

Furlong, M.J. and Zalucki, M.P. (2017) Climate change and biological control: The consequences of increasing temperatures on host–parasitoid interactions. *Current Opinion in Insect Science* 20, 39–44.

Ganesan, R., Janke, R.S., Kaltenpoth, M. and Flórez, L.V. (2023) Colonization dynamics of a defensive insect ectosymbiont. *Biology Letters* 19(5), 20230100.

Gimmi, E., Wallisch, J. and Vorburger, C. (2023) Defensive symbiosis in the wild: Seasonal dynamics of parasitism risk and symbiont-conferred resistance. *Molecular Ecology* 32(14), 4063–4077.

Girin, C. and Boulétreau, M. (1995) Microorganism-associated variation in host infestation efficiency in a parasitoid wasp, *Trichogramma bourarachae* (Hymenoptera: Trichogrammatidae). *Experientia* 51(4), 398–401.

Gruntenko, N.E., Ilinsky, Y.Y., Adonyeva, N.V., Burdina, E.V., Bykov, R.A. *et al.* (2017) Various *Wolbachia* genotypes differently influence host *Drosophila* dopamine metabolism and survival under heat stress conditions. *BMC Evolutionary Biology* 17(2), 252.

Gu, X., Ross, P.A., Gill, A., Yang, Q., Ansermin, E. *et al.* (2023) A rapidly spreading deleterious aphid endosymbiont that uses horizontal as well as vertical transmission. *Proceedings of the National Academy of Sciences USA* 120(18), e2217278120.

Guay, J.F., Boudreault, S., Michaud, D. and Cloutier, C. (2009) Impact of environmental stress on aphid clonal resistance to parasitoids: Role of *Hamiltonella defensa* bacterial symbiosis in association with a new facultative symbiont of the pea aphid. *Journal of Insect Physiology* 55(10), 919–926.

Hague, M.T.J., Caldwell, C.N. and Cooper, B.S. (2020) Pervasive effects of *Wolbachia* on host temperature preference. *mBio* 11(5). DOI: 10.1128/mbio.01768-20.

Hance, T., van Baaren, J., Vernon, P. and Boivin, G. (2007) Impact of extreme temperatures on parasitoids in a climate change perspective. *Annual Review of Entomology* 52(52), 107–126.

Harvey, J.A., Tougeron, K., Gols, R., Heinen, R., Abarca, M. *et al.* (2023) Scientists' warning on climate change and insects. *Ecological Monographs* 93(1), e1553.

Heimpel, G.E., Rosenheim, J.A. and Mangel, M. (1996) Egg limitation, host quality, and dynamic behavior by a parasitoid in the field. *Ecology* 77(8), 2410–2420.

Heyworth, E.R. and Ferrari, J. (2015) A facultative endosymbiont in aphids can provide diverse ecological benefits. *Journal of Evolutionary Biology* 28(10), 1753–1760.

Heyworth, E.R. and Ferrari, J. (2016) Heat stress affects facultative symbiont-mediated protection from a parasitoid wasp. *PLOS One* 11(11), e0167180.

Heyworth, E.R., Smee, M.R. and Ferrari, J. (2020) Aphid facultative symbionts aid recovery of their obligate symbiont and their host after heat stress. *Frontiers in Ecology and Evolution* 8.

Hilgenboecker, K., Hammerstein, P., Schlattmann, P., Telschow, A. and Werren, J.H. (2008) How many species are infected with *Wolbachia*? – A statistical analysis of current data. *FEMS Microbiology Letters* 281(2), 215–220.

Hosokawa, T. and Fukatsu, T. (2020) Relevance of microbial symbiosis to insect behavior. *Current Opinion in Insect Science* 39, 91–100.

Hudson, C.M., Stalder, D. and Vorburger, C. (2024) Clines of resistance to parasitoids: the multifarious effects of temperature on defensive symbioses in insects. *Current Opinion in Insect Science* 64, 101208.

Huigens, M.E., Luck, R.F., Klaassen, R.H.G., Maas, M.F., Timmermans, M.J. *et al.* (2000) Infectious parthenogenesis. *Nature* 405(6783), 178–179.

Hussain, M., Akutse, K.S., Ravindran, K., Lin, Y., Bamisile, B.S. *et al.* (2017) Effects of different temperature regimes on survival of *Diaphorina citri* and its endosymbiotic bacterial communities. *Environment Microbiology* 19(9), 3439–3449.

Kang, Z.W., Zhang, M., Cao, H.H., Guo, S.S., Liu, F.H. *et al.* (2022) Facultative endosymbiont *Serratia symbiotica* inhibits the apterization of pea aphid to enhance its spread. *Microbiology Spectrum* 10(6), e04066–22.

Kaur, R., Shropshire, J.D., Cross, K.L., Leigh, B., Mansueto, A.J. *et al.* (2021) Living in the endosymbiotic world of *Wolbachia*: A centennial review. *Cell Host & Microbe* 29(6), 879–893.

Kikuchi, Y., Tada, A., Musolin, D.L., Hari, N., Hosokawa, T. *et al.* (2016) Collapse of insect gut symbiosis under simulated climate change. *mBio* 7(5), e01578–16.

Koga, R., Meng, X.-Y., Tsuchida, T. and Fukatsu, T. (2012) Cellular mechanism for selective vertical transmission of an obligate insect symbiont at the bacteriocyte–embryo interface. *Proceedings of the National Academy of Sciences USA* 109(20), E1230–E1237.

Kovacs, J.L., Wolf, C., Voisin, D. and Wolf, S. (2017) Evidence of indirect symbiont conferred protection against the predatory lady beetle *Harmonia axyridis* in the pea aphid. *BMC Ecology* 17(1), 26.

Li, H., Shu, X., Meng, L., Zhou, X., Obrycki, J.J. *et al.* (2021) Prevalence of maternally-inherited bacteria in native and invasive populations of the harlequin ladybird beetle *Harmonia axyridis*. *Biological Invasions* 23(5), 1461–1471.

Libbrecht, R., Gwynn, D.M. and Fellowes, M.D.E. (2007) *Aphidius ervi* preferentially attacks the green morph of the pea aphid, *Acyrthosiphon pisum*. *Journal of Insect Behavior* 20(1), 25–32.

Ling, X., Guo, H., Di, J., Xie, L., Zhu-Salzman, K. *et al*. (2024) A complete DNA repair system assembled by two endosymbionts restores heat tolerance of the insect host. *Proceedings of the National Academy of Sciences USA* 121(51), e2415651121.

Liu, X.D., Lei, H.X. and Chen, F.F. (2019) Infection pattern and negative effects of a facultative endosymbiont on its insect host are environment-dependent. *Scientific Reports* 9(1), 4013.

López-Madrigal, S. and Duarte, E.H. (2019) Titer regulation in arthropod-*Wolbachia* symbioses. *FEMS Microbiology Letters* 366(23), fnz232.

Losey, J.E., Harmon, J., Ballantyne, F. and Brown, C. (1997) A polymorphism maintained by opposite patterns of parasitism and predation. *Nature* 388(6639), 269–272.

Łukasik, P. and Kolasa, M.R. (2024) With a little help from my friends: The roles of microbial symbionts in insect populations and communities. *Philosophical Transactions of the Royal Society B Biological Sciences* 379(1904), 20230122.

Martinez, J., Duplouy, A., Woolfit, M., Vavre, F., O'Neill, S.L. *et al*. (2012) Influence of the virus LbFV and of *Wolbachia* in a host-parasitoid interaction. *PLOS One* 7(4), e35081.

Martinez, J., Tolosana, I., Ok, S., Smith, S., Snoeck, K. *et al*. (2017) Symbiont strain is the main determinant of variation in *Wolbachia*-mediated protection against viruses across *Drosophila* species. *Molecular Ecology* 26(15), 4072–4084.

Martins, M., César, C.S. and Cogni, R. (2023) The effects of temperature on prevalence of facultative insect heritable symbionts across spatial and seasonal scales. *Frontiers in Microbiology* 14, 1321341.

McLean, A.H.C., Parker, B.J., Hrček, J., Henry, L.M. and Godfray, H.C.J. (2016) Insect symbionts in food webs. *Philosophical Transactions of the Royal Society B Biological Sciences* 371(1702), 20150325.

McLean, A.H.C., Hrček, J., Parker, B.J., Mathé-Hubert, H., Kaech, H. *et al*. (2020) Multiple phenotypes conferred by a single insect symbiont are independent. *Proceedings of the National Academy of Sciences USA* 287(1929), 20200562.

Mills, N. (2009) Parasitoids. In: Resh, V.H. and Cardé, R.T. (eds) *Encyclopedia of Insects*, 2nd edn. Academic Press, San Diego, California, pp. 748–751.

Min, K.T. and Benzer, S. (1997) *Wolbachia*, normally a symbiont of *Drosophila*, can be virulent, causing degeneration and early death. *Proceedings of the National Academy of Sciences USA* 94(20), 10792–10796.

Monticelli, L.S., Outreman, Y., Frago, E. and Desneux, N. (2019) Impact of host endosymbionts on parasitoid host range – from mechanisms to communities. *Current Opinion in Insect Science* 32, 77–82.

Montllor, C.B., Maxmen, A. and Purcell, A.H. (2002) Facultative bacterial endosymbionts benefit pea aphids *Acyrthosiphon pisum* under heat stress. *Ecological Entomology* 27(2), 189–195.

Moran, N.A. and Yun, Y. (2015) Experimental replacement of an obligate insect symbiont. *Proceedings of the National Academy of Sciences USA* 112(7), 2093–2096.

Moran, N.A., Munson, M.A., Baumann, P. and Ishikawa, H. (1993) A molecular clock in endosymbiotic bacteria is calibrated using the insect hosts. *Proceedings of the Royal Society B Biological Sciences* 253(1337), 167–171.

Moran, N.A., McCutcheon, J.P. and Nakabachi, A. (2008) Genomics and evolution of heritable bacterial symbionts. *Annual Review of Genetics* 42(1), 165–190.

Mouton, L., Henri, H., Bouletreau, M. and Vavre, F. (2006) Effect of temperature on *Wolbachia* density and impact on cytoplasmic incompatibility. *Parasitology* 132(1), 49–56.

Mouton, L., Henri, H., Charif, D., Boulétreau, M. and Vavre, F. (2007) Interaction between host genotype and environmental conditions affects bacterial density in *Wolbachia* symbiosis. *Biology Letters* 3(2), 210–213.

Müller, C.B., Williams, I.S. and Hardie, J. (2001) The role of nutrition, crowding and interspecific interactions in the development of winged aphids. *Ecological Entomology* 26(3), 330–340.

Murhububa, I.S., Bragard, C., Tougeron, K. and Hance, T. (2024) Preference of Pentalonia nigronervosa for infected banana plants tends to reverse after Banana bunchy top virus acquisition. *Scientific Reports* 14(1), 2993.

Nedoluzhko, A.V., Sharko, F.S., Tsygankova, S.V., Boulygina, E.S., Sokolov, A.S. *et al*. (2017) Metagenomic analysis of microbial community of a parasitoid wasp *Megaphragma amalphitanum*. *Genomics Data* 11, 87–88.

Oliver, K.M., Russell, J.A., Morant, N.A. and Hunter, M.S. (2003) Facultative bacterial symbionts in aphids confer resistance to parasitic wasps. *Proceedings of the National Academy of Sciences USA* 100(4), 1803–1807.

Oliver, K.M., Degnan, P.H., Hunter, M.S. and Moran, N.A. (2009) Bacteriophages encode factors required for protection in a symbiotic mutualism. *Science* 325(5943), 992–994.
Oliver, K.M., Degnan, P.H., Burke, G.R. and Moran, N.A. (2010) Facultative symbionts in aphids and the horizontal transfer of ecologically important traits. *Annual Review of Entomology* 55(1), 247–266.
Oliver, K.M., Smith, A.H. and Russell, J.A. (2014) Defensive symbiosis in the real world – advancing ecological studies of heritable, protective bacteria in aphids and beyond. *Functional Ecology* 28(2), 341–355.
Parker, B.J., Hrček, J., McLean, A.H.C., Brisson, J.A. and Godfray, H.C.J. (2021) Intraspecific variation in symbiont density in an insect–microbe symbiosis. *Molecular Ecology* 30(6), 1559–1569.
Patel, V., Chevignon, G., Manzano-Marín, A., Brandt, J.W., Strand, M.R. *et al.* (2019) Cultivation-assisted genome of candidatus *Fukatsuia symbiotica*; The enigmatic "X-type" symbiont of aphids. *Genome Biology and Evolution* 11(12), 3510–3522.
Perlmutter, J.I., Atadurdyyeva, A., Schedl, M.E. and Unckless, R.L. (2023) *Wolbachia* enhances the survival of *Drosophila* infected with fungal pathogens. *bioRxiv*.
Perreau, J., Patel, D.J., Anderson, H., Maeda, G.P., Elston, K.M. *et al.* (2021) Vertical transmission at the pathogen–symbiont interface: *Serratia symbiotica* and aphids. *mBio* 12(2), e00359–21.
Pintureau, B., Chapelle, L. and Delobel, B. (1999) Effects of repeated thermic and antibiotic treatments on a *Trichogramma* (Hym., Trichogrammatidae) symbiont. *Journal of Applied Entomology* 123(8), 473–483.
Pirotte, J.A.L.M., Lorenzi, A., Foray, V. and Hance, T. (2018) Impact of differences in nutritional quality of wingless and winged aphids on parasitoid fitness. *Journal of Experimental Biology* 221(20), jeb185645.
Pollmann, M., Moore, L.D., Krimmer, E., D'Alvise, P., Hasselmann, M. *et al.* (2022) Highly transmissible cytoplasmic incompatibility by the extracellular insect symbiont *Spiroplasma*. *iScience* 25(5), 104335.
Power, N.R., Rugman-Jones, P.F., Stouthamer, R., Ganjisaffar, F. and Perring, T.M. (2022) High temperature mortality of *Wolbachia* impacts the sex ratio of the parasitoid *Ooencyrtus mirus* (Hymenoptera: Encyrtidae). *PeerJ* 10, e13912.
Renoz, F. (2024) The nutritional dimension of facultative bacterial symbiosis in aphids: Current status and methodological considerations for future research. *Current Research in Insect Science* 5, 100070.
Renoz, F., Pons, I. and Hance, T. (2019) Evolutionary responses of mutualistic insect–bacterial symbioses in a world of fluctuating temperatures. *Current Opinion in Insect Science* 35, 20–26.
Russell, J.A. and Moran, N.A. (2006) Costs and benefits of symbiont infection in aphids: Variation among symbionts and across temperatures. *Proceedings of the Royal Society B Biological Sciences* 273(1586), 603–610.
Scarborough, C.L., Ferrari, J. and Godfray, H.C.J. (2005) Aphid protected from pathogen by endosymbiont. *Science* 310(5755), 1781–1781.
Scudder, G.G.E. (2017) The importance of insects. In: Foottit, G. and Adler, P.H. (eds) *Insect Biodiversity: Science and Society*. John Wiley & Sons Ltd, Chichester, UK, pp. 9–43.
Shan, H.W., Lu, Y.H., Bing, X.L., Liu, S.S. and Liu, Y.Q. (2014) Differential responses of the Whitefly *Bemisia tabaci* symbionts to unfavorable low and high temperatures. *Microbial Ecology* 68(3), 472–482.
Shan, H.W., Deng, W.H., Luan, J.B., Zhang, M.J., Zhang, Z. *et al.* (2017) Thermal sensitivity of bacteriocytes constrains the persistence of intracellular bacteria in Whitefly symbiosis under heat stress. *Environmental Microbiology Reports* 9(6), 706–716.
Shigenobu, S. and Stern, D.L. (2013) Aphids evolved novel secreted proteins for symbiosis with bacterial endosymbiont. *Proceedings of the Royal Society B Biological Sciences* 280(1750), 20121952.
Silva, I.M., Van Meer, M.M., Roskam, M.M., Hoogenboom, A., Gort, G. *et al.* (2000) Biological control potential of *Wolbachia*-infected versus uninfected wasps: Laboratory and greenhouse evaluation of Trichogramma cordubensis and T. deion strains. *Biocontrol Science and Technology* 10(3), 223–238.
Simon, J.C., Boutin, S., Tsuchida, T., Koga, R., Gallic, J.F.L. *et al.* (2011) Facultative symbiont infections affect aphid reproduction. *PLOS One* 6(7), e21831.
Simonet, P., Gaget, K., Balmand, S., Lopes, M.R., Parisot, N. *et al.* (2018) Bacteriocyte cell death in the pea aphid/*Buchnera* symbiotic system. *Proceedings of the National Academy of Sciences USA* 115(8), E1819–E1828.
Smith, A.H., Łukasik, P., O'Connor, M.P., Lee, A., Mayo, G. *et al.* (2015) Patterns, causes and consequences of defensive microbiome dynamics across multiple scales. *Molecular Ecology* 24(5), 1135–1149.

Smith, A.H., O'Connor, M.P., Deal, B., Kotzer, C., Lee, A. *et al.* (2021) Does getting defensive get you anywhere? Seasonal balancing selection, temperature, and parasitoids shape real-world, protective endosymbiont dynamics in the pea aphid. *Molecular Ecology* 30(10), 2449–2472.

Sochard, C., Bellec, L., Simon, J.C. and Outreman, Y. (2021) Influence of "protective" symbionts throughout the different steps of an aphid–parasitoid interaction. *Current Zoology* 67(4), 441–453.

Sokolova, M.I., Zinkevich, N.S. and Zakharov, I.A. (2002) Bacteria in ovarioles of females from maleless families of ladybird beetles *Adalia bipunctata* L. (Coleoptera: Coccinellidae) naturally infected with *Rickettsia*, *Wolbachia*, and *Spiroplasma*. *Journal of Invertebrate Pathology* 79(2), 72–79.

Stouthamer, R., Luck, R.F. and Hamilton, W.D. (1990) Antibiotics cause parthenogenetic *Trichogramma* (Hymenoptera/Trichogrammatidae) to revert to sex. *Proceedings of the National Academy of Sciences USA* 87(7), 2424–2427.

Stouthamer, R., Breeuwer, J.A.J., Luck, R.F. and Werren, J.H. (1993) Molecular identification of microorganisms associated with parthenogenesis. *Nature* 361(6407), 66–68.

Sudakaran, S., Kost, C. and Kaltenpoth, M. (2017) Symbiont acquisition and replacement as a source of ecological innovation. *Trends in Microbiology* 25(5), 375–390.

Svensson, E.I., Gomez-Llano, M. and Waller, J.T. (2020) Selection on phenotypic plasticity favors thermal canalization. *Proceedings of the National Academy of Sciences USA* 117(47), 29767–29774.

Taylor, G.P., Coghlin, P.C., Floate, K.D. and Perlman, S.J. (2011) The host range of the male-killing symbiont *Arsenophonus nasoniae* in filth fly parasitoids. *Journal of Invertebrate Pathology* 106(3), 371–379.

Tian, P.P., Zhang, Y.L., Huang, J.L., Li, W.Y. and Liu, X.D. (2023) *Arsenophonus* interacts with *Buchnera* to improve growth performance of aphids under amino acid stress. *Microbiology Spectrum* 11(3), e01792–23.

Tougeron, K., Iltis, C., Rampnoux, E., Goerlinger, A., Dhondt, L. *et al.* (2023) Still standing: The heat protection delivered by a facultative symbiont to its aphid host is resilient to repeated thermal stress. *Current Research in Insect Science* 3, 100061.

Truitt, A.M., Kapun, M., Kaur, R. and Miller, W.J. (2019) *Wolbachia Modifies Thermal Preference in Drosophila Melanogaster.Environmental Microbiology* 21(9), 3259–3268.

Tsuchida, T., Koga, R. and Fukatsu, T. (2004) Host plant specialization governed by facultative symbiont. *Science* 303(5666), 1989–1989.

Tsuchida, T., Koga, R., Horikawa, M., Tsunoda, T., Maoka, T. *et al.* (2010) Symbiotic bacterium modifies aphid body color. *Science* 330(6007), 1102–1104.

van Baaren, J., Le Lann, C. and van Alphen, J.J.M. (2010) Consequences of climate change for aphid-based multi-trophic systems. In: Kindlmann, P., Dixon, A.F.G. and Michaud, J.P. (eds) *Aphid Biodiversity under Environmental Change: Patterns and Processes*. Springer, Dordrecht, The Netherlands, pp. 55–68.

van Opijnen, T. and Breeuwer, J.A.J. (1999) High temperatures eliminate *Wolbachia*, a cytoplasmic incompatibility inducing endosymbiont, from the two-spotted spider mite. *Experimental & Applied Acarology* 23(11), 871–881.

Vavre, F., Mouton, L. and Pannebakker, B.A. (2009) *Drosophila*–parasitoid communities as model systems for host–*Wolbachia* interactions. *Advances in Parasitology* 70, 299–331.

Vorburger, C. (2018) Symbiont-conferred resistance to parasitoids in aphids – challenges for biological control. *Biological Control* 116, 17–26.

Vorburger, C. (2022) Defensive symbionts and the evolution of parasitoid host specialization. *Annual Review of Entomology* 67(67), 329–346.

Vorburger, C., Gehrer, L. and Rodriguez, P. (2010) A strain of the bacterial symbiont *Regiella insecticola* protects aphids against parasitoids. *Biology Letters* 6(1), 109–111.

Waage, J.K. and Hassell, M.P. (1982) Parasitoids as biological control agents – a fundamental approach. *Parasitology* 84(4), 241–268.

Wagner, S.M., Martinez, A.J., Ruan, Y.M., Kim, K.L., Lenhart, P.A. *et al.* (2015) Facultative endosymbionts mediate dietary breadth in a polyphagous herbivore. *Functional Ecology* 29(11), 1402–1410.

Walton, M.P., Loxdale, H.D. and Allen-Williams, L.J. (2011) Flying with a 'death sentence' on board: Electrophoretic detection of braconid parasitoid larvae in migrating winged grain aphids, *Sitobion avenae* (F.). *Bulletin of Entomological Research* 101(4), 443–449.

Wernegreen, J.J. (2012) Mutualism meltdown in insects: Bacteria constrain thermal adaptation. *Current Opinion in Microbiology* 15(3), 255–262.

Werren, J.H. (1997) Biology of *Wolbachia*. *Annual Review of Entomology* 42(42), 587–609.

Werren, J.H., Zhang, W. and Guo, L.R. (1997) Evolution and phylogeny of *Wolbachia*: Reproductive parasites of arthropods. *Proceedings of the Royal Society B Biological Sciences* 261(1360), 55–63.

Werren, J.H., Baldo, L. and Clark, M.E. (2008) *Wolbachia*: Master manipulators of invertebrate biology. *Nature Reviews Microbiology* 6(10), 741–751.

White, J.A., Kelly, S.E., Perlman, S.J. and Hunter, M.S. (2009) Cytoplasmic incompatibility in the parasitic wasp *Encarsia inaron*: Disentangling the roles of *Cardinium* and *Wolbachia* symbionts. *Heredity* 102(5), 483–489.

Whittle, M., Barreaux, A.M.G., Bonsall, M.B., Ponton, F. and English, S. (2021) Insect-host control of obligate, intracellular symbiont density. *Proceedings of the Royal Society B Biological Sciences* 288(1963), 20211993.

Zchori-Fein, E., Faktor, O., Zeidan, M., Gottlieb, Y., Czosnek, H. *et al.* (1995) Parthenogenesis-inducing microorganisms in *Aphytis* (Hymenoptera: Aphelinidae). *Insect Molecular Biology* 4(3), 173–178.

Zhang, B., Leonard, S.P., Li, Y. and Moran, N.A. (2019) Obligate bacterial endosymbionts limit thermal tolerance of insect host species. *Proceedings of the National Academy of Sciences USA* 116(49), 24712–24718.

Zhou, J.C., Li, Y.Y., Liu, Q.Q., Ning, S.F., Che, W.N. *et al.* (2019) Effects of temperature and superparasitism on quality and characteristics of thelytokous *Wolbachia*-infected *Trichogramma dendrolimi* Matsumura (Hymenoptera: Trichogrammatidae) during mass rearing. *Scientific Reports* 9(1), 18114.

Zientz, E., Dandekar, T. and Gross, R. (2004) Metabolic interdependence of obligate intracellular bacteria and their insect hosts. *Microbiology and Molecular Biology Reviews* 68(4), 745–770.

Zouache, K., Voronin, D., Tran-Van, V. and Mavingui, P. (2009) Composition of bacterial communities associated with natural and laboratory populations of *Asobara tabida* infected with *Wolbachia*. *Applied and Environmental Microbiology* 75(11), 3755.

Zytynska, S., Tighiouart, K. and Frago, E. (2021) The benefits and costs of hosting facultative symbionts in plant-sucking insects: A meta-analysis. *Molecular Ecology* 30, 2483–2494.

13 Challenges of Aflatoxin Biocontrol in Maize Under a Scenario of Climate Change

Maria S. Alaniz-Zanon, Maria L. Chiotta, Marianela Bossa, Paloma Rhein and Sofia N. Chulze*

Research Institute on Mycology and Mycotoxicology (IMICO) – National Scientific and Technical Research Council – Argentina (CONICET), National University of Río Cuarto (UNRC), Córdoba, Argentina

Abstract

Aflatoxins (AFs) are secondary metabolites produced by *Aspergillus* section *Flavi*, mainly *A. flavus*. This fungus is distributed worldwide and may infect maize and other crops at different stages of the agri-food chain. When environmental conditions are favourable (high temperature and drought), some strains may produce AFs, affecting the health of both humans and animals that consume contaminated food and feed, since they have been described as the most toxic natural compounds. The scenario of climate change (CC) that the world will have to face in the next decades, including an increase in global temperature, changes in precipitation patterns and increases in CO_2 levels in the atmosphere, is expected to increase levels of AF contamination, besides having a great impact on biodiversity and the behaviour of native micro-organisms. Furthermore, global population growth will necessitate more quantity and quality of food. Considering this context, it is urgent to develop and implement strategies to mitigate AF contamination by biological control based on non-toxigenic *A. flavus* strains. Several studies have revealed that this is a promising tool to prevent this problem. These biocontrol agents need to be efficient under CC. This chapter describes a case study in which a potential biocontrol agent was selected, characterized and evaluated with the aim of contributing to food safety.

13.1 Introduction

Food safety is one of the main current objectives in our world (WHO, 2019; Lee and Yoon, 2021). Only safe food can satisfy dietary needs and contribute to a healthy life, functional planet, profitable economy and secure future (UN, 2024). Additionally, the world population will have increased from 8.5 billion in 2030 to 10.4 billion in 2100, and 3 billion new middle-class consumers will appear in the coming years. Consequently, the demand for resources will increase exponentially and it is estimated that by 2030, the world will need at least 50% more food in comparison with current conditions (UN, 2022; Lalawmpuii, 2023).

Furthermore, climate change (CC) is also expected to impact food safety. The effects of CC can generate both changes in the nature of existing food safety risks and the emergence of new food safety risks (Camardo Leggieri *et al.*, 2019). The productivity of major field crops like maize and other cereals will be affected and the effect of CC on agriculture and food security will have direct and indirect impacts on nature, threatening biodiversity and increasing the risk

*Corresponding author: schulze@exa.unrc.edu.ar

DOI: 10.1079/9781800625099.0013

of consumer exposure to food-borne diseases (FAO, 2015; Bhattacharyya *et al.*, 2020). Solving and/or reducing this problem is a great challenge and responsibility for food producers and handlers, as well as for governments, industries and consumers around the world. In 2015 the United Nations proposed 17 sustainable development goals, which include eradicating hunger, achieving food security, guaranteeing a healthy life, promoting sustained economic growth and adopting urgent measures against CC, among others (UN, 2015).

Many surveys have demonstrated the ability of fungi to survive, grow and interact with a great diversity of macro- and micro-organisms. Also, they can produce mycotoxins, which are secondary metabolites produced by species within the genera *Aspergillus*, *Fusarium*, *Penicillium* and *Alternaria*. Since these toxins cause several diseases in both animals and humans, with immunotoxicity and teratogenic, hepatotoxic and nephrotoxic effects, it is imperative to reduce their accumulation in food and feed chains. Mycotoxin production is influenced by environmental factors, mainly temperature, relative humidity and insect presence. In this sense, food safety has acquired a leading role among the greatest concerns at global level. The effects on yield and quality of agricultural crops intended for food and feed production, especially regarding mycotoxins, have received particular attention during recent years, highlighting the importance of intensifying efforts to develop efficient mycotoxin management tools (Habschied *et al.*, 2021; Alaniz Zanon *et al.*, 2023).

Among the strategies suggested to mitigate mycotoxin contamination, the most promising to date is biological control, based on the use of micro-organisms with different methods of action. They can compete with toxigenic fungi by infection site or nutrients or produce antagonistic metabolites to avoid the infection, among other mechanisms. Currently, at a global level there is a trend to replace agrochemicals and implement more biological products. Biological control is gaining importance as it is an eco-friendly and economical tool. However, CC may affect biodiversity and this imbalance may generate mismatches between pests so their natural enemies may render biological control ineffective. Thus, CC can affect both beneficial and harmful organisms that may threaten both biodiversity and food safety.

Considering this situation, it is relevant to study the effectiveness and impact of biocontrol agents in a CC scenario, evaluating different conditions and taking action now to prevent problems in the future. In this chapter, more details about aflatoxins (AFs), one of the natural carcinogenic contaminants in maize, will be presented, as well as how CC can influence AF contamination and biocontrol strategies to mitigate the risk, including a case study that addresses the selection and evaluation of a biocontrol agent to reduce AF accumulation.

13.2 Aflatoxins in Maize

Maize (*Zea mays* L.) and other cereals are the main source of proteins, fibre, lipids, vitamins, minerals and calories for human and animal nutrition. Globally, annual cereal production has risen and now stands at 2800 million tonnes. This growth is primarily the result of a substantial increase in the forecast for world maize output during the 2023–24 harvest season (FAO, 2024). Simultaneously, every year cereal diseases cause economic losses of around US$40 trillion worldwide, and at least 20–40% of these losses are associated with infections with pathogenic micro-organisms (Ab Rahman *et al.*, 2018). Several factors, including environmental ones, especially water availability and temperature, influence infection with toxigenic fungi and consequently, mycotoxin contamination levels (Perrone *et al.*, 2020). In both pre- and postharvest stages, maize is vulnerable to infections with toxigenic fungi, which are adapted to a wide range of environmental conditions.

Among the fungal species that commonly contaminate maize, one of the most important is *Aspergillus flavus*, belonging to *Aspergillus* section *Flavi* (Camardo Leggieri *et al.*, 2021). The life cycle of this fungus in the maize environment can be divided into the colonization of plant debris in soil and the invasion of seeds and grains in actively growing crop plants (Horn, 2007). Conidia remain dormant in soil and germinate when appropriate nutritional sources are available. Wind and insects play an important role, since they disperse conidia to the infection site in developing maize ears. Additionally, both drought stress and high temperatures contribute to this infection. Infected cobs serve as an important source of infection in the next season. This species and *A. parasiticus* are the most important AF producers, and mycotoxin synthesis could happen at any stage of the production chain (Waliyar *et al.*, 2015). AFs are secondary metabolites synthesized by toxigenic strains of these species, mainly during water and thermal stress. The presence of AFs in agricultural products cannot be avoided. Kernels contaminated with AFs may be stored under diverse conditions and during storage as well as transportation, AF levels may increase (Alaniz Zanon *et al.*, 2022b).

Aflatoxins are the most toxic natural compounds known, highly carcinogenic and of great concern in food safety. Among all types of AFs, aflatoxin B_1 (AFB_1) is the most important in terms of occurrence and toxicity, and for this reason it has been classified as a group 1 human carcinogen by the International Agency for Research on Cancer (IARC, 2002). The human health implications of AF toxicity depend on the dose and duration of exposure, the individual's gender, age, health, immunity, environmental factors and diet. Patients with AF toxicity can have a range of non-specific signs and symptoms, but the most prominent features are those of hepatotoxicity (McMillan *et al.*, 2018). In animals, acute and chronic aflatoxicosis have also been described (Ingle *et al.*, 2020; Navale *et al.*, 2021).

Many countries have implemented regulations for these and other mycotoxins in food, feed and raw materials to ensure consumer safety, such as European (EC, 2023) and South American countries (Mercosur, 2002; ANVISA, 2011; CAA, 2021). Because of these restrictions, contamination with mycotoxins generates significant economic losses due to the rejection of products or lower prices during domestic and foreign trades (Ladeira *et al.*, 2017; Ojiambo *et al.*, 2018).

According to the last BIOMIN report (BIOMIN, 2023), 22% of samples of maize kernels from Europe were contaminated with AFs (average 17 ppm, maximum 1362 ppm), 9% from North America (average 40 ppm, maximum 327 ppm), 19% from Latin America (average 16 ppm, maximum 9846 ppm), 40% from Asia (average 55 ppm, maximum 1392 ppm), 10% from Sub-saharan Africa (average 210 ppm, maximum 2899 ppm) and 18% from the Middle East and North Africa (average 44 ppm, maximum 142 ppm). Moreover, as discussed in the next section, many of those regions would turn into hot spots due to CC.

13.3 Consequences of Climate Change on Aflatoxin Production

Abiotic factors, especially drought stress, can have a major impact on maize growth mainly during the critical silking period, which can cause both pest damage and increase in colonization by *A. flavus*, resulting in AFB_1 contamination. Currently, AF contamination of maize is prevalent in tropical areas, but unexpected AF contamination has been observed in several regions of Europe since 2000, most probably caused by increasingly drier and warmer summers. This contamination was related to a shift from *F. verticillioides* to *A. flavus*, leading to high AF content in maize kernels due to the climatic conditions. Fig. 13.1 summarizes the regions where meteorological conditions have been associated with AF contamination.

Regarding Europe, in Hungary AFs were not detected in a survey from 2009 to 2011,, although in 2012 AFB_1 was detected in 68.5% of samples evaluated. This situation was related to hot and dry weather with prolonged drought during spring and summer (Kos *et al.*, 2013). Subsequently, from 2013 to 2016, AFB_1 was frequently detected in Hungary (5–72% of samples analysed), with the exception of 2014 when cool and rainy weather conditions occurred. A similar situation was observed in Serbia, where AFs were also detected in several surveys (Janić Hajnal *et al.*, 2017; Kos *et al.*, 2018). In northern Italy, the first *A. flavus* outbreak occurred in 2003, probably due to very hot (>35°C) and dry conditions (Medina *et al.*, 2017). Using a predictive model based on *A. flavus* growth, AFB_1 production and crop phenology, Battilani *et al.* (2016) estimated that AF contamination of plant products, including maize, will also become a problem in north-west Europe. The areas of increased risk in Europe will be central and southern Spain, southern Italy, Greece, northern and south-eastern Portugal, Bulgaria,

Fig. 13.1. Regions where aflatoxin contamination risk was evaluated in relation to meteorological conditions.

Albania, Cyprus and European Turkey (Casu *et al.*, 2024).

In Asian countries, freshwater availability is projected to decrease by 2050 and coastal areas will face a higher risk of flooding. Rising temperatures are increasing the likelihood of heat waves across Asia, as well as droughts in arid and semi-arid areas of western, central and southern Asia, and leading to delays and weakening of the monsoon circulation in South-East Asia (IPCC, 2023).

In Latin America, there will be a significant increase in temperature and decrease in soil water that could reduce crop productivity. Increased risk of drought, floods and landslides is projected as well as large-scale changes and biome shifts in the Amazon (Fernández, *et al.*, 2023; IPCC, 2023).

In the USA, water resources will be stretched, and floods and drought are projected to be more frequent and intense, depending on the geographical area (IPCC, 2023). Kerry *et al.* (2021) investigated the future contamination risk of AFs in maize in southern Georgia, USA. That study showed that the main maize-growing counties are increasingly at risk of AF contamination, exceeding two weather risk factors (temperature >30°C and rainfall < 50 mm) in a greater proportion of years.

In African regions, hotter tropical climates and longer periods of drought stress would have a significant impact on the amount of food produced and their levels of mycotoxin contamination (Magan *et al.*, 2011; IPCC, 2023).

In Australia, an increase of temperature is predicted by 2100 and more hot days and heatwaves are expected. Indeed, extreme events, including marine heatwaves, major hail storms and fires, have been recorded in recent years (IPCC, 2023). These conditions can also impact AF contamination.

Considering the CC scenario, countries that at present are considered safe regions would be vulnerable to AF contamination in the future (Moretti *et al.*, 2019; Burdon and Zhan, 2020; Camardo Leggieri *et al.*, 2020; Magan and Medina, 2020). In this context, the use of different technologies such as breeding for resistance, adaptation of farming practices, predictive models and those that control pathogens and toxigenic fungi based on environmentally friendly tools such as biological control to reduce agrochemical inputs can be promising.

Future needs include research and scientific efforts to focus on gaining a better understanding of global CC and defining realistic strategies to mitigate future effects on agri-food systems worldwide. It is relevant that developing countries that will be affected by CC should have access to technological transfer and solutions through national authorities and policymakers (Casu *et al.*, 2024).

13.4 Biocontrol as an Effective Tool to Mitigate Aflatoxins

The urgent need to generate new technologies to mitigate the negative impact of CC has led to the global adoption of various strategies and action plans, such as the 2030 Agenda (Trivedi *et al.*, 2022; Sarrocco, 2023; UN, 2023). There is a great diversity of physical, chemical and biological strategies in the frame of integrated pest management. The integrated pest management adaptation to CC is addressed in Chapter 14. These strategies are based on reducing symptoms and, consequently, impact on the yield of crops affected by pathogenic and/or toxigenic fungi. Inputs based on biological control agents have been applied at both the pre- and postharvest stages, achieving a decrease in the use of agrochemicals by almost 50% (Magan, 2020).

Regarding AFs, research has focused on the development of predictive models to estimate the risks of AF accumulation in Europe, Australia, Africa and the USA, considering variables such as climatic conditions, maize cultivars and soil types (Chauhan *et al.*, 2015; Battilani *et al.*, 2016; Damianidis *et al.*, 2018). Recently, Yu *et al.* (2022) combined climatic conditions and maize growth stages in their model and predicted that around 90% of USA areas included in the study will be affected by high levels of AFs. However, every region around the world will have different characteristics so a model for each case would be necessary (Nji *et al.*, 2022).

The efficacy of applying *A. flavus* or *A. parasitucus* strains to soil at the preharvest stage has been demonstrated in different crops, including peanuts, cotton and maize (Dorner and Cole, 2002; Cotty and Antilla, 2003; Abbas *et al.*, 2006; Pitt and Hocking, 2006; Dorner and Horn, 2007; Alaniz Zanon *et al.* (2016, 2022a); Ortega-Beltran *et al.*, 2019). Regarding maize, this strategy has been evaluated in different countries around the world: United States (Abbas *et al.*, 2006), Nigeria (Atehnkeng *et al.*, 2008), Italy (Mauro *et al.*, 2018), Serbia (Savić *et al.*, 2020), Ghana (Agbetiameh *et al.*, 2020) and Argentina (Alaniz Zanon *et al.*, 2018, 2022b), among others. Many of these bioinputs are already commercially available (Moore, 2022). Many non-toxigenic *A. flavus* strains are the best biocontrol agents since they are more aggressive than *A. parasiticus* strains and they remain biologically active under identical environmental conditions to toxigenic strains, so they compete for the infection niche and for soil and plant nutrients with native strains (Cotty *et al.*, 2008). Other requirements for these bioinputs are mentioned in Fig. 13.2.

In relation to preparation of bioformulates for biological control of AFs, several matrices have been evaluated. Some formulations were designed for development of the biocontrol agent and its transportation to the field, while others utilized dry conidia grown in culture media and attached to a carrier. An important issue has been the physiological stage of the biocontrol agent when it is applied to the field, and the release ability. Therefore, multiple preparations such as alginate-based formulations (Daigle and Cotty, 1995) and the use of cereal kernels such as wheat (Bock and Cotty, 1999), barley (Abbas *et al.*, 2017) and long-grain rice (Dorner *et al.*, 2003; Alaniz Zanon *et al.*, 2013) have been considered. However, since these substrates are a source of food and feed, and therefore their use for bioinput manufacture would reduce their availability for human consumption, other strategies have been developed considering nutritional stability and a low cost of production. Some examples are the use of bioplastics (Abbas *et al.*, 2017), the design of water-dispersible granules that include tapioca starch (Weaver *et al.*, 2016), the reuse of plastic waste (Accinelli *et al.*, 2009; Accinelli and Abbas, 2011; Accinelli *et al.*, 2014), and the development of a biopolymer based on maize starch (Alaniz Zanon *et al.*, 2022a). Additionally, combinations of strains have been evaluated as potential formulations to reduce AF accumulation in crops (Atehnkeng *et al.*, 2014; Alaniz Zanon *et al.*, 2016). The integration of these technologies and strategies could contribute to food safety, although they still have to be evaluated under CC scenarios.

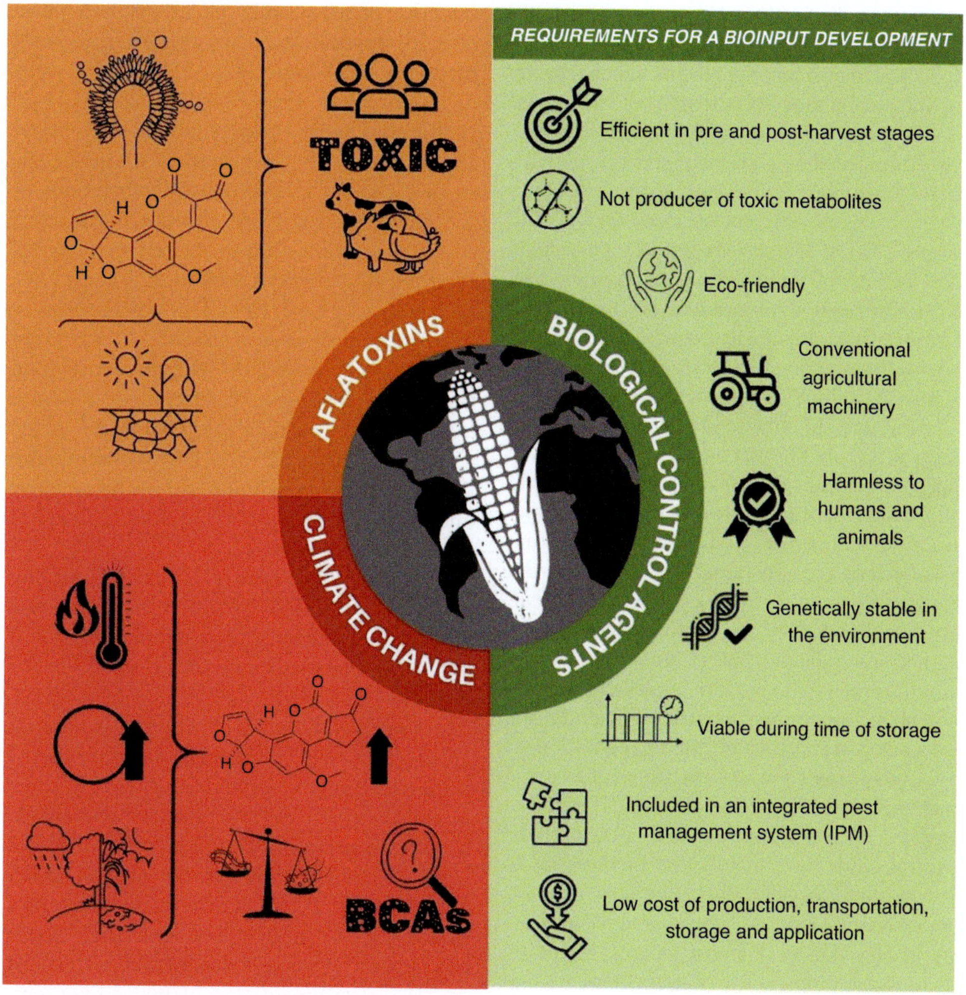

Fig. 13.2. Maize crops face aflatoxin contamination and climate change risks, and the promising alternative of biological control agents. Aflatoxins are produced by *Aspergillus* section *Flavi* mainly under stress conditions and cause important damage to human and animal health. Predicted climate change, including warmer conditions, increased CO_2 levels and modifications in precipitation patterns, would increase the risk of aflatoxin synthesis, modify microbial biodiversity and probably affect biocontrol agent efficiency. In this context, developing bioinputs based on biological control agents have to achieve several requirements.

13.5 Impact of Climate Change on Native Microbiota and Biological Control Agents

Maize productivity is threatened by global CC (Porter *et al.*, 2019). A reduction of about 3.8% in global maize production was reported between 1980 and 2008, countering some of the yield gains from breeding and other technological advances. In addition, maize is increasingly challenged by plant pathogens (Mueller *et al.*, 2020), insect pests (De Groote *et al.*, 2020) and other biotic and abiotic stress factors such as heat and drought (Cairns *et al.*, 2012).

Micro-organisms play an important role in the health and productivity of crops since they form complex associations with plants that contribute to agroecosystem stability, also determining the fertility of the soil, productivity of the crops and stress tolerance (Noman, *et al.*, 2021; Hartmann and Six, 2023). Furthermore, for the sustainable production of food and feed, healthy and productive soils are required. It is estimated that 1 gram of soil can contain up to 10 million micro-organisms belonging to thousands of different species (Torsvik and Øvreås, 2002). The soil microbiota provides a wide variety of ecosystem services including the decomposition of organic matter, release of nutrients and biological control of pests. In general, these ecosystem services are important for human well-being and economic prosperity.

However, in addition to the detrimental effects of CC on different agricultural crops, the approach that has been taken in recent years to soil management to improve yield and production in agriculture has had negative effects that threaten long-term sustainability of agricultural practices (Plaas *et al.*, 2019). Currently, agriculture occupies 40% of the Earth's surface. Some estimations indicate that at least 10–20% of semi-natural habitat per km^2 is needed to ensure ecosystem functioning, in particular pollination, biological pest control and climate regulation, as well as to prevent soil erosion, nutrient loss and water pollution (DeClerck *et al.*, 2021; Hartmann and Six, 2023). Nowadays, 18–33% of agricultural lands have insufficient biodiversity to provide these services, which implies an enormous security risk. Soil biodiversity decreases day by day due to different factors, including CC, intensive crops, abuse of agrochemicals and contamination of soils, air and water sources (DeClerck *et al.*, 2021). Furthermore, it is known that the incorporation of micro-organisms into the ecosystem can influence, at least temporarily, native microbial communities (Trabelsi and Mhamdi, 2013). The dynamics and evolution of various fungal species and biocontrol agents under different conditions are addressed in greater depth in Chapter 10. Additionally, a deeper analysis related to the impact of CC on bacteria and viruses is described in Chapters 11 and 12.

To date, there is scarce literature on the environmental impact of the incorporation of biocontrol agents into the ecosystem, and their effect on the microbial biodiversity of the soils where they are applied. A study conducted by Lewis *et al.* (2019) evaluated shifts in the genetic structure of native soil populations of *A. flavus* after the application of two different biocontrol products (Afla-Guard® and AF36). In this study Afla-Guard was more effective than AF36 in shifting the native soil populations of *A. flavus*, allowing a higher proportion of non-toxigenic strains. The authors demonstrated that one of the bioinputs was more effective in reducing AFs accumulation and more persistent in soil than the other. Besides their interesting results, no biodiversity analysis and no other species of micro-organisms were considered in the study. These aspects should be relevant considering that the effectiveness of biocontrol will ultimately depend on how the biodiversity in the ecosystem is affected and how, consequently, the balance between the positive and negative interactions of the present species is modified (Cardinale *et al.*, 2003).

Additionally, many studies regarding biological control of several diseases caused by pathogen and/or mycotoxigenic fungi have been published worldwide, including molecular and physiological interactions and demonstrating the mode of action of different biocontrol agents. However, few of them have considered resilience under CC conditions. When some of those studies were conducted, the abiotic factors under evaluation were temperature and water availability, but not CO_2 levels (Magan and Medina, 2020). Concerning biocontrol agents for AF mitigation, Rodriguez Sixtos Higuera (2017) and Gasperini *et al.* (2019) observed that inoculation of maize ears with spores of a non-toxigenic strain and a toxigenic strain, in the same proportion and under conditions of high CO_2, 27°C and water activity (a_w) of 0.958 and 0.976, did not show effectiveness in the control of AFB_1 contamination. These results suggest that it is extremely important to identify non-toxigenic *A. flavus* strains more resistant to CC and capable of controlling AF production in crops such as maize (Perrone *et al.*, 2020).

Since it is essential to implement strategies that enhance the beneficial effects of the native microbiota and prevent the loss of biodiversity, it is necessary to focus research on the environmental impact of bioinputs aiming to mitigate AF accumulation under CC conditions.

13.6 Case Study: Selection and Evaluation of a Biocontrol Agent to Reduce the Aflatoxin Accumulation

Regarding the biological control of AFs, the research group working at the Research Institute on Mycology and Micotoxicology (IMICO), Río Cuarto, Córdoba, Argentina, has focused for several years on the development of a strategy based on non-toxigenic *A. flavus* strains native to Córdoba province. During this study, the steps mentioned in Fig. 13.3 were followed as a general schedule for bioinput development.

It is worth mentioning that in many Argentinean regions, maize and peanut crops are included in the same crop rotation system, so they share soil microbiota and climate conditions, among other aspects. A biocontrol strategy based on competitive exclusion by non-toxigenic *A. flavus* strains was studied in both crops.

The first step was isolation of potential biocontrol agents within *Aspergillus* section *Flavi* from the agroecosystem where they were intended to be applied, particularly from soil and kernel samples. The isolation of *Aspergillus* section *Flavi* was performed using two different culture media: Dichloran–Bengal Rose–Chloramphenicol Agar (DRBC) and Dichloran–Glycerol 18% Agar (DG18) (Barros *et al.*, 2003; Alaniz Zanon *et al.*, 2018). Both *A. parasitucus* and *A. flavus* were morphologically and molecularly identified, selecting *A. flavus* strains for further characterization. Some strains of *A. flavus* synthesize cyclopiazonic acid (CPA), a toxic indole tetramic acid that commonly co-occurs with AFs in contaminated agricultural commodities such as maize and peanut. The potential biocontrol agents selected were not producers of either AFs nor CPA (Barros *et al.*, 2005; Alaniz Zanon, 2016; Alaniz Zanon *et al.*, 2016). Additionally, strain sclerotia morphotype was evaluated. In general, sclerotia type S are related to aflatoxigenic strains, while type L are associated with atoxigenic strains. For this reason, L strains were selected (Barros *et al.*, 2005; Alaniz Zanon *et al.*, 2016).

From the selected *A. flavus* strains, vegetative compatibility group (VCG) analysis was carried out. *A. flavus* communities from soils and crops are integrated by multiple VCGs. Members of the same VCG are considered to belong to the same clonal lineage and are different from other VCGs. This analysis was used to estimate the genetic diversity of populations and to monitor the incidence of potentially biocontrol strains applied in field trials. Non-toxigenic strains included in the most frequent VCG were selected (Barros *et al.*, 2006; Alaniz Zanon *et al.*, 2016). Mating type gene analysis was performed to select a potential biocontrol agent with low probability to recombine with toxigenic strain. As the population under study had shown an abundance of idiomorph *MAT1-1*, the potential biocontrol agent selected presented *MAT1-1* loci (Alaniz Zanon *et al.*, 2018).

In situ assays were conducted in the laboratory in maize kernels coinoculated with a toxigenic strain and one of the biocontrol candidates (Alaniz Zanon *et al.*, 2018). The highest AF reduction percentage observed was up to 68%. From these studies, the most effective strains were selected to conduct field trials. For this next stage, it was necessary to prepare a proper formulation, taking into account several considerations such as solid vs liquid preparation, physiological stage of the potential biocontrol agent, conditions of production, material costs, packaging, transportation, application machinery, etc. Therefore, for the first studies, bioformulates were prepared using long-grain rice. Simultaneously, the development of a maize starch biopolymer that works as substrate and carrier was set up and optimized through several studies (Alaniz Zanon *et al.*, 2022b). Also, data on viability and stability of the bioinput during the time of storage under different conditions were generated, confirming that the formulate remains unaltered for 6 months at 20–25°C and at 4°C. Furthermore, this bioformulation based on the AFCHG2 strain demonstrated safety in trials carried out in Wistar rats fed for 60 days, and in another group of animals inoculated intraperitoneally (Alaniz Zanon, 2016).

During field trials, different variables were considered for the design of the experiments: drought stress conditions vs natural conditions, phenological stage of the crop (for peanut: at planting, 50 days after planting and both inoculation stages; and for maize: at V4–V6 stage), number of applications (1 and 2), inoculum rates (25–50 kg/ha), agroecological regions with different soil, agricultural practices and climatic characteristics (areas with different

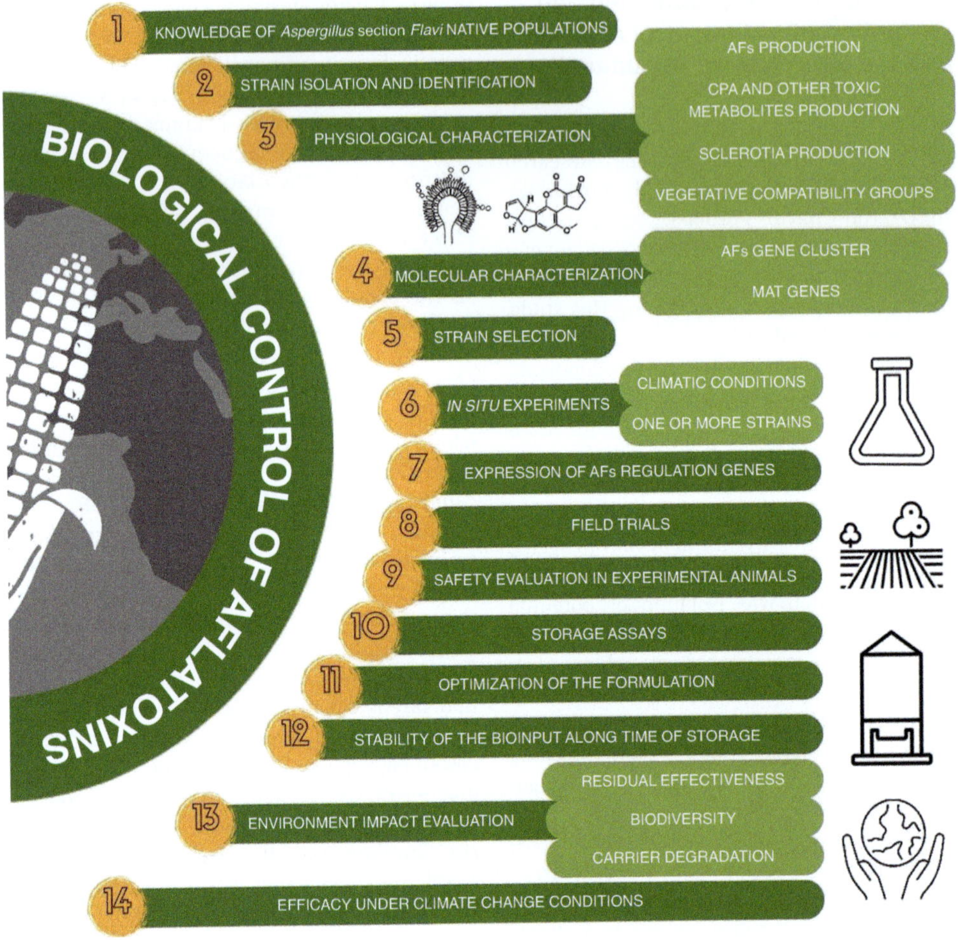

Fig. 13.3. Steps included in the development of a biological control agent. The first step is knowing the environment where the biocontrol agent will be applied. It is important to isolate strains from the same ecosystem, since they are well adapted to it, and identify the species of interest. Then, the pool of strains must be characterized regarding physiological and molecular criteria to select the best candidates as biocontrol agents. The chosen strains are then evaluated under *in situ* assays in the laboratory, considering different climatic conditions and strain mixtures. In those studies, in addition to determining AF content, it is relevant to analyse the expression of regulation genes within the AF gene cluster in the toxigenic strain/s in the presence of the potential biocontrol agents. Field trials are the next step that should be carried out during at least three crop seasons. Another interesting evaluation is the residual effect during storage of the biocontrol agent applied at field stage, as well as its safety for animals and viability during bioinput storage periods. Sometimes, it is necessary to optimize the matrix where the biocontrol agent grows and over which it is transported to the field. Finally, evaluation of the environmental impact, including biodiversity changes, carrier degradation in soil and residual effect for the following seasons, as well as evaluation of the biocontrol efficiency under CC are steps that will allow the bioinput to be successful.

rainfall patterns) (Alaniz Zanon *et al.*, 2013, 2016). The meteorological parameters evaluated were soil temperature, rainfall levels and environmental temperature. Also, the use of combined atoxigenic *A. flavus* strains within the same bioformulate was evaluated (Alaniz Zanon *et al.*, 2016, 2022a).

In these field trials, monitoring of biocontrol agent effectiveness was carried out by aflatoxigenic profile evaluation and VCG analysis within *A. flavus* isolates from both soil and kernel samples. Also, AFB_1 content in kernels at harvest stage was determined. In general, the most effective formulation in reducing AFB_1 accumulation was that based on the non-toxigenic *A. flavus* AFCHG2 strain used individually, achieving AFB_1 reduction up to 85%. During 2022 a storage assay was conducted in order to evaluate the effect of preharvest biocontrol treatments on AF accumulation in maize stored in silo bags, which are widely used in Argentina. Single and mixed inocula based on *A. flavus* AFCHG2 and ARG5/30 strains were evaluated. Hermetic bags were better than non-hermetic bags in preventing AF accumulation and both single and mixed inocula were effective during 3 and 6 months of storage, achieving AF reduction percentages up to 72% (Alaniz Zanon *et al.*, 2022b).

At present, environmental impact studies are ongoing to evaluate the effect of the biological control agent on the soil mycobiota (data not yet published). Further studies, including metagenomic analysis, are necessary to confirm that the introduction of this biocontrol agent, even if it is native to the same environment, does not cause an imbalance in the ecosystem microbiota. Also, under the scenario of CC and the increased risk of contamination with high levels of AFs, *in vitro* studies are being conducted to evaluate the resilience and effectiveness of the biocontrol agent selected. For this study, different conditions will be evaluated comparing current conditions vs simulated CC: 25°C vs 37°C, 400 ppm vs 1000 ppm CO_2, and a_w of 0.98 vs 0.94.

13.7 Conclusion

Growing evidence of the negative impacts of CC on agriculture as well as human and animal health requires more research in order to understand the underlying mechanisms of damage in a global context. Regarding AF accumulation in raw materials and several substrates intended for food and feed production, research around the world is focused on preventing and/or solving this problem. Countless studies have contributed to food safety through different tools that complement each other within an integrated pest management system. The effectiveness of applying atoxigenic *A. flavus* strains to control AF production has been confirmed worldwide. However, there is still much work to be done. Considering the CC scenario, it will be essential to combine efforts and resources in order to validate the efficacy of those inputs based on biological control strategies under conditions that will favour AF synthesis, mainly in regions of the world that are projected to turn into hot spots.

Finally, the coherent and conscious integration of knowledge developed to date regarding biological control in a holistic and globalized context would also provide the opportunity of generating new sustainable development policies that allow us to face the inevitable consequences of CC.

References

Ab Rahman, S.F.S., Singh, E., Pieterse, C.M. and Schenk, P.M. (2018) Emerging microbial biocontrol strategies for plant pathogens. *Plant Science* 267, 102–111.

Abbas, H.K., Zablotowicz, R.M., Bruns, H.A. and Abel, C.A. (2006) Biocontrol of aflatoxin in corn by inoculation with non-aflatoxigenic *Aspergillus flavus* isolates. *Biocontrol Science and Technology* 16(5), 437–449.

Abbas, H.K., Accinelli, C. and Shier, W.T. (2017) Biological control of aflatoxin contamination in US crops and the use of bioplastic formulations of *Aspergillus flavus* biocontrol strains to optimize application strategies. *Journal of Agricultural and Food Chemistry* 65(33), 7081–7087.

Accinelli, C. and Abbas, H.K. (2011) New perspectives for the application of bioplastic materials in the biocontrol of *Aspergillus flavus* in corn. *Toxin Reviews* 30(2–3), 71–78.

Accinelli, C., Saccà, M.L., Abbas, H.K., Zablotowicz, R.M. and Wilkinson, J.R. (2009) Use of a granular bioplastic formulation for carrying conidia of a non-aflatoxigenic strain of *Aspergillus flavus*. *Bioresource Technology* 100(17), 3997–4004.

Accinelli, C., Abbas, H.K., Vicari, A. and Shier, W.T. (2014) Aflatoxin contamination of corn under different agro-environmental conditions and biocontrol applications. *Crop Protection* 63, 9–14.

Agbetiameh, D., Ortega-Beltran, A., Awuah, R.T., Atehnkeng, J., Elzein, A. *et al.* (2020) Field efficacy of two atoxigenic biocontrol products for mitigation of aflatoxin contamination in maize and groundnut in Ghana. *Biological Control* 150, 104351.

Alaniz Zanon, M.S. (2016) Reducción de la entrada de aflatoxinas en la cadena agroalimentaria maní. Biocontrol basado en la exclusión competitiva. PhD thesis, Universidad Nacional de Río Cuarto, Argentina.

Alaniz Zanon, M.S., Chiotta, M.L., Giaj-Merlera, G., Barros, G. and Chulze, S. (2013) Evaluation of potential biocontrol agent for aflatoxin in Argentinean peanuts. *International Journal of Food Microbiology* 162(3), 220–225.

Alaniz Zanon, M.S., Barros, G.G. and Chulze, S.N. (2016) Non-aflatoxigenic *Aspergillus flavus* as potential biocontrol agents to reduce aflatoxin contamination in peanuts harvested in Northern Argentina. *International Journal of Food Microbiology* 231, 63–68.

Alaniz Zanon, M.S., Clemente, M.P. and Chulze, S.N. (2018) Characterization and competitive ability of non-aflatoxigenic *Aspergillus flavus* isolated from the maize agro-ecosystem in Argentina as potential aflatoxin biocontrol agents. *International Journal of Food Microbiology* 277, 58–63.

Alaniz Zanon, M.S., Bossa, M., Chiotta, M.L., Oddino, C., Giovanini, D. *et al.* (2022a) Pre-harvest strategy for reducing aflatoxin accumulation during storage of maize in Argentina. *International Journal of Food Microbiology* 380, 109887.

Alaniz Zanon, M.S., Oddino, C., Giovaninni, D., Barbero, C., Chiotta, M.L. *et al.* (2022b) Bioformulate to reduce the accumulation of aflatoxins in maize based on a biopolymer as a carrier and support for growth of the biocontrol agent. In: *Paper presented at the World Mycotoxin Forum 13th Conference*.

Alaniz Zanon, M.S., Pena, G., Yerkovich, N., Bossa, M., Chiotta, M.L. *et al.* (2023) Aflatoxins and fumonisins in maize under a climate change scenario. Biocontrol strategies at the pre-harvest stage. *European Journal of Plant Pathology* 167(4), 551–567.

ANVISA (2011) Agência Nacional de Vigilância Sanitária. Resolução RDC 07/2011.

Atehnkeng, J., Ojiambo, P.S., Ikotun, T., Sikora, R.A., Cotty, P.J. *et al.* (2008) Evaluation of atoxigenic isolates of *Aspergillus flavus* as potential biocontrol agents for aflatoxin in maize. *Food Additives and Contaminants* 25(10), 1264–1271.

Atehnkeng, J., Ojiambo, P.S., Cotty, P.J. and Bandyopadhyay, R. (2014) Field efficacy of a mixture of atoxigenic *Aspergillus flavus* Link: Fr vegetative compatibility groups in preventing aflatoxin contamination in maize (*Zea mays* L.). *Biological Control* 72, 62–70.

Barros, G., Torres, A., Palacio, G. and Chulze, S. (2003) *Aspergillus* species from section *Flavi* isolated from soil at planting and harvest time in peanut-growing regions of Argentina. *Journal of the Science of Food and Agriculture* 83(13), 1303–1307.

Barros, G., Torres, A. and Chulze, S. (2005) *Aspergillus flavus* population isolated from soil of Argentina's peanut-growing region. Sclerotia production and toxigenic profile. *Journal of the Science of Food and Agriculture* 85(14), 2349–2353.

Barros, G.G., Torres, A.M., Rodriguez, M.I. and Chulze, S.N. (2006) Genetic diversity within *Aspergillus flavus* strains isolated from peanut-cropped soils in Argentina. *Soil Biology and Biochemistry* 38(1), 145–152.

Battilani, P., Toscano, P., der Fels-Klerx, V., Moretti, A., Camardo Leggieri, M. *et al.* (2016) Aflatoxin B_1 contamination in maize in Europe increases due to climate change. *Scientific Reports* 6(1), 1–7.

Bhattacharyya, P., Pathak, H., Pal, S., Bhattacharyya, P., Pathak, H. *et al.* (eds) (2020) Impact of climate change on agriculture: Evidence and predictions. In: *Climate Smart Agriculture: Concepts, Challenges and Opportunities*. Springer, Cham, Switzerland, pp. 17–32.

BIOMIN (2023) DSM World Mycotoxin Survey. The Global Threat January–December 2022. Available at: www.dsm.com/anh/news/downloads/whitepapers-and-reports/dsm-world-mycotoxin-survey-2022-report.html (accessed 5 June 2025).

Bock, C.H. and Cotty, P.J. (1999) Wheat seed colonized with atoxigenic *Aspergillus flavus*: Characterization and production of a biopesticide for aflatoxin control. *Biocontrol Science and Technology* 9(4), 529–543.

Burdon, J.J. and Zhan, J. (2020) Climate change and disease in plant communities. *PLOS Biol* 18(11), e3000949.

CAA (2021) Código Alimentario Argentino. Capítulo III: De los productos alimenticios. Available at: www.argentina.gob.ar/sites/default/files/anmat_capitulo_iii_prod_alimenticios.pdf (accessed 5 June 2025).

Cairns, J.E., Sanchez, C., Vargas, M., Ordoñez, R. and Araus, J.L. (2012) Dissecting maize productivity: Ideotypes associated with grain yield under drought stress and well-watered conditions. *Journal of Integrative Plant Biology* 54(12), 1007–1020.

Camardo Leggieri, M., Giorni, P., Pietri, A. and Battilani, P. (2019) *Aspergillus flavus* and *Fusarium verticillioides* interaction: Modeling the impact on mycotoxin production. *Frontiers in Microbiology* 10, 2653.

Camardo Leggieri, M., Lanubile, A., Dall'Asta, C., Pietri, A. and Battilani, P. (2020) The impact of seasonal weather variation on mycotoxins: Maize crop in 2014 in Northern Italy as a case study. *World Mycotoxin Journal* 13, 25–36.

Camardo Leggieri, M., Toscano, P. and Battilani, P. (2021) Predicted aflatoxin B_1 increase in Europe due to climate change: Actions and reactions at global level. *Toxins* 13(4), 292.

Cardinale, B.J., Harvey, C.T., Gross, K. and Ives, A.R. (2003) Biodiversity and biocontrol: Emergent impacts of a multi-enemy assemblage on pest suppression and crop yield in an agroecosystem. *Ecology Letters* 6(9), 857–865.

Casu, A., Camardo Leggieri, M., Toscano, P. and Battilani, P. (2024) Changing climate, shifting mycotoxins: A comprehensive review of climate change impact on mycotoxin contamination. *Comprehensive Reviews in Food Science and Food Safety* 23, e13323.

Chauhan, Y., Tatnell, J., Krosch, S., Karanja, J., Gnonlonfin, B. *et al.* (2015) An improved simulation model to predict pre-harvest aflatoxin risk in maize. *Field Crops Research* 178, 91–99.

Cotty, P. and Antilla, L. (2003) *Managing Aflatoxins in Arizona*. Agricultural Research Service, US Department of Agriculture, New Orleans, Louisiana.

Cotty, P.J., Probst, C. and Jaime-Garcia, R. (2008) Etiology and management of aflatoxin contamination. In: Leslie, J.F., Bandyopadhyay, R. and Visconti, A. (eds) *Mycotoxins: Detection Methods, Management, Public Health And Agricultural Trade*. CABI, Wallingford, UK, pp. 287–299.

Daigle, D.J. and Cotty, P.J. (1995) Formulating atoxigenic *Aspergillus flavus* for field release. *Biocontrol Science and Technology* 5(2), 175–184.

Damianidis, D., Ortiz, B.V., Windham, G.L., Bowen, K.L., Hoogenboom, G. *et al.* (2018) Evaluating a generic drought index as a predictive tool for aflatoxin contamination of corn: From plot to regional level. *Crop Protection* 113, 64–74.

De Groote, H., Kimenju, S.C., Munyua, B., Palmas, S., Kassie, M. *et al.* (2020) Spread and impact of fall armyworm (Spodoptera frugiperda JE Smith) in maize production areas of Kenya. *Agriculture, Ecosystems & Environment* 292, 106804.

DeClerck, F.A.J., Koziell, I., Benton, T., Garibaldi, L.A., Kremen, C. *et al.* (2021) A whole earth approach to nature-positive food: Biodiversity and agriculture. In: von Braun, J., Afsana, K., Fresco, L.O. and Hassan, M.H.A. (eds) *Science and Innovations for Food Systems Transformation*. Springer, Cham, Switzerland, pp. 469–496.

Dorner, J.W. and Cole, R.J. (2002) Effect of application of nontoxigenic strains of *Aspergillus flavus* and *A. parasiticus* on subsequent aflatoxin contamination of peanuts in storage. *Journal of Stored Products Research* 38(4), 329–339.

Dorner, J.W. and Horn, B.W. (2007) Separate and combined applications of nontoxigenic *Aspergillus flavus* and *A. parasiticus* for biocontrol of aflatoxin in peanuts. *Mycopathologia* 163, 215–223.

Dorner, J.W., Cole, R.J., Connick, W.J., Daigle, D.J., McGuire, M.R. *et al.* (2003) Evaluation of biological control formulations to reduce aflatoxin contamination in peanuts. *Biological Control* 26(3), 318–324.

EC (2023) Commission Regulation (EU) 2023/915 of 25 April 2023 on maximum levels for certain contaminants in food and repealing Regulation (EC) No 1881/2006. *Official Journal of the European Union* 119, 103–157.

FAO (2015) Climate change and food security: Risks and responses. Available at: www.fao.org/3/i5188e/I5188E.pdf (accessed 5 June 2025).

FAO (2024) Global cereal supplies in 2023/24 remain comfortable; Early production prospects of wheat in 2024 favourable. Available at: www.fao.org/worldfoodsituation/csdb/en (accessed 5 June 2025).

Fernández, A., Franchini, M., Aragão, L., Carte, L. and Pacheco, D. (2023) Facing climate change across Latin America and the Caribbean. *Nature Climate Change* 13, 1014–1019.

Gasperini, A.M., Rodriguez-Sixtos, A., Verheecke-Vaessen, C., Garcia-Cela, E., Medina, A. *et al.* (2019) Resilience of biocontrol for aflatoxin minimization strategies: Climate change abiotic factors may affect control in non-GM and GM-maize cultivars. *Frontiers in Microbiology* 10, 493745.

Habschied, K., Krstanović, V., Zdunić, Z., Babić, J., Mastanjević, K. *et al.* (2021) Mycotoxins biocontrol methods for healthier crops and stored products. *Journal of Fungi* 7(5), 348.

Hartmann, M. and Six, J. (2023) Soil structure and microbiome functions in agroecosystems. *Nature Reviews Earth & Environment* 4(1), 4–18.

Horn, B.W. (2007) Biodiversity of *Aspergillus* section *Flavi* in the United States: A review. *Food Additives & Contaminants* 24(10), 1088–1101.

IARC (2002) IARC monographs on the evaluation of carcinogenic risks to humans. Some Traditional herbal medicines, mycotoxins, naphthalene and styrene. Lyon, France. Available at: https://publications.iarc.fr/Book-And-Report-Series/Iarc-Monographs-On-The-Identification-Of-Carcinogenic-Hazards-To-Humans/Some-Traditional-Herbal-Medicines-Some-Mycotoxins-Naphthalene-And-Styrene-2002 (accessed 5 June 2025).

Ingle, A.P., Gupta, I., Jogee, P. and Rai, M. (2020) Role of nanotechnology in the detection of mycotoxins: A smart approach. In: Rai, M. and Abd-Elsalam, K.A. (eds) *Nanomycotoxicology*. Academic Press, Cambridge, Massachusetts, pp. 11–33.

IPCC (2023) Climate change 2023: Synthesis report. Available at: www.ipcc.ch/report/ar6/syr/downloads/report/IPCC_AR6_SYR_LongerReport.pdf (accessed 5 June 2025).

Janić Hajnal, E., Kos, J., Krulj, J., Krstović, S., Jajić, I. *et al.* (2017) Aflatoxins contamination of maize in Serbia: The impact of weather conditions in 2015. *Food Additives & Contaminants: Part A* 34(11), 1999–2010.

Kerry, R., Ingram, B., Garcia-Cela, E., Magan, N., Ortiz, B.V. *et al.* (2021) Determining future aflatoxin contamination risk scenarios for corn in Southern Georgia, USA using spatio-temporal modelling and future climate simulations. *Scientific Reports* 11(6), 13522.

Kos, J., Mastilović, J., Hajnal, E.J. and Šarić, B. (2013) Natural occurrence of aflatoxins in maize harvested in Serbia during 2009–2012. *Food Control* 34(1), 31–34.

Kos, J., Janić Hajnal, E., Šarić, B., Jovanov, P., Mandić, A. *et al.* (2018) Aflatoxins in maize harvested in the Republic of Serbia over the period 2012–2016. *Food Additives & Contaminants: Part B* 11(4), 246–255.

Ladeira, C., Frazzoli, C. and Orisakwe, O.E. (2017) Engaging One Health for non-communicable diseases in Africa: Perspective for mycotoxins. *Frontiers in Public Health* 5, 266.

Lalawmpuii, P.K.R. (2023) Role of water-energy-food nexus in environmental management and climate action. *Energy Nexus* 11, 100230.

Lee, H. and Yoon, Y. (2021) Etiological agents implicated in foodborne illness world wide. *Food Science of Animal Resources* 41(1), 1–7.

Lewis, M.H., Carbone, I., Luis, J.M., Payne, G.A., Hagan, A.K. *et al.* (2019) Biocontrol strains differentially shift the genetic structure of indigenous soil populations of *Aspergillus flavus*. *Frontiers in Microbiology* 10, 477910.

Magan, N., Medina, A. and Aldred, D. (2011) Possible climate change effects onmycotoxin contamination of food crops pre- and postharvest. *Plant Pathology* 60(1), 150–163.

Magan, N. (2020) Importance of ecological windows for efficacy of biocontrol agents. In: De Cal, A., Melgarejo, P. and Magan, N. (eds) *How Research Can Stimulate Development of Comercial Biological Control against Plant Diseases. Progress in Biological Control*, Vol. 21. Springer, Cham, Switzerland, pp. 1–14.

Magan, N. and Medina, A. (2020) Climate change and resilience of biological control agents. In: De Cal, A., Melgarejo, P. and Magan, N. (eds) *How Research Can Stimulate Development of Comercial Biological Control against Plant Diseases. Progress in Biological Control*, Vol. 21. Springer, Cham, Switzerland, pp. 83–93.

Mauro, A., Garcia-Cela, E., Pietri, A., Cotty, P.J. and Battilani, P. (2018) Biological control products for aflatoxin prevention in Italy: Commercial field evaluation of atoxigenic *Aspergillus flavus* active ingredients. *Toxins* 10(1), 30.

McMillan, A., Renaud, J.B., Burgess, K.M.N., Orimadegun, A.E., Akinyinka, O.O. *et al.* (2018) Aflatoxin exposure in Nigerian children with severe acute malnutrition. *Food and Chemical Toxicology* 111, 356–362.

Medina, A., Gilbert, M.K., Mack, B.M., OBrian, G.R., Rodriguez, A. *et al.* (2017) Interactions between water activity and temperature on the *Aspergillus flavus* transcriptome and aflatoxin B1 production. *International Journal of Food Microbiology* 256, 36–44.

Mercosur (2002) MERCOSUR/GMC/Res. N° 25/02. Reglamento técnico MERCOSUR sobre límites máximos de aflatoxinas admisibles en leche, maní y maíz (Derogación de la Res. GMC N° 56/94).

Moore, G.G. (2022) Practical considerations will ensure the continued success of pre-harvest biocontrol using non-aflatoxigenic *Aspergillus flavus* strains. *Critical Reviews in Food Science and Nutrition* 62(15), 4208–4225.

Moretti, A., Pascale, M. and Logrieco, A.F. (2019) Mycotoxin risks under a climate change scenario in Europe. *Trends in Food Science & Technology* 84, 38–40.

Mueller, D.S., Wise, K.A., Sisson, A.J., Allen, T.W., Bergstrom, G.C. *et al.* (2020) Corn yield loss estimates due to diseases in the United States and Ontario, Canada, from 2016 to 2019. *Plant Health Progress* 21(4), 238–247.

Navale, V., Vamkudoth, K.R., Ajmera, S. and Dhuri, V. (2021) *Aspergillus* derived mycotoxins in food and the environment: Prevalence, detection and toxicity. *Toxicology Reports* 8, 1008–1030.

Nji, Q.N., Babalola, O.O. and Mwanza, M. (2022) Aflatoxins in maize: Can their occurrence be effectively managed in Africa in the face of climate change and food insecurity? *Toxins* 14(8), 574.

Noman, M., Ahmed, T., Ijaz, U., Shahid, M. and Azizullah, D. (2021) Plant–microbiome crosstalk: Dawning from composition and assembly of microbial community to improvement of disease resilience in plants. *International Journal of Molecular Sciences* 22(13), 6852.

Ojiambo, P.S., Battilani, P., Carry, J.W., Blum, B.H. and Carbone, I. (2018) Cultural and genetic approaches to manage aflatoxin contamination: Recent insights provide opportunities for improved control. *Phytopathology* 108(9), 1024–1037.

Ortega-Beltran, A., Moral, J., Picot, A., Puckett, R.D., Cotty, P.J. *et al.* (2019) Atoxigenic *Aspergillus flavus* isolates endemic to almond, fig and pistachio orchards in California with potential to reduce aflatoxin contamination in these crops. *Plant Disease* 103(5), 905–912.

Perrone, G., Ferrara, M., Medina, A., Pascale, M. and Magan, N. (2020) Toxigenic fungi and mycotoxins in a climate change scenario: ecology, genomics, distribution, prediction and prevention of the risk. *Microorganisms* 8(10), 1496.

Pitt, J.I. and Hocking, A.D. (2006) Mycotoxins in Australia: Biocontrol of aflatoxin in peanuts. *Mycopathologia* 162(3), 233–243.

Plaas, E., Meyer-Wolfarth, F., Banse, M., Bengtsson, J., Bergmann, H. *et al.* (2019) Towards valuation of biodiversity in agricultural soils: A case for earthworms. *Ecological Economics* 159, 291–300.

Porter, J.R., Challinor, A.J., Henriksen, C.B., Howden, S.M., Martre, P. *et al.* (2019) Invited review: Intergovernmental panel on climate change, agriculture and food – a case of shifting cultivation and history. *Global Change Biology* 25(8), 2518–2529.

Rodriguez Sixtos Higuera, A. (2017) Fungal interactions and control of aflatoxins in maize, pre-and post-harvest under different climate change scenarios. Doctoral dissertation. Cranfield University, Cranfield, UK. Available at: http://dspace.lib.cranfield.ac.uk/handle/1826/17884 (accessed 5 June 2025).

Sarrocco, S. (2023) Biological disease control by beneficial (micro)organisms: Selected breakthroughs in the past 50 years. *Phytopathology* 113(4), 732–740.

Savić, Z., Dudaš, T., Loc, M., Grahovac, M., Budakov, D. *et al.* (2020) Biological control of aflatoxin in maize grown in Serbia. *Toxins* 12(3), 162.

Torsvik, V. and Øvreås, L. (2002) Microbial diversity and function in soil: from genes to ecosystems. *Current Opinion in Microbiology* 5(3), 240–245.

Trabelsi, D. and Mhamdi, R. (2013) Microbial inoculants and their impact on soil microbial communities: A review. *BioMed Research International* 863240.

Trivedi, P., Batista, B.D., Bazany, K.E. and Singh, B.K. (2022) Plant–microbiome interactions under a changing world: Responses, consequences and perspectives. *New Phytologist* 234(6), 1951–1959.

UN (2015) The 17 goals. Available at: https://sdgs.un.org/goals (accessed 5 June 2025).

UN (2022) World Population Prospects 2022: Summary of results. Available at: www.un.org/development/desa/pd/sites/www.un.org.development.desa.pd/files/wpp2022_summary_of_results.pdf (accessed 5 June 2025).

UN (2023) Goal 2: Zero Hunger. Available at: www.un.org/sustainabledevelopment/hunger/#:~:text=Goal%202%20is%20about%20creating,climate%20change%2C%20and%20deepening%20inequalities (accessed 5 June 2025).

UN (2024) Why improving food safety is important. Available at: www.un.org/en/observances/food-safety-day (accessed 5 June 2025).

Waliyar, F., Osiru, M., Ntare, B.R., Kumar, K.V.K., Sudini, H. *et al*. (2015) Post-harvest management of aflatoxin contamination in groundnut. *World Mycotoxin Journal* 8(2), 245–252.

Weaver, M.A., Abbas, H.K., Jin, X. and Elliott, B. (2016) Efficacy of water-dispersible formulations of biological control strains of *Aspergillus flavus* for aflatoxin management in corn. *Food Additives & Contaminants: Part A* 33(2), 346–351.

WHO (2019) Food safety. Available at: www.who.int/news-room/fact-sheets/detail/food-safety (accessed 5 June 2025).

Yu, J., Hennessy, D.A., Tack, J. and Wu, F. (2022) Climate change will increase aflatoxin presence in US corn. *Environmental Research Letters* 17(5), 054017.

14 Adapting Integrated Pest Management to Climate Change

Joffrey Moiroux[1]*, Marie Perrin[2] and Myriam Siegwart[3]
[1]*Avignon Université, Aix Marseille Université, CNRS, IRD, IMBE, Avignon, France;*
[2]*Agroecology and Environment Research Unit, ISARA-Lyon, Avignon, France;*
[3]*INRAE, UR 1115 PSH, Plantes et Systèmes de culture Horticoles, Avignon, France*

Abstract

Integrated pest management (IPM) is a holistic approach to pest control that combines various climate-dependent tactics. This chapter first reviews how climate, mainly temperature, influences four of these tactics: mating disruption, biological control using entomopathogens and entomophagous arthropods, and insecticides. Reports of recent failures in these techniques worldwide highlight the urgent need for adaptations to climate change, which are explored in the second section. The third section discusses recent and ongoing research efforts to ensure that IPM remains a viable production strategy in the context of climate change, using the codling moth *Cydia pomonella* (Lepidoptera:Tortricidae) as a case study.

14.1 Introduction

Integrated Pest management (IPM) is a comprehensive approach to pest control that combines biological, biotechnological, chemical, physical and cultural methods, using chemical pesticides only when essential to keep pest populations below thresholds that would cause economically unacceptable damage or loss. This strategy, aimed at reducing risks to both environmental and human health, is expected to evolve in the future as climate change influences pest abundance (Deutsch *et al.*, 2018) (see Chapter 3) and affects the effectiveness of various control tactics used in IPM (Castex *et al.*, 2018).

Because pests and their natural enemies are predominantly ectotherms, rising temperatures could indeed significantly affect management methods that rely on direct actions on their biological processes – whether biochemical, neurological, physiological or behavioural. Various studies have thus in recent years anticipated or observed changes in the effectiveness of some IPM tactics. Notably, the use of chemical insecticides is expected to increase significantly in the coming decades due to climate change (Koleva and Schneider, 2009). Preparing IPM strategies for future conditions could help manage the anticipated rise in pest populations and reduce the growing reliance on insecticides.

In this chapter, the effects of climate on four commonly used IPM tactics – mating disruption, biological control using entomopathogens, biological control using entomophagous organisms, and insecticides – are first examined. In the second part, potential adaptations of these pest management tactics in response to climate change are discussed. Finally, a concrete example of adapting an IPM strategy to climate

*Corresponding author: joffrey.moiroux@univ-avignon.fr

 Biological Control Systems and Climate Change (eds Joan van Baaren *et al.*)
DOI: 10.1079/9781800625099.0014

change is presented, focusing on the codling moth *Cydia pomonella* (Lepidoptera: Tortricidae).

14.2 Influence of Temperature on Pest Management Methods

This first section examines the impact of climatic factors, primarily temperature given the extensive literature on this topic, on four major management methods used in IPM: pheromone-based mating disruption; biological control with entomopathogens and entomophagous organisms; and chemical insecticides and bioinsecticides. Although other tactics employed in IPM, such as plant extracts (e.g. neem), the sterile insect technique, crop rotation and intercropping, may also be impacted by climate change, they have been insufficiently studied in this context. Consequently, any predictions regarding their future effectiveness remain highly speculative; therefore, they are not discussed here.

14.2.1 Mating disruption

In IPM, industrially synthesized sex pheromones are commonly used to lure pests into traps, monitor their presence, disrupt their mating and control their populations (Witzgall *et al.*, 2010). While this method is still efficient to gauge pest presence and abundance, studies have reported recent failures to control some pests worldwide (El-Sayed *et al.*, 2021). It has been proposed that such lack of efficacy may first result from an increased pest density and advanced phenology with global warming, all crucial elements for the success of this strategy in IPM. Indeed, pheromone-based pest control frequently fails when initial pest population density is too high (Louis and Schirra, 2001). Second, these failures may result from decreased dispenser durability caused by an increased pheromone release rate (Tobin *et al.*, 2011) and/or accelerated pheromone degradation and increased isomeric and chemical impurities occurring at high temperatures (El-Sayed *et al.*, 2021). Tobin *et al.* (2011) deployed traps baited with disparlure, the synthetic form of the gypsy moth *Lymantria dispar* (Lepidoptera: Erebidae) sex pheromone, across a broad geographic gradient in the USA and observed that release rates over time linearly increased with increasing cumulative degree-days. Authors estimated that disparlure half-life time at some of the warmer locations did not exceed the length of time pheromone-baited traps were deployed in the field. In the future, dispensers may thus need to be replaced more frequently or contain more pheromone, but the increased cost may be at least partially compensated by a lower density, since increased release rate extends the distance downwind at which the concentration of the blend induces an insect response (Linn and Roelofs, 1991).

Finally, temperature's influence on the neural pathways of pests may disrupt pheromone attraction. However, higher diffusion temperatures often lead to increased electroantennogram responses to sex pheromones and enhanced expression of pheromone-mediated behaviours (Lemmen and Evenden, 2015), improved ability to track pheromone pulses (Baker *et al.*, 1988) and reduced specificity to odours. For instance, this was observed in the pink bollworm moth *Pectinophora gossypiella* (Lepidoptera: Gelechiidae), which exhibited peak responses to a broader range of blends and doses at the highest temperature in a wind tunnel (Linn and Roelofs, 1991). Therefore, an increase in diffusion temperature is expected to enhance the effectiveness of mating disruption. However, any negative impact of developmental temperature on male responses to pheromones, as observed in *Cydia pomonella*, could offset this positive effect (Perrin, 2022), and may partly explain the recent failures reported with this tactic.

14.2.2 Entomopathogens

Entomopathogens are commonly used to suppress pests in IPM, and climate is known to significantly affect their efficacy. This impact arises from its influence on the persistence of entomopathogens in agroecosystems, their pathogenicity, and the behaviour and immune response of host insects.

While ultraviolet radiations strongly affect the development and persistence in crops of all entomopathogenic taxa, relative humidity and temperature are strongly limiting for fungi (Quesada-Moraga *et al.*, 2024), and nematodes (Sharmila *et al.*, 2018), but less for viruses (Lacey, 2017), and bacteria (Leong *et al.*, 1980). These climatic factors also strongly influence the effectiveness of entomopathogenic organisms (see Chapters 10 and 11). A positive relationship between temperature and the toxicity of entomopathogenic bacteria or their toxins has thus been described for many pest species (e.g. He *et al.*, 2001). High temperature reduces time required for nematode penetration into hosts, resulting in higher pathogenicity (Chen *et al.*, 2003), although deleterious effects of high temperatures have also been reported (El Khoury *et al.*, 2018). Entomopathogenic fungi and viruses display strong inter- and intraspecific variations in their response to temperature, depending on their biotope of origin. Positive or negative relationships between temperature and their pathogenicity have thus been reported (Abbaszadeh *et al.*, 2011; Rizal *et al.*, 2022).

Humidity also determines the outcome of insect–pathogen interactions, especially in fungi that generally require at least 95% relative humidity for germination, penetration and infection (Hallsworth and Magan, 1999), although the microclimate within cuticular folds of insects may provide sufficient humidity (Acheampong *et al.*, 2020). Similarly, high humidity promotes the mortality of pests exposed to baculoviruses (Fuxa *et al.*, 1999) or nematodes (Radová and Trnková, 2010).

In addition to its direct influence on entomopathogens, climate may affect the success of this strategy by altering pest behaviour and immune responses. Infection by entomopathogens requires their ingestion by the insect or penetration through the cuticle or natural body openings, and global warming is expected to increase infection rates due to heightened feeding rates and mobility of pests (Elderd and Reilly, 2014). Climate also directly influences the insect immune system, which may be either weakened or strengthened as temperatures rise, depending on the specific immune component and species involved (Catalán *et al.*, 2012). A general assumption is that global warming will create more opportunities for pest insects to perform behavioural fever, a strategy where they move to microhabitats with temperatures that enhance their immune response (Catalán *et al.*, 2012) and/or inhibit pathogen growth, thereby reducing the negative consequences of infection (Valverde-Garcia *et al.*, 2018). However, this enhancement in immune response at higher temperatures could be compromised by the negative impact of climate change on the quality of food resources for pests (Zvereva and Kozlov, 2006). For instance, in the Indianmeal moth *Plodia interpunctella* (Lepidoptera: Pyralidae), the activity of phenoloxidase – an enzyme involved in the arthropod immune system – increased with rising temperatures when larvae were raised on suitable food but decreased when poor food was provided (Triggs and Knell, 2012).

14.2.3 Entomophagous arthropods

Central to the success of IPM is the promotion of biological control methods using predatory or parasitic organisms, either through introduction (i.e. classical and augmentative biological control) or conservation (i.e. conservation of native species as biological control agents via habitat manipulation and pesticide reduction). By fostering natural predator–prey relationships, biological control not only offers an environmentally friendly alternative to chemical pesticides but also contributes to the long-term stability of agroecosystems. Climate should impact pest control by directly affecting life-history traits, population dynamics, behaviours and community composition of natural enemies, and indirectly through cascading effects or spatial and temporal mismatches between trophic levels.

Climate change may initially influence the fitness of natural enemies and, consequently, their population dynamics through its effects on their life history traits. Development time, body size and longevity commonly decrease with increasing temperature in ectotherms (Perrin, 2022) (see Chapter 1), as well as with a decreased prey or host quality subsequent to an increase in temperature and/or CO_2 concentration. For example, aphids reared in warmer conditions were reported to be less suitable for parasitoid development than those reared in colder conditions due to their smaller size (Li and Mills, 2004). Additionally, increased concentrations of allelochemicals in plants, resulting from

CO_2 or O_3 enrichment, may be toxic to natural enemies, leading to higher mortality rates (Holton *et al.*, 2003).

Second, differential responses to climatic factors between pests and their natural enemies may alter spatial and temporal synchronization between the two trophic levels. This could potentially reduce biological control and even lead to the local extinction of specialist natural enemies (Barton, 2010; Wetherington *et al.*, 2017). Spatial or temporal mismatches have been predicted in some systems but not in others (Gerard *et al.*, 2013), as the thermal dependency of development is species specific. Moreover, increased asynchrony may not always decrease pest control. In some cases, an advanced phenology of entomophagous arthropods may help reduce pest populations, as early control is crucial to avoid high densities later in the season (Neuville *et al.*, 2016).

Third, the behaviours of both pests and natural enemies involved in biological control are known to change with climate. For example, high temperatures increase locomotion (Gilchrist, 1996) and feeding rates (Sentis *et al.*, 2012) of most natural enemies, thereby increasing their efficacy. However, high temperatures also increase aggressive defensive behaviours in aphids, reducing their susceptibility to natural enemies (Moiroux *et al.*, 2016). Contrary to common hypotheses, most experimental studies did not report any significant influence of climatic factors on the ability of natural enemies to locate their hosts or discriminate between pest-infected and control plant odours (Vuorinen *et al.*, 2004; Weldegergis *et al.*, 2015).

The various effects of climate on natural enemies and pests, which are highly species specific, can significantly influence the outcome of pest management by entomophagous arthropods. Experimentally increased temperatures or CO_2 concentrations often lead to higher pest abundance due to increased pest fecundity combined with decreased or unchanged parasitism and predation rates (e.g. Klaiber *et al.*, 2013). However, an increase in parasitism or predation rates can sometimes offset the increased pest fecundity, thereby maintaining stable pest control under these conditions (Chen *et al.*, 2007).

While experimental studies help disentangle the effects of various climatic factors, they often oversimplify agroecosystems and offer limited insights into how climate impacts the efficacy of biological control. Large-scale field approaches or meta-analyses of studies that quantify the impact of biological control across different latitudes provide valuable projections by directly examining the influence of climate on pest abundance in crops (e.g. Diehl *et al.*, 2013). Such studies can help predict changes in biological control due to climate change in specific cropping systems, and further analyses of this kind should be conducted in the future.

14.2.4 Insecticides

Although IPM advocates for their judicious use as a last resort, it is estimated that the application of chemical insecticides will greatly increase as a result of climate change, together with direct and externalized costs for growers and society (Koleva and Schneider, 2009). This overall need for increased application with global warming firstly results from a decreased concentration and persistence of products in plants and crops, caused by an increased plant growth rate and an accelerated volatilization, degradation or alteration of the components (Delcour *et al.*, 2015). Second, global warming is expected to enhance pest abundance and frequency of outbreaks, causing greater economic losses for growers (Deutsch *et al.*, 2018) (see Chapter 3) and prompting more insecticide applications. Additionally, climate change can hasten the evolution of insecticide resistance, as warmer conditions often result in shorter generation times for multivoltine pests or broaden their geographic distribution. By comparing spatialized presence-absence data of pyrethroid resistance in the mite pest *Halotydeus destructor* (Trombidiformes: Penthaleidae) with long-term climatic data in Australia, Maino *et al.* (2018) reported that insecticide resistance was positively correlated with climatic regions supporting increased voltinism of the pest. In China, global warming probably resulted in increased mean pesticide resistance in the diamondback moth *Plutella xylostella* (Lepidoptera: Plutellidae) through expanding its overwintering range (Ma *et al.*, 2021).

The efficacy of insecticides under climate change is further complicated by the temperature dependency of their toxicity, as observed in *P. xylostella* populations that display high resistance levels to methamidophos, an organophospate insecticide, during autumn and spring and low levels during summer in China (Wu *et al.*, 2004). Such relationship between temperature and toxicity varies widely depending on the chemical. Organophosphate insecticides, for example, typically show increased toxicity with rising temperatures, while pyrethroid insecticides often exhibit decreased toxicity (Satpute *et al.*, 2014). However, this trend is not uniform within insecticides. In a bioassay testing various insecticides against a Floridian population of the Asian citrus psyllid *Diaphorina citri* (Hemiptera: Psyllidae), two organophosphates, one carbamate and three neonicotinoids were found to be more toxic at higher temperatures. However, results varied within the pyrethroid group: one pyrethroid showed a positive temperature-dependent toxicity correlation while three others showed a negative correlation (Boina *et al.*, 2009).

The toxicity of bioinsectides may also vary with temperature. For example, a 32-fold increase in toxicity was observed between 15°C and 32°C in *Plutella xylostella* exposed to abamectin, a bioinsecticide derived from the bacteria *Streptomyces avermitilis* (Streptomycetales: Streptomycetaceae) (He *et al.*, 2001). The relationship between temperature and insecticide toxicity is also influenced by insect order (Li *et al.*, 2024), species and developmental stage. For example, the toxicity of deltamethrin increases with temperature for pests like the fall armyworm *Spodoptera frugiperda* (Lepidoptera: Noctuidae) and the tobacco budworm *Chloridea virescens* (Lepidoptera: Noctuidae) but decreases for the cabbage looper *Trichoplusia ni* (Lepidoptera: Noctuidae) and the boll weevil *Anthonomus grandis* (Coleoptera: Curculionidae) (Sparks *et al.*, 1982, 1983). In *Apis mellifera* (Hymenoptera: Apidae), adult workers reared at 33°C were significantly more susceptible to dimethoate compared to those reared at 35°C, whereas the opposite effect was observed in larvae (Medrzycki *et al.*, 2010). Factors such as developmental temperature (Medrzycki *et al.*, 2010), postapplication temperature (Boina *et al.*, 2009) and daily temperature fluctuations (Verheyen *et al.*, 2019) all play roles in influencing insecticide toxicity.

Other climatic factors influence insecticide toxicity and complexify projections. Among the rare studies available, Ge *et al.* (2013a) observed that the control of the brown planthopper *Nilaparvata lugens* (Hemiptera: Delphacidae) with an organophosphate insecticide decreased under elevated CO_2. Moreover, exposure to insecticides may result in better thermotolerance in pests, as observed in the same pest species (Ge *et al.*, 2013b). Consequently, making broad generalizations about climate-dependent insecticide toxicity is challenging, necessitating specific characterization of each system.

In IPM, chemical insecticides and biological control may be used simultaneously. As for pests, the toxicity of insecticides is commonly temperature dependent for both parasitoids (Abbes *et al.*, 2015) and predators (Mansoor *et al.*, 2015). Michalko and Košulič (2016) observed that temperature had no effect on the mortality of two *Philodromus* spider species (Araneae: Philodromidae) when exposed to spinosad or λ-cyhalothrin, except at the highest temperature. At this elevated temperature, nearly all the spiders exposed to cyhalothrin perished, while those exposed to spinosad had survival rates similar to the control group. Global warming therefore has the potential to indirectly reduce pest control in agricultural systems that rely on both entomophagous arthropods and insecticides by increasing the toxicity of insecticides to natural enemies. This is particularly concerning when insecticide toxicity decreases with rising temperatures for pests but increases for their predators or parasitoids (Perrin *et al.*, 2023), as detailed in the third section of this chapter.

14.3 Adaptations of Pest Management Methods to Global Warming

As highlighted in the previous section, climate change has the potential to affect the efficacy of various pest management techniques used in IPM. For a given pest–crop association, the frequency with which economic thresholds (i.e.the density of a pest that should trigger a specific management action) will be exceeded depends on the impact of climate change on pest abundance and pest management efficacy (Pedigo, 2002), as well as our ability to adapt control strategies to future climatic conditions. This second section emphasizes current and future developments that may enable growers using IPM to cope with climate change.

14.3.1 Agricultural practices

Although many avenues need to be developed by researchers and companies, growers may already adapt some of their practices to cope with climate change. First, IPM encourages the use and conservation of natural enemies, which are exposed to new climatic constraints. Various arrangements, such as hedgerows, flower strips, woodlots, sown vegetation and water points, are commonly promoted to enhance biological control (Albrecht *et al.*, 2020). These environmental elements also serve as crucial microclimatic refuges for agrobiodiversity during extreme climatic events (Harvey *et al.*, 2023). Therefore, increasing structural diversity in agricultural landscapes should be encouraged to maintain natural enemy populations and effective pest control in the face of climate change (Feit *et al.*, 2021).

Second, climate change is characterized by an average rise in temperatures as well as an increase in climatic variations. Using combinations of biological control agents suited to different climatic situations is thus a profitable approach to cope with a diversity of climatic factors, especially when synergy emerges from such combinations. For example, the fungus *Metarhizium flavoviride* (Hypocreales: Clavicipitaceae) is more efficient under warm temperatures than another fungus species, *Beauveria bassiana* (Hypocreales: Cordycipitaceae), in regulating the grasshopper *Melanoplus sanguinipes* (Orthoptera: Acrididae), while the opposite is observed in a cool temperature environment (Inglis *et al.*, 1997). However, simultaneous applications of the two pathogens result in better pest control than using *M. flavoviride* alone in hot temperatures and equal to using *B. bassiana* alone in cool temperatures. Coformulation of pathogens with broad overlapping temperature ranges may also help to counter behavioural fever used by insect pests to slow down an infection (Quesada-Moraga *et al.*, 2024). Some companies already offer combinations of natural enemies that cover different thermal ranges. For example, a mix containing the hymenopterous parasites *Encarsia formosa* (Hymenoptera: Aphelinidae) and *Eretmocerus eremicus* (Hymenoptera: Aphelinidae) is proposed to control the whiteflies (Hemiptera: Aleyrodidae) *Trialeurodes vaporariorum* and *Bemisia tabaci*. According to the manufacturer, the combination of the two species allows pest control throughout the season, with a predominance of *E. eremicus* during the hottest periods, as this species is more adapted than *E. formosa* to high temperatures and thermal fluctuations.

Finally, as stated in the previous section, the temperature dependency of insecticide toxicity – considered the last resort in IPM – varies by product and target species (Perrin *et al.*, 2023). When the effect of temperature on different products used against the same pest is known, it seems reasonable to adapt the seasonality of the products used to optimize their efficacy while mitigating potential risks to human health and the environment. Thus, products showing a negative relationship between temperature and toxicity should be reserved for early and late season, while those showing a positive relationship should be prioritized during the hottest periods. Of course, the effects of temperature on the toxicity of products against natural enemies should be considered. A concrete example is given in the section dedicated to *Cydia pomonella*. Alternative use of insecticides based on their sensitivity to temperature would optimize IPM and could contribute to postponing resistance development in pest populations. It is important to note, however, that limited data are available for the different pests and products, so further research is needed in this area, as there can be significant variations.

14.3.2 Research and development

Economic injury levels and economic thresholds

The calculation of economic thresholds (ET), which is central to IPM, is known to be climate dependent. Various studies have revealed significant geographical variations in these thresholds, highlighting the importance of validating them for ecoregions with distinct climates (Cárcamo *et al.*, 2024). Such findings emphasize the need to carefully recalibrate ETs for future climates. Depending on the region, certain climate conditions could allow plants to tolerate higher or lower levels of insect herbivory in the future (Wise and Lamb, 1998). Moreover, because of the positive relationship between temperature and metabolic rate in insects (Brown *et al.*, 2004), the rate of food consumption per individual is expected to increase significantly in the future (Deutsch *et al.*, 2018). Economic thresholds that have been established until now and are commonly based on pest density may thus need to be revised, as yield losses and associated costs may rise or decrease in the future for a similar pest density.

Insect phenology models

An essential aspect of IPM is the timely monitoring and application of control methods, particularly when pests resume their life cycle after diapause or when a new generation is expected to appear. This approach helps prevent a rapid increase in pest populations. Mathematical models based on the thermal requirements for development are frequently developed and made available to farmers to predict activity and emergence dates for various pests according to the year's climatic conditions (e.g. Gallardo *et al.*, 2009). However, due to the major selection pressure that temperature represents on the development cycle of ectotherms, evolutionary changes in the rate of development, phenology, or number of generations have already been observed or are predicted for various insects, whether they are phytophagous or predators and parasitoids (e.g. van Asch *et al.*, 2013; Tougeron *et al.*, 2020) (see Chapter 2). Therefore, it is essential to develop the most accurate models possible and to regularly readjust them based on the evolution of thermal requirements of species with climate change. This would allow growers to effectively use monitoring or control tactics at the right time and help to predict accurately the risks of temporal mismatch between pests and their natural enemies.

Combinations of products

As discussed above, several combinations of entomopathogens or entomophagous insects were identified to control pests under different climatic situations, and are even offered to growers by specialized companies. However, research has scarcely explored combinations that associate different control methods to address climate change, although such an approach could be highly relevant in IPM. Delnat *et al.* (2019) reported that an increased daily temperature variation, a common prediction under climate change, did not influence the toxicity of either the organophosphate insecticide chlorpyrifos or the entomopathogenic bacteria *Bacillus thuringiensis* (Bacillales: Bacillaceae) for the mosquito *Culex pipiens* (Diptera: Culicidae) when applied separately but, surprisingly, resulted in increased toxicity of a mixture of the two insecticides due to a synergistic effect. Using cocktails to overcome the constraints of future climate is thus an interesting approach that deserves further investigation, especially regarding side-effects that may occur on other insects, including natural enemies, and the environment.

Selection of climate-adapted strains of biological control organisms

Intraspecific variability in temperature, humidity or CO_2 concentration sensitivity has been described in several entomopathogenic fungi and nematodes, depending on their climate of origin (e.g.Borisade and Magan, 2015; Quesada-Moraga *et al.*, 2024) (see Chapters 10 and 11). Mukuka *et al.* (2010a) and Ulu and Susurluk (2014) thus reported a positive relationship between heat tolerance and, respectively, the mean annual temperature and the highest average annual temperature recorded at place of origin of differents strains of the nematode *Heterorhabditis bacteriophora* (Rhabditida: Heterorhabditidae). Interestingly, the last authors also observed a positive relationship between heat tolerance and efficiency of strains on *Galleria mellonella* (Lepidoptera: Pyralidae). Intraspecific variation was also reported for the reaction norm of life-history traits important for population dynamics of natural enemies (e.g. Moiroux *et al.*, 2010). For example, Traoré *et al.* (2006) observed in the egg parasitoid of weevils

Anaphes victus (Hymenoptera: Mymaridae) that individuals from a population sampled in Texas displayed faster development than those from Michigan or Québec when reared at high temperatures. Screening strains from different geographical regions, especially from warm and dry areas, may thus allow for the selection of biological control agents with the highest environmental competence in relation to climate change for a given region (Kalyebi *et al.*, 2006; Quesada-Moraga *et al.*, 2024). Moreover, cross-breeding tolerant strains may help to improve the ability of entomopathogenic or entomophagous organisms to overcome future climatic conditions, as observed in nematodes (Mukuka *et al.*, 2010b). Although such strains would be more resistant to extreme temperatures or drought, it also seems essential to improve the formulations of commercial products to increase the survival of these organisms in the face of climatic constraints. Encapsulation methods are, for example, being developed for nematodes to reduce mortality caused by insufficient soil moisture levels (Peters, 2016).

Genetic improvement of biological control agents

Combining the cross-breeding of tolerant strains with artificial selection can enhance the performance of commercially developed biological control agents. Mukuka *et al.* (2010b) successfully increased heat tolerance in the nematode *H. bacteriophora* by 3°C through cross-breeding tolerant parental strains from 60 populations, followed by an 11-step artificial selection process. Directional selection was also used to produce strains of the Mexican fruit fly *Anastrepha ludens* (Diptera: Tephritidae), with higher tolerance to desiccation compared to wild individuals, aiming to enhance the efficiency of the sterile insect technique in response to future climatic extremes (Tejeda *et al.*, 2017).

Moreover, artificial selection for traits like heat or desiccation resistance can lead to improvements in other valuable life-history traits. For instance, Salame *et al.* (2010) found that enhancing desiccation tolerance increased heat resistance in the nematode *Steinernema feltiae* (Rhabditida: Steinernematidae). Similarly, Madhulika and Singh (2015) reported a 60% increase in fecundity in the egg parasitoid *Trichogramma chilonis* (Hymenoptera: Trichogrammatidae) after 50 generations of selection for high-temperature resistance. However, such genetic enhancements require careful examination, as selecting for tolerance to one temperature might compromise tolerance to other temperatures (Lirakis and Magalhães, 2019) or other fitness-related traits. Carrière and Boivin (2001) observed that 15 generations of selecting for high-temperature egg-laying efficiency in *Trichogramma pretiosum* (Hymenoptera: Trichogrammatidae) led to reduced fecundity. Additionally, artificial selection is a lengthy and costly process, with some traits proving resistant to improvement in different species, such as desiccation resistance in *H. bacteriophora* (Mukuka *et al.*, 2010b). Integrating genetic and genomic approaches could help overcome these challenges and enhance the efficacy of biological control agents (Leung *et al.*, 2020). The knowledge derived from the engineering of the sterile insect technique, in which genetic sexing strains with temperature-sensitive lethality have been designed (Nguyen *et al.*, 2021), could be particularly valuable in the development of strains adapted to future climates.

Integrated pest management prioritizes biological control but may resort to insecticides when pest pressure becomes excessive. However, as mentioned in the *Insecticides* section, global warming could increase insecticide-induced mortality and sublethal effects on natural enemies. To address this issue, Kumar *et al.* (2008) developed a genetically improved strain of the egg parasitoid *Trichogramma chilonis* in India, which exhibited combined tolerance to both insecticides and high temperatures after 81 generations. Thus, selecting for natural resistance to future climatic conditions and insecticides appears to be a viable strategy in IPM to mitigate the impacts of climate change

14.4 Adapting Integrated Pest Management to Climate Change: A Case Study with the Codling Moth *Cydia Pomonella*

The fundamental principle of IPM is to address pest issues through a combination of methods tailored to the specific needs and characteristics of each agricultural system. Predicting a global influence of climate change on integrated pest management is thus presumptuous considering the diversity of climate-related effects on each method and pest species.

In an attempt to provide a guideline for adapting IPM to climate change, this chapter proposes a concrete example with the codling moth *Cydia pomonella*, a particularly relevant species in this context. Various methods are employed simultaneously to control this major apple pest worldwide, which has developed resistance or adaptations to most control measures. These methods include exclusion nets, mating disruption, chemical insecticides, bioinsecticides, entomopathogenic viruses, fungi and nematodes, classical and conservation biological control, and the sterile insect technique (Lacey *et al.*, 2005). Additionally, the influence of climate on the pest's phenology (e.g. Juszczak *et al.*, 2013) and life-history traits (Sæthre and Trond, 2002) is relatively well known, and the influence of temperature on the regulatory potential of some control strategies has been and is still being investigated (Perrin, 2022). This example is thus based on current knowledge that has already helped adapt pest control to climatic conditions, as well as ongoing research that should lead to the development of new adaptations to climate change. This reflection is grounded in the recommended integrated pest management strategy for south-eastern France, a region characterized by a Mediterranean climate, particularly affected by rising temperatures and decreasing precipitation (IPCC, 2022). Here, the codling moth is capable of producing three generations per year. Unlike other regions of the world where it has proven efficient (Thistlewood and Judd, 2019), the sterile insect technique is not used in France against this species, although research is currently under way. Therefore, this method is not discussed in this example, especially given the limited information available on the impact of climate on its efficiency.

One of the first steps in the integrated management of *C. pomonella* is the use of mating disruption at the beginning of the season to slow initial population growth, with this strategy being maintained throughout the year. However, this control method has become less effective in recent years. While early-season climate might not explain this decline, electroantennographic measurements have shown that high developmental temperatures significantly reduced male responses to pheromones, including codlemone, the main sex pheromone of *C. pomonella* (Perrin, 2022). Elevated temperatures could also decrease the longevity of dispensers, which are currently recommended for 150–180 days, as observed in other systems (Tobin *et al.*, 2011), though this hypothesis remains to be tested for this species to our knowledge. It is worth noting that, in addition to mating disruption products targeting only *C. pomonella*, other blends aimed at both this species and the oriental fruit moth *Grapholita molesta* (Lepidoptera: Tortricidae) are available to farmers. (Linn and Roelofs, 1991) reported that *G. molesta* exhibits peak responses to a broader range of pheromone blends and doses at the highest temperatures, both in a wind tunnel and in the field. The effectiveness of such products may thus increase in the future because of a decreased specificity to odours as temperatures rise. Nonetheless, there are limited data on the effect of climate on the efficacy of mating disruption, and much work remains to be done.

Another crucial early point in managing *C. pomonella* is establishing the biofix, which corresponds to the first flights of the season. This biofix can be empirically determined by capturing the first males in codlemone traps, which are monitored daily. In south-eastern France, the biofix is set when half of the traps in the trapping network capture the first moth. However, this method has certain limitations such as disruption of monitoring by mating disruption, or poor tracking. Therefore, mathematical models were developed and proven effective in predicting a fixed biofix for various regions worldwide. In the USA, Jones *et al.* (2013) found that sites at elevations less than 1300 ft fit a simple quadratic equation using only latitude, while for higher elevations, a multiple regression model using latitude and elevation was required. Thus, it is essential to develop targeted biofix models for each region, which will probably need to evolve with the climate to remain accurate. This biofix is crucial in IPM because it allows the prediction of the emergence dates of the first generation, whose larvae will cause the initial damage, thus enabling proactive pest control. Multiple predictive models were developed in different regions of the world and are currently offered to farmers (e.g. Boivin *et al.*, 2005; Graf *et al.*, 2018). Establishing the thermal requirements of local populations, which vary from region to region, and monitoring their potential evolution with climate warming would help to maintain and/or improve the accuracy of these models, which have sometimes failed to correctly predict emergence dates in recent years in south-eastern France, with discrepancies of up to 15 days reported. Moreover, Boivin *et al.* (2005) revealed that the emergence of insecticide resistance can be accompanied by changes in the thermal requirements and phenology of the codling moth, requiring the development of specific models to predict the emergence of these resistant populations. In light of the emergence of new resistances, potentially accelerated by climate warming, it is therefore crucial to frequently recalibrate existing models.

Those models help to estimate the timing for deploying nets, which are lowered after the apple trees have finished blooming. In south-eastern France, Alt'Carpo nets are increasingly used in both organic and integrated orchards, but some farmers have reported failures. It has been hypothesized that the common negative relationship between body size and temperature (Atkinson, 1994) could explain these failures, but experimental evidence does not support this assumption. Individuals from populations adapted to nets were indeed the same size, or even larger, than individuals from non-adapted populations. It is likely that behavioural adaptations independent of climate change have been selected (Siegwart *et al.*, 2013). Therefore, alternative methods should be used in conjunction with these nets, although they remain a major pest control strategy for the codling moth. In their absence, more frequent monitoring and treatments are required.

The calculation of the emergence date for the first generation of *C. pomonella* based on the biofix also determines the timing for the initial use of the codling moth granulovirus (CpGV), which is frequently used against this pest. However, there have been numerous cases of resistance to this entomopathogenic virus (Siegwart *et al.*, 2020). To counteract this resistance, various virus strains were developed and are used alternately between generations to prevent the selection of new ones. To our knowledge, no strain is currently marketed based on its thermal competence, although some have been identified by INRAE Avignon as being more effective at low or high temperatures, or over a broader range of temperatures, depending on their region of origin (Sauphanor *et al.*, unpublished data). It is therefore crucial for the industry to address this issue and develop virus strains that are effective under both current and future climatic conditions.

Although classical or conservation biological control relying on entomphagous organisms is not central to integrated management of the codling moth in south-eastern France, numerous natural enemies significantly limit its abundance. These include generalist predators such as the European earwig *Forficula auricularia* (Dermaptera: Forficulidae), which preys on *C. pomonella* eggs (Glen, 1975), or the great tit *Parus major* (Passeriformes: Paridae). Indigenous parasitoids, such as *Ascogaster quadridentata* (Hymenoptera: Braconidae) and *Pristomerus vulnerator* (Hymenoptera: Ichneumonidae), also play a role, as well as the exotic species *Mastrus ridens* (Hymenoptera: Ichneumonidae), which was recently introduced to France to control *C. pomonella* (Muru *et al.*, 2018).

To estimate the risks of a disruption in biological control due to climate change, it is necessary to study the thermal biology of the pest and its natural enemies for each system. Laboratory work conducted by Perrin (2022) on *C. pomonella*, *M. ridens* and *F. auricularia* revealed, for example, that while the codling moth can develop at a constant temperature of 35°C, the parasitoid and the predator are unable to complete their development at 32°C and 28°C respectively. Furthermore, while the parasitism rate of *M. ridens* remains stable between 20°C and 30°C, the predation rate of adult *F. auricularia* is drastically reduced when the ambient temperature reaches 30°C. Therefore, the risks of increased codling moth abundance during the hottest months due to the absence of natural enemies or a decrease in their regulatory potential are real. These results highlight the importance of taking appropriate measures to preserve auxiliaries in the face of climate change, such as planting flower strips (Damien *et al.*, 2017; Harvey *et al.*, 2023) or favouring spontaneous flowering vegetation (Laffon *et al.*, 2024), which provide insects with food sources while also serving as potential microclimatic refuges.

In autumn, it is recommended that farmers facing high pest pressures introduce entomopathogenic nematodes to reduce overwintering larvae number. In south-eastern France, two species are used to target the codling moth: *Steinernema carpocapsae* (Rhabditida: Steinernematidae) and *S. feltiae*, each with different temperature sensitivities. *S. feltiae* is preferred at lower temperatures, while *S. carpocapsae* has a broad range of activity and efficacy across a wider temperature spectrum. Research on other nematodes suggests that new species particularly effective at higher temperatures could be deployed during late-season heatwaves, such as *Steinernema riobrave* (Rhabditida: Steinernematidae) (Lacey and Unruh, 1998). Meanwhile, some species, like *Heterorhabditis indica* (Rhabditida: Heterorhabditidae) (Patil *et al.*, 2018), would be more suitable for increasing drought conditions. To our knowledge, no work has been done to select strains within a species that are adapted to greater thermal variations or persistent drought. However, Nimkingrat *et al.* (2013) successfully produced a cold-adapted hybrid strain of *S. feltiae* by crossing five naturally cold-active populations, followed by gradual directional selection down to 0°C. Notably, this strain did not exhibit any decrease in virulence against *C. pomonella*. Although autumn applications do not require strains that are effective at very high temperatures, it could be advantageous to develop, through cross-breeding and/or artificial selection, strains that are effective across a broad range of temperatures or drought-resistant, and to design appropriate formulations, to address future climate conditions.

When all those methods do not provide sufficient control, chemical insecticides may be needed. In south-eastern France, insecticides are applied when the infestation threshold reaches 1% of apples out of 1000 checked. Although rising temperatures might lead to higher metabolic needs (Brown *et al.*, 2004), the amount of damage caused by a single larva is not expected to change, as one apple is sufficient for its development. Therefore, it is likely that climate warming will not affect this economic threshold, although it may be reached more frequently if pest abundance increases. Moreover, elevated temperatures could reduce the duration of the wandering stage, which occurs between egg hatching and larval entry into the apple, by accelerating larval movement. It is therefore necessary to develop the most reliable predictive models possible to ensure that larvae are exposed to treatments before entering the apples.

When using insecticides, producers are advised to alternate between emamectin, chlorantraniliprole and spinosad, to combat the three annual generations of the codling moth, without considering temperature effects on their efficacy. However, Perrin *et al.* (2022) reported that the efficiency of emamectin and spinosad against the codling moth decreased with rising temperatures, while chlorantraniliprole's efficacy remained unaffected by temperature. Additionally, higher temperatures increased the toxicity of spinosad and chlorantraniliprole to the parasitoid *Mastrus ridens* (Perrin *et al.*, 2023), and emamectin to the European earwig *F. auricularia*. Based on these findings, emamectin should be reserved for early spring when cooler temperatures enhance its toxicity to the pest and minimize its impact on *Forficula auricularia*. Conversely, chlorantraniliprole should be used in summer, as rising temperatures do not influence its effectiveness against the pest or its toxicity against *F. auricularia*, although it may increase *M. ridens* mortality. Spinosad should be avoided due to its high toxicity to *Mastrus ridens* at all temperatures. Adapting insecticide use based on temperature sensitivity could thus enhance control effectiveness while delaying the development of resistance in pest populations and reducing their impact on natural enemies ans the environment.

This example, centred on a specific pest species and region, highlights the many opportunities to adapt agricultural practices to climate change (Fig. 14.1). However, the available data are still fragmented, and research on these topics needs to be encouraged.

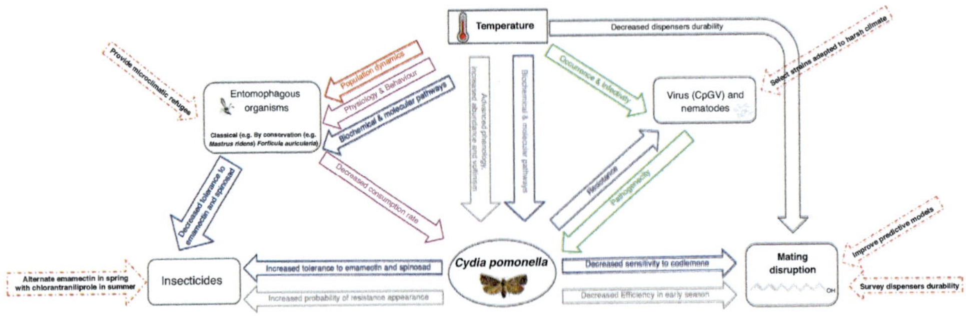

Fig. 14.1. Influence of increased temperature on tactics used in IPM against *Cydia pomonella* in south-eastern France, and potential adaptations. Solid line arrows with similar colours indicate logical links between elements, representing the indirect effects of temperature on pest management (e.g. increased temperature results in greater abundance and voltinism, which may increase the probability of resistance development). When 'increased' or 'decreased' is not indicated for a given arrow, a change due to temperature is expected but depends on the specific entomophagous or entomopathogenic species, making it difficult to indicate a general trend. Red dashed line arrows represent adaptations of the different tactics to cope with climate change.

Acknowledgement

We are grateful to Anahi Matteo for providing the photograph of *C. pomonella* used in our figure.

References

Abbaszadeh, G., Dhillon, M.K., Srivastava, C. and Gautam, R.D. (2011) Effect of climatic factors on bioefficacy of biopesticides in insect pest management. *Biopesticides International* 7(1), 1–14.

Abbes, K., Biondi, A., Kurtulus, A., Ricupero, M., Russo, A. *et al.* (2015) Combined non-target effects of insecticide and high temperature on the parasitoid *Bracon nigricans*. *PLOS One* 10(9), Article e0138411.

Acheampong, M.A., Coombes, C.A., Moore, S.D. and Hill, M.P. (2020) Temperature tolerance and humidity requirements of select entomopathogenic fungal isolates for future use in citrus IPM programmes. *Journal of Invertebrate Pathology* 174, Article 107436.

Albrecht, M., Kleijn, D., Williams, N.M., Tschumi, M., Blaauw, B.R. *et al.* (2020) The effectiveness of flower strips and hedgerows on pest control, pollination services and crop yield: A quantitative synthesis. *Ecology Letters* 23(10), 1488–1498.

Atkinson, D. (1994) Temperature and organism size-a biological law for ectotherms? In: Begon, M. and Fitter, A.H. (eds) *Advances in Ecological Research*, Vol. 25. Academic Press, New York, pp. 1–58.

Baker, T.C., Hansson, B.S., Löfstedt, C. and Löfqvist, J. (1988) Adaptation of antennal neurons in moths is associated with cessation of pheromone-mediated upwind flight. *Proceedings of the National Academy of Sciences* 85(24), 9826–9830.

Barton, B.T. (2010) Climate warming and predation risk during herbivore ontogeny. *Ecology* 91(10), 2811–2818.

Boina, D.R., Onagbola, E.O., Salyani, M. and Stelinski, L.L. (2009) Influence of posttreatment temperature on the toxicity of insecticides against *Diaphorina citri* (Hemiptera: Psyllidae). *Journal of Economic Entomology* 102(2), 685–691.

Boivin, T., Chadœuf, J., Bouvier, J.C., Beslay, D. and Sauphanor, B. (2005) Modelling the interactions between phenology and insecticide resistance genes in the codling moth *Cydia pomonella*. *Pest Management Science* 61(1), 53–67.

Borisade, O.A. and Magan, N. (2015) Resilience and relative virulence of strains of entomopathogenic fungi under interactions of abiotic stress. *African Journal of Microbiology Research* 9(14), 988–1000.

Brown, J.H., Gillooly, J.F., Allen, A.P., Savage, V.M. and West, G.B. (2004) Toward a metabolic theory of ecology. *Ecology* 85(7), 1771–1789.

Cárcamo, H., Herle, C., Schwinghamer, T., Robinson, S., Reid, P. *et al*. (2024) Revising economic injury levels for *Lygus* spp. in canola: The value of historical yield and insect data to improve decision making. *Crop Protection* 176, 106467.

Carrière, Y. and Boivin, G. (2001) Constraints on the evolution of thermal sensitivity of foraging in *Trichogramma*: Genetic trade-offs and plasticity in maternal selection. *American Naturalist* 157(5), 570–581.

Castex, V., Beniston, M., Calanca, P., Fleury, D. and Moreau, J. (2018) Pest management under climate change: The importance of understanding tritrophic relations. *Science of the Total Environment* 616, 397–407.

Catalán, T.P., Niemeyer, H.M., Kalergis, A.M. and Bozinovic, F. (2012) Interplay between behavioural thermoregulation and immune response in mealworms. *Journal of Insect Physiology* 58(11), 1450–1455.

Chen, F., Wu, G., Parajulee, M.N. and Ge, F. (2007) Impact of elevated CO_2 on the third trophic level: A predator *Harmonia axyridis* and a parasitoid *Aphidius picipes*. *Biocontrol Science and Technology* 17, 313–324.

Chen, S., Li, J., Han, X. and Moens, M. (2003) Effect of temperature on the pathogenicity of entomopathogenic nematodes (*Steinernema* and *Heterorhabditis spp.*) to *Delia radicum*. *BioControl* 48(6), 713–724. DOI: 10.1023/A:1026341325264.

Damien, M., Le Lann, C., Desneux, N., Alford, L., Al Hassan, D. *et al*. (2017) Flowering cover crops in winter increase pest control but not trophic link diversity. *Agriculture, Ecosystems and Environment* 247, 418–425.

Delcour, I., Spanoghe, P. and Uyttendaele, M. (2015) Literature review: Impact of climate change on pesticide use. *Food Research International* 68, 7–15.

Delnat, V., Tran, T.T., Janssens, L. and Stoks, R. (2019) Daily temperature variation magnifies the toxicity of a mixture consisting of a chemical pesticide and a biopesticide in a vector mosquito. *Science of the Total Environment* 659, 33–40.

Deutsch, C.A., Tewksbury, J.J., Tigchelaar, M., Battisti, D.S., Merrill, S.C. *et al*. (2018) Increase in crop losses to insect pests in a warming climate. *Science* 361(6405), 916–919.

Diehl, E., Sereda, E., Wolters, V. and Birkhofer, K. (2013) Effects of predator specialization, host plant and climate on biological control of aphids by natural enemies: A meta-analysis. *Journal of Applied Ecology* 50(1), 262–270.

El Khoury, Y., Oreste, M., Noujeim, E., Nemer, N. and Tarasco, E. (2018) Effect of temperature on the pathogenicity of Mediterranean native entomopathogenic nematodes (Steinernematidae and Heterorhabditidae) from natural ecosystems. *Redia* 101, 123–127.

Elderd, B.D. and Reilly, J.R. (2014) Warmer temperatures increase disease transmission and outbreak intensity in a host-pathogen system. *Journal of Animal Ecology* 83(4), 838–849.

El-Sayed, A.M., Ganji, S., Gross, J., Giesen, N., Rid, M. *et al*. (2021) Climate change risk to pheromone application in pest management. *Science of Nature* 108, Article 47.

Feit, B., Blüthgen, N., Daouti, E., Straub, C., Traugott, M. *et al*. (2021) Landscape complexity promotes resilience of biological pest control to climate change. *Proceedings of the Royal Society B* 288(1951), 20210547.

Fuxa, J.R., Sun, J.Z., Weidner, E.H. and LaMotte, L.R. (1999) Stressors and rearing diseases of *Trichoplusia ni*: Evidence of vertical transmission of NPV and CPV. *Journal of Invertebrate Pathology* 74(2), 149–155.

Gallardo, A., Ocete, R., López, M.A., Maistrello, L., Ortega, F. *et al*. (2009) Forecasting the flight activity of *Lobesia botrana* (Denis and Schiffermüller) (Lepidoptera, Tortricidae) in Southwestern Spain. *Journal of Applied Entomology* 133(8), 626–632.

Ge, L.-Q., Wu, J.C., Sun, Y.C., Ouyang, F. and Ge, F. (2013a) Effects of triazophos on biochemical substances of transgenic Bt rice and its nontarget pest *Nilaparvata lugens* Stål under elevated CO_2. *Pesticide Biochemistry and Physiology* 107(2), 188–199.

Ge, L.-Q., Huang, L.-J., Yang, G.-Q., Song, Q.-S., Stanley, D. *et al*. (2013b) Molecular basis for insecticide-enhanced thermotolerance in the brown planthopper *Nilaparvata lugens* Stål (Hemiptera: Delphacidae). *Molecular Ecology* 22(22), 5624–5634.

Gerard, P.J., Barringer, J.R.F., Charles, J.G., Fowler, S.V., Kean, J.M. *et al.* (2013) Potential effects of climate change on biological control systems: Case studies from New Zealand. *BioControl* 58(2), 149–162.

Gilchrist, G.W. (1996) A quantitative genetic analysis of thermal sensitivity in the locomotor performance curve of *Aphidius ervi*. *Evolution* 50(4), 1560–1572.

Glen, D. (1975) The effects of predators on the eggs of codling moth *Cydia pomonella*, in a cider-apple orchard in south-west England. *Annals of Applied Biology* 80, 115–119.

Graf, B., Höhn, H., Höpli, H.U. and Kuske, S. (2018) Predicting the phenology of codling moth, *Cydia pomonella*, for sustainable pest management in Swiss apple orchards. *Entomologia Experimentalis et Applicata* 166(8), 618–627.

Hallsworth, J.E. and Magan, N. (1999) Water and temperature relations of growth of the entomopathogenous fungi, *Beauveria bassiana*, *Metarhizium anisopliae*, and *Paecilomyces farinosus*. *Journal of Invertebrate Pathology* 74(3), 261–266.

Harvey, J.A., Tougeron, K., Gols, R., Heinen, R., Abarca, M. *et al.* (2023) Scientists' warning on climate change and insects. *Ecological Monographs* 93(1), Article e1553.

He, Y.X., Yang, X.J., Weng, Q.Y. and Huang, J. (2001) Biological toxicity of avermectin to *Plutella xylostella*. *Fujian Journal of Agricultural Science* 16(2), 24–27.

Holton, M.K., Lindroth, R.L. and Nordheim, E.V. (2003) Foliar quality influences tree-herbivore-parasitoid interactions: Effects of elevated CO2, O3, and plant genotype. *Oecologia* 137, 233–244.

Inglis, G.D., Johnson, D.L., Cheng, K.J. and Goettel, M.S. (1997) Use of pathogen combinations to overcome the constraints of temperature on entomopathogenic hyphomycetes against grasshoppers. *Biological Control* 8(2), 143–152.

IPCC (2022) *Climate Change 2022: Impacts, Adaptation and Vulnerability. Contribution of Working Group II to the Sixth Assessment Report of the Intergovernmental Panel on Climate Change*. Cambridge University Press, Cambridge, UK.

Jones, V.P.,R., Hilton, J.F., Brunner, W.J., Bently, D.G., Alston, B. *et al.* (2013) Predicting the emergence of the codling moth, *Cydia pomonella* (Lepidoptera: Tortricidae), on a degree-day scale in North America. *Pest Management Science* 69, 1393–1398.

Juszczak, R., Kuchar, L., Leśny, J. and Olejnik, J. (2013) Climate change impact on development rates of the codling moth (*Cydia pomonella* L.) in the Wielkopolska Region, Poland. *International Journal of Biometeorology* 57(1), 31–44.

Kalyebi, A., Overholt, W.A., Schulthess, F., Mueke, J.M. and Sithanantham, S. (2006) The effect of temperature and humidity on the bionomics of six African egg parasitoids (Hymenoptera: Trichogrammatidae). *Bulletin of Entomological Research* 96(3), 305–314.

Klaiber, J., Najar-Rodriguez, A.J., Dialer, E. and Dorn, S. (2013) Elevated carbon dioxide impairs the performance of a specialized parasitoid of an aphid host feeding on Brassica plants. *Biological Control* 66(1), 49–55.

Koleva, N.G. and Schneider, U.A. (2009) The impact of climate change on the external cost of pesticide applications in US agriculture. *International Journal of Agricultural Sustainability* 7(3), 203–216.

Kumar, G.A., Jalali, S.K., Venkatesan, T., Nagesh, M. and Lalitha, Y. (2008) Genetic improvement of egg parasitoid *Trichogramma chilonis* Ishii for combined tolerance to multiple insecticides and high temperature. *Journal of Biological Control* 22(2), 347–356.

Lacey, L.A. and Unruh, T.R. (1998) Entomopathogenic nematodes for control of codling moth, *Cydia pomonella* (Lepidoptera: Tortricidae): Effect of nematode species, concentration, temperature, and humidity. *Biological Control* 13(3), 190–197.

Lacey, L.A. and Unruh, T.R. (2005) Biological control of codling moth (*Cydia pomonella*, Lepidoptera: Tortricidae) and its role in integrated pest management, with emphasis on entomopathogens. *Vedalia* 12(1), 33–60.

Lacey, L.A. (ed.) (2017) *Microbial Control of Insect and Mite Pests: From Theory to Practice*. Academic Press, New York.

Laffon, L., Bischoff, A., Blaya, R., Lescourret, F. and Franck, P. (2024) Spontaneous flowering vegetation favours hoverflies and parasitoid wasps in apple orchards but has low cascading effects on biological pest control. *Agriculture, Ecosystems and Environment* 359, 108766.

Lemmen, J. and Evenden, M. (2015) Environmental conditions terminate reproductive diapause and influence pheromone perception in the long-lived moth *Caloptilia fraxinella*. *Physiological Entomology* 40, 30–42.

Leong, K.L.H., Cano, R.J. and Kubinski, A.M. (1980) Factors affecting *Bacillus thuringiensis* total field persistence. *Environmental Entomology* 9(5), 593–599.

Leung, K., Ras, E., Ferguson, K.B., Ariëns, S., Babendreier, D. *et al.* (2020) Next-generation biological control: The need for integrating genetics and genomics. *Biological Reviews* 95(6), 1838–1854.

Li, B. and Mills, N. (2004) The influence of temperature on size as an indicator of host quality for the development of a solitary koinobiont parasitoid. *Entomologia Experimentalis et Applicata* 110(3), 249–256.

Li, D., Jiang, K., Wang, X. and Liu, D. (2024) Insecticide activity under changing environmental conditions: A meta-analysis. *Journal of Pest Science* 97, 1711–1723.

Linn, C. and Roelofs, W. (1991) The effects of different blend ratios and temperature on the active space of the oriental fruit moth sex pheromone. *Physiological Entomology* 16(2), 211–222.

Lirakis, M. and Magalhães, S. (2019) Does experimental evolution produce better biological control agents? A critical review of the evidence. *Entomologia Experimentalis et Applicata* 167(7), 584–597.

Louis, F. and Schirra, K.J. (2001) Mating disruption of *Lobesia botrana* (Lepidoptera: Tortricidae) in vineyards with very high population densities. *IOBC-WPRS Bulletin* 24(2), 75–80.

Ma, C.S., Zhang, W., Peng, Y., Zhao, F., Chang, X.Q. *et al.* (2021) Climate warming promotes pesticide resistance through expanding overwintering range of a global pest. *Nature Communications* 12(1), Article 5351.

Madhulika, S. and Singh, A.K. (2015) Development of superior strain of *Trichogramma chilonis* Ishii for temperature tolerance. *Journal of Biological Control* 29(2), 68–74.

Maino, J.L., Umina, P.A. and Hoffmann, A.A. (2018) Climate contributes to the evolution of pesticide resistance. *Global Ecology and Biogeography* 27(2), 223–232.

Mansoor, M.M., Afzal, M., Raza, A.B.M., Akram, Z., Waqar, A. *et al.* (2015) Post-exposure temperature influence on the toxicity of conventional and new chemistry insecticides to green lacewing *Chrysoperla carnea* (Stephens) (Neuroptera: Chrysopidae). *Saudi Journal of Biological Sciences* 22(3), 317–321.

Medrzycki, P., Sgolastra, F., Bortolotti, L., Bogo, G., Tosi, S. *et al.* (2010) Influence of brood rearing temperature on honey bee development and susceptibility to poisoning by pesticides. *Journal of Apicultural Research* 49(1), 52–59.

Michalko, R. and Košulič, O. (2016) Temperature-dependent effect of two neurotoxic insecticides on predatory potential of *Philodromus* spiders. *Journal of Pest Science* 89, 517–527.

Moiroux, J., Lann, C.L., Seyahooei, M.A., Vernon, P., Pierre, J.S. *et al.* (2010) Local adaptations of life-history traits of a Drosophila parasitoid, *Leptopilina boulardi*: does climate drive evolution? *Ecological Entomology* 35(6), 727–736.

Moiroux, J., Abram, P.K., Louâpre, P., Barrette, M., Brodeur, J. and Boivin, G. (2016) Influence of temperature on patch residence time in parasitoids: Physiological and behavioural mechanisms. *Science of Nature* 103(3), Article 32.

Mukuka, J., Strauch, O., Waeyenberge, L., Viaene, N., Moens, M. and Ehlers, R.U. (2010a) Heat tolerance among different strains of the entomopathogenic nematode *Heterorhabditis bacteriophora*. *BioControl* 55, 423–434.

Mukuka, J., Strauch, O., Hoppe, C. and Ehlers, R.U. (2010b) Improvement of heat and desiccation tolerance in *Heterorhabditis bacteriophora* through cross-breeding of tolerant strains and successive genetic selection. *BioControl* 55, 511–521.

Muru, D., Auguste, A, Fauvergue, X., Malausa, T., Ris, N. *et al.* (2018) Un parasitoïde exotique pour lutter contre le carpocapse. *Phytoma* 710, 37–41.

Neuville, S., Le Ralec, A., Outreman, Y. and Jaloux, B. (2016) The delay in arrival of the parasitoid *Diaeretiella rapae* influences the efficiency of cabbage aphid biological control. *Biocontrol* 61, 115–126.

Nguyen, T.N., Choo, A. and Baxter, S.W. (2021) Lessons from *Drosophila*: Engineering genetic sexing strains with temperature-sensitive lethality for sterile insect technique applications. *Insects* 12(3), 243.

Nimkingrat, P., Khanam, S., Strauch, O. and Ehlers, R.U. (2013) Hybridisation and selective breeding for improvement of low temperature activity of the entomopathogenic nematode *Steinernema feltiae*. *BioControl* 58, 417–426.

Patil, J., Gowda, M.T., Vijayakumar, R. and Verghese, A. (2018) Fluctuating temperature: A cause for survival and development of entomopathogenic nematodes, *Heterorhabditis indica* and *Steinernema carpocapsae*. *Indian Journal of Experimental Biology* 56, 327–333.

Pedigo, L.P. (2002) *Entomology and Pest Management*, 4th edn. Prentice Hall, Upper Saddle River, New Jersey.

Perrin, M. (2022) Gestion du carpocapse des pommes: Impact des changements climatiques sur la lutte contre cydia pomonella dans les vergers de provence. Doctoral dissertation, Avignon Université.

Perrin, M., Moiroux, J., Maugin, S., Olivares, J., Rault, M. *et al.* (2022) Cross effects of heat stress and three insecticides on the survival of the codling moth *Cydia pomonella* (L.): Investigating the molecular and biochemical mechanisms. *Pesticide Biochemistry and Physiology* 185, Article 105139.

Perrin, M., Borowiec, N., Thaon, M., Siegwart, M., Delattre, T. *et al.* (2023) Differential influence of temperature on the toxicity of three insecticides against the codling moth *Cydia pomonella* (L.) and two natural enemies. *Journal of Pest Science* 97, 229–241.

Peters, A. (2016) Formulation of nematodes. *Methods in Molecular Biology* 1477, 121–135.

Quesada-Moraga, E., González-Mas, N., Yousef-Yousef, M., Garrido-Jurado, I. and Fernández-Bravo, M. (2024) Key role of environmental competence in successful use of entomopathogenic fungi in microbial pest control. *Journal of Pest Science* 97, 1–15.

Radová, S. and Trnková, Z. (2010) Effect of soil temperature and moisture on the pathogenicity of two species of entomopathogenic nematodes (Rhabditida: Steinernematidae). *Journal of Agrobiology* 27(1), 1–7.

Rizal, L.M., Furlong, M.J. and Walter, G.H. (2022) Responses of diamondback moth to diverse entomopathogenic fungi collected from non-agricultural habitats – effects of dose, temperature and starvation. *Fungal Biology* 126(10), 648–657.

Salame, L., Glazer, I., Chubinishvilli, M.T. and Chkhubianishvili, T. (2010) Genetic improvement of the desiccation tolerance and host-seeking ability of the entomopathogenic nematode *Steinernema feltiae*. *Phytoparasitica* 38(4), 359–368.

Satpute, N.S., Deshmukh, S.D., Rao, N.G.V., Tikar, S.N., Moharil, M.P. *et al.* (2014) Temperature-dependent variation in toxicity of insecticides against *Earias vitella* (Lepidoptera: Noctuidae). *Journal of Economic Entomology* 100(2), 357–360.

Sentis, A., Hemptinne, J.-L. and Brodeur, J. (2012) Using functional response modeling to investigate the effect of temperature on predator feeding rate and energetic efficiency. *Oecologia* 169(4), 1117–1125.

Sharmila, R., Priya, M.S., Subramanian, S., Poornima, K. and Pandiyan, M. (2018) Review on ecology of entomopathogenic nematodes. *Journal of Entomology and Zoology Studies* 6(4), 1086–1093.

Siegwart, M., Pierrot, M., Toubon, J.F., Maugin, S. and Lavigne, C. (2013) Adaptation to exclusion netting of the codling moth (*Cydia pomonella* L.) in apple orchards. *IOBC-WPRS Bulletin* 91, 127–131.

Siegwart, M., Maugin, S., Besse, S., Lopez-Ferber, M., Hinsberger, A. *et al.* (2020) Le carpocapse des pommes résiste au virus de la granulose. *Phytoma La Défense Des Végétaux* 738, 45–50.

Sparks, T.C., Shour, M.H. and Wellemeyer, E.G. (1982) Temperature-toxicity relationships of pyrethroids on three lepidopterans. *Journal of Economic Entomology* 75(4), 643–646.

Sparks, T.C., Pavloff, A.M., Rose, R.L. and Clower, D.F. (1983) Temperature-toxicity relationships of pyrethroids on *Heliothis virescens* (F.) (Lepidoptera: Noctuidae) and *Anthonomus grandis grandis* Boheman (Coleoptera: Curculionidae). *Journal of Economic Entomology* 76(2), 243–246.

Sæthre, M.G. and Trond, H. (2002) Effect of temperature on oviposition behavior, fecundity, and fertility in two northern European populations of the codling moth (Lepidoptera: Tortricidae). *Environmental Entomology* 31(5), 804–815.

Tejeda, M.T., Arredondo-Gordillo, J., Orozco-Dávila, D., Quintero-Fong, L. and Díaz-Fleischer, F. (2017) Directional selection to improve the sterile insect technique: Survival and sexual performance of desiccation resistant *Anastrepha ludens* strains. *Evolutionary Applications* 10(10), 1020–1030.

Thistlewood, H.M. and Judd, G.J. (2019) Twenty-five years of research experience with the sterile insect technique and area-wide management of codling moth *Cydia pomonella* (L.), in Canada. *Insects* 10(9), 292.

Tobin, P.C., Zhang, A., Onufrieva, K.S. and Leonard, D.S. (2011) Field evaluation of effect of temperature on release of disparlure from a pheromone-baited trapping system used to monitor gypsy moth (Lepidoptera: Lymantriidae). *Journal of Economic Entomology* 104(4), 1265–1271.

Tougeron, K., Brodeur, J., Le Lann, C. and Baaren, J. (2020) How climate change affects the seasonal ecology of insect parasitoids. *Ecological Entomology* 45(2), 167–181.

Traoré, L., Pilon, J.-G., Fournier, F. and Boivin, G. (2006) Adaptation of the developmental process of *Anaphes victus* (Hymenoptera: Mymaridae) to local climatic conditions across North America. *Annals of the Entomological Society of America* 99(6), 1121–1126.

Triggs, A. and Knell, R.J. (2012) Interactions between environmental variables determine immunity in the Indian meal moth *Plodia interpunctella*: Environmental interactions and immunity. *Journal of Animal Ecology* 81(2), 386–394.

Ulu, T.C. and Susurluk, I.A. (2014) Heat and desiccation tolerances of *Heterorhabditis bacteriophora* strains and relationships between their tolerances and some bioecological characteristics. *Invertebrate Survival Journal* 11(1), 4–10.

Valverde-Garcia, P., Santiago-Alvarez, C., Thomas, M.B., Garrido-Jurado, I. and Quesada-Moraga, E. (2018) Comparative effects of temperature and thermoregulation on candidate strains of entomopathogenic fungi for Moroccan Locust *Dociostaurus maroccanus* control. *BioControl* 63, 819–831.

Van Asch, M., Salis, L., Holleman, L.J., Lith, B. and Visser, M.E. (2013) Evolutionary response of the egg hatching date of a herbivorous insect under climate change. *Nature Climate Change* 3(3), 244–248.

Verheyen, J., Delnat, V. and Stoks, R. (2019) Increased daily temperature fluctuations overrule the ability of gradual thermal evolution to offset the increased pesticide toxicity under global warming. *Environmental Science and Technology* 53(8), 4600–4608.

Vuorinen, T., Nerg, A.-M. and Holopainen, J.K. (2004) Ozone exposure triggers the emission of herbivore-induced plant volatiles, but does not disturb tritrophic signalling. *Environmental Pollution* 131(2), 305–311.

Weldegergis, B.T., Zhu, F., Poelman, E.H. and Dicke, M. (2015) Drought stress affects plant metabolites and herbivore preference but not host location by its parasitoids. *Oecologia* 177(3), 701–713.

Wetherington, M.T., Jennings, D.E., Shrewsbury, P.M. and Duan, J.J. (2017) Climate variation alters the synchrony of host–parasitoid interactions. *Ecology and Evolution* 7(20), 8578–8587.

Wise, I.L. and Lamb, R.J. (1998) Economic threshold for plant bus, *Lygus* spp.(Heteroptera: Miridae), in canola. *Canadian Entomologist* 130(6), 825–836.

Witzgall, P., Kirsch, P. and Cork, A. (2010) Sex pheromones and their impact on pest management. *Journal of Chemical Ecology* 36(1), 80–100.

Wu, G., Jiang, S. and Miyata, T. (2004) Seasonal changes of methamidophos susceptibility and biochemical properties in *Plutella xylostella* (Lepidoptera: Yponomeutidae) and its parasitoid *Cotesia plutellae* (Hymenoptera: Braconidae). *Journal of Economic Entomology* 97(5), 1689–1698.

Zvereva, E.L. and Kozlov, M.V. (2006) Consequences of simultaneous elevation of carbon dioxide and temperature for plant–herbivore interactions: A metaanalysis. *Global Change Biology* 12(1), 27–41.

15 Biological Control and Integrated Pest Management in Greenhouse Crops and Adaptation to Climate Change

Carmelo Peter Bonsignore*

Dipartimento di Architettura e Design (DAeD), Laboratorio di Entomologia ed Ecologia applicata, Università degli studi "Mediterranea" di Reggio Calabria, Reggio Calabria, Italy

Abstract

Biological control is the use of natural enemies (predators, parasites, parasitoids and pathogens) or beneficial agents to manage a pest. Biological control is used as part of an integrated pest management (IPM) programme where evidence-based biological control measures are used in conjunction with other IPM measures. Greenhouse crops are high-value agricultural produce which are dependent on appropriate production factors, such as water, light, soil and temperature. Many specialized crops are grown in structurally evolved and well-managed greenhouse environments and they can be excellent indicators of the quality of production. The impact of climate change means that the management of greenhouse crops must undergo appropriate changes, both in terms of localization of production and in phytosanitary management, including biological control. This chapter explains in detail the possible effects of climate change on IPM, the application of traditional and biological control measures, and proposes strategies and potential technical means (orientation, whitening, resource management, choice of transplant, specific use of natural enemies) to mitigate the effects of global warming on greenhouse crops.

15.1 Introduction

Greenhouse crops, like all natural and agricultural environments, are subject to climate change. The use of greenhouses for agricultural production was originally developed to increase the average temperature of both soil and foliage to enable vegetables and fruit to be grown out of season or to protect crops from low temperatures. Climate change is generating elevated temperatures which, in some areas, may achieve a desired thermal increase in a greenhouse environment and extend the growing season in cool climates. However, climate change can cause both thermal and water stresses in greenhouses, and it is critical to quantify and mitigate these stresses to produce quality crops.

Preliminary studies have highlighted how the effects of climate change are causing substantial production losses (Sultan *et al.*, 2019). Climate shifts affect plants and animals, and particularly ectotherms, such as insects and

*Corresponding author: cbonsignore@unirc.it

 Biological Control Systems and Climate Change (eds Joan van Baaren *et al.*)
DOI: 10.1079/9781800625099.0015

mites, which cannot physiologically regulate their own body temperatures. Thermal increase because of elevated temperatures can cause an acceleration of metabolism, which is also capable of inducing stress. The effect can influence species interactions across trophic levels, particularly if members of different trophic levels show differential responses. In climate change processes, an increase in CO_2 as well as the frequency and intensity of extreme events such as drought, floods and heatwaves (IPCC, 2021) indirectly affects insect physiology by altering photosynthesis and plant health (Sharma *et al.*, 2020). The repercussions of climate change on plant–herbivore–natural enemy systems have been widely studied and evaluated in recent years (Tonnang *et al.*, 2022).

Global warming also affects soil and this is expected to have a stronger impact at high latitude where thermal conditions are the main limiting factor for the occurrence of pests and diseases. Nematode and fungal communities in soil can show rapid shifts in response to climate shifts and this can lead to a decline in species number, richness and diversity, and alter interrelationships such as predator–prey interactions. In turn, this can have significant effects on other system functions such as nutrient cycling in soil (Ruess *et al.*, 1999; Okulewicz, 2017). In hot arid areas, accentuated thermal stress on plant and faunal biosystems is aggravated by water shortage which can exacerbate problems associated with pest control, in particular the application of biological control measures. Greenhouse crop production is often located in hot, dry environments. Field and laboratory trials conducted by Romo and Tylianakis (2013) have demonstrated that the association between thermal increase and drought can lead to complex relationships between herbivores and their natural enemies. Even less predictable are the variations in activities of natural enemies and the subsequent adaptive management strategies adopted by farmers to mitigate the effects of climate change in greenhouse crops (Welch and Harwood, 2014). The implications of climate change related to biological control measures in greenhouses are therefore complex, and the adoption of IPM and biological control strategies is fundamental (Fig. 15.1).

15.2 Biological Control and Integrated Pest Management in Greenhouses

15.2.1 Greenhouse typology and thermal regulation

In general, the term 'greenhouse' refers to a cultivation environment protected by glass or synthetic material, the main purpose of which is to achieve a higher average temperature than the external environment (Choab *et al.*, 2019). Greenhouses are frequently used to extend seasonal production and have the benefits of creating a more controlled environment against pests and disease. Choice of greenhouse is closely linked to (i) country, (ii) external climatic conditions, (iii) crop type, (iv) available materials, and (v) the economic and sociocultural position of the human populations involved. Greenhouse structures are generally metal or wood with a covering that can be glass ('glasshouse') or, more commonly, plastic. Hemicylindrical-protected environments are known as tunnels or polytunnels. They are similar to plastic greenhouses but their thermo-hygrometric conditions are not always comparable to those of a plastic or glass greenhouse, although construction costs are lower and tunnels favour the rapid succession of crops, such as melon, zucchini and peppers.

Greenhouses are common in temperate and Mediterranean areas and scattered in other parts of the world (Cuesta Roble, 2019). Depending on greenhouse type, mulches may be used which are generally made of plastic and coloured white, black or silver to regulate soil temperature, reduce evaporation and restrain weeds. Thermal conditions achieved inside a greenhouse can favour poikilothermic herbivores, such as insects and mites, or create humid conditions favourable to fungal growth. These conditions may also favour the use of organisms as biological control agents or they may become pests (Vacante and Bonsignore, 2017). Adverse temperatures and excessive radiation can significantly reduce photosynthesis (Moreno-Teruel *et al.*, 2020), and natural ventilation in many cultivation environments is insufficient to reduce excess energy during hot, sunny days.

Temperature control includes both additional heating and cooling and, in recent years,

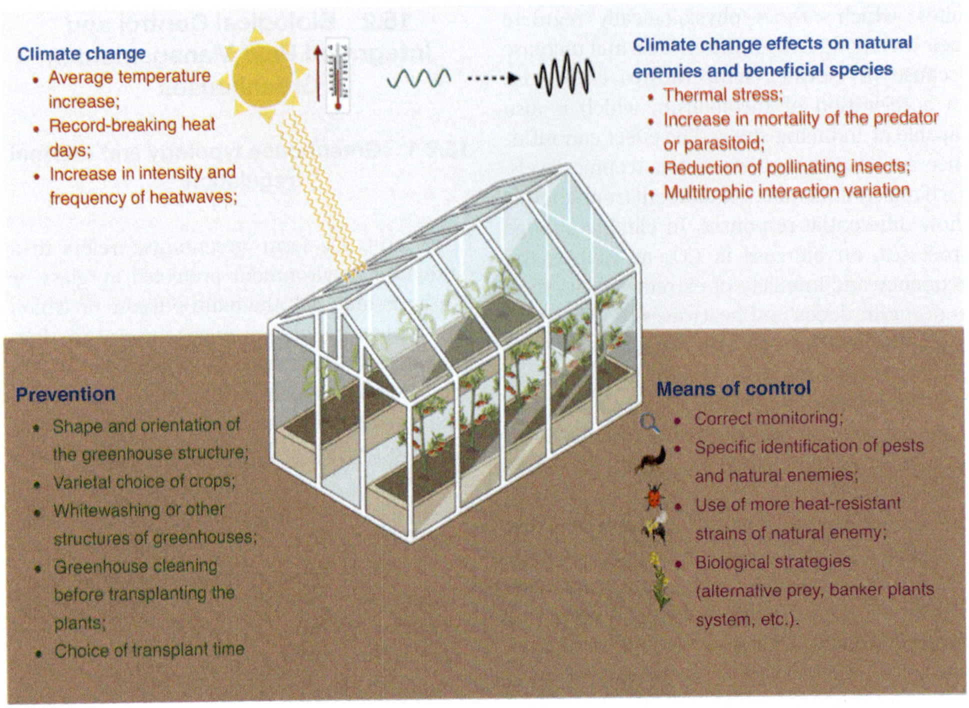

Fig. 15.1. General overview of preventive and control measures to mitigate the effects of climate change on biological control in greenhouses.

innovative greenhouses have been proposed which consider the need to reduce temperatures in the hottest months (Joudi and Farhan, 2015). Passive cooling systems include the installation of horizontal pipes at a certain depth in the ground, which are capable of exchanging heat (Mongkon *et al.*, 2013). Greenhouses may be equipped with adjustable and/or automatic ventilation (Costantino *et al.*, 2021).

Thermal regulation can be further controlled by a system of fans and plastic covers that are whitened on the external sides with carbonate of lime or slaked lime (calcium hydroxide – $Ca(OH)_2$). The concentration of calcium carbonate used varies according to the degree of transmissivity of the covering that is required. This temporary paint has liquid and powder shading agents to be used during summer to limit greenhouse temperature, to obtain a canopy-to-air temperature difference, and to create suitable growing conditions where the transpiration rate of crops is sufficiently low to avoid stress (Mashonjowa *et al.*, 2020). This traditional solution is designed to reflect sunlight, is low cost and deliberately designed not to interact with the plastic covering material. Some greenhouse producers mix iron-based products with hydrated lime to increase shading inside greenhouses. A fundamental aspect in order to optimize the treatment in whitening products is to use adhesive components to provide a high degree of protection against heavy rain without altering the shielding effect of the treatment (López-Martínez *et al.*, 2019).

Fixed-roof net or retractable roof production systems can be used to mitigate high temperatures and reduce light intensity (Nguyen and Lantzke, 2022). Side coverings with nets also ensure ventilation that allows cooling in warmer environments. The size and shape of netting can be designed to deter arthropod pests (Fatnassi *et al.*, 2006; Rigakis *et al.*, 2015). Moreover, different roofing materials can selectively filter incoming solar radiation (Schettini *et al.*, 2011).

15.2.2 Agrometeorology, modelling and forecasting in integrated pest management, biological control and pollination in greenhouse crops

An understanding of the interrelationships between greenhouse structures, the climate inside and outside greenhouses, and the species under cultivation is crucial for correct phytosanitary management in greenhouse crop production. Mathematical modelling has been used to quantify these relationships. Models that include systems theory and the control of climate parameters can improve the decision-making processes for producers, while also reducing the cost of production in terms of production inputs (Hemming *et al.*, 2019; Katzin *et al.*, 2022). Models can be used as decision aids in the tactical management and design of greenhouse systems (Lentz, 1998). This topic is discussed in detail in the specialized bibliography by Katzin *et al.* (2022).

The models developed for IPM and biological control for greenhouse crops are valuable. Modelling for biological control in greenhouses and other agroecosystems must consider the complex community-level interactions and this aspect has become fundamental in predicting the efficacy of using natural enemies to control pest populations (Laubmeier *et al.*, 2023). Future biological control modelling paths must be linked to a broader knowledge of the microfauna present in a greenhouse and the relationships between these and plants under cultivation. Model evaluation processes are also important for creating long-term datasets which can be used to improve the models (Deere *et al.*, 2023). The simplified predator–prey decision model for groups of cases, as recently developed by Moerkens *et al.* (2021), can be used in modelling for biological control. This model is aimed at predicting future pest population density and, most important, when to adopt a biological control programme to reduce a specific pest.

To date, theoretical models cannot be easily translated into real case scenarios, so it is important to study in detail and validate pre-existing models in relation to expected thermal variations. Also, it is crucial that the scientific community pays greater attention to the rapid changes in temperature that are occurring and considers the implications of conservation biological control strategies in the short and medium term from an economic perspective. Recently, Johnson *et al.* (2021) highlighted this general data deficiency across trophic levels and economic measures around the fundamental importance of the adoption of conservation biological control strategies.

High solar radiation that passes through glass or plastic greenhouses can significantly increase the temperature of the internal air, resulting in evident stress for cultivated greenhouse crops if it is associated with non-optimal water conditions. For this reason, in a climate change scenario, it is appropriate to adopt and optimize cooling ventilation using fans and other strategies such as evaporative cooling pad systems, whitening, external shading nets and internal thermal screens (Xu *et al.*, 2015; Costantino *et al.*, 2021). Careful monitoring of climatic conditions inside a greenhouse is critical to prevent fungal diseases that can develop in conditions which are optimal for plant growth. High humidity inside a greenhouse can cause botrytis and mildew, caused by *Botrytis cinerea* (Pers., 1794) and *Pseudoperonospora cubensis* (Berk. et Curt) Rostov, 1903, observed in Cucurbitaceae. *Pseudoperonospora cubensis* is a common and widely distributed fungal pathogen, inflicting heavy economic losses on growers due to its high treatment costs, particularly in the fruit-ripening period (Cohen *et al.*, 2015). It is necessary to reduce the humidity inside a greenhouse through a natural or forced increase in ventilation to prevent fungal damage (Cañadas *et al.*, 2017).

Pollination in greenhouses is another challenge. Crops that require or benefit from insect pollination are suffering because of the rapid decline in pollinating insects due to climate change in numerous areas investigated (Millard *et al.*, 2023). Bumblebees are the main providers of pollination ecosystem services in greenhouses and they are affected by increasing temperatures that lead to an increase in thermal stress in plant canopies with an associated reduction of pollen production (Whittington *et al.*, 2004). It is appropriate to investigate other pollinating insects that are capable of tolerating higher temperatures, such as afrotropical stingless bee species.

Always, it is critical to consider the problems of gene flow between managed greenhouse and wild insect populations (Kraus *et al.*, 2011; Kiatoko *et al.*, 2022; Gonzalez *et al.*, 2022).

15.2.3 Monitoring and identification of pests and their natural enemies

Monitoring, aimed at guiding decision-making processes in an IPM or biological control programme, is important for good agronomic practice. One of the fundamental aspects of a control programme's success is the correct identification of pests and their main natural enemies (Bonsignore and Vacante, 2017). Identification must occur as quickly as possible in order to proceed with localized control actions. Different monitoring techniques, such as sticky cards, pheromone traps and pitfall traps, are often specific to a taxonomic group (Marullo *et al.*, 2021) and do not always allow rapid identification of a pest. Morphological identification requires many years' experience in basic character states, discriminating different groups of taxa and referring to specialist collections for comparison (Marullo *et al.*, 2021). For newly introduced or difficult to identify herbivore species, molecular methods can be used, including real-time polymerase chain reaction restriction fragment polymorphism (Toda and Komazaki, 2002). The adoption of information technology (IT) platforms or information exchange groups, mainly based on photographs shared between producers or interested parties, has facilitated more rapid identification of pests, their natural enemies and pollinators, and this has helped to identify and document novel pathogens and pests (Meentemeyer *et al.*, 2015; Hulbert *et al.*, 2017). Structured monitoring of a crop conducted by technicians can facilitate identification of plant diseases, mites, wingless aphids or the immature stages of whiteflies and leaf miners. IPM strategies must consider these aspects. The most recent monitoring techniques, such as image analysis systems, smart traps and sensors, are already widespread for tree crops (Čirjak *et al.*, 2022) and these should be included in greenhouse monitoring programmes.

15.2.4 Biological control

In a greenhouse ecosystem, the successful containment of one or more herbivore pests is dependent on the stability of the IPM or biological control strategy. As greenhouse conditions often favour the proliferation of pests, it is critical that IPM or biological control measures work rapidly and effectively (Waage and Greathead, 1988). Natural enemies may not be able to enter and establish quickly enough in a greenhouse environment unless they are deliberately introduced. Predatory mites belonging to the family Phytoseiidae, such as *Amblyseilus swirskii* Athias-Henriot, *Neoseiulus californicus* (McGregor), *Neoseiulus cucumeris* (Oudemans) and *Phytoseiulus persimilis* Athias-Henriot, are widely used in greenhouse environments against tetranychoid and eriophyoid herbivore mites, whiteflies, thrips and free-living nematodes (McMurtry *et al.*, 2013). They are commercially available and widely used for biological control in greenhouses (Vangansbeke *et al.*, 2013; Knapp *et al.*, 2018). However, predatory mites tend to be adversely affected by high temperatures (Montserrat *et al.*, 2013).

As for other predators, recent studies suggest that the impacts of climate change on species distribution can vary among taxon and geographic scales (Liao *et al.*, 2023), so it is essential to investigate species or strains that can perform at high temperatures (Zhang *et al.*, 2018; Moradi *et al.*, 2023). An exploration of intraspecific variation in specific biocontrol traits by selective breeding could improve biological control measures in greenhouses and other agroecosystems (Nachappa *et al.*, 2011; Kruitwagen *et al.*, 2018; Dumont *et al.*, 2019). Parasitoids are insects whose larvae develop in or on the bodies of arthropods. They are more specific in their action than predatory mites but further research on their potential hosts and implications in biological control is necessary to improve their use as biological control agents (Heimpel *et al.*, 2021) even in the context of climate change.

The association between arthropods and fungi, bacteria and viruses has been well documented (Boucias and Pendland, 1998; van der Geest *et al.*, 2000). Viruses are ubiquitously present in insect populations but their impacts and dynamics are not well known (Myers and Cory, 2015), although they can be potential agents for controlling harmful insects, especially Lepidoptera (Miller and Ball, 1998; Myers and Cory, 2015). Among Bacteroidaceae, various strains of *Bacillus thuringiensis* Berliner (and other species) are sold for the control of various insect pests, such as moths (Lepidoptera), beetles (Coleoptera) and flies (Diptera). Entomopathogenic fungi can increase their growth rate and infection potential at higher temperatures, but a reduction in relative humidity associated with higher temperatures could reduce their capacity to act as effective control agents. The interaction between entomopathogens, herbivorous insects and host plants is complex. Plant primary and secondary metabolites can influence the outcomes of insect–pathogen interactions (Shikano, 2017). *Beauveria* species generally grow within moderate temperature ranges (Ferron, 1978), but some strains may be able to tolerate higher temperatures (Mishra *et al.*, 2015; Rehner and Kepler, 2017).

Currently, commercially widely available products for different crops containing *Trichoderma* spp., *Ampelomyces quisqualis*, *Bacillus* spp. and *Ulocladium* spp. are widely used to control botrytis and powdery mildew in greenhouses and field crops. In the future, biological control of pathogens in greenhouses could predominate over chemical control. There is some evidence that combinations of microorganisms can be used to promote resistance to abiotic stress in plants, although the mechanism of action is unclear (Poveda and Eugui, 2022). IPM is mandatory in the European Union (Helepciuc and Todor, 2022) and overall, the market for biocontrol agents is growing. In 2023 biological control measures were valued at $5.8 billion, and this figure is expected to reach $10.6 billion by 2032 (Research and Markets, 2024).

15.3 Implementation Protocol

15.3.1 Prevention

Total radiation can be partially controlled by a careful choice of greenhouse shape and orientation. East–west orientation is best for year-round greenhouse applications at all latitudes as this orientation receives greater total radiation in winter and less in summer except near the Equator (Sethi, 2009). Prevailing winds should be considered as they can help to remove hot air masses inside greenhouses. For tunnels built on sloping terrain, it is important to reduce their length to avoid anomalous and abundant flows of hot air towards one end of the tunnel. For climatic areas that suffer from high summer temperatures, it is also possible to introduce mixed structures capable of reducing sunlight in the hot season and accumulating heat in winter (see Section 15.2.1).

Whitening applications are widely used to reduce radiation. As described previously, calcium-based products can be distributed over the plastic or glass and this treatment is generally categorized according to greenhouse light transmissivity in relation to the density of whitening produced (Karaca, 2023). Some shade may be optimal to produce high-quality peppers in a greenhouse during summer and autumn months (Díaz-Pérez, 2013; Caruso *et al.*, 2020). Estimation of performance and effects beyond the measurement of transmitted light must consider the specific requirements of the plant species under cultivation and must be related to plant phenological phases. For example, sunburn can be induced in many kinds of fruit, flowers and vegetables, such as tomato, grown in warm or semi-arid climates if the sun's rays are direct. In other cases, such as peppers and aubergines, light influences floral induction (Thomas, 2006), particularly in semi-arid environments as in some areas in the Mediterranean (Nguyen *et al.*, 2022). Promotion of optimal plant growth can reduce the density and spread of pest populations of *Trialeurodes vaporariorum* (Westwood, 1856), such as in zucchini and tomato greenhouses (Bonsignore, 2015).

Control and mitigation of the climate inside a greenhouse are not always realistic. Sometimes, large greenhouses guarantee a greater air volume under one structure which can provide a large buffer space above the crop (Jewett and Jarvis, 2001), but the process of reducing or increasing temperature can require unsustainable energy expenditure. The microclimate around the surface of plant foliage most influences insects and mites and their predators, such as *Orius insidiosus* (Zhang and Shipp,

1998), and their possible control. Maintaining humidity around leaves is critical for establishing entomopathogenic fungi, such as *Beauveria bassiana* (Bals.-Criv.) Vuill. (Shipp *et al.*, 2003).

An effective cleaning or sanitizing programme at the end of the agricultural season before the next harvest, where economically feasible, will significantly reduce or eliminate pest problems (Ellouze *et al.*, 2020).

To obtain a better understanding of the effects of climate change and reduce unexpected consequences when biological control strategies are used, it is important to build a body of data for specific ecosystems. Only thus can successful interactions between pests, natural enemies and biological control measures be implemented (Thomson *et al.*, 2010).

15.3.2 Choices and cultural practices

There are many species of mirid bugs (Order Hemiptera, Family Miridae) which are widely distributed and considered to be excellent predators (Cassis and Schuh, 2012). They are phytophagous, which means they feed on live plant tissue during periods of prey scarcity, but they can also induce plant defence mechanisms by the activation of several metabolic pathways, which can lead to antixenosis in plants and the attraction of parasitoids (Pérez-Hedo *et al.*, 2021). Appropriate companion plants and the use of alternative foods can mitigate their negative behaviours (Konan *et al.*, 2021).

Thermal tolerance varies between species. *Nesidiocoris tenuis* Reuter is used in warmer climates while *Macrolophus pygmaeus* Rambur is used in temperate regions on tomato, aubergine, pepper and some ornamentals. *Macrolophus pygmeus* activated plant defence mechanisms against the two-spotted spider mite Tetranychus urticae Koch, but Pappas *et al.* (2015) did not observe similar activity in the greenhouse whitefly *Trialeurodes vaporariorum* infesting tomato plants.

Integrated pest management aims to keep the use of pesticides and other forms of intervention at economically and ecologically justified levels, and other measures such as crop rotation, selection of appropriate cultivars and cultivation techniques can be used. Producers should give priority to preventive actions, monitoring and biological plant protection alternatives at different plant growth stages before using chemical plant protection products. The control of *Thrips palmi* (Karny), which has expanded its distribution range in greenhouses (Cao *et al.*, 2019), is best achieved by the early introduction of *A. swirskii* and the use of metalized reflective mulching (Razzak *et al.*, 2019).

15.3.3 Sampling and population monitoring

Pests in greenhouses are often overlooked in ecological studies, regardless of their role in multitrophic interactions outside greenhouses. Most pests are opportunistic and, according to the stage of growth, occupy different niches, habitats and ecosystems, including soil and on different plant parts. It is essential to develop an effective sampling and monitoring plan to provide an evidence-based IPM or biological control plan. During cultivation, many operators choose to obtain a general picture of plant health, including the presence of insects, pathogens, nematodes and weeds. This type of monitoring is very detailed and well articulated and it helps small producers to better follow their crops. For large producers, a regular inspection is necessary to allow (i) identification of the pest, (ii) determination of the abundance of the populations present or of the symptom detected, and (iii) evaluation of the objectives achieved by the adoption of previous cultivation operations.

In greenhouses, as well as other agro-ecosystems, there are various techniques for pest monitoring. For some species such as whiteflies, aphids, spider mites or some psyllids, the methods are universal and involve direct observations of plant parts with counts or sticky traps. For other pests such as *Tuta absoluta* (Meyrick) or noctuide species such as *Spodoptera littoralis* (Boisduval) (Lepidoptera: Noctuidae), pheromone monitoring methods are preferred which are more species specific. Monitoring in greenhouses requires consideration of the climate region, the crop under cultivation and the type of pest. If the pests are flying insects, it is preferable to monitor during the hours of least activity. For signs of plant damage, such as chlorotic, stunted or malformed growth, feeding damage, honeydew, sooty mould and/or the presence of ants, the daily monitoring interval can be extended in line with the operator's schedule. Particular attention must be paid to the areas near the openings of greenhouses, as this is how most pests are introduced.

As with other ecosystems, adequate monitoring plans are critical to understand the specific ecosystem that has been created and to better understand the implications of climate change. Increasingly, climate shifts are leading to the introduction of new pests, which have only recently been detected and whose ecology is poorly understood (Kocarek *et al.*, 2015).

15.3.4 Economic injury levels, economic treshold and action treshold

The role of biological control measures in a shifting climate requires an understanding of the ecological interrelationships between different trophic levels in greenhouse systems, particularly in organic farming. From a practical perspective, the biological control of pests in greenhouse production must usually be integrated with other technical means. The aim is to develop effective control measures which have a minimum residual effect on the greenhouse ecosystem and include optimum growing condition for plants, favouring natural control mechanisms through the promotion of natural enemies, and the application of direct control measures to kill pests, diseases or weeds. The tools to manage pests and diseases include cultivation practices, habitat management and direct control measures, such as the biological control of pests by natural enemies and the use of insecticides of vegetal or mineral origin, pheromones, repellents, traps and physical barriers.

There are two aspects to an IPM strategy: the lowest population density of the pest that will cause economic damage, known as the economic injury level (EIL), and the economic threshold (ET), which is the population density at which control measures should be implemented to prevent a pest population from reaching the EIL. The ET for different crops and pests is beyond the scope of this chapter because there is substantial variation depending on the crop under cultivation, the pest type, the cost of control methods and time until harvest and market. However, it is evident that climate change, particularly higher temperatures and less predictable rainfall, and the arrival of new pests in response to climate shifts are increasing the rate at which EIL is reached. A further compounding factor is that, in greenhouses, several species are typically destructive insects feeding on a high-value crop which results in a low economic threshold value (Bonsignore and Vacante, 2017). Pests that require rapid control or preventive interventions include several phytophagous mites such as the broad mite *Polyphagotarsonemus latus* (Acari: Tarsonemidae) and the two-spotted spider mite *T. urticae* whose populations may increase rapidly under different thermal conditions.

15.3.5 Control measures

Appropriate knowledge of control measures available to growers in a greenhouse environment allows planned and effective implementation. Control measures include chemical, biological and physical methods and, if appropriately integrated, producers can achieve effective control of the pest or disease. Chemical pesticides are widely used. For the specific methods of action and/or type of formulation, please refer to the specialized bibliography (Stejskal *et al.*, 2021).

It is important to consider the effects of pesticides on both pests and host plants, particularly under elevated thermal conditions. Higher temperatures inside a greenhouse lead to greater volatility of the pesticide and a slower dispersion effect because of the covered environment, and this may lead to hazardous conditions for the operator. Among fumigants used in greenhouses, the use of 1,3-dichloropropene against nematodes should be accompanied by methods to reduce emissions into the atmosphere (Yates *et al.*, 2016).

Currently, pesticides represent the main means of control for plant pests and diseases in greenhouse environments. At a global level, although the noxious effects on human health and hazards are partly known, the quantities used are considerable, as are the number of compounds and formulations available (Choudhury and Saha, 2020). Climate change can have a direct and/or indirect impact on living organisms in an agricultural ecosystem, including crops and their associated pests. The performance of pesticides with the arrival of a new pest, or the greater rate of growth of different poikilotherms at higher temperatures, makes it difficult to predict the future. Recent studies have shown that at higher temperatures some active compounds in pesticides degrade more rapidly, by hydrolysis, photodegradation or greater volatilization (Hooper *et al.*, 2013; Otieno *et al.*, 2013; de Beeck *et al.*, 2017). Therefore, the performance of pesticides is likely to change because of thermal variations and with an increase in drought. Changes in the translocation and absorption of systemic insecticides under climate shifts are currently unknown (Khodaverdi *et al.*, 2016; Hubbard and Murillo, 2024). Therefore, the use of pesticide formulations that minimize the impact on beneficial insects (predators, parasitoids and pollinators) in the greenhouse is of fundamental importance.

It is essential to gauge control requirements after the correct identification of a pest or disease. A careful examination of natural enemies and population density must be made. In areas with strong thermal variations linked to the seasons, the lowering of temperature in the transition from autumn to winter suggests that the growth cycle of the pest species will decline and this may provide slightly greater time margins for phytocultural interventions. Within the decision-making process that leads to the use of a pesticide, it is of primary importance to use, among the pesticides allowed for that crop, those with the lowest effect on beneficial insects. Furthermore, it is essential in practice to improve the development, reproduction and efficacy of natural enemies within the greenhouse through alternative prey, banker plant systems, pollen and manipulated habitats (Pijnakker *et al.*, 2020). This would allow an increase in the permanence and survival of natural enemies in a greenhouse environment, especially in stressful conditions.

Mitigation of the adverse effects of climate change are related to the relationship between predators and prey and, for a serious pest such as *T. urticae*, artificial pollen provision can mitigate competition among its main natural enemies *Euseius stipulatus*, *N. californicus* and *P. persimilis* (Urbaneja-Bernat and Jaques, 2022). The habitat around a greenhouse can be modified to support predator populations, including spiders (Cotes *et al.*, 2018), or encourage wild populations of *N. tenuis* which can then move into a greenhouse environment (Bonsignore and van Baaren, 2024). This could help reduce pest populations in times of thermal stress.

Overall, it is important to involve greenhouse producers in large-scale experimental programmes to test which measures will work best to support biological control and lead to tangible improvements for production.

Climate shifts may benefit greenhouse crop production in cooler climates by extending the growing season, but detailed knowledge and monitoring of the climate within a greenhouse are essential to support an effective pest management programme (Shipp *et al.*, 2011). Appropriate breeding programmes to support the production of suitable cultivars and evidence-based planting regimes to mitigate climate shifts are important (Faralli *et al.*, 2019; Becker *et al.*, 2021).

15.4 Conclusion

Climate change is creating a gloomy scenario for greenhouse crop production. However, the adoption of biological control programmes based on natural enemies capable of reacting promptly to thermal stress, the correct choice of crops and the time of transplanting/sowing crops can be integrated with effective management of greenhouse microclimates by incorporating whitening measures and crop irrigation. The selection of natural enemies as biological control agents (predators and parasitoids) needs to consider the correct use of micro-organisms, such as entomopathogenic bacteria and fungi, and the prudent management of water resources to minimize crop stress. Greenhouse production must include a new vision in relation to the possible climate scenario for different areas. It will benefit from a transition towards food and agricultural systems that seek to safeguard biodiversity through programmes that provide ongoing technical assistance to support producers to make more sustainable choices. Comprehensive bio-ecological research on the natural enemies of pests is critical. Biological control in a greenhouse environment will play a fundamental role within an organic farming strategy, but it needs to be practical and cost-effective in its implementation. We conclude that natural enemies can be beneficial against crop pests and their action is the cornerstone to biological control.

References

Becker, C.C., Streck, N.A., Schwab, N.T., Uhlmann, L.O., Tomiozzo, R. *et al.* (2021) Climate risk zoning for gladiolus production under three climate change scenarios. *Revista Brasileira de Engenharia Agrícola e Ambiental* v25, 297–304.

Bonsignore, C.P. (2015) Effect of environmental factors on the flight activity of *Trialeurodes vaporariorum* (Westwood) under greenhouse conditions. *Entomological Science* 18, 207–216.

Bonsignore, C.P. and van Baaren, J. (2024) Complex habitats boost predator co-occurrence, enhancing pest control in sweet pepper greenhouses. *Horticulturae* 10, 614.

Bonsignore, C.P. and Vacante, V. (2017) Pest management in organic vegetable greenhouses. In: Vacante, V. and Kreiter, S. (eds) *Handbook of Pest Management in Organic Farming*. CABI, Wallingford, UK.

Boucias, D.G. and Pendland, J.C. (1998) *Principles of Insect Pathology*. Kluwer Academic Publ, Boston, Massachusetts.

Cao, L.J., Gao, Y.F., Gong, Y.J., Chen, J.C., Chen, M. *et al.* (2019) Population analysis reveals genetic structure of an invasive agricultural thrips pest related to invasion of greenhouses and suitable climatic space. *Evolutionary Applications* 12, 1868–1880.

Caruso, G., Formisano, L., Cozzolino, E., Pannico, A., El-Nakhel, C. *et al.* (2020) Shading affects yield, elemental composition and antioxidants of perennial wall rocket crops grown from spring to summer in southern Italy. *Plants* 9, 933.

Cassis, G. and Schuh, R.T. (2012) Systematics, biodiversity, biogeography, and host associations of the Miridae (Insecta: Hemiptera: Heteroptera: Cimicomorpha). *Annual Review of Entomology* 57, 377–404.

Cañadas, J., Sánchez-Molina, J.A., Rodríguez, F. and María del Águila, F.I. (2017) Improving automatic climate control with decision support techniques to minimize disease effects in greenhouse tomatoes. *Information Processing in Agriculture* 4, 50–63.

Choab, N., Allouhi, A., El Maakoul, A., Kousksou, T., Saadeddine, S. *et al.* (2019) Review on greenhouse microclimate and application: Design parameters, thermal modeling and simulation, climate controlling technologies. *Solar Energy* 191, 109–137.

Choudhury, P.P. and Saha, S. (2020) Dynamics of pesticides under changing climatic scenario. *Environmental Monitoring and Assessment* 192(1), 1–3.

Čirjak, D., Miklečić, I., Lemić, D., Kos, T. and Pajač Živković, I. (2022) Automatic pest monitoring systems in apple production under changing climatic conditions. *Horticulturae* 8, 520.

Cohen, Y., Van den Langenberg, K.M., Wehner, T.C., Ojiambo, P.S., Hausbeck, M. *et al.* (2015) Resurgence of *Pseudoperonospora cubensis*: The causal agent of cucurbit downy mildew. *Phytopathology* 105, 998–1012.

Costantino, A., Comba, L., Sicardi, G., Bariani, M. and Fabrizio, E. (2021) Energy performance and climate control in mechanically ventilated greenhouses: A dynamic modelling-based assessment and investigation. *Applied Energy* 288, 116583.

Cotes, B., González, M., Benítez, E., De Mas, E., Clemente-Orta, G. *et al.* (2018) Spider communities and biological control in native habitats surrounding greenhouses. *Insects* 9, 33.

Cuesta Roble (2019) Global greenhouse statistics. Available at: https://www.cuestaroble.com/statistics.html (accessed 7 June 2025).

de Beeck, L.O., Verheyen, J. and Stoks, R. (2017) Integrating both interaction pathways between warming and pesticide exposure on upper thermal tolerance in high-and low-latitude populations of an aquatic insect. *Environmental Pollution* 224, 714–721.

Deere, J.A., Janssen, A., Furlong, M.J. and Bonsall, M.B. (2023) Editorial: Integrating models into practice: The role of modelling in biocontrol and integrated pest management. *Frontiers in Ecology and Evolution* 11, 1243260.

Dumont, F., Réale, D. and Lucas, É. (2019) Can isogroup selection of highly zoophagous lines of a zoophytophagous bug improve biocontrol of spider mites in apple orchards? *Insects* 10, 303.

Díaz-Pérez, J.C. (2013) Bell pepper (Capsicum Annum L.) crop as affected by shade level: Microenvironment, plant growth, leaf gas exchange, and leaf mineral nutrient concentration. *HortScience* 48(2), 175–182.

Ellouze, W., Mishra, V., Howard, R.J., Ling, K.-S. and Zhang, W. (2020) Preliminary study on the control of cucumber green mottle mosaic virus in commercial greenhouses using agricultural disinfectants and resistant cucumber varieties. *Agronomy* 10(12), 1879.

Faralli, M., Matthews, J. and Lawson, T. (2019) Exploiting natural variation and genetic manipulation of stomatal conductance for crop improvement. *Current Opinion in Plant Biology* 49, 1–7.

Fatnassi, H., Boulard, T., Poncet, C. and Chave, M. (2006) Optimisation of greenhouse insect screening with computational fluid dynamics. *Biosystems Engineering* 93, 301–312.

Ferron, P. (1978) Biological control of insect pests by entomogenous fungi. *Annual Review of Entomology* 23, 409–442.

Gonzalez, V.H., Oyen, K., Aguilar, M.L., Herrera, A., Martin, R.D. *et al.* (2022) High thermal tolerance in high-elevation species and laboratory-reared colonies of tropical bumble bees. *Ecology and Evolution* 12, e9560.

Heimpel, G.E., Abram, P.K. and Brodeur, J. (2021) A phylogenetic perspective on parasitoid host ranges with implications for biological control. *Current Opinion in Insect Science* 44, 95–100.

Helepciuc, F.E. and Todor, A. (2022) EU microbial pest control: A revolution in waiting. *Pest Management Science* 78, 1314–1325.

Hemming, S., De Zwart, H.F., Elings, A., Righini, I. and Petropoulou, A. (2019) Remote control of greenhouse vegetable production with artificial intelligence – greenhouse climate, irrigation, and crop production. *Sensors* 19, 1807.

Hooper, M.J., Ankley, G.T., Cristol, D.A., Maryoung, L.A., Noyes, P.D. *et al.* (2013) Interactions between chemical and climate stressors: A role for mechanistic toxicology in assessing climate change risks. *Environmental Toxicology and Chemistry* 32, 32–48.

Hubbard, C.B. and Murillo, A.C. (2024) Behavioral resistance to insecticides: Current understanding, challenges, and future directions. *Current Opinion in Insect Science* 63, 101177.

Hulbert, J.M., Agne, M.C., Burgess, T.I., Roets, F. and Wingfield, M.J. (2017) Urban environments provide opportunities for early detections of Phytophthora invasions. *Biological Invasions* 19, 3629–3644.

IPCC (2021) Climate change 2021: The physical science basis. Available at: www.ipcc.ch/report/ar6/wg1/

Jewett, T.J. and Jarvis, W.R. (2001) Management of the greenhouse microclimate in relation to disease control: A review. *Agronomie* 21, 351–366.

Johnson, A.C., Liu, J., Reynolds, O., Furlong, M.J., Mo, J. *et al.* (2021) Conservation biological control research is strongly uneven across trophic levels and economic measures. *Pest Management Science* 77, 2165–2169.

Joudi, K.A. and Farhan, A.A. (2015) A dynamic model and an experimental study for the internal air and soil temperatures in an innovative greenhouse. *Energy Conversion and Management* 91, 76–82.

Karaca, C. (2023) Estimation of daily global solar radiation based on different whitening applications using temperature in Mediterranean type greenhouses. *Italian Journal of Agrometeorology* 1, 79–93.

Katzin, D., van Henten, E.J. and van Mourik, S. (2022) Process-based greenhouse climate models: Genealogy, current status, and future directions. *Agricultural Systems* 198, 103388.

Khodaverdi, H., Fowles, T., Bick, E. and Nansen, C. (2016) Does drought increase the risk of insects developing behavioral resistance to systemic insecticides? *Journal of Economic Entomology* 109, 2027–2031.

Kiatoko, N., Pozo, M.I., Van Oystaeyen, A., van Langevelde, F., Wäckers, F. *et al.* (2022) Effective pollination of greenhouse Galia musk melon (*Cucumis melo* L. var. reticulatus ser.) by afrotropical stingless bee species. *Journal of Apicultural Research* 61, 664–674.

Knapp, M., van Houten, Y., van Baala, E. and Groot, T. (2018) Use of predatory mites in commercial biocontrol: Current status and future prospects. *Acarologia* 58, 72–82.

Kocarek, P., Libor Dvorak, L. and Marketa Kirstov, M. (2015) *Euborellia annulipes* (Dermaptera: Anisolabididae), a new alien earwig in Central European greenhouses: Potential pest or beneficial inhabitant? *Applied Entomology and Zoology* 50, 201–206.

Konan, K.A.J., Monticelli, L.S., Ouali-N'goran, S.W.M., Ramirez-Romero, R., Martin, T. *et al.* (2021) Combination of generalist predators, *Nesidiocoris tenuis* and *Macrolophus pygmaeus*, with a companion plant, *Sesamum indicum*: What benefit for biological control of *Tuta absoluta*? *PLOS One* 16(9), e0257925.

Kraus, F.B., Szentgyörgyi, H. and Rożej, E. (2011) Greenhouse bumblebees (*Bombus terrestris*) spread their genes into the wild. *Conservation Genetics* 12, 187–192.

Kruitwagen, A., Beukeboom, L.W. and Wertheim, B. (2018) Optimization of native biocontrol agents, with parasitoids of the invasive pest *Drosophila suzukii* as an example. *Evolutionary Applications* 11, 1473–1497.

Laubmeier, A.N., Tabassum, N. and Tenhumberg, B. (2023) Temperature fluctuation alters optimal predator community composition for anticipated biological control. *Frontiers in Ecology and Evolution* 10, 998396.

Lentz, W. (1998) Model applications in horticulture: A review. *Scientia Horticulturae* 74, 151–174.

Liao, J.R., Chiu, M.C. and Kuo, M.H. (2023) Reassessing the presence of alien predatory mites and their prospects in the face of future climate change. *Pest Management Science* 79, 5186–5196.

López-Martínez, A., Valera-Martínez, D.L., Molina-Aiz, F.D., Moreno-Teruel, M.Á., Peña-Fernández, A. *et al.* (2019) Analysis of the effect of concentrations of four whitening products in cover transmissivity of Mediterranean greenhouses. *International Journal of Environmental Research and Public Health* 16, 958.

Marullo, R., Bonsignore, C.P. and Vono, G. (2021) Sampling methods according to thrips (Insecta, Thysanoptera) habitats. *Bulletin of Insectology* 74, 241–251.

Mashonjowa, E., Ronsse, F., Mhizha, T., Milford, J.R., Lemeur, R. *et al.* (2020) The effects of whitening and dust accumulation on the microclimate and canopy behaviour of rose plants (Rosa hybrida) in a greenhouse in Zimbabwe. *Solar Energy* 84, 10–23.

McMurtry, J.A., de Moraes, G.J. and Sourassou, N.F. (2013) Revision of the lifestyles of phytoseiid mites (Acari: Phytoseiidae) and implications for biological control strategies. *Systematic Applied Acarology* 18, 297–320.

Meentemeyer, R.K., Dorning, M.A., Vogler, J.B., Schmidt, D. and Garbelotto, M. (2015) Citizen science helps predict risk of emerging infectious disease. *Frontiers in Ecology and Environments* 13, 189–194.

Millard, J., Outhwaite, C.L., Ceausu, S., Carvalheiro, L.G., Silva, F. *et al.* (2023) Key tropical crops at risk from pollinator loss due to climate change and land use. *Science Advances* 9, eadh0756.

Miller, L.K.J. and Ball, L.A. (1998) *The Insect Viruses*. Plenum Press, New York.

Mishra, S., Kumar, P. and Malik, A. (2015) Effect of temperature and humidity on pathogenicity of native *Beauveria bassiana* isolate against *Musca domestica* L.. *Journal of Parasitic Disease* 39, 697–704.

Moerkens, R., Janssen, D. and Brenard, N. (2021) Simplified modelling enhances biocontrol decision making in tomato greenhouses for three important pest species. *Journal of Pest* 94, 285–295.

Mongkon, S., Thepa, S., Namprakai, P. and Pratinthong, N. (2013) Cooling performance and condensation evaluation of horizontal earth tube system for the tropical greenhouse. *Energy and Buildings* 66, 104–111.

Montserrat, M., Guzmán, C., Sahún, R.M., Belda, J.E. and Hormaza, J.I. (2013) Pollen supply promotes, but high temperatures demote, predatory mite abundance in avocado orchards. *Agriculture, Ecosystems and Environment* 164, 155–161.

Moradi, M., Joharchi, O., Döker, I., Khaustov, V.A., Salavatulin, V.M. *et al.* (2023) Effects of temperature on life table parameters of a newly described phytoseiid predator, *Neoseiulus neoagrestis* (Acari: Phytoseiidae) fed on *Tyrophagus putrescentiae* (Acari: Acaridae). *Acarologia* 63, 31–40.

Moreno-Teruel, M.Á., Valera, D., Molina-Aiz, F.D., López-Martínez, A., Peña, A. *et al.* (2020) Effects of cover whitening concentrations on the microclimate and on the development and yield of tomato (*Lycopersicon esculentum* Mill.) inside Mediterranean greenhouses. *Agronomy* 10, 237.

Myers, J.H. and Cory, J.S. (2015) Ecology and evolution of pathogens in natural populations of Lepidoptera. *Evolutionary Applications* 23, 231–247.

Nachappa, P., Margolies, D.C., Nechols, J.R. and Campbell, J.F. (2011) Variation in predator foraging behaviour changes predator–prey spatio-temporal dynamics. *Functional Ecology* 25, 1309–1317.

Nguyen, G.N. and Lantzke, N. (2022) Mitigating the adverse effects of semi-arid climate on capsicum cultivation by using the retractable roof production system. *Plants* 11, 2794.

Nguyen, G.N., Lantzke, N. and van Burgel, A. (2022) Effects of shade nets on microclimatic conditions, growth, fruit field, and quality of eggplant (*Solanum melongena* L.): A case study in Carnarvon, Western Australia. *Horticulturae* 8, 696.

Okulewicz, A. (2017) The impact of global climate change on the spread of parasitic nematodes. *Annals of Parasitology* 63(1), 15–20.

Otieno, P.O., Owuor, P.O., Lalah, J.O., Pfister, G. and Schramm, K.W. (2013) Impacts of climate-induced changes on the distribution of pesticides residues in water and sediment of Lake Naivasha, Kenya. *Environmental Monitoring and Assessment* 185, 2723–2733.

Pappas, M.L., Steppuhn, A., Geuss, D., Topalidou, N., Zografou, A. *et al.* (2015) Beyond predation: The zoophytophagous predator Macrolophus pygmaeus induces tomato resistance against spider mites. *PLOS One* 10(5), e0127251.

Pérez-Hedo, M., Riahi, C. and Urbaneja, A. (2021) Use of zoophytophagous mirid bugs in horticultural crops: Current challenges and future perspectives. *Pest Management Science* 77(1), 33–42.

Pijnakker, J., Vangansbeke, D., Duarte, M., Moerkens, R. and Wäckers, F.L. (2020) Predators and parasitoids-in-first: From inundative releases to preventative biological control in greenhouse crops. In: *Frontiers in Sustainable Food Systems* 4.

Poveda, J. and Eugui, D. (2022) Combined use of Trichoderma and beneficial bacteria (mainly Bacillus and Pseudomonas): Development of microbial synergistic bio-inoculants in sustainable agriculture. *Biological Control* 176, 105100.

Razzak, M.A., Seal, D.R., Stansly, P.A., Schaffer, B. and Liburd, O.E. (2019) A predatory mite, *Amblyseius swirskii*, and plastic mulch for managing melon thrips, *Thrips palmi*, in vegetable crops. *Crop Protection* 126, 104916.

Rehner, S.A. and Kepler, R.M. (2017) Species limits, phylogeography and reproductive mode in the *Metarhizium anisopliae* complex. *Annual Review of Phytopathology* 148, 60–66.

Research and Markets (2024) BioControl Agents – Global Market Outlook (2024–2032). Available at: www.researchandmarkets.com/research/z6hmnn/biocontrol_agents (accessed 7 June 2025).

Rigakis, N., Katsoulas, N., Teitel, N., Bartzanas, T. and Kittas, C. (2015) A simple model for ventilation rate determination in screenhouses. *Energy and Buildings* 87, 293–301.

Romo, C.M. and Tylianakis, J.M. (2013) Elevated temperature and drought interact to reduce parasitoid effectiveness in suppressing hosts. *PLOS One* 8(3), e58136.

Ruess, L., Michelsen, A., Schmidt, I.K. and Jonasson, S. (1999) Simulated climate change affecting microorganisms, nematode density and biodiversity in subarctic soils. *Plant and Soil* 212, 63–73.

Schettini, E., Salvador, F.R.D., Scarascia-Mugnozza, G. and Vox, G. (2011) Radiometric properties of photoselective and photoluminescent greenhouse plastic films and their effects on peach and cherry tree growth. *Journal of Horticultural Science and Biotechnology* 86, 79–83.

Sethi, V.P. (2009) On the selection of shape and orientation of a greenhouse: Thermal modeling and experimental validation. *Solar Energy* 83, 21–38.

Sharma, A., Kumar, V. and Shahzad, B. (2020) Photosynthetic response of plants under different abiotic stresses: A review. *Journal of Plant Growth Regulation* 39, 509–531.

Shikano, I. (2017) Evolutionary ecology of multitrophic interactions between plants, insect herbivores and entomopathogens. *Journal of Chemical Ecology* 43, 586–598.

Shipp, L., Zhang, Y., Hunt, D.W.A. and Ferguson, G. (2003) Influence of humidity and greenhouse microclimate on the efficacy of *Beauveria bassiana* (Balsamo) for control of greenhouse arthropod pests. *Environmental Entomology* 32, 1154–1163.

Shipp, L., Johansen, N., Vänninen, I. and Jacobson, R. (2011) Greenhouse climate: An important consideration when developing pest management programs for greenhouse crops. *Acta Horticulturae* 893, 133–143.

Stejskal, V., Vendl, T., Aulicky, R. and Athanassiou, C. (2021) Synthetic and natural insecticides: Gas, liquid, gel and solid formulations for stored-product and food-industry pest control. *Insects* 12, 590.

Sultan, B., Defrance, D. and Iizumi, T. (2019) Evidence of crop production losses in West Africa due to historical global warming in two crop models. *Scientific Reports* 9(1), 1.

Thomas, B. (2006) Light signals and flowering. *Journal of Experimental Botany* 57(13), 3387–3393.

Thomson, L.J., Macfadyen, S. and Hoffmann, A.A. (2010) Predicting the effects of climate change on natural enemies of agricultural pests. *Biological Control* 52, 296–306.

Toda, S. and Komazaki, S. (2002) Identification of thrips species (Thysanoptera: Thripidae) on Japanese fruit trees by polymerase chain reaction and restriction fragment length polymorphism of the ribosomal ITS2 region. *Bulletin of Entomological Research* 92, 359–363.

Tonnang, H.E., Sokame, B.M., Abdel-Rahman, E.M. and Dubois, T. (2022) Measuring and modelling crop yield losses due to invasive insect pests under climate change. *Current Opinion in Insect Science* 50, 100873.

Urbaneja-Bernat, P. and Jaques, J.A. (2022) Can pollen provision mitigate competition interactions between three phytoseiid predators of *Tetranychus urticae* under future climate change conditions? *Biological Control* 165, 104789.

Vacante, V. and Bonsignore, C.P. (2017) Natural enemies and pest control. In: Vacante, V. and Kreiter, S. (eds) *Handbook of Pest Management in Organic Farming*. CABI, Wallingford, UK.

van der Geest, L.P., Elliot, S.L., Breeuwer, J.A. and Beerling, E.A. (2000) Diseases of mites. *Experimental and Applied Acarology* 24, 497–560.

Vangansbeke, D., De Schrijver, L., Spranghers, T., Audenaert, J., Verhoeven, R. *et al.* (2013) Alternating temperatures affect life table parameters of Phytoseiulus persimilis, Neoseiulus californicus (Acari: Phytoseiidae) and their prey *Tetranychus urticae* (Acari: Tetranychidae). *Experimental and Applied Acarology* 61, 285–298.

Waage, J.K. and Greathead, D.J. (1988) Biological control: Challenges and opportunities. *Philosophical Transactions of the Royal Society B* 318, 111–128.

Welch, K.D. and Harwood, J.D. (2014) Temporal dynamics of natural enemy–pest interactions in a changing environment. *Biological Control* 75, 18–27.

Whittington, R., Winston, M.L., Tucker, C. and Parachnowitsch, A.L. (2004) Plant-species identity of pollen collected by bumblebees placed in greenhouses for tomato pollination. *Canadian Journal of Plant Science* 84, 599–602.

Xu, J., Li, Y., Wang, R.Z., Liu, W. and Zhou, P. (2015) Experimental performance of evaporative cooling pad systems in greenhouses in humid subtropical climates. *Applied Energy* 138, 291–301.

Yates, S.R., Ashworth, D.J., Zheng, W., Knuteson, J. and van Wesenbeeck, I.J. (2016) Effect of deep injection on field-scale emissions of 1,3-dichloropropene and chloropicrin from bare soil. *Atmospheric Environment* 137, 135–145.

Zhang, Y. and Shipp, J.L. (1998) Effect of temperature and vapour pressure deficit on the flight activity of *Orius insidiosus* (Hemiptera: Anthocoridae). *Environmental Entomology* 27, 736–742.

Zhang, G.H., Li, Y.Y., Tian, C.B., Xu, Y.J., Zhou, H.W. *et al.* (2018) Intraspecific variations on thermal susceptibility in the predatory mite *Neoseiulus barkeri* Hughes (Acari: Phytoseiidae): Responding to long-term heat acclimations and frequent heat hardenings. *Biological Control* 121, 208–215.

16 Population Growth of Biocontrol Agents and Prey Under Climate Change: A Case Study with Acarine Species

Matthew L. Meehan[1,2*] and Carlos Barreto[3]
[1]*University of Manchester, Manchester, UK;* [2]*Natural Resources Canada, Canadian Forest Service, Laurentian Forestry Centre, Quebec City, Canada;* [3]*Algoma University, Sault Ste Marie, Canada*

Abstract
Phytoseiid mite species are important biocontrol agents for several pest species, including the two-spotted spider mite (*Tetranychus urticae*), but climate change is expected to affect population growth of both phytoseiid mites and *T. urticae* (TSSM), with unknown consequences. In this chapter we leveraged 445 published intrinsic rate of growth (r_m) measurements of *Phytoseiulus persimilis*, *Neoseiulus californicus*, *Amblyseius swirskii* and TSSM to examine how temperature change affects their population growth to help inform pest management decisions and outcomes. Most measurements were conducted at 25°C, where on average we found that *P. persimilis* and *N. californicus* have greater population growth than TSSM at constant 25°C. But at 30°C, TSSM population growth was higher than that of all predators, suggesting their populations may begin to outpace predator populations. Despite limitations in the available data, our results point towards climate change adversely affecting the efficacy of these phytoseiid predators as biocontrol agents for TSSM.

16.1 Introduction

Phytoseiidae (Acari: Parasitiformes) are a hyperdiverse family within the order Mesostigmata with more than 2800 described species, equating to roughly one-fifth of described Mesostigmata (De Moraes *et al.*, 2004; Walter and Proctor, 2013; Demite *et al.*, 2023). Phytoseiidae are a well-described group, in part because of their economic importance as biocontrol agents to pest species, particularly in agroecosystems (Tixier, 2018). Several species are mass reared and sold as biocontrol agents, including *Phytoseiulus persimilis* Athias-Henriot, *Neoseiulus californicus* (McGregor) and *Amblyseius swirskii* Athias-Henriot. These three species are used across multiple countries and regions worldwide to control various pests' populations (Cock *et al.*, 2010; van Lenteren, 2012), including those of the two-spotted spider mite (TSSM) *Tetranychus urticae* Koch (Acari: Tetranychidae).

The TSSM is a cosmopolitan agricultural pest species (Migeon *et al.*, 2010) known to feed on more than 1100 plant species in 140 families (Grbić *et al.*, 2011). Specifically, TSSM can feed on economically important crops such as cotton, bean, soybean, eggplant, tomato, pepper, cucumber, maize, grape, citrus, strawberry, blueberry, blackberry, elderberry, apple, papaya,

*Corresponding author: matthew.meehan@nrcan-rncan.gc.ca

DOI: 10.1079/9781800625099.0016

pear and peach, in addition to ornamental plants including Barberton daisy, hibiscus, croton, mallow and rose (Bolland *et al.*, 1998; Navajas, 1998; Reddy and Baskaran, 2006; Silva *et al.*, 2009; Grbić *et al.*, 2011; Döker *et al.* 2020, 2021) where they feed by ingesting sap from plant tissue (Grbić *et al.*, 2007). The TSSM has large economic impacts on agriculture worldwide. For instance, measures to control TSSM in European hop cultures cost approximately 2 million euros annually (Weihrauch, 2004), whereas producers in the United States lost over 12,000 tons of cotton due to spider mites in 1 year, translating to more than $24 million in losses (Williams, 2012).

To mitigate economic and agricultural losses, phytoseiid species are used to reduce TSSM populations as numerous species feed on them, albeit to different degrees. *Phytoseiulus persimilis*, *N. californicus* and *A. swirskii* exhibit distinct feeding preferences and specializations, observable within their gnathosoma (mouthparts) (Liu *et al.*, 2017). As reported by McMurtry *et al.* (2013), *P. persimilis* is a specialist predator of *Tetranychus* species and has adapted to feeding on spider mites that produce complex web structures. *Neoseiulus californicus* is a selective predator of tetranychid mites and consumes spider mites across various webbing types and other non-spider mites, including Tarsonemidae and Tydeoidea species. Lastly, *Amblyseius swirskii* is a generalist predator that feeds on a diverse array of mites but is mainly used as a biocontrol agent for thrips and whiteflies. Although generalist predators (i.e. species that feed on a wide range of prey, including TSSM, other mites, insect larvae) from the Type III lifestyle, which includes *A. swirskii*, may struggle with the webbing produced by *Tetranychus* species (McMurtry *et al.*, 2013), *A. swirskii* is often studied concurrently with TSSM to observe population dynamics to various stimuli (e.g. temperature, pesticides, resource quality) (Riahi *et al.*, 2017; Seiedy *et al.*, 2017; Rahimi *et al.*, 2022) and remains an important secondary predator for TSSM.

Increasing temperatures with climate change may impact predator–prey interactions between phytoseiid mites and TSSM. As ectotherms, environmental temperatures are tightly correlated with mite metabolism and physiology, such as growth, feeding, movement and population growth rate (Gillooly *et al.*, 2001; Brown *et al.*, 2004; Savage *et al.*, 2004; Clarke, 2017). Ectotherm whole-organism physiological rates 'performance' follows a non-linear relationship where performance is lowest at cool and warm temperatures and highest at intermediate temperatures, i.e. their thermal optimum (T_{opt}) (Angilletta, 2009; Clarke, 2017). Due to this non-linear relationship, average performance differs when under fluctuating or constant temperatures (Paaijmans *et al.*, 2013; Kingsolver *et al.*, 2015; Marshall *et al.*, 2021). Here, the efficacy of these phytoseiid species as biocontrol agents for TSSM may be limited under climate change if responses to temperature differ between TSSM and their predators. For example, Coombs and Bale (2013) reported *P. persimilis* CT_{max} (critical thermal maximum, the upper thermal limit of performance) was, on average, 40°C. However, the reported average CT_{max} of TSSM individuals was 47.3°C, a ~ 7°C difference, suggesting extended heatwaves may be more detrimental to predators than to TSSM. Other studies have reported similar findings as short-term exposure to extreme temperatures can lower predator mite attack rate. For example, Li *et al.* (2022) found that the attack rate of the predator mite *Neoseiulus barkeri* Hughes was reduced by half on TSSM eggs, protonymphs and adults when held at 40°C, vs 25°C.Tscholl *et al.* (2022) observed that *P. persimilis* consumed more TSSM under an 'extreme' heatwave vs a 'mild' heatwave (28.6°C vs 22.6°C mean temperatures, respectively). However, faster development by TSSM under these 'extreme' heatwaves may limit predation as juvenile TSSM will develop faster into adults (Tscholl *et al.*, 2022). Ultimately, climate change has the potential to disrupt predator–prey interactions between these economically important species.

Intrinsic rate of natural increase (r) is calculated to gauge population growth (broadly, the number of births minus number of deaths) of a species. Several methods are used to calculate intrinsic rate of growth, but intrinsic rate of natural increase has historically been used in insect (and more generally, arthropod) population modelling (Birch, 1948; Howe, 1953; Smith, 1965; Laing, 1968). One such method is r_m, which represents maximum population growth under a specific stimulus, like temperature (Caughley and Birch, 1971). Notably, r_m and r notation are often used interchangeably to describe population growth of TSSM and phytoseiid predators. For consistency, we will use r_m when referring to intrinsic rate of natural increase (see Caughley and Birch (1971) and Cortés (2016) and references therein for more details on terminology and further perspectives on the topic).

Within the context of biocontrol, r_m is a useful indicator to determine if biological (food type), chemical (pesticides) or external (temperature and humidity change) factors affect predator and pest populations. For instance, Ali *et al.* (2013) tested how two types of watermelon cultivar affect r_m of *P. persimilis*, *N. californicus* and TSSM at 25°C and 30°C and found strong interactions in r_m between species, temperatures and cultivar types. Warming by 5°C increased population growth for all three species, but the average increase differed between species and watermelon cultivar. Typically, predator mite population growth was higher than that of TSSM at 30°C (Ali *et al.*, 2013). r_m has been used to assess the effect of climate change on population growth rate (Deutsch *et al.*, 2008; Vasseur *et al.*, 2014). Despite some limitations in using r_m for this purpose (see Amarasekare and Coutinho (2013) for more details), r_m may provide critical information into how increasing temperature affects population growth and dynamics for biocontrol agents and their prey.

In this chapter, we leveraged published intrinsic rate of natural increase (r_m) of TSSM and three of their phytoseiid predators (*P. persimilis*, *N. californicus*, and *A. swirskii*) to examine how temperature change affects population growth of these four species to inform how climate change may hinder the biological control of predators on TSSM. Specifically, this dataset may indicate temperatures where TSSM population growth is greater than that of their phytoseiid predators. This is information is critical to ascertain as TSSM outbreak is predicted to worsen under climate change (Litskas *et al.*, 2019)

16.2 Methods

We searched for published r_m values using the ISI Web of Science database using the following query in April 2023: ((TS=('*Phytoseiulus persimilis*' OR '*Neoseiulus californicus*' OR '*Amblyseius swirskii*' OR '*Tetranychus urticae*')) AND PY=(1998–2023)) AND AB=('intrinsic'). After this initial search, in February 2024 we found two additional papers that met our general criteria and were added into the study pool (Escudero and Ferragut, 2005; Abd El-Wahed and El-Halawany, 2012). We focused on articles published between 1998 and 2023, and specifically sought out articles with the keyword 'intrinsic' in the abstract. This was to limit the number of articles that conducted experiments with *P. persimilis*, *N. californicus* and *A. swirskii*, but especially with *T. urticae*, as these animals are routinely studied with r_m not always measured. Because of this decision, we recognize that some studies may be missing in our study pool, in addition to older studies that investigated the effects of temperature on mite population growth. Finally, we focused on how mean temperature change affected population growth, in lieu of other climate change-related impacts to growth, such as humidity (e.g. Le Hesran *et al.* 2019, 2020) as we focused on temperature–ectotherm performance relationships in this chapter. In total, we obtained an initial pool of 407 studies to extract data from.

After inspecting abstracts and methods, we removed any studies from the initial pool (i) that did not perform experiments on any of the four species, (ii) that did not measure population growth, and (iii) in which the article was not published in the English language. Furthermore, we enacted a dietary restriction for phytoseiid species to be included in our data review as resource quality and availability are subject to change under climate change. We removed studies and measurements in which predators did not consume *T. urticae* (TSSM) as a single prey source (i.e. here predators can only feed on TSSM to have their population growth measured), because we wanted to better compare the effects of temperature on population growth amongst phytoseiid predators. Diet can greatly affect population growth for these predators (Escudero and Ferragut, 2005; Kadkhodazadeh *et al.*, 2021), therefore we limited this variation by excluding other prey types and feeding sources. We did not have a dietary restriction for TSSM when extracting r_m values from studies.

After our initial screening process, we retained 141 studies from which to record r_m.

Conveniently, r_m values were solely collected from tables. For studies that examined the effect of pesticide or environmental stressors (like drought) on population growth, we only recorded the control population growth (when no pesticide or stressor was applied). We recorded the food source for TSSM during the experiments. Ideally, this was the binomial Latin name of the feeding resource for TSSM, but this level of detail was not always provided. As stated above, the sole feeding source for phytoseiid predators was TSSM. We also recorded the average experimentaltemperature(s) in which r_m was derived. If a range of experimental temperatures were provided instead of a single value, we recorded the average temperature within the given range (e.g.23–27°C with a recorded average of 25°C).

Methods to calculate r_m varied between retained studies but were predominantly calculated (and subsequently cited) with two formulas. The first, and older, method cited is Birch (1948) and is derived from the Euler–Lotka equation, which calculates r_m as seen in Eqn 16.1:

$$\sum_{x=0}^{\infty} e^{-rx} l_x m_x = 1 \qquad (16.1)$$

where age is indexed from zero, l_x is age-specific survival rate, and m_x is age-specific fecundity. Notably, only females are used to calculate population growth with this equation. Note that in Birch (1948), intrinsic rate of natural increase is denoted with r (as we have done here), while we added in age being indexed at zero when using sigma notation (the discrete summation of values). The second formula often cited and used to calculate r_m is from Chi and Liu (1985) and Chi (1988) (here, intrinsic rate of natural increase is denoted by r when citing this method). This method is an extension of Birch (1948) and includes age/stage-specific results and information from males and females, and is known as the age-stage, two-sex life table theory. This formula relies on additional calculations for age-specific survival and fecundity, as follows:

$$l_x = \sum_{j=1}^{k} s_{xj} \qquad (16.2)$$

$$m_x = \frac{\sum_{j=1}^{k} s_{xj} f_{xj}}{\sum_{j=1}^{k} s_{xj}} \qquad (16.3)$$

where s_{xj} is the age/stage-specific survival rate and f_{xj} is age/stage-specific fecundity rate. These parameters are used to subsequently calculate r_m. As stated earlier, for consistency within the chapter we will use r_m to denote population growth.

16.3 Results and Discussion

16.3.1 Predators and prey r_m values in function of temperature

Across the 140 studies (Table 16.1), we recorded 445 measurements from our three phytoseiid predators and TSSM. Seventy-five percent of r_m measurements (n= 334) were from TSSM with the remaining from the predators with 37, 52 and 22 measurements from *P. persimilis*, *N. californicus* and *A. swirskii*, respectively. Over half of r_m measurements (55.9%, from both phytoseiid predators and TSSM) were derived from a single temperature, 25°C, with a combined experimental temperature range of 12–35°C across all studies (Fig. 16.1).

We found that increasing temperatures corresponded with higher population growth for predators and TSSM. On average, the population growth for TSSM doubled from 20°C to 30°C, exceeding that of the closest predator, *N. californicus*, whose population growth rate only increased 1.6 times over the same temperature range. However, most experimental studies and measurements were conducted at a single temperature (25°C), restricting our ability to conclude how climate change will affect the population growth rate of predators and TSSM. Some notable exceptions include Rahimi *et al.* (2022), who examined the population growth of *A. swirskii* across six temperatures (20°C, 22°C, 25°C, 27°C, 30°C and 32°C) and found population growth was highest at 27°C before significantly declining at 30°C and 32°C. In addition, both Gotoh *et al.* (2015) and Osman *et al.* (2012) maintained TSSM at five temperatures (15°C, 20°C, 25°C, 30°C, 35°C) and found that r_m was highest at 30°C. It is worth noting that the *average* population growth of TSSM was higher at 35°C than 30°C across our study pool, but infrequent measurements at these warmer temperatures limit our ability to make definitive

Table 16.1. List of studies included in the literature review on temperature effects on the intrinsic rate of increase (r_m) of predatory and pest mite species.

In-text.citation	Paper Title	Journal.name	Year	Species	Temperature.C
Abd El-Wahed & El-Halawany	Effect of temperature degrees on the biology and life table parameters of *Tetranychus urticae* Koch on two pear varieties	egyptian Academic Journal of Biological Sciences	2012	T.urticae	15, 20, 25, 30
Abad-Moyano *et al.*	Comparative life-history traits of three phytoseiid mites associated with *Tetranychus urticae* (Acari: Tetranychidae) colonies in clementine orchards in eastern Spain: implications for biological control	Experimental and Applied Acarology	2009	P.persimilis, N.californicus	25
Abdel-Khalek *et al.*	Comparative biology and growth rate of the two predatory mites, *Cydnoseius negevi* and *Neoseiulus californicus (Acari: Phytoseiidae)*, reared on two pea cultivars	Persian Journal of Acarology	2019	N.californicus	27
Abdelwines & Ahmed	The effect of some fertilizer compounds as a resistance inducer in strawberry plants on life history parameters of *Tetranychus urticae* (Acari: Tetranychidae)	Persian Journal of Acarology	2022	T.urticae	28
Abo-Shnaf	Temperature-based life history and life table parameters of the two spotted-spider mite (Acari: Tetranychidae) on White Frangipani	egyptian Academic Journal of Biological Sciences	2017	T.urticae	20, 25, 30
Abou-Elella & Abdel-Khalek	Biology and life table analysis of *Tetranychus urticae* Acari: Tetranychidae) on different common pea and bean cultivars	Persian Journal of Acarology	2020	T.urticae	27
Ahmed *et al.*	Application of demographic analysis for assessing effects of pesticides on the predatory mite, *Phytoseiulus persimilis* (Acari: Phytoseiidae)	Persian Journal of Acarology	2021	P.persimilis	25

Continued

Table 16.1. Continued

In-text.citation	Paper Title	Journal.name	Year	Species	Temperature.C
Ali *et al.*	Biology and life table parameters of *Tetranychus urticae* Koch (Acari: Tetranychidae) and two phytoseiid predatory mites on two watermelon cultivars	Acarines	2013	P.persimilis, T.urticae, N.californicus	25, 30
Alinejad *et al.*	Sublethal effects of fenazaquin on biological performance of the two-spotted spider mite, *Tetranychus urticae* (Acari: Tetranychidae): application of age-stage, two-sex life tables	Acarina	2015	T.urticae	26
Alinejad *et al.*	Assessment of sublethal effects of spirodiclofen on biological performance of the predatory mite, *Amblyseius swirskii*	Systematic & Applied Acarology	2016	A.swirskii	26
Alinejad *et al.*	Demographic analysis of sublethal effects of propargite on *Amblyseius swirskii* (Acari: Phytoseiidae): Advantages of using age-stage, two sex life table in ecotoxicological studies	Systematic & Applied Acarology	2020	A.swirskii	26
Alinejad *et al.*	Sublethal effects of fenazaquin on life table parameters of the predatory mite *Amblyseius swirskii* (Acari: Phytoseiidae)	Experimental and Applied Acarology	2014	A.swirskii	26
Alipour *et al.*	Resistant rose cultivar affects life table parameters of two-spotted spider mite and its predators *Phytoseiulus persimilis* and *Amblyseius swirskii* (Phytoseiidae)	Systematic & Applied Acarology	2019	P.persimilis, T.urticae, A.swirskii	25
Alizade *et al.*	Effects of nitrogen fertilization on population growth of two-spotted spider mite	Systematic & Applied Acarology	2016	T.urticae	24

Continued

Table 16.1. Continued

In-text.citation	Paper Title	Journal.name	Year	Species	Temperature.C
Azadi-Qoort *et al.*	Biological responses of *Tetranychus urticae* (Acari: Tetranychidae) to different host plants: an investigation on bottom-up effects	Systematic & Applied Acarology		T.urticae	27
Bahari *et al.*	Long-term feeding on greenhouse cucumber affects life table parameters of two-spotted spider mite and its predator *Phytoseiulus persimilis*	Systematic & Applied Acarology	2018	P.persimilis, T.urticae	25
Barati & Hejazi	Reproductive parameters of *Tetranychus urticae* (Acari: Tetranychidae) affected by neonicotinoid insecticides	Experimental and Applied Acarology	2015	T.urticae	26
Bayu *et al.*	Impact of constant vs fluctuating temperatures on the development and life history parameters of *Tetranychus urticae* (Acari: Tetranychidae)	Experimental and Applied Acarology	2017	T.urticae	15, 20, 25, 30, 35
Belczewski & Harmsen	The effect of non-pathogenic phylloplane fungi on life-history traits of *urticae* (Acari: Tetranychidae)	Experimental and Applied Acarology	2000	T.urticae	20
Bounfour & Tanigoshi	Effect of temperature on development and demographic parameters of *Tetranychus urticae* and *Eotetranychus carpini borealis* (Acari: Tetranychidae)	Ecology and Population Ecology	2001	T.urticae	15, 20, 25, 30
Bozhgani *et al.*	Sublethal effects of chlorfenapyr on the life table parameters of two-spotted spider mite, *Tetranychus urticae* (Acari: Tetranychidae)	Systematic & Applied Acarology	2018	T.urticae	25

Continued

Table 16.1. Continued

In-text.citation	Paper Title	Journal.name	Year	Species	Temperature.C
Bozhgani *et al.*	The effects of Spiromesifen on life history traits and demographic parameters of predatory mite *Neoseiulus californicus* (Acari: Phytoseiidae) and its prey *Tetranychus urticae* Koch (Acari: Tetranychidae)	Systematic & Applied Acarology	2019	N.californicus, T.urticae	25
Bruessow *et al.*	Replacement of CTV-susceptible sour orange rootstock by CTV-tolerant ones may have triggered outbreaks of *Tetranychus urticae* in Spanish citrus	Agriculture, Ecosystems and Environment	2010	T.urticae	25
Canlas *et al.*	Biology and predation of the Japanese strain of *Neoseiulus californicus* (McGregor) (Acari: Phytoseiidae)	Systematic & Applied Acarology	2006	N.californicus	20, 25, 30
Castagnoli *et al.*	Tomato transgenic lines and *Tetranychus urticae*: changes in plant suitability and susceptibility	Experimental and Applied Acarology	2003	T.urticae	25
Chitgar *et al.*	Sublethal effects of *Thymus vulgaris* essential oil on life-table parameters of two-spotted spider mite, *Tetranychus urticae* Koch (Acari: Tetranychidae)	Archives of Phytopathology and Plant Protection	2013	T.urticae	25
Dana *et al.*	Demographic comparison of the *Tetranychus urticae* Koch. (Acari: Tetranychidae) reared on different cultivars of strawberry	Journal of Economic Entomology	2018	T.urticae	25
Dana *et al.*	Comparison of the life table and reproduction parameters of the *Tetranychus urticae* (Acari: Tetranychidea) on five strawberry varieties	International Journal of Acarology	2018	T.urticae	25
Dehghan *et al.*	Fitness of *Tetranychus urticae* Koch (Acari: Tetranychidae) on different soybean cultivars: biology and fertility life-tables	International Journal of Acarology	2009	T.urticae	25

Continued

Table 16.1. Continued

In-text.citation	Paper Title	Journal.name	Year	Species	Temperature.C
Dutton *et al.*	Uptake of Bt-toxin by herbivores feeding on transgenic maize and consequences for the predator *Chrysoperla carnea*	Ecological Entomology	2002	T.urticae	27
Eini et ak.	How pollen grains of 23 plant species affect performance of the predatory mite *Neoseiulus californicus*	BioControl	2022	N.californicus	25
El Taj *et al.*	Effects of host plants and seasons on the biology of the two-spotted spider mite, *Tetranychus urticae* (Koch)	African Entomology	2016	T.urticae	12, 31.5
El-Nahas *et al.*	Thermal requirements and life table parameters of the predatory mite, *Neoseiulus californicus* (McGregor) fed on *Tetranychus urticae* Koch (Acari: Phytoseiidae: Tetranychidae)	Acarines	2020	N.californicus	20, 27, 30
Elhalawany *et al.*	Influence of prey type on the biology and life-table parameters of *Neoseiulus californicus* (McGregor) (Acari: Phytoseiidae)	Acarines	2017	N.californicus	25
Elkholy *et al.*	Effect of different doses of gamma irradiation on biology and life table of *Tetranychus urticae* Koch (Acari: Tetranychidae) of two ornaments plants	egyptian Academic Journal of Biological Sciences	2022	T.urticae	25
Escudero & Ferragut	Life-history of predatory mites *Neoseiulus californicus* and *Phytoseiulus persimilis* (Acari: Phytoseiidae) on four spider mite species as prey, with special reference to *Tetranychus evansi* (Acari: Tetranychidae)	Biological Control	2005	P.persimilis, N.californicus, T.urticae	25
Fahim & El-Saiedy	Life table parameters of *Amblyseius swirskii* and *Neoseiulus californicus* (Acari: Phytoseiidae) reared on two strawberry cultivars	International Journal of Acarology	2021	N.californicus, A.swirskii	27

Continued

Table 16.1. Continued

In-text.citation	Paper Title	Journal.name	Year	Species	Temperature.C
Fathipour *et al.*	Spider mite host plant resistance traits improve the predatory performance of *Phytoseiulus persimilis* on cucumber, despite negative life history impacts	Biological Control	2019	P.persimilis	25
Ghaderi *et al.*	Demographic analysis of the effect of fenpyroximate on *Phytoseiulus persimilis* Athias-Henriot (Acari: Phytoseiidae)	Entomologia Generalis	2013	P.persimilis	25
Ghadim *et al.*	Demographic analysis of sublethal effects of spiromesifen on *Neoseiulus californicus* (Acari: Phytoseiidae)	Acarologia	2017	N.californicus	25
Ghasemzadeh & Qureshi	Demographic analysis of fenpyroximate and thiacloprid exposed predatory mite *Amblyseius swirskii* (Acari: Phytoseiidae)	PloS ONE	2018	A.swirskii	25
Golizadeh *et al.*	Comparative life table analysis of *Tetranychus urticae* Koch (Acari: Tetranychidae) on ten rose cultivars	Acarologia	2017	T.urticae	24
Gotoh *et al.*	Effect of temperature on life history of the predatory mite *Amblyseius* (*Neoseiulus*) californicus (Acari: Phytoseiidae)	Experimental and Applied Acarology	2004	N.californicus	20, 25, 30
Gotoh *et al.*	Development and reproduction of five *Tetranychus* species (Acari: Tetranychidae): Do they all have the potential to become major pests?	Experimental and Applied Acarology		T.urticae	15, 20, 25, 30, 35
Gotoh *et al.*	Effects of constant and variable temperatures on development and reproduction of the two-spotted spider mite *Tetranychus urticae* (Acari: Tetranychidae)	Experimental and Applied Acarology	2014	T.urticae	22

Continued

Table 16.1. Continued

In-text.citation	Paper Title	Journal.name	Year	Species	Temperature.C
Gotoh *et al.*	Influence of prey on developmental performance, reproduction and prey consumption of *Neoseiulus californicus* (Acari: Phytoseiidae)	Experimental and Applied Acarology	2006	N.californicus	25
Hadadi *et al.*	Long-term effects of cattail pollen on development, population growth potential, and predation capacity of *Amblyseius swirskii* (Acari: Phytoseiidae)	Biocontrol Science and Technology	2022	A.swirskii	25
Hashemi *et al.*	How does feeding on different diets affect the life history traits of *Neoseiulus californicus*?	International Journal of Acarology	2021	N.californicus	25
Havasi *et al.*	The effects of bifenazate on life history traits and population growth of *Amblyseius swirskii* Athias-Henriot (Acari: Phytoseiidae)	Systematic & Applied Acarology	2021	A.swirskii	25
Havasi *et al.*	Impact of hexythiazox on life table parameters of the *Amblyseius swirskii* (Acari: Phytoseiidae) and its prey *Tetranychus urticae*	Journal of Crop Protection	2021	A.swirskii	25
Havasi *et al.*	Life history traits and demographic parameters of *Neoseiulus californicus* McGregor (Acari: Phytoseiidae) treated with the Biomite (R)	Systematic & Applied Acarology	2020	N.californicus	25
Havasi *et al.*	Sublethal effects of diflovidazin on demographic parameters of the predatory mite, *Neoseiulus californicus* (Acari: Phytoseiidae)	International Journal of Acarology	2019	N.californicus	25
Havasi *et al.*	Influence of low -lethal concentrations of thiamethoxam on biological characteristics of *Neoseiulus californicus* (Acari: Phytoseiidae)	Journal of Crop Protection	2020	N.californicus	25

Continued

Table 16.1. Continued

In-text.citation	Paper Title	Journal.name	Year	Species	Temperature.C
Havasi *et al.*	Sublethal effects of diflovidazin on life table parameters of two-spotted spider mite *Tetranychus urticae* (Acari: Tetranychidae)	International Journal of Acarology	2018	T.urticae	25
Havasi *et al.*	Acaricidal activity of *Punica granatum* L. peels extract against Tetranychus urticae Koch (Acari: Tetranychidae)	Archives of Phytopathology and Plant Protection	2019	T.urticae	25
Havasi *et al.*	Sublethal effects of Dayabon (R) on biological parameters of two-spotted spider mite, *Tetranychus urticae* (Koch) (Acari: Tetranychidae)	Journal of Crop Protection	2020	T.urticae	25
Havasi *et al.*	The sublethal concentration of Cyflumetofen adversely affect demographic parameters of *Tetranychus urticae* (Acari: Tetranychidae): Using age-stage, two-sex life tables	International Journal of Acarology	2022	T.urticae	25
Hoffmann *et al.*	Arbuscular mycorrhizal symbiosis increases host plant acceptance and population growth rates of the two-spotted spider mite *Tetranychus urticae*	Oecologia	2009	T.urticae	25
Honarparvar *et al.*	Demographic parameters of two spotted spider mite, *Tetranychus urticae* Koch (Acari: Tetranychidae) on cotton	Archives of Phytopathology and Plant Protection	2011	T.urticae	28
Hoque *et al.*	Life tables of two-spotted spider mite *Tetranychus urticae* Koch (Acari: Tetranychidae) and its predator *Phytoseiulus persimilis* Athias-Henriot (Acari: Phytoseiidae).	Journal of Bio-Science	2008	P.persimilis, T.urticae	25, 26
Hosseini *et al.*	Plant growth-promoting *Rhizobacteria* enhance defense of strawberry plants against spider mites	Frontiers in Plant Science	2022	T.urticae	25

Continued

Table 16.1. Continued

In-text.citation	Paper Title	Journal.name	Year	Species	Temperature.C
Ibrahim *et al.*	Biological and ecological fitness of the two-spotted spider mite, *Tetranychus urticae* koch (Acari: Tetranychidae) on certain host plants	Acarines	2016	T.urticae	25
Kabiri *et al.*	Impact of different cotton (*Gossypium* spp.) cultivars, as host plant, on development and fertility life-table parameters of *Tetranychus urticae* (Acari: Tetranychidae)	International Journal of Acarology	2012	T.urticae	25
Kadkhodazadeh *et al.*	Suitability of different pollen grains and *Tetranychus urticae* as food for the predatory mite, *Amblyseius swirskii* (Acari: Phytoseiidae)	Persian Journal of Acarology	2021	A.swirskii	25
Kasap	Effect of apple cultivar and of temperature on the biology and life table parameters of the twospotted spider mite Tetranychus urticae	Phytoparasitica	2004	T.urticae	20, 25, 30, 35
Kavousi *et al.*	Demographic traits of *Tetranychus urticae* (Acari: Tetranychidae) on leaf discs and whole leaves	Journal of Economic Entomology	2009	T.urticae	25
Khanamani *et al.*	Effectiveness of factitious foods and artificial substrate in mass rearing and conservation of *Neoseiulus californicus* (Acari: Phytoseiidae)	International Journal of Acarology	2021	N.californicus	25
Khanamani *et al.*	Population growth response of *Tetranychus urticae* to eggplant quality: application of female age-specific and age-stage, two-sex life tables	International Journal of Acarology	2013	T.urticae	25
Khanamani *et al.*	How pollen supplementary diet affect life table and predation capacity of *Neoseiulus californicus* on two-spotted spider mite	Systematic & Applied Acarology	2017	N.californicus	25

Continued

Table 16.1. Continued

In-text.citation	Paper Title	Journal.name	Year	Species	Temperature.C
Khanamani *et al.*	Linking pollen quality and performance of *Neoseiulus californicus* (Acari: Phytoseiidae) in two-spotted spider mite management programmes	Pest Management Science	2017	N.californicus	25
Kheradmand *et al.*	Biological responses of *Tetranychus urticae* (Acari: Tetranychidae) to sub-lethal concentrations of the entomopathogenic fungus *Beauveria bassiana*	Bulletin of Entomological Research	2021	T.urticae	25
Khodayari & Abedini	Alfalfa responses to drought, salinity, and herbivory by *Tetranychus urticae* (Acari: Tetranychidae) and performance of the pest on water-stressed plants	Persian Journal of Acarology	2022	T.urticae	25
Khodayari *et al.*	The responses of cucumber plants subjected to different salinity or fertilizer concentrations and reproductive success of *Tetranychus urticae* mites on these plants	Experimental and Applied Acarology	2018	T.urticae	25
Khodayari *et al.*	The response of common bean (*Phaseolus vulgaris* L.) to salinity and drought stresses and life table parameters of *Tetranychus urticae* Koch reared on it	Systematic & Applied Acarology	2021	T.urticae	25
Krips *et al.*	Compatibility of host plant resistance and biological control of the two-spotted spider mite *Tetranychus urticae* in the ornamental crop gerbera	Biological Control	1999	P.persimilis	25
Krips *et al.*	Intrinsic rate of population increase of the spider mite *Tetranychus urticae* on the ornamental crop gerbera: Intraspecific variation in host plant and herbivore	Entomologia Experimentalis et Applicata	1998	T.urticae	25
Kumral *et al.*	Biological parameters and population development of *Tetranychus urticae* Koch, 1836 (Acari: Tetranychidae) on different pepper cultivars	Turkiye Entomoloji Dergisi - Turkish Journal of Entomology	2017	T.urticae	27

Continued

Table 16.1. Continued

In-text.citation	Paper Title	Journal.name	Year	Species	Temperature.C
Landeros *et al.*	Functional response and population parameters of *Phytoseiulus persimilis* (Acari: Phytoseiidae) Athias-Henriot on *Tetranychus urticae* Koch	Southwestern Entomologist	2001	P.persimilis	28
Landeros *et al.*	Effect of sublethal concentrations of avermectin on population parameters of *Tetranychus urticae* on strawberry	Southwestern Entomologist	2002	T.urticae	25
Leviticus *et al.*	Lethal and sublethal effects of fluralaner on the two-spotted spider mites, *Tetranychus urticae* Koch (Acari: Tetranychidae)	Pest Management Science	2019	T.urticae	27
Li *et al.*	Sublethal effects of bifenazate on life history and population parameters of *Tetranychus urticae* (Acari: Tetranychidae)	Systematic & Applied Acarology	2017	T.urticae	25
Maleknia *et al.*	How greenhouse cucumber cultivars affect population growth and two-sex life table parameters of *Tetranychus urticae* (Acari: Tetranychidae)	International Journal of Acarology	2015	T.urticae	25
Marcic	The effects of clofentezine on life-table parameters in two-spotted spider mite *Tetranychus urticae*	Experimental and Applied Acarology	2003	T.urticae	24.5, 25.5, 27
Marcic	Sublethal effects of tebufenpyrad on the eggs and immatures of two-spotted spider mite, *Tetranychus urticae*	Experimental and Applied Acarology	2005	T.urticae	27.5
Marcic	Sublethal effects of spirodiclofen on life history and life-table parameters of two-spotted spider mite (*Tetranychus urticae*)	Experimental and Applied Acarology	2007	T.urticae	28.5
Martinez-Villar *et al.*	Effects of azadirachtin on the two-spotted spider mite, *Tetranychus urticae* (Acari : Tetranychidae)	Experimental and Applied Acarology	2005	T.urticae	24

Continued

Table 16.1. Continued

In-text.citation	Paper Title	Journal.name	Year	Species	Temperature.C
Medo *et al.*	Acaricidal activity and sublethal effects of the microbial pesticide spinosad on *Tetranychus urticae* (Acari: Tetranychidae)	Systematic & Applied Acarology	2017	T.urticae	27
Miyazaki *et al.*	Lack of adaptation to a new host in a generalist herbivore: implications for host plant resistance to twospotted spider mites in cotton	Pest Management Science	2014	T.urticae	29.5
Moghadasi *et al.*	Life table and predation capacity of *Phytoseiulus persimilis* Athias-Henriot (Acari: Phytoseiidae) feeding on *Tetranychus urticae* Koch (Acari: Tetranychidae) on rose	Journal of Agricultural Science and Technology	2016	P.persimilis	25
Mohammed *et al.*	Effect of two e.g.yptian cotton varieties on development and life table of the two-spotted spider mite, *Tetranychus urticae* Koch (Acari: Tetranychidae) in relation to leaf chemical contents	e.g.yptian Academic Journal of Biological Sciences	2019	T.urticae	27
Mokhtari *et al.*	Lethal and sublethal impacts of cyflumetofen and bromopropylate on Tetranychus urticae Koch (Acari: Tetranychidae)	Persian Journal of Acarology	2022	T.urticae	25
Mollaloo *et al.*	Sublethal effects of pyridaben on life table parameters of the predatory mite *Neoseiulus californicus* (McGregor) (Acari: Phytoseiidae)	Zoology and Ecology	2018	N.californicus	25
Montazersaheb *et al.*	Effects of plant probiotic bacteria and herbivore-induced plant volatiles on life table parameters of *Tetranychus urticae* (Acari: Tetranychidae) on kidney bean's attached leaves	International Journal of Acarology	2021	T.urticae	25
Motahari *et al.*	The impact of cucumber nitrogen nutrition on life history traits of *Tetranychus urticae* (Koch) (Acari: Tetranychidae)	Acarologia	2013	T.urticae	25

Continued

Table 16.1. Continued

In-text.citation	Paper Title	Journal.name	Year	Species	Temperature.C
Mousavi *et al.*	Sublethal effects of milbemectin on biological parameters of *Amblyseius swirskii* (Acari: Phytoseiidae)	Systematic & Applied Acarology	2022	A.swirskii	25
Musa *et al.*	Acaricidal and sublethal effects of a chenopodium-based biopesticide on the two-spotted spider mite (Acari: Tetranychidae)	Experimental and Applied Acarology	2017	T.urticae	27
Musa *et al.*	Transovarial toxicity matters: lethal and sublethal effects of hexythiazox on the two-spotted spider mite (Acari: Tetranychidae)	Experimental and Applied Acarology	2022	T.urticae	26
Nachappa *et al.*	Response of a complex foraging phenotype to artificial selection on its component traits	Evolutionary Ecology	2010	P.persimilis	24
Nachman & Gotoh	Modeling the effects of constant and variable temperatures on the vital rates of an age-, stage-, and sex-structured population by means of the SANDY approach	Environmental Entomology	2015	T.urticae	22
Najafabadi	Comparative biology and fertility life tables of *Tetranychus urticae* Koch (Acari: Tetranychidae) on different common bean cultivars	International Journal of Acarology	2012	T.urticae	27
Najafabadi *et al.*	Life parameters of *Tetranychus urticae* (Acari: Tetranychidae) on six common bean cultivars	Journal of Economic Entomology	2014	T.urticae	27
Osman *et al.*	The impact of temperature on development and demographic parameters of *Tetranychus urticae* (Koch)	Acarines	2012	T.urticae	15, 20, 25, 30, 35
Osman *et al.*	Biological responses of the two-spotted spider mite, *Tetranychus urticae* to different host plant	Archives of Phytopathology and Plant Protection	2019	T.urticae	25

Continued

Table 16.1. Continued

In-text.citation	Paper Title	Journal.name	Year	Species	Temperature.C
Pietrosiuk *et al.*	Life history parameters of the two-spotted spider mite (*Tetranychus urticae* Koch) feeding on bean leaves treated with pyrrolizidine alkaloids	Journal of Applied Toxicology	2003	T.urticae	23
Puspitarini *et al.*	Host plant variability affects the development and reproduction of *Tetranychus urticae*	International Journal of Acarology	2021	T.urticae	26
Rahimi *et al.*	Impact of constant temperatures on population characteristics of *Amblyseuis swirskii* (Acari: Phytoseiidae)	Systematic & Applied Acarology	2022	A.swirskii	20, 22, 25, 27, 30, 32
Ramzmjou *et al.*	Effect of soybean cultivar on life history parameters of *Tetranychus urticae* Koch (Acari: Tetranychidae)	Journal of Pest Science	2009	T.urticae	25
Ramzmjou *et al.*	Life history traits of *Tetranychus urticae* Koch on three legumes (Acari: Tetranychidae)	Munis Entomology & Zoology	2009	T.urticae	25
Ramzmjou *et al.*	Comparative population growth parameters of the two-spotted spider mite, *Tetranychus urticae* Koch (Acari: Tetranychidae), on different common bean cultivars	Systematic & Applied Acarology	2009	T.urticae	25
Rencken & Pringe	Developmental biology of *Amblyseius californicus* (McGregor) (Acarina: Phytoseiidae), a predator of tetranychid mites, at three temperatures	African Entomology	1998	N.californicus	20, 25, 30
Riahi *et al.*	Population growth of the two-spotted spider mite, *Tetranychus urticae*, on three peach varieties in Iran	Acarologia	2011	T.urticae	27
Riahi *et al.*	Temperature effects on development and life table parameters of *Tetranychus urticae* (Acari: Tetranychidae)	Journal of Agricultural Science and Technology	2013	T.urticae	25, 27, 30, 33

Continued

Table 16.1. Continued

In-text.citation	Paper Title	Journal.name	Year	Species	Temperature.C
Riahi *et al.*	Linking life table and consumption rate of *Amblyseius swirskii* (Acari: Phytoseiidae) in presence and absence of different pollens	Annals of the Entomological Society of America	2017	A.swirskii	25
Riahi *et al.*	Natural diets vs factitious prey: comparative effects on development, fecundity and life table of *Amblyseius swirskii* (Acari: Phytoseiidae)	Systematic & Applied Acarology	2017	A.swirskii	25
Ribeiro *et al.*	Biological aspects of the two-spotted spider mite on strawberry plants under silicon application	Horticultura Brasileira	2021	T.urticae	25
Sa Argolo *et al.*	Compatibility of *Phytoseiulus persimilis* and *Neoseiulus californicus* (Acari: Phytoseiidae) with imidacloprid to manage clementine nursery pests	Crop Protection	2013	P.persimilis, N.californicus	25
Saber *et al.*	Sublethal effects of spirodiclofen, abamectin and pyridaben on life-history traits and life-table parameters of two-spotted spider mite, *Tetranychus urticae* (Acari: Tetranychidae)	Experimental and Applied Acarology	2018	T.urticae	25
Saemi *et al.*	Group-rearing did not affect the life table and predation rate of *Phytoseiulus persimilis* (Acari: Phytoseiidae) fed on *Tetranychus urticae*	Systematic & Applied Acarology	2017	P.persimilis	25
Sáenz-de-Cabezón *et al.*	Influence of sublethal exposure to triflumuron on the biological performance of *Tetranychus urticae* Koch (Acari: Tetranychidae)	Spanish Journal of Agricultural Research	2006	T.urticae	24
Sarbaz *et al.*	Side effects of spiromesifen and spirodiclofen on life table parameters of the predatory mite, *Neoseiulus californicus* McGregor (Acari: Phytoseiidae)	International Journal of Acarology	2017	N.californicus	25

Continued

Table 16.1. Continued

In-text.citation	Paper Title	Journal.name	Year	Species	Temperature.C
Savi *et al.*	Performance of *Tetranychus urticae* (Acari: Tetranychidae) on three hop cultivars (*Humulus lupulus*)	Experimental and Applied Acarology	2021	T.urticae	25
Sedaratian *et al.*	Comparative life table analysis of *Tetranychus urticae* (Acari: Tetranychidae) on 14 soybean genotypes	Insect Science	2011	T.urticae	28
Seiedy *et al.*	Interactions of two natural enemies of *Tetranychus urticae*, the fungal entomopathogen *Beauveria bassiana* and the predatory mite, *Phytoseiulus persimilis*	Biocontrol Science and Technology	2012	P.persimilis	25
Seiedy *et al.*	Development and reproduction of the predatory mite *Amblyseius swirskii* Athias-Henriot (Acari: Phytoseiidae) on *Tetranychus urticae* Koch (Acari: Tetranychidae) and *Bemisia tabaci* Gennadius (Heteroptera: Aleyrodidae)	International Journal of Acarology	2017	A.swirskii	25
Seki	Leaf-morphology-assisted selection for resistance to two-spotted spider mite *Tetranychus urticae* Koch (Acari: Tetranychidae) in carnations (*Dianthus caryophyllus* L)	Pest Management Science	2016	T.urticae	26
Seyed-Talebi *et al.*	Sublethal effects of *Beauveria bassiana* on life table parameters of two-spotted spider mite, *Tetranychus urticae* (Acari: Tetranychidae)	Biocontrol Science and Technology	2012	T.urticae	25
Shang *et al.*	Effects of B-azolemiteacrylic on life-history traits and demographic parameters of two-spotted spider mite, *Tetranychus urticae* (Acari: Tetranychidae)	Experimental and Applied Acarology	2022	T.urticae	25

Continued

Table 16.1. Continued

In-text.citation	Paper Title	Journal.name	Year	Species	Temperature.C
Shoorooei *et al.*	Antixenosis and antibiosis response of common bean (*Phaseolus vulgaris*) to two-spotted spider mite (*Tetranychus urticae*)	Experimental and Applied Acarology	2018	T.urticae	25
Soledad Pascua *et al.*	*Typha angustifolia* L. pollen as an alternative food for the predatory mite Neoseiulus californicus (McGregor) (Acari: Phytoseiidae)	Systematic & Applied Acarology	2020	N.californicus	25
Soltaniyan *et al.*	Suitability of pollen from different plant species as alternative food sources for *Neoseiulus californicus* (Acari: Phytoseiidae) in comparison with a natural prey	Journal of Economic Entomology	2018	N.californicus	25
Sulek & Cakmak	Performance of *Tetranychus urticae* (Acari: Tetranychidae) on six cotton varieties with varying degree of leaf pubescence	Systematic & Applied Acarology	2022	T.urticae	25
Toyoshima & Hinomoto	Intraspecific variation of reproductive characteristics of *Amblyseius californicus* (McGregor) (Acari: Phytoseiidae)	Applied Entomology and Zoology	2004	N.californicus	20
Tuan *et al.*	Survival and reproductive strategies in two-spotted spider mites: Demographic analysis of arrhenotokous parthenogenesis of *Tetranychus urticae* (Acari: Tetranychidae)	Journal of Economic Entomology	2016	T.urticae	25
Uddin *et al.*	Life table parameters of *Tetranychus urticae* Koch (Acari: Tetranychidae) on different bean varieties	African Entomology	2015	T.urticae	25
Uddin *et al.*	Life table parameters of an indigenous strain of *Neoseiulus californicus* McGregor (Acari: Phytoseiidae) when fed *Tetranychus urticae* Koch (Acari: Tetranychidae)	Entomological Research	2017	N.californicus	20, 25, 30

Continued

Table 16.1. Continued

In-text.citation	Paper Title	Journal.name	Year	Species	Temperature.C
Vangansbeke *et al.*	Alternating temperatures affect life table parameters of *Phytoseiulus persimilis*, *Neoseiulus californicus* (Acari: Phytoseiidae) and their prey *Tetranychus urticae* (Acari: Tetranychidae)	Experimental and Applied Acarology	2013	P.persimilis, N.californicus, T.urticae	15, 20
Wang *et al.*	Effects of sublethal concentrations of bifenthrin on the two-spotted spider mite, *Tetranychus urticae* (Acari: Tetranychidae)	Systematic & Applied Acarology	2014	T.urticae	26
Wang *et al.*	Sublethal effects of spinetoram on the two-spotted spider mite, *Tetranychus urticae* (Acari: Tetranychidae)	Pesticide Biochemistry and Physiology	2016	T.urticae	26
Xu *et al.*	The performance of *Tetranychus urticae* on five melon cultivars is correlated with leaf thickness	Systematic & Applied Acarology	2019	T.urticae	26
Yang *et al.*	Effect of photoperiods on development and acaricide susceptibility in the two-spotted spider mite, *Tetranychus urticae*	Experimental and Applied Acarology	2020	T.urticae	25

conclusions. Broadly, we found that TSSM individuals can persist under temperatures that exceed 30°C.

Fluctuating temperature regimes for these mites may provide further insights than constant temperatures alone when evaluating how warming affects mite population growth due to the non-linear relationship between temperature and ectotherm performance. For example, Bayu *et al.* (2017) investigated how constant and fluctuating temperatures affected TSSM across an average temperature gradient of 15–35°C, with temperatures fluctuating by 10°C. They found that r_m was lower when mites were maintained at 15°C than at fluctuating temperatures of 10–20°C. Interestingly, r_m was not significantly different between constant and fluctuating temperature treatments at the same mean temperature of 20°C, 25°C, 30°C and 35°C. However, Gotoh *et al.* (2014) found that fluctuating temperatures (varying from 0°C to 12°C in steps of 3°C, with an average temperature of 22°C) increased r_m for TSSM in most cases. Specifically, they found that temperature fluctuations of 3°C, 6°C and 9°C, but not 12°C, increased TSSM population growth rate due to shorter preoviposition periods at 3°C, 6°C and 9°C fluctuations. These results suggest that fluctuating temperatures can, at times, benefit, hinder or have minimal effect on TSSM population growth.

For phytoseiid predators, Vangansbeke *et al.* (2013) found that *P. persimilis* and *N. californicus* (but also TSSM) had greater r_m under fluctuating temperatures, relative to constant temperatures (15°C constant temperatures vs 5–20°C temperature regime with mean of 15°C). Notably, they found that temperature fluctuations

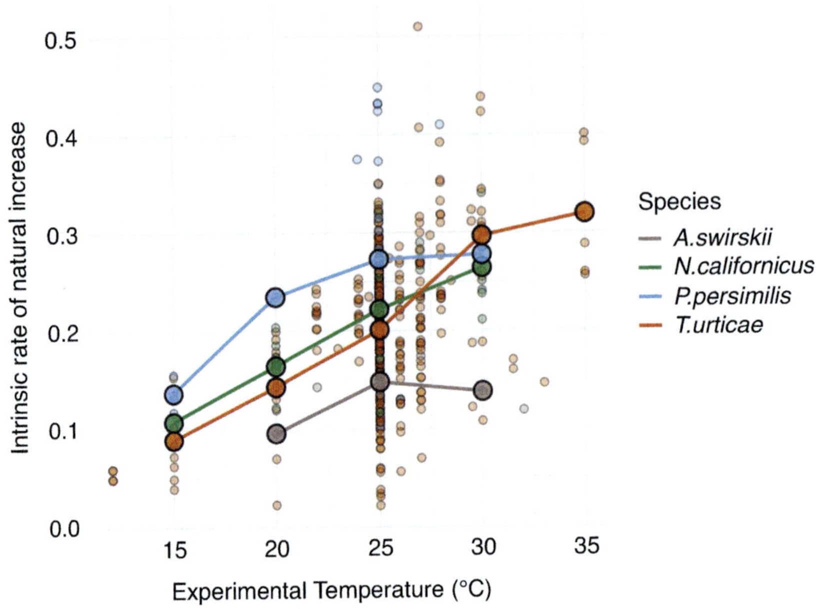

Fig. 16.1. Extracted intrinsic rate of natural increase (r_m) values from candidate studies with *Amblyseius swirskii*, *Neoseiulus californicus*, *Phytoseiulus persimilis* and *Tetranychus urticae* (TSSM) as test species. Average experimental temperatures ranged from 12°C to 35°C with most measurements occurring at 25°C (55.9% of 445 measurements). Small points represent extracted data values, while large points are average values at 15°C, 20°C, 25°C and 35°C of rate of natural increase for each species. Due to limited values at a given temperature, an 'average' may be represented by a single data point. Averages were calculated for *A. swirskii* at 20°C (1, number of r_m values), 25°C (12) and 30°C (1), for *N. californicus* at 15°C (2), 20°C (10), 25°C (28) and 30°C (7), for *P. persimilis* at 15°C (2), 20°C (1), 25°C (30) and 30°C (2), and for *T. urticae* at 15°C (9), 20°C (12), 25°C (179), 30°C (12) and 35°C (5).

increased average r_m more for predators than TSSM. These results stem from the non-linear relationship between temperature and ectotherm performance, as exposure to cooler or warmer temperatures can increase overall performance relative to the average constant temperature (Paaijmans *et al.*, 2013; Kingsolver *et al.*, 2015; Marshall *et al.*, 2021). As a result, climate warming along with increased temperature variability may result in novel outcomes, i.e. high temperature variability at cool temperatures may benefit ectotherms but be detrimental at warm temperatures (Kingsolver *et al.*, 2015). In all, further research is needed to investigate how varying temperature conditions will affect the population growth of phytoseiid predators and TSSM.

On average, we found that *P. persimilis* and *N. californicus* have greater population growth than TSSM at 25°C (see Fig. 16.1), but at 30°C, the average population growth of TSSM was higher than all predators. This indicates that under warmer temperatures, TSSM population growth begins to outpace predators. It is worth noting that Ali *et al.* (2013) studied TSSM with *P. persimilis* and *N. californicus* population growth concurrently at 30°C and found that phytoseiid predator population growth was typically greater than that of TSSM. However, the average population growth of TSSM across our pooled studies was higher at 30°C than to the two phytoseiid species. Various studies have also assessed the functional response of *P. persimilis*, *N. californicus* and *A. swirskii* when feeding on TSSM across a temperature gradient. Unsurprisingly, they found that temperature increases predator feeding rate until~25°C–35°C (depending on the species), whereafter feeding rate declines (Everson, 1980; Skirvin and Fenlon, 2003; Ahn *et al.*, 2010; Farazmand and Amir-Maafi, 2021).

Only a single study in our review measured predator feeding rate on TSSM alongside population growth. El-Nahas *et al.* (2020) found that *N. californicus* feeding rate on TSSM increased when temperatures changed from 20 to 30°C by both females and males on TSSM immatures. Based on previous feeding trials with these predators, higher predator feeding and attack rates under warming may potentially compensate for increased TSSM population growth with warming. Together, understanding the long-term impact of increasing temperatures on predator–prey populations (measured as abundances) for TSSM when predators are present requires further research. Similar studies include Meehan and Lindo (2023) and Thakur *et al.* (2017, 2018) who used Laelapidae predator mites (Acari: Parasitiformes: Mesostigmata) and springtail prey (Hexapoda: Collembola) to explore the effects of predation on prey populations across temperature gradients, where they have found non-linear relationships between temperature and predator–prey population size. Although r_m is solely measured for TSSM when predators are not present to gauge *maximum* population growth, this chapter (and studies within) have shown that temperature will affect phytoseiid predators and prey performance, potentially leading to novel predator–prey outcomes under higher temperatures.

Amblyseius swirskii had the lowest r_m out of the three predators (see Fig. 16.1). This can probably be attributed to *A. swirskii* being a generalist predator species (from the Type III lifestyle) as McMurtry *et al.* (2013) discussed their struggle with manoeuvering through webbing produced by TSSM. From our pool, one study investigated the effects of different temperatures on *A. swirskii*: Rahimi *et al.* (2022), who found that r_m was highest at 27°C. The authors pointed to limitations of *A. swirskii* as a biocontrol agent at higher temperatures due to an increased preoviposition period under warming. Furthermore, Kadkhodazadeh *et al.* (2021) observed mixed results on r_m when offering *A. swirskii* TSSM and seven pollen diets; the predator had higher r_m when reared on date palm, oak, pistachio and pomegranate compared to TSSM, which illustrates that TSSM is not the optimal prey for *A. swirskii* population growth. Past studies have shown that *A. swirskii* preferred other arthropod prey, like thrips, to TSSM (Xu and Enkegaard, 2010) as the reported thermal optimum of population growth for *A. swirskii* was higher when feeding on cattail pollen (*Typha latifolia* L.) than TSSM, here approximated by Rahimi *et al.* (2022) at 30.1°C (Lee and Gillespie, 2011). Having the lowest r_m of the three predators is not a positive aspect of *A. swirskii* in this scenario, with implications for using this species to control TSSM, as opposed to other biocontrol agents. As generalist predators such as *A. swirskii* also have difficulty manoeuvering through webbing produced by *Tetranychus* species, our findings suggest that *A. swirskii* should not be a primary predator for TSSM, particularly under warmer temperatures where population growth of TSSM greatly exceeds that of *A. swirskii*.

16.3.2 Influence of diet on r_m for predators and prey

We observed that resource type greatly affected population growth of both predators and prey and is particularly noticeable when looking at the range of r_m values for TSSM at 25°C (see Fig. 16.1). Azadi-Qoort *et al.* (2019), for instance, measured population growth of TSSM reared on cucumber, tomato, aubergine and watermelon and found its r_m was more than two times greater on mites feeding on cucumber than when reared on tomato plants. More specifically, Ibrahim *et al.* (2016) investigated TSSM rearing on different varieties of melon and found significantly higher r_m for TSSM when feeding on Sudanian melon compared to Musk and Snake melon. Other examples of indirect changes in food quality on r_m of TSSM include Hosseini *et al.* (2022), who reared TSSM on strawberry plants with different free-living plant growth-promoting rhizobacteria which negatively affected the life history and population dynamics of TSSM. Similarly, Hoffmann *et al.* (2009) observed an increase in r_m of TSSM reared on bean plants with arbuscular mycorrhizal symbiosis compared to bean plants without it. For predators, *N. californicus* displayed higher r_m when feeding on almond and maize pollen compared to other types of pollen; both almond and maize had the highest levels of protein and sucrose, making them better nutritional food sources for *N. californicus* (Khanamani *et al.*, 2017).

Differences in food quality for TSSM can also subsequently affect phytoseiid population growth rate. Ali *et al.* (2013) showed this as watermelon cultivar type strongly affected r_m of TSSM, alongside *P. persimilis* and *N. californicus* population growth when maintained on TSSM. Consistently, however, a temperature increase from 25°C to 30°C led to higher population growth for all three species across both cultivar types (Ali *et al.*, 2013). Furthermore, due to shifting metabolic and physiological demands with climate change, macromolecule intake (i.e. proteins, lipids, carbohydrates) for both phytoseiid mites and TSSM may change as well (for examples of such phenomena, see Lee *et al.*, 2015; Schmitz *et al.*, 2016; Zhang *et al.*, 2020). Because diet appears critical in shaping population growth (especially in TSSM), understanding how predator and TSSM diets may shift under climate change is necessary to predict their interactions in a warmer world.

16.4 Future Directions and Conclusions

Most studies calculated r_m using formulas derived from Birch (1948) along with Chi and Liu (1985) and Chi (1988). As noted above, differences between formulas stem from the inclusion of male demographics in measuring population growth for Chi and Liu (1985), as opposed to females exclusively for Birch (1948). Amongst our collection of studies, few studies used both equations to gauge population growth for their study species. Khanamani *et al.* (2013) tested the effects of seven aubergine cultivars on TSSM population growth. Across all seven cultivars, the female age-specific formula (Birch) had higher population growth for TSSM than the age-stage, two-sex formula (Chi and Liu, 1985), with the relative difference between r_m values being conditional on cultivar type. In our review we did not separate studies based on formula, but when comparing r_m between studies, one should consider, amongst other variables, how population growth was measured, even between seemingly similar methods.

In a future, warmer climate where the r_m of the prey is higher than that of its predators, predators released as biocontrol agents outside temperature-controlled facilities may not be efficient. It has been documented that TSSM CT_{max} temperature exceeds that of its predators (Coombs and Bale, 2013) while we found that the average population growth of TSSM based on r_m from extracted studies exceeds that of their predators at warmer temperatures (+30°C). More research is needed to conclusively identify the environmental temperatures in which TSSM population growth exceeds their predators. But in general, strategies will need to be developed to counteract higher population growth of TSSM if predators cannot control prey populations at higher temperatures. Alongside this, one could keep the temperature of the facilities at a level where predator, but not prey, populations are at maximum biocontrol efficacy. Incorporating temperature fluctuations to internal set-ups that limit TSSM population growth and maximize population growth and feeding of the predators may be beneficial. Overall, any biocontrol strategies to mitigate TSSM population will need to consider pest species, biocontrol agent and greenhouse plants and crops.

Phytoseiid mites are critically important biocontrol agents within agro-ecosystems (Tixier, 2018). As pest species, such as TSSM, become more resistant to pesticides (Van Leeuwen *et al.*, 2010; Xu *et al.*, 2018), these predators will potentially be more necessary to limit TSSM outbreaks.

Analysing 445 r_m measurements for three phytoseiid predators and TSSM from 141 studies pointed us to more knowledge gaps than answers. Specifically, understanding how r_m responds to temperatures other than 25°C, including males in life tables (Chi and Liu, 1985; Chi, 1988; El-Nahas *et al.*, 2020), and measuring predator feeding rates under warming using experimental arenas (like those of Barroso *et al.* (2023) and Ramachandran *et al.* (2021) who used other mesostigmatid mite species) are some examples of where data are lacking – here in this review and otherwise most current studies focus on the impact of constant temperature ranging from 25°C to 30°C. Nonetheless, from our collection of studies, we found that TSSM population growth may begin to outpace that of their predators under warming, specifically at temperatures greater than 30°C, where we detected higher r_m for TSSM, albeit with limited data. Unless higher predator feeding rates at warmer temperatures compensate for greater TSSM population growth, the efficacy of these species as biocontrol agents will probably be diminished, with large-scale economic consequences.

Ultimately, we call for studies using more realistic scenarios for these mites (e.g. Bayu *et al.*, 2017), for instance, by using fluctuating temperatures as treatments due to the non-linear relationship between temperature and ectotherm performance, or by considering asymmetrical day and night temperatures, extreme events and varying durations of heat events. In addition, incorporating diurnal temperature fluctuations, along with extreme warming and climate events (e.g. prolonged and intense droughts or precipitation) in multifactorial experiments will allow for better prediction of changes in species interactions. This is critical as studies have found interactive effects between drought and elevated CO_2 (Sinaie *et al.*, 2019) while both rainfall and humidity are strongly correlated with TSSM population size (Rao *et al.*, 2018). Together, this requires standard experimental designs to reduce confounding effects between studies, such as differences in feeding resources or r_m calculations. In addition, studies should closely examine how the nutritional or macromolecule content of food (e.g. proteins, lipids, carbohydrates) impact mite population growth. Finally, we recommend including both phytoseiid predators and TSSM within studies when investigating how temperature affects population growth of both predators and prey. Only then will our ability to conclude how climate change will affect the population growth rate of predators and TSSM be realized.

Acknowledgment

We would like to thank Alexandra Revynthi (University of Florida) for providing comments on an early version of this chapter.

References

Abd El-Wahed, N.M. and El-Halawany, A.S. (2012) Effect of temperature degrees on the biology and life table parameters of *Tetranychus urticae* Koch on two pear varieties. *Egyptian Academic Journal of Biological Sciences, B. Zoology* 4, 103–109.

Ahn, J.J., Kim, K.W. and Lee, J.-H. (2010) Functional response of *Neoseiulus californicus* (Acari: Phytoseiidae) to *Tetranychus urticae* (Acari: Tetranychidae) on strawberry leaves. *Journal of Applied Entomology* 134, 98–104.

Ali, F.S., Afifi, A.M., El-Saiedy, E.M.A. and Ahmed, M.M. (2013) Biology and life table parameters of *Tetranychus urticae* Koch (Acari: Tetranychidae) and two phytoseiid predatory mites on two watermelon cultivars. *Journal of the Egyptian Society of Acarology* 7, 25–30.

Amarasekare, P. and Coutinho, R.M. (2013) The intrinsic growth rate as a predictor of population viability under climate warming. *Journal of Animal Ecology* 82, 1240–1253.

Angilletta, M.J., Jr (2009) *Thermal Adaptation: A Theoretical and Empirical Synthesis*. Oxford University Press, Oxford, UK.

Azadi-Qoort, A., Sedaratian-Jahromi, A., Haghani, M. and Ghane-Jahromi, M. (2019) Biological responses of *Tetranychus urticae* (Acari: Tetranychidae) to different host plants: An investigation on bottom-up effects. *Systematic and Applied Acarology* 24, 659–674.

Barroso, G., Barreto, C., Padilha, A.C., Pazini, J.B., Santos, J.C. *et al.* (2023) The double-edged sword effect of warming on the soil predatory mite *Cosmolaelaps brevistilis* (Karg, 1978) (*Mesostigmata: Laelapidae*). *Entomological Communications* 5, ec05024.

Bayu, M.S.Y., Ullah, M.S., Takano, Y. and Gotoh, T. (2017) Impact of constant versus fluctuating temperatures on the development and life history parameters of *Tetranychus urticae* (Acari: Tetranychidae). *Experimental and Applied Acarology* 72, 205–227.

Birch, L.C. (1948) The intrinsic rate of natural increase of an insect population. *Journal of Animal Ecology* 17, 15–26.

Bolland, H.R., Gutierrez, J. and Flechtmann, C.H. (1998) *World Catalogue of the Spider Mite Family(Acari: Tetranychidae)*. Koninklijke Brill NV, Leiden, The Netherlands.

Brown, J.H., Gillooly, J.F., Allen, A.P., Savage, V.M. and West, G.B. (2004) Toward a metabolic theory of ecology. *Ecology* 85, 1771–1789.

Caughley, G. and Birch, L.C. (1971) Rate of increase. *Journal of Wildlife Management* 35, 658–663.

Chi, H. (1988) Life-table analysis incorporating both sexes and variable development rates among individuals. *Environmental Entomology* 17, 26–34.

Chi, H. and Liu, H. (1985) Two new methods for the study of insect population ecology. *Bulletin of the Institute of Zoology, Academia Sinica* 24, 225–240.

Clarke, A. (2017) *Principles of Thermal Ecology: Temperature, Energy and Life*. Oxford University Press, Oxford, UK.

Cock, M.J.W., Lenteren, J.C., Brodeur, J., Barratt, B.I.P., Bigler, F. *et al*. (2010) Do new access and benefit sharing procedures under the Convention on Biological Diversity threaten the future of biological control? *BioControl* 55, 199–218.

Coombs, M.R. and Bale, J.S. (2013) Comparison of thermal activity thresholds of the spider mite predators *Phytoseiulus macropilis* and *Phytoseiulus persimilis* (Acari: Phytoseiidae). *Experimental and Applied Acarology* 59, 435–445.

Cortés, E. (2016) Perspectives on the intrinsic rate of population growth. *Methods in Ecology and Evolution* 7, 1136–1145.

De Moraes, G.J., McMurtry, J.A., Denmark, H.A. and Campos, C.B. (2004) A revised catalog of the mite family phytoseiidae. *Zootaxa* 434, 1–494.

Demite, P.R., Moraes, G.J., McMurtry, J.A., Denmark, H.A. and Castilho, R.C. (2023) Phytoseiidae Database: A website for taxonomic and distributional information on phytoseiid mites (Acari). *Zootaxa* 3795, 571–577.

Deutsch, C.A., Tewksbury, J.J., Huey, R.B., Sheldon, K.S., Ghalambor, C.K. *et al*. (2008) Impacts of climate warming on terrestrial ectotherms across latitude. *Proceedings of the National Academy of Sciences USA* 105, 6668–6672.

Döker, İ., Revynthi, A.M., Mannion, C. and Carrillo, D. (2020) First report of acaricide resistance in *Tetranychus urticae* (Acari: Tetranychidae) from south Florida. *Systematic and Applied Acarology* 25, 1209–1214.

Döker, İ., Revynthi, A.M., Kazak, C. and Carrillo, D. (2021) Interactions among exotic and native phytoseiids (Acari: Phytoseiidae) affect biocontrol of two-spotted spider mite on papaya. *Biological Control* 163, 104758.

El-Nahas, R.A., Fouly, A.H., Yousef, A.E.-T.A. and Khalil, A.H. (2020) Thermal requirements and life table parameters of the predatory mite, *Neoseiulus californicus* (McGregor) fed on *Tetranychus urticae* Koch (Acari: Phytoseiidae: Tetranychidae). *Journal of the Egyptian Society of Acarology* 37–44.

Escudero, L.A. and Ferragut, F. (2005) Life-history of predatory mites *Neoseiulus californicus* and *Phytoseiulus persimilis* (Acari: Phytoseiidae) on four spider mite species as prey, with special reference to *Tetranychus evansi* (Acari: Tetranychidae). *Biological Control* 32, 378–384.

Everson, P. (1980) The relative activity and functional response of *phytoseiulus persimilis* (Acarina: Phytoseiidae) and *Tetranychus urticae* (Acarina: Tetranychidae): The effect of temperature. *Canadian Entomologist* 112, 17–24.

Farazmand, A. and Amir-Maafi, M. (2021) Use of functional response modeling to evaluate the effect of temperature on predation of *Amblyseius swirskii* (Acari: Phytoseiidae) adults preying on *Tetranychus urticae* (Acari: Tetranychidae) nymphs. *Journal of Economic Entomology* 114, 2271–2276.

Gillooly, J.F., Brown, J.H., West, G.B., Savage, V.M. and Charnov, E.L. (2001) Effects of size and temperature on metabolic rate. *Science* 293, 2248–2251.

Gotoh, T., Moriya, D. and Nachman, G. (2015) Development and reproduction of five *Tetranychus* species (Acari: Tetranychidae): Do they all have the potential to become major pests? *Experimental and Applied Acarology* 66, 453–479.

Gotoh, T., Saito, M., Suzuki, A. and Nachman, G. (2014) Effects of constant and variable temperatures on development and reproduction of the two-spotted spider mite *Tetranychus urticae* (Acari: Tetranychidae). *Experimental and Applied Acarology* 64, 465–478.

Grbić, M., Khila, A., Lee, K.-Z., Bjelica, A., Grbić, V. *et al.* (2007) Mity model: *Tetranychus urticae*, a candidate for chelicerate model organism. *BioEssays* 29, 489–496.

Grbić, M., Van Leeuwen, T., Clark, R.M., Rombauts, S., Rouzé, P. *et al.* (2011) The genome of *Tetranychus urticae* reveals herbivorous pest adaptations. *Nature* 479, 487–492.

Hoffmann, D., Vierheilig, H., Riegler, P. and Schausberger, P. (2009) Arbuscular mycorrhizal symbiosis increases host plant acceptance and population growth rates of the two-spotted spider mite *Tetranychus urticae*. *Oecologia* 158, 663–671.

Hosseini, A., Hosseini, M. and Schausberger, P. (2022) Plant growth-promoting rhizobacteria enhance defense of strawberry plants against spider mites. *Frontiers in Plant Science* 12, 783578.

Howe, R.W. (1953) The rapid determination of the intrinsic rate of increase of an insect population. *Annals of Applied Biology* 134–151.

Ibrahim, M.M.S., El-Esnawy, B.A. and El-Adawy, A.M. (2016) Biological and ecological fitness of the two-spotted spider mite, *Tetranychus urticae* Koch (Acari: Tetranychidae) on certain host plants. *Journal of the Egyptian Society of Acarology* 10, 31–35.

Kadkhodazadeh, F., Asadi, M. and Khanamani, M. (2021) Suitability of different pollen grains and *Tetranychus urticae* as food for the predatory mite, *Amblyseius swirskii* (Acari: Phytoseiidae). *Persian Journal of Acarology* 10, 321–334.

Khanamani, M., Fathipour, Y. and Hajiqanbar, H. (2013) Population growth response of *Tetranychus urticae* to eggplant quality: Application of female age-specific and age-stage, two-sex life tables. *International Journal of Acarology* 39, 638–648.

Khanamani, M., Fathipour, Y., Asghar Talebi, A. and Mehrabadi, M. (2017) Linking pollen quality and performance of *Neoseiulus californicus* (Acari: Phytoseiidae) in two-spotted spider mite management programmes. *Pest Management Science* 73, 452–461.

Kingsolver, J.G., Higgins, J.K. and Augustine, K.E. (2015) Fluctuating temperatures and ectotherm growth: Distinguishing non-linear and time-dependent effects. *Journal of Experimental Biology* 218, 2218–2225.

Laing, J.E. (1968) Life history and life table of *Phytoseiulus persimilis* Athias-Henriot. *Acarologia* 10, 578–588.

Le Hesran, S., Groot, T., Knapp, M., Bukovinszky, T., Forestier, T. *et al.* (2019) Phenotypic variation in egg survival in the predatory mite *Phytoseiulus persimilis* under dry conditions. *Biological Control* 130, 88–94.

Le Hesran, S., Groot, T., Knapp, M., Bukovinszky, T., Nugroho, J.E. *et al.* (2020) Maternal effect determines drought resistance of eggs in the predatory mite *Phytoseiulus persimilis*. *Oecologia* 192, 29–41.

Lee, H.-S. and Gillespie, D.R. (2011) Life tables and development of *Amblyseius swirskii* (Acari: Phytoseiidae) at different temperatures. *Experimental and Applied Acarology* 53, 17–27.

Lee, K.P., Jang, T., Ravzanaadii, N. and Rho, M.S. (2015) Macronutrient balance modulates the temperature-size rule in an ectotherm. *American Naturalist* 186, 212–222.

Li, W.-Z., Zhu, T., Li, H.-L. and Shang, S.-Q. (2022) The effects of short-term heat stress on functional response of *Neoseiulus barkeri* to *Tetranychus urticae*. *Journal of Applied Entomology* 146, 310–318.

Litskas, V.D., Migeon, A., Navajas, M., Tixier, M.-S. and Stavrinides, M.C. (2019) Impacts of climate change on tomato, a notorious pest and its natural enemy: Small scale agriculture at higher risk. *Environmental Research Letters* 14, 084041.

Liu, S., Lv, J., Wang, E. and Xu, X. (2017) Life-style classification of some *Phytoseiidae* (Acari: Mesostigmata) species based on gnathosoma morphometrics. *Systematic and Applied Acarology* 22, 629–639.

Marshall, K.E., Anderson, K.M., Brown, N.E.M., Dytnerski, J.K., Flynn, K.L. *et al.* (2021) Whole-organism responses to constant temperatures do not predict responses to variable temperatures in the ecosystem engineer *Mytilus trossulus*. *Proceedings of the Royal Society B Biological Sciences* 288, 20202968.

McMurtry, J.A., Moraes, G.J. and Sourassou, N.F. (2013) Revision of the lifestyles of phytoseiid mites (Acari:Phytoseiidae) and implications for biological control strategies. *Systematic and Applied Acarology* 18, 297–320.

Meehan, M.L. and Lindo, Z. (2023) Mismatches in thermal performance between ectothermic predators and prey alter interaction strength and top-down control. *Oecologia* 201, 1005–1015.

Migeon, A., Nouguier, E. and Dorkeld, F. (2010) Spider mites web: A comprehensive database for the Tetranychidae. In: Sabelis, M.W. and Bruin, J. (eds) *Trends in Acarology: Proceedings of the 12th International Congress*, Springer, pp. 557–560.

Navajas, M. (1998) Host plant associations in the spider mite *Tetranychus urticae* (Acari: Tetranychidae): Insights from molecular phylogeography. *Experimental and Applied Acarology* 22, 201–214.

Osman, M.A., Tawfik, A.A. and Abou-Elella, G.M. (2012) The impact of temperature on development and demographic parameters of *Tetranychus urticae* (Koch). *Journal of the Egyptian Society of Acarology* 6, 25–30.

Paaijmans, K.P., Heinig, R.L., Seliga, R.A., Blanford, J.I., Blanford, S. *et al.* (2013) Temperature variation makes ectotherms more sensitive to climate change. *Global Change Biology* 19, 2373–2380.

Rahimi, A., Moradi, F., Sadeghi, A., Fathipour, Y. and Maroufpoor, M. (2022) Impact of constant temperatures on population characteristics of *Amblyseius swirskii* (Acari: Phytoseiidae). *Systematic and Applied Acarology* 27, 1775–1786.

Ramachandran, D., Lindo, Z. and Meehan, M.L. (2021) Feeding rate and efficiency in an apex soil predator exposed to short-term temperature changes. *Basic and Applied Ecology* 50, 87–96.

Rao, K.S., Vishnupriya, R., Ramaraju, K. and Poornima, K. (2018) Effect of abiotic factors on the population dynamics of two spotted spider mite, *Tetranychus urticae* Koch and its predatory mite, *Neoseiulus longispinosus* (Evans) in brinjal ecosystem. *Journal of Experimental Zoology India* 21, 797–800.

Reddy, G.V.P. and Baskaran, P. (2006) Damage potential of the spider mite *Tetranychus ludeni* (Acari: Tetranychidae) on four varieties of eggplant. *International Journal of Tropical Insect Science* 26, 48–56.

Riahi, E., Fathipour, Y., Talebi, A.A. and Mehrabadi, M. (2017) Linking life table and consumption rate of *Amblyseius swirskii* (Acari: Phytoseiidae) in presence and absence of different pollens. *Annals of the Entomological Society of America* 110, 244–253.

Savage, V.M., Gillooly, J.F., Brown, J.H., West, G.B. and Charnov, E.L. (2004) Effects of body size and temperature on population growth. *American Naturalist* 163, 429–441.

Schmitz, O.J., Rosenblatt, A.E. and Smylie, M. (2016) Temperature dependence of predation stress and the nutritional ecology of a generalist herbivore. *Ecology* 97, 3119–3130.

Seiedy, M., Soleymani, S. and Hakimitabar, M. (2017) Development and reproduction of the predatory mite *Amblyseius swirskii* Athias-Henriot (Acari: Phytoseiidae) on *Tetranychus urticae* Koch (Acari: Tetranychidae) and *Bemisia tabaci* Gennadius (Heteroptera: Aleyrodidae). *International Journal of Acarology* 43, 160–164.

Silva, E.A., Reis, P.R., Carvalho, T.M.B. and Altoé, B.F. (2009) *Tetranychus urticae* (Acari: Tetranychidae) on *Gerbera jamesonii* Bolus and Hook (Asteraceae). *Brazilian Journal of Biology* 69, 1121–1125.

Sinaie, S., Sadeghi-Namaghi, H. and Fekrat, L. (2019) Effects of elevated CO_2 and water stress on population growth of the two-spotted spider mite, *Tetranychus urticae* Koch (Acari: Tetranychidae), on sweet pepper under environmentally controlled conditions. *Journal of Asia-Pacific Entomology* 22, 96–102.

Skirvin, D.J. and Fenlon, J.S. (2003) The effect of temperature on the functional response of *Phytoseiulus persimilis* (Acari: Phytoseiidae). *Experimental and Applied Acarology* 31, 37–49.

Smith, L.B. (1965) The intrinsic rate of natural increase of *Cryptolestes ferrugineus* (Stephens) (Coleoptera, Cucujidae). *Journal of Stored Products Research* 1, 35–49.

Thakur, M.P., Künne, T., Griffin, J.N. and Eisenhauer, N. (2017) Warming magnifies predation and reduces prey coexistence in a model litter arthropod system. *Proceedings of the Royal Society B Biological Sciences* 284, 20162570.

Thakur, M.P., Griffin, J.N., Künne, T., Dunker, S., Fanesi, A. *et al.* (2018) Temperature effects on prey and basal resources exceed that of predators in an experimental community. *Ecology and Evolution* 8, 12670–12680.

Tixier, M.S. (2018) Predatory mites (Acari: Phytoseiidae) in agro-ecosystems and conservation biological control: A review and explorative approach for forecasting plant-predatory mite interactions and mite dispersal. *Frontiers in Ecology and Evolution* 6, 192.

Tscholl, T., Nachman, G., Spangl, B. and Walzer, A. (2022) Heat waves affect prey and predators differently via developmental plasticity: Who may benefit most from global warming? *Pest Management Science* 78, 1099–1108.

Van Lenteren, J.C. (2012) The state of commercial augmentative biological control: Plenty of natural enemies, but a frustrating lack of uptake. *BioControl* 57, 1–20.

Vangansbeke, D., de Schrijver, L., Spranghers, T., Audenaert, J., Verhoeven, R. *et al.* (2013) Alternating temperatures affect life table parameters of *Phytoseiulus persimilis*, *Neoseiulus californicus* (Acari: Phytoseiidae) and their prey *Tetranychus urticae* (Acari: Tetranychidae). *Experimental and Applied Acarology* 61, 285–298.

Van Leeuwen, T., Vontas, J., Tsagkarakou, A., Dermauw, W. and Tirry, L. (2010) Acaricide resistance mechanisms in the two-spotted spider mite *Tetranychus urticae* and other important Acari: A review. *Insect Biochemistry and Molecular Biology* 40, 563–572.

Vasseur, D.A., Delong, J.P., Gilbert, B., Greig, H.S., Harley, C.D.G. *et al.* (2014) Increased temperature variation poses a greater risk to species than climate warming. *Proceedings of the Royal Society B* 281, 20132612.

Walter, D.E. and Proctor, H.C. (2013) *Mites: Ecology, Evolution and Behavior: Life at a Microscale*. Springer, New York.

Weihrauch, F. (2004) A new monitoring approach for the two-spotted spider mite *Tetranychus urticae* Koch (Acari: Tetranychidae) in hop culture. *Journal of Plant Diseases and Protection* 111, 197–205.

Williams, M.R. (2012) Cotton insect losses – 2011. In: *Proceedings of the 2012 Beltwide Cotton Conferences*. Available at: www.cotton.org/beltwide/proceedings/2005-2022/data/conferences/2012/papers/13657.pdf

Xu, X. and Enkegaard, A. (2010) Prey preference of the predatory mite, *Amblyseius swirskii* between first instar western flower thrips *Frankliniella occidentalis* and nymphs of the two-spotted spider mite *Tetranychus urticae*. *Journal of Insect Science* 10, 149.

Xu, D., He, Y., Zhang, Y., Xie, W., Wu, Q. *et al.* (2018) Status of pesticide resistance and associated mutations in the two-spotted spider mite, *Tetranychus urticae*, in China. *Pesticide Biochemistry and Physiology* 150, 89–96.

Zhang, P., Van Leeuwen, C.H.A., Bogers, D., Poelman, M., Xu, J. *et al.* (2020) Ectothermic omnivores increase herbivory in response to rising temperature. *Oikos* 129, 1028–1039.

17 Controlling Plant Disease with Plant Growth-Promoting Endophytes in the Context of Climate Change

Lorena Barra-Bucarei[1]*, Javiera Ortiz-Campos[1] and Hanna Cáceres Iparraguirre[2]

[1]Instituto de Investigaciones Agropecuarias, Chillán, Chile; [2]Universidad Autónoma de Ica, Chincha Alta, Perú

Abstract

Agriculture currently requires alternatives to synthetic chemicals for the phytosanitary and nutrient management of crops. There is growing consumer demand for products free of potentially toxic residues. Bioinputs produced from micro-organisms offer a concrete alternative that can replace conventional practices and advance sustainable agriculture. It has been widely demonstrated that micro-organisms support plants when they face with biotic and abiotic stresses. Within this group, endophytic micro-organisms are especially important because they internally colonize plants and perform multiple functions on their behalf.

This chapter demonstrates the strategic importance of endophytic micro-organisms in the nutrient and phytosanitary management of plants in the context of climate change, where environmental conditions are rapidly changing and crops have lost resilience due to conventional management. In addition, the diverse positive effects of endophytes on plant welfare and consequently on crop yield and quality are outlined.

17.1 Introduction

For many years, conventional agriculture has managed to increase productivity and, consequently, crop profitability, responding to the urgent task of feeding a growing world population. However, this type of farming has not adequately considered social and environmental concerns. Recently, synthetic agrochemicals that pollute the soil, water sources and air have been used intensively, causing environmental imbalances such as pest invasions due to insects and plant pathogen outbreaks due to climate change (Carvalho, 2017; Meena *et al.*, 2020).

Global climate has changed since preindustrial times; however, during the last few decades atmospheric CO_2 has increased by almost 30% and the temperature has increased by 0.3–0.6°C (Canadell *et al.*, 2007). While the increase in atmospheric CO_2 concentrations could boost plant growth and photosynthesis, increasing plant biomass and possibly improving plant water relations (Rajput *et al.*, 2021), warming also produces uncertainty regarding the consequences for crops, both at the phytosanitary and nutritional levels, and its impact on prim *Beauvericin* ary production in agricultural systems. The changing climate is most probably associated with changing

*Corresponding author: barra.bucarei@gmail.com

DOI: 10.1079/9781800625099.0017

patterns in climate variability, including extreme weather events such as precipitation anomalies and greater temperature variations. Regarding phytosanitation, climate change is expected to affect three areas: losses caused directly by plant diseases, the effectiveness of disease management strategies and the geographical distribution of plant diseases (Manoharachary and Kunwar, 2014).

Changes in average temperatures and precipitation have had a significant impact on agriculture, especially plant pathogens. Elevated CO_2 levels can enhance the aggressiveness of pathogens and alter the epidermal characteristics of plants, potentially increasing their susceptibility to infection (Lake and Wade, 2009). Increases in temperature allow pathogens and their vectors (such as disease-transmitting insects) to develop more rapidly because their life cycle is accelerated and their ability to infect new areas increases. In addition, warmer temperatures may lengthen plant growing cycles and thus provide more time for pathogens to attack plants, resulting in greater damage (Amari *et al.*, 2021). Higher humidity creates a favourable environment for the proliferation of many diseases that spread easily in stagnant water or damp conditions, while a lack of water could also favour pathogens by weakening plants and making them more susceptible to infection. In general, pathogens tend to adapt quickly to changing conditions and may take advantage while plants are vulnerable due to the stress caused by environmental conditions. Climate change has also impacted control strategies at field level, with heavy rainfall washing away plant protection products or reducing their efficacy due to rapid degradation.

On the other hand, climate change affects plant nutrition, since the increase in CO_2 and variations in temperature and precipitation modify the availability and general cycle of nutrients in the soil, both annually and seasonally (Koller and Phoenix, 2017). The relationship between plant nutrition and climate change is extremely complicated because climatic conditions affect all stages of plant growth, including plant development, metabolism, physiology and yield (Alshaal *et al.*, 2017). Higher CO_2 concentrations could have a positive effect by increasing plant biomass, but they could also negatively affect plant nutrient concentrations (Rajput *et al.*, 2021).

Regarding temperature, an increase could alter the availability of carbon (C), nitrogen (N) and phosphorus (P), greatly affecting plants since these are essential nutrients for their development. In addition, increased temperatures could affect the spatial variability of soil nutrients and the processes related to changes in habitat quality, resulting in a complicated scenario, as these factors influence soil microbial activity and therefore,the availability of nutrients for plants (Matias *et al.*, 2011). In terms of N, an increase in temperature could increase mineralization above the plant absorption rate, which would result in leaching of this nutrient (Lotze-Campen, 2011), accompanied by changes in the composition of the soil solution, including a decrease in pH and an increase in electrical conductivity (De Neve *et al.*, 2003). P mineralization in leaf litter has also proven to increase by an average of 48% with temperature increases of 0.3–0.6°C (Zalba *et al.*, 2007).

Climate change has also caused considerable variations in the volume, intensity, duration and distribution of rainfall. Significant variations in precipitation directly influence yields, especially in semi-arid and arid climates. Increases in precipitation increase the rate of nutrient leaching and soil erosion (Lotze-Campen, 2011). A 1.0% increase in precipitation is projected to result in a 1.5–2.0% increase in erosion, thereby increasing nutrient depletion (Cruse *et al.*, 2023). On the other hand, low water potential is a limiting factor for plant growth; fine roots can become dehydrated during a drought, which reduces nutrient uptake and leads to decreased root activity and nutrient mobility (Arndal *et al.*, 2013). In addition, when drought stress increases, relative leaf water content decreases. However, the impacts of drought on plant nutrition are complex and vary depending on the species, developmental stage, tissue type, drought severity and duration. The integrated effects of high temperatures and decreased rainfall could result in lower yields and an increased risk of erosion; consequently, the need for improved nutrient and water management is urgent (Howden and Crimp, 2011).

In the complex scenario of uncertainty associated with climate change, alternative solutions must be sought to increase agricultural production for the growing population. In addition to the above, there is currently a strong tendency towards sustainable agriculture, to reduce the negative externalities of conventional agriculture. The use of bioinputs is the most concrete sustainable alternative, as they have been well evaluated and increasingly explored among producers. Specifically, micro-organisms have frequently been used as bioinputs and new applications are increasingly being discovered to support crops. While micro-organisms may be affected by climate change, they are also fundamental in the process of plant adaptation to changing conditions. Several studies provide evidence that micro-organisms associated with plants can improve their tolerance to various abiotic stresses and their resistance to pests and diseases, as well as promoting their growth (Compant *et al.*, 2005; Mohanty and Swain, 2018; Dey and Raghuwanshi, 2020).

In recent decades, interest in the study of plant pathogen-controlling micro-organisms has increased significantly (Dhakal and Singh, 2019), as has the study of plant growth-promoting micro-organisms that exhibit various mechanisms based on the production of plant growth-promoting substances or biofertilization through the mobilization of mineral components in the soil (Grover *et al.*, 2021). Several studies have shown that under unfavourable conditions such as hot, dry and saline environments, conditions closely associated with climate change events, plants can grow and develop different adaptive capacities to these stresses, attributed, in part, to the micro-organisms associated with them (Compant *et al.*, 2021). Within this group, endophytic fungi are of increasing interest to the scientific community as allies in the complex scenario of climate change. These organisms can internally colonize plants, establishing a mutually beneficial relationship in which the micro-organisms provide the plants with resources for growth and survival, while the plant provides the endophytes with food and a suitable niche for their growth (Quesada-Moraga *et al.*, 2014).

This chapter highlights the importance of the use of micro-organisms, in particular endophytic micro-organisms, in crop phytosanitary and nutrient management strategies to cope with climate change conditions that may have a negative impact on food production.

17.2 Plant Growth-Promoting Endophytes: An Emerging Tool to Cope with Climate Change

It is well known that micro-organisms can promote plant growth through various mechanisms. They can be located on the outside of plants, such as the rhizosphere or phyllosphere, or inside them, known as the endosphere. A great diversity of beneficial micro-organisms (fungi and bacteria) lives in the endosphere and has received increasing attention due to their intimate interaction with plants and their effect on plant growth (Pérez-Montaño *et al.*, 2014; Hardoim *et al.*, 2015). Several studies have shown that endophytes can promote plant growth directly, usually by facilitating the acquisition of resources (nutrients), by modulating plant hormone levels or indirectly by relieving plants from biotic stresses through their action as biocontrollers (Tall and Meyling, 2018; Macedo-Raygoza *et al.*, 2019; Rana *et al.*, 2020). The association of plants with endophytes capable of promoting their growth has been perceived as beneficial even during stressful conditions. Inoculations with endophytic bacteria have been confirmed to increase plant growth and improve plant tolerance to abiotic stresses. For example, applications of *Bacillus subtilis* in maize (*Zea mays* L., Poaceae) stimulated the production of plant growth hormones, such as auxins, which increased root biomass and consequently water retention, improving their ability to tolerate drought (Vurukonda *et al.*, 2016). *Piriformospora indica* has also been shown to improve drought tolerance in crops, such as wheat (*Triticum aestivum* L., Poaceae) and barley (*Hordeum vulgare* L., Poaceae), by promoting root growth, increasing water and nutrient uptake, and stimulating the expression of stress-responsive genes (Varma *et al.*, 1999).

On the other hand, entomopathogenic fungi have received special attention due to their action as biological controllers and for the development of commercial products that have been relatively easy to scale, formulate and apply in the field (Litwin *et al.*, 2020). These fungi have a long history of evolutionary interaction with plants and have succeeded at internally colonizing them and promoting their growth through various mechanisms (Rivas-Franco *et al.*, 2020; Hu and Bidochka, 2021). This capacity may be provided by their structure, the formation of hyphae and mycelium, increasing the effective areas of root absorption (Chen *et al.*, 2020; García-Latorre *et al.*, 2021). Hyphae can extend their growth outwards from the roots, thus significantly increasing the volume of soil explored, resulting in a greater capacity to absorb water and nutrients, including N, P and potassium (K) (Behie *et al.*, 2012, 2017; Mohamed *et al.*, 2022; Moreira *et al.*, 2024).

Several studies have confirmed the contribution of entomopathogenic endophytic fungi to the successful growth and development of agricultural crops. For example, Jaber and Enkerli (2017) showed that foliar applications of *Beauveria bassiana*, *B. brongniartii* and *Metarhizium brunneum* strains increased the fresh weight of roots, shoots and height of faba bean (*Vicia faba* L., Fabaceae). Cabbage plants (*Brassica oleracea* L., Brassicaceae) colonized by *B. bassiana* and *M. brunneum* showed greater shoot length (123.0% for *B. bassiana* and 53.3% for *M. brunneum*) compared to control treatments (Dara *et al.*, 2017). Inoculations of *M. robertsii* and *M. brunneum* strains on coffee (*Coffea arabica* L., Rubiaceae) increased the dry and fresh mass of the shoot system (Martins *et al.*, 2024). Treating seeds and roots (irrigation) of sunflower (*Helianthus annuus L.*, Asteraceae) plants with endophytic *B. bassiana* strains increased plant height and chlorophyll contents (Sui *et al.*, 2024). In cereals, the growth promotion potential of *B. bassiana* has been clearly demonstrated; seed and soil applications increased ear production, root length and grain yield in bread wheat (Sánchez-Rodríguez *et al.*, 2018). Similarly, Afandhi *et al.* (2019) suggested a positive correlation between the *B. bassiana* inoculation method and the growth enhancement of bean (*Phaseolus vulgaris* L., Fabaceae). Tomato plants (*Solanum lycopersicum* L., Solanaceae) treated with *M. anisopliae* showed increased plant height, root length, shoots and root dry weight (Elena *et al.*, 2011). It is worth mentioning that several of these investigations included field trials, so their results provide evidence of the potential that endophytes have to cope with different climate change scenarios.

In addition to being a unique and highly specialized group of microbial agents that regulate insect pest populations and consequently possess desirable traits that favour their development as biopesticides (Lacey *et al.*, 2015), entomopathogenic fungi also can behave as saprophytes, living outside the plant, recycling nutrients from plant tissues (Saikkonen *et al.*, 2015).

The results presented above and in several other investigations confirm the capacity of endophytes to promote plant growth. Several mechanisms may be involved in this process, which may act independently or in combination. The most studied mechanisms are discussed below.

17.2.1 Phosphate solubilization

Phosphorus is considered an essential macronutrient because it plays important roles in all plant metabolic processes and its deficit could limit growth due to its structural, functional and metabolic properties (Plaxton and Tran, 2011). Phosphate solubilization can be achieved through several mechanisms and depends mainly on the ability of micro-organisms to produce organic acids in their respective environments (Bashan *et al.*, 2013; Ávila Martínez *et al.*, 2015; Restrepo-Franco *et al.*, 2015). Studies conducted by Matos *et al.* (2017) showed that the endophytic strain of the bacterium *Bacillus* EB-78 presented a P solubilization capacity in a solid medium, when $Ca_3(PO_4)_2$ and soy lecithin were used as P sources; this isolate significantly reduced the pH of the liquid medium and exhibited acid phosphatase activity. Similarly, Varga *et al.* (2020) determined that the endophytic strains *Rahnella* and *Burkholderia* had the ability to solubilize $Ca3(PO_4)_2$ in poplar (*Populus alba L.*, Salicaceae), which could be beneficial to plants in alkaline soils, in addition to the ability to partially solubilize aluminium (Al) phosphate and iron (Fe) phosphate, which could benefit plants in the most common acidic soils. Regarding fungi, they could be more efficient in nutrient solubilization than bacteria (Pradhan and Sukla, 2006; Khan *et al.*, 2010). Several genera of entomopathogenic endophytes have also shown their potential for P solubilization (Baron *et al.*, 2020; Barra-Bucarei *et al.*, 2020; Toscano-Verduzco *et al.*, 2020). Krell *et al.* (2018) demonstrated the ability of endophytic strains of *M. brunneum* to decrease nutritional stress by mobilizing phosphorus and transferring it to potato plants (*Solanum tuberosum* L., Solanaceae).

17.2.2 Plant hormone production

Phytohormones are versatile signalling molecules that affect various physiological processes of plant development, even at low concentrations (Khan *et al.*, 2012). Endophytes are a source of bioactive secondary metabolites such as phytohormones, including indole acetic acid, gibberellic acid and cytokinins (Khan *et al.*, 2008). Khan *et al.* (2014) determined that the endophytic bacterium *Sphingomonas* sp. strain LK11 produces indole-3-acetic acid and gibberellic acid, which promoted growth in tomato (*Solanum lycopersicum*). On the other hand, liquid media cultures of endophytic strains of *Pseudomonas resinovorans* and *Paenibacillus* showed positive results in the production of cytokinin-like compounds (Bhore *et al.*, 2010). Endophytic fungi also produce phytohormones. For example, the endophytic strains of *Oxalis corniculata* TS1 of *Aspergillus fumigatus* and BRL1 of *Fusarium proliferatum* secrete indole acetic acid and gibberellic acid in rice (*Oryza sativa* L., Poaceae) roots. The endophytic fungus *Falciphora oryzae* showed prominent stimulation of lateral root growth in *Arabidopsis thaliana* L. (Brassicaceae) plants by inducing the biosynthesis of indole derivatives (Sun *et al.*, 2020).

17.2.3 Increase in nitrogen availability

Nitrogen is essential for plant growth; its deficit often limits plant productivity. Atmospheric nitrogen (N_2) fixation is an ability that only bacteria exhibit; in this function rhizobacteria have been shown to be more competitive. However, there is evidence of N_2 fixation in foliage by the endophytic bacteria of *Pinus flexilis* (Moyes *et al.*, 2016). Similarly, Knoth *et al.* (2013) reported endophytic colonization of maize by diazotrophic strains isolated from dicotyledonous evergreen trees (*Populus trichocarpa* Torr. & A. Gray ex Hook., Salicaceae and *Salix sitchensis* Sanson ex Bong., Salicaceae); these strains produced substantial growth promotion, with increased leaf physiology. Although endophytic fungi do not have the capacity to fix atmospheric N, they have managed to make this element available to plants through other mechanisms. It has been demonstrated that strains of the genus *Metarhizium* (mainly root endophyte) transfer insect-derived N to the host plant and receive carbon from it. This has shown a significant increase in root growth, up to 31.8%, in *P. vulgaris* (Behie and Bidochka, 2014; Behie *et al.*, 2017).

17.2.4 Siderophore production

Siderophores are low molecular weight iron (Fe)-chelating ligands produced by almost all micro-organisms. These help plants absorb this element in Fe-limited soils. Fungi and endophytic bacteria have been reported to produce different types of siderophores. Maheshwari *et al.* (2019) analysed 84 endophytic bacterial isolates from nodules and roots of chickpea (*Cicer arietinum* L., Fabaceae) and pea (*Pisum sativum* L., Fabaceae) plants; ten isolates produced high amounts of Fe siderophore units (mainly hydroxamate siderophores) above 65.0%. Only one of the isolates (CPFR10) produced the catecholate type of siderophore. Studies with *Phialocephala fortinii* (a dark septate endophytic fungus) have indicated a potential for iron translocation in endophyte–plant mutualism (Bartholdy *et al.*, 2001). On the other hand, strains of the endophyte *Epichloë festucae* Leuchtm., Schardl & Siegel (Clavicipitaceae) have been observed to produce siderophores of a new type of fusarinin and ferricrocin, which are considered intracellular siderophores (Johnson, 2008; Koulman *et al.*, 2012). The production of Fe siderophores has also been recently reported for entomopathogenic endophytic fungi of the genera *Beauveria* and *Metarhizium* (Barra-Bucarei *et al.*, 2020; Siqueira *et al.*, 2020; Das *et al.*, 2023).

The potential of endophytic micro-organisms to promote plant growth has been widely demonstrated. This capacity could help plants alleviate the biotic and abiotic stresses produced by changing climate conditions, mainly due to the rapid adaptation of micro-organisms. This could not only increase yields at the field level but also reduce dependence on inorganic fertilizers.

17.3 Microbial Endophytes: Strategies and Mechanisms of Pathogen Control in the Context of Climate Change

Plant diseases, caused principally by pathogenic fungi and bacteria, annually produce important losses in agricultural crops. These losses threaten food security and human health, with serious economic implications for agriculture (Anderson *et al.*, 2004). The emergence of new or more aggressive diseases could be directly related to climate change (among other factors). This is because climate can influence all stages of the life cycle, not only those of the pathogen but also those of the host, consequently affecting the dynamics of many diseases. Beneficial micro-organisms, particularly those that can inhabit the plant endosphere, are a concrete alternative capable of increasing crop resilience.

Endophytic micro-organisms and their secondary metabolites can be effective tools for directly or indirectly reducing plant diseases. Endophytic bacteria have been reported to reduce the incidence and severity of various plant pathogens. Studies conducted by Barka *et al.* (2002) determined the antagonistic action of strains of endophytic bacteria of the genus *Pseudomonas* against *Botrytis cinerea* in grape vine (*Vitis vinifera* L., Vitaceae). Native endophytic bacteria have also been isolated from browntop millet (*Urochloa ramose* L., Poaceae) seeds, which promoted seedling development and defended against fungal diseases such as *Fusarium* (Verma and White, 2018). Studies conducted by Nejad and Johnson (2000) with bacteria isolated from oilseed rape (*Brassica napus* L., Brassicaceae) stems and roots determined that inoculation of these bacteria significantly reduced the symptoms of vascular wilt disease caused by the pathogenic fungi *Verticillium dahliae* and *F. oxysporum* f. sp. *lycopersici*, in *B. napus* and *S. lycopersicum*, respectively. Endophytic bacteria have even been reported to have an antagonistic effect against some viruses. For example, endophytes of the genus *Bacillus* have caused a significant reduction in banana bunchy top virus in banana (*Musa* spp., Musaceae) crops under field conditions (Harish *et al.*, 2009). In addition, the endophytic strains *Bacillus* sp. TS2, B. *subtilis* 26D and *B. thuringiensis* B-5351 have shown potential to control potato virus X in *S. tuberosum* (Sorokan *et al.*, 2020).

Several studies have shown the antagonistic action of endophytic entomopathogenic fungi against plant pathogens (Ownley *et al.*, 2008; Jaber and Salem, 2014; Jaber, 2018; Barra-Bucarei *et al.*, 2019a). This action could be due to different mechanisms that act individually or together (Vega *et al.*, 2009; Ownley *et al.*, 2010). Some researchers have even suggested that these fungi may have the ability to control pests and diseases simultaneously (Kim *et al.*, 2007; Ownley *et al.*, 2008). This could be associated with their ability to produce a diversity of bioactive metabolites, many of which have antimicrobial properties (Wagner and Lewis, 2000; Yu *et al.*, 2010).

The antagonistic effect of these micro-organisms, mainly of the *Beauveria* and *Metarhizium genera*, has been studied against foliar and root pathogenic fungi such as *Plasmopara viticola* in *V. vinifera* (Jaber, 2015); *Rhizoctonia solani* (Moreira *et al.*, 2024) and *Fusaroum solani* f. sp. *phaseoli* in *P. vulgaris* (Sasan and Bidochka, 2013); *Alternaria alternata*, *Botrytis cinerea* and *Rhizoctonia solani* in *S. lycopersicum* (Ownley *et al.*, 2004; Barra-Bucarei *et al.*, 2019a; Sinno *et al.*, 2021); *Pythium myriotylum* in cotton (*Gossypium hirsutum* L., Malvaceae) (Ownley *et al.*, 2010); *F. oxysporum* in sweet pepper (*Capsicum annuum* L., Solanaceae); *F. culmorum* in *T. aestivum* L. (Jaber, 2018); and *Sclerotinia sclerotiorum* in *H. annuus* (Sui *et al.*, 2024). In addition, the action of endophytic entomopathogenic fungi against bacteria such as *Xanthomonas axonopodis* pv. *malvacearum* in *G. hirsutum* has been reported (Ownley *et al.*, 2008). Similarly, their antagonistic action against viruses such as zuczhini yellow mosaic virus in zucchini (*Cucurbita pepo* L., Cucurbitaceae) (Jaber and Salem, 2014) and beet mild yellow virus in sugarbeet (*Beta vulgaris*

Fig. 17.1. Mechanisms of controlling plant diseases and plant growth promotion with endophyte micro-organisms in the context of climate change.

ssp. *vulgaris*, Amaranthaceae) (Dessauvages *et al*., 2024) has been demonstrated.

The ability of endophytic micro-organisms to act as antagonists of plant pathogens (fungi and bacteria) may result from the action of different mechanisms and their interactions; however, knowledge in this field is still scarce. Direct mechanisms such as competition, parasitism and antibiosis are produced by primary and secondary metabolites, enzymes or volatile compounds of endophytic micro-organisms, while indirect mechanisms such as the induction of resistance, are a plant response activated by endophytes (Fig. 17.1).

17.3.1 Competition

Competition for space and/or food occurs between the endophyte and the pathogen inside the plant. Endophytic micro-organisms have been shown to successfully colonize most plant tissues, from seed to fruit, providing a varied food source exploited by the endophytes, leaving less chance for pathogen colonization (Ownley *et al*., 2004). Early inoculation of these micro-organisms could provide advantages, as they probably exhaust the plant's resources, leaving few available for the pathogen. An example of this mechanism is the endophytic fungus of the genus *Trichoderma* that competes for sources of N, C, Fe and other growth factors to control plant pathogens, since these fungi are very competitive and take advantage of various nutrients (Vinale *et al*., 2008; Segarra *et al*., 2010).

17.3.2 Mycoparasitism

Mycoparasitism is an antagonistic interaction between a parasitic fungus and a beneficial host fungus that generally involves the action of extracellular lytic enzymes that break down the cell walls of the host. Although there are several studies on the action of endophytic fungi as mycoparasites, these only present results at the *in vitro* level, with no reports of their action on plants (Ownley *et al*., 2010). For example, species of the genus *Trichoderma* secrete enzymes that degrade the cell wall of pathogenic fungi, including β-glucosidase, endochitinases, proteases and mannosidases (Lima *et al*., 2024). Moreover, its mycopathogenic effect has been evaluated under *in vitro* conditions with positive results, such as those obtained by Bailey *et al*. (2008) in which some *Trichoderma* isolates showed aggressive mycoparasitism against *Moniliophthora roreri*, to the point that the pathogen could not recover.

17.3.3 Antibiosis

Antibiosis acts against phytopathogens and is produced by the direct action of toxic metabolites created by endophytes that act against other micro-organisms that are sensitive to them (Gowtham *et al.*, 2024). For example, beauvericin is a mycotoxin produced by *B. bassiana*, *Paecilomyces fumosoroseus* and *Fusarium* spp. (Logrieco *et al.*, 1998) which is of interest for agricultural crop protection because it has significant antibacterial, antifungal and antiviral activity (Xu *et al.*, 2010; Wang and Xu, 2012). Bassionalide is another metabolite produced by more than one genus of endophytic fungi and has been reported in *B. bassiana* and *Lecanicillium lecanii* (Kanaoka *et al.*, 1978). Fungi of the genus *Metarhizium* produce dextruxins (Roberts and St Leger, 2004; Ríos-Moreno *et al.*, 2016), which have been successfully evaluated in the control of the pathogens *F. oxysporum* and *Cladosporium herbarum* (Ravindran *et al.*, 2014).

Several strains of the genus *Beauveria* have been reported to have biocontrol properties attributed to antibiosis and their effects have been demonstrated on pathogens such as *B. cinerea*, *F. oxysporum* f. sp. *lycopersici* and *S. sclerotiorum* (Culebro-Ricaldi *et al.*, 2017; Barra-Bucarei *et al.*, 2019a, b). The endophyte *Penicillium pinophilum* from *Alloteropsis cimicina* is a rich source of alkaloids that have exhibited effectiveness against *E. coli*, *Enterococcus faecalis*, *Staphylococcus aureus*, *Klebsiella pneumoniae* and *Xanthomonas campestris* (Nischitha and Shivanna, 2021).

On the other hand, there are several reports of endophytic bacteria with antagonistic activity that result from the production of metabolites. This is the case of the genus *Bacillus*. Studies by Hasan *et al.* (2020) determined that metabolites produced by endophytic strains of *B. altitudinis* (HNH7) and *B. velezensis* (HNH9) showed a significant reduction in the severity of *Verticillium* wilt in plants of *Gossypium hirsutum*. Similar responses were found when using volatile organic compounds produced by *Bacillus* spp., which had adverse effects on fungal mycelial growth and *Sclerotinia sclerotia* production (Massawe *et al.*, 2018). Pavithra *et al.* (2020) suggested that opportunistic micro-organisms, such as phytopathogens, could be kept at bay by the antimicrobial substances produced by fungi. Similarly, the bactericidal and bacteriostatic capacity of different secondary metabolites has been demonstrated.

17.3.4 Induction of resistance

Induction of resistance was defined by Kloepper *et al.* (1992) as the process of active resistance dependent on the physical or chemical barriers of the host plant, activated by biotic or abiotic agents. This mechanism, used by endophytes to control pathogens, is considered an indirect mechanism because it can be stimulated in the plant by the presence of endophytes and allows reduction of disease symptoms in parts of the plant far from the site where the inducing agent is active (Kavroulakis *et al.*, 2007; Shoresh *et al.*, 2010). Endophytes can induce plant resistance by triggering systemic resistance through jasmonic acid, ethylene and/or salicylic acid-dependent signalling pathways (Bastias *et al.*, 2017; Vahabi *et al.*, 2018; Yousef-Yousef *et al.*, 2024). Plants that are endophytically colonized by micro-organisms may trigger a faster response, which may be associated with their ability to produce secondary metabolites that act as elicitors for phytoalexin production (Compant *et al.*, 2005; Gao *et al.*, 2010). Several bacteria reported as endophytes have also exhibited this mechanism of action.

Quite a few studies have shown that specific strains of *P. fluorescens* can induce systemic resistance in plants, providing enhanced defence against a wide range of pathogens. These studies have explained that these bacteria trigger plant immune responses without directly killing the pathogens, leading to more robust and sustained disease resistance (Bakker *et al.*, 2003, 2007 ; Maldonado-González *et al.*, 2011). Endophytic strains of the bacterium *B. halotolerans* (ATCC 25096T) have demonstrated antagonistic abilities that inhibit *Verticillium dahliae* spore germination and mycelial growth in cotton (*Gossypium hirsutum*), which has been associated with the overexpression of β-glucanase, possibly increasing the inhibition efficiency of this pathogen (Zhang *et al.*, 2019). The activation of resistance genes by combining fungi and endophytic bacteria has also been tested. Endophytic strains of *T. viride* and *P. fluorescens* were assessed for their synergistic ability to inhibit the growth of *Pythium aphanidermatum*, the causal agent of chili wilt. It was thus determined that the induction of defence-related enzymes, PR proteins and phenol accumulation occurred when the combination of strains were applied, which was not the case for individual applications (Muthukumar *et al.*, 2011). Likewise, symbiotic assemblages have proven to facilitate endophytic fungi to produce certain secondary metabolites that facilitate defence mechanisms by triggering induced systemic resistance and systemic acquired resistance. In the case of the endophytic fungus *B. bassiana*, Rondot and Reineke (2019) studied this fungus' ability to reduce the severity and incidence of downy mildew disease (*Plasmopara vitícola*) in grapevine plants through the activation of systemic resistance, where the expression level of defence genes encoding PR-2 (β−1,3-glucanases), PR-3 (chitinases), PR-5 (thaumatin-like proteins) and PR protein 10.3 increased after treatments with *B. bassiana*, within 24 hours post treatment.

As the number of studies addressing the action of endophytic microorganisms as control agents of phytopathogens increases, progress is also being made in elucidating the mechanisms involved in this process. However, many questions remain as to how endophytes modulate plant physiology and metabolism to activate their defence mechanisms.

17.4 Challenges and Limitations in the Use of Endophytic Micro-Organisms

While there is evidence of the roles that endophytic micro-organisms can play in supporting plant growth and adaptation to various climate change scenarios, it is necessary to:

- recognize the variability of climatic conditions, mainly aspects associated with temperatures, humidity and UV radiation, to establish trials and evaluate the effectiveness of these micro-organisms in different climatic and soil conditions
- further study and identify the specific mechanisms through which plant–endophyte interactions occur, particularly under variable environmental conditions
- determine the compatibility of these micro-organisms with other management and agronomic practices, to ensure their colonization, establishment and subsequent performance favour plant growth
- use strategies based on biostimulation of plants with endophytic micro-organisms, which could increase their adaptability to changing climate conditions and prepare plants to resist new or more aggressive phytopathogens
- use micro-organisms adapted to the agro-ecosystems where they will be used.

17.5 Conclusion

The plant endosphere represents a very diverse and dynamic environment for the micro-organisms that inhabit it, and as plants face changing environmental conditions, these micro-organisms support their adaptation processes. Endophytic micro-organisms therefore appear as a promising tool to combat the challenges of climate change. These organisms represent a natural resource which has not yet been fully explored and represents an immense potential for the discovery of new functions in favour of plant adaptation to changing environmental conditions. Endophytic micro-organisms not only enhance plant resilience through the various mechanisms associated with plant growth promotion and protection against phytopathogens, but also offer promising adaptive benefits to various biotic and abiotic stresses.

As research progresses, we expect to obtain a deeper understanding of the complex symbiotic relationships between plants and micro-organisms. These new discoveries will aid us in maximizing benefits for crops in the context of climate change. Further exploration of the ecological functions and specific applications of endophytic micro-organisms in agriculture could also lead to more sustainable and environmentally friendly practices.

References

Afandhi, A., Widjayanti, T., Emi, A.A.L., Tarno, H., Afiyanti, M. *et al.* (2019) Endophytic fungi *Beauveria bassiana* Balsamo accelerates growth of common bean (*Phaeseolus vulgaris* L.). *Chemical and Biological Technologies in Agriculture* 6(1), 1–6.

Alshaal, T., El-Ramady, H., Al-Saeedi, A.H., Shalaby, T., Elsakhawy, T. *et al.* (2017) The rhizosphere and plant nutrition under climate change. In: Naeem, M., Ansari, A. and Gill, S. (eds) *Essential Plant Nutrients: Uptake, Use Efficiency, And Management*. Springer, New York, pp. 275–308.

Amari, K., Huang, C. and Heinlein, M. (2021) Potential impact of global warming on virus propagation in infected plants and agricultural productivity. *Frontiers in Plant Science* 12, 649768.

Anderson, P.K., Cunningham, A.A., Patel, N.G., Morales, F.J., Epstein, P.R. *et al.* (2004) Emerging infectious diseases of plants: Pathogen pollution, climate change and agrotechnology drivers. *Trends in Ecology & Evolution* 19(10), 535–544.

Arndal, M.F., Merrild, M.P., Michelsen, A., Schmidt, I.K., Mikkelsen, T.N. *et al.* (2013) Net root growth and nutrient acquisition in response to predicted climate change in two contrasting heathland species. *Plant and Soil* 369, 615–629.

Ávila Martínez, E.G., Lizarazo Forero, L.M. and Cortés Pérez, F. (2015) Promoción del crecimiento de *Baccharis macrantha* (Asteraceae) con bacterias *solubilizadoras* de fosfatos asociadas a su *rizósfera*. *Acta Biológica Colombiana* 20(3), 121–131.

Bakker, P.A., Pieterse, C.M. and Van Loon, L.C. (2007) Induced systemic resistance by fluorescent *Pseudomonas* spp. *Phytopathology* 97(2), 239–243.

Bailey, B.A., Bae, H., Strem, M.D., Crozier, J., Thomas, S.E. *et al.* (2008) Antibiosis, mycoparasitism, and colonization success for endophytic *Trichoderma isolates* with biological control potential in *Theobroma cacao*. *Biological Control* 46(1), 24–35.

Bakker, P.A., Ran, L.X., Pieterse, C.M.J. and Van Loon, L.C. (2003) Understanding the involvement of rhizobacteria-mediated induction of systemic resistance in biocontrol of plant diseases. *Canadian Journal of Plant Pathology* 25(1), 5–9.

Barka, E.A., Gognies, S., Nowak, J., Audran, J.C. and Belarbi, A. (2002) Inhibitory effect of endophyte bacteria on *Botrytis cinerea* and its influence to promote the grapevine growth. *Biological Control* 24(2), 135–142.

Baron, N.C., de Souza Pollo, A. and Rigobelo, E.C. (2020) *Purpureocillium lilacinum* and *Metarhizium marquandii* as plant growth-promoting fungi. *PeerJ* 8, e9005.

Barra-Bucarei, L., France, A. and Millas, P. (2019b) Crossing frontiers: Endophytic entomopathogenic fungi for biological control of plant diseases. In: Hodkinson, T., Doohan, F., Saunders, M. and Murphy, B. (eds) *Endophytes for a Growing World*. Cambridge University Press, Cambridge, UK, pp. 67–92.

Barra-Bucarei, L., González, M.G., Iglesias, A.F., Aguayo, G.S., Peñalosa, M.G. *et al.* (2020) *Beauveria bassiana* multifunction as an endophyte: Rowth promotion and biologic control of *Trialeurodes vaporariorum*, (Westwood) (Hemiptera: Aleyrodidae) in tomato. *Insects* 11(9), 591.

Barra-Bucarei, L., France Iglesias, A., Gerding González, M., Silva Aguayo, G., Carrasco-Fernández, J. *et al.* (2019a) Antifungal activity of *Beauveria bassiana* endophyte against *Botrytis cinerea* in two solanaceae crops. *Microorganisms* 8(1), 65.

Bartholdy, B.A., Berreck, M. and Haselwandter, K. (2001) Hydroxamate siderophore synthesis by *Phialocephala fortinii*, a typical dark septate fungal root endophyte. *Biometals* 14, 33–42.

Bashan, Y., Kamnev, A.A. and de-Bashan, L.E. (2013) Tricalcium phosphate is inappropriate as a universal selection factor for isolating and testing phosphate-solubilizing bacteria that enhance plant growth: A proposal for an alternative procedure. *Biology and Fertility of Soils* 49(4), 465–479.

Bastias, D.A., Martínez-Ghersa, M.A., Ballaré, C.L. and Gundel, P.E. (2017) Epichloë fungal endophytes and plant defenses: Not just alkaloids. *Trends in Plant Science* 22(11), 939–948.

Behie, S.W. and Bidochka, M.J. (2014) Nutrient transfer in plant–fungal symbioses. *Trends in Plant Science* 19(11), 734–740.

Behie, S.W., Zelisko, P.M. and Bidochka, M.J. (2012) Endophytic insect-parasitic fungi translocate nitrogen directly from insects to plants. *Science* 336(6088), 1576–1577.

Behie, S.W., Moreira, C.C., Sementchoukova, I., Barelli, L., Zelisko, P.M. *et al.* (2017) Carbon translocation from a plant to an insect-pathogenic endophytic fungus. *Nature Communications* 8(1), 14245.

Bhore, S.J., Ravichantar, N. and Loh, C.Y. (2010) Screening of endophytic bacteria isolated from leaves of Sambung Nyawa [*Gynura procumbens* (Lour.) Merr.] for cytokinin-like compounds. *Bioinformation* 5(5), 191.

Canadell, J.G., Le Quéré, C., Raupach, M.R., Field, C.B., Buitenhuis, E.T. *et al.* (2007) Contributions to accelerating atmospheric CO_2 growth from economic activity, carbon intensity, and efficiency of natural sinks. *Proceedings of the National Academy of Sciences USA* 104(47), 18866–18870.

Carvalho, F.P. (2017) Pesticides, environment, and food safety. *Food and Energy Security* 6(2), 48–60.

Chen, Z., Jin, Y., Yao, X., Chen, T., Wei, X. *et al.* (2020) Fungal endophyte improves survival of *Lolium perenne* in low fertility soils by increasing root growth, metabolic activity and absorption of nutrients. *Plant and Soil* 452, 185–206.

Compant, S., Cambon, M.C., Vacher, C., Mitter, B., Samad, A. *et al.* (2021) The plant *endosphere* world–bacterial life within plants. *Environmental Microbiology* 23(4), 1812–1829.

Compant, S., Reiter, B., Sessitsch, A., Nowak, J., Clément, C. and Ait Barka, E. (2005) Endophytic colonization of *Vitis vinifera* L. by plant growth-promoting bacterium *Burkholderia* sp. strain PsJN. *Applied and Environmental Microbiology* 71(4), 1685–1693.

Cruse, R.M., Herzmann, D.E., Gelder, B.K., James, D.E. and Flanagan, D.C. (2023) Impacts of changing precipitation events on landscape soil erosion estimates. Presented at Soil Erosion Research Under a Changing Climate, January 8–13, American Society of Agricultural and Biological Engineers, Aguadilla, Puerto Rico, USA.

Culebro-Ricaldi, J.M., Ruíz-Valdiviezo, V.M., Rodríguez-Mendiola, M.A., Ávila-Miranda, M.E., Miceli, F.G. *et al.* (2017) Antifungal properties of *Beauveria bassiana* strains against *Fusarium oxysporum* f. sp. *lycopersici* race 3 in tomato crop. *Journal of Environmental Biology* 38(5), 821.

Dara, S.K., Dara, S.S. and Dara, S.S. (2017) Impact of entomopathogenic fungi on the growth, development, and health of cabbage growing under water stress. *American Journal of Plant Sciences* 8(6), 1224.

Das, J., Sahoo, B., Panigrahy, M. and Dangar, T.K. (2023) Genetic polymorphism and plant growth promotion traits of potent fungal entomopathogens of rice leaf folder. *Archives of Microbiology* 205(5), 216.

De Neve, S., Hartmann, R. and Hofman, G. (2003) Temperature effects on N mineralization: Changes in soil solution composition and determination of temperature coefficients by TDR. *European Journal of Soil Science* 54(1), 49–62.

Dessauvages, K., Scheifler, M., Francis, F. and Ben Fekih, I. (2024) A new isolate *Beauveria bassiana* GxABT-1: Efficacy against *Myzus persicae* and promising impact on the beet mild yellow virus-aphid association. *Insects* 15(9), 697.

Dey, R. and Raghuwanshi, R. (2020) Comprehensive assessment of growth parameters for screening endophytic bacterial strains in *Solanum lycopersicum* (Tomato). *Heliyon* 6(10), e05325.

Dhakal, R. and Singh, D.N. (2019) Biopesticides: A key to sustainable agriculture. *International Journal of Pure and Applied Bioscience* 7, 391–396.

Elena, G.J., Beatriz, P.J., Alejandro, P. and Lecuona, R.E. (2011) *Metarhizium anisopliae* (Metschnikoff) Sorokin promotes growth and has endophytic activity in tomato plants. *Advances in Biological Research* 5(1), 22–27.

Gao, F.K., Dai, C.C. and Liu, X.Z. (2010) Mechanisms of fungal endophytes in plant protection against pathogens. *African Journal of Micriobiological Research* 4(13), 1346–1351.

García-Latorre, C., Rodrigo, S. and Santamaría, O. (2021) Endophytes as plant nutrient uptake-promoter in plants. In: Maheshwari, D. and Dheeman, S. (eds) *Endophytes: Mineral Nutrient Management*, Vol. 3. Springer, Dordrecht, The Netherlands, pp. 247–265.

Gowtham, H.G., Hema, P., Murali, M., Shilpa, N., Nataraj, K. *et al.* (2024) Fungal endophytes as mitigators against biotic and abiotic stresses in crop plants. *Journal of Fungi* 10(2), 116.

Grover, M., Bodhankar, S., Sharma, A., Sharma, P., Singh, J. *et al.* (2021) PGPR mediated alterations in root traits: Way toward sustainable crop production. *Frontiers in Sustainable Food Systems* 4, 618230.

Hardoim, P.R., Van Overbeek, L.S., Berg, G., Pirttilä, A.M., Compant, S. *et al.* (2015) The hidden world within plants: Ecological and evolutionary considerations for defining functioning of microbial endophytes. *Microbiology and Molecular Biology Reviews* 79(3), 293–320.

Harish, S., Kavino, M., Kumar, N., Balasubramanian, P. and Samiyappan, R. (2009) Induction of defense-related proteins by mixtures of plant growth promoting endophytic bacteria against Banana bunchy top virus. *Biological Control* 51(1), 16–25.

Hasan, N., Farzand, A., Heng, Z., Khan, I.U., Moosa, A. *et al.* (2020) Antagonistic potential of novel endophytic *Bacillus* strains and mediation of plant defense against Verticillium wilt in upland cotton. *Plants* 9(11), 1438.

Howden, M.S. and Crimp, S.J. (2011) Regional impacts: Australia. In: Yadav, S., Redden, R., Hatfield, J., Lotze-Campen, H. and Hall, A. (eds) *Crop Adaptation to Climate Change*. John Wiley & Sons, Chichester, UK, pp. 143–155.

Hu, S. and Bidochka, M.J. (2021) Root colonization by endophytic insect-pathogenic fungi. *Journal of Applied Microbiology* 130(2), 570–581.

Jaber, L.R. and Salem, N.M. (2014) Endophytic colonisation of squash by the fungal entomopathogen *Beauveria bassiana* (Ascomycota: Hypocreales) for managing Zucchini yellow mosaic virus in cucurbits. *Biocontrol Science and Technology* 24(10), 1096–1109.

Jaber, L.R. and Enkerli, J. (2017) Fungal entomopathogens as endophytes: Can they promote plant growth? *Biocontrol Science and Technology* 27(1), 28–41.

Jaber, L.R. (2018) Seed inoculation with endophytic fungal entomopathogens promotes plant growth and reduces crown and root rot (CRR) caused by *Fusarium culmorum* in wheat. *Planta* 248, 1525–1535.

Jaber, L.R. (2015) Grapevine leaf tissue colonization by the fungal entomopathogen *Beauveria bassiana* sl and its effect against downy mildew. *BioControl* 60, 103–112.

Johnson, L. (2008) Iron and siderophores in fungal–host interactions. *Mycological Research* 112(2), 170–183.

Kanaoka, M., Isogai, A., Murakoshi, S., Ichinoe, M., Suzuki, A. *et al.* (1978) Bassianolide, a new insecticidal *Cyclodepsipeptide* from *Beauveria bassiana* and *Verticillium lecanii*. *Agricultural and Biological Chemistry* 42(3), 629–635.

Kavroulakis, N., Ntougias, S., Zervakis, G.I., Ehaliotis, C., Haralampidis, K. *et al.* (2007) Role of ethylene in the protection of tomato plants against soil-borne fungal pathogens conferred by an endophytic *Fusarium solani* strain. *Journal of Experimental Botany* 58(14), 3853–3864.

Khan, A.L., Hamayun, M., Kang, S.M., Kim, Y.H., Jung, H.Y. *et al.* (2012) Endophytic fungal association via gibberellins and indole acetic acid can improve plant growth under abiotic stress: An example of *Paecilomyces formosus* LHL10. *BMC Microbiology* 12, 1–14.

Khan, A.L., Waqas, M., Kang, S.M., Al-Harrasi, A., Hussain, J. *et al.* (2014) Bacterial endophyte *Sphingomonas* sp. LK11 produces gibberellins and IAA and promotes tomato plant growth. *Journal of Microbiology* 52, 689–695.

Khan, M.S., Zaidi, A., Ahemad, M., Oves, M. and Wani, P.A. (2010) Plant growth promotion by phosphate solubilizing fungi–current perspective. *Archives of Agronomy and Soil Science* 56(1), 73–98.

Khan, S.A., Hamayun, M., Yoon, H., Kim, H.Y., Suh, S.J. *et al.* (2008) Plant growth promotion and *Penicillium citrinum*. *BMC Microbiology* 8, 1–10.

Kim, H.Y., Choi, G.J., Lee, H.B., Lee, S.W., Lim, H.K. *et al.* (2007) Some fungal endophytes from vegetable crops and their anti-oomycete activities against tomato late blight. *Letters in Applied Microbiology* 44(3), 332–337.

Kloepper, J.W., Wei, G. and Tuzun, S. (1992) Rhizosphere population dynamics and internal colonization of cucumber by plant growth-promoting rhizobacteria which induce systemic resistance to *Colletotrichum orbiculare*. In: Tjamos, E., Papavizas, G. and Cook, R. (eds) *Biological Control of Plant Diseases: Progress and Challenges for the Future*. Springer, Dordrecht, The Netherlands, pp. 185–191.

Knoth, J.L., Kim, S.H., Ettl, G.J. and Doty, S.L. (2013) Effects of cross host species inoculation of nitrogen-fixing endophytes on growth and leaf physiology of maize. *Gcb Bioenergy* 5(4), 408–418.

Koller, E.K. and Phoenix, G.K. (2017) Seasonal dynamics of soil and plant nutrients at three environmentally contrasting sites along a sub-Arctic catchment sequence. *Polar Biology* 40, 1821–1834.

Koulman, A., Lee, T.V., Fraser, K., Johnson, L., Arcus, V. *et al.* (2012) Identification of extracellular siderophores and a related peptide from the endophytic fungus *Epichloë festucae* in culture and endophyte-infected *Lolium perenne*. *Phytochemistry* 75, 128–139.

Krell, V., Unger, S., Jakobs-Schoenwandt, D. and Patel, A.V. (2018) Endophytic *Metarhizium Brunneum mitigates* nutrient deficits in potato and improves plant productivity and vitality. *Fungal Ecology* 34, 43–49.

Lacey, L.A., Grzywacz, D., Shapiro-Ilan, D.I., Frutos, R., Brownbridge, M. *et al.* (2015) Insect pathogens as biological control agents: Back to the future. *Journal of Invertebrate Pathology* 132, 1–41.

Lake, J.A. and Wade, R.N. (2009) Plant–pathogen interactions and elevated CO_2: Morphological changes in favour of pathogens. *Journal of Experimental Botany* 60(11), 3123–3131.

Lima, P.C., Karimian, P., Johnston, E. and Hartley, C.J. (2024) The use of *Trichoderma* spp. for the bioconversion of agro-industrial waste biomass via fermentation: A review. *Fermentation* 10(9), 442.

Litwin, A., Nowak, M. and Różalska, S. (2020) Entomopathogenic fungi: Unconventional applications. *Reviews in Environmental Science and Bio/Technology* 19(1), 23–42.

Logrieco, A., Moretti, A., Castella, G., Kostecki, M., Golinski, P. *et al.* (1998) Beauvericin production by *Fusarium* species. *Applied and Environmental Microbiology* 64(8), 3084–3088.

Lotze-Campen, H. (2011) Climate change, population growth, and crop production: An overview. In: Yadav, S., Redden, R., Hatfield, J., Lotze-Campen, H. and Hall, A. (eds) *Crop Adaptation to Climate Change*. John Wiley & Sons, Chichester, UK, pp. 1–11.

Macedo-Raygoza, G.M., Valdez-Salas, B., Prado, F.M., Prieto, K.R., Yamaguchi, L.F. *et al.* (2019) Enterobacter cloacae, an endophyte that establishes a nutrient-transfer symbiosis with banana plants and protects against the black *Sigatoka pathogen*. *Frontiers in Microbiology* 10, 804.

Maheshwari, R., Bhutani, N. and Suneja, P. (2019) Screening and characterization of siderophore producing endophytic bacteria from *Cicer arietinum* and *Pisum sativum* plants. *Journal of Applied Biology and Biotechnology* 7(5), 7–14.

Maldonado-González, M., Bakker, P.A. and Mercado-Blanco, J. (2011) Involvement of induced systemic resistance in control of verticillium wilt by fluorescent *Pseudomonas* spp. *IOBC/Wprs Bulletin* 71, 73–76.

Manoharachary, C. and Kunwar, I.K. (2014) Host–pathogen interaction, plant diseases, disease management strategies, and future challenges. In: Goyal, A. and Manoharachary, C. (eds) *Future Challenges in Crop Protection Against Fungal Pathogens*. Springer, New York, pp. 185–229.

Martins, J.L.A., Franzin, M.L., Ferreira, D.D.S., Magina, L.C.R., Martins, E.F. *et al.* (2024) Metarhizium-inoculated coffee seeds promote plant growth and biocontrol of coffee leaf miner. *Microorganisms* 12(9), 1845.

Massawe, V.C., Hanif, A., Farzand, A., Mburu, D.K., Ochola, S.O. *et al.* (2018) Volatile compounds of endophytic *Bacillus* spp. have biocontrol activity against *Sclerotinia sclerotiorum*. *Phytopathology* 108(12), 1373–1385.

Matias, L., Castro, J. and Zamora, R. (2011) Soil-nutrient availability under a global-change scenario in a Mediterranean mountain ecosystem. *Global Change Biology* 17(4), 1646–1657.

Matos, A.D., Gomes, I.C., Nietsche, S., Xavier, A.A., Gomes, W.S. *et al.* (2017) Phosphate solubilization by endophytic bacteria isolated from banana trees. *Anais Da Academia Brasileira de Ciências* 89(4), 2945–2954.

Meena, R.S., Kumar, S., Datta, R., Lal, R., Vijayakumar, V. *et al.* (2020) Impact of agrochemicals on soil microbiota and management: A review. *Land* 9(2), 34.

Mohamed, A.H., Abd El-Megeed, F.H., Hassanein, N.M., Youseif, S.H., Farag, P.F. *et al.* (2022) Native *rhizospheric* and endophytic fungi as sustainable sources of plant growth promoting traits to improve wheat growth under low nitrogen input. *Journal of Fungi* 8(2), 94.

Mohanty, S. and Swain, C.K. (2018) Role of microbes in climate smart agriculture. In: Panpatte, D., Jhala, Y., Shelat, H. and Vyas, R. (eds) *Microorganisms for Green Revolution: Volume 2: Microbes for Sustainable Agro-Ecosystems*. Springer, Dordrecht, The Netherlands, pp. 129–140.

Moreira, F.M., Machado, T.I., Torres, C.A.R., Souza, H.R.D., Celestino, M.F. *et al.* (2024) *Purpureocillium lilacinum* SBF054: Endophytic in *Phaseolus vulgaris*, Glycine max, and *Helianthus annuus*; antagonistic to *Rhizoctonia solani*; and virulent to *Euschistus heros*. *Microorganisms* 12(6), 1100.

Moyes, A.B., Kueppers, L.M., Pett-Ridge, J., Carper, D.L., Vandehey, N. *et al.* (2016) Evidence for foliar endophytic nitrogen fixation in a widely distributed subalpine conifer. *New Phytologist* 210(2), 657–668.

Muthukumar, A., Eswaran, A. and Sangeetha, G. (2011) Induction of systemic resistance by mixtures of fungal and endophytic bacterial isolates against *Pythium aphanidermatum*. *Acta Physiologiae Plantarum* 33, 1933–1944.

Nejad, P. and Johnson, P.A. (2000) Endophytic bacteria induce growth promotion and wilt disease suppression in oilseed rape and tomato. *Biological Control* 18(3), 208–215.

Nischitha, R. and Shivanna, M.B. (2021) Antimicrobial activity and metabolite profiling of endophytic fungi in *Digitaria bicornis* (Lam) Roem. and Schult and *Paspalidium flavidum* (Retz.) *A. Camus*. *3 Camus* (2), 53.

Ownley, B.H., Gwinn, K.D. and Vega, F.E. (2010) Endophytic fungal entomopathogens with activity against plant pathogens: Ecology and evolution. *BioControl* 55, 113–128.

Ownley, B.H., Griffin, M.R., Klingeman, W.E., Gwinn, K.D., Moulton, J.K. *et al.* (2008) *Beauveria bassiana*: Endophytic colonization and plant disease control. *Journal of Invertebrate Pathology* 98(3), 267–270.

Ownley, B.H., Pereira, R.M., Klingeman, W.E., Quigley, N.B. and Leckie, B.M. (2004) *Beauveria bassiana*, a dual purpose biocontrol organism, with activity against insect pests and plant pathogens. In: Lartey, R. and Caesar, A. (eds) *Emerging Concepts in Plant Health Management. Research Signpost*. Auburn Hills, Michigan, pp. 255–269.

Pavithra, G., Bindal, S., Rana, M. and Srivastava, S. (2020) Role of endophytic microbes against plant pathogens: A review. *Asian Journal of Plant Science* 19(1), 54–62.

Pérez-Montaño, F., Alías-Villegas, C., Bellogín, R.A., Del Cerro, P., Espuny, M.R. *et al.* (2014) Plant growth promotion in cereal and leguminous agricultural important plants: From microorganism capacities to crop production. *Microbiological Research* 169(5–6), 325–336.

Plaxton, W.C. and Tran, H.T. (2011) Metabolic adaptations of phosphate-starved plants. *Plant Physiology* 156(3), 1006–1015.

Pradhan, N. and Sukla, L.B. (2006) Solubilization of inorganic phosphates by fungi isolated from agriculture soil. *African Journal of Biotechnology* 5, 10.

Quesada-Moraga, E., Herrero, N. and Zabalgogeazcoa, Í. (2014) Entomopathogenic and nematophagous fungal endophytes. *Advances in Endophytic Research* 85–99.

Rajput, V.D., Harish Singh, R.K., Verma, K.K. and Sharma, L. (2021) Recent developments in enzymatic antioxidant defence mechanism in plants with special reference to abiotic stress. *Biology* 10(4), 267.

Rana, K.L., Kour, D., Kaur, T., Devi, R., Yadav, A.N. *et al.* (2020) Endophytic microbes: Biodiversity, plant growth-promoting mechanisms and potential applications for agricultural sustainability. *Antonie Van Leeuwenhoek* 113(8), 1075–1107.

Ravindran, K., Chitra, S., Wilson, A. and Sivaramakrishnan, S. (2014) Evaluation of antifungal activity of Metarhizium anisopliae against plant phytopathogenic fungi. In: Kharwar, R., Upadhyay, R., Dubey, N. and Raghuwanshi, R. (eds) *Microbial Diversity and Biotechnology in Food Security*. Springer, New York, pp. 251–255.

Restrepo-Franco, G.M., Marulanda-Moreno, S., Fe-Pérez, Y.D.L., Díaz-de la Osa, A., Baldani, V.L.D. *et al.* (2015) Bacterias solubilizadoras de fosfato y sus potencialidades de uso en la promoción del crecimiento de cultivos de importancia económica. *Revista CENIC Ciencias Biologicas* 46, 63–76.

Rivas-Franco, F., Hampton, J.G., Narciso, J., Rostás, M., Wessman, P. *et al.* (2020) Effects of a maize root pest and fungal pathogen on entomopathogenic fungal rhizosphere colonization, endophytism and induction of plant hormones. *Biological Control* 150, 104347.

Roberts, D.W. and St Leger, R.J. (2004) Metarhizium spp., cosmopolitan insect-pathogenic fungi: Mycological aspects. *Advances in Applied Microbiology* 54(1), 1–70.

Ríos-Moreno, A., Garrido-Jurado, I., Resquín-Romero, G., Arroyo-Manzanares, N., Arce, L. *et al.* (2016) Destruxin A production by *Metarhizium brunneum* strains during transient endophytic colonisation of Solanum tuberosum. *Biocontrol Science and Technology* 26(11), 1574–1585.

Rondot, Y. and Reineke, A. (2019) Endophytic *Beauveria bassiana* activates expression of defence genes in grapevine and prevents infections by grapevine downy mildew *Plasmopara viticola*. *Plant Pathology* 68(9), 1719–1731.

Saikkonen, K., Mikola, J. and Helander, M. (2015) Endophytic phyllosphere fungi and nutrient cycling in terrestrial ecosystems. *Current Science* 109, 121–126.

Sánchez-Rodríguez, A.R., Raya-Díaz, S., Zamarreño, Á.M. and García-Mina, J.M. (2018) An endophytic *Beauveria bassiana* strain increases spike production in bread and durum wheat plants and effectively controls cotton leafworm (Spodoptera littoralis) larvae. *Biological Control* 116, 90–102.

Sasan, R.K. and Bidochka, M.J. (2013) Antagonism of the endophytic insect pathogenic fungus *Metarhizium robertsii* against the bean plant pathogen *Fusarium solani* f. sp. *phaseoli*. *Canadian Journal of Plant Pathology* 35(3), sp. .

Segarra, G., Casanova, E., Avilés, M. and Trillas, I. (2010) Trichoderma asperellum strain T34 controls fusarium wilt disease in tomato plants in soilless culture through competition for iron. *Microbial Ecology* 59, 141–149.

Shoresh, M., Harman, G.E. and Mastouri, F. (2010) Induced systemic resistance and plant responses to fungal biocontrol agents. *Annual Review of Phytopathology* 48(1), 21–43.

Sinno, M., Ranesi, M., Di Lelio, I., Iacomino, G., Becchimanzi, A. *et al.* (2021) Selection of endophytic *Beauveria bassiana* as a dual biocontrol agent of tomato pathogens and pests. *Pathogens* 10(10), 1242.

Siqueira, A.C.O., Mascarin, G.M., Gonçalves, C.R., Marcon, J., Quecine, M.C. *et al.* (2020) Multi-trait biochemical features of *Metarhizium species* and their activities that stimulate the growth of tomato plants. *Frontiers in Sustainable Food Systems* 4, 137.

Sorokan, A., Cherepanova, E., Burkhanova, G., Veselova, S., Rumyantsev, S. *et al.* (2020) Endophytic *Bacillus* spp. as a prospective biological tool for control of viral diseases and non-vector *Leptinotarsa decemlineata* Say. in *Solanum tuberosum* L. *Frontiers in Microbiology* 11, 569457.

Sui, L., Lu, Y., Yang, H., Liu, Z., Wang, L. *et al.* (2024) Endophytic *Beauveria bassiana* promotes sunflower growth and confers resistance against *Sclerotinia disease*. *BioControl* 70, 119–130.

Sun, X., Wang, N., Li, P., Jiang, Z., Liu, X. *et al.* (2020) Endophytic fungus *Falciphora oryzae* promotes lateral root growth by producing indole derivatives after sensing plant signals. *Plant, Cell & Environment* 43(2), 358–373.

Tall, S. and Meyling, N.V. (2018) Probiotics for plants? Growth promotion by the entomopathogenic fungus *Beauveria bassiana* depends on nutrient availability. *Microbial Ecology* 76(4), 1002–1008.

Toscano-Verduzco, F.A., Cedeño-Valdivia, P.A., Chan-Cupul, W., Hernández-Ortega, H.A., Ruiz-Sánchez, E. *et al.* (2020) Phosphates solubilization, indol-3-acetic acid and siderophores production by *Beauveria brongniartii* and its effect on growth and fruit quality of *Capsicum chinense*. *Journal of Horticultural Science and Biotechnology* 95(2), 235–246.

Vahabi, K., Reichelt, M., Scholz, S.S., Furch, A.C., Matsuo, M. *et al.* (2018) *Alternaria brassicae* induces systemic jasmonate responses in *Arabidopsis* which travel to neighboring plants via a *Piriformsopora indica* hyphal network and activate abscisic acid responses. *Frontiers in Plant Science* 9, 626.

Varga, T., Hixson, K.K., Ahkami, A.H., Sher, A.W., Barnes, M.E. *et al.* (2020) Endophyte-promoted phosphorus solubilization in Populus. *Frontiers in Plant Science* 11, 567918.

Varma, A., Verma, S., Sahay, N., Bütehorn, B. and Franken, P. (1999) *Piriformospora indica*, a cultivable plant-growth-promoting root endophyte. *Applied and Environmental Microbiology* 65(6), 2741–2744.

Vega, F.E., Goettel, M.S., Blackwell, M., Chandler, D., Jackson, M.A. *et al.* (2009) Fungal entomopathogens: New insights on their ecology. *Fungal Ecology* 2(4), 149–159.

Verma, S.K. and White, J.F. (2018) Indigenous endophytic seed bacteria promote seedling development and defend against fungal disease in browntop millet (*Urochloa ramosa* L.). *Journal of Applied Microbiology* 124(3), 764–778.

Vinale, F., Sivasithamparam, K., Ghisalberti, E.L., Marra, R., Woo, S.L. *et al.* (2008) Trichoderma–plant–pathogen interactions. *Soil Biology and Biochemistry* 40(1), 1–10.

Vurukonda, S., Vardharajula, S., Shrivastava, M. and Sk, Z.A. (2016) Enhancement of drought stress tolerance in crops by plant growth promoting rhizobacteria. *Microbiological Research* 184, 13–24.

Wagner, B.L. and Lewis, L.C. (2000) Colonization of corn, Zea mays, by the entomopathogenic fungus *Beauveria bassiana*. *Applied and Environmental Microbiology* 66(8), 3468–3473.

Wang, Q. and Xu, L. (2012) Beauvericin, a bioactive compound produced by fungi: A short review. *Molecules* 17(3), 2367–2377.

Xu, L., Wang, J., Zhao, J., Li, P., Shan, T. *et al.* (2010) Beauvericin from the endophytic fungus, *Fusarium redolens*, isolated from *Dioscorea zingiberensis* and its antibacterial activity. *Natural Product Communications* 5(5), 1934578X1000500527.

Yousef-Yousef, M., García-Espinoza, F., García del Rosal, M.J. and Quesada-Moraga, E. (2024) Guardians within: Entomopathogenic fungi-driven antibiosis and compensatory growth combines to reduce the damage of herbivores to melon plants. *Journal of Pest Science* 98, 291–308.

Yu, H., Zhang, L., Li, L., Zheng, C., Guo, L. *et al.* (2010) Recent developments and future prospects of antimicrobial metabolites produced by endophytes. *Microbiological Research* 165(6), 437–449.

Zalba, P., Bravo, O.A., Garay Schiebelbein, M.M. and Peinemann, N. (2007) Moisture and temperature effect on soil phosphorus availability. *Agrochimica* 51, 55–63.

Zhang, L., Li, W., Tao, Y., Zhao, S., Yao, L. *et al.* (2019) Overexpression of the key virulence factor 1, 3–1, 4-β-D-glucanase in the endophytic bacterium *Bacillus halotolerans* Y6 to improve Verticillium resistance in cotton. *Journal of Agricultural and Food Chemistry* 67(24), 6828–6836.

18 Challenges and Opportunities for Conservation Biological Control of Arthropods Under Climate Change

Armando Alfaro-Tapia[1], Kévin Tougeron[2], Enrique Maldonado-Santos[3], Cécile Le Lann[4], Jacques Brodeur[5], Joan van Baaren[4] and Blas Lavandero[3]*

[1]Centro Regional de Investigación e Innovación para la Sostenibilidad de la Agricultura y los Territorios Rurales, Centro Ceres, Pontificia Universidad Católica de Valparaíso, Quillota, Chile; [2]Écologie des Interactions et Changements Globaux, Institut de Recherche en Biosciences, Université de Mons, Mons, Belgium; [3]Laboratorio de Control Biológico, Instituto de Ciencias Biológicas, Universidad de Talca, Talca, Chile; [4]Université de Rennes, CNRS, ECOBIO (Ecosystèmes, Biodiversité, Évolution), UMR, Rennes, France; [5]Département de Sciences Biologiques, Institut de Recherche en Biologie Végétale, Université de Montréal, Montréal, Canada

Abstract

This chapter investigates the potential impacts of climate change on conservation biological control (CBC), focusing on pest–natural enemy interactions and ecosystem resilience. CBC is a commonly employed strategy to promote an environment that supports the persistence of diverse natural enemy populations in agro-ecosystems over both time and space, to regulate pest species. CBC utilizes habitat manipulation to enhance natural enemy populations, promoting pest control through SNAP (Shelter, Nectar, Alternative Prey). Climate change threatens these strategies by altering species distribution, phenology and abundance, exacerbated by land use changes. Such disruptions may lead to mismatches in spatial and temporal interactions between pests and natural enemies, challenging CBC effectiveness. However, there are opportunities for adaptation, including exploiting new resource subsidies due to desynchronization. This chapter highlights the dual role of habitat management in fostering trophic interactions while mitigating antagonistic impacts among biological control agents, amidst ongoing challenges and opportunities posed by climate change.

18.1 Introduction

This chapter examines how climate change may impact conservation biological control (CBC). CBC is the deliberate modification of habitats through management practices that facilitate natural enemies present in ecosystems to achieve greater control of pests (Landis *et al.*, 2000; Heimpel and Mills, 2017). CBC programmes include habitat manipulation tactics such as plant diversification within and at the periphery of

*Corresponding author: blavandero@utalca.cl

DOI: 10.1079/9781800625099.0018

crop fields to achieve SNAP (Shelter, Nectar, and Alternative Prey or hosts) (Gurr *et al.*, 2017). CBC habitat management also includes tactics to limit pesticide-induced mortality and fitness reduction of natural enemies of pests (Roubos *et al.*, 2014). Sometimes, CBC is confused with habitat management to reduce pest infestation, which consists of limiting pest reproduction and attractiveness on cultivated plants. This is generally achieved through the use of repellents or plant species that change the behaviour of the pest species, thereby reducing target crop colonization or diluting the visual and chemical stimuli to the pest (Price, 1997).

The fitness of both pests and natural enemies depends on their tolerance to abiotic factors, mainly temperature, and variations in species interactions within a community (Tylianakis *et al.*, 2008). Current populations of many arthropod taxa (e.g. pollinators, natural enemies of agricultural pests) are in serious decline across many geographic zones (Harvey *et al.*, 2023). Climate change is expected to induce further shifts in their distribution, phenology and abundance (Renner and Zohner, 2018; Harvey *et al.*, 2023). As global climate conditions change continuously with the increase in mean temperature and the occurrence of unpredictable extreme events (e.g. heatwaves, droughts, fires, unseasonal cold spells and floods) (Bathiany *et al.*, 2018), arthropods will be affected, destabilizing the composition, functioning and resilience of ecosystems (Sentis *et al.*, 2022). Land use change and agricultural intensification have exacerbated the detrimental effects of climate change worldwide, Land use change, agricultural intensification and climate change have detrimental effects that are deeply linked together in their causes (e.g.change in energy fluxes or biogeochemical cycles) and consequences (e.g. change in realized niche, change in microclimatic conditions) (Dale, 1997; Ostberg *et al.*, 2015). Those pressures are additively or synergistically affecting to varying degrees biodiversity and the services ensured by arthropods in agroecosystems (pollination, herbivory, remineralization, predation) (Prather *et al.*, 2013; Marta *et al.*, 2021; Sohlström *et al.*, 2022).

Even in simplified agro-ecosystems, the nature and magnitude of climate change impacts remain difficult to predict because of many ecological and evolutionary intermingling factors such as the sensitivity of individual species, the diversity of climatic factors, the complexity of guild and trophic interactions, the unpredictability of extreme climatic events and the potential of arthropods to mitigate and adapt to climate change. Nevertheless, pest diversity and crop production losses are predicted to increase overall with change (Harvey *et al.*, 2023) (see Chapter 3 of this book). Climate change is therefore expected to further impact CBC. A major concern relates to expected spatial and temporal mismatches between pests and their natural enemies, as well as a loss of synchrony between foraging biocontrol agents and the availability of plant resources they may use (e.g. pollen, nectar) (Hance *et al.*, 2007; Gérard *et al.*, 2020) (see Chapter 2 of this book). Nevertheless, many opportunities may arise due to the positive impact of desynchronization by promoting the exploitation of new resources, including greater effects of resource subsidies (Tylianakis *et al.*, 2004).

We first describe how habitat management and pesticide selectivity may, on the one hand, promote trophic interactions and, on the other hand, reduce antagonistic interactions among biological control agents. Next, we explore the impact that climatic variation may have on ecosystem services provided by different types of natural enemies within the specific context of CBC. The last two sections identify and discuss the challenges and opportunities for efficient implementation of CBC strategies under climate change.

18.2 Conservation Biological Control: Why Species Interactions are Important

Several ecosystem services, such as pest regulation, depend on multitrophic interactions (Wood *et al.*, 2015). A desirable goal of CBC is to preserve or re-establish self-sustaining agro-ecosystems that ensure the persistence of natural enemies favouring trophic interactions that reduce pest populations (Begg *et al.*, 2017). CBC history may be seen as a progressive advancement in our knowledge of the biology and ecology of interactions between plants, pests and their natural enemies (Begg *et al.*, 2017; Heimpel and Mills, 2017). This advancement has been achieved by increasing our knowledge of functional complementarity among natural enemy communities (Holland *et al.*, 2016, 2020).

These trophic interactions play a vital role in maintaining the balance and stability of the ecosystem, as well as enhancing the effectiveness of natural enemies in controlling pest populations (Holland *et al.*, 2016). CBC originates from the understanding that agricultural landscapes are challenging environments for natural enemies due to frequent and intensive disturbance regimes that may reduce the availability of vital resources at crucial developmental stages. Therefore, CBC efforts tend to focus on maintaining viable populations of focal species by deploying species-specific management tools to achieve SNAP (Schmitz and Barton, 2014; Laws, 2017), as well as weakening the intensity of antagonistic interactions among natural enemies to enhance pest suppression (Wood *et al.*, 2015; Holland *et al.*, 2016). The maintenance of the integrity of interactions under climate change has become a high priority (Chidawanyika *et al.*, 2019) since climate change is likely to have severe impacts on the composition and functioning of agro-ecosystems, which could alter the degree of success of conservation biological control (Furlong and Zalucki, 2017; Skendžić *et al.*, 2021).

The dynamics of pest populations are influenced by climatic conditions, food resources (bottom-up forces) and natural enemy–pest interactions (top-down forces) (Vidal and Murphy, 2017; Le Gal *et al.*, 2020). Increasing arthropod abundance and functional diversity by enhancing biodiversity can lead to lower pest populations by the direct and indirect effect on natural enemies (Gurr *et al.*, 2017; Sentis *et al.*, 2022; Tougeron *et al.*, 2022). Predation, including parasitism, has a variety of important and far-reaching effects on structuring trophic network composition and influencing ecosystem processes, including the biocontrol of pests.

Traditionally, the direct consumption of pests has been regarded as the primary mechanism for reducing pests in natural ecosystems (Stenberg *et al.*, 2021). However, the indirect effects of predators, such as inducing anti-predator behaviour in prey, may significantly impact pest population dynamics (Lima, 1998; Preisser and Bolnick, 2008), leading to trophic cascade (Beckerman *et al.*, 1997; Nelson *et al.*, 2004). These non-consumptive effects can sometimes even outweigh the direct impact of predation through consumption. This intricate network of relationships among pests and their natural enemies is constantly changing; each individual/species in a node has a range of offensive and defensive tactics and counterstrategies to respond to their interacting individuals/species (Chidawanyika *et al.*, 2019). For example, through the provision of non-crop resources using cover crops, the early recruitment of natural enemies can translate to a decrease in pest populations via the provision of alternative hosts (Peñalver-Cruz *et al.*, 2020; Alvarez-Baca *et al.*, 2023) and/or food resources (Judt *et al.*, 2023). Theoretical models have suggested that CBC can alter a predator's probability of encountering prey, which may determine its ability to suppress herbivore densities (Kean *et al.*, 2003). Although the provision of alternative hosts and food resources (e.g. nectar) benefits natural enemies, future studies should also include the interaction of these management practices at the local level with the structural complexity of a habitat. Shelter and niche partitioning may have an important role in maintaining natural enemy abundance, and the overall seasonal effect on pest populations depends on landscape complexity, as local and landscape effects may interact in positive or negative ways.

As interactions occur not only among species at different trophic levels but also within trophic levels (i.e. intraguild predation and competition) (Jeavons *et al.*, 2023), direct and indirect interactions need to be studied to grasp the full extent of how CBC can promote trophic cascades to the benefit of farmers (Fig. 18.1). In contrast to direct interactions, indirect interactions occur between two species that are separated in time or space and require the involvement of at least one additional mediating species (Begg *et al.*, 2017; Luliano and Gratton, 2020). Natural enemies may rely on alternative prey/host availability to succeed when their main pest target is scarce (Luliano and Gratton, 2020). For instance, it has been suggested to incorporate alyssum *Lobularia maritima* (L.) Desv. (Brassicaceae) as a companion plant and use *Ephestia kuehniella* Zeller (Lepidoptera: Pyralidae) e.ggs as an alternative prey source to promote populations of *Orius laevigatus* (Fieber) (Hemiptera: Anthocoridae), a biocontrol agent of *Macrosiphum euphorbiae* Thomas (Hemiptera:

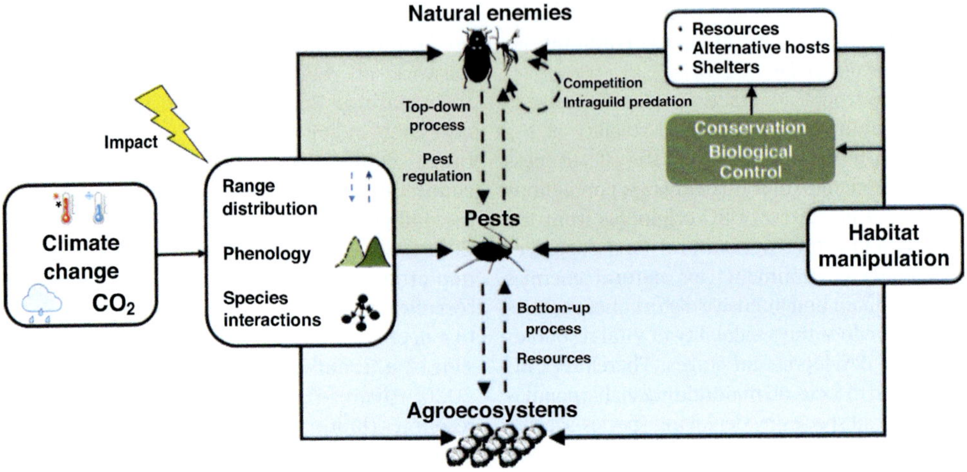

Fig. 18.1. Effect of climate change on CBC. The impacts on the distribution, phenology and interactions of different system components in a food chain are shown. Solid lines indicate direct effects, and dashed lines indicate each level's direct and indirect interactions.

Aphididae) in strawberry cropping systems (Zuma *et al.*, 2023).

18.3 Interactions and Climatic Variation

Species interactions are crucial to community structure and function, and climate change can affect the nature and intensity of intra- and interspecific interactions and trophic cascades (Pekár *et al.*, 2015) by affecting species characteristics (Abrego *et al.*, 2021). For instance, climate-induced changes in phenology may not equally affect all individuals, species or trophic levels within a system (Tougeron *et al.*, 2019). As such, species with fast dispersal or more plastic behaviours will have a better chance to disperse and track their habitats under new conditions than those with low dispersal rates, which may lag (Barton and Ives, 2014). These scenarios result in disrupted population dynamics (Renner and Zohner, 2018), in which food web structures and interactions will be modified.

Under warmer climates, trophic interactions may be more intense because of higher activity rates of ectotherms (Pekár *et al.*, 2015) or even less intense as, for example, temperature could influence the emission of info-chemicals (Sentis *et al.*, 2015). Therefore, the intensity of interactions and consequently the biocontrol efficacy of some predators (e.g. spiders) are expected to differ between areas at different scales with more climatic variation (Schemske *et al.*, 2009). However, studies have shown that the effects on survival, development and reproduction could be negative for herbivores, affecting natural enemy fitness indirectly and potentially hindering biocontrol services (Han *et al.*, 2019).

18.3.1 Predation

Some predators can have a significant effect on pest regulation (Petráková *et al.*, 2016), but sometimes, generalist predators have been considered to have limited value in the biological control of insects (Cuff *et al.*, 2022). This might be particularly true for specialist predators (stenophagy) such as coccinellids and predatory mites, but generalist predators (euryphagy), which are often much more abundant in agroecosystems, also contribute to the suppression of pests as they may act in synergy (Losey and Denno, 1999). For example, spiders have been found to suppress agricultural pests in most studies. Results from a recent meta-analysis suggest that in 79% of cases, a large mean effect size of increased spider density on pest suppression can be found (see Chapter 6). There is also some evidence that there is an increase in the predation pressure on herbivores toward the tropics in terrestrial ecosystems (Roslin *et al.*, 2017). A few meta-analyses (Michalko *et al.*, 2019) have even shown that generalist predators can indeed trigger trophic cascades and improve crop performance.

Nevertheless, the effect of temperature in these studies may be confounded by other factors related to the latitude, such as a more stable environment, longer time for coevolution and higher specialization (Schemske *et al.*, 2009). In any case, predatory interactions may change, and therefore the outcome from extreme climatic variability could be very context dependent. In a general review study synthesizing broad trends in the impacts of global change drivers on over 1000 biotic interactions from 688 separate published studies, Tylianakis *et al.* (2008) found that climate change had variable effects on predation. Recently, in a study investigating the combined effect of heatwaves and risk of predation (simulated as an exposure to scent from black widow spiders *Latrodectus hesperus* Chamberlin and Ivie (Araneae: Theridiidae)) on resource foraging of field cricket *Gryllus lineaticeps* Stål (Orthoptera: Gryllidae), the authors concluded that crickets invested less energy in self-maintenance, which could impair key physiological processes (Stahlschmidt *et al.*, 2024), adding to some evidence that temperature may benefit predators. Other non-additive effects, with mutual inhibition between predation and heat shocks, on prey phenotype (e.g. body size) and behavioural responses have also been suggested (e.g. Sentis *et al.*, 2017), demonstrating further the difficulty of identifying general patterns among different study systems. In any case, these results highlight the high variability of climate variation effect on the interactions between some arthropods important for biological control, as plant damage and crop yield respond to herbivory with different intensities (Halaj and Wise, 2001).

18.3.2 Parasitism

For trophic systems such as plant–herbivore–parasitoid, this is highly critical, as different responses to thermal effects may lead to decoupled phenological cycles (e.g. Hance *et al.*, 2007; Furlong and Zalucki, 2017) (see Chapter 2 of this book). Parasitoids may be more vulnerable to climate warming than their hosts, as the estimated optimum temperature for various parasitoids is consistently lower (Furlong and Zalucki, 2017). Deleterious effects of warming may also be disproportionate up the food chain, with more physiological damage on predators or parasitoids than on their prey or hosts (Tougeron *et al.*, 2019). This widespread phenomenon suggests that trophic interactions under conditions of extreme temperatures may be weakened because of a reduction in the strength of top-down or bottom-up processes (Rosenblatt and Schmitz, 2016). Many parasitoids have limited host ranges and may be more sensitive to heat than plants and their herbivore hosts (Harvey *et al.*, 2020). This is because they are affected both directly and indirectly through the effects of high temperature on the insect host and the host plant (Hance *et al.*, 2007; van Baaren *et al.*, 2010).

Extreme temperature events can also modify the outcome of intrinsic competition in parasitoids (Chen *et al.*, 2019), as parasitoids of the same and of different species often lay eggs in the same individual host (Harvey *et al.*, 2013), altering the dynamics and structure of the host–parasitoid multitrophic interactions. Specifically for parasitoids, climate change has the potential to generate additional evolutionary pressures (Tougeron *et al.*, 2019) (see more below in the section entitled 'Competition).

18.3.3 Competition

Ecological competition arises when two or more organisms strive to acquire limited resources (e.g. food and space), leading to a negative impact on their fitness (Ode *et al.*, 2021). Competition between natural enemies of pests may have both beneficial and detrimental effects on pest management (Li *et al.*, 2021).

Competition can take place in two ways: through interference, where one natural enemy directly hinders another's ability to control pests; or indirectly through exploitative competition, when one species is more efficient at locating and exploiting limited resources than a competing species (Jeavons *et al.*, 2023). Effective CBC requires an understanding of these competitive interactions in the short term, so a balance between natural enemy coexistence and pest population suppression can exist (Muller and Brodeur, 2002; Ode *et al.*, 2021). Interference competition among natural enemies in CBC is a critical interaction where natural enemies engage in confrontations over prey or habitat resources. This type of competition can influence the control of pest populations by affecting their foraging behaviours (Mills and Heimpel, 2018) and the reproductive capacity of the predatory species involved. For instance, larger or more aggressive predators may outcompete and displace smaller or less aggressive predators. This is the case of *Harmonia axyridis* (Pallas) (Coleoptera: Coccinellidae), which has been shown to outcompete other smaller species of coccinellids, including native species in its invasive range, with different types of evidence in different geographical regions (Lucas *et al.*, 2002; Michaud, 2002; Nóia *et al.*, 2008; Kenis *et al.*, 2020; Ortiz-Martínez *et al.*, 2020). These shifts can lead to situations where the natural enemies, that CBC relies upon, face increased competition for limited resources, such as prey/host or suitable habitats (Begg *et al.*, 2017). Therefore, if the natural enemies that are most effective at controlling the target pest are outcompeted by others, the overall effectiveness of CBC strategies could be compromised.

18.3.4 Intraguild predation and competition

Intraguild predation (IGP), as prey/host resources, may be mediated by both predator–predator or predator–parasitoid interactions and habitat structure and can reduce pest suppression (Finke and Denno, 2002). Predators sharing a common extraguild prey (in the context of CBC, a pest) may engage in predation of each other (Sentis *et al.*, 2022; Jeavons *et al.*, 2023) or by feeding on a parasitized host (Langellotto *et al.*, 2006). In addition, IGP may be especially detrimental when generalist predators, who typically have broad feeding habits, target specialized natural enemies, thereby hindering the control of a particular pest species (Ortiz-Martínez *et al.*, 2020).

Several studies have demonstrated that IGP often arises among members of the same species (cannibalism) or between different species (interspecific IGP), particularly when food resources are inadequate (Straub *et al.*, 2007), with some traits of enemies (e.g. hunting mode) and their prey (e.g. catchability and profitability) promoting this interaction (Hambäck *et al.*, 2021). However, CBC may limit IGP, as increasing the complexity of an agroecosystem may benefit predators and thus promote the biocontrol of pests. For instance, the presence of floral resources has been observed to reduce IGP by increasing larvae survival of two sympatric ladybird species and enhancing pest control in a greenhouse setting (Liang *et al.*, 2022). In a different study, landscape complexity did not affect the overall frequency of IGP interactions (estimated through molecular methods) among four coccinellid species, although these were highly common, independent of the surrounding landscape and seasonal abundance of these species and their prey (Ortiz-Martínez *et al.*, 2020).

Therefore, the impact of IGP must be considered when developing biological control programmes (Boivin and Brodeur, 2006), and strategies that might include enhancing habitat diversity may render IGP less likely by reducing the probability of encounters between species. As conventional agriculture may promote weak omnivory coupled with strong intraguild predation due to monocultures, these could lead to the loss of an intraguild prey and increase dominance of one species, thereby shortening food chains (Takimoto *et al.*, 2012). This situation would consequently promote the population abundance of pests with high growth rates as a result of the loss of control by their natural enemies. As a consequence of changes in their foraging behaviour while interacting with more intraguild predators or with less strict interspecific competitors, the biocontrol potential of natural enemies can further be disrupted (Finke and Denno, 2002).

On the other hand, IGP can theoretically improve pest suppression if one meso-predator is less effective than another at target pest suppression. IGP would improve the performance of the top predator compared to a monotypic pest diet (Michalko *et al.*, 2019). However, the use and effect of IGP for generalist predators remain controversial. IGP can be an adaptive strategy for several reasons and some of them are not mutually exclusive. (i) IGP can help cope with energy limitation because it is an additional source of energy during periods or in environments with scarce prey (Haghani *et al.*, 2019). (ii) A meso-predator may represent a food of high nutritional quality from which a top predator can extract all the nutrients needed for growth, development and reproduction (Guo *et al.*, 2016). (iii) Some predators, although probably a minority, are protein limited (Toft *et al.*, 2019). As well as the above, there is some evidence that habitat modification can affect biological control services by many predators/parasitoids (Tylianakis and Romo, 2010), as heterogeneous habitats can provide more shelters from IGP, enhance niche complementarity among predators, and consequently increase the predation pressure on specific pest species (Tylianakis and Romo, 2010; Michalko *et al.*, 2019). IGP is especially noticeable among spiders in agro-ecosystems with low heterogeneity compared to most natural habitats (Marshall and Rypstra, 1999).

The consequences of climate change for intraguild interactions and pest control were modelled by Laubmeier *et al.* (2023) based on empirical studies, demonstrating that the foraging behaviour of arthropod predators is influenced by an interaction between temperature and predator body size. The model predicts that optimal communities would often include predators with complementary foraging activity (based on the temperature at which they are most active, indicating that biological control could occur over a wide temperature range). Under an increase in average temperature by 2.5°C and an increase in daily variability by 10%, the optimal community changed drastically but at 20°C and 35°C, foraging activity reduced the negative effects of intraguild interference on pest consumption. The authors suggest that favouring some predators, such as Lycosidae spiders (Aranea: Arachnida) and large ground beetles (e.g. *Pterostichus* Bonelli and *Poecilus* Bonelli (Coleoptera: Carabidae)), through reduced tilling and increased quantities of refuge habitats may promote CBC, even with increased IGP.

18.4 Challenges for Conservational Biological Control

As climate change increases mean temperatures throughout the globe, many negative aspects of biological control may appear as a consequence (Table 18.1), such as the lack of synchronization between a parasitoid and its hosts due to differences in thermal requirements or host shifts. If the life cycle of a pest is shortened and the number of generations increases, natural enemies may fail to maintain pest populations under the economic threshold. The creation and loss of phenological mismatches could become common given the substantial variation in phenological responses to climate change among taxa (Miller-Rushing *et al.*, 2010). However, more evidence of loss of synchronization is required to foresee general patterns, if any, identify the consequences for population dynamics (Damien and Tougeron, 2019), and understand how these may change species composition and lead to variance in trophic cascades, which could impact pest control under climatic variation.

The negative impact of desynchronization between interacting species is expected to be higher for temperate species, occurring either very early in the spring or very late in autumn when the probability of emergence of a given natural enemy in the absence of any potential prey is the highest (Forrest and James, 2011). On the other hand, a positive impact of desynchronization would result in more prey for generalist natural enemies as shifts in phenology could alleviate existing mismatches and promote the exploitation of newly available resources (Miller-Rushing *et al.*, 2010). It is also worth mentioning that many species have evolved different strategies for coping with the loss of temporal synchronization. It has been observed that plasticity and local adaptation could be involved in the response of aphid parasitoid overwintering strategies to different thermal environments (Tougeron *et al.*, 2019). For instance, diapause incidence in populations of harsh winter conditions is higher than in populations of mild winter conditions (Tougeron *et al.*, 2018), showing that non-diapause insects may be winter-active (Alfaro-Tapia *et al.*, 2022). Therefore, aphid parasitoid populations might possess genetic variation that allows them

Table 18.1. Major challenges and opportunities for CBC under climate change.

Major challenges	Major opportunities	Examples where CBC can harness these opportunities
Lack of synchronization between a parasitoid and its hosts due to differences in thermal requirements or host shifts.	Understanding the factors that affect the dynamics of life cycles of prey/host and natural enemies to better synchronize population cycles.	Use or exploitation of symbionts that may provide adaptation, as well as providing additional microclimatic and buffering conditions to improve phenological and thermal synchronization through habitat manipulation.
Targeted habitat manipulation strategies involving increasing nectar and pollen delivery may be required for biological control in general.	Managing plants and ground covers to supplement natural enemies with alternative resources, providing shelter during extreme weather conditions.	Using an alternative plant host or flower strip to achieve early colonization and abundance and shelter during extreme heat or cold.
Sublethal concentrations of pesticides significantly impacting pests and their natural enemies, reducing the chances for biological control under climate change.	Exploring temperature ranges of species to understand how they respond to climate change and sublethal doses of insecticides.	Selection and combination of augmentative BC of organisms that can maintain their populations even after extreme weather events.
The degradation of forest and shrub ecosystems, driven by climate change and the expansion of agriculture, significantly diminishes the potential for implementing CBC strategies on a larger scale.	Exploring how CBC strategies operate in relation to the life traits and functional ecology of species, along geographical and climatic gradients.	Landscape features in combination with local management that help to mitigate extreme weather impacts on agent fitness and population stability
Local extinction of plant resources, host and prey availability modified by climatic variation and land use change.	Effectiveness of natural enemies through natural selection to face local extinctions due to climate change.	Including more resilient agents to varying environmental conditions, higher resistance to toxins or increased fitness when feeding on non-prey food.

to adjust the synchrony or asynchrony of their phenology under climate change events (Damien and Tougeron, 2019). On the other hand, in Hymenopteran parasitoids, another strategy in unfavourable environments, characterized by smaller hosts and host scarcity, is to switch the sex ratio of their offspring towards males, the less costly sex (Moiroux *et al.*, 2014). However, how it may affect the control of a pest in the longer term remains unclear (Ode and Hardy, 2008).

Habitat manipulation strategies involving the extended provision of nectar and pollen may be required not only to increase resource availability but also to provide shelter during extreme weather conditions and therefore have the potential for reducing the effect of potential desynchronizations, as natural enemies with additional SNAP may be able to arrive earlier and sustain populations for larger periods. On the other hand, as global warming increases the distribution of herbivores, this may be accompanied by host plants shifting (Thomas *et al.*, 2001), and herbivores may then escape predation or parasitism. CBC may be pivotal to maintaining more natural enemies, including species foraging on non-crop or other crop resources, where target pest herbivores may have host-shifted.

Combinations of agrochemicals with other stressors, such as climate change, are characteristic of agro-ecosystems (Van den Brink *et al.*, 2018). Pesticide effectiveness is strongly influenced by the active ingredients and environmental factors, which impact the insecticide concentrations at a spatial and temporal scale (Müller, 2018) (see Chapter 14 of this book). Sublethal concentrations of pesticides significantly impact pests and their natural enemies (Müller, 2018) and therefore the timing of key life stages, changes in reproductive strategies, changes in behaviour and migration patterns, and even population declines, as well as modifications in orientation and foraging patterns of natural enemies (Bartling *et al.*, 2024). As an example, *Aphidius ervi* Haliday (Hymenoptera: Braconidae) exposed to pyrethroid insecticides exhibited an irreversible, unco-ordinated ovipositor extrusion and consequently failed to lay eg.gs (Desneux *et al.*, 2004), and predators such as *Coccinella septempunctata* Linnaeus (Coleoptera, Coccinellidae) modified their foraging behaviour and tended to stay on plant parts known to receive fewer pesticides during treatment, thus demonstrating a clear repellent effect (Wiles and Jepson, 1994).

Declines in insect richness have been found to be greater in semi-natural grassland areas embedded in a landscape with a high number of arable fields compared with areas surrounded by forest (Seibold *et al.*, 2019). These effects, although not yet studied, may change the availability of hosts and prey, as well as reduce the availability of other trophic resources such as nectar and pollen. Evidence of landscape effects on the overwintering active parasitoids has shown that landscape features could mitigate extreme weather (Tougeron *et al.*, 2016). In this study, smaller insects from simple landscapes were more cold-resistant. Since parasitoid size is linked to fecundity and longevity, under stressful temperature conditions fitness could be reduced. If climate change exacerbates the loss of these important habitats, natural enemies will not be able to benefit from the mitigating properties of these more complex landscapes (see also, for carabids, Alford *et al.*, 2023). Local extinction of plant resources and host and prey availability may also be modified by both climatic variation and land use change. Recent evidence of the reduction of abundant prey host resources due to climate change corroborates this (van Klink *et al.*, 2023). Fundamental changes occurring in insect assemblages may have broad repercussions for food webs and ecosystem functioning, although more data are needed to understand the role of climate change.

Climatic variation may also challenge the effectiveness of natural enemies through natural selection. As an example, Lavandero and Tylianakis (2013) present some evidence of the loss of phylogenetic congruence among populations of hosts and parasitoids due to a reduction in the mean number of cold days in the central valley of Chile. In this study, the genotypes were determined for host and parasitoid *Aphelinus mali* (Haldeman) (Hymenoptera: Aphelinidae), a highly specialist parasitoid of the woolly apple aphid *Eriosoma lanigerum* Hausmann (Hemiptera: Eriosomatinae), which was introduced in the 1920s and under climatic selection (with 9–11 generations per year for the aphid). The climatic conditions of central Chile (17–37°S) have significantly changed, warming in the central valley and western Andes to 0.25°C/decade (Falvey and Garreaud, 2009). The narrowing of host–genotype–niche breadth observed here for the parasitoids could reduce the resilience of the network to changes in host genetic structure or invasion by novel host genotypes, thereby inhibiting biological control. Similarly, selection could occur at the holobiont level. Aphids notoriously harbour obligate and secondary endosymbionts (see Chapter 12), some of which relate to host defences as well as heat tolerance (Tougeron and Iltis, 2022), but their potential implication for biological control, including how CBC can mitigate and reduce natural enemy mortality due to extreme weather, has not yet been explored.

18.5 Opportunities

Several solutions can be implemented to sustain CBC strategies in the context of climate change (see Table 18.1). They necessarily include a better understanding of how each individual species responds to climate change and associated change drivers, of the complex ecological mechanisms underlying their responses, and of the functioning of agricultural exploitations and solutions put in place as part of CBC (Lehmann *et al.*, 2020).

One of the major obstacles to the success of biological control is the fact that the life cycle of crop pests and their potential natural enemies is not necessarily synchronized (Thomson *et al.*, 2010), especially when the pests establish before their natural enemies in the fields (Gómez-Marco *et al.*, 2016). Even in cases where the arrival of pests and natural enemies is almost simultaneous, pest populations tend to develop faster than those of their natural enemies, due to reproductive strategies such as parthenogenesis, faster development rates, or because they take better advantage of environmental warming, therefore limiting the regulatory impact of natural enemies (Thomson *et al.*, 2010). In contrast, there may not be enough pests at suitable stages available early in the season to support foraging activity, reproduction and population persistence of natural enemies, leading to potential high mortality rates or dispersal out of the crop (Grabenweger *et al.*, 2007).

To remove these caveats, the management of plants in and around the crop with selected ground covers to supplement natural enemies with alternative resources can favour the settlement of significant natural enemy populations on alternative hosts/prey before the arrival of the pests (Alvarez-Baca *et al.*, 2020). For example, in *Citrus* orchards in Spain, Gómez-Marco *et al.* (2016) showed that ground covers with grasses (Poaceae) attract stenophagous aphids, which are used as alternative prey by predators before the main *Citrus* pest, the aphid *Aphis spiraecola* Patch (Hemiptera: Aphididae), arrives. In apple orchards of New Zealand, Irvin *et al.* (2006) showed that covers of buckwheat (*Fagopyrum esculentum* Moench (Polygonaceae)) and alyssum (*L. maritima*) below the trees significantly increased parasitism of the light-brown apple moth *Epiphyas postvittana* (Walker) (Lepidoptera: Tortricidae), by the parasitoid *Dolichogenidea tasmanica* (Cameron) (Hymenoptera: Braconidae). The longevity of the parasitoids increases when they can feed on the flowers of these two plant species, allowing a better synchronization with the aphid host at the beginning of the infestation.

Developing landscapes and field margins that allow the survival and retention of many natural enemy species necessarily increases their capacity as biological agents, in particular through high functional redundancy, extended activity windows for pest suppression, a wider range of thermal refuges and the overall diversity of beneficial species (Shields *et al.*, 2019). A more diverse natural enemy community, favoured by efficient CBC strategies, also constitutes a reservoir of potential solutions against new pests. In addition, one needs to take into account how landscape and crop management affect the different life stages of pests and their natural enemies (Ma *et al.*, 2025).

Synchronization between natural enemies and their prey/hosts generally results from a long coevolutionary process (Visser and Gienapp, 2019). However, climate change may alter this equilibrium because it now occurs over a very short period (Thomson *et al.*, 2010; Tougeron *et al.*, 2019; Harvey *et al.*, 2023). Such an issue arises frequently in importation (classical) biological control when an introduced natural enemy is not adapted to local climatic conditions (Harms *et al.*, 2021), in terms of both phenology and thermal tolerance range. If such thermal mismatches occur between trophic levels, it may allow plants to escape from their pests, but for most agricultural pests studied to date, climate change would favour them over their natural enemies, which may arrive too late (i.e. a mismatch in phenological advance) (Furlong and Zalucki, 2017). Conservation biological control approaches can facilitate this by providing additional microclimatic and buffering conditions to improve phenological and thermal synchronization between natural enemies and their prey or hosts, and refuges to resist extreme climatic conditions, through the manipulation of vegetation cover and plant species composition (Alford *et al.*, 2018).

However, the effects of extreme climatic events such as heatwaves or droughts, as well as fluctuations in temperature or humidity, are still poorly considered in understanding the dynamics between plants, phytophagous pests and natural enemies (Hance *et al.*, 2007). As a result, we need to better explore the temperature ranges of species, as eurythermic or stenothermic natural enemies respond differently to climate change, and some species will not be able to track their hosts or prey as easily, as their generation time and adaptation capacities are more reduced (Thomson *et al.*, 2010).

In the system involving the European larch *Larix decidua* Mill. (Pinaceae) and its phytophagous insect, the large bud moth *Zeiraphera diniana* Guenée (Lepidoptera: Tortricidae) in the Alps, long-term data of the frequency of pest outbreaks, either by direct observation from 1953 to now or by dendrochronology for more than 1000 generations (Büntgen *et al.*, 2020), revealed a frequency of pest outbreak every 9.3 years on average, during 1173 years. The pest outbreaks are linked to particular overwintering conditions since warm winters are followed by an arrest of the occurrence of pest outbreaks because larval diapause of the large bud moth is altered and induces a loss of synchronization with its host at spring emergence (Büntgen *et al.*, 2020).

In Brittany (western France), the aphid species *Sitobion avenae* (Fabricius) and *Rhopalosiphum padi* (Linnaeus) (Hemiptera: Aphididae) stopped reproducing sexually in cereal fields around 50 years ago, allowing them to maintain activity the whole winter instead of being in diapause at the eg.g stage (Dedryver *et al.*, 2001). Moreover, following winter warming, these parthenogenetic populations now reproduce throughout the winter as long as the temperature is above 4°C, resulting in new host resources available for natural enemies. In 2010–2011, a third aphid species, *Metopolophium dirhodum* (Walker) (Hemiptera: Aphididae), originally present in cereal fields only during the spring, was observed in the fields during the winter (Tougeron *et al.*, 2017). The increasing presence of parthenogenetic aphids during the winter, both in abundance and species richness, when wheat plants are highly sensitive to virus transmission enhances the risks of barley yellow dwarf virus transmission (Van den Eynde *et al.*, 2020) at this period. However, in this region, winter warming also modified the expression of diapause in aphid parasitoids (Tougeron *et al.*, 2017). By benefiting from winter crop covers and flowering strips, the parasitoids succeeded in developing a population on alternative hosts inside the cover and parasitizing aphids as soon as they arrived in the field, inducing a strong biocontrol (Damien *et al.*, 2017) and preventing further pest outbreaks the following spring. This strong biocontrol is also facilitated by phenological changes in plants; climate warming-induced flowering during the winter provides nectar resources for the parasitoids, which contribute to increasing their fecundity and longevity (Damien *et al.*, 2020). This example demonstrated the rapid response of insects to climate warming, as it showed how they can cope with phenological change.

How does knowledge of the impacts of climate change on the development of plants, phytophagous insects and their natural enemies help limit pest outbreaks? Data on the population dynamics of pests and their natural enemies in different climatic conditions can help identify the most critical periods when pest populations increase rapidly without being controlled by their natural enemies. Following the same crop for several years can determine the climatic conditions, landscape structure and agricultural practices most favourable to either pest outbreaks or CBC by natural enemies (Jego *et al.*, 2023). In cases where the level of synchronization between pest and natural enemy populations is a significant ecological driver of their interactions, the next step is to identify which type of management of the crop or its surroundings can help to increase CBC efficiency. More attention should be devoted to generalist biocontrol species, as they are most likely to be able to cope with climate variations and attack a wider range of hosts in the context of species redistribution (Schmitz and Barton, 2014). In landscape management and the implementation of CBC strategies, it is crucial that plants are resistant and resilient to changing temperatures and can continue to provide nectar, shelter, pollen, etc. in the future (Gurr *et al.*, 2017).

Changes in temperature/precipitation/CO_2 level in each location can favour the natural enemy, the phytophagous insect or the plant, but also involve other organisms of the system. For instance, each species of a trophic web coexists with several species of symbiotic bacteria or fungi, which can influence their resistance to hot or cold thermal stresses caused by climate change as well as other biotic stressors (Heyworth *et al.*, 2020; Tougeron *et al.*, 2023). We still need a better understanding of how environmental resources can be managed as part of CBC. In this context, future research should focus on gaining a better understanding of the role of abiotic pressures on symbiotic communities (Tougeron and Iltis, 2022), the role of food resources on gut microbiota (Kamenova *et al.*, 2018) and how certain microbes are transmitted horizontally through the environment

(Pons *et al.*, 2022). The net result is an escape of the plant from the phytophagous insect, a pest outbreak, better biocontrol by natural enemies or even a disappearance of the whole system if the plants or the symbionts are not supported by the new climatic conditions (van Baaren *et al.*, 2010; Ma *et al.*, 2024).

Solutions for CBC under the influence of climate change should therefore strongly consider the functional ecology of organisms within the agro-ecosystem (Begg *et al.*, 2017). This is particularly true because the response of organisms is never constant in the face of variations in the environment. Complex mechanisms are at play concerning distribution and movement between habitats, exploitation of microclimates, response to environmental filters, availability of food resources and interactions between species (Chaplin-Kramer *et al.*, 2011; Tena *et al.*, 2016). Exploring how CBC strategies operate in relation to the life traits and functional ecology of species along geographical and climatic gradients provides useful avenues for improvement in the context of climate change (Alfaro-Tapia *et al.*, 2022; Jego *et al.*, 2023). It is necessary to investigate the spatiotemporal dynamics of agricultural landscapes to understand this functional aspect of CBC (Landis *et al.*, 2000), and in particular the hidden dynamics (beyond the structure and composition of elements) of agricultural practices from local scales to functional landscapes (Murrell, 2017; Brusse *et al.*, 2024).

Management strategies need to consider not only the production and release of natural enemies but also their resilience and adaptive capacity. The genetic improvement of natural enemies has become a useful strategy to identify and enhance desired traits of natural enemies and thereby overcome some constraints (Bielza *et al.*, 2020; Leung *et al.*, 2020). For example, the use of microsatellite markers revealed genetic differentiation between laboratory-reared strains of the zoophytophagous *Nesidiocoris tenuis* Reuter (Hemiptera: Miridae). One strain identified to have lower levels of phytophagy and higher levels of zoophagy was then used in a selective breeding programme to improve biocontrol traits of interest (Pérez-Hedo *et al.*, 2024).

Several traits have been suggested in the literature to enhance the effectiveness of a natural enemy, and identifying the appropriate traits to be prioritized may be the first step to increasing our knowledge and harnessing significant changes in research and natural enemy production. Factors limiting the performance of key natural enemies should be the main focus. For example, for generalist predators, traits such as pesticide compatibility or prey-density dependence, the use of alternative non-suitable host plants and resistance to extreme environmental conditions are also relevant for CBC. Combining CBC and augmentative BC (the rearing and release of natural enemies) and including more resilient agents to varying environmental conditions (temperature, humidity, cold resistance) may become a novel strategy for the future management of pests. The benefits of selection for higher resistance to toxins, whether artificially applied (pesticides) or plant-produced (plant defences), or increased fitness when feeding on non-prey food (supplemented or plant-derived), should also be considered in insect breeding for biological control.

Under climatic extreme variability, augmentative biological control, even when using very effective organisms selected for better fitness, will not be able to function properly without considering CBC strategies. For instance, the combined use of flower strips of *F. esculentum* and release of the parasitic wasp *Trichogramma ostriniae* Pang et Chen (Hymenoptera: Trichogrammatidae) decreased the percentage of fruits infested by the European corn borer as a result of the synergy of both tactics (Russell and Bessin, 2009). A concentrated and stable source of nectar, such as those offered by the flower strips, under the ever hotter, dry summers for crops under temperate climates, can surely make the difference in terms of extending the life and fitness of released natural enemies, as well as providing shelter for these and other natural enemies during times of extreme heat.

18.6 Conclusion

There is a strong complementarity between habitat manipulation for CBC and the use of organisms in augmentative biological control, both to compensate for the inadequacies of effective pest control in the event of extreme climatic variability and at the same time to select relevant traits under these conditions. CBC leverages natural enemies to manage pest populations, reducing reliance on chemical pesticides and fostering ecological balance. However, the efficacy of CBC faces significant challenges as climate change alters the dynamics of agro-ecosystems.

To overcome these challenges, a deeper understanding of the ecological responses of both pests and their natural enemies to climate change is essential. This includes studying their thermal tolerances, phenological and distribution shifts, and the impact of altered climatic and meteorological patterns. Additionally, habitat management strategies, such as the use of ground covers, hedges and flowering strips, can enhance the resilience of natural enemy populations by providing alternative resources and microclimatic refuges. Genetic improvements in natural enemies, through selective breeding for traits such as temperature resilience and pesticide resistance, offer promising avenues to bolster CBC under changing climatic conditions. It is, however, important to note that, with our current knowledge, simple conclusions about the impact of climate change on biological control are not straightforward.

Ultimately, successful CBC in the face of climate change requires a multifaceted approach, integrating ecological and organismal biology research, landscape management and genetic knowledge.

References

Abrego, N., Roslin, T., Huotari, T., Ji, Y.Q., Schmidt, N.M. *et al.* (2021) Accounting for species interactions is necessary for predicting how arctic arthropod communities respond to climate change. *Ecography* 44(6), 885–896.

Alfaro-Tapia, A., Alvarez-Baca, J.K., Tougeron, K., Lavandero, B., Le Lann, C. *et al.* (2022) Overwintering strategies and life-history traits of different populations of *Aphidius platensis* along a latitudinal gradient in Chile. *Entomologia Generalis* 42(1), 127–145.

Alford, L., Tougeron, K., Pierre, J.S., Burel, F. and van Baaren, J. (2018) The effect of landscape complexity and microclimate on the thermal tolerance of a pest insect. *Insect Science* 25(5), 905–915.

Alford, L., Roudine, S., Pierre, J.S., Burel, F. and van Baaren, J. (2023) Landscape effects on the thermotolerance of carabid beetles and the role of behavioral thermoregulation. *Insect Science* 30(1), 251–263.

Alvarez-Baca, J.K., Alfaro-Tapia, A., Lavandero, B., Le Lann, C. and van Baaren, J. (2020) Suitability and profitability of a cereal aphid for the parasitoid *Aphidius platensis* in the context of conservation biological control of *Myzus persicae* in Orchards. *Insects* 11(6), 381.

Alvarez-Baca, J.K., Montealegre, X., Alfaro-Tapia, A., Zepeda-Paulo, F., van Baaren, J. *et al.* (2023) Composition and food web structure of aphid-parasitoid populations on plum Orchards in Chile. *Insects* 14(3), 288.

Bartling, M.T., Brandt, A., Hollert, H. and Vilcinskas, A. (2024) Current insights into sublethal effects of pesticides on insects. *International Journal of Molecular Sciences* 25(11), 6007.

Barton, B.T. and Ives, A.R. (2014) Species interactions and a chain of indirect effects driven by reduced precipitation. *Ecology* 95(2), 486–494.

Bathiany, S., Dakos, V., Scheffer, M. and Lenton, T.M. (2018) Climate models predict increasing temperature variability in poor countries. *Science Advances* 4(5), aar5809.

Beckerman, A.P., Uriarte, M. and Schmitz, O.J. (1997) Experimental evidence for a behavior-mediated trophic cascade in a terrestrial food Chain. *Proceedings of the National Academy of Sciences USA* 94(20), 10735–10738.

Begg, G.S., Cook, S.M., Dye, R., Ferrante, M., Franck, P. *et al.* (2017) A functional overview of conservation biological control. *Crop Protection* 97, 145–158.

Bielza, P., Balanza, V., Cifuentes, D. and Mendoza, J.E. (2020) Challenges facing arthropod biological control: Identifying traits for genetic improvement of predators in protected crops. *Pest Management Science* 76(11), 3517–3526.

Boivin, G. and Brodeur, J. (2006) Intra-and interspecific interactions among parasitoids: Mechanisms, outcomes and biological control. In: Boivin, G. and Brodeur, J. (eds) *Trophic and Guild in Biological Interactions Control*. Springer, Cham, Switzerland, pp. 123–144.

Brusse, T., Tougeron, K., Barbottin, A., Henckel, L., Dubois, F. *et al.* (2024) Considering farming management at the landscape scale: Descriptors and trends on biodiversity. A review. *Agronomy for Sustainable Development* 44(3), 30.

Büntgen, U., Arseneault, D., Boucher, E., Churakova, O., Gennaretti, F. *et al.* (2020) Prominent role of volcanism in common era climate variability and human history. *Dendrochronologia* 64, 125757.

Chaplin-Kramer, R., O'Rourke, M.E., Blitzer, E.J. and Kremen, C. (2011) A meta-analysis of crop pest and natural enemy response to landscape complexity. *Ecology Letters* 14(9), 922–932.

Chen, C., Donner, S.H., Biere, A., Gols, R. and Harvey, J.A. (2019) Simulated heatwave conditions associated with global warming affect development and competition between hyperparasitoids. *Oikos* 128(12), 1783–1792.

Chidawanyika, F., Mudavanhu, P. and Nyamukondiwa, C. (2019) Global climate change as a driver of bottom-up and top-down factors in agricultural landscapes and the fate of host-parasitoid interactions. *Frontiers in Ecology and Evolution* 7, 80.

Cuff, J.P., Tercel, M.P.T.G., Drake, L.E., Vaughan, I.P., Bell, J.R. *et al.* (2022) Density-independent prey choice, taxonomy, life history, and web characteristics determine the diet and biocontrol potential of spiders (Linyphiidae and Lycosidae) in cereal crops. *Environmental DNA* 4(3), 549–564.

Dale, V.H. (1997) The relationship between land-use change and climate change. *Ecological Applications* 7(3), 753–769.

Damien, M. and Tougeron, K. (2019) Prey–predator phenological mismatch under climate change. *Current Opinion in Insect Science* 35, 60–68.

Damien, M., Le Lann, C., Desneux, N., Alford, L., Al Hassan, D. *et al.* (2017) Flowering cover crops in winter increase pest control but not trophic link diversity. *Agriculture, Ecosystems & Environment* 247, 418–425.

Damien, M., Llopis, S., Desneux, N., van Baaren, J. and Le Lann, C. (2020) How does floral nectar quality affect life history strategies in parasitic wasps? *Entomologia Generalis* 40(2), 147–156.

Dedryver, C.A., Hullé, M., Le Gallic, J.F., Caillaud, M.C. and Simon, J.C. (2001) Coexistence in space and time of sexual and asexual populations of the cereal aphid *Sitobion avenae*. *Oecologia* 128(3), 379–388.

Desneux, N., Pham-Delègue, M.H. and Kaiser, L. (2004) Effects of sub-lethal and lethal doses of lambda-cyhalothrin on oviposition experience and host-searching behaviour of a parasitic wasp, *Aphidius ervi*. *Pest Management Science* 60(4), 381–389.

Falvey, M. and Garreaud, R.D. (2009) Regional cooling in a warming world: Recent temperature trends in the southeast Pacific and along the west coast of subtropical South America (1979–2006). *Journal of Geophysical Research: Atmospheres* 114, D4.

Finke, D.L. and Denno, R.F. (2002) Intraguild predation diminished in complex-structured vegetation: Implications for prey suppression. *Ecology* 83(3), 643–652.

Forrest, J.R.K. and James, D.T. (2011) An examination of synchrony between insect emergence and flowering in Rocky Mountain meadows. *Ecological Monographs* 81(3), 469–491.

Furlong, M.J. and Zalucki, M.P. (2017) Climate change and biological control: The consequences of increasing temperatures on host-parasitoid interactions. *Current Opinion in Insect Science* 20, 39–44.

Gérard, M., Vanderplanck, M., Wood, T. and Michez, D. (2020) Global warming and plant–pollinator mismatches. *Emerging Topics in Life Sciences* 4(1), 77–86.

Gómez-Marco, F., Tena, A., Jaques, J.A. and García, A.U. (2016) Early arrival of predators controls *Aphis spiraecola* colonies in citrus clementines. *Journal of Pest Science* 89(1), 69–79.

Grabenweger, G., Hopp, H., Jaeckel, B., Balder, H., Koch, T. *et al.* (2007) Impact of poor host-parasitoid synchronisation on the parasitism of *Cameraria ohridella* (Lepidoptera: Gracillariidae). *European Journal of Entomology* 104(1), 153–158.

Guo, F., Kainz, M.J., Sheldon, F. and Bunn, S.E. (2016) The importance of high-quality algal food sources in stream food webs – current status and future perspectives. *Freshwater Biology* 61(6), 815–831.

Gurr, G.M., Wratten, S.D., Landis, D.A. and You, M.S. (2017) Habitat management to suppress pest populations: Progress and prospects. *Annual Review of Entomology* 62, 91–109.

Haghani, S., Zahedi-Golpayegani, A., Saboori, A. and Allahyari, H. (2019) The effect of con/heterospecific diet on predation, oviposition and longevity of *Amblyseius swirskii*, *Neoseiulus californicus* and *Phytoseiulus persimilis* (Acari: Phytoseiidae). *Systematic and Applied Acarology* 24(11), 2240–2252.

Halaj, J. and Wise, D.H. (2001) Terrestrial trophic cascades: How much do they trickle? *American Naturalist* 157(3), 262–281.

Hambäck, P.A., Cirtwill, A.R., García, D., Grudzinska-Sterno, M., Miñarro, M. *et al.* (2021) More intraguild prey than pest species in arachnid diets may compromise biological control in apple Orchards. *Basic and Applied Ecology* 57, 1–13.

Han, P., Becker, C., Sentis, A., Rostás, M., Desneux, N. *et al.* (2019) Global change-driven modulation of bottom–up forces and cascading effects on biocontrol services. *Current Opinion in Insect Science* 35, 27–33.

Hance, T., van Baaren, J., Vernon, P. and Boivin, G. (2007) Impact of extreme temperatures on parasitoids in a climate change perspective. *Annual Review of Entomology* 52(1), 107–126.

Harms, N.E., Knight, I.A., Pratt, P.D., Reddy, A.M., Mukherjee, A. *et al.* (2021) Climate mismatch between introduced biological control agents and their invasive host plants: Improving biological control of tropical weeds in temperate regions. *Insects* 12(6), 549.

Harvey, J.A., Poelman, E.H. and Tanaka, T. (2013) Intrinsic inter-and intraspecific competition in parasitoid wasps. *Annual Review of Entomology* 58, 333–351.

Harvey, J.A., Heinen, R., Gols, R. and Thakur, M.P. (2020) Climate change-mediated temperature extremes and insects: From outbreaks to breakdowns. *Global Change Biology* 26(12), 6685–6701.

Harvey, J.A., Tougeron, K., Gols, R., Heinen, R., Abarca, M. *et al.* (2023) Scientists' warning on climate change and insects. *Ecological Monographs* 93(1), e1553.

Heimpel, G.E. and Mills, N.J. (2017) *Biological Control – Ecology and Applications*. Cambridge University Press, Cambridge, UK.

Heyworth, E.R., Smee, M.R. and Ferrari, J. (2020) Aphid facultative symbionts aid recovery of their obligate symbiont and their host after heat stress. *Frontiers in Ecology and Evolution* 8, 56.

Holland, J.M., Bianchi, F.J.J.A., Entling, M.H., Moonen, A.C., Smith, B.M. *et al.* (2016) Structure, function and management of semi-natural habitats for conservation biological control: A review of European studies. *Pest Management Science* 72(9), 1638–1651.

Holland, J.M., Jeanneret, P., Moonen, A.C., van der Werf, W., Rossing, W.A.H. *et al.* (2020) Approaches to identify the value of seminatural habitats for conservation biological control. *Insects* 11(3), 195.

Irvin, N.A., Scarratt, S.L., Wratten, S.D., Frampton, C.M., Chapman, R.B. *et al.* (2006) The effects of floral understoreys on parasitism of leafrollers (Lepidoptera: Tortricidae) on apples in New Zealand. *Agricultural and Forest Entomology* 8(1), 25–34.

Jeavons, E., Le Lann, C. and van Baaren, J. (2023) Interactions between natural enemies and pollinators: Combining ecological theory with agroecological management. *Entomologia Generalis* 43(2), 243–259.

Jego, L., Li, R.N., Roudine, S., Ma, C.S., Le Lann, C. *et al.* (2023) Parasitoid ecology along geographic gradients: Lessons for climate change studies. *Current Opinion in Insect Science* 57, 101036.

Judt, C., Korányi, D., Zaller, J.G. and Batáry, P. (2023) Floral resources and ground covers promote natural enemies but not pest insects in apple Orchards: A global meta-analysis. *Science of the Total Environment* 903, 166139.

Kamenova, S., Mayer, R., Rubbmark, O.R., Coissac, E., Plantegenest, M. *et al.* (2018) Comparing three types of dietary samples for prey DNA decay in an insect generalist predator. *Molecular Ecology Resources* 18(5), 966–973.

Kean, J., Wratten, S., Tylianakis, J. and Barlow, N. (2003) The population consequences of natural enemy enhancement, and implications for conservation biological control. *Ecology Letters* 6(7), 604–612.

Kenis, M., Nacambo, S., Van Vlaenderen, J., Zindel, R. and Eschen, R. (2020) Long term monitoring in Switzerland reveals that *Adalia bipunctata* strongly declines in response to *Harmonia axyridis* invasion. *Insects* 11(12), 883.

Landis, D.A., Wratten, S.D. and Gurr, G.M. (2000) Habitat management to conserve natural enemies of arthropod pests in agriculture. *Annual Review of Entomology* 45(1), 175–201.

Langellotto, G.A., Rosenheim, J.A. and Williams, M.R. (2006) Assessing trophic interactions in a guild of primary parasitoids and facultative hyperparasitoids: Stable isotope analysis. *Oecologia* 150(2), 291–299.

Laubmeier, A.N., Tabassum, N. and Tenhumberg, B. (2023) Temperature fluctuation alters optimal predator community composition for anticipated biological control. *Frontiers in Ecology and Evolution* 10, 998396.

Lavandero, B. and Tylianakis, J.M. (2013) Genotype matching in a parasitoid–host genotypic food web: An approach for measuring effects of environmental change. *Molecular Ecology* 22(1), 229–238.

Laws, A.N. (2017) Climate change effects on predator–prey interactions. *Current Opinion in Insect Science* 23, 28–34.

Le Gal, A., Robert, C., Accatino, F., Claessen, D. and Lecomte, J. (2020) Modelling the interactions between landscape structure and spatio-temporal dynamics of pest natural enemies: Implications for conservation biological control. *Ecological Modelling* 420, 108912.

Lehmann, P., Ammunét, T., Barton, M., Battisti, A., Eigenbrode, S.D. *et al.* (2020) Complex responses of global insect pests to climate warming. *Frontiers in Ecology and the Environment* 18(3), 141–150.

Leung, K., Ras, E., Ferguson, K., Ariens, S., Babendreier, D. *et al.* (2020) Next-generation biological control: The need for integrating genetics and genomics. *Biological Reviews* 95(6), 1838–1854.

Li, H.R., Li, B.P., Lövei, G.L., Kring, T.J. and Obrycki, J.J. (2021) Interactions among native and non-native predatory Coccinellidae influence biological control and biodiversity. *Annals of the Entomological Society of America* 114(2), 119–136.

Liang, Y.Y., Chen, X., Dai, H.J., Wang, J., Guo, X.J. *et al.* (2022) Flower provision reduces intraguild predation between predators and increases aphid biocontrol in tomato. *Journal of Pest Science* 95(1), 461–472.

Lima, S.L. (1998) Nonlethal effects in the ecology of predator-prey interactions. *Bioscience* 48(1), 25–34.

Losey, J.E. and Denno, R.F. (1999) Factors facilitating synergistic predation: The central role of synchrony. *Ecological Applications* 9(2), 378–386.

Lucas, É., Gagné, I. and Coderre, D. (2002) Impact of the arrival of *Harmonia axyridis* on adults of *Coccinella septempunctata* and *Coleomegilla maculata* (Coleoptera: Coccinellidae). *European Journal of Entomology* 99(4), 457–463.

Luliano, B. and Gratton, C. (2020) Temporal resource (dis)continuity for conservation biological control: From field to landscape scales. *Frontiers in Sustainable Food Systems* 4, 127.

Ma, C.S., Wang, B.X., Wang, X.J., Lin, Q.C., Zhang, W. *et al.* (2025) Crop pest responses to global changes in climate and land management. *Nature Reviews Earth & Environment* 6, 264–283.

Ma, G., Ma, C.S., Le Lann, C. and van Baaren, J. (2024) Effects of climate change on insect phenology. In: Gonzalez-Tokman, D. and Datillo, W. (eds) *Effects of Climate Change on Insects: Physiological, Evolutionary, and Ecological Responses*. Oxford University Press, Oxford, pp. 89–109.

Marshall, S.D. and Rypstra, A.L. (1999) Patterns in the distribution of two wolf spiders (Araneae: Lycosidae) in two soybean agroecosytems. *Environmental Entomology* 28(6), 1052–1059.

Marta, S., Brunetti, M., Manenti, R., Provenzale, A. and Ficetola, G.F. (2021) Climate and land-use changes drive biodiversity turnover in arthropod assemblages over 150 years. *Nature Ecology & Evolution* 5(9), 1291–1300.

Michalko, R., Pekár, S., Dul'a, M. and Entling, M.H. (2019) Global patterns in the biocontrol efficacy of spiders: A meta-analysis. *Global Ecology and Biogeography* 28(9), 1366–1378.

Michaud, J.P. (2002) Invasion of the Florida citrus ecosystem by *Harmonia axyridis* (Coleoptera: Coccinellidae) and asymmetric competition with a native species, *Cycloneda sanguinea*. *Environmental Entomology* 31(5), 827–835.

Miller-Rushing, A.J., Hoye, T.T., Inouye, D.W. and Post, E. (2010) The effects of phenological mismatches on demography. *Philosophical Transactions of the Royal Society B: Biological Sciences* 365(1555), 3177–3186.

Mills, N.J. and Heimpel, G.E. (2018) Could increased understanding of foraging behavior help to predict the success of biological control? *Current Opinion in Insect Science* 27, 26–31.

Moiroux, J., Brodeur, J. and Boivin, G. (2014) Sex ratio variations with temperature in an egg parasitoid: Behavioural adjustment and physiological constraint. *Animal Behaviour* 91, 61–66.

Müller, C. (2018) Impacts of sublethal insecticide exposure on insects – facts and knowledge gaps. *Basic and Applied Ecology* 30, 1–10.

Muller, C.B. and Brodeur, J. (2002) Intraguild predation in biological control and conservation biology. *Biological Control* 25, 216–223.

Murrell, E.G. (2017) Can agricultural practices that mitigate or improve crop resilience to climate change also manage crop pests? *Current Opinion in Insect Science* 23, 81–88.

Nelson, E.H., Matthews, C.E. and Rosenheim, J.A. (2004) Predators reduce prey population growth by inducing changes in prey behavior. *Ecology* 85(7), 1853–1858.

Nóia, M., Borges, I. and Soares, A.O. (2008) Intraguild predation between the aphidophagous ladybird beetles *Harmonia axyridis* and *Coccinella undecimpunctata* (Coleoptera: Coccinellidae): The role of intra- and extraguild prey densities. *Biological Control* 46(2), 140–146.

Ode, P.J. and Hardy, I.C.W. (2008) Parasitoid sex ratios and biological control. In: Wajnberg, É., Bernstein, C. and van Alphen, J. (eds) *Behavioral Ecology of Insect Parasitoids: From Theoretical Approaches to Field Applications*. Blackwell, Oxford, pp. 229–252.

Ode, P.J., Vyas, D.K. and Harvey, J.A. (2021) Extrinsic inter-and intraspecific competition in parasitoid wasps. *Annual Review of Entomology* 67(1), 305–328.
Ortiz-Martínez, S., Staudacher, K., Baumgartner, V., Traugott, M. and Lavandero, B. (2020) Intraguild predation is independent of landscape context and does not affect the temporal dynamics of aphids in cereal fields. *Journal of Pest Science* 93(1), 235–249.
Ostberg, S., Schaphoff, S., Lucht, W. and Gerten, D. (2015) Three centuries of dual pressure from land use and climate change on the biosphere. *Environmental Research Letters* 10(4), 044011.
Pekár, S., Michalko, R., Loverre, P., Líznarová, E. and Černecká, L. (2015) Biological control in winter: Novel evidence for the importance of generalist predators. *Journal of Applied Ecology* 52(1), 270–279.
Pérez-Hedo, M., Ortells-Fabra, R., Alonso-Valiente, M., Ruíz-Rivero, O. and Urbaneja, A. (2024) Enhancing the biocontrol potential of the predator *Nesidiocoris tenuis* through genetic selection. *Biological Control* 188, 105413.
Petráková, L., Michalko, R., Loverre, P., Sentenská, L., Korenko, S. *et al.* (2016) Intraguild predation among spiders and their effect on the pear psylla during winter. *Agriculture, Ecosystems & Environment* 233, 67–74.
Peñalver-Cruz, A., Alvarez, D. and Lavandero, B. (2020) Do hedgerows influence the natural biological control of woolly apple aphids in Orchards? *Journal of Pest Science* 93(1), 219–234.
Pons, I., Scieur, N., Dhondt, L., Renard, M.E., Renoz, F. *et al.* (2022) Pervasiveness of the symbiont *Serratia symbiotica* in the aphid natural environment: Distribution, diversity and evolution at a multitrophic level. *FEMS Microbiology Ecology* 98(1), fiac012.
Prather, C.M., Pelini, S.L., Laws, A., Rivest, E., Woltz, M. *et al.* (2013) Invertebrates, ecosystem services and climate change. *Biological Reviews* 88(2), 327–348.
Preisser, E.L. and Bolnick, D.I. (2008) The many faces of fear: Comparing the pathways and impacts of nonconsumptive predator effects on prey populations. *PLOS One* 3(6), e2465.
Price, P.W. (1997) *Insect Ecology*. John Wiley & Sons, Hoboken, New Jersey.
Renner, S.S. and Zohner, C.M. (2018) Climate change and phenological mismatch in trophic interactions among plants, insects, and vertebrates. *Annual Review of Ecology, Evolution, and Systematics* 49(1), 165–182.
Rosenblatt, A.E. and Schmitz, O.J. (2016) Climate change, nutrition, and bottom-up and top-down food web processes. *Trends in Ecology & Evolution* 31(12), 965–975.
Roslin, T., Hardwick, B., Novotny, V., Petry, W., Andrew, N. *et al.* (2017) Higher predation risk for insect prey at low latitudes and elevations. *Science* 356(6339), 742–744.
Roubos, C.R., Rodriguez-Saona, C. and Isaacs, R. (2014) Mitigating the effects of insecticides on arthropod biological control at field and landscape scales. *Biological Control* 75, 28–38.
Russell, K. and Bessin, R. (2009) Integration of *Trichogramma ostriniae* releases and habitat modification for suppression of European corn borer (*Ostrinia nubilalis* Hübner) in bell peppers. *Renewable Agriculture and Food Systems* 24(1), 19–24.
Schemske, D.W., Mittelbach, G.G., Cornell, H.V., Sobel, J.M. and Roy, K. (2009) Is there a latitudinal gradient in the importance of biotic interactions? *Annual Review of Ecology, Evolution, and Systematics* 40(1), 245–269.
Schmitz, O.J. and Barton, B.T. (2014) Climate change effects on behavioral and physiological ecology of predator-prey interactions: Implications for conservation biological control. *Biological Control* 75, 87–96.
Seibold, S., Gossner, M., Simons, N., Bluthgen, N., Muller, J. *et al.* (2019) Arthropod decline in grasslands and forests is associated with landscape-level drivers. *Nature* 574(7780), 671–674.
Sentis, A., Ramon-Portugal, F., Brodeur, J. and Hemptinne, J.L. (2015) The smell of change: Warming affects species interactions mediated by chemical information. *Global Change Biology* 21(10), 3586–3594.
Sentis, A., Hemptinne, J.L. and Brodeur, J. (2017) Non-additive effects of simulated heat waves and predators on prey phenotype and transgenerational phenotypic plasticity. *Global Change Biology* 23(11), 4598–4608.
Sentis, A., Hemptinne, J.L., Magro, A. and Outreman, Y. (2022) Biological control needs evolutionary perspectives of ecological interactions. *Evolutionary Applications* 15(10), 1537–1554.
Shields, M.W., Johnson, A.C., Pandey, S., Cullen, R., González-Chang, M. *et al.* (2019) History, current situation and challenges for conservation biological control. *Biological Control* 131, 25–35.
Skendžić, S., Zovko, M., Živković, I.P., Lešić, V. and Lemić, D. (2021) The impact of climate change on agricultural insect pests. *Insects* 12(5), 440.

Sohlström, E.H., Brose, U., van Klink, R., Rall, B.C., Rosenbaum, B. *et al.* (2022) Future climate and land-use intensification modify arthropod community structure. *Agriculture, Ecosystems & Environment* 327, 107830.

Stahlschmidt, Z.R., Joura, H., Makarem, J.R. and Sun, J.L. (2024) Hot and scared: How do heatwaves and predation risk impact resource acquisition and allocation? *Biology Letters* 20(4), 20240009.

Stenberg, J.A., Sundh, I., Becher, P.G., Björkman, C., Dubey, M. *et al.* (2021) When is it biological control? A framework of definitions, mechanisms, and classifications. *Journal of Pest Science* 94(3), 665–676.

Straub, C.S., Finke, D.L. and Snyder, W.E. (2007) Are the conservation of natural enemy biodiversity and biological control compatible goals? *Biological Control* 45(2), 225–237.

Takimoto, G., Post, D.M., Spiller, D.A. and Holt, R.D. (2012) Effects of productivity, disturbance, and ecosystem size on food-chain length: Insights from a metacommunity model of intraguild predation. *Ecological Research* 27(3), 481–493.

Tena, A., Wäckers, F.L., Heimpel, G.E., Urbaneja, A. and Pekas, A. (2016) Parasitoid nutritional ecology in a community context: The importance of honeydew and implications for biological control. *Current Opinion in Insect Science* 14, 100–104.

Thomas, C.D., Bodsworth, E.J., Wilson, R.J., Simmons, A.D., Davies, Z.G. *et al.* (2001) Ecological and evolutionary processes at expanding range margins. *Nature* 411(6837), 577–581.

Thomson, L.J., Macfadyen, S. and Hoffmann, A.A. (2010) Predicting the effects of climate change on natural enemies of agricultural pests. *Biological Control* 52(3), 296–306.

Toft, S., Cuende, E., Olesen, A.L., Mathiesen, A., Larsen, M.M. *et al.* (2019) Food and specific macronutrient limitation in an assemblage of predatory beetles. *Oikos* 128(10), 1467–1477.

Tougeron, K. and Iltis, C. (2022) Impact of heat stress on the fitness outcomes of symbiotic infection in aphids: A meta-analysis. *Proceedings of the Royal Society B Biological Sciences* 289(1971), 20212660.

Tougeron, K., van Baaren, J., Burel, F. and Alford, L. (2016) Comparing thermal tolerance across contrasting landscapes: First steps towards understanding how landscape management could modify ectotherm thermal tolerance. *Insect Conservation and Diversity* 9(3), 171–180.

Tougeron, K., Le Lann, C., Brodeur, J. and van Baaren, J. (2017) Are aphid parasitoids from mild winter climates losing their winter diapause? *Oecologia* 183(3), 619–629.

Tougeron, K., van Baaren, J., Llopis, S., Ridel, A., Doyon, J. *et al.* (2018) Disentangling plasticity from local adaptation in diapause expression in parasitoid wasps from contrasting thermal environments: A reciprocal translocation experiment. *Biological Journal of the Linnean Society* 124(4), 756–764.

Tougeron, K., Brodeur, J., Le Lann, C. and Baaren, J. (2019) How climate change affects the seasonal ecology of insect parasitoids. *Ecological Entomology* 45(2), 167–181.

Tougeron, K., Couthouis, E., Marrec, R., Barascou, L., Baudry, J. *et al.* (2022) Multi-scale approach to biodiversity proxies of biological control service in European farmlands. *Science of the Total Environment* 822, 153569.

Tougeron, K., Iltis, C., Rampnoux, E., Goerlinger, A., Dhondt, L. *et al.* (2023) Still standing: The heat protection delivered by a facultative symbiont to its aphid host is resilient to repeated thermal stress. *Current Research in Insect Science* 3, 100061.

Tylianakis, J.M. and Romo, C.M. (2010) Natural enemy diversity and biological control: Making sense of the context-dependency. *Basic and Applied Ecology* 11(8), 657–668.

Tylianakis, J.M., Didham, R.K. and Wratten, S.D. (2004) Improved fitness of aphid parasitoids receiving resource subsidies. *Ecology* 85(3), 658–666.

Tylianakis, J.M., Didham, R.K., Bascompte, J. and Wardle, D.A. (2008) Global change and species interactions in terrestrial ecosystems. *Ecology Letters* 11(12), 1351–1363.

van Baaren, J., Le Lann, C. and van Alphen, J.J.M. (2010) Consequences of climate change for aphid-based multi-trophic systems. In: Kindlmann, P., Dixon, A. and Michaud, J. (eds) *Aphid Biodiversity Under Environmental Change: Patterns and Processes*. Springer, Dordrecht, The Netherlands, pp. 55–68.

Van den Brink, P.J., Boxall, A., Maltby, L., Brooks, B., Rudd, M. *et al.* (2018) Toward sustainable environmental quality: Priority research questions for Europe. *Environmental Toxicology and Chemistry* 37(9), 2281–2295.

Van den Eynde, R., Leeuwen, T. and Haesaert, G. (2020) Identifying drivers of spatio-temporal dynamics in barley yellow dwarf virus epidemiology as a critical factor in disease control. *Pest Management Science* 76(8), 2548–2556.

Van Klink, R., Bowler, D.E., Gongalsky, K.B., Shen, M.H., Swengel, S.R. *et al.* (2023) Disproportionate declines of formerly abundant species underlie insect loss. *Nature* 628(8007), 359–364.
Vidal, M.C. and Murphy, S.M. (2017) Bottom-up vs. top-down effects on terrestrial insect herbivores: A meta-analysis. *Ecology Letters* 21(1), 138–150.
Visser, M.E. and Gienapp, P. (2019) Evolutionary and demographic consequences of phenological mismatches. *Nature Ecology & Evolution* 3(6), 879–885.
Wiles, J.A. and Jepson, P.C. (1994) Sub-lethal effects of deltamethrin residues on the within-crop behaviour and distribution of *Coccinella septempunctata*. *Entomologia Experimentalis et Applicata* 72(1), 33–45.
Wood, S.A., Karp, D.S., DeClerck, F., Kremen, C., Naeem, S. *et al.* (2015) Functional traits in agriculture: Agrobiodiversity and ecosystem services. *Trends in Ecology* 531–539.
Zuma, M., Njekete, C., Konan, K.A.J., Bearez, P., Amiens-Desneux, E. *et al.* (2023) Companion plants and alternative prey improve biological control by *Orius laevigatus* on strawberry. *Journal of Pest Science* 96(2), 711–721.

19 Emerging Insect Invasions and Climate Change: Which Biocontrol Strategies for Effective Prediction and Early Detection of Invaders?

Marie-Anne Auger-Rozenberg* **and Alain Roques**
INRAE, URZF, Orléans, France

Abstract
This chapter documents the triggering effect of global warming on the potential establishment of non-native insect pests transported with human activities, including those originating from subtropical and tropical areas. With globalization, continuous changes in trade routes and transported goods result in the increasing arrival of unanticipated emerging plant pests, which are not targeted by the quarantine lists. Following the presentation of representative case studies, the chapter details novel methods proposed for tackling such species: sentinel plantings in other continents to identify potential invaders prior to their arrival and trappings using lures with generic attractiveness at ports of entry for detecting them as early as possible upon arrival.

19.1 Introduction: More and More 'Emerging' Species on the Road

Climate change and biological invasions are interacting key processes affecting global biodiversity, and must not be considered separately (Walther *et al.*, 2009). Worldwide, the establishment of non-native insect species on a continent outside their original range has exponentially increased since the beginning of the 20th century, without any signs of saturation (Seebens *et al.*, 2017). For instance, the number of non-native insect species newly establishing per year in Europe almost doubled between the periods 1950–1974 and 2000–2020, passing from 9.7 to 17 to reach a total of 1559 species in 2022 (extracted from the International Non-native Insect Establishment Database; Turner *et al.*, 2024). Similar trends have been observed in North America (Isitt *et al.*, 2024), Australasia (Isitt *et al.*, 2024) and China (Roques *et al.*, 2020). The vast majority of these new introductions are accidental but linked to human activities, mostly associated with the plant trade or as hitchhikers in containers or goods that have no particular biological significance for them (Fenn-Moltu *et al.*, 2023). In particular, the recent increase of the trade in live plants all over the world has triggered the transportation of insects associated with such plants (Liebhold *et al.*, 2012; Bonnamour *et al.*, 2023).

A growing number of these insect species, which arrive with the opening of new trade routes, are considered as 'emerging' since they have never previously been identified as invasive in a region outside their original range (Seebens

*Corresponding author: marie-anne.auger-rozenberg@inrae.fr

DOI: 10.1079/9781800625099.0019

et al., 2018), and often, they do not cause any damage in their native area due to both coevolution with their host plants and control provided by their natural enemies. The recent arrivals in Europe of species such as the box tree moth (*Cydalima perspectalis*), the Asian hornet (*Vespa velutina*) and the North American seed bug (*Leptoglossus occidentalis*) provide significant examples of such processes since none of them had previously invaded any other regions. The same consideration could be applied to major tree pests such as the emerald ash borer (*Agrilus planipennis*) and the Asian long-horned beetle (*Anoplophora glabripennis*) before they invaded North America in the late 1990s.

Moreover, global warming has already enabled a number of non-native insect species to establish and expand into regions in which they previously could not survive and reproduce (Walther *et al.*, 2009). Insect species originating from subtropical and even tropical regions are thus establishing in growing numbers in regions with more temperate climates as a direct result of climate change. In 2009, Walther *et al.* estimated the number of such species present in Europe as 400. At present, this number appears to have largely been underestimated. The International Non-native Insect Establishment Database (Turner *et al.*, 2024) revealed that 643 insect species native to subtropical/tropical regions were recorded as established in Europe, their temporal records following the same dynamics of increase as that the species originating from temperate regions (Fig. 19.1A). Tropical Africa was the main supplier of these species during the 19th and 20th centuries but since 2000, the Neotropics and tropical Asia (South-East Asia and Indomalaya) have become the dominant sources (Fig. 19.1B).

Usually, these non-native insects from subtropical/tropical regions remained in their site of introduction, most often in the Mediterranean area. The situation has changed during recent years for a number of these species as evidenced by pest insects associated with palm trees. While until the mid-2000s the red palm weevil (*Rhynchophorus ferrugineus*) and the palm moth (*Paysandisia archon*) remained confined to their respective areas of introduction in southern Spain and the Canary Islands, they colonized the major part of the Mediterranean region between 2004 and 2006 (Kontodimas *et al.*, 2006), and then even more northern areas such as Brittany and the south of England for the moth (EPPO, 2003). This rapid spread has been attributed to a combined effect of global warming, allowing better development of these tropical species, and increasing trade in ornamental palms. The arrival of the weevil in Greece was thus concomitant with massive imports of infested palm trees from Spain for the 2004 Olympic Games in Athens (Roques, 2010).

An indirect estimation of the triggering effect of climate change on insect invasions is also given by the temporal changes in colonization by non-native insect species of subtropical and tropical trees and shrubs planted in Europe. Whereas during the 19th and most of the 20th century only a few non-native insect species followed non-native plant species originating from tropical/subtropical regions and planted in Europe, such as palms, eucalypts and exotic legume trees, these plants showed a sharp acceleration in the recruitment of non-native insects during the last 20 years. The converse was observed for conifers native to Europe, on which the rate of arrival of non-native insect species decreased (Fig. 19.2). However, it remains difficult to disentangle the effects of climate change from those of the growing trade in such plants for ornament and the development of new commercial routes offering transport possibilities to additional insect species.

More generally, for a large number of insects specific temperature thresholds delimit barriers constraining their geographical distribution. In temperate latitudes, low temperatures are usually a key factor limiting the range of species, including native pests (Bellard *et al.*, 2013). Minimum temperature thresholds exist for both insect winter survival and the different development stages (egg, larval instar, pupa). Climate change, by removing or moving these barriers in space, allows expansion into areas previously unsuitable for establishment. Even a slight increase in winter temperatures thus allows survival in geographical areas previously inaccessible due to harsh climates. Thus, many species respond quickly to rising temperatures by moving their range northward or to higher elevations (Hickling *et al.*, 2006). The northwards expansion in North America of an important pest bark beetle, the southern pine beetle (*Dendroctonus frontalis*), is one of the

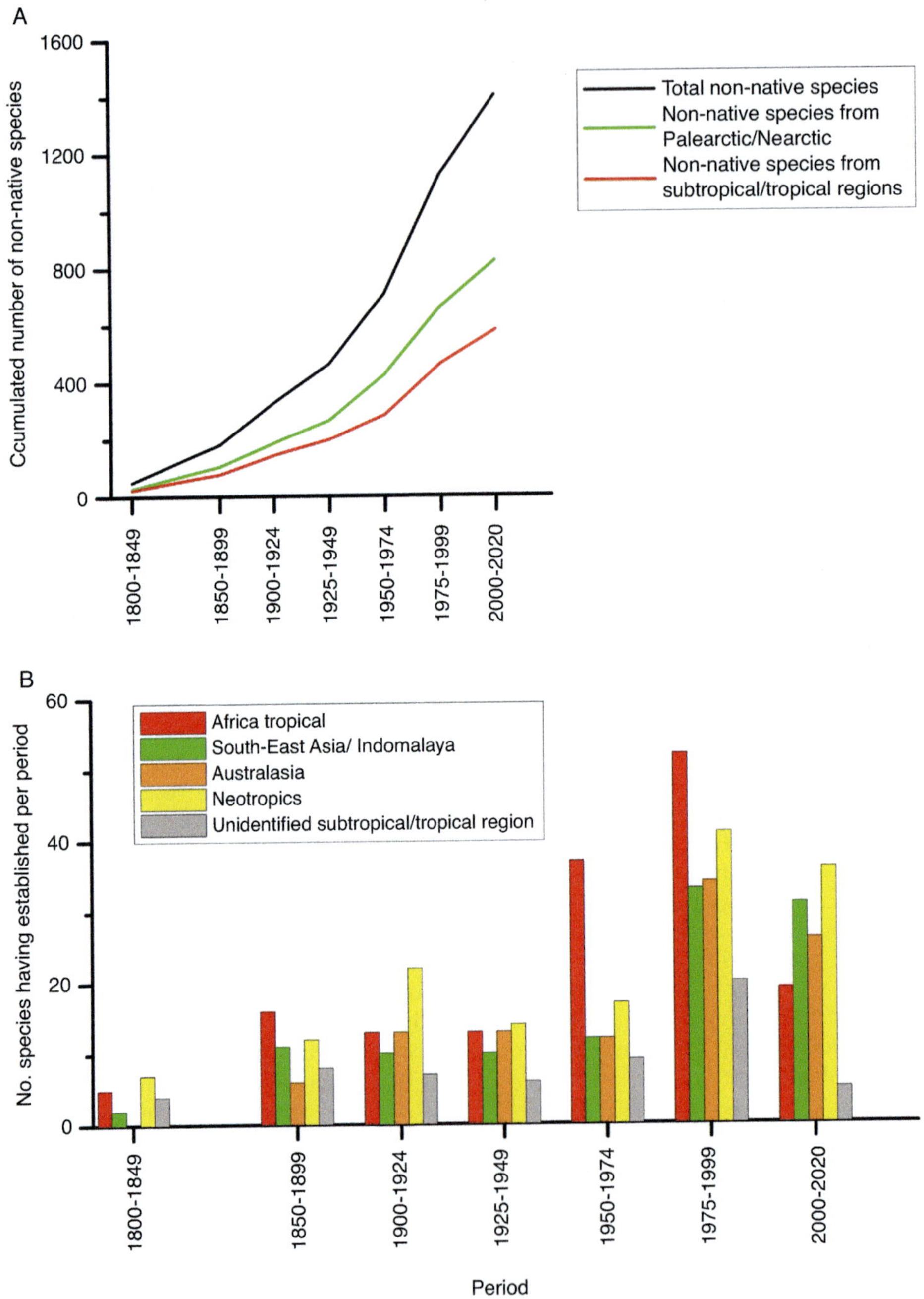

Fig. 19.1. A. Establishments in Europe of non-native insect species from the 19th century to 2020 based on date of first records of 1404 species, 141 others having no precise date of first record. B. Change over time of the native regions of insects from subtropical/tropical areas establishing in Europe (Roques, unpublished data; partly extracted from the Zenodo International Non-native Insect Establishment Data; Turner *et al*., 2024). The intervals on the *x*-axis are proportional to the duration of the considered time periods.

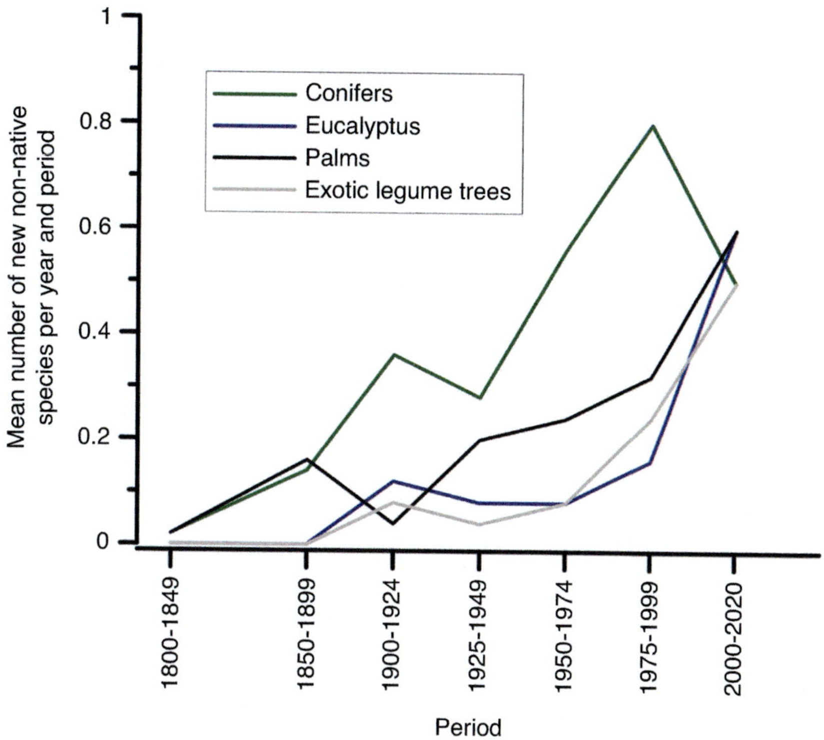

Fig. 19.2. Temporal changes in the recruitment of non-native insects by native conifers and some non-native tree species (adapted and updated from Eschen *et al.*, 2015). The intervals on the *x*-axis are proportional to the duration of the considered time periods.

best examples. Until the late 1990s, its survival during winter was only possible in the south-eastern part of the USA, due to an estimated lethal winter temperature of −16°C delimiting the northern limit of the insect's range (Ungerer *et al.*, 1999). Latitudinal northwards movement of the −16°C isotherm with winter warming has then allowed the beetle to colonize areas at much more northerly latitudes in the north-eastern United States (Tran *et al.*, 2007) as well as higher altitudes in more southerly areas such as Mexico (Sáenz-Romero *et al.*, 2023). This species is highly polyphagous and is therefore not constrained by host resources in its area of expansion. Early detection with pheromone traps is used to monitor the pest across the United States, especially in the newly invaded areas, but also outside infested areas (Dodds *et al.*, 2018) in order to detect its arrival as quickly as possible and try to manage affected forest ecosystems.

Similarly to non-native species introduced from subtropical and tropical areas, native pest species can also be unintentionally transported beyond, even far beyond, their natural ranges through human activities. Such translocations have probably occurred for a long time without previously leading to the establishment of the species due to unfavourable climatic conditions in the new environments. However, with rising temperatures, many of these translocated pests are now able to establish well beyond their natural range. The pine processionary moth (*Thaumetopoea pityocampa*) serves as a notable example (Roques, 2015; see Section 19.2.1). However, climate warming can produce contradictory effects within the same species, site and seasonal year. For instance, increased winter survival due to milder temperatures may be offset by higher mortality during summer of the same year caused by excessive heat.

Additionally, climate change affects the entire ecosystem, including hosts (plants), competing insects and natural enemies, simultaneously (Skendžic *et al.*, 2021). The responses of each of these components may differ for the same temperature increase and the interactions may be significantly altered in one direction or another. For example, studies suggest that the effectiveness of some predators, such as ladybirds, may increase in a warmer environment, resulting in a reduction of the expansion of their aphid prey (Cannon, 1998). By contrast, global warming allowed the pine processionary moth to expand northwards much faster than its natural enemies (Auger-Rozenberg *et al.*, 2015).

In this context, a critical challenge is to detect these unanticipated insect pests as early as possible before their introduction/expansion into new areas. Early detection is essential to increase the likelihood of successful eradication upon arrival in non-infested sites, and to provide more accurate estimates of invasion/expansion frequencies. In Europe, plant health control systems at borders are based on quarantine lists drawn up by the European Plant Protection Organisation (OEPP-EPPO). These lists (A1 for non-native quarantine species currently absent from Europe; A2 for non-native quarantine species locally present in Europe; Alert list for other non-native species at risk) mainly target non-native organisms on the basis of damage observed in their region of origin and/or on the continents they have invaded. Emerging species, unknown before their arrival in Europe, generally slip through the net and are therefore often absent from the quarantine lists which regulate the controls carried out by the phytosanitary services in European ports of entry.

19.2 Case Studies of Pest Insects Simultaneously Affected by Climate Change and Globalization

19.2.1 Native pests behaving as invaders

Pine processionary moth Thaumetopoea pityocampa *(Lepidoptera: Notodontidae)*

This European species was cited in the 4th IPCC report as one of the rare insects for which the role of climate change has clearly been demonstrated to result in range expansion (Rosenzweig *et al.*, 2007). The larval development of this species native to the Mediterranean basin occurs during autumn and winter, the larvae developing gregariously within a silk tent. The lower threshold for larval survival has been estimated at −16°C, a value that is rarely observed in most of Western and Central Europe at present, whereas nocturnal feeding on pine needles is possible as soon as the daily temperature in the tent exceeds 9°C before a night with temperature above 0°C (Battisti *et al.*, 2005). Winter warming has thus increased the potential survival of the larvae in more northern and mountainous areas. Therefore, from the late 1970s on, winter warming has allowed processionary moth populations to colonize previously unfavourable areas and progressively extend their range in both northward latitudes (at a rate of ca. 5.5 km/year; Roques *et al.*, 2015) and altitudes (Battisti *et al.*, 2006) in Europe. In contrast, a retraction is observed at the southern edge of the range in North Africa due to rising higher temperatures affecting the survival of the eggs and young larvae in summer (Bourougaaoui *et al.*, 2024).

However, more and more observations are revealing that the expansion indeed results from a combined effect of climate warming and accidental human-mediated translocations. Winter warming is progressively making most of Western Europe favourable to the survival of the moth larvae if introduced (Robinet *et al.*, 2012). Thus, since 2003 establishments of isolated moth colonies have regularly been detected in northern and eastern France, dozens to hundreds of kilometres away from the natural front of expansion and which therefore could not result from natural adult dispersal. Genetic analyses showed that these colonies were the result of transporting and planting pine trees with soil containing moth pupae from southern areas (Robinet *et al.*, 2012). Translocated moth larvae thus behave like non-native invaders. Early detection of such long-distance jumps, for example through pheromone trapping networks, is essential to allow eradication before the colonies expand too much in space.

19.2.2 Invasive pests affected by global change

Asian tiger mosquito Aedes albopictus *(Diptera: Culicidae)*

The Asian tiger mosquito, native to South-East Asia, was initially introduced into Albania in 1979 as hitchhiking eggs in used tyres (Adhami and Murati, 1987). It rapidly spread around the world (Del Lesto *et al.*, 2022) and has been recorded in more than 25 European countries (Lührsen *et al.*, 2023). Environmental factors, including weather conditions, affect both the potential presence and seasonal activity of the species which represents a serious threat as it transmits vector-borne diseases such as dengue and chikungunya (Lühken *et al.*, 2023). In its native tropical region, the larvae develop in tree holes filled with water. Its establishment in invaded areas is favoured by a warmer climate. It is clear that the tiger mosquito has thrived due to a number of factors including its strong adaptability, but also increased global trade and travel, and hotter and wetter conditions due to climate change, which shorten the incubation period for its eggs and means winters are no longer cold enough to kill them off (Benelli *et al.*, 2020).

Oriental fruit fly Bactrocera dorsalis *(Diptera: Tephritidae)*

The Oriental fruit fly is native to tropical and subtropical regions of Asia. It has emerged as one of the most invasive and destructive insect pests of tropical and subtropical fruits and vegetables over the last two decades and has invaded most of Africa and the Indian Ocean islands and now threatens the United States, Australia and the European Union, where it is classified as a priority quarantine pest (Nugnes *et al.*, 2018). This highly polyphagous pest belongs to a complex of about 90 species with several sympatric and morphologically cryptic species, some of which are regularly intercepted at European borders. As a result of climate change and changes in cold stress boundaries, the potential global range for *B. dorsalis* is predicted to expand northwards, particularly in Mediterranean Europe (Stephens *et al.*, 2007; Qin *et al.*, 2019) where it has already invaded more temperate agricultural areas such as southern Italy (Nugnes *et al.*, 2018). The recurrent incursions and the evidence of reproduction events on European territory suggest that *B. dorsalis* may be in its establishment phase in Europe, and that its permanent establishment would be extremely damaging (Charbonnel *et al.*, 2023).

Southern green stink bug Nezara viridula *(Hemiptera: Pentatomidae)*

The southern green stink bug is a worldwide pest of many crops and cropping systems. It is thought to be of southern origin but occurs throughout tropical, subtropical and warm temperate regions. In Japan, the northern limit of its distribution was delimited by the mean temperature isothermal line of +5°C for the coldest month of the year during adult overwintering. Since 2000 the bug has been observed expanding northward, beyond the previous range. This northward range expansion was promoted by an increase in the newly colonized areas of the mean and lowest winter month temperatures by 1–2°C from the 1950s to the 1990s. The resulting winter temperature of more than 5°C allowed adult survival during overwintering. Global warming also accelerates the development of immature bugs in the autumn. However, contradictory effects of seasonal warming were observed, the increase in summer temperature having a negative impact on diapause and nymphal development. In the newly colonized areas, adult diapause is induced after mid-September, resulting in late-season ineffective reproduction, the progeny having no chance of attaining adulthood and surviving winter (Musolin, 2007). The northward expansion of this bug is therefore the result of a balance between adverse effects of global warming.

Asian ambrosia beetle Xylosandrus crassiusculus *(Coleoptera: Curculionidae)*

This polyphagous ambrosia beetle originates from south-eastern Asia and is invasive on several continents. As a fungus-farming borer, it can infest many of the trees and shrubs that make up Mediterranean scrubland. Established in Italy since 2003 at least, the species is rapidly spreading in southern Europe. However, *X. crassiusculus* includes two very diverging clusters, mostly allopatric and exhibiting different climate niches. One cluster invaded the Americas, Europe, Africa and Oceania, with several independent introductions from multiple sources, whereas the other only invaded the Pacific islands and Africa. Nevertheless, modelling studies revealed that most of southern and western Europe was favourable to the establishment of these two clusters since the early 20th century (Urvois *et al.*, 2024). Since large climatically suitable regions still remain uncolonized, the recent colonization of Europe is probably due to an increase in propagule pressure via international trade rather than to recent environmental changes.

Eucalypt pests

This is another example of the interactions between climate warming and changes in plant trade. A total of 20 species of arthropods, including 18 insects and are mites, are presently recorded on eucalyptus trees planted in Europe (Saintonge *et al.*, 2025). All these species are non-native, originating from Australasia along with their host plants. They essentially established in the Mediterranean area where climate allows the growth of eucalypts.

The temporal sequence of introductions reveals that two wood-boring species of cerambycids (*Phoracantha semipunctata* and *P. recurva*) and two large weevil species (*Gonipterus* spp.) arrived in Europe during the second half of the 20th century, probably with the importation of eucalyptus logs. However, the export of eucalyptus wood from Australasia became limited, and no further introductions of cerambycids or large weevils have been observed since then, even though the Australian fauna includes many other eucalypt-related species in these two genera (Hurley *et al.*, 2016). By contrast, 13 of the 14 species that have arrived since 2000 are tiny insects associated with foliage (Saintonge *et al.*, 2025). This change can be attributed to the development of the trade in young ornamental plants and cut flowers surrounded by eucalyptus foliage. Trade between Mediterranean countries has probably also allowed rapid dissemination throughout the region. As soon as a country was invaded by one of these species, it thus served as a 'bridgehead', as observed for the eulophid gall-maker *Leptocybe invasa* (Saintonge *et al.*, 2025). However, these foliage insects generally cause only minor damage and are often well controlled by their natural enemies, imported with them or released by humans for biological control (Mendel *et al.*, 2017). On the other hand, the threats from xylophages remain very high and even more important in the context of climate change, generating long and severe droughts.

19.3 New Tools for Identifying Potential Invaders Prior to Introduction and Early Detection Upon Arrival

The case studies above illustrate the need to consider climate change and globalization simultaneously as the driving forces behind the expansion of non-native insects into previously uninfested areas (Robinson *et al.*, 2020). In this context, the anticipation of arrivals and their early detection at ports of entry are essential to improve the chances of eradication before they establish in the wild, as well as to obtain a better estimate of introduction frequencies. Several approaches have already been developed such as the anticipation of areas at risk using species distribution models (SDM) (Li *et al.*, 2023; Urvois *et al.*, 2024), the establishment of lists of species at risk by 'horizon scanning' (Roy *et al.*, 2019) or by 'hierarchical clustering' comparing assemblages of invasive species between regions (Eschen *et al.*, 2014). All these approaches nevertheless present the bias of considering species already identified as possibly at risk.

Thus, two significant challenges remain: (i) how to identify potential invaders before their arrival, particularly the 'emerging' ones that are not considered pests in their native range; and (ii) which devices should be deployed at potential entry points to detect unanticipated species, not on the quarantine lists, upon arrival. These challenges have led to the proposal of new biocontrol strategies which have been tentatively applied to a key group of invaders: insects associated with woody plants.

19.3.1 Sentinel plantings as a tool for identifying potential invaders before their introduction

To provide earlier warning of risks to our main forest species by identifying harmful non-native insects before they are introduced and become established, the settlement of sentinel plantings has been proposed. Such designs are to be set up in continents/regions that are the primary sources of live plant and wood product trade. Sentinel plantings can be categorized into two main approaches differing in the origin of the plants and the objectives: sentinel plantations *sensu stricto* and sentinel nurseries (Eschen *et al.*, 2019).

Sentinel plantations *sensu stricto*

Also called '*ex-patria* sentinel plantings', these correspond to the planting of plants outside their native range, especially in the continent/region identified as a possible source of invaders. For example, the planting of European woody plants in China aims at identifying Chinese insects capable of colonizing such plants and thus providing information about the potential impact of these species if they are introduced to Europe.

Such designs have been tested in the field since the late 2000s (Roques *et al.*, 2015). By planting seedlings of seven European tree species, five broad-leaved (*Quercus petraea*, *Q. suber*, *Q. ilex*, *Fagus sylvatica* and *Carpinus betulus*) and two conifer species (*Abies alba* and *Cupressus sempervirens*), on two distant Chinese sites, it has been possible to identify over a period of 3 years more than 100 Chinese insect species capable of colonizing them. The observation of recurrent damage affecting several plants from 1 year to the next then enabled the creation of a shortlist of 29 species considered as regular colonizers. The next step was to establish, through *in situ* rearing, the capacity of these species to complete their entire cycle on the European plant. Their potential for survival during transport was then estimated, as well as the possibilities of detection, taking into account their cycle and biological traits. In the end, four of these species were considered to represent a serious potential threat if introduced into Europe (Roques *et al.*, 2015).

Three years were enough to obtain a plateau of colonization by species affecting foliage as well as by root feeders. However, due to the too small diameter of the stems of the plants, a much longer time (>8 years) would have been necessary to observe a colonization by xylophagous insects. This requires funding and long-term monitoring, especially if the plantation is grown from seed, which is easier to export than seedlings but develops less quickly. Taxonomic difficulties also represent a major obstacle in these new epidemiological surveillance approaches. While adult insects are generally easier to identify, apart from the possible presence of species new to science, the majority of individuals collected were in immature stages, and in most cases no identification key for these stages was available. Systematic sequencing of larvae and eggs was expected to help by comparison with databases such as BOLD (www.boldsystems.org/) but in this case it only allowed identification of a very small number of additional species, mainly lepidopteran moths. Genetic databases therefore need to be strengthened.

Sentinel nurseries

Also called '*in-patria* sentinel plantings', these correspond to plantings of commonly exported plants in their native range, avoiding any phytosanitary treatment; for example, plantings in China of the tree species most exported to Europe. The objectives therefore differ from those of the previous approach. Surveys and identification of pest–host associations in these plantations aim at identifying insects that could be introduced into the importing country through the live plant trade, and at estimating the probability of their presence on these plants.

This approach was also tested in the field during 2010–2020. Two sentinel nurseries were set up in China, using five species of trees and shrubs corresponding to the most exported to Europe since 2000 (*Acer palmatum*, *Buxus microphylla*, *Fraxinus chinensis*, *Ilex cornuta*, *Zelkova schnederiana*). In just 2 years, surveys revealed 95 new native-to-native insect–host connections, which had not been mentioned previously in the Chinese or international literature, and generated a list of species that had never before been identified as likely to be introduced with these plants (Kenis *et al.*, 2018). The surveys also revealed the high probability of introduction of certain species present on most of the plants, with the box tree moth *Cydalima perspectalis* as a striking example. At each of the sites, the box tree moth defoliated nearly all the planted box shrubs, even though it was not considered a pest in these regions. Setting up such a system in the early 2000s could certainly have alerted us to the danger of uncontrolled imports of Chinese box seedlings before the insect became established in Europe (2007). Similarly, as in sentinel plantations, the entomological fauna in the first few years mainly consisted of defoliators (Kenis *et al.*, 2018). However, sentinel nurseries gave much more information about other insect guilds because older plants available on the local market could be used for plantations, unlike non-native ones which are subject to stricter regulatory constraints in their use. Actually, only 5 years after the sentinel nurseries were set up, the first woodborers (cerambycids, bark beetles, buprestids) were observed on fast-growing species (*Acer*, *Fraxinus*, *Zelkova*) (Roques, unpublished).

The international community has begun to generalize these sentinel plantations, adopting a standard methodology for planting (Morales-Rodríguez *et al.*, 2019), monitoring and identifying damage with the publication of a guide that can be used on all continents (www.cabidigitallibrary.org/doi/book/10.1079/9781786394415.0000) (Roques *et al.*, 2017). New sentinel plantations of forest species on a larger scale have been set up in South Africa, the United States and China, as well as in Europe using Chinese plants, as part of a European Horizon 2020 project (HOMED project – Holistic Management of Emerging Forest Pests and Diseases 2018–2023). Currently, these plantations only concern forest species or ornamental shrubs, but they could tentatively be extended to plants of agronomic interest.

19.3.2 Trapping with generic broad-spectrum attractants, a tool for early detection of invasive species upon arrival

Even if they are identified before they arrive, some of these non-native species will inevitably be introduced. It is therefore important to be able to detect them as early as possible, preferably as soon as they arrive at potential points of entry so that effective eradication measures can be applied. In the case of non-native forest insects, the vast majority of first records (89%) did not occur in forests but in urban areas at short distance from ports and airports likely to serve as points of entry (Branco *et al.*, 2019). In order to maximize detection possibilities while minimizing operational costs, research has therefore focused on defining generic attractants capable of simultaneously capturing the largest possible number of species, including known and unknown, unanticipated species.

A group of wood borers, including numerous invasive species such as the cerambycids or long-horned beetles (Cerambycidae), appeared particularly well adapted to this objective. These species pose significant risks due to their long life cycles and larvae that develop concealed within wood (e.g. logs, trunks, packaging materials, pallets and wood products), facilitating their unintentional transport across continents through international trade. Substantial progress has been made over the last 10 years in the chemical ecology of cerambycids. At present, sex pheromones (produced by females to attract males) or sex aggregation pheromones (produced by males to attract both sexes) are known for more than 400 species in this family (Millar and Hanks, 2017). More importantly, the components of these pheromones appear to be relatively well conserved at world level between related species at the generic, tribal or even subfamily level (Millar *et al.*, 2018). Thus, the pheromone of a European insect could be attractive to a Chinese or American insect of the same genus. Combining several pheromones into a single blend was therefore expected to provide a generic lure allowing the simultaneous trapping of cerambycid species from different subfamilies, tribes and genera.

After a successful preliminary test in France which did not show any antagonistic effects between compounds (Fan *et al.*, 2019), a world trapping programme was deployed from 2019 to 2022 to check the potential generic attractiveness of a blend combining eight pheromones of cerambycids. The trial was based on the principle that a species trapped by a blend in large numbers on its native continent has a significant probability of being trapped by the same blend when it arrives at a port of entry on another continent. More than 1250 traps, mainly black multifunnel traps, were placed in 168 sites covering most European countries, many parts of China, Siberia, the

USA, Canada, Australia and the Caribbean. This network allowed trapping of 376 species from eight subfamilies and 60 tribes (Roques *et al.*, 2023). The substantial capture of species in different continents for each of the most important subfamilies or tribes confirmed the generic attractiveness of the blend.

However, cerambycids are not the only non-native wood-boring pests of concern; bark and ambrosia beetles (Curculionidae Scolytinae) and jewel beetles (Buprestidae) are also significant threats. The use of traps containing plant-emitted volatiles, ethanol and [-] α-pinene in addition to the cerambycid multipheromone blend resulted in satisfactory generic attractiveness for trapping non-native bark and ambrosia beetles (Barnouin *et al.*, 2020). In contrast, no compounds or blends with generic attractiveness were identified for jewel beetles, for which the colour of the trap, green or yellow, seems to be important (Rassati *et al.*, 2019).

On the basis of these results, an annual survey programme for early detection of non-native wood-borers was launched in 2021 by the French Ministry of Agriculture in 29 potential ports of entry, including sea and fluvial ports, aiports, sawmills, national plant markets and truck parkings at borders (Roques *et al.*, 2023). At each site, a set of two traps, one black (intended for cerambycids and bark beetles) and one green (intended for jewel beetles), is deployed within the perimeter of the port to allow early detection upon arrival, both traps being baited with the multipheromone blend for cerambycids, ethanol and [-] α-pinene. Another set of two traps, also one black and one green, is settled in a wooded area located within 1 km from the port in order to check possible dispersal of the non-native insects. Highly significant results have already been obtained. In 2023, non-native wood-borers were captured in 24 of the 29 monitored ports of entry, for a total of 22 species including four cerambycids and 12 bark and ambrosia beetles, and almost 3000 specimens (Goudet and Roques, 2025). Most species were only captured inside the port but the external traps showed that an Asian cerambycid threatening fruit trees was beginning to disperse outside. This monitoring design is being tested in several countries of southern Europe (Spain, Italy, Portugal) and Switzerland (Roques *et al.*, 2022) and also in North America (Santoiemma *et al.*, 2024).

Currently, the process of sorting trapped insects and having taxonomists identify them to detect potential non-native species takes several weeks. As a result, immediate eradication upon detection is unrealistic. Ongoing research aims to develop automated traps capable of remotely transmitting real-time photos of daily captures, coupled with artificial intelligence for species identification.

References

Adhami, J. and Murati, N. (1987) Presence of the mosquito *Aedes albopictus* in Aalbania (in Albanian, French summary). *Revista Mjekësore* 1, 13–16.

Auger-Rozenberg, M.-A., Barbaro, L., Battisti, A., Blache, S., Charbonnier, Y. *et al.* (2015) Ecological responses of parasitoids, predators and associated insect communities to the climate-driven expansion of pine processionary moth. In: Roques, A. (ed.) *Processionary Moths and Climate: An Update*. Springer/Quae Editions, Change, pp. 311–358.

Barnouin, T., Soldati, F., Roques, A., Faccoli, M., Kirkendall, L.R. *et al.* (2020) Bark beetles and pinhole borers recently or newly introduced to France (Coleoptera: Curculionidae, Scolytinae and Platypodinae). *Zootaxa* 4877(1), 051–074.

Battisti, A., Stastny, M., Netherer, S., Robinet, C., Schopf, A. *et al.* (2005) Expansion of geographic range in the pine processionary moth caused by increased winter temperatures. *Ecological Applications* 15(6), 2084–2096.

Battisti, A., Stastny, M., Buffo, E. and Larsson, S. (2006) A rapid altitudinal range expansion in the pine processionary moth produced by the 2003 climatic anomaly. *Global Change Biology* 12, 662–671.

Bellard, C., Bertelsmeier, C., Leadley, P., Thuiller, W., Leroy, B. *et al.* (2013) Will climate change promote future invasions? *Global Change Biology* 19, 3740–3748.

Benelli, G., Wilke, A. and Beier, J. (2020) *Aedes albopictus* (Asian tiger mosquito). *Trends in Parasitology* 36(11), 942–943.

Bonnamour, A., Blake, R.E., Liebhold, A.M., Nahrung, H.F., Roques, A. *et al.* (2023) Historical plant introductions predict current insect invasions. *Proceedings of the National Academy of Sciences USA* 120, e2221826120.

Bourougaaoui, A., Robinet, C., Ben Jamâa, M.L. and Laparie, M. (2024) A retrospective analysis on the effects of climate warming on the pine processionary moth at the southern edge of its range. *Oikos* 12, e10989.

Branco, M., Nunes, P., Roques, A., Fernandes, M.R., Orazio, C. *et al.* (2019) Urban trees facilitate the establishment of non-native forest insects. *NeoBiota* 52, 25–46.

Cannon, R.J.C. (1998) The implications of predicted climate change in the UK, with emphasis on non-indigenous species. *Global Change Biology* 4, 785–796.

Charbonnel, E., Ouvrard, D., Benoit, L. and Chapuis, M.-P. (2023) Identifying the specific status and geographical origin of european incursions of an invasive and cryptic pest, *Bactrocera dorsalis* (Diptera: Tephritidae). ANSES Scientific and Doctoral Days, Maisons-Alfort, France, 2 Octobre. Available at: h ttps://agritrop.cirad.fr/607493/ (accessed 11 June 2025).

Del Lesto, I., Liberato, C., Casini, R., Magliano, A., Ermenegildi, A. *et al.* (2022) Is Asian tiger mosquito (*Aedes albopictus*) going to become homodynamic in southern Europe in the next decades due to climate change? *Royal Society Open Science* 9(12), 220967.

Dodds, K.J., Aoki, C.F., Arango-Velez, A., Cancelliere, J., D'Amato, A.W. *et al.* (2018) Expansion of southern pine beetle into northeastern forests: Management and impact of a primary bark beetle in a new region. *Journal of Forestry* 116, 178–191.

EPPO (2003) First report of *Paysandisia archon* in United Kingdom. *Reporting Services* 8, 2003/121.

Eschen, R., Holmes, T., Smith, D., Roques, A., Santini, S. *et al.* (2014) Likelihood of establishment of tree pests and diseases based on their worldwide occurrence as determined by hierarchical cluster analysis. *Forest Ecology and Management* 315, 103–111.

Eschen, R., Roques, A. and Santini, A. (2015) Taxonomic dissimilarity in patterns of interception and establishment of alien arthropods, nematodes and pathogens affecting woody plants in Europe. *Diversity and Distributions* 21(1), 36–45.

Eschen, R., O'Hanlon, R., Santini, A., Vannini, A., Roques, A. *et al.* (2019) Safeguarding global plant health: The rise of sentinels. *Journal of Pest Science* 92, 29–36.

Fan, J.T., Denux, O., Courtin, C., Bernard, A., Javal, M. *et al.* (2019) Multicomponent blends for trapping native and exotic longhorn beetles at potential points-of-entry and in forests. *Journal of Pest Science* 92, 281–297.

Fenn-Moltu, G., Ollier, S., Caton, B., Liebhold, A.M., Nahrung, H. *et al.* (2023) Alien insect dispersal mediated by the global movement of commodities. *Ecological Applications* 33, e2721.

Goudet, M. and Roques, A. (2025) Surveillance actuelle des exotiques envahissants par le Département de la Santé des Forêts, et contrôle aux frontières. In: Robinet, C., Saintonge, F.-X., Tassus, X. and Brault, S. (eds) *Invasion et Expansion d'insectes Bioagresseurs Forestiers. Quels Risques Pour La Forêt Française Dans Le Contexte Des Changements Globaux?* Quae Editions, Versailles, France, pp. 203–219.

Hickling, R., Roy, D.B., Hill, J.K., Fox, R. and Thomas, C.D. (2006) The distributions of a wide range of taxonomic groups are expanding polewards. *Global Change Biology* 12(3), 450–455.

Hurley, B.P., Garnas, J., Wingfield, M.J., Branco, M., Richardson, D.M. *et al.* (2016) Increasing numbers and intercontinental spread of invasive insects on eucalypts. *Biological Invasions* 18, 921–933.

Isitt, R., Liebhold, A.M., Turner, R.M., Battisti, A., Bertelsmeier, C. *et al.* (2024) Asymmetrical insect invasions between three world regions. *NeoBiota* 90, 35–51.

Kenis, M., Li, H.M., Fan, J.T., Courtial, C., Auger-Rozenberg, M.-A. *et al.* (2018) Sentinel nurseries to assess the phytosanitary risks from insect pests on importations of live plants. *Scientific Reports* 8, 11217.

Kontodimas, D.C., Milonas, P.G., Vassiliou, V., Thymakis, N. and Economou, D. (2006) The occurrence of *Rhynchophorus ferrugineus* in Greece and Cyprus and the risk against the native greel palm tree *Phoenix theophrasti*. *Entomologia Hellenica* 16, 11–15.

Li, D.X., Li, Z.X., Wang, X.X., Wang, L., Khoso, A.G. *et al.* (2023) Climate change and international trade can exacerbate the invasion risk of the red imported fire ant *Solenopsis invicta* around the globe. *Entomologia Generalis* 43, 315–323.

Liebhold, A.M., Brockerhoff, E., Garrett, L., Parke, J. and Britton, K. (2012) Live plant imports: The major pathway for the forest insect and pathogen invasions of the US. *Frontiers in Ecology and the Environment* 10, 135–143.

Lühken, R., Brattig, N. and Becker, N. (2023) Introduction of invasive mosquito species into Europe and prospects for arbovirus transmission and vector control in an era of globalization. *Infectious Diseases of Poverty* 12, 109.

Lührsen, D.S., Zavitsanou, E., Cerecedo, C., Pardo-Araujo, M., Palmer, J. *et al*. (2023) Adult *Aedes albopictus* in winter: Implications for mosquito surveillance in southern Europe. *Lancet Planetary Health* 9, e729–e731.

Mendel, Z., Protasov, A., La Salle, J., Blumberg, D., Brand, D. *et al*. (2017) Classical biological control of two eucalyptus gall wasps: Main outcome and conclusions. *Biological Control* 105, 66–78.

Millar, J.G. and Hanks, L.M. (2017) Chemical ecology of cerambycids. In: Wang, Q. (ed.) *Cerambycidae of the World: Biology and Pest Management*. CRC Press/Taylor and Francis, Boca Raton, Florida, pp. 161–208.

Millar, J.G., Richards, A.B., Halloran, S., Zou, Y., Boyd, E. *et al*. (2018) Pheromone identification by proxy: Identification of aggregation-sex pheromones of North American cerambycid beetles as a strategy to identify pheromones of invasive Asian congeners. *Journal of Pest Science* 92, 213–220.

Morales-Rodríguez, C., Anslan, S., Auger-Rozenberg, M.-A., Augustin, S., Baranchikov, Y. *et al*. (2019) Forewarned is forearmed: Harmonized approaches for early detection of potentially invasive pests and pathogens in sentinel plantings. *NeoBiota* 47, 95–123.

Musolin, D.L. (2007) Insects in a warmer world: Ecological, physiological and life-history responses of true bugs (Heteroptera) to climate change. *Global Change Biology* 13, 1565–1585.

Nugnes, F., Russo, E., Viggiani, G. and Bernardo, U. (2018) First record of an invasive fruit fly belonging to *Bactrocera dorsalis* Complex (Diptera: Tephritidae) in Europe. *Insects* 9(4), 182.

Qin, Y., Wang, C., Zhao, Z., Pan, X. and Li, Z. (2019) Climate change impacts on the global potential geographical distribution of the agricultural invasive pest, *Bactrocera dorsalis* (Hendel) (Diptera: Tephritidae). *Climatic Change* 155, 145–156.

Rassati, D., Marini, L., Marchioro, M., Rapuzzi, P., Magnani, G. *et al* (2019) Developing trapping protocols for wood-boring beetles associated with broadleaf trees. *Journal of Pest Science* 92, 267–279.

Robinet, C., Imbert, C.E., Rousselet, J., Sauvard, D., Garcia, J. *et al*. (2012) Warming up combined with the trade of large trees allowed long-distance jumps of pine processionary moth in Europe. *Biological Invasions* 14, 1557–1569.

Robinson, T.B., Martin, N., Loureiro, T.G., Matikinca, P. and Robertson. M.P. (2020) Double trouble: The implications of climate change for biological invasions. *NeoBiota* 62, 463–487.

Roques, A. (2010) Alien forest insects in a warmer world and a globalized economy: Impacts of changes in trade, tourism and climate on forest biosecurity. *New Zealand Journal of Forestry* 40, 77–94.

Roques, A., Fan, J.-T., Courtial, B., Zhang, Y.-Z., Yart, A. *et al*. (2015) Planting sentinel European trees in eastern Asia as a novel method to identify potential insect pest invaders. *PLOS One* 10(5), 0120864.

Roques, A., Shi, J., Auger-Rozenberg, M.-A., Ren, L., Augustin, S. *et al*. (2020) Are invasive patterns of non-native insects related to woody plants differing between Europe and China? *Frontiers in Forests and Global Change* 2, 91.

Roques, A., Feddern, N., Garcia, A., Marchioro, M., Mas, H. *et al*. (2022) Early detection of exotic xylophagous species using generic lures at potential ports-of-entry in Europe. In: *IUFRO Conference Division 7 – Forest Health Pathology and Entomology*, Lisbon, September 6–9, .

Roques, A., Ren, L., Rassati, D., Shi, J., Akulov, E. *et al*. (2023) Worldwide tests of generic attractants, a promising tool for early detection of non-native cerambycid species. *NeoBiota* 84, 169–209.

Roques, A. (ed.) (2015) *Processionary Moths and Climate Change: An Update*. Springer, Berlin.

Roques, A., Cleary, M., Matsiakh, I. and Eschen, R. (eds) (2017) *Field Guide for the Identification of Damage on Woody Sentinel Plants*. CABI, Wallingford, UK.

Rosenzweig, C., Casassa, G., Karoly, D.J., Imeson, C., Liu, A. *et al*. (2007) Assessment of observed changes and responses in natural and managed systems. In: Parry, M.L., Canziani, O., Palutikof, J. and Linden, P. (eds) *Climate Change 2007: Impacts, Adaptation and Vulnerability. Contribution of Working Group II to the Fourth Assessment Report of the Intergovernmental Panel on Climate Change*. Cambridge University Press, Cambridge, pp. 79–131.

Roy, H.E., Bacher, S., Essl, F., Adriaens, T., Aldridge, D.C. *et al*. (2019) Developing a list of invasive alien species likely to threaten biodiversity and ecosystems in the European union. *Global Change Biology* 25, 1032–1048.

Sáenz-Romero, C., Cambrón-Sandoval, V.H., Hammond, W., Méndez-González, J., Luna-Soria, H. *et al.* (2023) Abundance of *Dendroctonus frontalis* and *D. mexicanus* (Coleoptera: Scolytinae) along altitudinal transects in Mexico: Implications of climatic change for forest conservation. *PLOS One* 18(7), e0288067.

Saintonge, F.-X., Bailly, A., Ben Jamâa, M.H., Branco, M., Dhahri, S. *et al.* (2025) Ravageurs sur eucalyptus. In: Robinet, C., Saintonge, F.X., Tassus, X. and Brault, S. (eds) *Invasion et Expansion d'insectes Bioagresseurs Forestiers. Quels Risques Pour La Forêt Française Dans Le Contexte Des Changements Globaux?* Quae Editions, Versailles, France, pp. 167–179.

Santoiemma, G., Battisti, A., Courtin, C., Curletti, G., Faccoli, M. *et al.* (2024) Testing a trapping protocol for generic surveillance of wood-boring beetles in heterogeneous landscapes. *NeoBiota* 95, 77–95.

Seebens, H., Blackburn, T.M., Dyer, E.E., Genovesi, P., Hulme, P.E. *et al.* (2017) No saturation in the accumulation of alien species worldwide. *Nature Communications* 8, 14435.

Seebens, H., Blackburn, T.M., Dyer, E.E., Genovesi, P., Hulme, P.E. *et al.* (2018) Global rise in emerging alien species results from increased accessibility of new source pools. *Proceedings of the National Academy of Sciences USA* 115(10), e2264–e2273.

Skendžic, S., Zovko, M., Pajac Živkovic, I., Lešie, V. and Lemić, D. (2021) Effect of climate change on introduced and native agricultural invasive insect pests in Europe. *Insects* 12, 985.

Stephens, A.E.A., Kriticos, D.J. and Leriche, A. (2007) The current and future potential geographical distribution of the oriental fruit fly, *Bactrocera dorsalis* (Diptera: Tephritidae). *Bulletin of Entomological Research* 97, 369–378.

Tran, J.K., Ylioja, T., Billings, R.F., Régnière, J. and Ayres, M.P. (2007) Impact of minimum winter temperatures on the population dynamics of *Dendroctonus frontalis*. *Ecological Applications* 17, 882–899.

Turner, R., Blake, R. and Liebhold, A.M. (2024) International non-native insect establishment data. Version 2.0. Available at: https://zenodo.org/records/5245302 (accessed 11 June 2025).

Ungerer, M.J., Ayres, M.P. and Lombardero, M.J. (1999) Climate and the northern distribution limits of *Dendroctonus frontalis* Zimmerman (Coleoptera: Scolytidae). *Journal of Biogeography* 26, 1133–1145.

Urvois, T., Auger-Rozenberg, M.-A., Roques, A., Kerdelhué, C. and Rossi, J.-P. (2024) Intraspecific niche models for the invasive ambrosia beetle *Xylosandrus crassiusculus* suggest contrasted responses to climate change. *Oecologia* 204(4), 761–774.

Walther, G.R., Roques, A., Hulme, P.E., Sykes, M.T., Pyšek, P. *et al.* (2009) Alien species in a warmer world: Risks and opportunities. *Trends in Ecology and Evolution* 24, 686–692.

20 The Influence of Microclimates on the Success of Biological Control

Sylvain Pincebourde*
UMR CNRS 7261 Institut de Recherche sur la Biologie de l'Insecte, Université de Tours, Tours, France

Abstract

Biological control is widely used to improve crop yields and food security with environmentally friendly approaches. This chapter discusses the extent to which the microclimate of crop systems can modulate the efficiency of biological control strategies. Fundamental concepts related to the thermal biology of insect pests and control agents are provided to comprehend the importance of microclimatic fluctuations in the population dynamics of both pests and their biological control agents. Agricultural landscapes are largely heterogeneous from a microclimate perspective, and even monocultures of small crops display thermal diversity. The relationship between microclimatic conditions and the performance of biological control agents is reviewed, generating the conclusion that there is room for improving biological control by considering the mosaic of microclimates that is often crop specific. However, generalizing remains difficult given the paucity of studies at pertinent scale. Mechanistic approaches are mandatory to comprehend the relationships and increase our predictive capacities.

20.1 Introduction

The concept of a microclimate is intuitive for us humans. During a heatwave, people tend to move to woods and forests, or near rivers and fountains, to seek for cooler environments. The same concept applies to most animals. The microclimate (i.e. climate at microscale) has been defined in slightly different ways depending on the discipline or the context. A general, all-purpose definition states that microclimates are fine-scale climate variations that deviate from the background atmosphere at least temporarily (Stoutjesdijk and Barkman, 2014; Bramer *et al.*, 2018). Note that the term 'deviate' is used instead of the word 'decoupled' in the original version of this definition (Bramer *et al.*, 2018). The temperature in the forest understorey can deviate strongly from atmospheric temperature (De Frenne *et al.*, 2013, 2019), but this deviation does not necessarily imply a full decoupling between the two.

The role of microclimate in buffering extreme atmospheric conditions (e.g. high temperature, drought, stormy winds) or in modulating environmental conditions (e.g. availability of microhabitats with optimal conditions) is an important aspect for species maintenance, distribution, coexistence and adaptation (Cloudsley-Thompson, 1962; Angilletta, 2009; Kemppinen *et al.*, 2024). This applies not only to natural ecosystems but also to anthropogenic systems such as urban and cultivated fields (Pincebourde *et al.*, 2016, 2017), and therefore the ecology of microclimates should be included in biological control strategies as well.

*Corresponding author: sylvain.pincebourde@univ-tours.fr

 Biological Control Systems and Climate Change (eds Joan van Baaren *et al.*)
DOI: 10.1079/9781800625099.0020

Yet, the concept of microclimate necessitates consideration of several aspects, from which microclimatic variables and how to measure them, to questions about the scale at which we should measure these variables (Briscoe *et al.*, 2023; Kemppinen *et al.*, 2024). This chapter provides the main conceptual keys to tackle the role of microclimates in the success of biological control, before reviewing current knowledge on agricultural microclimates and the evidence of their influence on the success of agents used to control for pests.

The contribution of trait-based approaches to understanding plant–arthropod interactions and improving conservation biological control was reviewed recently to reveal that (i) functional studies on plant–herbivore interactions identified several traits as important for trophic resource provision, that relate to resource quantity, quality, accessibility and availability, but that (ii) the traits associated with microclimate modifications by the plant and the insect have been little explored (Gardarin *et al.*, 2018). This gap clearly limits our ability to assess the role of microclimate in the fate of biological control plans. Yet, microclimatic variables such as temperature can vary tremendously across agricultural fields and crop units such as greenhouses. The question now is how microclimates modulate the outcome of biological control programmes. Therefore, this chapter proposes a conceptual framework to link mechanistically the performance of pests (or pathogens) and control agents, and the microclimate mosaic by integrating the problem of spatial and temporal scales and the influence of environmental modifications that are part of the agricultural practices into a biophysical modelling approach to obtain quantitative assessment of agent efficiency (Fig. 20.1).

20.2 Fundamental Concepts in the Link Between Microclimate and Biological Control

Two components are fundamental to understanding the influence of microclimates on the fate of biological control strategies. The thermal biology of both the pest and the introduced control agent should be determined, as well as the drivers of the microclimatic conditions for these species (see Fig. 20.1). Conceptually, assessing the vulnerability of a species through only one of these two aspects is likely to fail to provide accurate outputs (Pincebourde and Casas, 2015, 2019).

20.2.1 The thermal biology of ectotherms

As ectotherms, the body temperature of arthropods varies largely depending on the microclimatic conditions around them. Ultimately, the body temperature of an ectotherm is the physiologically relevant temperature, i.e. the temperature at which the whole organism, tissues, organs and cells are functioning. Temperature drives virtually all physiological and biochemical rates including enzyme activity, metabolic rate, growth rate and feeding rate (Hochachka and Somero, 2002; Chown and Nicolson, 2004). The relationship between (body) temperature and these traits (or rates) can be described using thermal performance curves (TPCs) which are basically non-linear models fitted to trait measurements at various temperatures (Angilletta, 2006; Sinclair *et al.*, 2016). For most traits (for example, developmental rate of an insect larva), the TPC often is asymmetrical (see Fig. 20.1): the trait value starts from zero at low temperature and slowly increases as temperature increases, up to an optimal temperature which maximizes the trait value, and then performance drops rapidly (faster than the rate of increase from cold to optimal temperature) until it again reaches zero at high temperature. The TPC allows us to determine the temperature range that maximizes the performance of the ectotherm, including population growth rate of an insect predator, for example. The shape and temperature range of TPCs can vary considerably between species, between populations of the same species, and between life stages or sex from the same population (Sinclair *et al.*, 2016). In natural systems, the complementarity between different predators displaying contrasting TPCs attenuates the negative impacts of warming on food webs (Pepi and McMunn, 2021).

Thermal limits are traits of particular importance as they can be used to assess

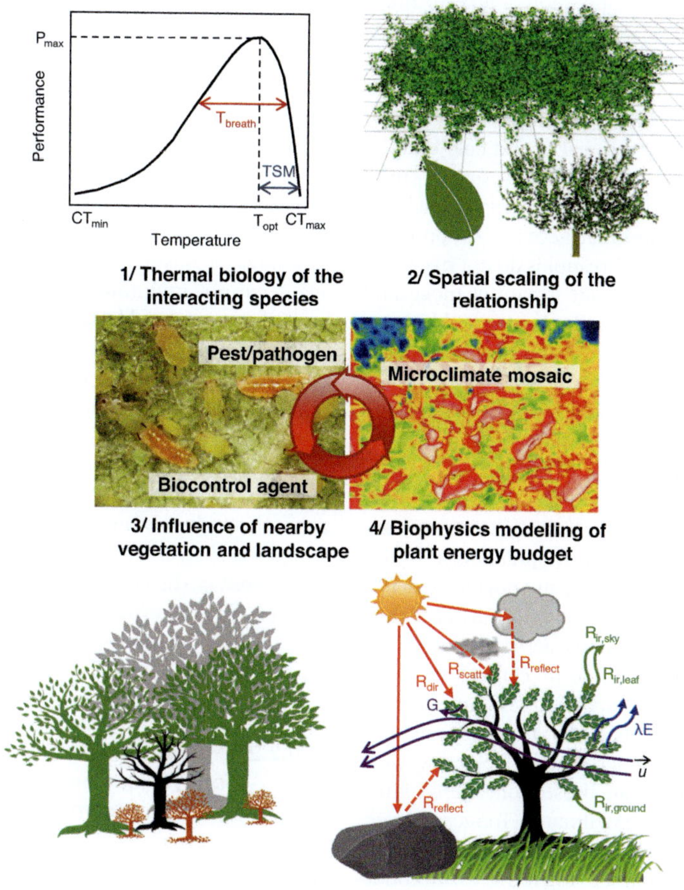

Fig. 20.1. Conceptual view of the mechanistic framework proposed to analyse the relationship between the pest (or pathogen), the control agent (predator or pathogen) and the microclimate mosaic. (1) Know the thermal biology of the different species. Typically, the thermal sensitivity can be described by determining the thermal performance curve (TPC) of traits such as developmental or feeding rates. These TPCs contain information that is critical to understanding the thermal sensitivity: the critical temperatures (CTmin and CTmax), the optimal temperature (Topt) at which performance is maximized (Pmax), the thermal breadth (Tbreadth) which is the temperature range producing a high performance, and the thermal safety margin (TSM) that is the temperature range between optimal and maximal temperatures. (2) Determine the spatial scale at which the relationship operates. In the case of plant–insect–pathogen multitrophic interactions, species may interact at the level of the leaf which can contain some levels of thermal heterogeneity, but additional thermal variance arises at the upper scales of the single plant up to the crop field or orchard level. (3) Consider the influence of the landscape elements surrounding the relationship. In agricultural systems, one can manipulate these elements to positively influence the impacts of the biological control strategy (planting trees, hedges, flowering strips, etc.) by providing alternative food sources and thermal refuges. (4) Finally, all these elements can be integrated into a biophysical model of the plant–arthropod energy budget to predict quantitatively the relative performances of the pest and the agent within the microclimate mosaic generated by the interaction between the crop field and the local climate. These models integrate the portion of incoming solar energy that is absorbed by the plant surface. Rdir, direct solar radiation; Rscatt and Rreflect, the radiation scattered and reflected by the atmosphere or nearby elements; Rir, the infrared radiation emitted by the sky, the leaf and the ground surface; u, wind speed and direction; G, the heat exchanged via convection with ambient air; λE, latent heat lost during evapotranspiration.

whether a species or population of an ectotherm can maintain itself under given (micro)climatic conditions. Thermal limits are normally used as proxies of survival and several metrics exist (Angilletta, 2009). For example, the critical thermal limits, at both warm and cold ends of the temperature range (CTmax and CTmin, respectively), are the temperatures at which co-ordinated neuromuscular activity is suppressed – these temperature thresholds do not necessarily kill the individual, but it cannot move and therefore cannot escape the heat or cold. Other metrics more often used in agronomical studies are the lower and upper lethal temperatures (LLT and ULT, respectively), which correspond to the temperature causing mortality of a given portion of a group (e.g. 50%) during a given exposure time.

Thermal limits, whatever the metrics, vary widely between species and populations (Sunday *et al.*, 2011), although they are phylogenetically constrained (Hoffmann *et al.*, 2013) and their level of plasticity remains perhaps insufficient to tolerate the amplitude of climate warming (Gunderson and Stillman, 2015). For example, aphid parasitoids collected in the winter in open landscapes (cooler microclimate) were 1°C more resistant to cold than those from closed landscapes (warmer microclimate) (Tougeron *et al.*, 2016). Hence, the vegetation cover provides shelter for parasitoids during the stressful conditions of winter (Suggitt *et al.*, 2011; Tougeron *et al.*, 2016). Also, these thermal limits can differ between interacting species such as host–parasitoid relationships, with host pest herbivores often being more tolerant to higher temperatures than their parasitoids, thus suggesting a thermal mismatch under warming conditions (Malinski *et al.*, 2024).

The identification of the consequences of this thermal mismatch for the success of biological control remains extremely rare and challenging (see also Chapters 2 and 16). Ideally, the best scenario would be to use a control agent that displays high tolerance levels for heat to ensure its persistence and the control of the undesirable species during the growth season. For example, the thermal physiology of a fly used as a biological control agent of aquatic weeds indicated that both larval and adult stages were sufficiently heat tolerant to potentially control Brazilian water weed in all invaded areas in South Africa (Smith *et al.*, 2022). However, the thermal thresholds remain to be compared to microclimatic temperature of the plant both underwater (for larvae) and in the air (for adults) during summer period. A similar case is observed between a macrophyte and a tropical coleopteran species (Goddard *et al.*, 2022). Nevertheless, the use of tropical species (highly heat tolerant) to control for aquatic weeds often generates a mismatch during winter as the agent cannot maintain itself in some invaded areas that are too cold (Harms *et al.*, 2021). Here again, surveying microclimate temperatures near the aquatic weed together with thermal physiology should provide crucial information to better understand the non-establishment of the control agent in these areas. Looking at the problem of non-establishment of the biological agent through only one lens – e.g. thermal physiology – is likely to fail to provide comprehensive assessment of the potential success of biological control (Pincebourde and Casas, 2015; Nqayi *et al.*, 2023).

20.2.2 Fundamentals of environmental biophysics: what generates microclimates?

Microclimates exist because transfers of heat, water and more generally energy occur through multiple paths and directions between the various elements of the environment and the atmosphere (see Fig. 20.1). Briefly, radiative (solar energy, shading, etc.), latent (heat exchange during evapotranspiration) and sensible (heat exchange via conduction and convection) heat transfers all modulate the amount of energy transiting through a microhabitat, translating into a temperature variation relative to the atmospheric conditions around the given microhabitat (Gates, 1980; Campbell and Norman, 1998; Nobel, 1999). The field of biophysical ecology emerged to model and quantify these heat transfers and the resulting temperature of a given microhabitat (Briscoe *et al.*, 2023). Solar radiation is the primary source of (heat) energy in the environment. As such, the level of exposure of a microhabitat to solar radiation often directly drives the temperature excess (i.e. difference between microhabitat temperature and ambient temperature): a shaded microhabitat could be cooler than atmosphere (e.g. the understorey of dense forests; Zellweger *et al.*, 2019) while a microhabitat directly exposed to the sun can be warmer (e.g. herbaceous vegetation in grasslands; Scherrer and Körner, 2010). Thus, the shading effect of above-ground vegetation directly drives soil temperature patterns (Suggitt *et al.*, 2011). In addition, humidity plays a great role in contrasting microhabitat temperature relative to atmospheric conditions. Well-watered plants that evapotranspirate lose latent energy into the atmosphere, contributing to maintaining temperatures below the temperature of the air layer above (Nobel, 1999).

20.2.3 Microclimates and the eternal question of scales

The notion of a microclimate includes a spatial extent scale. For example, microclimate corresponds to horizontal distances of 0.001–100 m while mesoclimate lies in the range 1–200 km and macroclimate corresponds to distances > 200 km (Bramer *et al.*, 2018). However, this scale is largely biased by our own (human) vision. A distance of 100 m does not mean the same for an aphid and a large reptile. Considering this body size perspective, recent studies have proposed that the microclimate dimension is below 1 cm and 10 cm for a spider mite (0.2 mm in size) and an aphid (2 mm in size), respectively (Pincebourde and Woods, 2020). Indeed, for organisms with very small body size, the term nano-climate was sometimes employed, referring to the climate within surface boundary layers such as a leaf surface or air cavity in the soil (Stoutjesdijk and Barkman, 2014). However, this term was never really used by ecologists and physiologists, and indeed is no longer needed if we scale the microclimate relative to the body size of the studied organism (Pincebourde and Woods, 2020).

Indeed, for most arthropods that affect agronomical and agricultural systems, their small body size means that their heat budget is mostly driven by conduction processes with the vegetation surface on which they live rather than by radiation or convection with atmospheric conditions (Pincebourde *et al.*, 2021). The bodies of small herbivore insects can be largely influenced by the leaf surface temperature to the point that tiny species exactly match leaf surface temperature even when exposed to solar radiation (Pincebourde *et al.*, 2021). Paradoxically, this small-scale perspective opens novel avenues for herbivores living, and potentially thermoregulating, at the leaf surface because most plant species display large surface temperature gradients across their individual leaves, up to 10–15°C in both cultivated and wild plants (Caillon *et al.*, 2014; Pincebourde and Suppo, 2016; Leigh *et al.*, 2017; Saudreau *et al.*, 2017). Therefore, we should focus on the smallest scales when studying the potential role of plant microclimates in the success of releasing predatory arthropods as biological control agents.

20.3 The Mosaic of Microclimates of Agricultural Landscapes

From the concepts developed above, we understand that the architectural complexity of plants within agricultural landscapes determines the microclimates available for arthropods. Therefore, even monocultures are not thermally homogeneous. The term 'mosaic' illustrates the extent to which microclimates, at the pertinent scale for a given arthropod, vary in space and time within a given agricultural system. Below, this microclimatic heterogeneity is illustrated in several representative agricultural systems (Fig. 20.2).

20.3.1 Microclimate mosaics in crop open fields

Crop open fields are often characterized by a short layer of dense vegetation above the soil, implying exposure to solar radiation of the top surface of crop canopies and a variable shading effect near the soil surface, depending on the canopy density. High-resolution measurements of potato field surface temperatures with thermography demonstrated the high thermal heterogeneity available for insects to thermoregulate: insect pests encounter within a few centimetres (10–20 cm) most of the thermal

Fig. 20.2. The microclimate mosaic of two agricultural systems contrasting in their architectural complexity. (Left panel) A winter wheat field, with a focus on the portion of about 1 × 1 m (the metal marks help to match the visible and infrared images). The two infrared images display the spatial heterogeneity of surface temperatures in the morning and the afternoon for this field located near Paris, France (spring). Note that the thermal gradient between soil surface exposed to solar radiation and in the shade and the vegetation surface is reversed between morning and afternoon, with a temperature range of ~4°C and ~14°C, respectively. (Right panel) An apple orchard with a focus on a single leaf that is partially shaded by the other leaves around it, generating a temperature gradient of about 7°C across a single leaf (infrared image).

microenvironments (temperature range of up to 20°C) recorded at the whole potato field level (Faye *et al.*, 2017). In addition, the microclimate heterogeneity increases with solar radiation level and pests encounter the same range of microclimates over even shorter distances (Faye *et al.*, 2017). This study also indicated that the diversity (but not abundance) of insect herbivores on potatoes positively correlates with the level of thermal heterogeneity, and it is anticipated that the same would apply to the diversity of insect predators and parasitoids.

Finally, the temporal dynamics of crop systems probably matters; alternating crops in the field may produce sudden microclimate shifts, although only indirect cues are available. For instance, Fox *et al.* (2016) indicated that planting a winter crop after maize stover harvest, instead of leaving bare ground, allows maintenance of the abundance and diversity of predatory insects. In addition, cover crops in general increase per capita seed predation relative to bare soil, suggesting that cover crops facilitate weed biocontrol (Blubaugh *et al.*, 2016), and microclimatic modification could be a driver.

20.3.2 The microclimate mosaic in agroforestry systems

From a biophysical point of view, agroforestry systems play with the benefit that crops gain from being in the shade of trees. The distance to which a single tree influences crop yield and ecosystem services was estimated to be up to 17 m in an African system with relatively tall trees (Roupsard *et al.*, 2020), suggesting indirectly that this single tree canopy can potentially modify the local microclimate at this relatively large scale. In coffee agriculture, the amount of shade directly determines the ability to mitigate microclimate temperature fluctuations as well as soil moisture variability, improving tolerance of the crop to thermal extremes, especially during the dry season (Lin, 2007; de Carvalho *et al.*, 2020). Again, the forest canopy cover is key, and this includes the presence of forest patches surrounding the coffee plantation; biological control and richness and abundance of natural enemies increased with increasing forest cover at multiple spatial scales in Brazil (Medeiros *et al.*, 2019).

Nevertheless, very few studies have attempted to determine the microclimate effects of trees within the crop field on yield. The presence of poplar intercropping produced a globally cooler and wetter microenvironment in a winter wheat field compared to open crop field, improving the grain yield by 16% (Kanzler *et al.*, 2018). It is unclear, however, if this positive effect on yield is a direct effect on the crop ecophysiology and/or an indirect effect via improvement of biological control of pests and diseases by natural enemies.

Finally, the 'abiotic edge effect' might matter in agroforestry systems. In fragmented forest patches, during summer, the forest edges were warmer and more fluctuating, drier and windier than the forest interior, whereas higher insolation was observed in north-facing edges (Bernaschini *et al.*, 2019). The amplitude of this effect is enough to accelerate the development of pest insects (Bernaschini *et al.*, 2019).

20.3.3 The microclimate in greenhouses

Greenhouses generally produce distinct microclimates compared to open fields, often with warmer and wetter conditions depending on the type of greenhouse (Jewett and Jarvis, 2001). These conditions are favourable for hyphomycete-based mycoinsecticides for whitefly control in greenhouse tomato crops in the Mediterranean area, even though moderate wind speed (ventilation) conditions within greenhouses can greatly decrease the wetness of the leaf boundary layer (Fargues *et al.*, 2005). The microclimate of greenhouses can be more easily engineered than open fields, for instance allowing an increase in the relative humidity to improve infection rates by the entomopathogenic fungus *Beauveria bassiana* (Cordycipitaceae) on various pest insects and mites (Shipp *et al.*, 2003). In addition, soil solarization within greenhouses relies on solar radiation to heat up the bare soil to reach lethal temperatures for most pests between two cultures (Lindquist, 1999). Generally, the microclimate of greenhouses is expected to be particularly heterogeneous due to the complex air flows that develop, depending on the structure of the greenhouse (Fatnassi *et al.*, 2003).

20.3.4 Fine -scale microclimates in agriculture

Weather stations provide climatic data at a scale that is not pertinent for small organisms such as insects (Potter *et al.*, 2013; Pincebourde and Woods, 2020; Mosedale *et al.*, 2024). However, among the few studies that directly compared the phenological predictions of insects when using coarse data from weather stations vs more pertinent microclimatic data, some reported that coarse climatic data can still be useful (Coetzee, 2012), probably because the comparison was made within the range of temperature that generates a linear response in biological traits (Mosedale *et al.*, 2024). However, most agronomical studies unequivocally demonstrated that fine-grain microclimate data better predict biological responses such as insect developmental times (Faye *et al.*, 2014; Rebaudo *et al.*, 2016). Indeed, the question of scale is central when attempting to determine if microclimatic conditions allow more accurate predictions on insect dynamics than macroscale climatic conditions. Given that most insect pest and biocontrol agents are small, this author advocates for adjusting the scale of microclimate studies to the most pertinent scale for these species: fine-scale microclimates that one can find at the surface of vegetation and soil (see Fig. 20.2) (Pincebourde and Woods, 2020).

As explained in the first section of this chapter, small insects (less than ~5 mm) are largely influenced by the temperature of the surface on which they live (Ma *et al.*, 2021; Pincebourde *et al.*, 2021). Whatever the agricultural system (agroforestry, open fields, orchards, greenhouses, etc.), leaf surfaces can display significant surface temperature gradients. For instance, apple shows a temperature range of 7–12°C within its individual leaves (Caillon *et al.*, 2014; Cahon *et al.*, 2018), and up to 20°C under extreme conditions of high solar radiation and low wind speed (Saudreau *et al.*, 2017). Comparable surface temperature ranges were also observed on non-crop species (Leigh *et al.*, 2017) or ornamental plants (Roth-Nebelsick, 2001).

The lifestyle of the herbivore arthropod can also complicate the picture as the species can induce modifications in plant tissue when interacting with it. For example, the atmosphere within leaf mines can warm up to 12–15°C above ambient air when exposed to solar radiation (Pincebourde and Casas, 2006). This warmer mine atmosphere can be beneficial by accelerating larval development during moderate summer conditions but alternatively becomes risky during heatwaves as the mine temperature approaches the ULT of the caterpillar (Pincebourde *et al.*, 2007; Saudreau *et al.*, 2013). Warmer microclimates are also expected for other insects living inside plant tissues such as galling insects and borers (Pincebourde and Casas, 2016). The feeding activity of ectophagous herbivores can induce either warmer or cooler leaf microenvironments compared to intact leaf tissues (Pincebourde and Casas, 2019).

In addition, plants have evolved various structures across their leaves, such as trichomes and domatia, which can interact with the leaf microclimate to enrich the set of thermal microenvironments for natural or released enemies. The presence of trichomes at the leaf surface can influence leaf surface temperature (by increasing or decreasing it depending on environmental conditions) and also the amount of damaging UV-B radiation (Bickford, 2016), which can shelter biocontrol agents from adverse conditions although this has never been tested directly. Domatia are often evoked to provide a more humid microenvironment for very small arthropods (Rowles and O'Dowd, 2009). Nevertheless, this hypothesis is only weakly supported by the available data (Schmidt, 2014).

20.4 The Relationship Between Microclimates and Pests/Biological Control Agents

Agricultural systems are characterized by particular microclimates that deviate sometimes strongly from the regional climate. Both pests and biocontrol agents can potentially exploit these fine-scale temperature gradients to thermoregulate behaviourally, either to approach their optimal temperature for larval development (Caillon *et al.*, 2014) or to avoid overheating and improve survival (Leith *et al.*, 2024; Pincebourde, 2024). In addition, the microclimate of herbivore-induced structures (galls, leaf mines, etc.) is also important for the higher trophic levels including parasitoids that attack these herbivores and that therefore develop within the warmer-than-air atmosphere of these structures (Woods *et al.*, 2021).

This section reviews the available evidence of a link between microclimates and biocontrol agents, either direct or indirect.

20.4.1 Direct evidence of a relationship

There is little direct evidence that the microclimate at crop scale (relatively fine scale) modulates the success of biological control. The presence of vegetation (e.g. rye grass) planted with seed corn, instead of bare soil, decreased the maximal surface temperature experienced by parasitoids when within their capsule employed for augmentative release, thereby improving emergence success (Orr *et al.*, 1997). Microclimatic refuges may appear as the crop plant grows and develops, thereby generating non-permanent refuges. For example, whorl-stage maize plants offer suitable microclimatic conditions for foraging ladybirds, buffering temperature extremes at midday by more than 3°C and providing free water, thereby acting as a refuge during hot and dry season in agro-ecosystems (Pan *et al.*, 2020). Intercropping maize with other crops can improve biological control by conservation of natural enemies. More generally, even if the idea of accounting for microclimatic variable to refine the estimate of efficacy of a biological control agent has been evoked for decades, researchers have been somewhat clumsy in their choice of microclimatic variable and scale (Morandi *et al.*, 2007).

The timing of 'inoculation' of the biological control agent is theoretically important when considering microclimatic conditions. For instance, the release of a fungus (*Gliocladium roseum*: Bionectriaceae) to control the grey mould (fungal pathogen) on strawberries provided best results (i.e. higher number of colonies) when performed at midday or early afternoon when high temperatures, low humidity and absence of a water film on the foliage prevailed (Carvalho, 1994). Similarly, the release of mycoparasites to manage powdery mildew disease on melon provided more effective results if done around sunset to ensure a long period with high humidity that allows germination of the mycoparasite's spores (Romero *et al.*, 2007). Seasonal timing is also relevant. For instance, the pathogenicity of mitosporic fungus against locust species, including the deadly desert locust (*Schistocerca gregaria*: Acrididae), varies tremendously across and between seasons in relation to environmental temperature but also the thermoregulation behaviour of the locusts (Blanford and Thomas, 1999), although the link with agricultural microclimates remains to be determine.

Integrating both the thermal biology of the biological control agent and the microclimate of the target is of primary importance. The example of honey bees illustrates this very well: the microclimate temperature at brood nest ranges from 32°C to 36°Cdue to the thermoregulatory behaviours of bees (Jones *et al.*, 2004) and therefore any agent released against one of the many bee pests (arthropods and pathogens) should be able to endure a constant temperature in this range (Abou-Shaara and Staron, 2019). Temperature drives the efficiency of entomopathogenic fungi against varroa mites (Davidson *et al.*, 2003), but only a few isolates can tolerate up to 35°C (Rodríguez *et al.*, 2009).

20.4.2 Indirect evidence: habitat manipulations

Habitat manipulation is an efficient way to create suitable conditions – including microclimates – for conserving natural enemies. Some of the most common forms of shelters include beetle banks, shelterbelts, hedgerows, flower strips, intercropping and selective conservation of arable weeds. These habitat manipulations can generate favourable microclimate conditions to help maintain biological control agents in or near a crop field, especially during weather extremes in summer and winter (Gontijo, 2019). Overwintering shelters are particularly important in annual crop systems because natural enemies cannot complete their life cycle in transient cultivated fields and need to seek refuge in the vegetation around the field before the end of autumn (Gontijo, 2019).

Artificial shelters are used sometimes for specific pests/predators. Although their microclimate has never been surveyed, these shelters may provide a microclimate refuge against dry and hot environments in greenhouses (e.g. in a Coccinellidae; Pérez-Rodriguez and Messelink, 2023). Alternatively, these shelters may turn into heat traps if they are affected by solar radiation under open-field situations, since these structures cannot evacuate the excess of radiation energy by evapotranspiration. Shading and insect-proof screens are also widely used in agriculture for both microclimate control and insect exclusion. From a microclimate point of view, this strategy is efficient for crop production in otherwise adverse environmental conditions (Mahmood *et al.*, 2018). Since screens are made of a porous material, they provide only moderate microclimatic modifications depending on the type and configuration of the screen (Mahmood *et al.*, 2018). However, it remains unclear to what extent these microclimate modifications influence the insect pests and natural enemies that are naturally present within the crop.

20.5 New Tools for Surveying Microclimates When Applying Biocontrol Strategies

Precision agriculture has progressed rapidly in recent years but curiously, it has only just started to tackle the question of biocontrol agent release (see also Chapters 21 and 22). The use of unmanned aerial vehicles (UAV), typically drones, to map crop microclimates at a fine spatial scale was made possible when technological advancements allowed the installation of infrared cameras on these drones (Faye *et al.*, 2015). Remote infrared thermography allows mapping of surface temperatures and application of landscape metric analyses to infer the spatiotemporal dynamics of crop microclimates and their relationships with insect pest abundance (Faye *et al.*, 2015, 2017). Landscape analyses should also be used to infer the spatial association between the pest and the agent (Karimzadeh and Sciarretta, 2022). Thermography can also determine the amount of infrared radiation emitted by tree leaves onto crops in agroforestry systems (Karvatte *et al.*, 2020) and even estimate the tree density needed to obtain a significant microclimatic effect (Villani *et al.*, 2020).

Nevertheless, the use of UAVs for thermographic acquisitions comes with challenges related to spatial scaling and the biases in temperature assessment produced by the distance between the drone and the surface being measured (Faye *et al.*, 2016). A powerful approach can be to associate remote sensing thermography with sensor networks to benefit from high spatial and temporal resolution of microclimatic measurements. For example, wireless systems now exist to alleviate the challenge of deploying multiple sensors by permitting access to the data in real time (Jawad *et al.*, 2017). Miniaturization of temperature sensors is also progressing such that systems can be applied onto leaf surfaces to record the finest temperatures that matter for insect pests and biocontrol agents (e.g. RFID-based wireless sensor systems; (Palazzi *et al.*, 2019). Interestingly, UAVs can also be used to release biocontrol agents in a crop (e.g. mass release of the egg parasitoid *Trichogramma*: Trichogrammatidae; Martel *et al.*, 2021).

Future developments should include the use of UAVs to both measure microclimates and release the agent simultaneously and in real time for a powerful and efficient control strategy against pests. Ideally, thermography can be combined with other remote sensing approaches (multispectral) to better link detection of damage, microclimatic conditions and ecophysiological responses of the plant to assist in decision support systems for various pest management programmes (Abd El-Ghany *et al.*, 2020). Such an integrative approach (including machine learning) was applied recently to assess the efficiency of biological control of the fungal pathogen *Fusarium* spp. by the antagonistic fungi *Trichoderma* spp. (Manganiello *et al.*, 2024). Indeed, a machine-learning model parameterized with hyperspectral data enabled early detection of the pathogen, exemplifying an effective artificial intelligence-based decision support method for crop monitoring and biocontrol strategies (Manganiello *et al.*, 2024).

20.6 Conclusion and Perspectives

In our minds, the term 'microclimate' often refers to a microhabitat that serves as a refuge against adverse (macro)climatic conditions. For instance, the understorey of dense forests buffers extreme temperatures above the canopy, potentially offering a way to escape the heat for species that move toward the forest interior (Zellweger *et al.*, 2019). Nevertheless, it should be stressed that the microclimate at the vegetation surface can act as both a buffer and a magnifier of climatic conditions (Woods *et al.*, 2021). Under solar radiation, leaf surface temperature can be much warmer than the surrounding air layer, exposing organisms at the leaf surface to potentially challenging abiotic conditions (Pincebourde and Woods, 2012). Under temperate cropping systems, we can expect that most crops are magnifiers of the macroscale conditions. This property concerns the portion of vegetation that is exposed to incoming solar energy. By contrast, we can expect the soil surface underneath the crop vegetation to act more as a buffer of the environmental extremes, with soil temperatures often cooler than ambient air, particularly in irrigated cultures (Monteith and Unsworth, 2008). This contrast between top-vegetation layer/magnifier and soil level/buffer can also be exploited by organisms, as demonstrated for some aphids that can escape overheating at the wheat leaf surface by dropping onto the cooler soil surface (Ma *et al.*, 2018, 2021). Some natural enemies can also be partitioned between these two types of microhabitats, and this property should be exploited in pest management programmes.

To conclude, although we have solid indices of a strong relationship between microclimate mosaics in agricultural fields and the efficacy of control agents (or natural enemies), we are still unable to draw the multiple links between the numerous drivers of these microclimates, the architectural and ecophysiological complexities that are almost unique to each cropping system, and the integrative responses of both the pest (or pathogen) and the control agent. Usually, mechanistic modelling is a powerful approach to describe these connections (see Fig. 20.1), and biophysical ecology modelling should be adequate to predict microclimate patterns from crop properties including plant architecture and gas exchange (Nobel, 1999; Monteith and Unsworth, 2008; Pincebourde and Woods, 2012). Future studies should probably focus on designing a conceptual framework to better guide experimental research, as was done for natural ecological systems (Briscoe *et al.*, 2023). Such a conceptual framework would allow the community to shift from case-by-case studies to general concepts. Ultimately, this mechanistic approach offers a predictive tool to study the effects of climate change on these agricultural microclimate mosaics and their consequences for biological control under climate change.

References

Abd El-Ghany, N.M., Abd El-Aziz, S.E. and Marei, S.S. (2020) A review: Application of remote sensing as a promising strategy for insect pests and diseases management. *Environmental Science and Pollution Research* 27, 33503–33515.

Abou-Shaara, H.F. and Staron, M. (2019) Present and future perspectives of using biological control agents against pests of honey bees. *Egyptian Journal of Biological Pest Control* 29, 24.

Angilletta, M.J. (2006) Estimating and comparing thermal performance curves. *Journal of Thermal Biology* 31, 541–545.

Angilletta, M.J. (2009) *Thermal Adaptation – A Theoretical and Empirical Synthesis*. Oxford University Press, New York.

Bernaschini, M.L., Trumper, E., Valladares, G. and Salvo, A. (2019) Are all edges equal? Microclimatic conditions, geographical orientation and biological implications in a fragmented forest. *Agriculture Ecosystems & Environment* 280, 142–151.

Bickford, C.P. (2016) Ecophysiology of leaf trichomes. *Functional Plant Biology* 43, 807–814.

Blanford, S. and Thomas, M.B. (1999) Host thermal biology: The key to understanding host-pathogen interactions and microbial pest control? *Agricultural and Forest Entomology* 1, 195–202.

Blubaugh, C.K., Hagler, J.R., Machtley, S.A. and Kaplan, I. (2016) Cover crops increase foraging activity of omnivorous predators in seed patches and facilitate weed biological control. *Agriculture Ecosystems & Environment* 231, 264–270.

Bramer, I., Anderson, B.J., Bennie, J., Bladon, A.J., De Frenne, P. *et al.* (2018) Advances in monitoring and modelling climate at ecologically relevant scales. *Advances in Ecological Research* 58, 101–161.

Briscoe, N.J., Morris, S.D., Mathewson, P.D., Buckley, L.B., Jusup, M. *et al.* (2023) Mechanistic forecasts of species responses to climate change: The promise of biophysical ecology. *Global Change Biology* 29, 1451–1470.

Cahon, T., Caillon, R. and Pincebourde, S. (2018) Do aphids alter leaf surface temperature patterns during early infestation? *Insects* 9, 34.

Caillon, R., Suppo, C., Casas, J., Woods, H.A. and Pincebourde, S. (2014) Warming decreases thermal heterogeneity of leaf surfaces: Implications for behavioural thermoregulation by arthropods. *Functional Ecology* 28, 1449–1458.

Campbell, G.S. and Norman, J.M. (1998) *An Introduction to Environmental Biophysics*. Springer, New York.

Carvalho, R.A. (1994) Effect of microclimate on biological control of grey mould (*Botrytis cinerea*) by Gliocladium roseum in strawberries. MSc thesis, University of Guelph, Ontario, Canada.

Chown, S.L. and Nicolson, S.W. (2004) *Insect Physiological Ecology: Mechanisms and Patterns*. Oxford University Press, New York.

Cloudsley-Thompson, J.L. (1962) Microclimates and the distribution of terrestrial arthropods. *Annual Review of Entomology* 7, 199–222.

Coetzee, J.A. (2012) Meteorological weather station data can be used in climate matching studies of biological control agents. *Biocontrol Science and Technology* 22, 419–427.

Davidson, G., Phelps, K., Sunderland, K.D., Pell, J.K., Ball, B.V. *et al.* (2003) Study of temperature-growth interactions of entomopathogenic fungi with potential for control of Varroa destructor (Acari: Mesostigmata) using a nonlinear model of poikilotherm development. *Journal of Applied Microbiology* 94, 816–825.

de Carvalho, A.F., Fernandes-Filho, E.I., Daher, M., Gomes, L.C., Cardoso, I.M. *et al.* (2020) Microclimate and soil and water loss in shaded and unshaded agroforestry coffee systems. *Agroforestry Systems* 95, 119–134.

De Frenne, P., Rodriguez-Sanchez, F., Coomes, D.A., Baeten, L., Verstraeten, G. *et al.* (2013) Microclimate moderates plant responses to macroclimate warming. *Proceedings of the National Academy of Sciences USA* 110, 18561–18565.

De Frenne, P., Zellweger, F., Rodríguez-Sánchez, F., Scheffers, B.R., Hylander, K. *et al.* (2019) Global buffering of temperatures under forest canopies. *Nature Ecology & Evolution* 3, 744–749.

Fargues, J., Smits, N., Rougier, M., Boulard, T., Ridray, G. *et al.* (2005) Effect of microclimate heterogeneity and ventilation system on entomopathogenic hyphomycete infection of Trialeurodes vaporariorum (Homoptera: Aleyrodidae) in Mediterranean greenhouse tomato. *Biological Control* 32, 461–472.

Fatnassi, H., Boulard, T. and Bouirden, L. (2003) Simulation of climatic conditions in full-scale greenhouse fitted with insect-proof screens. *Agricultural and Forest Meteorology* 118, 97–111.

Faye, E., Herrera, M., Bellomo, L., Silvain, J.F. and Dangles, O. (2014) Strong discrepancies between local temperature mapping and interpolated climatic grids in tropical mountainous agricultural landscapes. *PLoS ONE* 9, e105541.

Faye, E., Rebaudo, F., Yánez-Cajo, D., Cauvy-Fraunié, S., Dangles, O. *et al.* (2015) A toolbox for studying thermal heterogeneity across spatial scales: From unmanned aerial vehicle imagery to landscape metrics. *Methods in Ecology and Evolution* 7, 437–446.

Faye, E., Dangles, O. and Pincebourde, S. (2016) Distance makes the difference in thermography for ecological studies. *Journal of Thermal Biology* 56, 1–9.

Faye, E., Rebaudo, F., Carpio, C., Herrera, M. and Dangles, O. (2017) Does heterogeneity in crop canopy microclimates matter for pests? Evidence from aerial high-resolution thermography. *Agriculture Ecosystems & Environment* 246, 124–133.

Fox, A.F., Kim, T.N., Bahlai, C.A., Woltz, J.M., Gratton, C. *et al.* (2016) Cover crops have neutral effects on predator communities and biological control services in annual cellulosic bioenergy cropping systems. *Agriculture Ecosystems & Environment* 232, 101–109.

Gardarin, A., Plantegenest, M., Bischoff, A. and Valantin-Morison, M. (2018) Understanding plant–arthropod interactions in multitrophic communities to improve conservation biological control: Useful traits and metrics. *Journal of Pest Science* 91, 943–955.

Gates, D.M. (1980) *Biophysical Ecology*. Springer, New York.

Goddard, M., Owen, C.A., Martin, G. and Coetzee, J.A. (2022) The thermal physiology of *Lysathia* sp. (Coleoptera: Chrysomelidae), a biocontrol agent of parrot's feather in South Africa, supports its success. *Biocontrol Science and Technology* 32, 837–846.

Gontijo, L.M. (2019) Engineering natural enemy shelters to enhance conservation biological control in field crops. *Biological Control* 130, 155–163.

Gunderson, A.R. and Stillman, J.H. (2015) Plasticity in thermal tolerance has limited potential to buffer ectotherms from global warming. *Proceedings of the Royal Society of London Series B-Biological Sciences* 282, 20150401.

Harms, N.E., Knight, I.A., Pratt, P.D., Reddy, A.M., Mukherjee, A. *et al.* (2021) Climate mismatch between introduced biological control agents and their invasive host plants: Improving biological control of tropical weeds in temperate regions. *Insects* 12, 549.

Hochachka, P.W. and Somero, G.N. (2002) *Biochemical Adaptation: Mechanism and Process in Physiological Evolution*. Oxford University Press, New York.

Hoffmann, A.A., Chown, S.L., Clusella-Trullas, S. and Fox, C. (2013) Upper thermal limits in terrestrial ectotherms: How constrained are they? *Functional Ecology* 27, 934–949.

Jawad, H.M., Nordin, R., Gharghan, S.K., Jawad, A.M. and Ismail, M. (2017) Energy-efficient wireless sensor networks for precision agriculture: A review. *Sensors* 17, 1781.

Jewett, T.J. and Jarvis, W.R. (2001) Management of the greenhouse microclimate in relation to disease control: A review. *Agronomie* 21, 351–366.

Jones, J.C., Myerscough, M.R., Graham, S. and Oldroyd, B.P. (2004) Honey bee nest thermoregulation: Diversity promotes stability. *Science* 305, 402–404.

Kanzler, M., Böhm, C., Mirck, J., Schmitt, D. and Veste, M. (2018) Microclimate effects on evaporation and winter wheat (*Triticum aestivum* L.) yield within a temperate agroforestry system. *Agroforestry Systems* 93, 1821–1841.

Karimzadeh, R. and Sciarretta, A. (2022) Spatial patchiness and association of pests and natural enemies in agro-ecosystems and their application in precision pest management: A review. *Precision Agriculture* 23, 1836–1855.

Karvatte, N.J., Miyagi, E.S., de Oliveira, C.C., Barreto, C.D., Mastelaro, A.P. *et al.* (2020) Infrared thermography for microclimate assessment in agroforestry systems. *Science of the Total Environment* 731, 139252.

Kemppinen, J., Lembrechts, J.J., Van Meerbeek, K., Carnicer, J., Chardon, N.I. *et al.* (2024) Microclimate, an important part of ecology and biogeography. *Global Ecology and Biogeography* 33, e13834.

Leigh, A., Sevanto, S., Close, J.D. and Nicotra, A.B. (2017) The influence of leaf size and shape on leaf thermal dynamics: Does theory hold up under natural conditions? *Plant Cell and Environment* 40, 237–248.

Leith, N.T., Miller, E.A. and Fowler-Finn, K.D. (2024) Thermoregulation enhances survival but not reproduction in a plant-feeding insect. *Functional Ecology* 38, 1344–1356.

Lin, B.B. (2007) Agroforestry management as an adaptive strategy against potential microclimate extremes in coffee agriculture. *Agricultural and Forest Meteorology* 144, 85–94.

Lindquist, R.K. (1999) Temperature in the management of insect and mite pests in greenhouses. In: Hallman, G.J. and Denlinger, D.L. (eds) *Temperature Sensitivity in Insects and Application in Integrated Pest Management*. CRC Press, Boca Raton, pp. 287–299.

Ma, C.S., Ma, G. and Pincebourde, S. (2021) Survive a warming climate: Insect responses to extreme high temperatures. *Annual Review of Entomology* 66, 163–184.

Ma, G., Bai, C.-M., Wang, X.-J., Majeed, M.Z. and Ma, C.-S. (2018) Behavioural thermoregulation alters microhabitat utilization and demographic rates in ectothermic invertebrates. *Animal Behaviour* 142, 49–57.

Mahmood, A., Hu, Y., Tanny, J. and Asante, E.A. (2018) Effects of shading and insect-proof screens on crop microclimate and production: A review of recent advances. *Scientia Horticulturae* 241, 241–251.

Malinski, K.H., Moore, E.M. and Kingsolver, J.G. (2024) Heat stress and host-parasitoid interactions: Lessons and opportunities in a changing climate. *Current Opinion in Insect Science* 64, 101225.

Manganiello, G., Nicastro, N., Ortenzi, L., Pallottino, F., Costa, C. and Pane, C. (2024) Trichoderma biocontrol performances against baby-lettuce fusarium wilt surveyed by hyperspectral imaging-based machine learning and infrared thermography. *Agriculture* 14, 307.

Martel, V., Johns, R.C., Jochems-Tanguay, L., Jean, F., Maltais, A. *et al.* (2021) The use of UAS to release the egg parasitoid *Trichogramma* spp. (Hymenoptera: Trichogrammatidae) against an agricultural and a forest pest in Canada. *Journal of Economic Entomology* 114, 1867–1881.

Medeiros, H.R., Grandinete, Y.C., Manning, P., Harper, K.A., Cutler, G.C. *et al.* (2019) Forest cover enhances natural enemy diversity and biological control services in Brazilian sun coffee plantations. *Agronomy for Sustainable Development* 39, 50.

Monteith, J.L. and Unsworth, M.H. (2008) *Principles of Environmental Physics*. Academic Press, San Diego, California.

Morandi, M.A.B., Maffia, L.A., Mizubuti, E.S.G., Alfenas, A.C., Barbosa, J.G. *et al.* (2007) Relationships of microclimatic variables to colonization of rose debris by Botrytis cinereaand the biocontrol agent Clonostachys rosea. *Biocontrol Science and Technology* 16, 619–630.

Mosedale, J.R., Eyre, D., Korycinska, A., Everatt, M., Grant, S. *et al.* (2024) Mechanistic microclimate models and plant pest risk modelling. *Journal of Pest Science* 97, 1749–1766.

Nobel, P.S. (1999) *Physicochemical and Environmental Plant Physiology*. Academic Press, New York.

Nqayi, S., Zachariades, C., Coetzee, J., Hill, M., Chidawanyika, F. *et al.* (2023) Do thermal requirements of *Dichrorampha odorata*, a shoot-boring moth for the biological control of *Chromolaena odorata*, explain its failure to establish in South Africa? *African Entomology* 31, e13597.

Orr, D.B., Landis, D.A., Mutch, D.R., Manley, G.V., Stuby, S.A. *et al.* (1997) Ground cover influence on microclimate and *Trichogramma* (Hymenoptera: *Trichogrammatidae*) augmentation in seed corn production. *Environmental Entomology* 26, 433–438.

Palazzi, V., Gelati, F., Vaglioni, U., Alimenti, F., Mezzanotte, P. *et al.* (2019) Leaf-compatible autonomous RFID-based wireless temperature sensors for precision agriculture. In: *IEEE Topical Conference on Wireless Sensors and Sensor Networks (WiSNet)*, IEEE, Orlando, FL, USA, pp. 1–4.

Pan, H., Xiu, C., Liu, B., Wyckhuys, K.A.G. and Lu, Y. (2020) Whorl-stage maize provides a microclimate refuge for predatory ladybeetles. *Biological Control* 142, 104162.

Pepi, A. and McMunn, M. (2021) Predator diversity and thermal niche complementarity attenuate indirect effects of warming on prey survival. *American Naturalist* 198, 33–43.

Pérez-Rodriguez, J. and Messelink, G.J. (2023) Artificial shelters enhance the establishment of the aphidophagous predator Scymnus interruptus on sweet pepper plants. *Biological Control* 177, 105110.

Pincebourde, S. and Casas, J. (2006) Multitrophic biophysical budgets: Thermal ecology of an intimate herbivore insect-plant interaction. *Ecological Monographs* 76, 175–194.

Pincebourde, S. and Woods, H.A. (2012) Climate uncertainty on leaf surfaces: The biophysics of leaf microclimates and their consequences for leaf-dwelling organisms. *Functional Ecology* 26, 844–853.

Pincebourde, S. and Casas, J. (2015) Warming tolerance across insect ontogeny: Influence of joint shifts in microclimates and thermal limits. *Ecology* 96, 986–997.

Pincebourde, S. and Casas, J. (2016) Hypoxia and hypercarbia in endophagous insects: Larval position in the plant gas exchange network is key. *Journal of Insect Physiology* 84, 137–153.

Pincebourde, S. and Suppo, C. (2016) The vulnerability of tropical ectotherms to warming is modulated by the microclimatic heterogeneity. *Integrative and Comparative Biology* 56, 85–97.

Pincebourde, S. and Casas, J. (2019) Narrow safety margin in the phyllosphere during thermal extremes. *Proceedings of the National Academy of Sciences USA* 116, 5588–5596.

Pincebourde, S. and Woods, H.A. (2020) There is plenty of room at the bottom: Microclimates drive insect vulnerability to climate change. *Current Opinion in Insect Science* 41, 63–70.

Pincebourde, S., Sinoquet, H., Combes, D. and Casas, J. (2007) Regional climate modulates the canopy mosaic of favourable and risky microclimates for insects. *Journal of Animal Ecology* 76, 424–438.

Pincebourde, S., Murdock, C.C., Vickers, M. and Sears, M.W. (2016) Fine-scale microclimatic variation can shape the responses of organisms to global change in both natural and urban environments. *Integrative and Comparative Biology* 56, 45–61.

Pincebourde, S., van Baaren, J., Rasmann, S., Rasmont, P., Rodet, G. *et al.* (2017) Plant–insect interactions in a changing world. *Advances in Botanical Research* 81, 289–332.

Pincebourde, S., Dillon, M.E. and Woods, H.A. (2021) Body size determines the thermal coupling between insects and plant surfaces. *Functional Ecology* 35, 1424–1436.

Pincebourde, S. (2024) Has behavioural thermoregulation evolved solely to stay alive in insects, nothing more? *Functional Ecology* 38, 1342–1343.

Potter, K.A., Woods, H.A. and Pincebourde, S. (2013) Microclimatic challenges in global change biology. *Global Change Biology* 19, 2932–2939.

Rebaudo, F., Faye, E. and Dangles, O. (2016) Microclimate data improve predictions of insect abundance models based on calibrated spatiotemporal temperatures. *Frontiers in Physiology* 7, 139.

Rodríguez, M., Gerding, M. and France, A. (2009) Selection of entomopathogenic fungi to control *Varroa destructor* (Acari: Varroidae). *Chilean Journal of Agricultural Research* 69, 534–540.

Romero, D., De Vicente, A., Zeriouh, H., Cazorla, F.M., Fernández-Ortuño, D. *et al.* (2007) Evaluation of biological control agents for managing cucurbit powdery mildew on greenhouse-grown melon. *Plant Pathology* 56, 976–986.

Roth-Nebelsick, A. (2001) Computer-based analysis of steady-state and transient heat transfer of small-sized leaves by free and mixed convection. *Plant Cell and Environment* 24, 631–640.

Roupsard, O., Audebert, A., Ndour, A.P., Clermont-Dauphin, C., Agbohessou, Y. *et al.* (2020) How far does the tree affect the crop in agroforestry? New spatial analysis methods in a Faidherbia parkland. *Agriculture Ecosystems & Environment* 296, 106928.

Rowles, A.D. and O'Dowd, D.J. (2009) Leaf domatia and protection of a predatory mite *Typhlodromus doreenae Schicha* (Acari: Phytoseiidae) from drying humidity. *Australian Journal of Entomology* 48, 276–281.

Saudreau, M., Pincebourde, S., Dassot, M., Adam, B., Loxdale, H.D. *et al.* (2013) On the canopy structure manipulation to buffer climate change effects on insect herbivore development. *Trees – Structure and Function* 27, 239–248.

Saudreau, M., Ezanic, A., Adam, B., Caillon, R., Walser, P. *et al.* (2017) Temperature heterogeneity over leaf surfaces: The contribution of the lamina microtopography. *Plant Cell and Environment* 40, 2174–2188.

Scherrer, D. and Körner, C. (2010) Infra-red thermometry of alpine landscapes challenges climatic warming projections. *Global Change Biology* 16, 2602–2613.

Schmidt, R.A. (2014) Leaf structures affect predatory mites (Acari: Phytoseiidae) and biological control: A review. *Experimental and Applied Acarology* 62, 1–17.

Shipp, J.L., Zhang, Y., Hunt, D.W.A. and Ferguson, G. (2003) Influence of humidity and greenhouse microclimate on the efficacy of *Beauveria bassiana* (Balsamo) for control of greenhouse arthropod pests. *Environmental Entomology* 32, 1154–1163.

Sinclair, B.J., Marshall, K.E., Sewell, M.A., Levesque, D.L., Willett, C.S. *et al.* (2016) Can we predict ectotherm responses to climate change using thermal performance curves and body temperatures? *Ecology Letters* 19, 1372–1385.

Smith, R., Coetzee, J.A. and Hill, M.P. (2022) Best of both worlds: The thermal physiology of *Hydrellia egeriae*, a biological control agent for the submerged aquatic weed, Egeria densa in South Africa. *BioControl* 67, 365–374.

Stoutjesdijk, P.H. and Barkman, J.J. (2014) *Microclimate Vegetation & Fauna*. KNNV Publishing, Zeist, The Netherlands.

Suggitt, A.J., Gillingham, P.K., Hill, J.K., Huntley, B., Kunin, W.E. *et al.* (2011) Habitat microclimates drive fine-scale variation in extreme temperatures. *Oikos* 120, 1–8.

Sunday, J., Bates, A. and Dulvy, N. (2011) Global analysis of thermal tolerance and latitude in ectotherms. *Proceedings of the Royal Society of London Series B-Biological Sciences* 278, 1823–1830.

Tougeron, K., van Baaren, J., Burel, F., Alford, L., Stewart, A. *et al.* (2016) Comparing thermal tolerance across contrasting landscapes: First steps towards understanding how landscape management could modify ectotherm thermal tolerance. *Insect Conservation and Diversity* 9, 171–180.

Villani, L., Castelli, G., Sambalino, F., Oliveira, L.A.A. and Bresci, E. (2020) Integrating UAV and satellite data to assess the effects of agroforestry on microclimate in Dodoma region, Tanzania. In: *IEEE International Workshop on Metrology for Agriculture and Forestry (MetroAgriFor)*. pp. 338–342.

Woods, H.A., Pincebourde, S., Dillon, M.E. and Terblanche, J.S. (2021) Extended phenotypes: Buffers or amplifiers of climate change?*Trends in Ecology and Evolution* 36, 889–898.

Zellweger, F., Coomes, D., Lenoir, J., Depauw, L., Maes, S.L. *et al.* (2019) Seasonal drivers of understorey temperature buffering in temperate deciduous forests across Europe. *Global Ecology and Biogeography* 28, 1774–1786.

21 Advanced Crop Protection Strategies in Response to Climate Change: Integrating Technology and Innovation for Sustainable Agriculture

Darija Lemić*, Ivana Pajač Živković, Sandra Skendžić and Helena Viric Gasparic

Department of Agricultural Zoology, University of Zagreb Faculty of Agriculture, Zagreb, Croatia

Abstract

In response to the challenges of climate change, this chapter addresses the critical aspects of adapting to and mitigating the challenges of crop protection in modern agriculture. It focuses on the use of advanced technologies and innovative methods to detect and control pests to ultimately increase crop yields. Techniques such as automated monitoring systems, meteorological data for pest assessment and integrated pest management are examined. The chapter also looks at the use of machine learning and artificial intelligence (AI) in pest control. It also explores the integration of remote sensing through satellite imagery and drones, with a focus on modern methods such as image processing and machine learning algorithms for early pest detection. These strategies represent a departure from conventional approaches to establish a new standard in agriculture. This development promotes site-specific cultivation methods, supports sustainable crop production and reduces the use of pesticides, offering a comprehensive approach to crop protection in times of climate change.

21.1 Pests under Evolving Climate

In the field of agriculture, climate change in Europe poses a complex challenge, especially with regard to the management of important agricultural pests (Skendžić *et al.*, 2021a). The dynamic interplay of rising temperatures, changing precipitation patterns and changing ecosystems influences the distribution, abundance and behaviour of pest species. Scientific evidence shows that climate change is a catalyst for the spread of pests that disrupt the ecological balance and threaten agricultural systems (Skendžić *et al.*, 2021b). Increased temperatures extend the geographic range of pest species and allow them to thrive in regions once considered inhospitable. Changing precipitation patterns contribute to the emergence of new habitats or the modification of existing habitats and create favourable conditions for the spread of pests (Skendžić *et al.*, 2021a). In addition, climate-related shifts in pest life cycles and phenology make pest control strategies even more complex (Heeb *et al.*, 2019; Pajač Živković *et al.*, 2025). Favourable weather conditions, such as warmer temperatures in spring, which favour the growth of pest populations, or mild winters,

*Corresponding author: dlemic@agr.hr

 Biological Control Systems and Climate Change (eds Joan van Baaren *et al.*)
DOI: 10.1079/9781800625099.0021

which favour the survival of pests, have led to an increased use of chemical pesticides in agriculture (Skendžić *et al.*, 2021a). Although this practice is intended to reduce pest pressure, it has led to increased environmental and economic costs. Based on future climate projections, the increase in extreme weather events is expected to directly and indirectly increase pest damage (Lemić *et al.*, 2024).

Climate change affects pests and the use of pesticides through both direct and indirect mechanisms. Directly, rising temperatures can extend the growing season and increase pest reproduction rates, leading to higher pest populations. In addition, extreme weather events can disrupt the natural enemies of pests, reducing their effectiveness and further increasing pest pressure (Heeb *et al.*, 2019). Indirectly, climate change is altering ecosystems and agricultural practices. Changing precipitation patterns and increased CO_2 levels can affect plant health and resistance and make crops more susceptible to pest infestations. These changing conditions can also shift the distribution and phenology of pests, making pest control more difficult (Heeb *et al.*, 2019; Pajač Živković *et al.*, 2025). Consequently, there is a likelihood that the use of chemical pesticides will continue to increase as a strategy to manage increased pest pressure (Olatinwo and Hoogenboom, 2014). However, this reliance on chemical solutions is unsustainable and potentially harmful to the environment and human health, emphasizing the urgency of developing new methods such as integrated pest management, remote sensing and AI-driven pest control. Given the urgency, scientists are advocating for interdisciplinary research, inventive pest control methods and the inclusion of climate-resilient strategies in agricultural policies to ensure food security in the face of climate change challenges while protecting the environment (Wijerathna-Yapa and Pathirana, 2022; Harvey *et al.*, 2023).

21.2 Traps and Monitoring Systems: From Traditional to Modern Approaches

Pests require constant monitoring for sustainable crop management, especially considering the impact of climate change uncertainties, such as changing temperature and precipitation patterns, on pest dynamics and behaviour. In this context, pest monitoring is an important component for farmers to understand and respond effectively to these changes.

Since the establishment of the International Organization for Biological and Integrated Control (OILB/IOBC), particular attention has been paid to the assessment of pest and beneficial insect populations. The first manual on pest monitoring methods in apple orchards was published in 1969, followed by a second edition in 1974. The traditional methods for monitoring pests and beneficial insects have remained unchanged to this day, apart from improvements due to technological development. Depending on the agricultural production, some traditional monitoring methods are specific to fruit growing, viticulture, arable farming and horticulture. However, they all have the common goal of determining the pest population and deciding whether such a population is causing economic damage (Barić and Pajač Živković, 2020).

These methods, which are often characterized by their simplicity and accessibility, play an important role in providing valuable insights to farmers. Visual inspections, manual traps and community-based observations are common elements of traditional monitoring. Traditional pest monitoring remains a fundamental and widely used approach in agriculture, bridging the gap between local practices and the evolving landscape of modern technologies (Barić and Pajač Živković, 2020).

Traditional monitoring methods offer several advantages in agriculture. They are cost-effective and thus accessible to farmers with limited resources, and their simple implementation requires minimal training. They also rely on local knowledge, promote community-based approaches to pest control and generally have a low impact on the environment. However, these methods also have their limitations (Fig. 21.1). They may lack the precision and accuracy of modern technologies, which can lead to pest populations being underestimated or detected too late. The time-consuming nature of manual monitoring processes combined with susceptibility to human error poses a challenge, especially in large agricultural areas. In addition, traditional methods may not provide real-time data, making it difficult to respond immediately

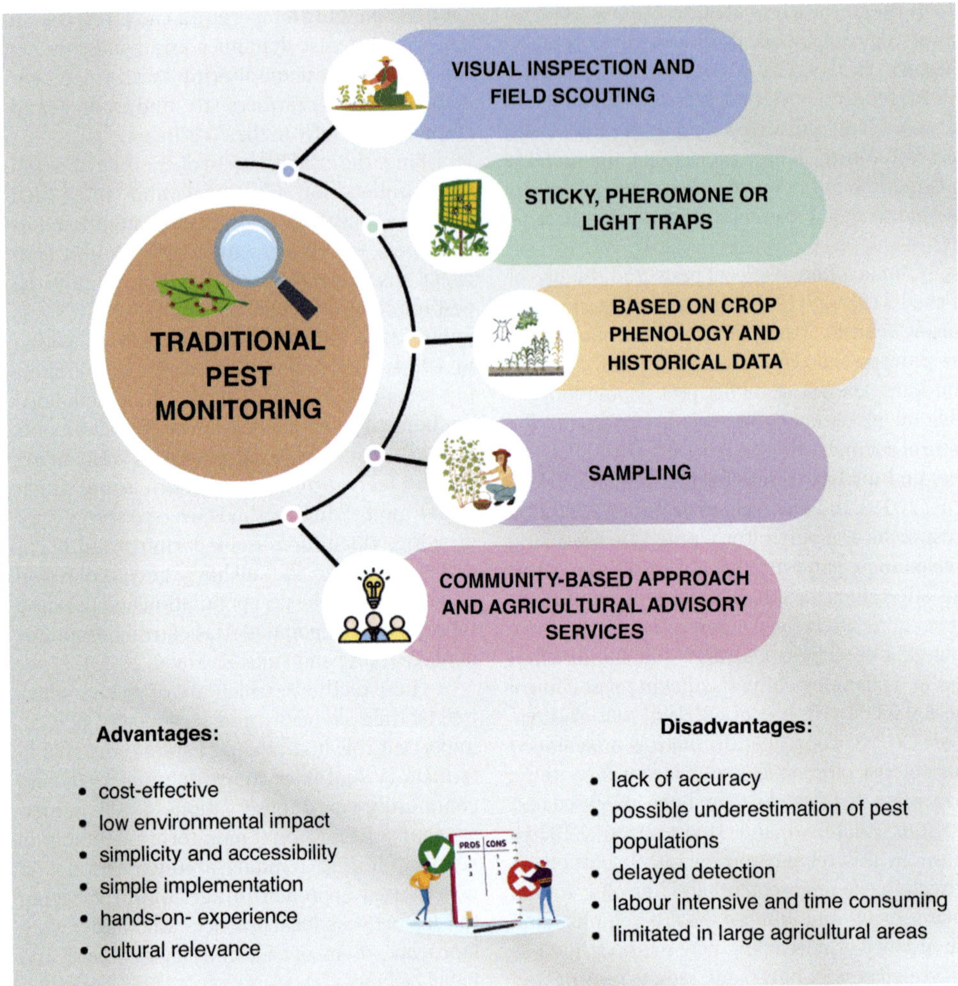

Fig. 21.1. Traditional pest monitoring techniques and its advantages and disadvantages.

to rapidly changing pest dynamics or weather conditions. Scalability can be an issue, especially on large or industrial farms, where more efficient and scalable monitoring approaches may be required for effective pest control (Preti *et al.*, 2021a).

The integration of new technologies into agricultural monitoring provides farmers with accurate, real-time data for adaptive and informed decision making in the face of changing climatic conditions and agricultural challenges, as well as facilitating accurate and timely pest control. This is more relevant for large farms due to their greater capacity to invest in these technologies and scale them effectively, whereas smaller farms face significant limitations due to the high costs and technical expertise required.

21.3 Modern Techniques: A Focus on Automation and AI

In recent years, the development of artificial intelligence (AI), remote sensing and autonomous agricultural vehicles such as weeding robots has created excellent opportunities to drastically reduce the use of pesticides (Iost Filho *et al.*, 2020). Through the use of technologies such as automated monitoring systems and the integration of machine learning, agriculture can be one step ahead in detecting and understanding pest populations. Early detection is important in a changing climate where the frequency and intensity of pest outbreaks can become unpredictable. Continuous monitoring allows farmers to take timely and targeted pest control measures to minimize the risk of crop damage and optimize the use of resources. In addition, the data obtained from pest monitoring help in the development of adaptation strategies, allowing farmers to develop preventive strategies, such as optimizing crop rotation, selecting resistant plant varieties and adjusting planting dates and locations.

One promising strategy in the new era of surveillance is automated pest monitoring, which relies on the use of electronic traps known as e-traps. These devices allow entomologists and farmers to continuously assess the presence and abundance of pests in real time by taking photos remotely, reducing the need for physical field visits (Ehler, 2006; Sutherst *et al.*, 2011; Barzman *et al.*, 2015; Hellin and Fisher, 2019; Sciarretta and Calabrese, 2019; Dong *et al.*, 2020; Faria *et al.*, 2021). These technological advances play a role in improving pest control strategies.

Automated pest monitoring systems driven by data and AI in modern agriculture offer benefits such as reduced pesticide use, less labour in difficult areas and fewer trips to remote fields, resulting in lower fuel consumption and carbon emissions (Blommers, 1994; Preti *et al.*, 2021b; Rydhmer *et al.*, 2022). These innovations ensure site-specific and environmentally friendly crop protection with fewer pesticide residues. Modern agriculture utilizes technologies such as multispectral drones for remote sensing observations, intelligent decision support systems, the Internet of Things (IoT)[1] and automated traps. These traps utilize methods such as photo interruption, wing beat analysis and vibration detection to efficiently monitor pests (Potamitis *et al.*, 2009Potamitis *et al.*, 2017; Chen *et al.*, 2014; Goldshtein *et al.*, 2017). Monitoring systems can be categorized into fully automated systems with species identification software or semi-automated systems where human experts view the images captured by camera-equipped traps (Sciarretta and Calabrese, 2019). Pattern recognition techniques based on artificial neural networks (ANNs) are increasingly used in the development of digital systems, with a focus on optimizing image quality for accurate pest detection in traps (Zhang *et al.*, 2013; Hernández-Serna and Jiménez-Segura, 2014; Čirjak *et al.*, 2022, Čirjak *et al.*, 2023a, b; Schrader *et al.*, 2022).

The design of traps influences their effectiveness in capturing target organisms. Factors such as colour, size and shape influence the attraction of target pests and the deterrence of non-target pests, as demonstrated by the use of yellow sticky traps in agriculture. These traps effectively attract aphids, whiteflies and certain thrips, which are visually attracted to the colour yellow. This makes them a practical tool for pest control without affecting non-target species that do not respond to yellow in a similar way (Vernon and Gillespie, 1995; Holguin *et al.*, 2010).

[1]The Internet of Things (IoT) refers to a network of interconnected physical devices embedded with sensors, software and other technologies that enable them to collect and exchange data over the internet.

21.4 The Future of Pest Monitoring in Agriculture

Monitoring pests in agriculture requires a dual approach that includes both traditional methods and modern technologies. Traditional methods, characterized by cost-effectiveness and community-based practices, offer accessible solutions based on local knowledge (see Fig. 21.1). However, they come with limitations, such as the possible underestimation of pest populations, lack of precision and the inability to provide real-time data.

In response to the challenges posed by climate change, the integration of advanced technologies becomes imperative. Satellite imagery, multispectral drones, machine learning and artificial intelligence are technologies that provide precision and real-time data. This information can be used to understand dynamic environmental conditions and facilitate data-driven decision making in the formulation of crop protection strategies. AI-powered automated pest monitoring systems contribute to timely interventions, reducing pesticide use and supporting environmentally sustainable pest control. They contribute to sustainable agriculture by providing real-time insights and promoting the precision of pest control strategies. While traditional methods are community-inclusive and cost-effective, modern technologies provide the adaptability and precision needed to cope with climate variability.

However, the integration of these advanced technologies also comes with a number of drawbacks and limitations. The high costs associated with the purchase and maintenance of modern equipment can be prohibitive, especially for smallholder farmers. In addition, the need for technical expertise to operate and interpret the data from these technologies can be a barrier, as not all farmers have access to the necessary training and support. Furthermore, reliance on technology can lead to concerns about data privacy and the risk of technology failure, which can disrupt pest monitoring and control (Heeb *et al.*, 2019).

To summarize, while traditional methods provide a solid foundation for pest monitoring, the evolving climate requires the precision and adaptability of modern technologies. The key to sustainable and effective pest management in agriculture lies in balancing these approaches and overcoming their respective limitations.

21.5 Integrated Pest Management and Threshold Levels

Integrated pest management (IPM) involves challenges such as managing multiple pests, implementation time and cost, and the requirement for continuous monitoring. Addressing these challenges is necessary for effective application (Rajotte *et al.*, 2009). IPM is a systematic approach to crop protection that aims to achieve high yields with minimal reliance on chemical inputs (Deguine *et al.*, 2021). A key component of this approach is the crop protection decision ladder, which is based on six fundamental pillars: biology, ecology, models, sampling, taxonomy and, most importantly, economic decision levels (EDLs), including economic threshold levels (ETLs) (Pedigo and Rice, 2009). Accurate pest population predictions and effective signalling systems are necessary to achieve balance in pest management. Only through detailed monitoring and understanding of EDLs can the timing and type of control measures be determined (Bažok *et al.*, 2018; Kadoić Balaško and Virić Gašparić, 2022).

The integration of ETLs into the monitoring framework adds an economic dimension to the decision-making process in crop protection. The ETL serves as a key indicator of the level of pest populations at which intervention is economically justified. This differentiated approach allows farmers to optimize their responses to pest pressure, especially in the context of dynamic environmental conditions caused by climate change. The synergy of monitoring practices and ETL gives farmers a strategic advantage to make well-informed decisions tailored to the evolving challenges of climate change. This integrative approach not only facilitates early detection of pests but also enables farmers to intervene precisely when necessary. The result is a more effective and targeted pest control strategy that minimizes the economic impact on agricultural production. This comprehensive approach goes beyond simply detecting and controlling pests – it represents a broader commitment to sustainable agriculture. Armed with real-time data and the economic threshold perspective, farmers can take pest control measures that are not only economically viable but also environmentally sustainable. This dual focus on economic efficiency and environmental responsibility puts agriculture on a path that makes it resilient and adaptable in the face of climate-related uncertainties (Pedigo and Rice, 2009).

21.5.1 Understanding threshold levels

The threshold is an important node in the decision ladder as it determines when pesticides should be applied based on the density of pest populations that represent potential economic damage. Without these thresholds, agriculture runs the risk of incurring unnecessary costs and causing unwarranted harm to the environment. By introducing economic decision levels, agriculture benefits from the targeted use of pesticides, cost efficiency, increased profitability, biodiversity conservation and environmental integrity (Knipling, 1979; Dhaliwal *et al.*, 2006).

Understanding how thresholds are calculated and the assumptions on which they are based can facilitate their application and increase confidence in them as valuable tools for IPM. Thresholds can be based on yield, crop quality or defoliation, depending on the crop and pest specificity (Murray *et al.*, 2020).

Yield-based thresholds are based on measured losses due to pest infestation, which has a direct impact on yield, while *quality-based thresholds* take into account a quality standard of the crop. Crop quality standards are defined by a combination of factors, including physical characteristics, chemical composition and safety parameters. These standards ensure that crops meet certain requirements for consumer health, safety and satisfaction (FAO, 2024). The thresholds for leaf loss are related to the percentage of crop defoliation and yield loss (Moffitt, 1986). The pest does not necessarily have to feed on the specific plant part that is the target of the crop. For example, insects that feed on the leaves of cereals, such as leafhoppers or cereal aphids, can cause yield losses in cereal grains. Similar damage is caused by other pests such as Colorado potato beetles, flower thrips, cabbage maggots, wireworms, etc. The yield loss can be calculated as the difference between the yield of healthy and infested plants. In addition to direct feeding damage, the pests also cause indirect damage by transmitting viruses or secreting honeydew, which impairs product quality, for example by reducing the hectolitre weight and protein content of the grains (Bažok *et al.*, 2018).

A *traditional numerical threshold* refers to a specific level of pest population or damage that triggers a decision to implement control measures. It is a predetermined numerical value that helps farmers or pest controllers to decide when intervention is required to prevent economic damage or yield loss. Once the pest population or damage exceeds the set numerical threshold, this signals the need for timely and targeted action to control the pest and mitigate potential agricultural losses (Möhring *et al.*, 2020). Numerical thresholds are a key component of IPM strategies as they provide a quantitative basis for decision making. Traditional numerical thresholds for pests are still used but can be inflexible and often lack an accurate link between pest density and resulting yield losses. The traditional approach of using numerical thresholds for pests has its limitations. Furthermore, the traditionally established thresholds are not flexible enough in situations with fluctuating crop values and control costs (Onstad, 1987).

In contrast, *economic thresholds* are defined as the level of pest population density at which the resulting damage corresponds to the cost of implementing control measures (including pesticides and their application). The recommendation to spray pesticides depends on crossing the economic threshold, i.e. a point at which the value of the potential damage exceeds the cost of control. Factors such as yield loss are often taken into account when deciding when to spray, and in certain cases quality losses can also influence decision thresholds (Onstad, 1987). Economic thresholds critically evaluate the balance between the costs associated with pest control and the value of potential losses (Onstad, 1987). The use of control measures based on these thresholds offers several advantages: (i) it prevents negative financial outcomes where control costs exceed the initial pest damage; (ii) it minimizes the risk of unnecessary control of non-target pest populations, thereby conserving beneficial organisms and positively impacting biodiversity conservation; (iii) it mitigates the risk of pests developing resistance to pesticides (Lande *et al.*, 1997; Naranjo, 2001; Liang *et al.*, 2012). Economic thresholds determine pest control decisions and regulate pesticide use (Ferron, 1999). The development and application of these thresholds require a comprehensive understanding of the pests, the resulting damage, crop responses to the damage (including yield reduction) and the assessment of crop value and control costs (Hueth and Regev, 1974).

The benefit–cost ratio (BCR) influences decisions on pesticide use. In IPM programmes, a BCR ratio of 1.5:1 is often used, i.e. measures are only taken if the potential damage avoided is at least 1.5 times higher than the control costs (Deguine *et al.*, 2021). Not every occurrence of pests requires intervention. Trying to prevent all damage is not only uneconomical but also incompatible with the principles of integrated farming (Silvie *et al.*, 2013).

21.6 Revolutionizing Pest Management: Machine Learning and Advanced Algorithms in Precision Pest Control

In recent years, the application of machine learning algorithms in the field of decision making for crop protection or pest control has become increasingly important (Li *et al.*, 2023).

Traditional approaches based on expert experience have shown certain limitations, especially in situations where crop values and spraying costs vary. Machine learning algorithms offer the possibility of a more advanced approach that uses data on pests, crops, costs and values to better rationalize pesticide use and achieve more profitable outcomes (Jha *et al.*, 2019; Tian *et al.*, 2020). An increasingly researched approach is the use of machine learning models to calculate economic thresholds. These models use data on control costs, crop values and damage caused by pests to determine the optimal pest population density that leads to an economically viable solution. The use of such models would allow farmers to make informed decisions about spraying, taking into account specific conditions and variations in costs and values (Bondad *et al.*, 2023).

Another approach currently being investigated is the use of analytical tools and algorithms to assess losses caused by pests (Bondad *et al.*, 2023). The combination of sensor data, geospatial information, geodata and machine learning algorithms allows the monitoring and identification of pests, as well as the estimation of their abundance, frequency, localization and impact on crops. Based on this information, farmers can define control thresholds, considering the risk of yield loss and the value of the crop (Ewert *et al.*, 2015). This approach enables better monitoring and more precise spraying decisions, reducing the need for unnecessary crop treatments.

In addition, the application of machine learning algorithms enables the monitoring and prediction of pest population dynamics, which is important for timely spraying decisions (Caffarra *et al.*, 2012; Khaliq *et al.*, 2014). Based on historical data on pest populations, weather conditions and other relevant factors (environmental conditions), algorithms can create models and predictions that help determine the optimal time for spraying. This reduces the need for routine spraying and enables targeted interventions only when they are really necessary.

21.6.1 Optimizing pest control through strategic utilization of weather stations

Weather stations are important for monitoring meteorological parameters that influence the behaviour of various pests and the susceptibility of plants. The use of weather stations in agriculture to improve crop protection through real-time weather monitoring is still an innovative approach. However, as technology advances, it is becoming increasingly important (Javaid *et al.*, 2022). These meteorological data can be obtained directly from individual farms, either within the farm premises or in close proximity to a particular crop of interest (Gleason *et al.*, 2008). They are equipped with sensors that measure parameters such as temperature, humidity, wind speed and precipitation. These data provide valuable insight into the prevailing weather conditions at a particular location. By utilizing data from weather stations, farmers can better understand how these factors influence pest activity. In addition, these systems aim to develop pest models and risk thresholds by using field phenology data together with weather data and growing degree days (Crimmins *et al.*, 2020; Damos *et al.*, 2021). By helping to establish early warning systems for pest outbreaks, they could contribute to the development of targeted pest control strategies. By continuously monitoring weather conditions, farmers can receive alerts and notifications when conditions are favourable for a potential pest infestation or disease development. This type of on-the-ground data gives farmers the ability to make decisions based on facts, rather than relying solely on their intuition or short-term weather forecasts for a larger area. This facilitates proactive decision making and ensures precise application of pest control measures exactly when they are needed. Some weather stations are equipped with leaf wetness duration (LWD) sensors that are used exclusively for disease warning systems to help farmers determine the right time to apply fungicides or bactericides (Gleason *et al.*, 2008).

As technology advances, the integration of meteorological data with machine learning algorithms offers the opportunity to improve agricultural pest control methods. This paves the way for a future where farmers can make informed decisions based on accurate predictions to effectively control pests and ensure optimal crop yields and quality.

Table 21.1. List of predictive models with brief description.

Model	Features	Reference
Machine learning for crop yield prediction	Machine learning models are used to forecast crop yields accurately by analysing data on soil quality, climate conditions and water availability	Chlingaryan *et al.*, 2018 Elavarasan *et al.*, 2018
Optimizing resource management	These models help farmers plan their planting and harvesting schedules, allocate resources more efficiently, and maximize overall productivity	Abbas *et al.*, 2020 Wei *et al.*, 2020
Plant disease detection	Machine learning models are employed for the early detection and classification of plant diseases by analysing images of crops and leaves, allowing for timely interventions	Demilie, 2024 Naeem *et al.*, 2021
Weed management	Precision agriculture utilizes machine learning algorithms to identify and classify weed species in real time, reducing the need for herbicides and improving crop health	Behmann *et al.*, 2015 Espejo-Garcia *et al.*, 2021
Pest insect detection	Remote sensing devices and machine learning techniques are integrated to detect and monitor pest insects automatically, aiding in pest management decisions and protecting crops	Sagar *et al.*, 2017 Skawsang *et al.*, 2019
Data-driven decision making	Enables farmers to make data-driven decisions regarding irrigation, fertilization and pesticide application, leading to better resource utilization while limiting environmental damage	Balducci *et al.*, 2018; Kamir *et al.*, 2020
Improved crop quality and yields	By implementing machine learning-based predictions and interventions, farmers can achieve higher crop quality and increased yields while minimizing losses	Portal Prognozno-Izveštajne Službe Zaštite Bilja (2024)

21.6.2 Predictive models and machine learning in modern farming practices

In modern agriculture, predictive modelling using machine learning and data analytics is integrated into agricultural practices. These advanced models are invaluable, enabling farmers to predict crop yields, optimize resource allocation and manage the challenges posed by crop diseases, weeds and insect pests with unprecedented accuracy and efficiency (Behmann *et al.*, 2015). In Table 21.1, we review the predictive models in modern agriculture, highlighting their diverse applications and the transformative impact they have on today's agricultural practices.

Predictive modelling has significantly changed the agricultural landscape, bringing both economic and environmental benefits. *FarmLogs*, for example, integrates data from satellites, weather stations and soil sensors to provide real-time insights that enable farmers to optimize their cropping plans and irrigation methods, resulting in higher crop yields, reduced water consumption and minimized fertilizer wastage, improving profitability and sustainability. Similarly, the *Climate Corporation* uses historical weather data and machine learning to create accurate yield forecasts and help farmers make strategic decisions (dataforest.ai). *Blue River Technology's 'See & Spray' technology*, which uses machine learning for real-time weed detection and precise herbicide application, reduces chemical use by up to 90%, highlighting its significant environmental benefits. In addition, *Ceres Imaging* uses aerial imagery analysed with data science methods to detect stress and nutrient imbalances in crops, leading to healthier harvests and less waste of resources (Dataforest, 2023).

Despite the transformative potential of predictive modelling, its implementation faces a number of challenges. Issues of data privacy, data quality and limited access to technology, particularly in rural areas and developing countries, are significant barriers. Ensuring data quality and integration is necessary, as inconsistencies can compromise the effectiveness of models. Overcoming these challenges is necessary for wider adoption and impact of predictive modelling in agriculture (Dataforest, 2023).

21.6.3 Adopting precision agriculture in pest management challenges

In response to the challenges posed by pest management in modern agriculture, the demand for advanced precision technologies is growing. Precision agriculture (PA) is a strategic approach to agricultural management that focuses on spatial precision to optimize both productivity and sustainability in crop production. The main objective of this approach is to improve the effectiveness of pest management through the strategic and precise use of pesticides and other inputs in agro-ecosystems. To accomplish this difficult task, the integration of cutting-edge technologies such as remote sensing and variable rate technology (VRT) into pest management strategies is a compelling solution. These technologies are expected to improve crop yields, enable data-driven decision making in the field, optimize pesticide use to reduce costs, increase profit margins and minimize environmental impact (Du *et al.*, 2008; Méndez-Vázquez *et al.*, 2019).

21.6.4 Role of remote sensing technology in PA

Remote sensing technology provides site-specific data on plant health, supporting its application in this context. By continuously monitoring and assessing plant health and identifying areas of pest infestation in real time, remote sensing enables precise and targeted intervention. This is done by detecting signs of pest infestation, such as defoliation of plants, changes in plant colour and geometric deformations of plant tissue. The use of precise spectral data collected by remote sensing sensors enables the detection, investigation and prediction of field pests (Zhang *et al.*, 2023).

Remote sensing in agriculture is generally a non-invasive approach and remote sensing systems rely primarily on information obtained from the reflection (or transmission) of visible near-infrared and short-wave infrared (VIS-NIR-SWIR) radiation by the plant. The radiation reflected by plants is quantified as spectral reflectance and analysed across different wavelengths in remote sensing. The most important spectral ranges for remote sensing of vegetation changes are therefore the visible (VIS; 400–700 nm), near-infrared (NIR; 700–1300 nm) and short-wave infrared (SWIR; 1300–2500 nm) regions within the electromagnetic spectrum. Currently, cost-effective multi- and hyperspectral VIS-NIR-SWIR sensors are available on various remote sensing platforms, including ground-based, air-borne and satellite-based systems.

In recent decades, research into the spectral reflectance of plants has greatly increased thanks to advances in multispectral and hyperspectral cameras, sensors and field spectroradiometers that can accurately measure the entire electromagnetic spectrum (350–2500 nm). These sensors analyse the spectral reflectance of plants and enable the calculation of vegetation indices, which serve as indicators for biotic and abiotic stress parameters and enable yield predictions. Which vegetation index to use for pest control depends on several factors, including the type of crop, the pest in question and the availability of remote sensing data. Integrating

Table 21.2. List of vegetation indices used for pest detection in fields, including calculation formulas and specific features.

Vegetation index	Calculation	Features
Normalized Difference Vegetation Index (NDVI)	$NDVI = \frac{NIR - RED}{NIR + RED}$	Assessing overall plant health; detecting pest-induced stress
Normalized Difference Water Index (NDWI)	$NDWI = \frac{R860 - R1240}{R860 + R1240}$	Detecting changes in water content; aiding in the identification of areas affected by pests like root-feeding nematodes, which can impact water uptake
Enhanced Vegetation Index (EVI)	$EVI = G\frac{NIR - RED}{NIR + C1 * RED - C2 * BLUE + L}$	Improved version of NDVI; more precise assessment of vegetation health, including stress caused by pests
Soil Adjusted Vegetation Index (SAVI)	$SAVI = \frac{NIR - RED}{NIR + RED + L} * (1 + L)$	Accounts for variations in soil brightness; suitable for areas with diverse soil types where pests may affect crops differently
Chlorophyll Index (CI)	$CI = \frac{R750 - R705}{R750 + R705}$	Early detection of stress caused by pests, which disrupt chlorophyll production
Photochemical Reflectance Index (PRI)	$PRI = \frac{R530 - R570}{R530 + R570}$	Focuses on changes in plant pigments and photosynthetic activity, serving as a sensitive indicator of pest-induced stress
Carotenoid Reflectance Index 1 (CRI1)	$CRI1 = \frac{1}{R510} - \frac{1}{R550}$	Monitors pigment-related stress in crops; particularly useful for early pest detection

multiple indices and data sources can provide a more comprehensive picture of crop health and pest impact. In addition, machine learning and remote sensing techniques are increasingly being used to automate the analysis of vegetation indices and detect pest-induced stress in crops.

There are more than 500 known vegetation indices that serve as indicators of different types of plant stress. In practice, however, a few well-established indices are generally used to detect pests in the field. These indices include the Normalized Difference Vegetation Index (NDVI), the Normalized Difference Water Index (NDWI), the Enhanced Vegetation Index (EVI), the Soil Adjusted Vegetation Index (SAVI), the Chlorophyll Index (CI), the Photochemical Reflectance Index (PRI) and the Carotenoid Reflectance Index 1 (CRI1) (Skendžić *et al.*, 2023). More detailed information on each of these indices can be found in Table 21.2.

21.6.5 Advancements in sensing technologies and site-specific pest management

Vegetation indices are used in PA for site-specific management and on-site screening for various plant breeding objectives, such as the evaluation of yield potential, adaptability to stress parameters and overall plant quality (Skendžić *et al.*, 2023). The above-mentioned indices are usually provided by satellite-based multispectral sensors such as Landsat, Sentinel-2, SPOT, etc. and a variety of air-borne multispectral sensors (e.g. multispectral drones). Variable rate technology enhances this approach by enabling the precise application of pesticides only where and when they are needed, reducing the unnecessary use of chemicals. In conventional orchards and market gardens, radial air-assisted sprayers are favoured, while tractor-mounted boom sprayers are typically used in field crops (Zhu *et al.*, 2017).

The conventional use of these sprayers often leads to an inefficient distribution of pesticides. This inefficiency results from the tendency of the sprayers to distribute the pesticides evenly across the entire field, regardless of factors such as the presence or absence of plants in the rows, the condition of the plant structures or the growth stage of the plants. In addition, inconsistent application can encourage the development of pesticide-resistant pests, while high production costs, regulatory compliance issues, erosion, runoff and impacts on non-target organisms compound the disadvantages.

To overcome these challenges, modern agriculture increasingly relies on precision farming. This involves using advanced technologies and techniques to optimize the use of pesticides and minimize their negative impact on the environment and human health (Alam *et al.*, 2020). Through the integration of remote sensing data, which includes vegetation indices, and the use of geographic information systems (GIS) together with global positioning system (GPS) technologies, the creation of application maps is becoming a central aspect of modern PA, especially for the targeted use of pesticides or site-specific pest control. By analysing the spatial variability of vegetation health and stress levels within a field, farmers can create precise application maps that highlight areas where pest pressure is higher or where plants are obviously stressed. These maps optimize pesticide application by ensuring efficient resource use and reducing environmental impact. Using GIS and GPS, farmers can navigate their equipment to specific locations in the field and apply pesticides with precision (Campos *et al.*, 2020). Currently, the focus of site-specific pesticide management is primarily on herbicides but as advances are made, there is a growing trend to apply this approach to fungicides and insecticides as well (Zanin *et al.*, 2022). The introduction of site-specific pest management in the early phenological stages of soya and maize crops resulted in a 56.16% reduction in pesticide costs, while crop yields remained consistent with historical data (Zanin *et al.*, 2022).

However, it is also important to recognize some limitations and challenges. The initial investment in remote sensing equipment and technology for PA can be significant, making it less accessible to smallholder farmers. Considering the size of crops, smaller crops in more complex agricultural landscapes might pose different challenges (Atta-Aidoo *et al.*, 2022). Smaller farms may have fewer pest problems due to their diverse cropping strategies, which can naturally reduce pest infestations. However, this does not eliminate the need for pest control. In fact, pest control in a complex landscape can still benefit from advanced technologies, although the scope and methods of application may vary. IPM and site-specific management are also necessary for smaller farms, as they help optimize resources and improve crop health without relying heavily on chemical inputs. Even though smaller farms are less likely to face large-scale pest outbreaks, they can still benefit from PA tools to effectively monitor and manage pest pressure (Atta-Aidoo *et al.*, 2022). In addition, the accuracy of pest detection and prescription maps depends on the quality of the remote sensing data and the ability to interpret it correctly, which may require specialized knowledge and training.

The prospects for remote sensing in pest control are promising. Advances in sensor technology, including hyperspectral and multispectral cameras, will provide even more detailed and accurate data for pest detection and crop monitoring. Data analysis in the form of machine learning and artificial intelligence algorithms will play a greater role in automating decision making. In addition, the integration of remote sensing with other innovative techniques such as Internet of Things sensors and autonomous sprayers has the potential to further improve the efficiency and sustainability of pest control in agriculture. In a world where sustainable agriculture is paramount, the synergy between remote sensing, variable rate spraying and IPM will continue to be a valuable tool to both protect the environment and maintain agricultural productivity (UNDP, 2021) (Fig. 21.2).

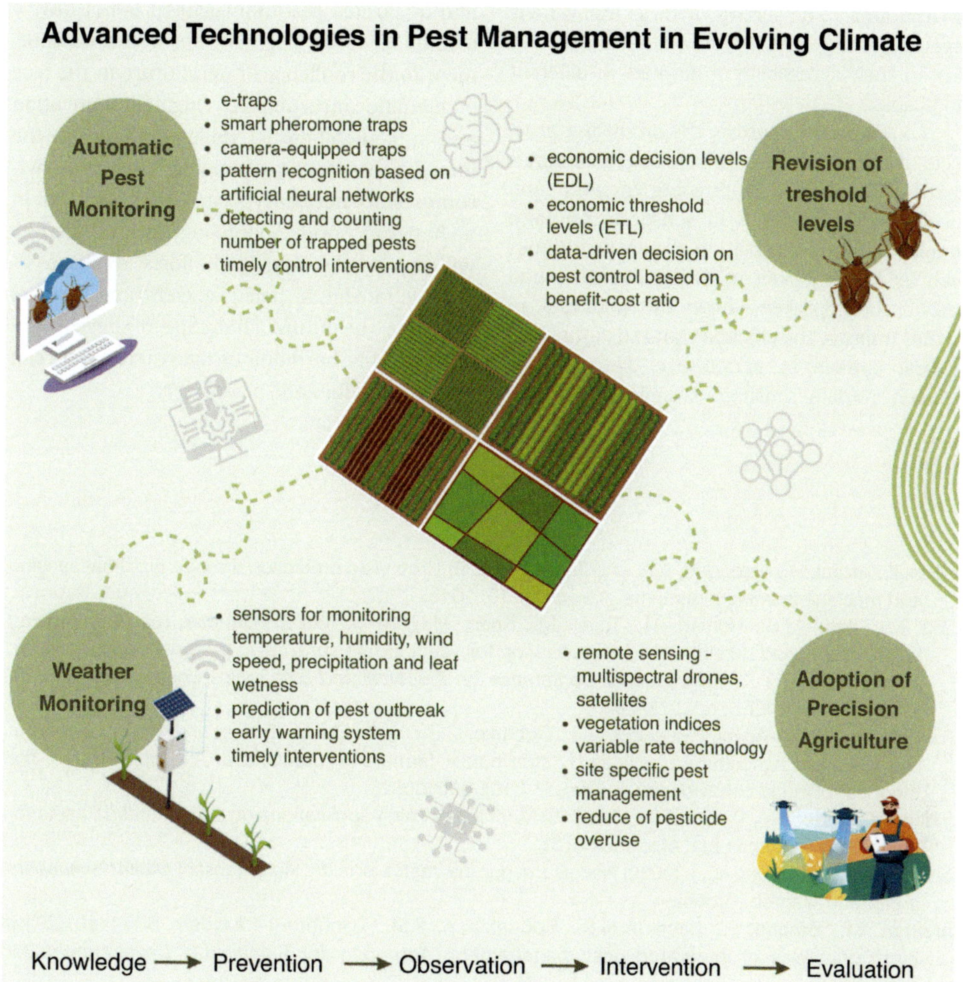

Fig. 21.2. Advanced technologies in pest management in evolving climate.

21.7 Conclusion

As climate change progresses, adaptation measures in agriculture are necessary to maintain food production and minimize environmental impact. The multiple challenges posed by climate change require a holistic and forward-looking approach to crop protection. The comprehensive integration of traditional and modern monitoring techniques combined with the incorporation of economic threshold levels provides farmers with strategic tools to deal with uncertainty and make informed decisions tailored to dynamic and rapidly changing environmental conditions.

The transformative impact of predictive modelling, machine learning and remote sensing technologies represents a paradigm shift in modern agriculture. These technological advances are enabling farmers to predict crop yields with unprecedented accuracy, optimize resource use and effectively manage crop diseases, weeds and pests. The impact of these innovations goes beyond the economic sphere and contributes significantly to the overall goal of sustainable agriculture and environmental protection. However, harnessing these technological advances also requires overcoming challenges such as data protection, ensuring data quality and bridging the access gap in rural areas. The success of these innovations depends not only on their technical capabilities but also on their accessibility to farmers in different landscapes.

Looking to the future, the promising prospects of remote sensing in pest control emphasize the importance of ongoing research and development. Advances in sensor technology, coupled with the integration of new techniques such as IoT (Internet of Things) sensors and autonomous sprayers, have the potential to further improve the efficiency and sustainability of pest control in agriculture. The synergy between remote sensing, variable technology and integrated pest management is not only a technological development but also a commitment to the resilience of agriculture in the face of climatic uncertainties. Ongoing innovation and a comprehensive understanding of the interaction between technology and the environment are necessary to sustain agriculture in a changing world. Addressing these challenges will require the combined efforts of researchers, technologists, policy makers and farmers to shape the future landscape of agriculture and promote harmonious coexistence between human activities and biodiversity.

References

Abbas, F., Afzaal, H., Farooque, A.A. and Tang, S. (2020) Crop yield prediction through proximal sensing and machine learning algorithms. *Agronomy* 10, 1046.

Alam, M., Alam, M.S., Roman, M., Tufail, M., Khan, M.U. *et al.* (2020) Real-time machine-learning based crop/weed detection and classification for variable-rate spraying in precision agriculture. In: *Presented at the 7th International Conference on Electrical and Electronics Engineering (ICEEE)*, Antalya, Turkey. IEEE, pp. 273–280.

Atta-Aidoo, J., Antwi-Agyei, P., Dougill, A.J., Ogbanje, C.E., Akoto-Danso, E.K. *et al.* (2022) Adoption of climate-smart agricultural practices by smallholder farmers in rural Ghana: An application of the theory of planned behavior. *PLOS Climate* 1(10), e0000082.

Balducci, F., Impedovo, D. and Pirlo, G. (2018) Machine learning applications on agricultural datasets for smart farm enhancement. *Machines* 6, 38.

Barić, B. and Pajač Živković, I. (2020) Načela integrirane zaštite bilja. In: *Agronomski Fakultet Sveučilišta u Zagrebu*. Zagreb, Croatia.

Barzman, M., Bàrberi, P., Birch, A.N.E., Boonekamp, P.M., Dachbrodt-Saaydeh, S. *et al.* (2015) Eight principles of integrated pest management. *Agronomy for Sustainable Development* 35, 1199–1215.

Bažok, R., Lemić, D., Čačija, M. and Drmić, Z. (2018) Ekonomski prag štetnosti I prag odluke (Kritični Broj) na primjeru žičnjaka I lisnih sovica. *Glasilo Biljne Zaštite* 18(5), 500–513.

Behmann, J., Mahlein, A.K., Rumpf, T., Römer, C. and Plümer, L. (2015) A review of advanced machine learning methods for the detection of biotic stress in precision crop protection. *Precision Agriculture* 16, 239–260.

Blommers, L.H. (1994) Integrated pest management in European apple orchards. *Annual Review of Entomology* 39, 213–241.

Bondad, J., Harrison, M.T., Whish, J., Sprague, S. and Barry, K. (2023) Integrated crop-disease models: New frontiers in systems thinking. *Farming Systems* 1(1), 100004.

Caffarra, A., Rinaldi, M., Eccel, E., Rossi, V. and Pertot, I. (2012) Modelling the impact of climate change on the interaction between grapevine and its pests and pathogens: European grapevine moth and powdery mildew. *Agriculture, Ecosystems & Environment* 148, 89–101.

Campos, J., Gallart, M., Llop, J., Ortega, P., Salcedo, R. *et al.* (2020) On-farm evaluation of prescription map-based variable rate application of pesticides in vineyards. *Agronomy* 10(1), 102.

Chen, Y., Why, A., Batista, G., Mafra-Neto, A. and Keogh, E. (2014) Flying insect classification with inexpensive sensors. *Journal of Insect Behavior* 27, 657–677.

Chlingaryan, A., Sukkarieh, S. and Whelan, B. (2018) Machine learning approaches for crop yield prediction and nitrogen status estimation in precision agriculture: A review. *Computers and Electronics in Agriculture* 151, 61–69.

Crimmins, T.M., Gerst, K.L., Huerta, D.G., Marsh, R.L., Posthumus, E.E. *et al.* (2020) Short-term forecasts of insect phenology inform pest management. *Annals of the Entomological Society of America* 113(2), 139–148.

Damos, P., Tsikos, E., Louta, M. and Papathanasiou, F. (2021) Towards the development of a smart plant protection solution for improved pest management of dry beans (*Phaseolus vulgaris* L.) in northern Greece. In: *Proceedings of the 1st International Electronic Conference on Entomology*. Available at: https://sciforum.net/paper/view/10568 (accessed 13 June 2025).

Dataforest (2023) Data science in agriculture: From predictive analytics to precision farming. Available at: https://dataforest.ai/blog/revolutionizing-farming-through-data-science-exploring-the-role-and-applications-of-data-science-in-agriculture (accessed 13 June 2025).

Deguine, J.P., Aubertot, J.N., Flor, R.J., Lescourret, F., Wyckhuys, K.A.G. *et al.* (2021) Integrated pest management: Good intentions, hard realities. A review. *Agronomy for Sustainable Development* 41, 38.

Demilie, W.B. (2024) Plant disease detection and classification techniques: A comparative study of the performances. *Journal of Big Data* 11(1), 5.

Dhaliwal, G.S., Singh, R. and Chhillar, B.S. (2006) *Essentials of Agricultural Entomology*. Kalyani Publishers, New Delhi, India.

Dong, Y., Xu, F., Liu, L., Du, X., Ren, B. *et al.* (2020) Automatic system for crop pest and disease dynamic monitoring and early forecasting. *IEEE Journal of Selected Topics in Applied Earth Observations and Remote Sensing* 13, 4410–4418.

Du, Q., Chang, N.B., Yang, C. and Srilakshmi, K.R. (2008) Combination of multispectral remote sensing, variable rate technology and environmental modeling for citrus pest management. *Journal of Environmental Management* 86(1), 14–26.

Ehler, L.E. (2006) Integrated Pest Management (IPM): Definition, historical development and implementation, and the other IPM. *Pest Management Science* 62, 787–789.

Elavarasan, D., Vincent, D.R., Sharma, V., Zomaya, A.Y. and Srinivasan, K. (2018) Forecasting yield by integrating agrarian factors and machine learning models: A survey. *Computers and Electronics in Agriculture* 155, 257–282.

Espejo-Garcia, B., Malounas, I., Vali, E. and Fountas, S. (2021) Testing the suitability of automated machine learning for weeds identification. *AI* 2, 34–47.

Ewert, F., Rötter, R., Bindi, M., Webber, H., Trnka, M. *et al.* (2015) Crop modelling for integrated assessment of risk to food production from climate change. *Environmental Modelling & Software* 72, 287–303.

FAO (2024) Ensuring food quality and safety. Available at: www.fao.org/4/W9474T/W9474T03.htm (accessed 13 June 2025).

Faria, P., Nogueira, T., Ferreira, A., Carlos, C. and Rosado, L. (2021) AI-powered mobile image acquisition of vineyard insect traps with automatic quality and adequacy assessment. *Agronomy* 11, 731.

Ferron, P. (1999) Protection intégrée des cultures: Évolution du concept et de son application. *Cahiers Agricultures* 8, 389–396.

Gleason, M.L., Duttweiler, K.B., Batzer, J.C., Taylor, S.E., Sentelhas, P.C. *et al.* (2008) Obtaining weather data for input to crop disease-warning systems: Leaf wetness duration as a case study. *Scientia Agricola* 65, 76–87.

Goldshtein, E., Cohen, Y., Hetzroni, A., Gazit, Y., Timar, D. *et al.* (2017) Development of an automatic monitoring trap for Mediterranean fruit fly (*Ceratitis capitata*) to optimize control applications frequency. *Computers and Electronics in Agriculture* 139, 115–125.

Harvey, J.A., Tougeron, K., Gols, R., Heinen, R., Abarca, M. *et al.* (2023) Scientists' warning on climate change and insects. *Ecological Monographs* 93(1), e1553.

Heeb, L., Jenner, E. and Cock, M.J.W. (2019) Climate-smart pest management: Building resilience of farms and landscapes to changing pest threats. *Journal of Pest Science* 92, 951–969.

Hellin, J. and Fisher, E. (2019) Climate-smart agriculture and non-agricultural livelihood transformation. *Climate* 7, 48.

Hernández-Serna, A. and Jiménez-Segura, L.F. (2014) Automatic identification of species with neural networks. *PeerJ* 2, e563.

Holguin, G.A., Lehman, B.L., Hull, L.A., Jones, V.P. and Park, J. (2010) Electronic traps for automated monitoring of insect populations. *IFAC Proceedings* Volumes 43, 49–54.

Hueth, D. and Regev, U. (1974) Optimal agricultural pest management with increasing pest resistance. *American Journal of Agricultural Economics* 56, 543–552.

Iost Filho, F.H., Heldens, W.B., Kong, Z. and de Lange, E.S. (2020) Drones: Innovative technology for use in precision pest management. *Journal of Economic Entomology* 113(1), 1–25.

Javaid, M., Haleem, A., Singh, R.P. and Suman, R. (2022) Enhancing smart farming through the applications of Agriculture 4.0 technologies. *International Journal of Intelligent Networks* 3, 150–164.

Jha, K., Doshi, A., Patel, P. and Shah, M. (2019) A comprehensive review on automation in agriculture using artificial intelligence. *Artificial Intelligence in Agriculture* 2, 1–12.

Kadoić Balaško, M. and Virić Gašparić, H. (2022) Prognoza I pragovi odluke suzbijanja važnijih štetnika strnih žitarica. *Glasilo Biljne Zaštite* 22(3), 373–388.

Kamir, E., Waldner, F. and Hochman, Z. (2020) Estimating wheat yields in Australia using climate records, satellite image time series and machine learning methods. *ISPRS Journal of Photogrammetry and Remote Sensing* 160, 124–135.

Khaliq, M., Javed, M., Sohail, M. and Sagheer, M. (2014) Environmental effects on insects and their population dynamics. *Journal of Entomology and Zoology Studies* 2, 1–7.

Knipling, E.F. (1979) *The Basic Principles of Insect Population Suppression and Management*. US Government Printing Office, Department of Agriculture, Washington, DC.

Lande, R., Saether, B.E. and Enger, S. (1997) Threshold harvesting for sustainability of fluctuating resources. *Ecology* 78, 1341–1350.

Li, S., Yuan, Z., Peng, R., Leybourne, D., Xue, Q. *et al*. (2023) An effective farmer-centred mobile intelligence solution using lightweight deep learning for integrated wheat pest management. Available at: https://ssrn.com/abstract=4410392 (accessed 13 June 2025).

Lemić, D., Kriticos, D.J., Virić Gasparić, H., Pajač Živković, I., Duffy, C. *et al*. (2024) Global change and adaptive biosecurity: Managing current and emerging *Aleurocanthus woglumi* threats to Europe. *Current Opinion in Insect Science* 62, 101164.

Liang, J., Tang, S. and Cheke, R.A. (2012) An integrated pest management model with delayed responses to pesticide applications and its threshold dynamics. *Nonlinear Analysis: Real World Applications* 13(5), 2352–2374.

Méndez-Vázquez, L.J., Lira-Noriega, A., Lasa-Covarrubias, R. and Cerdeira-Estrada, S. (2019) Delineation of site-specific management zones for pest control purposes: Exploring precision agriculture and species distribution modeling approaches. *Computers and Electronics in Agriculture* 167, 105101.

Moffitt, L.J. (1986) Risk-efficient thresholds for pest control decisions. *Journal of Agricultural Economics* 37, 69–75.

Möhring, N., Wuepper, D., Musa, T. and Finger, R. (2020) Why farmers deviate from recommended pesticide timing: The role of uncertainty and information. *Pest Management Science* 76(8), 2787–2798.

Murray, K., Jepson, P., Bouska, C., Scherr, M. and Walenta, D. (2020) Integrated pest management summit reveals barriers, needs, and goals for agricultural extension. *Journal of Extension* 58(3), Article .

Naeem, S., Ali, A., Chesneau, C., Tahir, M.H., Jamal, F. *et al*. (2021) The classification of medicinal plant leaves based on multispectral and texture feature using machine learning approach. *Agronomy* 11, 263.

Naranjo, S.E. (2001) Conservation and evaluation of natural enemies in IPM systems for *Bemisia tabaci*. *Crop Protection* 20, 835–885.

Olatinwo, R. and Hoogenboom, G. (2014) Weather-based pest forecasting for efficient crop protection. In: Abrol, D.P. (ed.) *Integrated Pest Management: Current Concepts and Ecological Perspective*. Elsevier, Amsterdam, The Netherlands, pp. 59–78.

Onstad, D.W. (1987) Calculation of economic-injury levels and economic thresholds for pest management. *Journal of Economic Entomology* 80(2), 297–303.

Pajač Živković, I., Vilizzi, L., Piria, M., Čirjak, D., Miklečić, I. *et al*. (2025) Invasion risk of non-native insects to agricultural lands: A screening for croatia. *Management of Biological Invasions* 16(1), 167–185.

Pedigo, L.P. and Rice, M.E. (2009) *Entomology and Pest Management*, 6th edn. PHI Learning Private Limited, New Delhi, India.

Portal Prognozno-Izveštajne Službe Zaštite Bilja (2024). Available at: www.pissrbija.com/default.aspx (accessed 13 June 2025).

Potamitis, I., Ganchev, T. and Kontodimas, D. (2009) On automatic bioacoustic detection of pests: The cases of *Rhynchophorus ferrugineus* and *Sitophilus oryzae*. *Journal of Economic Entomology* 102, 1681–1690.

Potamitis, I., Eliopoulos, P. and Rigakis, I. (2017) Automated remote insect surveillance at a global scale and the Internet of Things. *Robotics* 6, 19.

Preti, M., Verheggen, F. and Angeli, S. (2021a) Insect pest monitoring with camera-equipped traps: Strengths and limitations. *Journal of Pest Science* 94, 203–217.

Preti, M., Favaro, R., Knight, A.L. and Angeli, S. (2021b) Remote monitoring of *Cydia pomonella* adults among an assemblage of nontargets in sex pheromone-kairomone-baited smart traps. *Pest Management Science* 77, 4084–4090.

Rajotte, E.G., Norton, G.W., Kazmierczak, R.F., Lambur, M. and Allen, W. (2009) The role of knowledge, attitudes and perceptions in the uptake of agricultural and agroforestry innovations among smallholder farmers in sub-Saharan Africa. *International Journal of Agricultural Sustainability* 7(1), 40–50.

Rydhmer, K., Bick, E., Still, L., Strand, A., Luciano, R. *et al.* (2022) Automating insect monitoring using unsupervised near-infrared sensors. *Scientific Reports* 12, 2603.

Sagar, D., Nebapure, S.M. and Chander, S. (2017) Development and validation of weather based prediction model for *Helicoverpa armigera* in chickpea. *Journal of Agrometeorology* 19, 328–333.

Schrader, M.J., Smytheman, P., Beers, E.H. and Khot, L.R. (2022) An open-source low-cost imaging system plug-in for pheromone traps aiding remote insect pest population monitoring in fruit crops. *Machines* 10, 52.

Sciarretta, A. and Calabrese, P. (2019) Development of automated devices for the monitoring of insect pests. *Current Agricultural Research Journal* 7, 19.

Silvie, P.J., Renou, A., Vodounnon, S., Bonni, G., Adegnika, M.O. *et al.* (2013) Threshold-based interventions for cotton pest control in West Africa: What's up 10 years later? *Crop Protection* 43, 157–165.

Skawsang, S., Nagai, M., Tripathi, N.K. and Soni, P. (2019) Predicting rice pest population occurrence with satellite-derived crop phenology, ground meteorological observation, and machine learning: A case study for the central plain of Thailand. *Applied Sciences* 9, 4846.

Skendžić, S., Zovko, M., Pajač Živković, I., Lešić, V. and Lemić, D. (2021a) Effect of climate change on introduced and native agricultural invasive insect pests in Europe. *Insects* 12, 985.

Skendžić, S., Zovko, M., Živković, I.P., Lešić, V. and Lemić, D. (2021b) The impact of climate change on agricultural insect pests. *Insects* 12(5), 440.

Skendžić, S., Zovko, M., Lešić, V., Pajač Živković, I. and Lemić, D. (2023) Detection and evaluation of environmental stress in winter wheat using remote and proximal sensing methods and vegetation indices – A review. *Diversity* 15(4), 481.

Sutherst, R.W., Constable, F., Finlay, K.J., Harrington, R., Luck, J. *et al.* (2011) Adapting to crop pest and pathogen risks under a changing climate. *Wiley Interdisciplinary Reviews: Climate Change* 2, 220–237.

Tian, H., Wang, T., Liu, Y., Qiao, X. and Li, Y. (2020) Computer vision technology in agricultural automation – A review. *Information Processing in Agriculture* 7(1), 1–19.

United Nations Development Programme (UNDP) (2021) Precision agriculture for smallholder farmers. UNDP Global Centre for Technology, Innovation and Sustainable Development, Singapore. Available at: www.undp.org/publications/precision-agriculture-smallholder-farmers (accessed 13 June 2025).

Vernon, R.S. and Gillespie, D.R. (1995) Influence of trap shape, size, and background contrast on capture of Frankliniella occidentalis (Thysanoptera: Thripidae) in yellow sticky traps. *Journal of Economic Entomology* 88(2), 288–293.

Wei, M.C.F., Maldaner, L.F., Ottoni, P.M.N. and Molin, J.P. (2020) Carrot yield mapping: A precision agriculture approach based on machine learning. *Ai* 1(2), 229–241.

Wijerathna-Yapa, A. and Pathirana, R. (2022) Sustainable agro-food systems for addressing climate change and food security. *Agriculture* 12(10), 1554.

Zanin, A.R.A., Neves, D.C., Teodoro, L.P.R., da Silva Júnior, C.A., da Silva, S.P. *et al.* (2022) Reduction of pesticide application via real-time precision spraying. *Scientific Reports* 12(1), 5638.

Zhang, M., Chen, T.E., Gu, X., Chen, D., Wang, C, *et al.* (2023) Hyperspectral remote sensing for tobacco quality estimation, yield prediction, and stress detection: A review of applications and methods. *Frontiers in Plant Science* 14, 1073346.

Zhang, X., Yang, Y.H., Han, Z., Wang, H. and Gao, C. (2013) Object class detection: Survey. *ACM Computing Surveys* 46, 1–53.

Zhu, H., Liu, H., Shen, Y., Liu, H. and Zondag, R.H. (2017) Spray deposition inside multiple-row nursery trees with a laser-guided sprayer. *Journal of Environmental Horticulture* 35(1), 13–23.

Čirjak, D., Aleksi, I., Lemić, D. and Pajač Živković, I. (2023a) EfficientDet-4 deep neural network-based remote monitoring of codling moth population for early damage detection in apple orchard. *Agriculture* 13(5), 961.

Čirjak, D., Miklečić, I., Lemić, D., Kos, T. and Pajač Živković, I. (2022) Automatic pest monitoring systems in apple production under changing climatic conditions. *Horticulturae* 8(6), 520.
Čirjak, D., Aleksi, I., Miklečić, I., Antolković, A., Vrtodušić, R. *et al*. (2023b) Monitoring system for *Leucoptera malifoliella* (O. Costa, 1836) and its damage based on artificial neural networks. *Agriculture* 13(1), 67.

22 How Could Digital Tools Help IPM and Biocontrol Methods to Mitigate the Impacts of Climate Change?

Yelitza C. Colmenarez[1]*, Carlos Vásquez[2], Steve Edgington[3] and Bryony Taylor[3]
[1]CABI Latin America, FEPAF, Botucatu, Brazil; [2]Universidad Técnica de Ambato, Facultad de Ciencias Agropecuarias, Cevallos, Ecuador; [3]CABI, Bakeham Lane, Egham, UK

Abstract
The Latin America and Caribbean region plays a crucial role in global food security but faces significant challenges due to climate change and the threat of extreme weather events. In this context, the adoption of climate-smart agricultural practices is essential. The Plantwise Plus programme supports these efforts by promoting integrated pest management (IPM) and digital agriculture, offering farmers expert advice and resources through a global network of plant clinics. In parallel, early warning systems (EWS), integrating information and communication technology (ICT) tools like Internet of Things (IoT) devices and data analytics are transforming farming practices and enhancing pest and disease management. These technologies enable proactive decision making, adopting sustainable practices and improving resilience to climate change. In coffee production, climate-induced challenges are mitigated through advanced technologies, including event processing systems and expert-based prediction models. The integration of these systems is vital for fostering sustainable agricultural practices and ensuring long-term food security.

22.1 Importance of High Biodiversity in the Neotropical Region and Its Potential for Biocontrol

The Neotropical region, encompassing Latin America and parts of the Caribbean, is one of the world's most biodiverse areas, spanning South America, Central America, Mexico and the Caribbean (Colmenárez and Vasquez, 2024; Ojeda and Novillo, 2024). Renowned for its exceptional richness, this region has evolved through intricate biological processes that have generated unparalleled diversity across all levels of life (Antonelli, 2022). The Neotropics harbour approximately 37% of the world's seed plants (90,000 to 110,000 species), over 4600 bird species, more than 3000 amphibian and reptile species and around 1700 mammal species – exceeding the combined biodiversity of tropical Africa and South-East Asia in several taxa, such as vascular plants, butterflies, amphibians and mammals (Raven *et al.*, 2020; Túnez *et al.*, 2021).

Tropical forests, the most biodiverse habitats on Earth, dominate the Neotropics, with the Amazon Basin alone covering about 5,500,000 km². Additional key regions include the Chocó forests, the severely diminished Mata Atlântica

*Corresponding author: y.colmenarez@cabi.org

 Biological Control Systems and Climate Change (eds Joan van Baaren *et al.*)
DOI: 10.1079/9781800625099.0022

(now only 7% of its original 1,500,000 km²) and the highly diverse Andean forests. In comparison, African rainforests cover roughly 3,900,000 km² (Raven *et al.*, 2020). Despite having a smaller land area than the Afrotropical region, Latin America supports unparalleled species richness, partly due to its tropical forests' vast extent and ecological diversity (León-Cortés and Córdoba-Aguilar, 2024).

This biodiversity is not merely an ecological marvel but also a critical resource for sustainable practices such as biological control. Globally, more than 2000 species of insects and other organisms have been introduced to manage invasive species, achieving partial or complete control of at least 226 invasive insect species and 57 invasive weed species (Heimpel and Cock, 2017; Heimpel *et al.*, 2024). These biocontrol initiatives have substantially benefited environmental conservation, food security, human health and economic well-being (Mason *et al.*, 2023; Colmenárez and Vasquez, 2024).

However, the Neotropics face increasing threats from biodiversity loss, driven by large-scale deforestation, habitat degradation and land conversion for agriculture, cattle ranching and exotic tree plantations (Barlow *et al.*, 2018; Song *et al.*, 2021; Baeza *et al.*, 2022). These pressures are compounded by climate change, which disrupts ecosystems while reducing their ability to mitigate its effects, including carbon sequestration and climate regulation (Bakhtary *et al.*, 2024). Despite misconceptions of tropical forests as pristine wildernesses, human activity has long shaped these ecosystems, and their continued degradation risks a cascade of ecological, economic and social consequences (Fletcher *et al.*, 2021).

This chapter underscores the critical importance of preserving Neotropical biodiversity, for both its intrinsic ecological value and its potential to underpin sustainable biological control practices. It addresses pressing global challenges, including food security, pest management and climate resilience, while also emphasizing the role of ICT tools in enhancing crop management and supporting informed decision making (Kumar *et al.*, 2024).

22.2 Climate-smart Agriculture Approaches: Impact of the Plantwise Plus Programme in the Dissemination of Practices Among Farming Communities

Latin America and the Caribbean (LAC) are known for their diverse agricultural production (including forestry), which has grown at an average annual rate of 2.7%. The region, which holds a third of the world's freshwater resources and a significant portion of arable land, plays a crucial role in global food security. It accounts for 15% of global agricultural exports and is the largest net food exporter worldwide (Soria-Ruiz *et al.*, 2021). However, the Latin America and Caribbean agricultural sector faces many challenges since, on the one hand, chemical compounds are widely used in the region for pest and pathogen control (Colmenárez and Vasquez, 2024) and on the other hand, agricultural economies in the region are threatened by climate change and variability, including rising temperatures, changing rainfall patterns and more intense, more frequent extreme weather events that could affect yields and food security in the region (Prager *et al.*, 2020).

Since climate impacts on crop yields vary depending on crop type, production systems and availability of certain technologies, these impacts are being studied with particular interest in the agricultural sector because of its dependency on climatic stability and its importance in contributing to low-income livelihoods and food security (Banerjee *et al.*, 2021). Thus, LAC faces the challenge of producing enough food while conserving its natural resources and ecosystems and adapting to climate change.

From the analysis presented by Prager *et al.* (2020), adaptation to climate change can be addressed via five points of action:

- inclusion of climate change scenarios as a key element in agricultural ministry and research institute decision-making processes
- support research and adoption of drought- and heat-tolerant crop varieties
- promotion of sustainable irrigation as an effective adaptive strategy

- recovery of degraded lands and sustainable intensification to prevent further deforestation, one of the main causes of climate change in the first place
- implementation of climate-smart practices and technologies to increase productivity while improving adaptability to climate change.

These action points not only allow decision makers and farmers to adapt to changing climates, but practices such as climate-smart agriculture (which can encompass agroecological approaches, integrated pest management) can also improve resilience, yields and incomes. For example, a study in Kenya found that farmers who adopted climate-smart practices such as integrated pest management (IPM), biological control tactics and crop rotation reduced their pesticide costs by up to 70%, leading to higher net incomes (Ahmed *et al.*, 2024). Biological control, a key component of IPM, uses natural predators, parasitoids, pathogens, cultural practices and natural substances to manage pests, thereby minimizing chemical use (van Lenteren and Cock, 2020; Colmenárez and Vasquez, 2024).

According to Rodrigues *et al.* (2023), digital technologies can transform the economy, including food systems, by enhancing productivity and strengthening connections with markets for products and services. In agriculture, digital innovations are revolutionising production models through advanced methods of gathering, storing, handling, transferring and analysing large volumes of data. Farmer adoption of digital agriculture is key to the transition to a productive, sustainable and resilient agriculture, hence the development of telecommunications infrastructure is crucial for inclusive rural development and the provision of public goods (Gemtou *et al.*, 2024). These elements underpin progress in creating public information, online communities and technological applications essential for decision making, technology transfer, market access and the mechanisation and automation of tasks (Sotomayor *et al.*, 2021).

In this context, CABI (Centre for Agriculture and Bioscience International) has established the Plantwise Plus Programme to try to change farmers' appreciation of the IPM approach and at the same time promote farmer adoption of digital agriculture. This programme has been conceived as an initiative to establish an easy-access tool where farmers struggling with plant pests and diseases can get help from local agricultural advisory officers, known as plant doctors, in meeting places known as plant clinics (www.cabi.org/projects/plantwise/).

Since its creation, Plantwise Plus has set up a sustainable network of over 3700 plant clinics operated by national partners in 34 countries worldwide providing diagnoses and management advice for any problem and crop. Plantwise Plus has trained over 10,000 plant doctors and connects them with knowledge resources such as the Plantwise Plus Knowledge Bank, Plantwise Plus Online Management System and national research centres to provide local diagnostic support.

According to Colmenárez *et al.* (2020), the Plantwise Plus approach is based on three interlinked components.

- An ever-growing network of locally run plant clinics, where farmers can find advice to manage and prevent crop problems. Agricultural advisory staff can identify any problem on any crop brought to the clinics, with the support of a national and international network of diagnostic laboratories, and provide appropriate recommendations guided by national and international best practice standards.
- Improved information flows between everyone whose work supports farmers (e.g. extension, research, input suppliers and regulators). Collaboration within national plant health systems enables these actors to be more effective in their work to improve plant health, with concrete benefits for farmers.
- The Plantwise knowledge bank, a database with online and offline resources for pest diagnostic and advisory services, provides both locally relevant, comprehensive plant health information for everyone and a platform for collaboration and information sharing between plant health stakeholders.

Since farmer decision making is a complex and multifaceted process that is influenced by personal, technological, organisational, institutional and political factors (Verburg *et al.*, 2022), the Plantwise Plus Toolkit (https://

Plantwiseplustoolkit.org/) has focused efforts on addressing the real challenges faced by agricultural advisory services and ensures the best possible impact, taking in account the Principles for Digital Development.

- *Understand the existing ecosystem*: digital ecosystems are defined by culture, gender and social norms, political environment, economy, technology infrastructure and other factors that can affect an individual's ability to access and use a technology or to participate in an initiative.
- *Share, reuse and improve*: to share, reuse and improve is, in essence, to collaborate. Collaboration is essential to achieving our shared vision of a more equitable world. We have the most impact when we share information, insights, strategies and resources across silos related to geographies, focus areas and organisations. By sharing, reusing and improving existing initiatives, we pool our collective resources and expertise and avoid costly duplication and fragmentation. Ideally, this leads to streamlined services for people.
- *Design with people*: this means inviting those who will use or be affected by a given technology policy, solution or system to lead or otherwise meaningfully participate in the design of those initiatives.
- *Design for inclusion*: when leveraged intentionally and to its fullest potential, technology can overcome, rather than exacerbate, existing inequality. To design for inclusion is to seize the opportunity for digital initiatives to drive social progress by dismantling systemic barriers related to gender, disability, income, geography and other factors.
- *Build for sustainability*: building for sustainability means thinking about leveraging the inherent scalability of digital technology solutions early on.
- *Establish people-first data practices*: this principle emphisised the need to avoid collecting data that are used to create value (financial or otherwise) for a company or organisation, without delivering any direct value back to those people from whom the data are derived.
- *Create open and transparent practices*: to establish and maintain trust in the digital ecosystem, all people – whether or not they are directly impacted by a given initiative – must have confidence in digital policies, services and systems and the associated data handling.
- *Anticipate and mitigate harms*: several potential harms may arise from any given digital initiative, so to live up to the commitment to do no harm, policymakers and practitioners need to anticipate and work to mitigate harms, even those that originate outside a given initiative.
- *Use evidence to improve outcomes*: over time, practices in monitoring and evaluating technology initiatives have shifted to focus on their impact on people and communities, rather than merely measuring access and usage.

The main driver of the Plantwise Plus concept is its provision of information that farmers need to manage pests on their crops. As a result, Plantwise Plus puts as much emphasis on building capacity for correct pest diagnosis as it does on promoting alternative effective and safe methods that plant doctors can recommend to a farmer to reduce crop losses to pests through interconnected activities in the following areas (Otieno *et al.*, 2021).

- *Plantwise Plus training*: with increased use of digital devices, training on symptom-based field diagnosis of pests and diseases and giving good IPM-based advice has integrated elements on the use of tablet computers and smartphones to access information and capture data. The training on use of digital devices involves a generalised introduction to handling the devices, how the devices can be deployed in specific activities of Plantwise Plus work and their practical use to collect pest data through the Plantwise Plus Data Collection app or to access pest management information through the Plantwise Plus Factsheet Library app.
- *Plant clinics*: plant clinics have been established to show farmers new sustainable ways of managing pests and diseases while maintaining crop productivity (Colmenárez *et al.*, 2022; Colmenárez and Vasquez, 2024). Networks of plant clinics have contributed to the detection of new pest introductions, for example tomato

leafminer (*Tuta absoluta*), fall armyworm (FAW) (*Spodoptera frugiperda*) and maize lethal necrosis disease (MLND), in Africa, Asia and Latin America.

- *Developing open-access information resources*: one of the tools developed for this purpose is the Plantwise Plus Knowledge Bank (www. Plantwise Plus.org/knowledgebank), an interactive tool created to support diagnosis of pests through its extensive content of images and information materials in the form of factsheets that enables users to easily filter the information to prompt decision and action on any pest of interest.
- *Building the capacity of in-country experts in the development of extension material*: one of the uses of pest data by countries in the coming years is likely to be in broader applications in pest forecasting and advisories to farmers. The main kinds of data collected at plant clinics are those on the crops, places of production and the farmers, the diagnosed problems and the recommendations given to the farmers for managing the problems. The data are fed into a repository called the Plantwise Plus Online Management System, which was developed specifically to help countries manage these data coming from the field.

As summarised by Cartmell (2021), social media and ICT tools have become integral to plant health advisory services, enabling faster diagnosis, knowledge sharing and farmer outreach. Platforms like WhatsApp, Telegram and WeChat support horizontal communication among plant doctors and vertical connections with researchers, improving diagnostic accuracy and response times. In Uganda, plant doctors effectively use platforms such as Instagram and Snapchat, while in Ghana Telegram groups provide diagnostic support and training through expert-led sessions. Nicaragua and Honduras highlight how ICTs improve communication among decentralised networks, especially in remote areas. In China and Costa Rica, WeChat and WhatsApp have facilitated mass outreach and diagnostics, respectively. Studies show that these tools reduce the time to deliver pest management advice from weeks to days, preventing major crop losses. However, challenges include poor image quality, high participation costs and unmoderated information, underscoring the need for validation and integration with structured resources like the Knowledge Bank.

Some countries have taken steps to institutionalise Plantwise Plus elements, such as plant clinics and data management, by integrating them into national policies and plans for agricultural extension and crop protection, with dedicated budgets at national and local levels (e.g. Kenya's County Integrated Development Plan) (Otieno *et al.*, 2021). However, the successful scaling of Plantwise Plus innovations depends significantly on the level of ownership by programme beneficiaries.

In Latin America, the implementation of the Plantwise Plus programme has shown farmers from Bolivia and Costa Rica new ways of managing pests in a sustainable way while maintaining crop productivity. This has resulted in a reduction in pesticide use via the incorporation of less toxic active ingredients and sustainable pest management strategies, such as biological control and the use of cultural and ethological practices, by smallholder farmers, in both countries (Colmenárez *et al.*, 2022).

In conclusion, Plantwise Plus has served as both a knowledge base and an intelligence-gathering system for farmers. As a knowledge tool, it provides farmers with guidance on managing pests and diseases affecting their crops and as an intelligence tool, it collects local news and reports on emerging threats where plant doctors, through plant clinics, play a crucial role in utilising and contributing to the system (Beverley and Thakur, 2021). Because all Plantwise Plus innovations were developed with donor funding, they remain freely accessible to current and potential beneficiaries. This ensures equitable access for stakeholders in the agriculture sector, fostering widespread adoption and sustained impact.

22.3 Early Warning Systems in Agriculture: Importance and Examples

Climate change poses a significant challenge to ensuring an adequate food supply to support a growing global population, as it exacerbates the prevalence and impact of pests and diseases in agriculture (Elsayed *et al.*, 2022). Smallholder crop and livestock production systems are particularly vulnerable to climate change and are increasingly experiencing the devastating effects of more frequent and severe extreme weather events (Nakalembe *et al.*, 2021). Addressing these challenges requires innovative solutions, and one such approach is the integration of advanced information and communications technologies (ICT) into farming practices, known as 'smart farming' (Thongnim *et al.*, 2024). This modern agricultural paradigm leverages ICT tools and systems, such as precision sensors, IoT devices, data analytics and real-time monitoring, to enhance crop management and improve decision making (Kumar *et al.*, 2024).

Although these tools may introduce innovations that alter traditional extension approaches to agricultural development by enabling information- and knowledge-based approaches, they must be considered part of the evolving needs for sustainable agriculture (Otieno *et al.*, 2021). By enabling more efficient pest and disease control, smart farming offers a transformative method to adapt to the effects of climate change on agricultural productivity (Kumar *et al.*, 2024; Thongnim *et al.*, 2024).

The early warning system (EWS) is a critical tool for efficiently lowering risks from hazards to agricultural productivity and pests and diseases as it can acquire relevant and timely information in areas where this information or data is unavailable (Elsayed *et al.*, 2022). Early warning systems are essential for mitigating risks from various hazards by predicting or detecting early signs of emerging food safety risks. Systems are then able to issue timely warnings which have become important decision-making aids for authorities and organisations worldwide. Modern systems, powered by real-time, diverse data and advanced AI and machine learning, have shifted from reactive approaches, which address existing incidents, to proactive models that predict future risks (FAO, 2023). Raising awareness and supporting the uptake of these innovative digital tools are vital to strengthening food safety measures.

The transition from agri-enterprises to ICT-based tools should increasingly anchor plant health management as a priority issue in regular practices for sustainable agriculture (Otieno *et al.*, 2021). Accordingly, effective ICT-based tools for pest and disease management should:

- *enhance efficiency and accuracy*: facilitate timely monitoring, enable reliable pest detection and identification, and integrate seamlessly with existing data sources
- *support informed decision making*: provide farmers with access to real-time information on weather, soil, pests and crops, enabling them to make optimal decisions through geo-mapping, crop planning and tailored farm plans
- *improve accessibility*: ensure user-friendly interfaces for field operations, facilitate communication of pest advisories and responses, and enhance access to pest management resources and information
- *promote adaptability*: the ICT tool should be suitable for various cropping systems and environmental conditions and allow for easy interoperability and data integration.

Applying the Internet of Things (IoT) technology in agriculture presents numerous opportunities to revolutionise farming practices. Through smart irrigation, precision farming, crop and soil monitoring, smart greenhouses, supply chain management, livestock monitoring, agricultural drones and pest and disease management, farmers can enhance efficiency, productivity and sustainability. Key findings from this review highlight the transformative potential of IoT technology in agriculture. By enabling informed decision making and optimising resource usage, IoT facilitates precision agriculture, allowing for tailored adjustments to maximise crop yield while maximising inputs. Moreover, cost savings are achieved through efficient resource utilisation and predictive maintenance practices. Enhanced sustainability is promoted through reduced environmental impact and the conservation of resources. Data-driven insights generated by IoT devices enhance farmers' ability to plan and make good decisions. Additionally, supply chain optimisation reduces waste and ensures product quality, while remote management offers convenience and flexibility for farmers, allowing them to monitor and adjust operations from anywhere. These key findings underscore the significant benefits of integrating IoT technology into agricultural practices. In the face of challenges posed by climate change, population growth and resource limitations, leveraging IoT technology in agriculture becomes increasingly imperative. By embracing the opportunities presented by IoT, farming can mitigate risks, optimise processes and establish a more stable food chain.

In eastern Africa, there are two primary early warning systems for invasive flying pests funded by the United Nations Food and Agriculture Organisation (FAO), the eLocust system and the FAW Monitoring and Early Warning System (FAMEWS), both of which rely on a mobile application or dedicated satellite-connected tablet that allows for collecting, processing, storing and disseminating observational data on pest presence and impact across wide areas (Cressman *et al.*, 2016; Buchaillot *et al.*, 2020; Brown *et al.*, 2022). Measurements are made by farmers, agronomists, technicians and others working in agriculture, such as national survey and control officers.

The *eLocust* system is a good example of how the latest technology can be adapted for practical use in developing countries where the valuable data collected, recorded and transmitted provide the foundation for early warning, forecasting and preventive control strategy, allowing intervention to be undertaken in a more efficient and timely matter (Cressman *et al.*, 2016). On the other hand, the *FAMEWS* was developed to gather data on the impacts of FAW on maize crops and intercropping effects. However, the quality of its results was limited when applied only to limited areas in Kenya and some other countries (Buchaillot *et al.*, 2020).

Other prediction tools have been developed, such as the community-based fall armyworm monitoring and early warning system, which has been piloted in five East African countries, funded by the United States via the FAO; the use of species-specific pheromone traps and light lures to warn of a specific moth's presence; static pest distribution maps for the probability of pest presence from various sources; and the Pest Risk Information Service (PRISE) (see next section), supported by CABI, piloted in Ghana, Kenya, Malawi and Zambia (Day *et al.*, 2024).

Although several ICT-based tools are available, governments and international organisations should prioritise the uptake of technology (technology acceptance) by users, taking into account the influence of factors such as technology moderators (complexity, purpose), user's circumstances (individual moderators) and organisational moderators (voluntariness of technology use) (Sun and Zhang, 2006). Concomitantly, some challenges must be overcome, including a lack of ICT infrastructures through which internet-based services can be provided and accessed, low-level interest in utilising agricultural information among rural farmers, inadequate knowledge of rural farmers' agricultural information needs, level of ICT literacy and lack of information on providing agency in rural areas to meet the information needs of rural farmers through internet-based services (Mbagwu *et al.*, 2018).

Additionally, user engagement throughout the entire mobile application development process is critical in the identification of the main system requirements, thus ensuring that mobile applications are effective, about user acceptance and sustained usage, for the entire target audience and not just those who are easiest to reach (Ochilo *et al.*, 2019). As a result, farmers would become more resilient to climate change with positive benefits also contributing to sustainable production practices that are climate-smart, while consumers would have improved access to safer food as a result of production practices that minimise risks such as contamination of agricultural products with pesticides (Otieno *et al.*, 2021).

22.3.1 PRISE: what it is and its impact

The *Pest Risk Information SErvice* (PRISE) is an innovative crop pest risk forecasting product designed for smallholders and commercial producers in developing countries. Bringing together a broad range of stakeholders, including plant protection authorities, space experts, private sector companies and the farmers themselves, PRISE uses state-of-the-art pest modelling techniques to provide users with advanced information on the most appropriate times to act against key pests, allowing them to prepare suitable interventions and mitigate crop losses (Taverner, 2019; Day *et al.*, 2024) (Fig. 22.1).

This EWS unites the ability of the UK partners (CABI and Assimila Ltd) to collect and combine disparate datasets, manipulate data using computational and modelling expertise, and leverage well-established international development networks. PRISE leverages

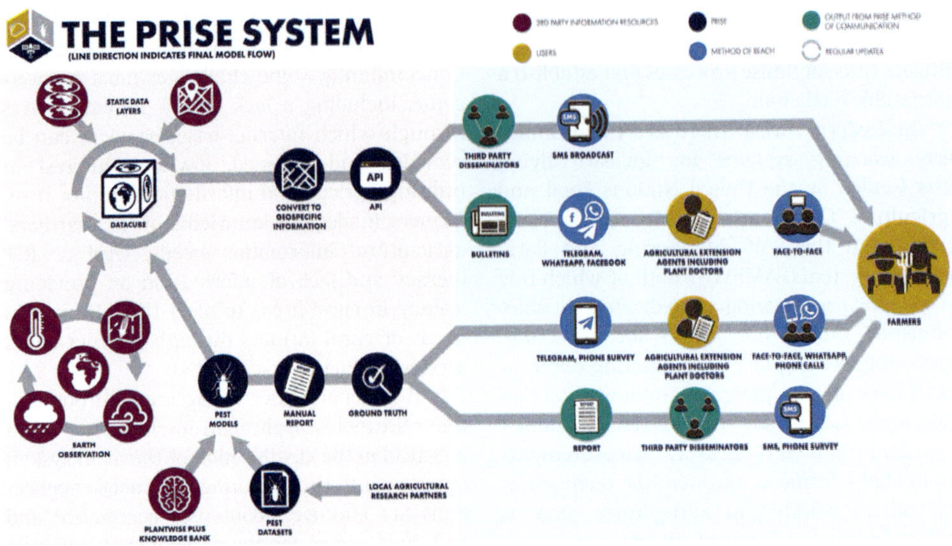

Fig. 22.1. A schematic of the PRISE system. Available at: https://academic.oup.com/view-large/figure/437312261/pmad026_fig2.jpg (accessed 16 June 2025).

advanced space infrastructure and sophisticated algorithms to address real-time pest issues effectively. Pest field data collected locally are used to build and validate models, providing robust information and enabling iterative system improvements. Time-to-act messages (TSA) are combined with information relating to Good Agricultural Practices information and communicated to users through tailored channels, while there is an option to collect user feedback simultaneously to refine the process.

Monitoring and evaluation assess the system's impact against baseline conditions, ensuring its effectiveness. Additionally, capacity-building initiatives and the active involvement of stakeholders from both private and public sectors support the creation of sustainable business plans, ensuring PRISE's long-term viability and widespread adoption (Fig. 22.2).

According to Tambo *et al.* (2023), dissemination of early warning pest alerts in combination with actionable IPM information to smallholder farmers can contribute to the adoption of sustainable crop protection technologies and improve the standard of living of farm households. The use of PRISE has benefited smallholder farmers in Ghana by improving FAW management in maize, combining a FAW time-to-act message with Good Agricultural Practices (Lowry *et al.*, 2022). Exposure to the pest risk information campaign significantly improved the timing of control actions and adoption of multiple IPM practices although it did not affect pesticide use. In addition, households receiving early warning pest alerts achieved a 4% increase in maize yield and were 38% less likely to report hunger. Finally, authors stated out that outcomes were better when women, alone or with their spouses, received the information compared to male-only recipients.

22.4 Coffee Case Studies

Coffee production is highly vulnerable to environmental changes, and reports from the Intergovernmental Panel on Climate Change (IPCC) predict that climate change will significantly reduce global coffee yields and shrink the amount of land suitable for cultivation by 2050 (Bilen *et al.*, 2023). A recent review examined the effects of climate change and variability on coffee production across countries in America, Africa and Asia, which were categorised and described by specific impact groups (Bilen *et al.*, 2023). Analyses revealed a general decline in coffee yields across America, Africa and Asia,

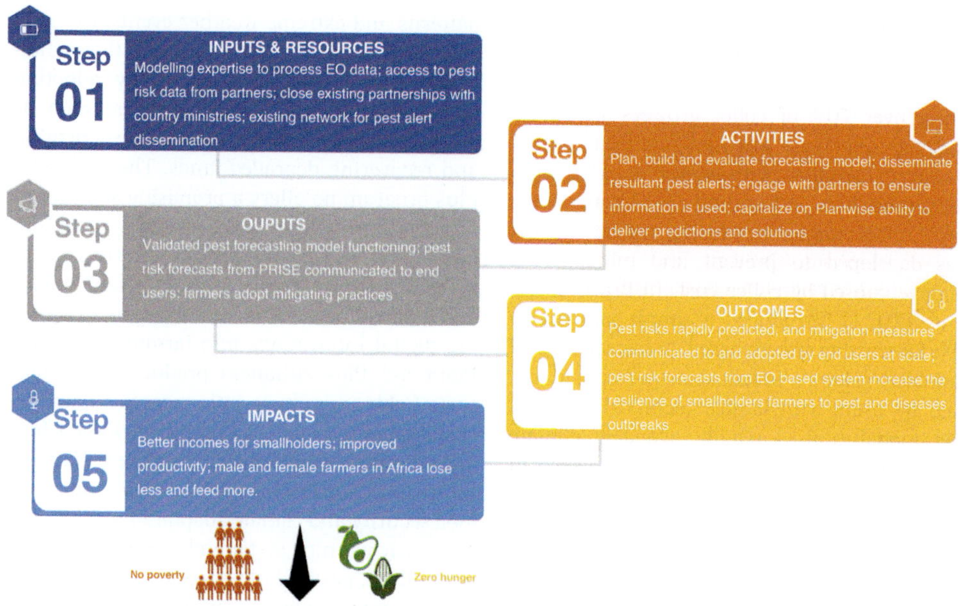

Fig. 22.2. Theory of change of the PRISE system. Reproduced from: www.cabi.org/Uploads/CABI/work-for-us/Pest-Risk-Information-SErvice-(PRISE)-Consultancies.pdf (accessed 16 June 2025).

with losses reaching up to 70% in the Americas. Climate-related factors have indirectly increased the distribution and reproduction rates of pests like the coffee berry borer, white stem borer and nematodes such as *Meloidogyne incognita*, as well as leaf miners in Brazil. Diseases like coffee rust have severely affected regions in Colombia, Central America and Nicaragua, with shorter incubation periods predicted to intensify epidemics.

Coffee rust is the most significant disease affecting coffee crops worldwide since it causes substantial yield losses, mainly in regions with favourable meteorological conditions, such as Colombia and Brazil where production is reduced by approximately 30% and 35%, respectively. To mitigate the effects of coffee rust, numerous studies in computational science have explored the use of expert systems and machine learning techniques to develop effective management and prediction strategies.

Recent research efforts have emphasised detecting coffee rust through climate monitoring and crop management parameters. However, many existing monitoring systems struggle to integrate and analyse data from multiple sources, limiting their ability to effectively identify anomalies and validate generated warnings. Plazas *et al.* (2016) proposed the development of event processing methodologies based on event analysis tools such as the complex event processing (CEP) engine and prediction system integration for EWS applied to coffee rust detection, capable of analysing multiple incoming events from the monitoring system and validating the warnings detection. They evaluated an experimental prototype in a field test, and the results demonstrated a successful integration of the CEP component with the prediction system, validating coffee rust incidence predictions with satisfactory performance in both evaluations. However, upgrading the server hardware could enhance results, as limitations in handling high event volumes caused some reports to be missed during prototype testing.

On the other hand, a rule-based expert system was proposed, where the rules are created based on the expert knowledge of specialists and technical reports about the behaviour of the disease during a crop year. According to the authors, this is a novel approach in this field using not only expert knowledge but also technical reports on the coffee rust problem. This system, designed using the Buchanan methodology, achieved an average accuracy of

66.67% in correctly detecting coffee rust warnings (Buitrón *et al.*, 2019).

In Central America and the Caribbean, where over 70% of coffee growers are smallholders, coffee rust has significantly impacted producer incomes, with losses of 15%–20% during the 2012–13 and 2013–14 campaigns. Thus, a regional early warning system (REWS) was developed to prevent and mitigate the damage caused by coffee rust. In Brazil, Pozza *et al.* (2021) introduced a simple and accessible forecasting system based on meteorological variables making it practical for widespread use, even by users with limited technical training. The model utilises readily available climate data which can be forecasted up to 15 days in advance. This lead time is sufficient for planning fungicide applications, helping producers spray at optimal times, reduce unnecessary treatments, and enhance disease control. The model was validated in two major coffee-producing areas (Carmo do Rio Claro and Nova Resende, Minas Gerais), and demonstrated high accuracy in predicting outbreaks of coffee rust.

In Colombia, Cafexport, in collaboration with scientists from CABI and other UK-based partners, is undertaking a 24-month initiative to support coffee growers (ends March 2026). The project involves collecting data, including temperature, rainfall, coffee berry borer (CBB) population levels and berry maturity, from coffee plantations. This is achieved through a combination of remote sensing technologies and on-site analyses conducted alongside farmers (www.cabi.org/projects/enabling-safe-and-climate-smart-coffee-production-in-colombia/). The collected data will be integrated with satellite-based earth observation data and historical climate trends to develop an early warning alert system for CBB. This system aims to assist farmers in optimising the timing of biopesticide applications, thereby minimising crop losses.

22.5 Conclusion

The LAC region plays a critical role in global food security, but its agricultural sector faces significant challenges due to climate change and overreliance on chemical pesticides. Climate change impacts, including changing rainfall patterns and extreme weather events, threaten yields and food security. To address these challenges, the region must adapt by adopting climate-smart agricultural practices, such as drought-resistant crops, sustainable irrigation and recovering degraded lands. The Plantwise Plus programme offers a promising solution to improve farmers' resilience by promoting IPM and digital agriculture, providing farmers with access to expert advice through plant clinics and a global network of resources. By integrating digital innovations into farming practices, Plantwise Plus enhances productivity, fosters sustainable agriculture, influences the uptake of sustainable methods of control, including biological control, and supports rural development, ultimately contributing to the region's improved food security and climate adaptation.

Early warning systems play a crucial role in agriculture, particularly in addressing the challenges posed by climate change, which exacerbates pest and disease outbreaks, affecting crop productivity and food security. Integrating ICT tools such as IoT devices, real-time monitoring and data analytics is transforming farming practices by enabling proactive decision making and enhancing pest and disease management. Systems like PRISE and eLocust exemplify how technology can empower farmers with timely, actionable information to mitigate risks. However, widespread adoption of these tools is contingent on overcoming barriers such as limited ICT infrastructure, low technology acceptance and insufficient training among rural farmers. Additionally, user engagement and feedback during the development and implementation of these systems are essential to ensure they meet farmers' needs and are sustainable in the long term. By leveraging digital tools for early pest detection and management, farmers can improve resilience to climate change, increase productivity and contribute to sustainable agricultural practices. As these technologies evolve, their integration into national and regional agricultural systems holds the potential to significantly enhance food security and climate adaptability in developing regions.

Coffee production faces significant challenges due to climate change, with coffee rust emerging as one of the most destructive diseases globally. The impact of climate change on coffee yields is exacerbating the spread and severity of pests and diseases, particularly in regions like Colombia, Brazil and Central America. Efforts to manage and mitigate the effects of coffee rust have increasingly incorporated advanced technologies, such as event processing systems and machine learning techniques, to predict and monitor the disease. These systems, including CEP engines and expert systems, have shown promising results in improving prediction accuracy and enabling early intervention. However, challenges remain, such as integrating multiple data sources and system scalability.

In response to these challenges, initiatives such as REWS have been developed to provide a more comprehensive solution by integrating climate, phytosanitary and economic data. These systems aim to reduce rust-related damage significantly, thereby safeguarding the livelihoods of smallholder coffee producers who are particularly vulnerable. The success of such systems will depend on their ability to effectively gather and analyse diverse data and on farmers' adoption of these technologies. By advancing monitoring and early warning systems, coffee production can be better protected against the growing threat of coffee rust and other climate-induced challenges, ensuring more sustainable and resilient agricultural practices for the future.

References

Ahmed, M.M., Sharif, U., Raza, A. and Safdar, M. (2024) Climate-smart agriculture for resilience and profitability. In: Garwi, J., Masengu, R. and Chiwaridzo, O.T. (eds) *Sustainable Practices for Agriculture and Marketing Convergence*. IGI Global, Hershey, Pennsylvania, pp. 27–50.

Antonelli, A. (2022) The rise and fall of Neotropical biodiversity. *Botanical Journal of the Linnean Society* 199(1), 8–24.

Baeza, S., Velez-Martin, E., de Abelleyra, D., Banchero, S., Gallego, F. *et al.* (2022) Two decades of land cover mapping in the Río de la Plata grassland region: The MapBiomas Pampa initiative. *Remote Sensing Applications: Society and Environment* 28, 100834.

Bakhtary, H., Palmegiani, I., Ayestas, M. and Hahn, G. (2024) Biodiversity and climate change action in agriculture and opportunities for building synergies. Available at: https://wwfint.awsassets.panda.org/downloads/biodiversity-and-climate-action-in-agriculture-and-food-systems.pdf (accessed 14 June 2025).

Banerjee, O., Cicowiez, M., Rios, A. and de Lima, C. (2021) Climate change impacts on agriculture in Latin America and the Caribbean: An application of the Integrated Economic-Environmental Modeling (IEEM) Platform. Available at: https://publications.iadb.org/en/climate-change-impacts-agriculture-latin-america-and-caribbean-application-integrated-economic (accessed 14 June 2025).

Barlow, J., França, F., Gardner, T.A., Hicks, C.C., Lennox, G.D. *et al.* (2018) The future of hyperdiverse tropical ecosystems. *Nature* 559, 517–526.

Beverley, C. and Thakur, M. (2021) Plantwise: A knowledge and intelligence tool for food security through crop protection. In: Scott, P., Strange, R., Korsten, L. and Gullino, M.L. (eds) *Plant Diseases and Food Security in the 21st Century*. Springer, Cham, Switzerland, pp. 231–248.

Bilen, C., El Chami, D., Mereu, V., Trabucco, A., Marras, S. *et al.* (2023) A systematic review on the impacts of climate change on coffee agrosystems. *Plants* 12, 102.

Brown, M.E., Mugo, S., Petersen, S. and Klauser, D. (2022) Designing a pest and disease outbreak warning system for farmers, agronomists and agricultural input distributors in East Africa. *Insects* 13, 232.

Buchaillot, M.L., Cairns, J., Hamadziripi, E., Wilson, K., Hughes, D. *et al.* (2020) Multi-scale remote sensing for fall armyworm monitoring and early warning systems. In: *Presented at International Geoscience and Remote Sensing Symposium (IGARSS)*, Waikoloa, pp. 4886–4889.

Buitrón, E.J.G., Corrales, D.C., Avelino, J., Iglesias, J.A. and Corrales, J.C. (2019) Rule-based expert system for detection of coffee rust warnings in colombian crops. *Journal of Intelligent and Fuzzy Systems* 36(5), 4765–4775.

Cartmell, S. (2021) Sustainability of Plantwise: An assessment after 10 years of the programme. Available at: www.cabi.org/wp-content/uploads/Plantwise-Sustainability-Assessment-synthesis-report-_-FINAL.pdf (accessed 14 June 2025).

Colmenárez, Y.C. and Vasquez, C. (2024) Benefits associated with the implementation of biological control programmes in Latin America. *BioControl* 69, 303–320.

Colmenárez, Y.C., Vásquez, C., Fidelis, E.G. and Corniani, N. (2020) Biological control as a key tool for the management of invasive species in Latin America and the Caribbean. In: Chong, P., Newman, D. and Steinmacher, D. (eds) *Agricultural, Forestry and Bioindustry Biotechnology and Biodiscovery*. CABI, Wallingford, UK, pp. 357–386.

Colmenárez, Y.C., Vásquez, C., de Freitas Bueno, A., Cantor, F., Hidalgo, E. *et al.* (2022) Sustainable management of the invasive Tuta absoluta (Lepidoptera: Gelechiidae): An overview of case studies from Latin American countries participating in Plantwise. *Journal of Integrated Pest Management* 13(1), 1–16.

Cressman, K., Van Der Elstraeten, A. and Pedrick, C. (2016) eLocust3: An innovative tool for crop pest control. Available at: www.fao.org/3/a-i6058e.pdf (accessed 14 June 2025).

Day, C., Murphy, S.T., Styles, J., Taylor, B., Beale, T. *et al.* (2024) Forecasting the population development of within-season insect crop pests in sub-Saharan Africa: The Pest Risk Information Service. *Journal of Integrated Pest Management* 15(1), 1–16.

Elsayed, A., Change, C. and Moawad, N. (2022) Framework for intelligent early warning systems of crop diseases. *International Journal of Advanced Trends in Computer Science and Engineering* 11(5), 191–195.

FAO (2023) *Early Warning Tools and Systems for Emerging Issues in Food Safety: Technical Background*. FAO, Rome.

Fletcher, M.S., Hamilton, R., Dressler, W. and Palmer, L. (2021) Indigenous knowledge and the shackles of wilderness. Proceedings of the National Academy of Sciences USA 118(40), 1–7.

Gemtou, M., Guillén, B.C. and Anastasiou, E. (2024) Smart farming technologies and sustainability. In: Lynn, T., Rosati, P., Kreps, D. and Conboy, K. (eds) *Digital Sustainability: Leveraging Digital Technology to Combat Climate Change*. Palgrave Macmillan, Oxford, UK, pp. 99–120.

Heimpel, G.E. and Cock, M.J.W. (2017) Shifting paradigms in the history of classical biological control. BioControl 63(1), 27–37.

Heimpel, G.E., Abram, P.K., Causton, C.E., Celis, S.L., Coll, M. *et al.* (2024) A benefit–risk analysis for biological control introductions based on the protection of native biodiversity. Ecological Applications 34(6), 1–21.

Kumar, V., Sharma, K.V., Kedam, N., Patel, A., Kate, T.R. and Rathnayake, U . (2024) A comprehensive review on smart and sustainable agriculture using IoT technologies. Smart Agricultural Technology 8, 100487.

León-Cortés, J.L. and Córdoba-Aguilar, A. (2024) Conservation and threats that insects face in the neotropics. In: León-Cortés, J.L. and Córdoba-Aguilar, A. (eds) *Insect Decline and Conservation in the Neotropics*. Springer Nature, Cham, Switzerland, pp. 3–13.

Lowry, A., Durocher-Granger, L., Oronje, M.-L., Mutisya, D., Mfune, T. *et al.* (2022) Optimizing the timing of management interventions against fall armyworm in African smallholder maize: Modelling the pattern of larval population emergence and development. Crop Protection 157, 105966.

Mason, P.G., Barratt, B.I.P., McKay, F., Klapwijk, J.N., Silvestri, L.C. *et al.* (2023) Impact of access and benefit sharing implementation on biological control genetic resources. BioControl 68(3), 235–251.

Mbagwu, F.C., Benson, O.V. and Onuoha, C.O. (2018) Challenges of meeting information needs of rural farmers through internet-based services: Experiences from developing countries in Africa. In: *Presented at Transform Libraries, Transform Societies*. World Library and Information Congress, Kuala Lumpur Malaysia, pp. 1–9.

Nakalembe, C., Becker-Reshef, I., Bonifacio, R., Hu, G., Humber, M.L. *et al.* (2021) A review of satellite-based global agricultural monitoring systems available for Africa. Global Food Security 29, 100543.

Ochilo, W.N., Ruffhead, H., Rumsey, A., Chege, F., Lusweti, C. *et al.* (2019) Can you ensure that ICT for development apps are downloaded and used? A case study of the Plantwise data collection app for Plant Health in Kenya. Journal of Agricultural and Food Information 20(3), 237–253.

Ojeda, R.A. and Novillo, A. (2024) Diversity and conservation of Neotropical mammals. In: Scheiner, S.M. (ed.) Encyclopedia of Biodiversity, 3rd edn. Elsevier, New York, pp. 204–222.

Otieno, W., Ochilo, W., Migiro, L., Jenner, W. and Kuhlmann, U. (2021) *Tools for Pest and Disease Management by Stakeholders: A case study on plantwise*. Burleigh Dodds Science Publishing, Cambridge, UK.

Plazas, J.E., Rojas, J.S., Corrales, D.C. and Corrales, J.C. (2016) Validation of coffee rust warnings based on complex event processing. In: Gervasi, O., Murgante, B., Misra, S., Rocha, A. and Torre, M.C. (eds) *Computational Science and Its Applications*. Springer, Cham, Switzerland, pp. 684–699.

Pozza, E.A., dos Santos, E., Gaspar, N., de Souza Vilela, X., de Carvalho Alves, M. *et al.* (2021) Coffee rust forecast systems: Development of a warning platform in a minas gerais state, Brazil. *Agronomy* 11, 2284.

Prager, S., Rios, A., Schiek, B., Almeida, J. and Gonzalez, C. (2020) Vulnerability to climate change and economic impacts on the agriculture sector in Latin America and the Caribbean. Available at: https://publications.iadb.org/en/vulnerability-to-climate-change-and-economic-impacts-in-the-agriculture-sector-in-latin-america-and-the-caribbean (accessed 14 June 2025).

Raven, P.H., Gereau, R.E., Phillipson, P.B., Chatelain, C., Jenkins, C.N. *et al.* (2020) The distribution of biodiversity richness in the tropics. *Science Advances* 6(37), 5–9.

Rodrigues, M., Nunes, S. and Sotomayor, O. (2023) Digitalization in the Agricultural Sector in Latin America and the Caribbean. Available at: https://agriculturadigital.cepal.org/sites/default/files/2024-08/Digitalization%20in%20the%20agricultural%20sector%20in%20Latin%20America%20and%20the%20Caribbean%20-%20Heterogeneity%20and%20public%20policy.pdf (accessed 14 June 2025).

Song, X.P., Hansen, M.C., Potapov, P., Adusei, B., Pickering, J. *et al.* (2021) Massive soybean expansion in South America since 2000 and implications for conservation. *Nature Sustainability* 4(9), 784–792.

Soria-Ruiz, J., Fernández-Ordoñez, Y.M., Medina-García, G., Quijano-Carranza, J.A., Ramírez-Guzmán, M.E. *et al.* (2021) Agriculture in Latin America: Recent advances and food demands by 2050. In: Bochtis, D.D., Pearson, S., Lampridi, M., Marinoudi, V. and Pardalos, P. (eds) *Information and Communication Technologies for Agriculture-Theme IV:Actions*. Springer, Cham, Switzerland, pp. 139–154.

Sotomayor, O., Ramírez, E. and Martínez, H. (2021) Digitalización y cambio tecnológico en las mipymes agrícolas y agroindustriales en América Latina. Available at: www.cepal.org/es/publicaciones/46965-digitalizacion-cambio-tecnologico-mipymes-agricolas-agroindustriales-america (accessed 14 June 2025).

Sun, H. and Zhang, P. (2006) The role of moderating factors in user technology acceptance. *International Journal of Human Computer Studies* 64, 53–78.

Tambo, J.A., Mbugua, F., Duah, S.A., Oppong-Mensah, B., Ocloo, C.Y. *et al.* (2023) Pest risk information, agricultural outcomes and food security: Evidence from Ghana. *Food Security* 15(6), 1667–1683.

Taverner, D. (2019) Pest risk information service (PRISE). Adaptation and resilience. Available at: www.spacefordevelopment.org/catalogue/pest-risk-information-service-prise-cabi/ (accessed 14 June 2025).

Thongnim, P., Yuvanatemiya, V. and Srinil, P. (2024) Smart agriculture: Transforming agriculture with technology. In: Hassan, F., Sunar, N., Basri, M.A., Mahmud, M.S., Ishak, M.H. *et al.* (eds) *Methods and Applications for Modeling and Simulation of Complex Systems*. Springer, Singapore, pp. 362–376.

Túnez, J.I., Nardelli, M., Ibañez, E. and Peralta, D. (2021) A review of the conservation status of Neotropical mammals. In: Nardelli, M. and Túnez, J.I. (eds) *Molecular Ecology and Conservation Genetics of Neotropical Mammals*. Springer, Cham, Switzerland, pp. 335–371.

van Lenteren, J.C. and Cock, M.J.W. (2020) The uptake of biological control in Latin America and the Caribbean. In: Lenteren, J.C., Bueno, V., Luna, M. and Colmenárez, Y.C. (eds) *Biological Control in Latin America and the Caribbean: Its Rich History and Bright Future*. CABI Publishing, Wallingford, UK, pp. 473–508.

Verburg, R.W., Verberne, E. and Negro, S.O. (2022) Accelerating the transition towards sustainable agriculture: The case of organic dairy farming in the Netherlands. *Agricultural Systems* 198, 103368.

23 Involvement of Companies in Biocontrol in the Context of Climate Change

Joan van Baaren*
Université de Rennes, CNRS, ECOBIO (Ecosystèmes, Biodiversité, Évolution), UMR, Rennes, France

Abstract
This chapter reviews products marketed by companies for the biocontrol of pests and pathogens. These products include biostimulants and biofertilizers, biopesticides, semiochemicals and the micro- and macro-organisms involved in biocontrol. They are used in various biological control methods, including classical biological control, biological control by augmentation (inoculative or inundative releases) and conservation biological control. The way in which climate change may affect the production or use of these different products, depending on the biological control method, is then presented. Next, the research avenues developed by companies, or which could be developed in the future, are listed. The results show that it is in the field of conservation biocontrol that the greatest amount of research remains to be developed.

23.1 Introduction

Industrial research into the control of pests, weeds and plant diseases has been based mainly on pesticides since the 1920s, with an explosion following the discovery of DDT (dichlorodiphenyltrichloroethane) in 1939. The period 1950–1970 saw the development of organophosphate products such as malathion and parathion, as well as carbamates, and the rapid growth of multinational agrochemical companies, which were initially based mainly in North America and Europe, followed by a rapid expansion in Asia (Perkins, 1982). The pesticide market, which includes insecticides (to control pests), herbicides (to control weeds) and fungicides (to control crop pathogens), was worth US$62 billion in 2020, 40% of which was spent on herbicides, mainly for field crops such as maize, soya and wheat (IHS Markit, 2020).

However, research is highlighting the dangers of pesticide use for both biodiversity and human health. In Europe, for example, insect populations have declined by 70–80% in recent decades, mainly as a result of intensive farming and the use of pesticides (Jactel *et al.*, 2020). In France, the agency for food, environmental and occupational health and safety (ANSES) has recently recognized that pesticide residues in food increase the risk of cancer and neurological disorders, particularly in children (ANSES, 2021). Therefore, pesticides used in agriculture are under increasing regulatory pressure from new national and international laws aiming at reducing their environmental and human health impact.

*Corresponding author: joan.van-baaren@univ-rennes.fr

 Biological Control Systems and Climate Change (eds Joan van Baaren *et al.*)
DOI: 10.1079/9781800625099.0023

From the 1990s onwards, research focused on developing alternative solutions, which included the development of GMOs and more targeted pesticides for conventional agriculture, marketed by multinationals. More sustainable solutions are also being developed in parallel, including three types of alternative biological control methods: those using biopesticides (products derived from living organisms or their components), those using micro-organisms (viruses, bacteria, entomopathogenic fungi) and finally those using macro-organisms (mainly parasitoid and predatory insects and mites, although it is also possible to use fish, birds and bats) (see parts I and II of this book for more detail). These markets are developing rapidly; the market for biocontrol agents (biopesticides, micro- and macro-organisms) reached US$5.23 billion in 2023 and is expected to reach US$20.4 billion by 2032, with a compound annual growth rate (CAGR) of over 16.4% between 2024 and 2032. However, the global biopesticide market represents only around 8% of the global chemical pesticide market, which is valued at US$62 billion (IHS Markit, 2020). This new market will be accompanied by the creation of a large number of companies specializing in biopesticides, micro- and macro-organisms, for their use in organic agriculture or integrated pest management (IPM) (IHS Markit, 2020).

In addition, during this same period, research showed that the fight against pests, weeds and plant diseases should not consist of a single method but a combination of solutions aimed at strengthening plant resistance (e.g. adding biostimulants, selecting varieties, for example), detecting the arrival of pathogens and pests at an early stage, boosting populations of natural enemies, and acting both locally at field level (above and below ground level) and more widely at farm, landscape and territory levels.

As a result, the range of products offered by companies is constantly growing. They are listed in different categories by organizations involved in food safety and sustainable agriculture (International Code of Conduct on Pesticide Management of the FAO) (EPA, 2021; IHS Markit, 2020) (Table 23.1).

The first category is biostimulants and biofertilizers, which come from natural products with a non-toxic mechanism of action for the environment or human health. These substances are mainly of organic or mineral origin and are applied to either the soil or the foliage of plants. They promote the growth and/or health of plants by stimulating natural processes. Biostimulants are growth regulators including plant hormones (phytohormones) or gas and microbial stimulants. For instance, *Trichoderma* fungi protect plants against fungal diseases by inhibiting the growth of pathogens or stimulating plant natural defences. Biofertilizers directly supply essential nutrients (phosphorus, potassium and nitrogen) to plants or increase their availability. They can be non-microbial, such as plant extracts (seaweed extracts, for example, which contain trace elements, are very commonly used), or humic and fulvic acids, compounds extracted from the breakdown of organic matter that facilitate the absorption of nutrients. They are most often symbiotic microbes, such as mycorrhizal fungi, or bacteria, such as *Rhizobium*, which fix nitrogen. Rhizospheric micro-organisms that act at root level can also improve nutrient absorption by encouraging root development.

The second category is composed of biochemical biopesticides, which are natural substances derived from plants, insects or other biological materials that act by affecting the behaviour, growth or reproduction of pests. Essential oils fall into this category, as do pyrethrins, a natural insecticide derived from chrysanthemums, and pelargonic acid derived from plants in the geranium family. The third category includes semiochemicals, which are substances that provoke behavioural or physiological responses in receptor organisms. These signals can be intra- (pheromones) or interspecific (allomones, kairomones, synomones), for example to lure pests. They can be used in prophylaxis to detect the arrival of pests at an early stage or to control a pest population (sterile male technique, for example). The fourth category consists of living micro-organisms (biocontrol agents, BCA), either natural or obtained by genetic manipulation, which are introduced into the environment to control a population of another living organism considered to be harmful. This category includes microbial agents such as bacteria, fungi, viruses or protozoa, one of the best known being *Bacillus thuringiensis* (Bt). The final last category of products marketed by companies to replace synthetic pesticides is macro-organisms (insects,

Table 23.1. Range of products offered by companies in the field of biological control.

Type of products	Examples	Method of biocontrol	Impact of climate change	Research avenues
Biostimulants (improve plant growth and health), microbiome	Non-microbial: growth regulators, plant hormones or amino acids Microbial: *Trichoderma*, crop microbiome engineering	NA	Duration of product action Quality of product, which can deteriorate rapidly	Evaluation of product resistance to different types of thermal stress Formulation, dosage application methods Encapsulation of active ingredients or microbial agents
Biofertilizers (enable nitrogen to be fixed or nutrients to be better absorbed), microbiome	Non-microbial: seaweed extracts, humic and fulvic acids Microbial: *Rhizobium*, mycorrhizal fungi, crop microbiome engineering	NA		
Biochemical biopesticides	Essential oils, pyrethrins, pelargonic acid, etc.	NA		
Semiochemicals	Pheromones (used in traps), allomones, kairomones, synomones	NA	Duration of release, dispersion Modification of the behaviour of individuals who have to perceive these odours	Study of behaviour at different temperatures and humidity levels
Micro-organisms	Bacteria, viruses, entomopathogenic fungi, yeasts	Inoculation and Inundative biocontrol	Survival, transmission, all micro- and macro-organism life-history traits after release	Discovery of resistance sources to various thermal stresses Selection of resistant strains
Macro-organisms	Insects, moths, predators Parasitoids			
Mix of non-cultivated plants	Intercropping (companion plants), including those favouring plant microbiota, covers, flowering strips, hedgerows, spontaneous or sown grassy margins, etc.	Conservation biological control	Survival, phenology	Growth under different climatic conditions Mix targeted for the natural enemies of pests that should be attracted, and/or for the targeted microbiome
Mix crop plant varieties	Mix of associational cultivars		Different maturity time Plant defences	Thermal stress resistance genes Ancient varieties Adapted tools for sowing and harvesting: drones, sensors, etc.
Mix crop species	Agro-forestry, mixed cropping (legumes, cereals , etc.)			

mites, predators or parasitoids) used in various biocontrol techniques to limit populations of pests, disease vectors and weeds.

Biocontrol using micro- or macro-organisms comprises four main methods. Traditional biocontrol involves introducing natural enemies or exogenous pathogens to control an invasive pest. For example, the introduction of the vedalia beetle *Rodolia cardinalis* (Coccinellidae) to control mealybugs in the United States has been a historic success (DeBach and Rosen, 1991). Inoculative biocontrol involves the periodic introduction of biological agents to maintain their populations at a sufficient level. This method is commonly used in greenhouses for predatory mites such as *Phytoseiulus persimilis*, which control spider mites (van Lenteren, 2012). Inundative biocontrol, on the other hand, is characterized by the massive release of biological agents for an immediate effect, such as the use of entomopathogenic nematodes to control beetle larvae in crops (Ehlers, 2001) or the use of *Trichogramma brassicae* oophagous parasitoids to control the European corn borer (*Ostrinia nubilalis*). Finally, conservation biocontrol focuses on ecological habitat management (such as hedges and flower strips) to preserve and naturally increase populations of indigenous natural enemies by providing them with favourable conditions such as food resources or alternative prey and hosts (Landis *et al.*, 2000).

However, all these alternative solutions are affected by climate change, as has been detailed in the various chapters of this book, whether it be plant defences (Chapter 5), micro-organisms (Chapters 10, 11 and 12) or macro-organisms, including their behaviour and their perception of semiochemicals (Chapters 1, 2, 4, 6–9, 14). In addition, Chapters 3–5 highlight predictions showing that the problems of pests, weeds and plant diseases will very probably be more prevalent in the future than they are at present due to biodiversity loss caused by climate change.

The aim of the following discussion is to identify the research areas and methodologies which companies could develop to promote biocontrol in the context of climate change. In the first part, we cover the impacts of climate change according to the different products marketed and the different biocontrol methods, and in the second, the different research areas in which companies could develop new products.

23.2 Impact of Climate Change on the Types of Products Marketed as Alternatives to Pesticides and Impact on Biocontrol Methods

The impacts of climate change on the various products on the market are summarized in Table 23.1. With regard to biostimulants and biofertilizers used to promote plant growth and resistance, their production will not be directly affected by climate change, although their use may be. Just as the effects of chemical pesticides are affected by climatic conditions (see Chapter 14), natural products, whether applied in the soil, on the roots or on the foliage of plants, and whatever their formulations (liquid or powder), will be affected by climatic conditions (temperature, humidity and CO_2 levels). For example, they can be washed out by heavy rain or be less persistent in the event of a heatwave (Carbó *et al.*, 2018) (see Chapter 14). The same applies to biochemical biopesticides, which will not be affected in their production but in their use, with dosages and application methods that will have to take account of the new conditions linked to climate change. Semiochemicals, whose production should not be affected either, will disperse at different distances or be perceived differently depending on the temperature or humidity of the air (Tobin *et al.*, 2011) (see Chapter 14).

Micro-organisms used as biopesticides will be subject to the same hazards, with their effectiveness and/or persistence affected in the field (see Chapters 10 and 11). It has been shown, for example, that the activity of micro-organisms used in biological control is strongly affected by humidity levels; in the control of thrips such as *Frankliniella occidentalis*, laboratory trials were conducted with three entomopathogenic fungi and their efficacy was maximum for humidity levels above 80%, a level that must be maintained for 25 hours after spraying on the leaves and for even longer periods when the thrips are on the flowers (Wraight *et al.*, 2016).

Finally, several chapters in this book have dealt with the effects of climate change on macro-organisms (see Chapters 1, 2, 4, 6–9) and these effects will be the same for macro-organisms produced for biocontrol when they are released in the field or in greenhouses.

Of the various biocontrol methods using micro- or macro-organisms, classical biological control is the least likely to be affected by climate change. The number of new species introduced outside their area of origin is constantly increasing as a result of the development of world trade and the intensification of intercontinental travel. These introductions are often unintentional, via sea or air transport, but can also be intentional (Hulme, 2009) and the resulting biological invasions pose serious threats to local biodiversity, human health and the economy (Pyšek *et al.*, 2020). Introduced organisms are becoming established in increasing numbers (Seebens *et al.*, 2017), particularly as a result of climate change which creates favourable conditions for these organisms in the area of introduction, but the actual method of introducing a natural enemy into a region where the invasive species is established is unlikely to be affected. Indeed, when choosing species to introduce, it is necessary to carry out studies to check whether the climatic conditions in the area of introduction will be favourable to the natural enemy both at the time of introduction and in the future. It is possible to study the plasticity and adaptive potential of the natural enemy to be introduced (see Chapter 9, for example) and to compare these with future conditions in the area of introduction, as one of the conditions for choosing the species.

Inoculative and inundative biological control are methods that generally involve the marketing of biological control agents, which are mass-produced by companies on alternative prey or hosts or on artificial diets. These methods are mainly used in glasshouses, and more rarely in fields, as the macro-organisms introduced disperse easily, limiting their local action in open fields. Species with low mobility, such as parasitoids belonging to the genus *Trichogramma*, are used successfully in open fields but this is the case for only a few species. On the other hand, their effectiveness has been demonstrated in closed environments such as greenhouses. Climate change may affect this method, as the environment (greenhouses or fields) is subject to the climate and local conditions may become less favourable to the agents released. In particular, heatwaves can reduce the effectiveness of introduced natural enemies, by reducing their reproductive capacities, which is particularly harmful in the case of inoculative releases in which the released organisms must reproduce on their own to increase their populations and play their regulatory role, or by limiting their mobility, attack rate or longevity (Vayssade *et al.*, 2012; Vuarin *et al.*, 2012). In addition, this can affect several generations (Roux *et al.*, 2010), which is problematic because in the strategy of inundative releases, it is often expected that the natural enemies will reproduce and establish themselves for several generations in the release zones.

Conservation biological control will be greatly affected by climate change, as it is an integrative method and all its components may be affected (Chapter 18). Similarly, IPM (see Chapter 14), which uses some of the same methods as conservation biological control, will also be affected by climate change. In particular, conservation biological control relies primarily on crop diversification on a field scale (mix of varieties, mixed crops), and the plants will be affected by climate change, which will have indirect consequences on pests/pathogens and their natural enemies (see Chapter 5). At the level of the natural enemies, research has highlighted differences between natural enemy populations in heat stress resistance capacities (see Chapter 9) (Sun *et al.*, 2017), which could be used in a selection process for heat stress-resistant strains that can be marketed by companies. Besides, this method often uses habitat management by setting up habitats that favour natural enemies (hedges, grassy strips, flower strips). These semi-natural habitats increase natural enemy populations by providing them with nutritive resources, alternative prey/hosts and shelters (Damien *et al.*, 2017; Roudine *et al.*, 2025). These habitats are suffering the consequences of climate change and research is needed to adapt these environments. However, the products marketed as part of this method do not often come from companies specialized in biocontrol.

23.3 Development of Biocontrol Products in the Context of Climate Change

The research avenues for adapting the various products linked to biocontrol on the market to the context of climate change are summarized in Table 23.1.

For biostimulants, biofertilizers and non-microbial or microbial biopesticides, before being marketed or used in a given area, resistance to intense climatic stresses (heatwaves, cold spells, extraordinary levels of humidity or drought) must be measured, and application doses or methods must be assessed including these extreme conditions. In fact, it has been shown that extreme events will affect pests/pathogens and their natural enemies such as micro- or macro-organisms more than an increase in average temperatures (Sunday *et al.*, 2019). This is theorized by the 'climate extremes hypothesis', which highlights the importance of extreme climatic events compared to the mean temperatures in shaping the thermal limits of the species and thus determining their range (Sunday *et al.*, 2019).

One line of research focuses on the development of new biostimulants or biofertilizers to enhance plant resistance to climatic stresses, such as drought or flooding, which has been highlighted for various species of fungi of the genus *Trichoderma*, used as microbial biostimulants (Kashyap *et al.*, 2017). Several traits and genes have been identified that could be modified to improve the beneficial action of these fungi (Kashyap *et al.*, 2017). Another major avenue for development is linked to the holobiont concept, which states that plants and their microbiota should be considered together as a level of biological organization rather than separately (Vandenkoornhuyse *et al.*, 2015). In symbiotic plant–arbuscular mycorrhizal fungi interactions, the fungi supply the plant with up to 80% of the phosphorus it needs and up to 20% of the nitrogen. Plants can select microbial communities that enable them to limit the effects of stresses, including climatic stresses (Begum *et al.*, 2019). A new technology, called 'microbiome engineering', could make it possible to improve adaptation to climate change (Rodriguez and Durán, 2020). By focusing on the microbiomes that exist in extreme conditions, it is also possible to carry out research into transferring these microbiomes to regions affected by climate change in order to improve yields (Rodriguez and Durán, 2020).

Another major avenue of research is the development of nanotechnologies and, in particular, the development of encapsulation (do Nascimento Junior *et al.*, 2021) of biocontrol agents (plant extracts, essential oils, etc.) in nanoparticles, which limits their degradation under abiotic stress conditions, prolongs their duration of action and allows better control of their release into the environment.

Regarding semiochemicals, since the release of odours by plants, whether infested or not, is affected by temperature and humidity conditions, as is the ability of pests and their natural enemies to perceive these odours, the whole system will be affected (Tobin *et al.* 2011). Semiochemicals are mainly used by companies to produce pheromone traps to attract insect pests and thus detect their presence at an early stage. Chapter 14 describes in detail how the effectiveness of these traps will change in the context of climate change. The research avenues concern the adaptation of trap dispensers to increase the duration of odour release into the environment and the resistance of these odours to harmful abiotic conditions.

With regard to the adaptation of micro- and macro-organisms used in biocontrol, a major avenue of research involves the selection of species or, more often, populations/strains of a species adapted to more variable climatic conditions. For example, in the case of entomopathogenic fungi of the *Beauveria* genus, used in biological control of a wide variety of insect pests such as Coleoptera, Lepidoptera and Hemiptera, it has been possible to select lines better able to withstand periods of drought and higher temperatures than the lines initially marketed (Quesada-Moraga *et al.*, 2024). Regarding macro-organisms, parasitoids of the *Trichogramma* genus have been studied the most in order to select populations adapted to different climatic conditions (see Chapter 9).

In the context of biocontrol by inoculation or inundative release, a major area of research has been the improvement of mass production methods and their automation, which has reduced the cost of using these biological control agents. This research has led to the development of large-scale bioreactors for the industrial production of biological control agents. The selection of more productive lines of BCAs was necessary but in the context of climate change, the production of climatic stress-resistant lines should become a priority. This has been the case for both micro-organisms, with work on the rapid multiplication of insecticidal bacteria or entomopathogenic fungi, and macro-organisms, with the organization of mass production lines for generalist predators such as ladybirds or bugs, or more specialist parasitoids such as *Trichogramma* (Grenier, 2009). *Trichogramma* are mass-produced by companies and then mass-released once or several times a season, at a rate of 300,000 individuals per hectare. In Europe, 50,000 ha of maize are treated each year with oophagous parasitoids (*Trichogramma brassicae*) to limit populations of the European corn borer (*Ostrinia nubilalis*).

Still in the context of inoculation or inundative biocontrol, current and future research is focusing on the use of drones to target release sites for biological control agents, so as to optimize the quantity of individuals released (Del Pozo-Valdivia *et al.*, 2021). This approach also requires the development of geolocation technologies. In the context of climate change, they could help to locate the least climatically stressful microhabitats for release in order to reduce mortality.

In the context of conservation biological control, the main factor is plant diversification, which can take place within the crop itself (mixtures of varieties or mixed crops), in the immediate vicinity of the crop (field margins, hedgerows, flower strips, grass strips) or on a wider farm or landscape scale, according to the 13 principles of agroecology of A. Wezel. In all cases, the selection of varieties and species to be planted inside or outside the field must account for local biotic and abiotic conditions and focus as much as possible on indigenous species for greater effectiveness and adaptation to climatic conditions.

Intra-field plant diversification involves, first and foremost, varietal selection, which has been based on a number of criteria including productivity and resistance to pests and diseases but which should now include adaptation to climate change. Research is now focusing on the selection of varieties that can withstand heat and drought while maintaining economically acceptable yields. In the case of wheat, for example, it has been possible to identify variability in resistance to heat stress between lines. More resistant varieties can be selected but current knowledge shows that only temperature increases under 1°C do not lead to significant yield reductions, which shows the limits of this method (Tack *et al.*, 2016). For wheat, maize and rice, drought resistance could be based on varieties with deeper roots that enable them to exploit water contained in the soil at greater depths (Zheng *et al.*, 2022). Research is also focusing on genes regulating stomatal closure to limit water loss during dry periods (Matkowski and Daszkowska-Golec, 2023). With the frequency of fungal and bacterial diseases expected to increase as a result of climate change, research of varieties that are resistant to disease in wetter or more variable climatic conditions is also a major challenge. For example, Xiao *et al.* (2024) recently discovered genes for wheat resistance to drought and cold temperatures in the cultivar G29, which also has genes for resistance to fungal diseases such as powdery mildew and rusts.

Technically, the varietal selection of plants adapted to climate change can be developed thanks to new technologies that make it possible to modify specific genes. For example, genes for resistance to high temperatures and drought can be directly modified using the CRISPR-Cas9 technique (Lu *et al.*, 2024). However, in their review, Daniel *et al.* (2023) highlighted the still unresolved methodological aspects of these techniques and their proven and potential off-target effects, including, for example, the delivery and transformation efficiency or potential off-target effect in the non-coding regions (even if it is the protein-coding genes that are the primary target). In fine, these off-target effects can be at the origin of unintended genetic modifications, which can have unpredictable consequences on crops. Another method is the research of resistance genes, which can be sought in ancient cultivars that are better adapted to water and heat stress (Ceccarelli *et al.*, 2010).

At the intra-field scale, the production of mixed crops, such as wheat with legumes or clover, is developing in various regions. However, these mixed crops require the development of new tools, particularly for harvesting the crops. In the same field, different varieties or different crops can develop at variable rates that are difficult to predict following the increase in climate variability, leading to heterogeneous maturity periods within the field, further accentuating the harvesting problem (Altieri and Nicholls, 2017). Tools that are beginning to be marketed include tractors and combine harvesters equipped with sensors capable of detecting stages of maturity and harvesting only those that are at the right stage or of the right species in the case of mixed crops, even if this requires several passes to harvest. Harvesting robots allow continuous harvesting over a shorter period, which is very useful when the optimum harvesting period is shortened by increasingly variable weather conditions. In the same vein, precision agriculture, based on a whole range of sensors, including air and soil climate sensors, will help mitigate the effects of climate change (Bukhamsin *et al.*, 2025). However, it should be noted that precision agriculture is fossil energy intensive for both its production and its use. In addition, data storage also has a negative impact on the carbon footprint of these new tools. The total carbon footprint of these technologies (production and use) must therefore be studied more precisely in order to determine its negative action on the climate compared to the benefits in terms of adaptation to climate change.

Different types of vegetation can be planted around fields to increase natural enemy populations, by providing them with a stable refuge area during times of ploughing and harvesting, alternative hosts or prey when there are none or not enough on the crop, and food for certain stages, such as nectar or pollen. These types of vegetation include flower strips, herbaceous strips and hedgerows (Heimpel and Mills, 2017).

In the case of flower strips, the selection of plant species to be sown around a field must be adapted to the target crop so as not to allow the recruitment of organisms harmful to the crop (Baggen and Gurr, 1998). For example, it has been shown that mustard planted around a wheat crop can increase the rate of parasitism of cereal aphids located in the wheat, without attracting other aphid species harmful to wheat (Damien *et al.*, 2017). On the other hand, planting mustard near a beet crop would give rise to a high probability of attracting aphid species that can develop on both mustard and beet, such as Myzus persicae Sulzer and Aphis fabae Scopoli (Aphididae). In addition, the species planted around fields should encourage the maintenance and development of the natural enemies of the crop's target pests, by meeting their specific needs, for example by providing nectar, pollen, shelter or more buffered temperatures than in the field (Tscharntke *et al.*, 2005; Damien *et al.*, 2017).

Regarding nectar, its nutritional quality depends on both the flower species, as not all plants produce nectar of identical composition, and the biocontrol agent, as the needs of different species are not always the same. Nectar can be used by parasitoids and is known to increase fecundity and/or longevity in many species (Wäckers and van Rijn, 2012; Damien *et al.*, 2020). For some species, it is the only food resource used in adulthood. Nectar is essentially composed of sugars (sucrose, glucose and fructose) and water, but the volume produced and the concentration, composition and proportion of the different sugars vary greatly between flower species and environmental conditions (Heil, 2011). Depending on the ratio between the quantity of disaccharides and monosaccharides, a distinction is made between nectars known as 'dominant hexose', 'rich hexose', 'rich sucrose' and 'dominant sucrose' (Baker and Baker, 1983), with flowers producing 'dominant sucrose' nectar appearing to have a better effect on parasitoid populations. For example, buckwheat (*Fagopyrum esculentum*, Polygonacea) or cornflower (*Centaurea cyanus*, Asteracea) nectar of the 'dominant sucrose' type has a greater effect on the fitness of many parasitoid species than that of white mustard (*Synapis alba*, Brassicacea), which is more 'dominant hexose' (Damien *et al.*, 2020).

Apart from the quality of the nectar itself, there is the question of its accessibility, as many parasitoids do not have a proboscis (tube-like mouthpart used to suck up the nectar) like bees, and the morphology of the flower may prevent them from accessing the nectar, given the morphology of their mouthparts (short mandibles). The plant species mentioned above have a morphology that allows parasitoids access to the nectar (Russell, 2015; Damien *et al.*, 2020), which is not the case for lacy phacelia (*Phacelia tanacetifolia*), for example. Other macro-organisms used in biocontrol, such as hoverflies and lacewings, feed on nectar when they are adults (Jervis *et al.*, 2005; Lundgren, 2009), and adding flowers to the environment can increase their effectiveness and abundance, whereas their regulatory action is carried out by the larval stages.

In addition, the nectar of a particular flower species can have different effects on different natural enemies. For example, the nectar of *Ammi majus* (L.) (Apiaceae) has a positive effect on the longevity of hoverflies (*Episyrphus balteatus*) and lacewings (*Chrysoperla carnea*) but not on that of the parasitoid *Microplitis mediator* (Haliday).

Finally, it is important to use flower mixtures to spread nectar production over time, to favour natural enemies throughout the growing season, and between crops to keep natural enemies in place before the arrival of the targeted pests. These mixtures must also be adapted to the context of climate change, to the local and future climatic conditions. It is also important that the flowering mixtures chosen must be inexpensive if they are to be adopted by farmers, and must not include plants considered to be weeds, harmful to the crop to be protected.

In addition to flower strips, grass strips can be refuges for predatory natural enemies such as ground beetles, rove beetles and spiders (Tougeron *et al.*, 2022). These stable microhabitats allow overwintering in a general environment disturbed by agricultural practices or thermal stress, in the immediate vicinity of fields. For example, 'beetle banks' are installed in agro-ecosystems with uncultivated grass strips around the edges of fields (Thomas and Marshall, 1999). They may be composed of rye grass (*Lolium perenne* L.) or spontaneous grassy vegetation. This type of habitat is more favourable to ground beetles than hedgerows, which are home to species other than those that predate arable crop pests (Tougeron *et al.*, 2022). Because of the changes in phenology brought about by climate change in both pests and their natural enemies, which, despite an increase in average temperature, also includes extreme events such as sudden, short-lived periods of frost or extreme heat, the importance of these thermal refuge areas is increasing sharply. In western France, for example, a loss of diapause in aphid parasitoids has been observed since 2010. Individuals active in winter can withstand the new average temperatures but remain very sensitive to cold spells (Tougeron *et al.*, 2017) (see Chapter 2).

23.4 Conclusion

A wide range of companies have begun to consider the adaptation of biological control to climate change. These companies are marketing many new products, adapting their production to the new climatic conditions or taking future climatic conditions into account in their production. Not all these companies are necessarily listed as 'biocontrol companies', especially in the field of conservation biological control. Most of the other chapters in this book have shown that adaptation to climate change must combine several different methods to maintain a stable and healthy environment, such as agro-forestry, habitat management to favour biocontrol agents and mass production of these agents in cases where conservation biological control proves insufficient (Altieri and Nicholls, 2017). There is still a wide range of possibilities for companies to develop new products adapted to climate change in this area (see also Chapters 21 and 22). However, when designing these new products, it is important to assess the carbon footprint linked to their production and use to ensure that they have no or little effect on climate change and that they can actually be taken into account for adaptation to climate change.

Acknowledgements

This chapter is produced within the framework of the Biodiversity Climate Chair, of which Joan van Baaren is the holder, funded by the University of Rennes Foundation, with the patronage of the companies OKwind, Giboire, Veolia, Samsic and the Fonds de Dotation Roullier.

References

Altieri, M.A. and Nicholls, C.I. (2017) The adaptation and mitigation potential of traditional agriculture in a changing climate. *Climatic Change* 140, 33–45.

ANSES (2021) Rapport d'activité 2021. Available at: www.anses.fr/fr/system/files/ANSES-RA2021-Rapport2021.pdf (accessed 16 June 2025).

Baggen, L.R. and Gurr, G.M. (1998) The influence of food on *Copidosoma koehleri* (Hymenoptera: Encyrtidae), and the use of flowering plants as a habitat management tool to enhance biological control of potato moth, *Phthorimaea operculella* (Lepidoptera: Gelechiidae). *Biological Control* 11, 9–17.

Baker, H.Q. and Baker, I. (1983) Floral nectar sugar constituents in relation to pollinator type. In: Jones, C.E. and Little, R.J. (eds) *Handbook of Experimental Pollination Biology*. Van Nostrand Reinhold, New York, pp. 117–141.

Begum, N., Qin, C., Ahanger, M.A., Raza, S., Khan, M.I. *et al.* (2019) Role of arbuscular mycorrhizal fungi in plant growth regulation: Implications in abiotic stress tolerance. *Frontiers in Plant Science* 10.

Bukhamsin, A., Kosel, J., Blilou, I. and Salama, K.N. (2025) Accelerating adoption of species-agnostic plant sensors for precision farming. *Nature Reviews Electrical Engineering* 2, 58–70.

Carbó, A., Torres, R., Teixidó, N., Usall, J., Medina, A. *et al.* (2018) Impact of climate change environmental conditions on the resilience of different formulations of the biocontrol agent *Candida* sake CPA-1 on grapes. *Letters in Applied Microbiology* 67(1), 2–8.

Ceccarelli, S., Grando, S., Maatougui, M., Michael, M., Slash, M. *et al.* (2010) Plant breeding and climate changes. *Journal of Agricultural Science* 148(6), 627–637.

Damien, M., Le Lann, C., Desneux, N., Alford, L., Al-Hassan, D. *et al.* (2017) Change in plant phenology during winter increases pest control but not trophic link diversity. *Agriculture Ecosystems and Environment* 247, 418–425.

Damien, M., Llopis, S., Desneux, N., Van Baaren, J. and Le Lann, C. (2020) How does floral nectar quality affect life history strategies in parasitic wasps? *Entomologia Generalis* 40(2), 147–156.

Daniel, M.A., Sebastin, R., Yu, J.-K., Soosaimanickam, M.P. and Chung, J.W. (2023) Enhancing horticultural crops through genome editing: Applications, benefits, and considerations. *Horticulturae* 9, 884.

DeBach, P. and Rosen, D. (1991) *Biological Control by Natural Enemies*. Cambridge University Press, Cambridge, UK.

Del Pozo-Valdivia, A., Morgan, E. and Chandler, B. (2021) In-field evaluation of drone-released lacewings for aphid control in California organic lettuce. *Journal of Economic Entomology* 114(5), 1882–1888.

do Nascimento Junior, D.R., Tabernero, A., Cabral Albuquerque, E.C. d. M. and Vieira de Melo, S.A.B. (2021) Biopesticide encapsulation using supercritical CO2: A comprehensive review and potential applications. *Molecules* 26, 4003.

Ehlers, R.U. (2001) Mass production of entomopathogenic nematodes for plant protection. *Applied Microbiology and Biotechnology* 56(5–6), 623–633.

EPA (2021) Biopesticide Active Ingredients. Available at: gov/ingredients-used-pesticide-products/biopesticide-active-ingredients (accessed 16 June 2025).

Grenier, S. (2009) In vitro rearing of entomophagous insects – past and future trends: A minireview. *Bulletin of Insectology* 62(1), 1–6.

Heil, M. (2011) Nectar: Generation, regulation and ecological functions. *Trends in Plant Science* 16(4), 191–200.

Heimpel, G.E. and Mills, N.J. (2017) *Biological Control: Ecology and Applications*. Cambridge University Press, New York.

Hulme, P.E. (2009) Trade, transport and trouble: Managing invasive species pathways in an era of globalization. *Journal of Applied Ecology* 46(1), 10–18.

IHS Markit (2020) Global Crop Protection Market: Overview and Trends. Available at: com/research-library/global-crop-protection-market.html (accessed 18 June 2025).

Jactel, H., Imler, J.L., Lambrechts, L., Failloux, A.B. and Lebreton, J.D. (2020) Insect decline: Immediate action is needed. *Comptes Rendus Biologies* 343(3), 267–293.

Jervis, M.A., Boggs, C.L. and Ferns, P.N. (2005) Egg maturation strategy and its associated trade-offs: A synthesis focusing on Lepidoptera. *Ecological Entomology* 30(4), 359–375.

Kashyap, P.L., Rai, P., Srivastava, A.K. and Kumar, S. (2017) Trichoderma for climate resilient agriculture. *World Journal of Microbiology and Biotechnology* 33, 155.

Landis, D.A., Wratten, S.D. and Gurr, G.M. (2000) Habitat management to conserve natural enemies of arthropod pests in agriculture. *Annual Review of Entomology* 45, 175–201.

Lu, X., Zhang, F., Zhang, C., Li, G., Du, Y. *et al.* (2024) TaTPS11 enhances wheat cold resistance by regulating source-sink factor. *Plant Physiology and Biochemistry* 211, 108695.

Lundgren, J.G. (2009) *Relationships of Natural Enemies and Non-Prey Foods*. Springer Science and Business Media, Dordrecht, The Netherlands.

Matkowski, H. and Daszkowska-Golec, A. (2023) Update on stomata development and action under abiotic stress. *Frontiers in Plant Science* 14, 1270180.

Perkins, J.H. (1982) *Insects, Experts, and the Insecticide Crisis: The Quest for New Pest Management Strategies*. Springer, New York.

Pyšek, P., Hulme, P.E., Simberloff, D., Bacher, S., Blackburn, T.M. *et al*. (2020) Scientists' warning on invasive alien species. *Biological Reviews* 95(6), 1511–1534.

Quesada-Moraga, E., González-Mas, N., Yousef-Yousef, M., Garrido-Jurado, I. and Fernandez-Bravo, M. (2024) Key role of environmental competence in successful use of entomopathogenic fungi in microbial pest control. *Journal of Pest Science* 97, 1–15.

Rodriguez, R. and Durán, P. (2020) Natural holobiome engineering by using native extreme microbiome to counteract the climate change effects. *Frontiers in Bioengineering and Biotechnology* 8, 568.

Roudine, S., Le Ralec, A., Bouvaine, S., Alford, L., Duval, F. *et al*. (2025) Conservation biological control in winter reduces vector-borne virus incidence in cereal crops. *Agriculture Ecosystems and Environment* 378, 109275.

Roux, O., Le Lann, C., Van Alphen, J.J.M. and Van Baaren, J. (2010) How does heat shock affect the life history traits of adults and progeny of the aphid parasitoid *Aphidius avenae* (Hymenoptera: Aphidiidae). *Bulletin of Entomological Research* 100, 543–549.

Russell, M. (2015) A meta-analysis of physiological and behavioral responses of parasitoid wasps to flowers of individual plant species. *Biological Control* 82, 96–103.

Seebens, H., Blackburn, T., Dyer, E., Genovesi, P., Hulme, P. *et al*. (2017) No saturation in the accumulation of alien species worldwide. *Nature Communications* 8(1), 14435.

Sun, Y., Brönnimann, O., Roderick, G.K., Poltavsky, A., Lommen, S.T. *et al*. (2017) Climatic suitability ranking of biological control candidates: A biogeographic approach for ragweed management in Europe. *Ecosphere* 8(4), e01731.

Sunday, J., Bennett, J.M., Calosi, P., Clusella-Trullas, S., Gravel, S. *et al*. (2019) Thermal tolerance patterns across latitude and elevation. *Philosophical Transactions of the Royal Society B: Biological Sciences* 374(1778), 20190036.

Tack, J., Barkley, A., Rife, T.W., Poland, J. and Lanier Nalley, L. (2016) Quantifying variety-specific heat resistance and the potential for adaptation to climate change. *Global Change Biology* 22, 2904–2912.

Thomas, C.F. and Marshall, E.J. (1999) Arthropod abundance and diversity in differently vegetated margins of arable fields. *Agriculture, Ecosystems and Environment* 72, 131–144.

Tobin, P.C., Zhang, A., Onufrieva, K.S. and Leonard, D.S. (2011) Field evaluation of effect of temperature on release of disparlure from a pheromone-baited trapping system used to monitor gypsy moth (Lepidoptera: Lymantriidae). *Journal of Economic Entomology* 104(4), 1265–1271. DOI: 10.1603/EC11063.

Tougeron, K., Le Lann, C., Brodeur, J. and Baaren, J. (2017) Are aphid parasitoids from mild winter climates losing their winter diapause? *Oecologia* 183(3), 619–629.

Tougeron, K., Couthouis, E., Hecq, F., Barascou, L., Baudry, J. *et al*. (2022) Multi-scale approach to biodiversity proxies of biological control service in European farmlands. *Science of the Total Environment* 822, 153569.

Tscharntke, T., Klein, A.M., Kruess, A., Steffan-Dewenter, I. and Thies, C. (2005) Landscape perspectives on agricultural intensification and biodiversity – ecosystem service management. *Ecology Letters* 8, 857–874.

van Lenteren, J.C. (2012) The state of commercial augmentative biological control: Plenty of natural enemies, but a frustrating lack of uptake. *BioControl* 57, 1–20.

Vandenkoornhuyse, P., Quaiser, A., Duhamel, M., Le Van, A. and Dufresne, A. (2015) The importance of the microbiome of the plant holobiont. *New Phytologist* 206(4), 1196–1206.

Vayssade, C., Martel, V., Moiroux, J., Fauvergue, X., van Alphen, J.J.M. *et al*. (2012) The response of life-history traits to a new species in the community: A story with Drosophila parasitoids from the Rhône and Saone valleys. *Biological Journal of the Linnean Society* 107, 153–165.

Vuarin, P., Allemand, R., Moiroux, J., van Baaren, J. and Gibert, P. (2012) Geographic variations of life history traits and trade-offs in different populations of the parasitoid *Leptopilina heterotoma*. *Naturwissenschaften* 99(11), 903–912.

Wäckers, F.L. and van Rijn, P.C.J. (2012) Pick and mix: Selecting flowering plants to meet the requirements of target biological control insects. In: Gurr, G.M., Wratten, S.D., Snyder, W.E. and Read, D.M.Y. (eds) *Biodiversity and Insect Pests*. John Wiley and Sons, Chichester, UK, pp. 139–165.

Wraight, S.P., Ugine, T.A., Ramos, M.E. and Sanderson, J.P. (2016) Efficacy of spray applications of entomopathogenic fungi against western flower thrips infesting greenhouse impatiens under variable moisture conditions. *Biological Control* 97, 31–47.

Xiao, B., Qie, Y., Jin, Y., Yu, N., Sun, N. *et al.* (2024) Genetic basis of an elite wheat cultivar Guinong 29 with harmonious improvement between multiple diseases resistance and other comprehensive traits. *Scientific Reports* 14, 14336.

Zheng, X., Yu, Z., Shi, Y. and Liang, P. (2022) Differences in water consumption of wheat varieties are affected by root morphology characteristics and post-anthesis root senescence. *Frontiers in Plant Science* 12, 814658.

24 Looking for Methods to Value Ecosystem Services? An Economic Toolkit with a Focus on Biocontrol

Marie-Hélène Hubert* and Julie Ing
Université de Rennes, CNRS, UMR, Rennes, France

Abstract
In the context of climate change and the alarming loss of biodiversity, there is an urgent need for policy makers to consider the environmental impact of any policy. As a result, economic environmental valuation is playing an increasing role in decision making. The aim of this chapter is to discuss why and how we value ecosystem services. We first review the theoretical background of environmental valuation. We then explore the empirical literature on environmental valuation, which can be divided into two general approaches: revealed preference and stated preference. While revealed preference approaches examine the actual behaviour of users of ecosystem services, stated preference approaches are survey based and elicit people's intended future behaviour. Finally, we discuss research perspectives and the difficulties that need to be overcome in order to economically value the ecosystem service of biocontrol.

24.1 Introduction

A variety of factors, including climate change, overuse of natural resources, changes in land and sea use, pollution, intensification of agriculture and the introduction of invasive species, have led to disruptions in ecosystem functioning. Van Klink *et al.* (2020) indicate a 10.56% decline in terrestrial insect abundance per decade. This observation is particularly concerning given the vital role insects play in providing, regulating, and maintaining numerous ecosystem services, including pollination, biological control and food provisioning (Noriega, *et al.*, 2018). A key strategy to address this issue is to reduce intensive agriculture and pesticide use. A recent study in Martinique demonstrated an average 21% reduction in biodiversity in banana plots frequently treated with glyphosate (El jaouhari, *et al.*, 2023).

Beyond the concerns about biodiversity losses for their own sake, the global decline of biodiversity and ecosystem services is also an economic issue. According to Losey and Vaughan (2006), insects provide ecosystem services worth at least $57 billion per year in the United States. Karp *et al.* (2013) estimated the economic benefit of natural biological control of coffee pests at US$75–310 per hectare per year, focusing on pest control services provided by native birds and bats to tropical coffee plantations. This underscores the importance of

*Corresponding author: marie-helene.hubert@univ-rennes.fr

DOI: 10.1079/9781800625099.0024

implementing public policies to safeguard biodiversity. To develop such policies, it is essential to assess their costs and benefits and assign a monetary value to ecosystem services. However, economically valuing these services raises a number of issues.

As De Groot *et al.* (2012) noted, the majority of the value of ecosystem services lies outside the market and should be regarded as non-tradable public benefits. This suggests the potential absence of a market price for these assets. Furthermore, in instances where market prices are present for ecosystem services, their full value may not be reflected in these prices. Ecosystem services can be regarded as positive externalities, meaning that they provide a positive benefit to social and economic activities but the cost of maintaining or enhancing them is not considered in their direct costs. A classic example of positive externalities is the pollination process created by a honey producer's bees. Another example is the use of biocontrol agents that decrease the need for pesticides, which has a positive impact on ecosystems (Nicholson *et al.*, 2024) and human health. The absence of a price, or the existence of a market price that does not reflect the full value of ecosystem services, may imply that these services are not fully considered by decision makers. For instance, to implement augmentative biocontrol programmes, natural enemies can be purchased directly from various suppliers, but the majority of natural enemies are not included in the market. Conversely, conservation biological control does not depend on a market for natural enemies. Instead, it relies on a variety of conservation measures, such as hedges or flowering strips, to increase the abundance and diversity of natural enemies. Even if a price is established for natural enemies, it does not accurately reflect the positive benefits of their use on biodiversity, water and soil quality, adjacent crops or worker health.

Economic value can be used to reflect the importance of ecosystem services in monetary terms (Small *et al.*, 2017). Despite the criticisms and limitations of this approach (McCauley, 2006), assigning economic value to these services is crucial. As Munns and Rea (2015) explain, when values are quantified in monetary terms, the benefits of competing alternatives can be compared directly with the costs of implementing those alternatives.

Economic value is derived from individual preferences, which are influenced by individual needs, perceptions and worldviews, as well as the scarcities imposed by nature. Preferences in economics refer to the way in which individuals rank different outcomes based on their personal values and criteria. These preferences, along with factors like income, price, and technology, influence production and consumption decisions. For instance, individual preferences for an agricultural product may be influenced by various attributes, such as production techniques (organic or intensive farming), visual quality or product origin. Similarly, farmers' decisions regarding crop protection methods, such as biocontrol vs pesticides, are influenced by a variety of constraints, including income, labour availability, technological factors and time. Farmers' choices are also influenced by their personal beliefs. For instance, farmers' decisions regarding diversified farming practices, including reduced tillage or the use of flower strips, are influenced by their perceptions of how these practices affect the gross margin (Rosa-Schleich *et al.*, 2024). Likewise, farmers' decisions to reduce pesticides may be influenced by their perceptions of environmental and health impacts (Bakker *et al.*, 2021; Finger and Möhring, 2022).

By its very nature, economic value is an anthropocentric value. However, this does not imply a disregard of other species or future generations, as humans have the capacity to integrate their well-being into their own preferences. This approach emphasizes an instrumental value defined exclusively in terms of its impact on the well-being of individuals. In contrast, the concept of intrinsic value posits that its worth is inherent in environmental goods, independent of human preferences. In this chapter, we will use the anthropocentric approach.

As environmental economic values are typically defined in monetary terms, they can be incorporated into a cost–benefit analysis.

This analysis has been employed by public authorities in making decisions regarding major investment projects (e.g. dam construction). There are numerous discussions on the application of cost–benefit analysis to biodiversity (see Wegner and Pascual, 2011; Nyborg, 2014). Nevertheless, only a few studies have applied it to ecosystem services, and even fewer have applied it to biological control, with the exceptions of De Groote *et al.* (2003), Daniels *et al.* (2017), Valente *et al.* (2018) and Chakir *et al.* (2016). In cost–benefit analysis, gains and losses are defined in terms of increments and decrements of human well-being, which are measured as follows: individuals' willingness to pay for an improvement or willingness to pay to avoid a change; or willingness to accept compensation to tolerate a change (Pearce, 1998). For instance, one can estimate individuals' willingness to pay for a theoretical bee protection policy (Mwebaze *et al.*, 2018), landowners' willingness to accept pesticide reduction (Neves do Prado and Abildtrup, 2022) or consumers' willingness to pay for pesticide reduction (Bazoche *et al.*, 2014). To our knowledge, willingness to accept the implementation of flower strips has never been estimated.[1]

The objective of this chapter is to understand why and how economists value ecosystem services. Before presenting the methods used to value ecosystem services, it is essential to understand why public authorities need to promote ecosystem services and to present the various policies they may use to that purpose. The outline of this chapter is as follows. The first section presents the economic approach to ecosystem services. This section presents the characteristics of ecosystem services that make the intervention of public authorities necessary, and briefly explains key concepts such as total economic value, opportunity costs and discounting that are central in economic valuation of ecosystem services. The second section will present different economic valuation methods, using the classical distinction between revealed preference methods and stated preference methods. Each method will be briefly described and illustrated with one case study.

24.2 Why Should We Assign an Economic Value to Ecosystem Services?

24.2.1 Environmental valuation: a tool for public policies

Theoretical background for environmental policies

The valuation of ecosystem services is related to two complementary approaches in economics. The positive approach aims to explain the existing world, while the normative approach provides prescriptions or recommendations based on value judgements and describes the world as it should be. This second approach defines the optimal outcome, that is, the allocation chosen by a social planner who is constrained only by the endowment of resources. When applied to ecosystem services, this process involves first assessing the value of these services using a positive approach and then defining appropriate public policies using a normative approach. Since the economic value of ecosystem services is typically not reflected in markets, government intervention is essential to ensure decision makers act in the common interest by internalizing externalities. Economists usually classify goods and services based on whether people can be prevented from consuming them (excludability) and whether one person's consumption reduces their availability for others (rivalry or subtractability). Based on these two characteristics, four categories of goods are defined, as shown in Table 24.1.

This classification clarifies the role of public authorities in providing or safeguarding specific types of goods. Economists concur that public goods tend to be underprovided. Indeed, there is an incentive to free ride. Free riding occurs when individuals benefit from a resource without paying for it or by paying only a portion of the cost. It is now well established that common-pool resources are generally overused and subject to the tragedy of the commons, as described by Hardin (1968). This tragedy illustrates the tension between short-term self-interest and long-term collective well-being. It asserts that decisions made by individuals acting in their own self-interest are likely to result in the depletion of the resource.

Table 24.1. Classification of goods and services.

	Excludable	Non-excludable
Rivalrous	Private goods (agricultural products from private land)	Common pool resources (fisheries, groundwater basins)
Non-rivalrous	Club goods (private wildlife parks and conservation reserves)	Public goods (biological control, pollination, biodiversity conservation)

Some ecosystem services fall into the categories of public and common goods, and thus require public intervention to provide or protect them. According to Costanza (2008), the categorization of ecosystem services can be outlined as follows: the majority of provisioning services are classified as private goods, or common goods in the absence of clearly defined property rights. Most regulatory and cultural services are public goods, while some recreational services are club goods. Biocontrol falls into the category of public goods, which can hinder its development. The cost of implementing biocontrol programmes, such as purchasing biocontrol agents or investing in agroecological infrastructure, is borne by farmers, while most benefits accrue to society as a whole. In contrast, hedges and field edges are private goods when located on private land, yet they generate positive externalities by supporting multiple ecosystem services and enhancing the aesthetic appeal of the landscape.

Examples of environmental policies

Environmental valuation is necessary for formulating effective public policies. It is essential for determining the optimal level of ecosystem services (e.g. conservation and land use policies) prior to designing a mechanism that will ensure this level is implemented (e.g. payment for ecosystem services). Furthermore, the valuation of ecosystem services is crucial for determining appropriate actions in the event of environmental damage, as well as for preventing such damage from occurring (e.g. environmental damage compensation).[2]

CONSERVATION AND LAND USE POLICIES. Land abandonment, urbanization and agricultural intensification are the most significant drivers of ecosystem changes (Pereira *et al.*, 2022). It is imperative for land use planning to quantify the changes in ecosystem services resulting from different policy options and the impacts of these changes on society. Defining land use policies requires comparing the value of certain ecosystem services against trade-offs with other valuable land uses. Bateman *et al.* (2013) demonstrate that land use decisions focused solely on agriculture reduce overall ecosystem service values. Integrating a broader range of services, such as greenhouse gas emissions and sequestration, open-access recreational visits, urban green spaces and biodiversity (including natural pest enemies and pollinators), into land use planning can significantly enhance total value, encompassing both market and non-market goods and services. Preserving biodiversity also depends on effective agricultural land management. Using a bio-economic model, Kirchweger *et al.* (2020) show that increasing field sizes and loss of permanent green edges substantially reduce both pollination and agricultural gross margins. However, this loss is more than offset by the economic gains from field enlargement.

PAYMENT FOR ECOSYSTEM SERVICES. As Heal (2000) explained, to conserve systems, we must make conservation more attractive than any other uses. To achieve this, it is essential to translate the social importance of ecosystem services into financial benefits and ensure that these benefits accrue to the owners of the ecosystems as a reward for their conservation efforts. The valuation of ecosystem services can serve as a foundation for designing payment for ecosystem services, such as agri-environmental schemes. These programmes serve as policy instruments, aiming to achieve conservation objectives through incentive mechanisms. Specifically, agri-environment schemes are government programmes designed to assist farmers in managing their land in an environmentally friendly manner. These programmes are crucial for preserving farmed environments of high nature value, fostering improved genetic diversity and safeguarding agro-ecosystems.

To develop effective agri-environment schemes, it is crucial to understand farmers' preferences regarding payment stability (Bougherara *et al.*, 2021), the adoption of action- or result-based schemes (Canessa *et al.*, 2023) and the role of collective bonuses (Kuhfuss, *et al.*, 2016). By comprehending farmers' preferences and the impact of contracts on the extensification and provision of ecosystem services, policy makers can design more effective contracts and enhance farmers' participation (Bartolini *et al.*, 2021). Furthermore, understanding public preferences regarding these schemes (types of agro-environmental measures, conditions of payment, responsible institutions, intervention sites) is necessary to identify solutions to finance them (Lévesque *et al.*, 2021).

ENVIRONMENTAL DAMAGE COMPENSATION. Environmental valuation can be a useful tool for compensating environmental damage or preventing further damage. The EU's Environmental Liability Directive (Directive 2004/35/CE) provides a precise definition of environmental damage, categorizing it as 'a damage to protected species and natural habitats, water and land, if its contamination threatens human health'. Kermagoret *et al.* (2016) categorize compensation options into three types: monetary compensation, investment in publicly owned assets or ecological restoration. However, biodiversity compensation can take the form of either or both monetary and environmental units (Gastineau and Taugourdeau, 2014). The Environmental Liability Directive is based on the polluter pays principle, meaning that polluters are responsible for implementing the necessary preventive and remediation actions and paying for their costs. Determining the appropriate compensation for environmental damage necessitates the valuation of ecosystem services. This compensation can be provided either *ex ante*, in advance of an action's potential impact, or *ex post*, in response to an accident's aftermath.[3]

24.2.2 Key concepts for valuing ecosystem services

Total economic value

To value ecosystem services, economists use the concept of total economic value. This concept, pioneered by the London School of Economics 30 years ago, offers a comprehensive synthesis of environmental economics' efforts to define the various values associated with the environment (Plottu and Plottu, 2007). Total economic value is divided into use values, comprising direct, indirect and option values, and non-use values, which include existence and bequest values (Fig. 24.1).

The use value incorporates tangible elements. Direct use value refers to the value of all goods and services derived from the direct use of ecosystems, consumptive use of resources (e.g. provisioning services such as agricultural products) or non-consumptive use of services (e.g. landscape amenity). Indirect use value is derived from the functioning of ecosystems underlying direct use activities, such as regulating services (e.g. pollination and natural enemies

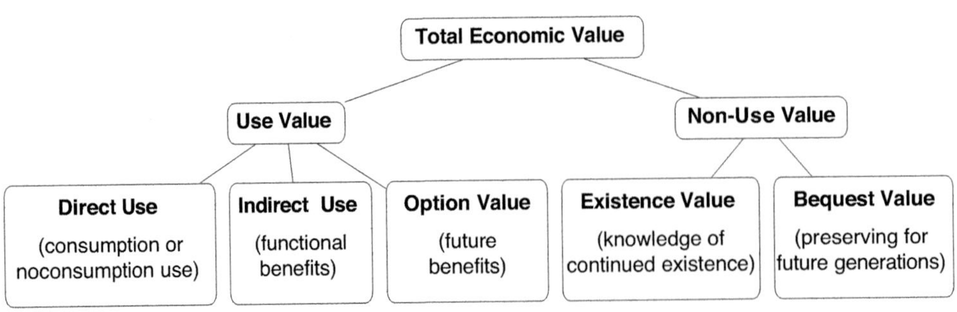

Fig. 24.1. Different components of the total economic value. Adapted from Pearce (1994).

of crop pests). The concept of option value was introduced by Weisbrod (1964). It represents the value that people place on the option to use, directly or indirectly, a resource or service in the future, even if not currently in use, for instance, the benefit from knowing that a field is being preserved from pesticides so that natural enemies of pests that live there can be used later and soil quality can be preserved.

Non-use value, on the other hand, is associated with the mere existence of a resource or service, irrespective of its direct or indirect use. Notable non-use values include the existence value developed by Krutilla (1967) and the bequest value. Existence value reflects the satisfaction derived from the knowledge that a natural element will be preserved, irrespective of its present or future utilization. For many animals, existence value constitutes a significant portion of their total economic value. For instance, Kontoleon and Swanson (2003) estimate that the existence value of the giant panda, defined as the willingness to pay (WTP) for the conservation of 500 pandas within their natural habitat, represents 73% of the total WTP. In contrast, bequest value refers to the value that individuals gain from being able to pass a resource on to future generations. For instance, a farmer's willingness to pay to preserve a pesticide-free field for future generations exemplifies bequest value.

Opportunity cost

Determining the value of ecosystem services requires consideration of the interactions between them, as a single natural asset can meet numerous needs. For instance, in addition to producing food, fibre and fuel for human needs, agriculture has a significant impact on carbon sequestration, flood control, groundwater recharge, soil conservation, biodiversity preservation, scenic vistas and purifying water, soil and air (Lv *et al.*, 2010).

The concept of opportunity cost, a central tenet in economic thinking, can provide a useful framework for evaluating ecosystem services. In economics, the opportunity cost of a choice is defined as the value of the best alternative foregone when a choice must be made between several mutually exclusive alternatives given limited resources. To illustrate, the opportunity cost of implementing flower strips in conservation biological control includes the benefits from farmlands that would have been received if those farmlands had been used for other purposes. To ensure accurate valuation, it is crucial to distinguish the full range of complementary and competitive ecosystem services before any aggregation (Turner *et al.*, 2003). To assess opportunity costs, it is necessary to consider the spatial aspects of ecosystem services (Costanza, 2008). Indeed, the benefits and costs associated with ecosystem services can occur at different scales. For instance, the benefits of ecosystem services such as carbon sequestration occur at regional and global scales, while the opportunity cost of ecosystem services, such as foregone agricultural expansion, is experienced at the local level (Ruijs *et al.*, 2017). Similarly, while some benefits of biological control are local, such as health and landscape improvements, others are global, like the contribution of pesticides to climate change. Some of the costs associated with biological control are also local, such as the cost of planting a flower strip or the foregone revenue from farming. However, other costs are more broadly felt, such as the introduction of invasive species during augmentative biological control.

Discounting

Time and its valuation are essential to the calculation of ecosystem services values and to define public policies. Most environmental policies involve a trade-off between short-term costs and longer-term benefits. The environmental benefits of these policies depend on the lifespan of pollutants, and these benefits may extend to several generations, as is the case for climate change or soil pollution.

Economists employ the concept of discounting to make benefits and costs occurring in different time periods comparable by expressing their values in present terms. The selection of discount rate can have far-reaching implications in cost–benefit analysis and public policy evaluation, particularly in the context of long-term policies such as climate policy. Costanza *et al.* (2021) propose a strategy of adjusting discount rates based on the type of capital involved, categorizing it as built (e.g. infrastructure, machinery, transportation), human (e.g. human beings and their attributes), social (e.g. social networks) and natural (e.g. natural environment and biodiversity). Utilizing this approach, they estimate the total net present value of global ecosystem services to be within the range of US$3.8–5.0 quadrillion ($\times 10^{15}$), which is approximately equivalent to twice the net present value of Gross Domestic Product.[4]

Discounting is a critical component of the valuation of biocontrol programmes. This is due to the fact that some costs are incurred in the present, such as investments in agro-ecological infrastructures, while benefits are realized over time, such as improved soil quality.

24.3 How Should We Assign an Economic Value to Ecosystem Services?

The objective of this section is to present the different methods developed by economists to characterize the economic value of ecosystem services in terms of willingness to pay for a gain in biodiversity or environmental quality, to avoid a loss in environmental quality or biodiversity, or to accept compensation to tolerate a loss or go without a benefit. The willingness to pay (WTP) is the minimum amount individuals are willing to pay for an additional unit of biodiversity or environmental quality, or to preserve natural resources. The willingness to accept (WTA) refers to the minimum amount of money individuals would be willing to allocate for more environmentally friendly behaviours.

The environmental valuation literature can be divided into two general approaches: revealed preference and stated preference. The unifying characteristic of the former (also referred to as market-based approaches) is that the valuation of non-market goods and services is inferred by observing individuals' actual behaviours and, in particular, purchases in actual markets (Mendelsohn, 2019). For instance, paying a price premium for a house close to a park reveals higher preferences for ecosystem services. These approaches are used to estimate the direct and indirect use values of ecosystem services (see Fig. 24.1). As these approaches are based on observing individuals' behaviours, they are particularly well suited for conducting post-policy evaluations.[5]

Conversely, stated preference approaches entail the formulation of a hypothetical scenario, delineating the ramifications of an environmental policy, such as the establishment of protected areas or the implementation of an environmental policy (Hanley and Czajkowski, 2019). Individuals express their intended future behaviour by participating in a survey. These approaches are particularly useful in estimating the values of ecosystem services for which observational data are not available. This ability to estimate a broad spectrum of ecosystem service values, ranging from use values to non-use values, is a key strength of stated preference methods (Johnston *et al.*, 2017). Furthermore, they are the only viable approach to run both *ex ante* and *ex post* policy evaluations.[6]

24.3.1 Revealed preference approaches

The revealed preference approaches include three main methods: (i) hedonic pricing, (ii) travel cost, and (iii) production function.[7] Each of these methods involves observing individuals' behaviours to infer the value of ecosystem services. For instance, the hedonic pricing method involves observing the purchase of durable goods, such as property, while the travel cost method focuses on travel expenditures. Each method estimates the WTP or WTA by conducting an econometric analysis with the objective of quantifying the marginal effect of a change to ecosystem services on the observed behaviour.[8] In the following paragraphs, we provide a general description of each of these methods and offer some examples of domains of application. We then illustrate the methodology with one case study, focusing on examples related to climate change, biodiversity and the adoption of sustainable pesticide use practices.

Hedonic pricing

The hedonic pricing method is typically, though not exclusively, employed in the property market, where a significant portion of ecosystem services is implicitly traded. For instance, households may be willing to pay a premium for a property located in a suburb with biodiversity, such as the former beltline railbed in Paris, which has been gradually converted into an urban walkway. This approach aims to assess the environmental quality of housing by conducting an econometric analysis of property markets using two comprehensive databases. The first database provides information on house property prices and characteristics. The second database describes ecosystem services, such as biodiversity levels, the presence of invasive species and exposure to flood and storm risks. The econometric analysis aims to identify the marginal impact of changes in ecosystem services on property prices. This marginal effect is interpreted as the WTP or the WTA.

A key strength of hedonic pricing lies in its ability to capture both direct and indirect values (see Fig. 24.1), making it a highly versatile tool. For instance, it has been used to estimate the WTP for preserving natural resources such as marine resources (Gopalakrishnan *et al.*, 2011; Asche *et al.*, 2021) or endangered species (Kontoleon and Swanson, 2003). Other studies have estimated the WTP for air quality (Carriazo *et al.*, 2013; Nam *et al.*, 2022) or for amenities in urban areas (Mei *et al.*, 2017; Jensen *et al.*, 2021). Other studies have estimated the external cost of invasive species in urban areas. For instance, Isely *et al.* (2017) have quantified the external costs of invasive *Phragmites* in Michigan's lower Grand River watershed (or, in other words, the WTP of the individuals for biodiversity). Their findings indicate that property values rise by $3.90 for each metre a property is situated farther from *Phragmites*. Finally, some studies have estimated the social cost of extreme events, such as storms or floods (Zivin *et al.*, 2023). This can be interpreted as the WTP of households to avoid climate damages or their WTP for climate policies.

To illustrate this methodology, let us examine the study by Cobián Álvarez *et al.* (2019). The primary objective of this study was to estimate the WTP for flood protection in Jakarta, Indonesia. To this end, the authors collected data from the 2007 Jakarta flood from the United Nations Department of Safety and Security (UNDSS). The data on housing prices and other housing characteristics (number of rooms, house size, wall, roof and floor materials) were extracted from the Indonesian Family Life Survey. The study also incorporated additional data to contextualize the neighbourhood characteristics, including transport access and unemployment rates. The econometric analysis revealed that a 1% rise in floodwater levels was associated with a 0.124% decrease in monthly housing rental prices. Based on this finding, the researchers calculated the total WTP of all households in Jakarta for eliminating flooding on an annual basis, which is approximately 8.1% of Jakarta's Gross Domestic Product. This suggests that households in Jakarta would be willing to allocate up to 8% of Jakarta's GDP to invest in climate policies.

Travel cost method

The travel cost method is a cost–benefit analysis that quantifies the value of ecosystem services. This method is used to determine the value of the recreational benefits of ecosystem services. While access to these areas is typically free, users still incur a cost in the form of travel expenses. For instance, natural areas such as beaches, woodlands, forests and heritage sites are often the target of recreational trips. The travel cost method utilizes costs associated with travelling to and within a site, such as the cost of time, as a proxy for the price of recreation. Specifically, individuals expend resources on travel and other expenses to access these sites, providing insights into the perceived value they place on the recreational experience. The econometric analysis conducted to estimate WTP or WTA leverages survey responses. The travel cost method is primarily employed to estimate the direct use value for a wide range of natural areas. For instance, it has been used to estimate the recreational benefit of beaches (Bonhomme and Schaffar, 2021). Other studies have focused on the value of biodiversity through birdwatching (Czajkowski *et al.*, 2019; Jayalath *et al.*, 2023).

We illustrate this method by drawing on the study by Cetin *et al.* (2021). The authors investigated the recreational potential and monetary value of urban catchments with rich biodiversity in Istanbul. Rather than using observational data from different sources, the travel cost method collects data by conducting a survey.[9] The survey questions can be divided into two categories. The first category includes information specific to travel cost methods, such as travel cost[10], travel distance and travel time. Both travel distance and travel time are used as proxies for opportunity cost of time. The second category focuses on the respondents' socio-economic characteristics (e.g. age, gender, education). By leveraging this information, the researchers can identify the type of respondents with the highest WTP to visit the site. For instance, do families with children have the highest WTP for this site? Following the collection of data, the researchers conducted an econometric study where the explained variable was the number of visits per site and the explanatory variables belonged to the two categories described above. The study's findings, as reported in Cetin *et al.* (2021), indicate that the benefit from visiting the site is eight times higher than the average travel cost per year and per person. This indicates that visitors derive greater benefits from recreational visits than the average travel cost associated with visiting the catchment area.

The current practices for assessing urban catchments focus on evaluating their monetary value via water-related services, such as water provision and water purification, without considering their potential recreational benefits. By neglecting the recreational benefits, policy makers may underestimate the total benefit of periurban catchments in a cost–benefit analysis, potentially leading to underprovision of this service.

Production function method

To date, the revealed preference approaches described value the impact of changes in ecosystem services by observing households' behaviour. However, environmental conditions also play a significant role in certain productive activities. For instance, wild pollination has been shown to enhance agricultural yields. The objective of the production function method is to observe the effects of a change in the provision of ecosystem services on either the level of production function or the production cost or a firm's revenues. The production function method has a strong interdisciplinary influence, especially from the field of ecology. The selection of economic and environmental data is based on economic theoretical frameworks and ecological models.

This method has been widely used to capture the indirect value of a wide range of ecosystem services (see Fig. 24.1). For instance, Singh and Mishra (2014) estimated the effect of forest cover on water quality and cost of water treatment using data from the Western Ghats of peninsular India. Vincent *et al.* (2016) quantified the value of forest water purification services in the form of ensuing cost savings of municipal drinking water treatment in Malaysia, and McCauley (2006) estimated this ecosystem service in China's Sichuan province. Other studies have focused on quantifying the income-stabilizing role of forests and natural biodiversity in developing countries (Noack *et al.*, 2019).

To illustrate the production function method, we rely on the study by Tibesigwa, *et al.* (2019). The primary objective of this study was to quantify the changes in crop revenue of smallholders who farm in Tanzania. These changes were associated with the reduction of natural habitats of wild pollinators. The pollination service was measured by the distance of the field to the natural habitats. The other ecological variables included climate and soil characteristics. The socio-economic vector included variables such as labour cost, seedling cost and household characteristics. The econometric analysis revealed that the decline in natural habitats in Tanzania between 2008 and 2013 led to a significant reduction in crop revenue for smallholder farmers, with an average loss of 29% and a median loss of 4%.

Advantages and limitations of revealed preference approaches

These approaches have been successfully implemented in a variety of contexts. The primary strength of these approaches lies in their foundation on actual (or observed) decisions made by individuals, households or firms. This inherent characteristic enhances the reliability of the outcomes, making them more reliable indicators of people's preferences.

However, it is important to note that there are several challenges associated with the practical implementation of these methods. A significant challenge is that they rely on data from uncontrolled settings. The valuation may be biased due to the many variables that could be correlated with the good being valued. Each method also has its own limitations. For instance, in the case of hedonic pricing, households/individuals may not have perfect information, which leads to potential bias in the estimation of the WTP or WTA. The travel cost method is particularly challenging when it comes to multiple-purpose trips. For instance, individuals may visit a site and visit their family on the same trip. In such cases, the WTP or WTA estimation may not be accurate.

Revealed preference approaches and biocontrol programmes

These approaches empower researchers to estimate the external impacts of biocontrol programs by leveraging observational data. For instance, data on property values in urban or periurban areas can be used to conduct hedonic analysis to estimate the positive effects of planting biodiversity-generating grass strips in urban areas. The travel cost method can be used to estimate the positive effects of flower bands on biodiversity along hiking trails (recreational sites). Finally, exhaustive data on farmers' revenues and crop yields could be useful in estimating the external impacts of biological control, such as their potential role in curbing carbon emissions. According to Wyckhuys *et al.* (2022), the aggregate carbon footprint of pesticides is approximately 73 million tons of CO_2 equivalent, which constitutes 10% of Germany's greenhouse gas emissions in 2022 (WRI, 2024).

24.3.2 Stated preference approaches

Two primary methods are used to determine stated preference values: contingent valuation and choice experiment (Hanley and Czajkowski, 2019). While revealed preference approaches value the non-market impacts of the change in the provision of ecosystem services by observing individuals' behaviour, stated preference approaches ask individuals directly to express their preferences for gaining potential environmental benefits. As for revealed preference approaches, the values for ecosystem services can be uncovered through the use of econometric analysis. The subsequent paragraphs offer a general description of each method and present some examples of its applications. We then illustrate the methodology with one case study, focusing on examples related to climate change, biodiversity and the adoption of sustainable pesticide use.

Contingent valuation methods

The contingent valuation is based on a hypothetical survey in which two alternative scenarios are proposed: the status quo scenario vs the pro-environmental scenario. The pro-environmental scenario outlines the ecosystem services in question, the institutional context in which they will be provided and the financing mechanisms. A random sample of people is then directly asked to express their potential willingness to pay for (WTP) or willingness to accept (WTA) the proposed hypothetical change in the level of provision of the good. The scenario is further enriched by additional questions that enquire about the respondents' sociodemographic characteristics. The econometric analysis utilizes the responses to estimate the mean WTA or WTP of the sample.

The contingent valuation method can be applied to assign a value to a wide range of ecosystem services. This includes the valuation of benefits associated with improving air and water quality (Orgill-Meyer *et al.*, 2018) and assessing the benefits of forest conservation (Gelo and Koch, 2015; Abdeta *et al.*, 2023) or the preservation of a wide range of natural resources, such as wetlands (Ghanian *et al.*, 2022) and fish (Danley, *et al.*, 2021). Additionally, this method has been utilized to estimate the WTP for climate adaptation policies (Vàsquez *et al.*, 2022) and clean air policies (Blackman *et al.*, 2020). A key strength of this method is its ability to capture a wide range of benefits from ecosystem services, including use values (Danley, *et al.*, 2021), existence values (Kontogianni *et al.*, 2012; Landry *et al.*, 2020) and bequest values (O'Garra, 2009).

We use as an illustrative case the study by da Costa and Santos (2016). The objective of this study is to assess the probability of consumers purchasing food items produced in farming systems that adopt sustainable pesticide use, as a function of the price premium level (or the additional price consumers are willing to pay compared to a conventional item), household income and knowledge about environmentally friendly and healthier production systems. The study was conducted in urban areas of Portugal. The contingent valuation method necessitates the establishment of a clear and credible hypothetical scenario. In da Costa and Santos (2016), the contingent valuation scenario details the environmental and health benefits of consuming food items produced with fewer pesticides. Additionally, the respondents are presented with a hypothetical scenario in which an agricultural policy is implemented to promote the sustainable use of pesticides. This policy would impose additional costs on farmers and result in higher prices for consumers. The respondents are asked to determine the maximum increase in their average monthly expenditure in fruits and vegetables they are willing to pay to accept a low-pesticide policy. This scenario is one component of a comprehensive survey. To identify the factors that positively impact the WTA, the following policy questions are posed: (i) the household's habits regarding food consumption, (ii) consumer awareness and perceptions of pesticide and fertilizer use risks, (iii) the socio-economic characteristics of the respondents. The responses to the questionnaire are then analysed statistically. The results indicate that consumers are willing to increase their average expenditures on food and vegetables produced using sustainable pesticide practices by 163%. This price premium is lower for households in the first decile of income, leaving them more exposed to the risks of pesticide use.

Choice modelling

This approach represents a different type of survey-based valuation technique. Unlike the contingent valuation method, which values the total change, choice modelling can value environmental changes that are multidimensional. Indeed, policies usually affect the different attributes of ecosystem services. This approach enables the quantification of marginal values for each attribute, providing a comprehensive assessment of the ecosystem's worth. This approach is widely encountered in marketing, transport economics and public health literature; only in recent years has it gained popularity among environmental economists. In environmental economics, respondents have the choice between competing environmental projects. In transport economics, they choose between different transport modes.

The choice modelling method has been used to estimate people's preferences and WTP for non-market goods (Carson *et al.*, 1990; Adamowicz *et al.*, 1994). For instance, Enriquez-Acevedo *et al.* (2018) estimated the WTP for maintaining beach ecosystem services in the Colombian Caribbean region. In a similar vein, Brey *et al.* (2007) applied a choice modelling method to estimate the WTP for carbon sequestration, while Diafas *et al.* (2017) estimated the bequest value of forest services. This method has been extended to value pro-environmental policies. For instance, Nthambi, *et al.* (2021) examined the WTA public adaptation measures, such as preventing flood damage on cropland through the construction of dams. They employed choice modelling methods to estimate farmers' preferences regarding various attributes of a sand storage dam project, including organizational governance and its impact on deforestation. Finally, Sokolowski (2024) aims to understand households' preferences for adopting different behaviours to improve air quality.

To illustrate the choice modelling method, we will use the study developed by Chèze *et al.* (2020) as an example. The primary objective of this study is to assess the relative importance of various factors that influence conventional farmers' decisions to reduce or maintain their use of synthetic pesticides. The researchers conducted a survey in which farmers could choose between maintaining current agricultural practices (*status quo scenario*) and adopting alternative practices that reduce their use of pesticides (*low-use-pesticide scenarios*).[11] Each scenario is described using four attributes: impact on yearly profit, risk of production loss, administrative commitment, and health and environmental impacts. To ascertain which type of respondents are more willing to move away from the status quo, the survey contains two additional parts. One part is dedicated to general questions on farmers' activities (e.g. size of the farm, use of pesticides). The other section focuses on the farmer's socio-economic situation, including income level, gender, age and education level. An econometric analysis of the respondents' choices is conducted to estimate the WTA to lower their pesticide use. The analysis revealed that, under stable conditions, farmers in the sample would be willing to accept an average annual payment of €90 per hectare to reduce their pesticide use, given the associated risk of production losses. It is important to note that farmers with low and insecure incomes are more reluctant to move away from high-pesticide practices.

As discussed earlier, it is critical to place an economic value on the environmental benefits of reducing pesticide use in favour of biological control in order to incentivize farmers. However, the environmental benefits should be balanced against the environmental costs, such as invasive species. The Asian ladybird, an aphid predator, was voluntarily released in Europe in the 1990s as a biological control agent due to its voracious appetite. However, it has become highly invasive on four continents. Employing a choice modelling approach enables us to obtain a comprehensive understanding of how respondents assess various attributes of integrated pest management programmes. For instance, Chakir *et al.* (2016) provide an economic valuation of the benefits and the costs of the Asian ladybird. The benefits are measured by the individuals' WTP for a reduction in pesticide use, and the costs are measured in terms of WTP for biodiversity losses (preservation of the two-spot ladybird) and human nuisances due to the invasion of the Asian ladybird. Their findings indicate that the total WTP for enhanced biodiversity and reduced nuisances exceeds the WTP for reduced pesticides.

Advantages and limitations of stated preference approaches

The contingent valuation and choice modelling methods are survey based, and the advantage of asking hypothetical questions is that one can ask about any value or about behaviour in new and unusual circumstances that are otherwise hard to measure. Consequently, both methods are capable of capturing the full range of values presented in Fig. 24.1. However, each method has its own particular strengths and limitations. Choice modelling can address multidimensional changes. Contingent valuation can ascertain the value of the total change in the provision of ecosystem services.

Despite the prevalence of stated preference approaches, they continue to be a subject of substantial controversy. Both contingent valuation and choice modelling methods are sensitive to study design. The contingent valuation method has limitations and potential biases, such as hypothetical bias (individuals may not act in accordance with stated preferences), strategic behaviour (such as free riding), scope problems (the valuation of ecosystem services is not proportional to its size), anchoring bias (the valuation depends on the first piece of information presented) and information bias (the framing of the question may influence the answer). Choice modelling possesses some disadvantages. The primary challenge lies in the cognitive complexity involved in handling multiple, intricate choice scenarios. Additionally, choice modelling methods face greater challenges in deriving values for a sequence of elements implemented by a policy when compared to a contingent valuation alternative.

Stated preference approaches and biocontrol programmes

Stated preferences approaches can be implemented *ex ante* in areas where biocontrol programmes have not yet been adopted. Researchers can estimate the price premiums consumers are willing to pay for reduced exposure to pesticides using contingent valuation. They can also estimate farmers' willingness to accept potential reduced profits due to biocontrol programmes in order to obtain current or future environmental benefits, such as the preservation of soil quality. Using choice modelling, researchers can quantify the marginal values of the different attributes of ecosystem services. For instance, they can measure farmers' willingness to accept biocontrol programmes while isolating different attributes, such as their awareness of environmental quality or the impact of reduced exposure to pesticides on human, animal and ecosystem health.

24.4 Future Directions for Research on Economic Valuation for Biocontrol

This chapter has discussed how ecosystem services can be valued and why they should be. While there is a growing body of economic literature on the economic valuation of biodiversity or endangered species preservation, low pesticide use programmes and pollination (Lee *et al.*, 2024), the number of articles evaluating the social value of biocontrol is still relatively scarce within the field of economics. However, to promote bio-based pest management practices and develop relevant agricultural policies, policy makers must assess their full costs and benefits by leveraging the findings of environmental valuation methods.

To the best of our knowledge, most existing studies estimate the private costs and benefits of biocontrol. These studies, which are based on agro-ecological economic models that combine models of pest population dynamics, pest movement and land use economic decisions (Naranjo *et al.*, 2015), succeed in quantifying the change in crop yields or farmers' profit. However, they do not consider the external effects of biocontrol programmes. Some of these models, such as those of Martinet and Roques (2022), incorporate feedback effects of farmer decisions on ecological dynamics. However, these models are limited in their ability to fully capture the external impacts of biocontrol programmes, as noted by Nicholson *et al.* (2024) and Roy *et al.* (2023). There is a need for further research on the economic valuation of biocontrol programmes.

The revealed preference and stated preference approaches discussed above are able to place a monetary value on the external impacts of biocontrol programmes. These approaches complement agro-ecological models. To conduct accurate economic valuation, it is essential to define and quantify the impacts on ecosystem services, and interdisciplinary collaboration between ecology and economics is crucial (Naranjo *et al.*, 2015). For instance, detailed information on the beneficial effects of integrated pest management on restoring ecosystem functions, curbing energy needs and carbon emissions is essential to establish hypothetical scenarios for contingent valuation or choice modelling methods. They can also serve as a basis to inform individuals if the travel cost method is employed. Additionally, researchers can utilize data on ecosystem services to implement revealed preference methods, such as hedonic pricing or production function.

Future economic studies may face the following challenge: how can results from existing and planned biocontrol programmes be used to better inform individuals? Indeed, the assumption of informed preferences is central to estimate the willingness to pay or the willingness to accept for the preservation of species, so citizens' unfamiliarity with the species being valued is an important issue to consider. Indeed, individual choices are highly correlated with group average choices (Bramoullé *et al.*, 2009). Individuals tend to emulate and learn from the behaviour of others, particularly those in their close circle such as friends, neighbours and colleagues. Economic studies have demonstrated the significance of peer effects in promoting pro-environmental behaviours (Allcott, 2011).

In light of these findings, the following question arises: how should uncertainty and risk behaviour be considered when designing public policies? There is considerable uncertainty in evaluating the risks associated with the use of pesticides on health and biodiversity. Different attitudes towards risk may imply different reactions to the same public policy. Another key issue is accurately considering the private effects of biocontrol and social benefits within a single study. Private costs and benefits can be estimated at the local level, while social benefits should be estimated at a larger scale. Furthermore, these costs and benefits may materialize over different time frames.

Notes

[1] Using an online survey and a qualitative approach, Serée *et al.* (2023) evaluated the perception of the aesthetic and biodiversity value of cropped fields and wildflower strips on the basis of their visual appearance. Morandin *et al.* (2016) assessed the combined economic benefits of pest control and pollination ecosystem services in adjacent crops, at sites with and without hedgerows on field edges. Considering installation costs of hedgerow planting, benefits from the potential pesticide reduction and potential crop seed set increases from native bee enhancement, they determine the net benefit of hedgerow plantings over time.

[2] Laurans, *et al.* (2013) distinguish three main categories of ecosystem service economic valuation based on its primary purpose: decisive, where valuation informs a specific decision *ex ante*, aiding a decision maker in choosing among alternatives; technical, where valuation is applied after a policy or project is selected to refine the economic instrument used for implementation; and informative, where valuation provides insights intended to indirectly influence decision making.
[3] In practice, individuals who violate pesticide regulations are typically sued for their direct impacts on human health and water contamination rather than for harming biodiversity. For example, in 2012, the agricultural chemical company Syngenta paid $105 million in fines to cities and towns across the Midwest for contaminating water supplies with the herbicide atrazine. In 2013, a landscaping company in Oregon was fined just $2886 for using a pesticide that caused the death of 50,000 bumblebees. Although monetary valuation of biodiversity losses is recommended for determining the fines, it is not always enforced.
[4] Some authors also consider spatial discounting. Indeed, empirical studies indicate that people tend to discount ecosystem services across space (Yamaguchi and Shah, 2020). In other words, they give a lower value to ecosystem services that are far away.
[5] An *ex post* policy evaluation focuses on the realized costs and benefits of a policy.
[6] An *ex ante* policy evaluation attempts to predict future costs and benefits of a policy.
[7] The averting behaviour and the costs of illness are also classified as revealed preference approaches (Mendelsohn, 2019). But since they are less used in the economic literature, we will not develop them in this chapter.
[8] The econometric method employed to estimate the WTP and WTA is specific to each method. However, we will not discuss it in this chapter.
[9] Participants are selected in order to get a representative sample of Istanbul's residents' characteristics such as household size, education level and living environment.
[10] The travel cost includes fuel costs, toll roads fee, tickets fee, entrance or parking fee, vehicle rental prices, taxi fee, etc.
[11] The survey is conducted among a representative sample, here, a set of French farmers (field crops, vegetable farming, wine growing among others).

References

Abdeta, D., Ayana, A.N. and Bekele, Y. (2023) Willingness to pay for forest conservation: Evidence from a contingent valuation survey analysis in Southwest Ethiopia. *Global Ecology and Conservation* 46, e02551.

Adamowicz, W., Louviere, J. and Williams, M. (1994) Combining revealed and stated preference methods for valuing environmental amenities. *Journal of Environmental Economics and Management* 26(3), 271–292.

Allcott, H. (2011) Social norms and energy conservation. *Journal of Public Economics* Special Issue(9), 1082–1095.

Asche, F., Bronnmann, J. and Cojocaru, A.L. (2021) The value of responsibly farmed fish: A hedonic price study of ASC-certified whitefish. *Ecological Economics* 188, 107135.

Bakker, L., Sok, J., Van Der Werf, W. and Bianchi, F. (2021) Kicking the habit: What makes and breaks farmers' intentions to reduce pesticide use? *Ecological Economics* 180, 106868.

Bartolini, F., Vergamini, D., Longhitano, D. and Povellato, A. (2021) Do differential payments for agri-environment schemes affect the environmental benefits? A case study in North-Eastern Italy. *Land Use Policy* 107, 104862.

Bateman, I.J., Harwood, A.R., Mace, G.M., Watson, R.T., Abson, D.J. *et al.* (2013) Bringing ecosystem services into economic decision-making: Land use in the United Kingdom. *Science* 341(6141), 45–50.

Bazoche, P., Combris, P., Giraud-Héraud, E., Seabra Pinto, A., Bunte, F. *et al.* (2014) Willingness to pay for pesticide reduction in the EU: Nothing but organic? *European Review of Agricultural Economics* 41(1), 87–109.

Blackman, A., Qin, P. and Yang, J. (2020) How costly are driving restrictions? Contingent valuation evidence from Beijing. *Journal of Environmental Economics and Management* 104, 102366.

Bonhomme, K. and Schaffar, A. (2021) L'évaluation économique des plages par la méthode des coûts de déplacement: Une démarche exploratoire. *Économie Rurale* 375, 61–79.

Bougherara, D., Lapierre, M., Préguet, R. and Sauquet, A. (2021) Do farmers prefer increasing, decreasing, or stable payments in agri-environmental schemes? *Ecological Economics* 183(C), 106946.

Bramoullé, Y., Djebbari, H. and Fortin, B. (2009) Identification of peer effects through social networks. *Journal of Econometrics* 150(1), 41–55.

Brey, R., Riera, P. and Mogas, J. (2007) Estimation of forest values using choice modeling: An application to Spanish forests. *Ecological Economics* 64(2), 305–312.

Canessa, C., Venus, T.E., Wiesmeier, M., Mennig, P. and Sauer, J. (2023) Incentives, rewards or both in payments for ecosystem services: Drawing a link between farmers' preferences and biodiversity levels. *Ecological Economics* 213, 107954.

Carriazo, F., Ready, R. and Shortle, J. (2013) Using stochastic frontier models to mitigate omitted variable bias in hedonic pricing models: A case study for air quality in Bogotá, Colombia. *Ecological Economics* 91, 80–88.

Carson, R., Hanemann, M. and Steinberg, D. (1990) A discrete choice contingent valuation estimates of the value of Kenai King salmon. *Journal of Behavioral Economics* 19(1), 53–68.

Cetin, N.I., Bourget, G. and Tezer, A. (2021) Travel-cost method for assessing the monetary value of recreational services in the Ömerli Catchment. *Ecological Economics* 190(C), 107192.

Chakir, R., David, M., Gozlan, E. and Sangare, A. (2016) Valuing the impacts of an invasive biological control agent: A choice experiment on the Asian ladybird in France. *Journal of Agricultural Economics* 67(3), 619–638.

Chèze, B., David, M. and Martinet, V. (2020) Understanding farmers' reluctance to reduce pesticide use: A choice experiment. *Ecological Economics* 167, 106349.

Cobián Álvarez, J., Armando, J. and Resosudarmo, B. (2019) The cost of floods in developing countries' megacities: A hedonic price analysis of the Jakarta housing market, Indonesia. *Environmental Economics Policy Studies* 21(4), 555–577.

Costanza, R. (2008) Ecosystem services: Multiple classification systems are needed. *Biological Conservation* 141(2), 250–252.

Costanza, R., Kubiszewski, I., Stoeckl, N. and Kompas, T. (2021) Pluralistic discounting recognizing different capital contributions: An example estimating the net present value of global ecosystem services. *Ecological Economics* 183, 106961.

Czajkowski, M., Giergiczny, M., Kronenberg, J. and Englin, J. (2019) The individual travel cost method with consumer-specific values of travel time savings. *Environmental and Resource Economics* 74, 961–984.

da Costa, C.A. and Santos, J. (2016) Estimating the demand curve for sustainable use of pesticides from contingent-valuation data. *Ecological Economics* 127, 121–128.

Daniels, S., Witters, N., Beliën, T., Vrancken, K., Vangronsveld, J. *et al.* (2017) Monetary valuation of natural predators for biological pest control in pear production. *Ecological Economics* 134, 160–173.

Danley, B., Sandorf, E.D. and Campbell, D. (2021) Putting your best fish forward: Investigating distance decay and relative preferences for fish conservation. *Journal of Environmental Economics and Management* 108, 102475.

De Groot, R., Brander, L., van der Ploeg, S., Costanza, R., Bernard, F. *et al.* (2012) Global estimates of the value of ecosystems and their services in monetary units. *Ecosystem Services* 1(1), 50–61.

De Groote, H., Ajuonu, O., Attignon, S., Djessou, R. and Neuenschwander, P. (2003) Economic impact of biological control of water hyacinth in southern Benin. *Ecological Economics* 45(1), 105–117.

Diafas, I., Barkmann, J. and Mburu, J. (2017) Measurement of bequest value using a non- monetary payment in a choice experiment – the case of improving forest ecosystem services for the benefit of local communities in rural Kenya. *Ecological Economics* 140, 157–165.

El jaouhari, M., Damour, G., Tixier, P. and Coulis, M. (2023) Glyphosate reduces the biodiversity of soil macrofauna and benefits exotic over native species in a tropical agroecosystem. *Basic and Applied Ecology* 73, 18–26.

Enriquez-Acevedo, T., Botero, C.M., Cantero-Rodelo, R., Pertuz, A. and Suarez, A. (2018) Willingness to pay for beach ecosystem services: The case study of three Colombian beaches. *Ocean & Coastal Management* 161, 96–104.

Finger, R. and Möhring, N. (2022) The adoption of pesticide-free wheat production and farmers' perceptions of its environmental and health effects. *Ecological Economics* 198, 107463.

Gastineau, P. and Taugourdeau, E. (2014) Compensating for environmental damages. *Ecological Economics* 97, 150–161.

Gelo, D. and Koch, S.F. (2015) Contingent valuation of community forestry programs in Ethiopia: Controlling for preference anomalies in double-bounded CVM. *Ecological Economics* 114, 79–89.

Ghanian, M., Ghoochani, O.M., Noroozi, H. and Cotton, M. (2022) Valuing wetland conservation: A contingent valuation analysis among Iranian beneficiaries. *Journal for Nature Conservation* 66, 126140.

Gopalakrishnan, S., Smith, M.D., Slott, J.M. and Murray, A.B. (2011) The value of disappearing beaches: A hedonic pricing model with endogenous beach width. *Journal of Environmental Economics and Management* 61(3), 297–310.

Hanley, N. and Czajkowski, M. (2019) The role of stated preference valuation methods in understanding choices and informing policy. *Review of Environmental Economics and Policy* 13(2), 248–266.

Hardin, G. (1968) The tragedy of the commons. *Science* 162(3859), 1243–1248.

Heal, G. (2000) Valuing ecosystem services. *Ecosystems* 3(1), 24–30.

Isely, P., Nordman, E.E., Howard, S. and Bowman, R. (2017) Phragmites removal increases property values in Michigan's lower grand river watershed. *Journal of Ocean and Coastal Economics* 4(1), 5.

Jayalath, T.A., Lloyd-Smith, P. and Becker, M. (2023) Biodiversity benefits of birdwatching using citizen science data and individualized recreational demand models. *Environmental and Resource Economics* 86(1), 83–107.

Jensen, C.U., Panduro, T.E., Lundhede, T.H., von Graevenitz, K. and Thorsen, B.J. (2021) Who demands peri-urban nature? A second stage hedonic house price estimation of household's preference for peri-Urban nature. *Landscape and Urban Planning* 207, 104016.

Johnston, R.J., Boyle, K.J., Adamowicz, W.V., Bennett, J., Brouwer, R. *et al.* (2017) Contemporary guidance for stated preference studies. *Journal of the Association of Environmental and Resource Economists* 4(2), 319–405.

Karp, D.S., Mendenhall, C.D., Sandí, R.F., Chaumont, N., Ehrlich, P.R. *et al.* (2013) Forest bolsters bird abundance, pest control and coffee yield. *Ecology Letters* 16(11), 1339–1347.

Kermagoret, C., Levrel, H., Carlier, A. and Dachary-Bernard, J. (2016) Individual preferences regarding environmental offset and welfare compensation: A choice experiment application to an offshore wind farm project. *Ecological Economics* 129, 230–240.

Kirchweger, S., Clough, Y., Kapfer, M., Steffan-Dewenter, I. and Kantelhardt, J. (2020) Do improved pollination services outweigh farm-economic disadvantages of working in small-structured agricultural landscapes? Development and application of a bio-economic model. *Ecological Economics* 169, 106535.

Kontogianni, A., Tourkolias, C., Machleras, A. and Skourtos, M. (2012) Service providing units, existence values and the valuation of endangered species: A methodological test. *Ecological Economics* 79, 97–104.

Kontoleon, A. and Swanson, T. (2003) The willingness to pay for property rights for the giant panda: Can a charismatic species be an instrument for nature conservation? *Land Economics* 79(4), 483–499.

Krutilla, J.V. (1967) Conservation reconsidered. *American Economic Review* 57, 777– 786.

Kuhfuss, L., Pr´eget, R., Thoyer, S. and Hanley, N. (2016) Nudging farmers to enrol land into agri-environmental schemes: The role of a collective bonus. *European Review of Agricultural Economics* 43(4), 609–636.

Landry, C.E., Shonkwiler, J.S. and Whitehead, J.C. (2020) Economic values of coastal erosion management: Joint estimation of use and existence values with recreation demand and contingent valuation data. *Journal of Environmental Economics and Management* 103, 102364.

Laurans, Y., Rankovic, A., Bill´e, R., Pirard, R. and Mermet, L. (2013) Use of ecosystem services economic valuation for decision making: Questioning a literature blindspot. *Journal of Environmental Management* 119, 208–219.

Lee, K., McDermott, S. and Fernandez, L. (2024) Using economics to inform and evaluate biological control programs: Opportunities, challenges, and recommendations for future research. *BioControl* 69, 237–252.

Lévesque, A., Kermagoret, C., Poder, T.G., L'Ecuyer-Sauvageau, C., He, J. *et al.* (2021) Financing on-farm ecosystem services in southern Quebec, Canada: A public call for pesticides reduction. *Ecological Economics* 184, 106997.

Losey, J.E. and Vaughan, M. (2006) The economic value of ecological services provided by insects. *BioScience* 56(4), 311–323.

Lv, Y., Zhong Gu, S. and Mei Guo, D. (2010) Valuing environmental externalities from rice–wheat farming in the lower reaches of the Yangtze river. *Ecological Economics* Special Section(7), 1436–1442.
Martinet, V. and Roques, L. (2022) An ecological-economic model of land-use decisions, agricultural production and biocontrol. *Royal Society Open Science* 9(10), 220169.
McCauley, D.J. (2006) Selling out on nature. *Nature* 443(7107), 27–28.
Mei, Y., Hite, D. and Sohngen, B. (2017) Demand for urban tree cover: A two-stage hedonic price analysis in California. *Forest Policy and Economics* 83, 29–35.
Mendelsohn, R. (2019) An examination of recent revealed preference valuation methods and results. *Review of Environmental Economics and Policy* 13.
Morandin, L.A., Long, R.F. and Kremen, C. (2016) Pest control and pollination cost–benefit analysis of hedgerow restoration in a simplified agricultural landscape. *Journal of Economic Entomology* 109(3), 1020–1027.
Munns, Jr, W.R. and Rea, A.W. (2015) Ecosystem services: Value is in the eye of the beholder. *Integrated Environmental Assessment and Management* 11(2), 332–333.
Mwebaze, P., Marris, G.C., Brown, M., MacLeod, A., Jones, G. *et al.* (2018) Measuring public perception and preferences for ecosystem services: A case study of bee pollination in the UK. *Land Use Policy* 71, 355–362.
Nam, K.M., Ou, Y., Kim, E. and Zheng, S. (2022) Air Pollution and housing values in Korea: A hedonic analysis with long-range transboundary pollution as an instrument. *Environmental and Resource Economics* 82(2), 383–407.
Naranjo, S.E., Ellsworth, P.C. and Frisvold, G.B. (2015) Economic value of biological control in integrated pest management of managed plant systems. *Annual Review of Entomology* 60(1), 621–645.
Neves do Prado, L. and Abildtrup, J. (2022) Landowners' willingness to accept pesticide reduction in the Pipiripau river basin (Brazil). *Environmental Challenges* 9, 100607.
Nicholson, C.C., Knapp, J., Kiljanek, T., Albrecht, M., Chauzat, M.P. *et al.* (2024) Pesticide use negatively affects bumble bees across European landscapes. *Nature* 628(8007), 355–358.
Noack, F., Riekhof, M.C. and Di Falco, S. (2019) Droughts, biodiversity, and rural incomes in the tropics. *Journal of the Association of Environmental and Resource Economists* 6(4), 823–852.
Noriega, J.A., Hortal, J., Azc´arate, F.M., Berg, M.P. and Bonada, N. (2018) Research trends in ecosystem services provided by insects. *Basic and Applied Ecology* 26, 8–23.
Nthambi, M., Markova-Nenova, N. and W¨atzold, F. (2021) Quantifying loss of benefits from poor governance of climate change adaptation projects: A discrete choice experiment with farmers in Kenya. *Ecological Economics* 179, 106831.
Nyborg, K. (2014) Project evaluation with democratic decision-making: What does cost–benefit analysis really measure? *Ecological Economics* 106, 124–131.
O'Garra, T. (2009) Bequest values for marine resources: How important for indigenous communities in less-developed economies? *Environmental and Resource Economics* 44, 179–202.
Orgill-Meyer, J., Jeuland, M., Albert, J. and Cutler, N. (2018) Comparing contingent valuation and averting expenditure estimates of the costs of irregular water supply. *Ecological Economics* 146, 250–264.
Pearce, D. (1998) Cost–benefit analysis and environmental policy. *Oxford Review of Economic Policy* 14(4), 84–100.
Pearce, D. (ed.) (1994) *The Economic Value of Biodiversity*. Earthscan, London.
Pereira, P., Inacio, M., Kalinauskas, M., Bogdzevi˘c, K., Bogunovic, I. *et al.* (2022) Land-use changes and ecosystem services. In: Pereira, P., Gomes, E. and Rocha, J. (eds) *Mapping and Forecasting Land Use*. Elsevier, St Louis, Missouri, pp. 1–27.
Plottu, E. and Plottu, B. (2007) The concept of total economic value of environment: A reconsideration within a hierarchical rationality. *Ecological Economics* 61(1), 52–61.
Rosa-Schleich, J., Loos, J., Ferrante, M., Mußhoff, O. and Tscharntke, T. (2024) Mixed farmers' perception of the ecological-economic performance of diversified farming. *Ecological Economics* 220, 108174.
Roy, H., Pauchard, A., Stoett, P., Renard Truong, T., Bacher, S. *et al.* (2023) *Summary for Policymakers of the Thematic Assessment Report on Invasive Alien Species and Their Control*. IPBES Secretariat, Bonn, Germany.
Ruijs, A., Kortelainen, M., Wossink, A., Schulp, C. and Alkemade, R. (2017) Opportunity cost estimation of ecosystem services. *Environmental and Resource Economics* 66(4), 717–747.
Serée, L., Legras, S., Chiron, F., Valantin-Morison, M. and Gardarin, A. (2023) People prefer arable fields and flower strips with continuous soil cover and diversified vegetation. *Ecology and Society* 28(2), 9.

Singh, S. and Mishra, A. (2014) Deforestation-induced costs on the drinking water supplies of the Mumbai metropolitan, India. *Global Environmental Change* 27, 73–83.

Small, N., Munday, M. and Durance, I. (2017) The challenge of valuing ecosystem services that have no material benefits. *Global Environmental Change* 44, 57–67.

Sokolowski, J. (2024) Assessing individual environmental capital and pro-climate behaviour: A residential sector choice experiment on heating investment preferences. *Journal of Cleaner Production* 447, 141123.

Tibesigwa, B., Siikam¨aki, J., Lokina, R. and Alvsilver, J. (2019) Naturally available wild pollination services have economic value for nature dependent smallholder crop farms in Tanzania. *Scientific Reports* 9(1), 3434.

Turner, R., Paavola, J., Cooper, P., Farber, S., Jessamy, V. *et al.* (2003) Valuing nature: Lessons learned and future research directions. *Ecological Economics* 46(3), 493–510.

Valente, C., Goncalves, C., Monteiro, F., Gaspar, J., Silva, M. *et al.* (2018) Economic outcome of classical biological control: A case study on the eucalyptus snout beetle, gonipterus platensis, and the parasitoid anaphes nitens. *Ecological Economics* 149, 40–47.

Van Klink, R., Bowler, D., Gongalsky, K., Swengel, A., Gentile, A. *et al.* (2020) Meta-analysis reveals declines in terrestrial but increases in freshwater insect abundances. *Science* 368(6489), 417–420.

Vàsquez, W.F., Beaudin, L., Murray, T.J., Pedlowski, M.A. and de Rezende, C.E. (2022) Preferences for sea level rise adaptation: A contingent valuation study in Rio de Janeiro State, Brazil. *Ocean & Coastal Management* 218, 106020.

Vincent, J.R., Ahmad, I., Adnan, N., Burwell, W., Pattanayak, S. *et al.* (2016) Valuing water purification by forests: An analysis of Malaysian panel data. *Environmental and Resource Economics* 64, 59–80.

Wegner, G. and Pascual, U. (2011) Cost-benefit analysis in the context of ecosystem services for human well-being: A multidisciplinary critique. *Global Environmental Change* Special Issue(2), 492–504.

Weisbrod, B.A. (1964) Collective-consumption services of individual-consumption goods. *Quarterly Journal of Economics* 78(3), 471–477.

WRI (2024) Data Lab, Climate Platform. World Resources Institute. Available at: www.wri.org/data (accessed 17 June 2025).

Wyckhuys, K.A., Furlong, M.J., Zhang, W. and Gc, Y.D. (2022) Carbon benefits of enlisting nature for crop protection. *Nature Food* 3(5), 299–301.

Yamaguchi, R. and Shah, P. (2020) Spatial discounting of ecosystem services. *Resource and Energy Economics* 62, 101186.

Zivin, J.G., Liao, Y. and Panassie, Y. (2023) How hurricanes sweep up housing markets: Evidence from Florida. *Journal of Environmental Economics and Management* 118, 10.

25 Biocontrol Seen from Innovation Management: A Diversity of Living Technologies Facing Global Change

Aura Parmentier-Cajaiba[1]*, Manuel Boutet[1] and Thibaut Malausa[2]
[1]Université Côte d'Azur, CNRS, UMR GREDEG, Sophia Antipolis, France; [2]INRAE, Université Côte d'Azur, CNRS, ISA, Sophia Antipolis, France

Abstract
This chapter aims to demonstrate that links between climate change and biocontrol depend on the type of biological control strategies considered here and their putative associated business models. So far, the inundation strategy has been favoured over conservation, inoculation or acclimatization. It shows that each type of biocontrol strategy may contribute in contrasting ways to the transformation of agriculture to fit planetary limits. It also shows that these strategies display contrasting resistance levels towards global changes. The main message is that having an organizational perspective on biocontrol innovation in a climate change context calls for new analytical criteria to design innovations, such as adaptability, complexity, interdependencies and types of knowledge, beyond the traditionally used cost–benefit or benefit–risk balance.

25.1 Introduction

The first section of this chapter sets the scene by considering biocontrol as an innovation that should fit planetary limits. To do so, we introduce the concepts of zombie technology and living technology (Halloy, 2019; Arnoud *et al.*, 2020; Monnin *et al.*, 2020), a typology to characterize technologies related to their impact on climate change and other planetary limits. It invites us to conceive technological innovation to make economy and production more coherent with contemporary societal needs.

In the second section, we apply the zombie/living technology framework to analyse biological control strategies' technical and organizational characteristics: inundative, inoculative, classical and conservation. This theoretical framework provides a lens through which to organize and adapt to climate change. We notably raise whether living technologies and associated business models could contribute less and be more resistant to climate change.

In the third section, we analyse which innovation and diffusion processes could be promoted to widely implement all the available biological control methods and increase their contribution to the agro-ecological transition and greater resistance to global change. To start with a simple example before presenting our argument, for a long time, the modernization of farming was synonymous with industrialization, with the critical and recurring question of expanding scale. Biocontrol has been and is still being thought through isn't that necessary that frame today. But from now on, that frame is a significant

*Corresponding author: aura.parmentier@univ-cotedazur.fr

 Biological Control Systems and Climate Change (eds Joan van Baaren *et al.*)
DOI: 10.1079/9781800625099.0025

obstacle to considering climate change because, as demonstrated here, sustainability and scaling collide and are often contradictory.

All in all, our analysis provides a theoretical and practical underpinning of organizational robustness that aligns both the impact of climate on biocontrol and the effect of biocontrol and its deployment on climate change.

25.2 Farming Innovation and Ecological Transition: From Zombie to Living Technology

25.2.1 The agro-ecological transition from an innovation management point of view

Farming is impacted by climate change and contributes to climate change. Year after year, the IPCC reports the responsibility of agriculture for global warming (IPCC, 2023). The IPBS also points out the effects of agricultural inputs on biodiversity. In turn, the impacts of climate change and biodiversity loss on agriculture are increasingly visible worldwide.

Adaptation to climate change can be seen as a simple question of infrastructures or changing a few parameters of cultural systems to adapt to new conditions. However, this displaces the problem without solving it. For example, infrastructures like greenhouses could be seen as solutions, but when the difference between the target greenhouse microclimate and the external climate increases, so do costs, dependency on networks (especially energy networks), energy consumption and eventually environmental impacts. From an innovation management point of view, implementing agro-ecological transitions calls for innovations coming less from engineering and more from living entities. This means a shift from chemistry to biology, from scaling to anchoring, by adopting an organizational perspective. This is consistent with the concept of robustness that Hamant (2023, 2024) proposes as an alternative view to performance and with his call for adapting productive systems towards more robustness.

The first issue to overcome is the overvalorization of technological innovation, to include new parameters that already exist but need to be noticed, i.e. organizational parameters. From an economic perspective, innovation has become a must-do for organizations because of its assumed relationship to economic growth (Aggeri, 2023). Innovation is often limited to technological innovation in terms of a base of technology (e.g. artificial intelligence, internet, nanotechnology) that has the potential to provide new or alternative products and services and confer an advantage to users, such as simplification, increase in speed, etc. Other forms of innovation should be advertised more. We seldom hear about organizational, social or architectural innovation; these innovations are not necessarily based on techniques and can be leveraged to create social changes. They can be new ways of doing business, providing a simpler product architecture, but also in relating actors and prompting them to operate an activity in a new way. . These innovations are less visible and less spectacular, but not less creative and valuable. The reduction of innovation to its technological side has ecological consequences. For instance, the multiplication of objects requires energy and resources such as rare earths and metals; their accumulation over time leads to a problem we do not know how to solve – recyclability.

A second issue to overcome is the standardization paradigm in agricultural innovation. Innovation in agriculture (including biocontrol) is still primarily an old form of industrialization. It relies on standardization and scalability, very similar to Fordism in the car industry in the first part of the 20th century. Farming has been profoundly transformed worldwide during the 20th century, and the two pillars of this modernization were research and industrialization, working hand in hand. The aim of the associated innovation in public policy was generalization through standardization and scaling (only considered from the perspective of *enlarging* scale), etc. This standardization – decreasing the dependence on local pedoclimatic, biological and socio-economic conditions – was supported by technologies, machines and chemistry in particular. It was a product of research, understood as a process of technological innovation organized from the labs to the fields, labelled as technology-push. Climate change jeopardizes these systems, and local cultivation conditions resurface as a core question, bringing the land back to centre stage.

25.2.2 Agricultural innovation: from zombie technologies to living technologies

Our proposition is a take on how to innovate in biocontrol to help face climate change and enhance the capacities of farming to adapt to these changes. This approach is defined as an 'ecological redirection' (Bonnet *et al.*, 2021; Monnin, 2021a) targeting strategies and moves that make daily activities fit a development trajectory compatible with planetary limits. In this approach, there are some ecologically problematic activities that we inherit, that will inevitably continue, and that we will have to find collective ways to manage – called 'negative commons'. Recent work on the sustainability of technology contributes to that framework and offers a stimulating new perspective. The researcher in physics, José Halloy, has developed a 'strong sustainability' approach (Halloy, 2019; Arnoud *et al.*, 2020) and proposes two notions: living vs zombie technologies that characterize technologies based on their potential of ecologization. Technologies are classified according to their capacity to ensure human and other living entities can cohabit on Earth and must be analysed within the framework of the 'Anthropocene'.

The question then is whether a technical system is sustainable from an ecological perspective over a long period, in opposition to technical systems that have detrimental effects due to their impact on non-renewable resources and natural settings. That typology is characterized by three dimensions: the nature of the resources used (whether they are renewable or not), the lifetime of the technology (its duration in working order), and the end of life (meaning the duration of the technology as waste). In this regard, living technologies have consequently a lower impact than zombie ones (Table 25.1) because their duration in working order lasts over a long period and they have a short life span as waste. The opposite holds true for zombie technology.

Chemical pesticides and fertilizers can be considered zombie technologies: they stay present and active longer than the timespan of their use, and there are multiple cases of them creating 'negative commons' such as polluted rivers left for communities to manage.

25.3 Biocontrol: Fifty Shades of Living Technologies

The above-mentioned living/zombie analysis underlines the importance of distinguishing between the different methods making up 'biocontrol' (Eilenberg *et al.*, 2001). They do not display the same properties regarding their resilience to climate change and ecological impact. Moreover, the resilience and sustainability of each biocontrol method strongly depend on the organization model of innovation deployment. Indeed, a reciprocal relationship exists between a given biocontrol strategy and the organization developed to implement that strategy. Business models as a concept can be a lever (Malausa *et al.*, 2024); the type of strategy chosen has an impact on the kind of business model (Boutet and Parmentier-Cajaiba, 2022), and a variety of organizational settings can fit the same strategy (Parmentier-Cajaiba, 2024). These differences in biocontrol strategies and the diversity and quantity of agents used eventually lead to various products and/or services with multiple extents of resilience and sustainability.

25.3.1 Inundation biocontrol

Inundation biocontrol is 'using living organisms to control pests when control is achieved

Table 25.1. Properties of zombie and living technologies. From Monnin (2021b, p.21).

	Resources	Sustainability	End of life
Zombie technologies	Finite (long-term exhaustion)	Minimum durability in working order	Maximum life span as waste
Living technologies	Renewable (strong sustainability)	Maximum durability in working order	Minimum life span as waste

exclusively by the released organisms themselves' (Eilenberg *et al.*, 2001, p. 395). Inundation strategies imply repeated releases of biocontrol agent organisms to regulate pest densities at a level consistent with farming production objectives.

Its logic of operation – repeated application of large quantities – is the closest to the use of chemical pesticides. It can be sold to farmers in the same way, as a product defined by a stable list of characteristics (passivated product: Callon, 2011), applied with machines, distributed by resellers, circulating along the same logistical chains, and produced in factories. Economies of scale are at the heart of this organizational model, and the stability of chemicals, which enables them to circulate in this system, has been the target of years of effort in formulation, conditioning and packaging.

The change from chemicals to biocontrol agents has been the implicit programme of almost the entire biocontrol industry for over 30 years, with minimal success observed so far, especially in greenhouse systems. Biocontrol agents are systematically less able to enter, at each level, systems built around another technology. Products based on biocontrol agents are less stable, especially when they are alive. Storage is complicated on production sites, during transportation and on farms. Regarding production, furnishing agents regularly and on a large scale is a challenge – with the necessity to redesign the production line for each one entirely. Research efforts have been concentrated on the reduction of that gap.

In other words, inundation biocontrol has been an integral part of the redeployment of the dynamic of industrialization in agriculture over the past decades. Based on the most sustainable technology, living agents, the deployment of that strategy has been developed mainly in an organizational frame with low sustainability and low adaptability.

On the side of sustainability, mass production has consequences on planetary limits because of land use, raw materials consumed, building and maintenance of production facilities, and the continuous use of energy. The transport and storage of these products also have organizational consequences. Potential consequences are different for micro- and macro-organisms. Macro-organisms require maintaining a chain of temperature, especially during transport, which has an environmental cost. Micro-organisms can be formulated in various ways that make them more or less sensitive to storage (Buitink *et al.*, 2022). However, in some instances, their use implies respecting temperature conditions not only on the production site but also on the shelves of co-operatives and even maybe on farm shelves.

On the side of adaptability, mass production is based on economies of scale and is related to product standardization and centralization of production. These are the logics of industrialization, which are a burden to adaptation. Furthermore, in this techno-push dynamic of innovation, more knowledge is needed about the fit of biocontrol to local geographical, biological and environmental conditions.

25.3.2 Inoculation biocontrol

Inoculation biocontrol, 'the intentional release of a biological control agent with the expectation that it will multiply and control the pest for an extended period, but not permanently' (Eilenberg *et al.*, 2001, p. 393), highlights the possibility of medium-term equilibrium. An example of the use of this strategy is the inoculation in greenhouses of the mirid bug *Macrolophus pygmaeus* (Rambur). Miridae transient establishment is favoured by the provision of alternative foods such as eggs of *Ephestia kuhniella* (Zeller) Pyralidae *and Artemia franciscana* (Kellog) Artemiidae (Moerkens *et al.*, 2017). Compared to inundative biocontrol, production can be done in smaller quantities (Eilenberg *et al.*, 2001) and renewed less frequently because agents must be able to reproduce in the target environment.

This strategy requires skill and knowledge to optimize the establishment of the biocontrol agent according to its phenotypic traits and environmental conditions. Knowledge of the identification and development cycle would help growers identify the number of agents and their stage of development. Knowledge about their reproduction would help assess the number of agents to produce and, if necessary, call in for a new release.

On the side of adaptation, we must first note that it is a strategy to address changing ecosystems. It is knowledge-intensive and based on living technology. It requires a comprehensive understanding of the local context with specific attention to consequences. It is, however, generally centralized, intending to introduce one or a few biocontrol agents over the agrosystems of an entire country from one or a few mass-rearing units owned by one or a few actors.

Regarding adaptability, the inoculation strategy can rely partly – more than inundation – on living organisms' adaptation capacity. But, more importantly, it is based on solid knowledge about biocontrol agents and the target environment. Where the inundation strategy leads to concentrated attention and efforts to optimize the chain of production of insects in the factory, the inoculation strategy encourages the development of knowledge about the conditions of establishment and the installation of solid lines of information on the site. Here, adaptability results from relying on the capacity of living organisms and their communities to adapt and develop knowledge about farming conditions. A system with a clear information line from the deployment site to the strategic centre of decision and knowledge development provides a chance for resilience.

25.3.3 Classical biocontrol

Classical biological control is the 'intentional introduction of an exotic, usually co-evolved, biological control agent for permanent establishment and long-term pest control' (Eilenberg *et al.*, 2001, p. 391). Achieving this biocontrol is primarily based on analytical work aimed at identifying an agent that naturally controls the pest in its native habitat, developing a mass production process, ensuring the biocontrol agent will not impair the ecosystem, and setting field release modalities to optimize establishment success. These activities are research centred and temporary (programmes of around 4–10 years). Once the work is achieved, in theory, no activity is required by agricultural practitioners. The classical biocontrol approach might be complemented by further inoculative releases to increase the density of the biocontrol agent if it is considered insufficient to control the pest after the classical biocontrol programme.

In the classical biocontrol approach, the delivery form is thus one-shot, and the related competencies are meta-scientific knowledge that must be redeployed for each case. Therefore, these techniques' diffusion modes are mainly based on services and cannot be developed as a product. This particularity of classical biocontrol makes it a low-impact technique related to climate change and planetary limits. The more significant risk of these techniques is unbalancing local biodiversity, but scientific work and regulatory processes aim to decrease this risk.

So far, in France and many other countries, this form of biocontrol has been operated mainly by public institutions. The investment and risk taking of different value chain actors are variable among countries and classical biocontrol programmes. For example, in some cases, stakeholders strongly finance the programmes operated by public institutions when they expect highly positive financial effects. When pests have substantial adverse effects on biodiversity and health, local public organizations may be essential actors in the financing and co-ordination of the programmes.

On the side of adaptation, we must first note that it is a strategy to address changing ecosystems. It is knowledge-intensive and based on living technology. It requires a comprehensive understanding of the local context with specific attention to consequences. It is, however, generally centralized, intending to introduce one or a few biocontrol agents over the agrosystems of an entire country from one or a few mass-rearing units owned by one or a few actors.

25.3.4 Conservation biocontrol

Conservation biocontrol is a 'modification of the environment or existing practices to protect and enhance specific natural enemies or other organisms to reduce the effect of pests' (Eilenberg *et al.*, 2001, p. 396). This strategy implies using various products and services that enhance and adapt the production system. Reflection is more holistic; it is oriented towards elaborating and maintaining productive environments based on natural equilibria, even when a significant pest is targeted and when such equilibria can be reached in the short term.

This strategy is decentralized, place-based and has little potential for scalability since it is not standardized. Here lies a difficulty: these properties are seen as weaknesses in industrial agricultural systems. However, sustainability and adaptability will remain out of reach if we do not overcome this fixation on industrial agrarian standards. Industrialization may still be present (e.g. producing biomass to feed resident natural enemies: seeds for flower mixes, pollens, *Ephestia* eggs, *Artemia* cysts, sugars, nests for birds, etc.) but not necessarily centralized. Production can be local and may follow a custom-made output logic.

Because of the nature of conservation biocontrol, the development of a professional branch could propose a mix of products and services and provide counselling and orientations to users on a variety of techniques favouring biocontrol services provided by resident organisms. These might include use and mix of crop and non-crop plant species, choice of seeds to produce and to set up flower mixes, production of supplementary foods such as pollens or *Artemia*, spatial arrangement of ecological infrastructures (hedges, landscape management), etc.

Storage and transport are also less standardized and organized in more flexible ways. Products such as plants, seeds and other vegetables participate as part of the productive activity or contribute to the overall landscape. The farming productive activity involves buying products such as seeds or small plants. Those products can be either sent by mail for seeds or grown and marketed locally. Other products, like nesting boxes, can be made out of organic matter to limit their footprint; their transportation is not recurrent but happens once for an extended period. As such, transportation and storage have a lower impact on that strategy.

Overall, conservation biocontrol appears highly sustainable in relation to to production and delivery. We also suggest that this strategy has tremendous potential for adaptability since counselling and producing elements for conservation strategy could be modified easily in light of new needs related to climate change. Because industrialization would be more custom based, it would help to adapt to abiotic changes.

25.3.5 Not all biocontrols are equal in the face of planetary limits

Looking at biocontrol strategies as living technology and thinking about their impact on climate invites us to think of biocontrol not only as a final object but also as a process. If biocontrol organic agents are more favourable than chemicals in carrying out an ecological transition, one cannot forget the process of producing and releasing them in the environment. Taking that whole process into consideration shows that biocontrol strategies are not equal in terms of technological aliveness, and their impact on planetary limits and climate change is different whether we consider one strategy or another.

Thinking about the deployment of biocontrol in light of the burden of planetary limits and, in particular, climate change also shows that biocontrol strategies have different qualities in terms of sustainability and adaptability and that a diversity of users and crops can use those strategies following their level of ecologization (Lamine, 2011).

Figure. 25.1 shows that public policies and investment have been mainly focused on biocontrol strategies oriented toward crop protection products as alternatives to chemical pesticides. The other biocontrol strategies, oriented toward other forms (service, co-operation, territorial logistics), have not so far lead to a massive diffusion.

The inundation strategy is low in terms of sustainability and adaptability; however, that strategy is easy to deliver in the form of products and can help the more conventional practitioners to engage in different types of products with various effects but through the same practices. Inoculation and classical biocontrol strategies are higher in terms of adaptability and sustainability but require, for the former, more observation to follow the level of pests and for the latter to act at a place-based level, hence to carry transsectoral and collaborative actions. Both observation and collaboration require major moves for conventional farmers. Therefore, conservation strategy is seldom used for productive agriculture as it implies a drastic change in users' practices. We argue that less complex biocontrol

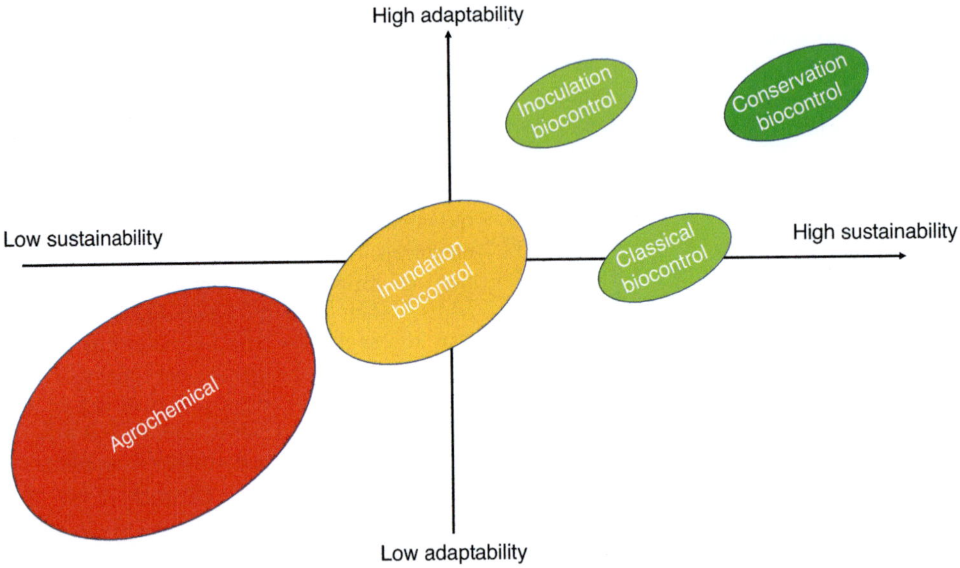

Fig. 25.1. Representation of biocontrol strategy positioning regarding sustainability vs adaptability potential (size of ovals approximates the level of private and public research and innovation investment for each strategy).

strategies provide steps toward agro-ecological crop systems by integrating new sets of practices into the paraphernalia of practitioners. Due to their organic properties, biocontrol strategies are robust; they also have the ability to adapt to the variations of pedoclimatic conditions inherent in climate change. Because of their nature and orientation to sustainability and adaptability of crop systems, biocontrol strategies offer a range of techniques to contribute to an agro-ecological transition of systems.

25.4 Innovation Strategies in Biocontrol to Combat Global Change

In this section, we will delve into the changes to be made in terms of practices, knowledge creation and reappropriation at the different stages of biocontrol elaboration and deployment to target higher sustainability and resistance. From this perspective, agronomic knowledge can be complemented with other knowledge to fully integrate the range of biocontrol strategies into habits and practices. This thinking is carried out at three levels of the value chain: research and development, production and commercialization, and final use.

This implies breaking with the implicit organizational model of conventional agriculture that has driven the development of biocontrol so far. For the past 30 years, biocontrol has been developed with the idea of perfect pesticide substitutes. However, substitutes have limited success since biocontrol is limited to products presented with the same properties as those of chemical products: immediate agronomic efficacy, availability, modes of application, storage and transportation. When it comes to technological change, it is common to view the new technology in light of the current one's properties even before considering the qualities of the latest technology. This approach favoured biocontrol strategies that fit in conventional systems, i.e. industrialized, deterritorialized, standardized and scalable. Because of these organizational similarities with agrochemistry, the inundative strategy has received much attention and effort from public and private sectors. On the side of sustainability and adaptability, these same qualities are somewhat detrimental, since they imply the use of a vast amount of fossil resources, and will therefore have a negative impact on climate change.

25.4.1 Research and development: change of habits and knowledge perception

Scientific breakthroughs and innovations in biocontrol are currently mainly implemented through a techno-push process (Schumpeter, 1912; Evangelista, 2018). As such, scientific research comes first in the process; science is expected to produce academic knowledge, which remains mainly abstract and distinct from the domain it applies to. Science is evaluated from the academic point of view; this means that biocontrol solutions are mainly intended (and financed) as the product of scientific knowledge and are not yet linked to the practical needs of users (or as an additional secondary objective). Following this techno-push model, users' interests are considered only when the product is on the market.

This innovation production model has drawbacks because knowledge production is separated from the development of solutions. It implies thinking about knowledge abstractly, and without links to actual problems and actors. It also means that the organizational dimension of biocontrol is not addressed explicitly, whereas using biocontrol implies a drastic modification of the organization of productive systems.

Producing abstract knowledge has great value and fuels the serendipity of science. However, the current development and deployment of biocontrol are related to the need for an agro-ecological transition to further food production in the broader context of climatic change.

This requires both the elaboration of new academic knowledge and the consideration of a practical end-in-view linked to agro-ecological transition. This change in academic knowledge type production aims to articulate more academic knowledge elaboration by developing solutions adapted to users' needs.

So far, the development of solutions is separated from the production of academic knowledge. In France, that logic can be seen throughout the different institutions contributing to each stage. The production of academic knowledge is the responsibility of INRAE (National Institute for the Research on Agriculture, Food and Environment). At the same time, the development of solutions and the tests that go with them are the responsibility of the Technical Institutes for Agriculture (ITA), which specialize in specific crops or broader farming sectors.

Besides this artificial separation, producing knowledge and solutions toward implementing an ecological transition implies also stepping back from the current idea that the market will take any innovation as long as it works and responds to a need. However, as was already the case with chemical pesticides, niche markets are rarely addressed this way and given the diversity of agriculture in France, a large part of agriculture is niche from the point of view of any industry, whether it be thatinputs or agrifood. Controversies reveal diverse points of view on a technique based on different knowledge about that technology; the current landscape depicted does not let the possibility of controversies appear, which generally happen when a new technology arises (i.e. nuclear for electricity, GMOs in biology, etc.) (Chateauraynaud and Debaz, 2017).

As an emerging technology, biocontrol must find its way to the market and provide agriculture with new means to produce without harming the environment. Understanding how the public receives a technology implies associating sciences with social sciences. Hence, we invite readers to consider seriously the development of multidisciplinary teams right from the start of a biocontrol solution development.

Despite years of scientific work in the domain, the deployment of biocontrol solutions remains confidential. The development of chemistry in the 1980s and genetics in the 1990s sidelined the massive development of biocontrol. Because of the environmental and climatic crises, alternatives to chemicals have only recently appeared as a necessity. This urges the development of solutions that, in return, call for reducing the time between development and deployment on territories. To more quickly integrate academic knowledge in the development of solutions, we believe that different kinds of users and actors in the supply chain can be better involved. This move can help create quicker adaptations to better fit with organizational deployment.

At this stage of the development of innovation in biocontrol, we argue that one must integrate more closely academic knowledge creation and the associated development of solutions to fuel interdisciplinary research and rely more on sectoral users to find solutions to fit users' needs and existing organizations of the domain. This can be referred to as co-innovation when it relies on various actors. It can be implemented by relying on the collaborative research perspective and belongs to the domain of open innovation (Chesbrough, 2003). Also, we invite the biocontrol research community to consider farmers and other agricultural actors as lead users or even professional users (Shah and Tripsas, 2007) and, as such, to take their perspectives and considerations seriously.

25.4.2 Production and distribution of biocontrol: From a product-based to a service-based agriculture

Bringing biocontrol to users implies at least two activities: the first is ensuring the production of biocontrol solutions, and the second is ensuring their distribution. The availability of farming intrants has so far mainly occurred in the form of products. In this section, we shall explore how the variety of biocontrol agents and strategies impact the final form to be implemented and the associated organizational forms. Besides products, an extensive range of biological control techniques can be delivered through a service.

Production and business processes: implications for biocontrol

Before delving into the consequences for biocontrol, we first provide the reader with generic elements on product and service production and consumption. The process of development of products and services is different. The production of a product is referred to as a production process, whereas the production of a service is referred to as a business process. In a production process, the outcome (i.e. product) must be produced before it is distributed whereas in a business process, the result (i.e. service) is achieved during its production. As such, a production process implies flows of raw materials and final products and the need for production facilities and storage. These production flows are necessarily associated with the need for storage at each phase of the process. In the case of a business process, there are no raw materials, storage or production facilities. In relation to sustainability, the generic differentiation between these two processes has environmental impacts. Facilities, raw materials and storage have a footprint, whereas services have a lesser effect. In relation to adaptability, a production process is detached from the contexts of use. Any adaptation will be delayed; a market signal takes a long time to interpret and strategically answer, and in the rare cases where dedicated lines of information are created between producers and users, they are complicated and costly to set up and maintain. At the same time, a service implies having contact with beneficiaries and can be adapted more easily and quickly.

For biocontrol, differentiating between products and services makes sense because of the nature of biocontrol agents and the variety of strategies that can be carried out. However, the current organization of the farming arena is principally based on products rather than services. The supply chain of farming intrants was created and developed around the circulation of material artefacts to fuel and support farming productivity. Ultimately, developing a service-based organization means structural change to achieve better sustainability and adaptability.

Specificities of biocontrol and impact on its distribution

In the current farming context, releasing a product has the advantage of providing a form of biocontrol that fits farming habits and appears to be a substitute, such as biopesticides. It can also favour the development of new farming routines and habits by integrating an agent that does not have the same behaviour and whose use form is identical to the products farmers are used to. The production of living-agent-base biocontrol, such as macro-organisms and, to a lesser extent, micro-organisms, is sensitive to industrial processes. As such, the production of biocontrol intrants under the conventional product form limits the type of agents and strategies that can be transmitted to users since most biocontrol strategies are not product based and imply a service-based approach.

In the short term, the use of biocontrol products can bring users to consider effectiveness in a new light that would lead them to change their practice (this point is developed in the next section); choosing to develop biocontrol product then can have an advantage in terms of the sameness it shares with chemical base products.

Biocontrol is diverse but its diffusion as a product is only possible for one strategy – inundation – and is limited to some micro-organisms and their metabolites. We will elaborate further on the three other biocontrol strategies.

So far, the inundation strategy based on micro-organisms is the most prominent biological control diffusion. The diffusion of other strategies and agents is still in its infancy. Based on the previous analysis of biocontrol in terms of living/zombie technology, we showed that this strategy is less ecologically compatible than other strategies. However, using it offers a way to reduce chemical pesticide use.

Integrating other strategies, such as the inoculation strategy, generally based on macro-organism agents, impacts the overall supply chain in at least two ways. First, using macro-organisms tends to influence the distance between the macro-organisms' production and the users' geographical distribution. Second, relying on inoculation as a biocontrol strategy impacts the type of knowledge available to users. Inoculation strategy implies at least being able to follow the native population in the production fields and assess when it is time to inoculate that population again. That knowledge can be developed by final users (i.e. farmers), but intermediate users can also be trained to support farmers in these new uses. Typically, counsellors from intrant co-operatives, extension services or technical advice services could be responsible for transmitting that knowledge to leverage the use of the technology.

The conservation strategy is more complex than inoculation or inundation in terms of the variety of artefacts that can be used, whether biological material such as seeds, plants or organisms, or inert artefacts such as attractive colours or nesting boxes. Conservation strategy thus requires an even more complex understanding of interactions between plants, auxiliaries, pests and the physical environment. That strategy can be supported by furthering services and developing counselling to structure conservation territories; it can be at various scales, that of the plot, of the farm or even a region. These new forms of services could be reinforced by providing knowledge services on the dynamic of the insect population developed for the inoculation strategy.

Eventually, classical biocontrol aims to establish an exotic organism to regulate an exotic pest. That last strategy is already consistent at the territorial level but mainly developed as an extension service delivered through public service. Those services must be reinforced if one wants to extend them, and one can also imagine the proposition of private services; it will, however, entail training more people who can take care of parts of the process or the whole process.

Delving into the consequences of expanding the type of biocontrol into the supply chain shows that it impacts the organisation of the production system regarding, at least, the distance between the production facility and the field. This calls for a regional or territorial way of designing organizational structures to support an agro-ecological turn. It also implies a change in communication habits and information transmission between distributors and farmers. The following section identifies the changes in practices that using the broader range of biocontrol implies and the knowledge needed to achieve this turn.

25.4.3 Major transformation of farming practices when embracing the whole range of biocontrol

Changes to introducing biocontrol into the field

Whether using one strategy or another, biocontrol implies investing in new practices. The first and more evident change in behaviour is related to the preventive nature of biocontrol, which opposes the use of chemicals as a curative means. This is typically the case for biocontrol products made of micro-organisms or micro-organisms, and also the case of the sterile insect technique (SIT).

Biocontrol acts more as a prevention whereas chemicals are applied as a curative means of control. To ensure that professionals use biocontrol so that it performs as well as possible, it should be supported by an application service to ensure use of the product at the right time. This implies changing deeply rooted habits as when users maintain those habits, the consequences for biocontrol are negative, which makes it appear inefficient. Explaining to users is not sufficient to make sure that habits will change. Besides the specificity of the inundation strategy, developed through products with few services, every biocontrol strategy implies anticipating more than when using chemicals.

When engaging in inoculation strategy, the complexity increases; observing and recognizing living organisms becomes necessary to optimize the use of biocontrol agents in the long run. The inoculation strategy, whose objective is to improve the presence of living organisms, implies observation and recognition of these organisms to assess the need for inoculation, the level of organisms to inoculate and when the inoculation must be repeated. Observing, recognizing and differentiating one organism from others is part of the work.

The conservation strategy is the more complex biocontrol strategy; it requires knowledge already described in previous paragraphs and knowledge about plant–insect interactions: using plants as sentinels to help fight pests and weeds, as a trap for pests. Understanding and testing interactions and using plants and auxiliaries are at the core of that strategy.

Efficiency should not be assessed as one solution for one problem; it becomes an overall assessment of plots or territories, and agronomics comes back into the picture.

When taken seriously, biocontrol implies a diversity of relations that are sometimes simplified by being presented as a product. However, whatever the strategy, biocontrol suggests a new manner of transmission to users. The passivation (Eymard-Duvernay, 1986; Callon, 2011) of biocontrol is different from the use of chemicals. Instead of an object that is ready to use with a clear procedure (mode of use), the diversity of biocontrol agents and strategies implies a new way of knowing, acting, communicating and disseminating it to users.

From changing intrant strategy to knowledge shifting

In this chapter, we follow the journey of biocontrol along the supply chain. By doing so, we sketch the effects of biocontrol on pests in the fields and human structure in the various areas of the supply chain needed to achieve productive farming.

The journey's starting point is the lab; research is primarily separated from the field. This level is more than a change in the kind of knowledge developed. The need is to widen the type of knowledge to be considered. Scientific knowledge is the Holy Grail of researchers, which is good. However, this does not prevent researchers from considering farmers' knowledge about their practices in order to produce academic knowledge that elaborates on practical needs (Avenier and Parmentier Cajaiba, 2012).

At the field level, the fundamental change in knowledge with the inundation strategy is about anticipating. This means treating the pests even when they are not yet visible which will be easier if the capacity to recognize them increases. Agronomics must also be reintegrated into production routines. Biocontrol, whatever the strategy, is more efficient when basic prophylaxis principles are practised daily.

Counsellors and intrant sellers also must change their knowledge base. Their paraphernalia changes and they must adapt to it; they must know their properties and the different strategies required to adjust production and be able to explain the differences that producers have to expect and adapt to. They should adopt a new logic and integrate those intrants' organic and living nature. Counselling requires covering a variety of biocontrol strategies and agents and contributing to diffuse agro-ecological principles. This means that knowledge must not only cover products but also the design of plots to further conservative biocontrol and biological interactions between the various living beings on the plot. They should also be able to adapt the proposition to territorial specificities.

For field workers, the dawn of productive agro-ecology can make them valuable intermediaries between what is happening in the field and the production. Seeing and assessing pests and auxiliary populations and providing a first daily evaluation would help decision making in choosing and refining subsequent actions. It could contribute to changing the future organization of plots when developing a targeted conservation strategy.

25.5 Conclusion

The organizational dimension is already present in the deployment of all biocontrol strategies. Both climate change and planetary boundary issues require this organizational dimension to be considered to address them effectively. However, this consideration is challenging because the organizational dimension is often taken for granted and conflated with the model derived from the modernization of agriculture.

The model arising from agricultural modernization is one of large-scale industrial development, characterized by standardization and replication. It typically results in centralized organizations where production sites are far removed from consumption areas. Innovation in this context is viewed as a process driven by research laboratories, aiming for generalization and thus disconnected from field conditions. This model leads to production and innovation structures that are not sustainable or adaptable and therefore fail to address the challenges associated with climate change and planetary boundaries.

If this model's constraints and limitations can be overcome, biocontrol appears highly promising. It can even be seen as a way to expand the use of living technologies (the most sustainable) in agriculture while reducing reliance on zombie technologies. However, achieving this requires integrating the organizational dimension into the deployment process.

We have discussed the organizational strengths and limitations of the various biocontrol strategies from the perspective of sustainability and adaptability to climate change. These strategies are not equal and depart more or less from the principles promoted by industrial agriculture.

We then outlined the necessary transformation pathways to integrate organizational dimensions into biocontrol deployment projects, enabling them to achieve sustainability and adaptability objectives. The scope of work required is vast.

Implementing an ecological transition in agriculture calls for integrating all products and services that have a lower impact on climate change and fit other planetary limits. Paradoxically, biocontrol research and innovation whether public or private, invests too little in strategies that display the lowest environmental negative impacts (such as conservation biocontrol; see Table 25.2). Investing in and diffusing all the possible biocontrol strategies, in particular the most environment friendly, is not less economically sustainable (Bale *et al.*, 2007; Naranjo *et al.*, 2015; 2019) but requires organizational changes along value chains and innovation–diffusion pathways.

Table 25.2. Summary of the properties of each biological control strategy.

Type of strategy	Key biological properties	Human intervention	Innovation model	Type of biomass needed	Quantity of biomass to be produced	Efficiency of biomass production technology	Storage or transportation efficiency
Inundative biocontrol							
Micro-organisms	Population growth rate in industrial conditions Short-term predation/ parasitism/ competition in the field Phenological stage with low or absent activity (for storage or transport)	Industrial production, formulation	Scaled, centralized, standardized (deterritorialized), from the lab	Micro-organisms	++	++	++
Macro-organisms		Industrial production		Macro-organisms	+++	+	+
Inundative biocontrol							
Micro-organisms	Short-term establishment capacity in the target ecosystem Short- to mid-term predation/ parasitism/ competition in the field Phenological stage with low or absent activity (for storage or transport)	Industrial production	Scaled, centralized, anchored, from the lab and driven by local information	Micro-organisms	+	++	++
Macro-organisms		Industrial production		Macro-organisms	++	+	+

Continued

Table 25.2. Continued

Type of strategy	Key biological properties	Human intervention	Innovation model	Type of biomass needed	Quantity of biomass to be produced	Efficiency of biomass production technology	Storage or transportation efficiency
Classical biocontrol							
Using insects	Adaptation capacities to target ecosystems Long-term predation/ parasitism impact on the target pest	Laboratory evaluation and production	One-shot (sustainable problem solving), centralized, anchored, knowledge-intensive (lab and outdoor science)	Macro-organisms	+	+	+
Conservation biocontrol							
Ecological infrastructures favouring beneficial micro- or macro-organisms	Short- to long-term ecological equilibrium with high regulation of target pests by resident natural enemies.	Establishment of ecological infrastructures, landscape management	Adaptive process, territorialized, anchored, collective knowledge	Permanently established biomass (hedges)	N/A	N/A	N/A
Agronomic practices favouring macro- or micro-organisms		Agronomic practices		N/A	N/A	N/A	N/A
Use of flower mixes, service plants, pollens, etc.		Industrial production		Seeds, pollen, plants	+	++	++

References

Aggeri, F. (2023) *L'Innovation, Mais Pour Quoi Faire? Essai Sur un Mythe Économique, Social et Managérial*. Seuil, Paris.

Arnoud, M., Goupil, C., Halloy, J. and Herbert, E. (2020) *Trois Questions à*. Letter published by the Institut National de SHS du CNRS.

Avenier, M.J. and Parmentier Cajaiba, A. (2012) The dialogical model: Developing academic knowledge for and from practice. *European Management Review* 9, 199–212.

Bale, J.S., van Lenteren, J.C. and Bigler, F. (2007) Biological control and sustainable food production. *Philosophical Transactions of the Royal Society B. Biological Science* 363, 761–776.

Bonnet, E., Landivar, D. and Monnin, A. (2021) *Héritage et Fermeture. Une Écologie Du Démantèlement*. Editions Divergences, Paris.

Boutet, M. and Parmentier-Cajaiba, A. (2022) Biocontrol in France: Prospects for structuring a developing sector. In: Fauvergue, X., Rusch, A., Barret, M., Bardin, M. and Jacquin-Joly, E. (eds) *Extended Biocontrol*. Springer, Dordrecht, The Netherlands, pp. 219–232.

Buitink, J., Douzals, J.P., Duliège, É., Lebeau, F., Marchi, M. *et al.* (2022) New technologies for the deployment of extended biocontrol. In: Fauvergue, X., Rusch, A., Barret, M., Bardin, M. and Jacquin-Joly, E. (eds) *Extended Biocontrol*. Springer, Dordrecht, The Netherlands, pp. 245–255.

Callon, M. (2011) Qu'est-ce qu'un agencement marchand? In: Callon, M., Akrich, M., Dubuisson-Quellier, S., Grandclement, C. and Hennion, A. (eds) *Sociologie Des Agencements Marchands*. Presse des Mines, Paris, pp. 325–440.

Chateauraynaud, F. and Debaz, V. (2017) *Aux bords de l'irréversible. Sociologie pragmatique des transformations*. Pétra, Paris.

Chesbrough, H. (2003) The era of open innovation. Available at: http://sloanreview.mit.edu/article/the-era-of-open-innovation/ (accessed 18 June 2025).

Eilenberg, J., Hajek, A. and Lomer, C. (2001) Suggestions for unifying the terminology in biological control. *BioControl* 46, 387–400.

Evangelista, R. (2018) Technology and economic development: The Schumpeterian legacy. *Review of Radical Political Economies* 50, 136–153.

Eymard-Duvernay, F. (1986) La qualification des produits. In: Salais, R. and Thevenot, L. (eds) *Le Travail: Marchés, Règles, Conventions*. INSEE, Paris, pp. 239–247.

Halloy, J. (2019) La durabilité des machines vivantes. *Angle Mort* 13, 9–23.

Hamant, O. (2023) *Antidote au culte de la performance. La robustesse du vivant*. Gallimard, Les Riceys, France.

Hamant, O. (2024) Is incoherence required for sustainability? *Anthropocene Review* 12.

IPCC (2023) AR6 Synthesis Report: Climate Change 2023. Available at: www.ipcc.ch/report/sixth-assessment-report-cycle/ (accessed 18 June 2025).

Lamine, C. (2011) Transition pathways towards a robust ecologization of agriculture and the need for system redesign. Cases from organic farming and IPM. *Journal of Rural Studies* 27, 209–219.

Malausa, T., Parmentier-Cajaiba, A., Brisset, M.-N. and Boutet, M. (2024) Biocontrol(s) from a pesticide-free agricultural perspective. In: Jacquet, F., Jeuffroy, M.H., Jouan, J., Latruffe, L. and Le Cadre, E. (eds) *Towards Pesticide-Free Agriculture – Research and Innovations in a Future Crop Protection Paradigm*. Synthèses, Paris, pp. 111–138.

Moerkens, R., Berckmoes, E., Van Damme, V., Wittemans, L., Tirry, L. *et al.* (2017) Inoculative release strategies of Macrolophus pygmaeus Rambur (Hemiptera: Miridae) in tomato crops: Population dynamics and dispersal. *Journal of Plant Diseases and Protection* 124, 295–303.

Monnin, A. (2021a) Planetary negative commons. *Multitudes* 85, 117–125.

Monnin, A., Halloy, J. and Nova, N. (2020) Au-delà du low tech: Technologies zombies, soutenabilité et inventions. Interview croisée de José Halloy et Nicolas Nova par Alexandre Monnin. In: Attard, I., Charles, E., Fergus, K., Monnin, A. and Pigeon, L. (eds) *Low Tech: Face Au Tout-Numérique, Se Réapproprier Les Technologies*. Ritimo, Paris, pp. 120–128.

Monnin, A. (2021b) Les communs négatifs de l'Anthropocène. In: Bonnet, E., Landivar, D. and Monnin, A. (eds) *Héritage et Fermeture. Une Écologie Du Démantèlement*. Editions Divergences, Paris, pp. 10–55.

Naranjo, S.E., Ellsworth, P.C. and Frisvold, G.B. (2015) Economic value of biological control in Integrated Pest Management of managed plant systems. *Annual Review of Entomology* 60, 621–645.

Naranjo, S.E., Frisvold, G.B. and Ellsworth, P.C. (2019) Economic value of arthropod biological control. In: Onstad, D. and Crain, P. (eds) *The Economics of Integrated Pest Management. CABI*. Wallingford, UK, pp. 49–85.

Parmentier-Cajaiba, A. (2024) Towards new business models for the deployment and growth of biocontrol in agriculture. *Innovations Agronomiques* 91, 1.

Schumpeter, J.A. (1912) *The Theory of Economic Development*. Routledge, London.

Shah, S.K. and Tripsas, M. (2007) The accidental entrepreneur: The emergent and collective process of user entrepreneurship. *Strategic Entrepreneurship Journal* 1, 123–140.

Index

Printed and bound by CPI Group (UK) Ltd, Croydon, CR0 4YY

13/09/2026

14961117-0006